# THIRD
# ASIAN MATHEMATICAL
# CONFERENCE 2000

Proceedings of the

# THIRD ASIAN MATHEMATICAL CONFERENCE 2000

University of the Philippines
Diliman, Philippines

23-27 October 2000

**Editors**

**Toshikazu Sunada**

*Tohoku University, Sendai, Japan*

**Polly W Sy**

*University of the Philippines, Diliman, Philippines*

**Yang Lo**

*The Chinese Academy of Sciences, Beijing, China*

World Scientific
*New Jersey • London • Singapore • Hong Kong*

*Published by*

World Scientific Publishing Co. Pte. Ltd.

P O Box 128, Farrer Road, Singapore 912805

*USA office:* Suite 1B, 1060 Main Street, River Edge, NJ 07661

*UK office:* 57 Shelton Street, Covent Garden, London WC2H 9HE

**British Library Cataloguing-in-Publication Data**
A catalogue record for this book is available from the British Library.

**PROCEEDINGS OF THE THIRD ASIAN MATHEMATICAL CONFERENCE 2000**



ISBN 981-02-4947-0

Printed in Singapore by FuIsland Offset Printing

# PREFACE

The Third Asian Mathematical Conference (AMC2000) was held on 23-27 October 2000 at the University of the Philippines, Diliman Campus. This event was organized by the Southeast Asian Mathematical Society in collaboration with many mathematical societies in Asia. This Conference covered the main fields of mathematics; namely, algebra and number theory, analysis, geometry and topology, applied mathematics, and mathematics education.

During the 5-day conference, there were 8 plenary lectures, 47 invited lectures and over100 short communications. Most of the plenary and invited addresses were delivered by mathematicians from Asia. There were about 500 participants in the Conference. This proceedings volume is a collection of papers presented in the Conference.

The AMC2000 has achieved its main purposes of providing a forum for Asian mathematicians to meet and discuss mathematical problems on a wide range of topics and of supporting the objectives of the World Mathematical Year 2000. Moreover, it has fostered friendship among citizens of the participating countries and contributed to the development of mathematics research in the host country as well as participating countries.

The first Asian Mathematical Conference (AMC) was held in the University of Hong Kong in 1990. The Second was held in the Suranaree University of Technology, Thailand in 1995 and this Conference was the third in the series. Because of the success of these three AMC conferences, it has been decided that the AMC will be held once every four years. During the third AMC, Singapore has agreed to host the fourth AMC in the year 2004.

The Southeast Asian Mathematical Society (SEAMS) is a society formed in 1972 whose members are mathematicians mostly from Southeast Asian countries including Brunei, Hong Kong, Indonesia, Malaysia, the Philippines, Singapore, Thailand and Vietnam. Its primary objectives are

(1) To promote the advancement of mathematical sciences in Southeast Asia

(2) To facilitate the exchange of information about the current research work and teaching methods not only among mathematicians in Southeast Asia but also with mathematicians from outside the region.
(3) To encourage and foster friendly collaboration with the members of mathematical and scientific institutions.

Since 1972, the promotion of mathematical research has been done through the sponsorship of mathematical conferences, symposia, meetings of mathematicians, lecture tours and other academic and social activities among its members. The facilitation of the exchange of research information has been realized through the publication of the Southeast Asian Bulletin of Mathematics starting in 1977. We believe that the SEAMS will continue to play its important role of promoting cross cultural and international links for mathematicians in the region as well as the whole of Asia.

On behalf of the AMC2000 Steering Committee and AMC2000 Scientific Committee, we would like to express our gratitude and appreciation to all sponsors, donors and friends for their help and support for the Conference. We would like to extend our thanks to all plenary speakers, invited speakers and guests for having kindly accepted our invitation and to all participants for their enthusiastic response. Finally, we would like to acknowledge and express our thanks for the help and support of the faculty, staff, alumni and friends of the Department of Mathematics, University of the Philippines, Diliman Campus in the preparation for and during the Conference.

Toshikazu Sunada
Polly W. Sy
Yang Lo
November 2001

# ACKNOWLEDGMENTS

## The Conference Organizers would like to acknowledge and thank the support given by the following Sponsors and Donors:

## SPONSORS:

International Mathematical Union
Abdus Salam International Center for Theoretical Physics

UP Diliman Mathematics Foundation, Inc.
Ellen Francisco-Fajardo Foundation, Inc.
Tan Yan Kee Foundation, Inc
Philamlife Foundation, Inc.
National Natural Science Foundation of China

Chinese Mathematical Society
Indonesian Mathematical Society
Korean Mathematical Society
Mathematical Society of Japan
Mathematical Society of the Philippines
Mathematical Society of Thailand
Singapore Mathematical Society
Vietnam Mathematical Society
Southeast Asian Mathematical Society

Ateneo de Manila University
De La Salle University
AMA Education System

Philippine Social Security System
Philippine Health Insurance Corporation
Philippine Government Service Insurance System
UCPB General Insurance Company, Inc.
Aegon Life Insurance (Philippines), Inc.
First Guarantee Life Assurance Company, Inc.
The Insular Life Assurance Company, Limited
CAP Family of Companies
Nippon Life Philippines
Actuarial Advisers, Inc.
CMG Life Insurance Company, Inc.

Aloha Mental Arithmetic Philippines, Inc.
Canon Information Technologies Philippines, Inc.
First Metro Investment Corporation
Medrano Law Office
Nestle (Philippines)
Pilipinas Shell Petroleum Corporation
SoftTech Advantage, Inc.
Q.C.D. Ventures Corporation

LandBank of the Philippines
Anchor Bank

## Donors ( Friends and Alumni of the Department )

Joaquin B. Bamba
Annalizza A. Bernardo
Rizaldy T. Capulong
Eleonora Y. Cinco
Melchor V. Ciubal
Peter Godfrey C. Darvin
Coralyn G. del Rosario
Ronald L. Evangelista
Arnold P. Frias
Lih Ko-Wei
Hiroko and Mitsuo Morimoto
Allan R. Santos
Judy Frances Aquino See
Edgar B. Solilapsi
Toshikazu Sunada
Horacio T. Templo
Kenji Ueno
Anonymous Donors

## Special Thanks to:

HIS Sounds Philippines (Dr. Romulo G. Pizaña, Musical Director/Conductor)
The University of the Philippines Filipiniana Dance Group (Mr. Van Manalo, Artistic Director)

# AMC2000 Committees

**The AMC2000 Scientific Committee :**

Yang Lo, The Chinese Academy of Sciences, Beijing (Chairman)
Dinh Dung, Hanoi Institute of Mathematics, Hanoi
Kim Do Han, Seoul National University, Seoul
Lee Seng Luan, National University of Singapore
Liu Fon-Che, Academia Sinica, Taipei
Ma Zhi Ming, The Chinese Academy of Sciences, Beijing
Masaki Maruyama, Kyoto University, Kyoto
Milagros P. Navarro, University of the Philippines, Diliman
M.S. Raghunathan, Tata Institute of Fundamental Research
Vladimir Romanov, Sobolev Institute of Mathematics Novosibirsk, Koptyuga
Aner Shalev, The Hebrew University of Jerusalem
Toshikazu Sunada, Tohoku University, Sendai

**The AMC2000 Steering Committee:**

Polly W. Sy, University of the Philippines, Diliman (Chairperson)
Do Long Van, Hanoi Institute of Mathematics, Hanoi
Li Daqian, Fudan University, Shanghai
Lih Ko-Wei, Academia Sinica, Taipei
Ann Chi Kim, Dolsan Mathematical Institute, Pusan
Koh KheeMeng, National University of Singapore
Mitsuo Morimoto, International Christian University, Tokyo
Bienvenido F. Nebres, Ateneo de Manila University
Kazuo Okamoto, Tokyo University, Tokyo
Shum Kar-Ping, The Chinese University of Hong Kong
Kenji Ueno, Kyoto University, Kyoto

**The AMC2000 International Organizing Committee :**

Lorna I. Paredes, University of the Philippines, Diliman (Chairperson)
Eisa Al-Said, King Saud University, Riyadh
Lorna S. Almocera, University of the Philippines, Cebu
Yolando B. Beronque, De La Salle University, Manila
Sergio R. Canoy, Mindanao State University-Iligan Institute of Technology
Wanida Hemakul, Chulalongkorn University, Bangkok
Wenlin Li, The Chinese Academy of Sciences, Beijing
Ling San, National University of Singapore
Maslina, Universiti Kebangsaan Malaysia, Selangor
Sri Wahyuni, Gadjah Mada University, Yogyakarta
Teofina Rapanut, University of the Philippines, Baguio
Walter Roth, Universiti Brunei Darussalam

## The AMC2000 Local Organizing Committee :

Polly W. Sy (Chairperson)
Luz R. Nochefranca (Secretary)
Maritina T. Castillo
Ricardo D.H. Del Rosario
Jose Maria L. Escaner IV
Augusto Y. Hermosilla
Ricardo Enrico C. Namay II
Caridad M. Natividad
Nenita C. Ocampo
Lorna I. Paredes
Romulo G. Pizaña
Dakila A. Reyes
Ivy C. Suan
Evelyn L. Tan
Ma. Vivien V. Visaya

**ASSISTING OFFICERS & STAFF of the Philippine Social Security System**

Chairman & Pres. Carlos A. Arellano
EVP Leopoldo S. Veroy
EVP Horacio T. Templo
SVP Marla M. Laurel
SVP Edgar B. Solilapsi
VP Marissu Bugante
VP Alfredo S. Villasanta
AVP Rizaldy T. Capulong
Amando P. Montano
Matias C. Defensor
Ronaldo W. Recio
Jesse Revecho
Jannet V. Villacorta
Arnold Frias
Melchor Vergel L. Ciubal
Nemesiano Mogello

**ASSISTING FACULTY and STAFF of the Department of Mathematics, University of the Philippines, Diliman**

Ma. Nerissa M. Abara
Tresio P. Acuzat
Julius Caesar C. Agapito
Ma. Sonia J. Andres
Christine Abegail A. Antonio
Carlene Perpetua P. Arceo
Joel N. Bangalan
Julius M. Basilla
Noemi R. Barcial
Angela Faith P. Bernardo
Rowena Alma L. Betty
Carolina H. Briones
Rex R. Briones
Grace D. Cabading
Chona M. Caguintuan
Adora A. Calaor
Ronel J. Calugay
Eloisa S. Cantuba
Kristine Joy E. Carpio
Don G. Castronuevo
Tom N. Chu
Filipinas L. Cristobal
Laarni dela Cruz
Angelito M. delos Reyes
Nenita A. delos Santos
Danilo C. Dematera
Carolina M. Dimasayao
Mace I. Eclevia
Jeopista C. Espina
Rene P. Felix
Boogin C. Gamboa
Michelle P. Guevara
Nelson C. Gomintong
Ma. Celeste F. Ignacio
Lagrimas D. Iloreta
Catherine N. Laniog
Glolibeth T. Lapay
Roosevelt Y. Loto
Neddie T. Lozano
Benjie L. Luces
Frances P'Ann I. Macabuhay
Kimberly May E. Magsajo
Rhodora D. Monte
Augusto Morales J r.
Manuel I. Musngi
Ann Marie E. Natividad
Fidel R. Nemenzo
Cherryl N. Ortega
Raul O. Padilla
Gilbert M. Parrocha
Ma. Garnet J. Porciuncula
Tricia Caroline M. Quijano
Reiko C. Rabe
Cherry Rose A. Ramos
Crispin R. Rance
Reynaldo Romualdo M. Rey
Josephine F. Reyes
Gaudella O. Ruiz
Virgina G. Saban
Gil Sadorra
Remedios D. Santos
Maranatha Consuelo L. Serrano
Anna Marie S. Tiro
Natividad A. Valencia
Michelle C. Vallejos
Cecille V. Verzola
Ma. Redina M. Victoria
Jesie G. Villanueva
Jimmy V. Viloria

**Other ASSISTING FACULTY Members :**

Jaime D.L. Caro, Dept. of Computer Science, University of the Philippines,Diliman
Felix P. Muga II, Dept. of Mathematics, Ateneo de Manila University
Andrew Vizcarra, Dept. of Mathematics, Ateneo de Manila University

# CONTENTS

# REMARKS ON THE SOLUTIONS OF $\Delta_P U + |U|^{Q-2}U = 0$

JULIUS CAESAR C. AGAPITO*, LORNA I. PAREDES†,
REYNALDO M. REY AND POLLY W. SY

*Department of Mathematics, College of Science, University of the Philippines, Diliman, Quezon City, Philippines*

In this paper, the positive solution of the equation $\Delta_p u + |u|^{q-2}u = 0$, where $p \neq q$, and its behavior relative to a certain limiting conditions on $p$ and $q$ shall be discussed.

For $p, q \in (1, \infty)$, Ôtani[3] studied the following one dimensional equation

$$(E^{p,q}) \qquad \begin{cases} \Delta_p u + |u|^{q-2}u = 0 & \text{on} \quad (0,1) & (1) \\ u(0) = u(1) = 0 & & (2) \end{cases}$$

where $\Delta_p u = (|u_x|^{p-2}u_x)_x$.

A function $u$ is said to be a *solution* of $(E^{p,q})$ if $u \in W_0^{1,p}(0,1)$ and $u$ satisfies (1) in the distribution sense.

In Ôtani's paper, he proved the following two theorems.

**Theorem 1** [3] Suppose that $p, q > 1$ and $p \neq q$. If $u$ is a solution of $(E^{p,q})$, then $u$ satisfies the following:

i) $u \in C^{\alpha}([0,1]) \cap C^{<q>}([0,1]]\backslash Z(u))$, where

$$Z(u) = \{x \in [0,1] \mid u_x(x) = 0\}, \quad \alpha = \min\{< \frac{2-p}{p-1} > +1, < q >\}$$

and

$$< r >= \begin{cases} \infty & \text{if } r \text{ is an even integer} \\ \min\{n \mid n \geq r,\ n \text{ nonnegative integer}\} & \text{otherwise.} \end{cases}$$

ii) $\frac{p-1}{p}|u_x(x)|^p + \frac{1}{q}|u_x(x)|^q = \text{constant for all } x \in [0,1]$

iii) $\lim_{t\to 0^+} u_x(t) = \lim_{t\to 1^-}[-u_x(t)]$

---

*MACARTHUR AND JOSEFINA DELOS REYES PROFESSORIAL CHAIR AWARDEE

†PCI BANK FOUNDATION PROFESSORIAL CHAIR AWARDEE

iv) $||u_x||^p_{L^p(0,1)} = ||u||^q_{L^q(0,1)} = \frac{q(p-1)}{pq-q+p}|\lim_{t\to 0+} u_x(t)|^p$.

**Theorem 2** [3] Suppose that $p, q > 1$ and $p \neq q$. Then $(E^{p,q})$ has a unique positive solution. Furthermore, for the functional $R$ defined by

$$R(v) = \frac{||v||_{L^q(0,1)}}{||v_x||_{L^p(0,1)}}$$

we have $R(u_{p,q}) = \sup\{R(v)| \quad v \in W_0^{1,p}(0,1) \quad \text{and} \quad v \neq 0\}$, where $u_{p,q}$ denotes the unique positive solution of $(E^{p,q})$.

**Remark:** The solution $u_{p,q}$, in addition, satisfy the following:
i) $u_{p,q}(x) = u_{p,q}(1-x)$ for any $x \in [,1]$
ii) $(u_{p,q})_x$ is positive and decreasing on $[0, \frac{1}{2})$ with $(u_{p,q})_x(\frac{1}{2}) = 0$
iii) $u_{p,q}(\frac{1}{2}) = \max_{x\in[0,1]} u_{p,q}(x)$

In APRS[1], an explicit formula for $u_{p,q}$ was obtained.

**Theorem 3** If $u_{p,q}$ is the unique positive solution of $(E^{p,q})$, then for any $x \in [0, \frac{1}{2}]$,

$$u_{p,q}(x) = \left(2q^{-1}B(q^{-1}, 1-p^{-1})\right)^{\frac{p}{q-p}} \left(\frac{q(p-1)}{p}\right)^{\frac{1}{q-p}} f_{p,q}^{-1}(2xq^{-1}B(q^{-1}, 1-p^{-1})),$$

where $B$ is the Beta function and $f_{p,q}(s) = \int_0^s (1-t^q)^{\frac{-1}{p}} dt$ for $s \in [0,1]$.

**Corollary 3.1** The best possible constant for the Sobolev-Poincare type inequality

$$||v||_{L^q(0,1)} \leq C_{p,q}||v_x||_{L^p(0,1)} \quad \text{for all } v \in W_0^{1,p}(0,1)$$

is given by

$$C_{p,q} = \frac{p^{\frac{1}{q}} q^{1-\frac{1}{p}} (pq-q+p)^{\frac{1}{p}-\frac{1}{q}}}{2(p-1)^{\frac{1}{p}} B(q^{-1}, 1-p^{-1})} \quad .$$

**Corollary 3.2** Let $\alpha(p)$ be a function defined on $(1, \infty)$ such that $1 < p\alpha(p) \neq p$ for all $p \in (1, \infty)$ and $\lim_{p\to\infty} \alpha(p) = a$. If $q = p\alpha(p)$, then

$$\lim_{p\to\infty} u_{p,q}(x) = 2^{\frac{a}{a-1}} \left(\frac{1}{2} - |x - \frac{1}{2}|\right) \qquad \text{for all} \quad x \in [0,1].$$

**Corollary 3.3** Let $\alpha(q)$ be a function defined on $(1,\infty)$ such that $1 < q\alpha(q) \neq q$ for all $q \in (1,\infty)$ and $\lim_{q\to\infty}\alpha(q) = b$. If $p = q\alpha(q)$, then

$$\lim_{q\to\infty} u_{p,q}(x) = 2^{\frac{a}{1-b}}\left(\frac{1}{2} - |x - \frac{1}{2}|\right) \qquad \text{for all} \quad x \in [0,1].$$

Idogawa[2] showed that,

**Theorem 4** Let $D_{p,q} = \max_{x\in[0,1]} u_{p,q}(x)$. Then

i) for fixed $q > 1$, we have

$$\lim_{p\to 1+} D_{p,q} = 2^{\frac{1}{q-1}}$$

ii) for fixed $p > 1$, we have

$$\lim_{q\to 1+} D_{p,q} = 2^{\frac{-p}{p-1}}\left(\frac{p-1}{p}\right)$$

And in addition to this, we have

**Proposition 5** For some $\delta_0 > 1$, let $\alpha(p)$ be a function defined on $(1, 1+\delta_0]$ such that $p \neq \alpha(p) > 1$ for all $p \in (1, 1+\delta_0]$ and $\lim_{p\to 1+}\alpha(p) = a > 1$. If $q = \alpha(p)$, then, for any $x \in (0, \frac{1}{2}]$

$$\lim_{p\to 1+} u_{p,q}(x) = 2^{\frac{1}{a-1}}.$$

**Proof:** From Theorem 3, we have, for any $x \in [0, \frac{1}{2}]$

$$u_{p,q}(x) = 2^{\frac{p}{q-p}}\left(\frac{1}{q}B(\frac{1}{q}, 1 - \frac{1}{p})\right)^{\frac{p}{q-p}}\left(\frac{q(p-1)}{p}\right)^{\frac{1}{q-p}} f_{p,q}^{-1}\left(\frac{2x}{q}B(\frac{1}{q}, 1 - \frac{1}{p})\right).$$

Let $q = \alpha(p)$ and $\lim_{p\to 1+}\alpha(p) = a > 1$. For any fixed $s \in [0,1)$, we have

$$\begin{aligned}\lim_{p\to 1+} f_{p,q}(s) &= \lim_{p\to 1+}\int_0^s (1 - t^q)^{-\frac{1}{p}}dt \\ &= \int_0^s (1 - t^a)^{-1}dt < \infty\end{aligned}$$

which implies that $\lim_{p\to 1+} f_{p,q}(s) = \infty$ if and only if $s = 1$.

Since, by definition, $B(\frac{1}{q}, 1-\frac{1}{p}) = \frac{\Gamma(\frac{1}{q})\Gamma(1-\frac{1}{p})}{\Gamma(\frac{1}{q}+1-\frac{1}{p})}$, where $\Gamma(\cdot)$ is the gamma function, then

$$\lim_{p\to 1^+} B(\frac{1}{q}, 1-\frac{1}{p}) = \lim_{p\to 1^+} \frac{\Gamma(\frac{1}{q})}{\Gamma(\frac{1}{q}+1-\frac{1}{p})} \cdot \lim_{p\to 1^+} \Gamma(1-\frac{1}{p})$$
$$= \frac{\Gamma(\frac{1}{a})}{\Gamma(\frac{1}{a})} \cdot \infty = \infty.$$

Therefore, for any $x \in (0, \frac{1}{2}]$, we have $\lim_{p\to 1^+} \frac{2x}{q} B(\frac{1}{q}, 1-\frac{1}{p}) = \infty$, and hence,

$$\lim_{p\to 1^+} f_{p,q}^{-1}(\frac{2x}{q} B(\frac{1}{q}, 1-\frac{1}{p})) = 1.$$

Note that

$$[\tfrac{1}{q}B(\tfrac{1}{q}, 1-\tfrac{1}{p})]^{\frac{p}{q-p}} [\tfrac{q(p-1)}{p}]^{\frac{1}{q-p}} =$$
$$(\tfrac{1}{q})^{\frac{p}{q-p}} (\tfrac{q}{p})^{\frac{1}{q-p}} (\frac{\Gamma(\frac{1}{q})}{\Gamma(\frac{1}{q}+1-\frac{1}{p})})^{\frac{p}{q-p}} (\Gamma(1-\tfrac{1}{p}))^{\frac{p}{q-p}} (p-1)^{\frac{1}{q-p}}.$$

Since $\frac{1}{p}$ is not an integer, we see that

$$(\Gamma(1-\frac{1}{p}))^{\frac{p}{q-p}} (p-1)^{\frac{1}{q-p}} = ((\frac{\pi}{(\sin\frac{\pi}{p})\Gamma(\frac{1}{p})})^p (p-1))^{\frac{1}{q-p}}.$$

From

$$\lim_{p\to 1^+} \frac{p-1}{(\sin\frac{\pi}{p})^p} = \frac{1}{\pi},$$

it follows easily that

$$\lim_{p\to 1^+} [\frac{1}{q}B(\frac{1}{q}, 1-\frac{1}{p})]^{\frac{p}{q-p}} [\frac{q(p-1)}{p}]^{\frac{1}{q-p}} = 1.$$

Hence, for each $x \in (0, \frac{1}{2}]$, we have $\lim_{p\to 1^+} u_{p,q}(x) = 2^{\frac{1}{a-1}}$.

**Proposition 6** For some $\delta_0 > 1$, let $\beta(q)$ be a function defined on $(1, 1+\delta_0]$ such that $q \neq \beta(q) > 1$ for all $q \in (1, 1+\delta_0]$ and $\lim_{q\to 1^+} \beta(q) = b > 1$. If $p = \beta(q)$, then for any $x \in (0, \frac{1}{2}]$,

$$\lim_{q\to 1^+} u_{p,q}(x) = 2^{\frac{-b}{b-1}} (\frac{b-1}{b})(1-(1-2x)^{\frac{b}{1-b}}).$$

**Proof:** Since $p = \beta(q)$ and $\lim_{q\to 1^+} \beta(q) = b > 1$, we have for each $s \in [0, 1]$

$$\lim_{q\to 1+} f_{p,q}(s) = \int_o^s (1-t)^{\frac{-1}{b}}\,dt$$
$$= (\frac{b}{b-1})(1-(1-s)^{\frac{b-1}{b}}) =: g(s).$$

Now if $y = g(s)$ where $y \in [0, \frac{b}{b-1})$, then $g^{-1}(y) = 1-(1-(\frac{b-1}{b}))^{\frac{b}{b-1}}$, and thus

$$\lim_{q\to 1+} f_{p,q}^{-1}(y) = 1-(1-(\frac{b-1}{b})y)^{\frac{b}{b-1}}.$$

Hence, for each $x \in (0, \frac{1}{2}]$, Theorem 3 yields,

$\lim_{q\to 1+} u_{p,q}(x)$
$$= \left(2B(1,1-\tfrac{1}{b})\right)^{\frac{b}{1-b}} \left(\tfrac{b-1}{b}\right)^{\frac{1}{1-b}} \left(1-(1-2x(\tfrac{b-1}{b})B(1,1-\tfrac{1}{b}))^{\frac{b}{1-b}}\right)$$
$$= 2^{\frac{-b}{b-1}}(\tfrac{b-1}{b})(1-(1-2x)^{\frac{b}{1-b}}).$$

This proves the proposition.

**Acknowledgment:** This research is supported by the **Natural Sciences Research Institute** of the University of the Philippines, Diliman.

## References

[1] Agapito, J.C.C., L.I. Paredes, R.M. Rey, P.W. Sy, A Note on the Positive Solution of $\Delta_p u + |u|^{q-2}u = 0$, Matimyas Matematika 22(1999) pp1-8.

[2] Idogawa, T., Lecture Notes

[3] Ôtani, M., "On Certain Second Order Ordinary Differential Equations Associated with Sovolev-Poincare-type Inequalities", Nonlinear Analysis, Theory, Method and Applications, 8 (11) 1984, 1225-1270.

# ON ONE-PHASE STEFAN PROBLEMS FOR SUBLINEAR HEAT EQUATIONS

TOYOHIKO AIKI

*Department of Mathematics, Faculty of Education, Gifu University, Gifu 501-1193, Japan.*

HITOSHI IMAI

*Department of Applied Physics and Mathematics, Faculty of Engineering, University of Tokushima, Tokushima 770-8506, Japan.*

NAOYUKI ISHIMURA

*Department of Mathematics, Hitotsubashi University, Kunitachi, Tokyo 186-8601, Japan.*

YOSHIO YAMADA

*Department of Mathematical Sciences, Waseda University, Ohkubo, Shinjyuku-ku, Tokyo 169-8555, Japan.*

We are interested in the one-phase Stefan problem for sublinear heat equations with Dirichlet boundary condition. We survey our recent establishments concerning the unique solvability and the asymptotic behavior of solutions.

## 1 Introduction

This article surveys our recent studies on the one-phase Stefan problems for sublinear heat equations with Dirichlet boundary condition [5,6]: Find a free boundary $x = l(t) > 0$ on $[0, T]$ with $T > 0$ and a function $u = u(t, x)$ on $Q(T; l(t)) := \{(t, x) \,|\, 0 < t < T, 0 < x < l(t)\}$, which satisfy

$$\begin{cases} u_t = u_{xx} + u^{1-\alpha} & \text{in } Q(T; l(t)) \\ u(t, 0) = 0 & \text{for } 0 \le t < T \\ u(t, l(t)) = 0 & \text{for } 0 \le t < T \\ l'(t) = -u_x(t, l(t)) & \text{for } 0 < t < T \\ u(0, x) = u_0(x) \ge 0 & \text{on } 0 \le x \le l_0 := l(0), \end{cases} \tag{1}$$

where $l_0 > 0$ is a constant taken to be $l_0 = 1$ hereafter without loss of generality. In (1) $u_0$ is a given smooth function satisfying

$$u_0 > 0 \quad \text{in } 0 < x < l_0,$$
$$u_0 = 0 \quad \text{and} \quad (u_0)_x \neq 0 \quad \text{at } x = 0, l_0.$$

The exponent $\alpha$ is assumed to be $0 < \alpha < 1$, which makes our nonlinear parabolic equations sublinear.

Existence and uniqueness for (1) is a first important problem. Due to the lack of the Lipschitz regularity of nonlinear terms, which is inherent in sublinearity, careful treatment is necessary. For example, we recall works [1,8,12]. Our first results [5], on the other hand, state that (1) is wellposed for every $T > 0$; existence and uniqueness hold true also in the free boundary problem for sublinear heat equations with certain class of nonzero initial data.

Our next aim is to investigate the asymptotic behavior of solutions. It is well known that if $\alpha < 0$ there exist blowing-up solutions for (1) [2,3]. It is a fascinating project to undertake whether a similar grow-up pattern takes place or not if $0 < \alpha < 1$. We remark that in the sublinear case no solution blows up within finite time. Our concern is focused on the asymptotic behavior of the solution $\{u(t,x), l(t)\}$ if $t \to \infty$. Roughly speaking, the sublinear nonlinearity makes the free boundary be apt to grow simultaneously with the solution itself. Note that, in (1) under $\alpha < 0$, if the Dirichlet boundary condition is replaced by the Neumann boundary condition $u_x(t, 0) = 0$ for $0 \leq t < T$, then the boundedness follows as established in [2].

*Acknowledgements.* This work is partially supported by Grants-in-Aids for Scientific Research (Nos.09740139, 10354001), from the Japan Ministry of Education, Science, Sports and Culture.

## 2 Unique Solvability

First we clarify the definition of a solution to the problem (1).

*Definition.* We call that a pair $\{u, l\}$ with $u \in W^{1,2}([0,T]; L^2(0,\infty))$ and $l \in C^{1+\nu}([0,T])$ $(0 < \nu < 1)$ is a solution of (1) on $[0,T]$, if the next conditions (S1)(S2) hold.
(S1): $u$ is non-negative on $[0,T] \times (0,\infty)$, $u_x \in C^\nu(\mathrm{Cl}.Q(T;l))$, $u_{xx} \in C(\mathrm{Cl}.Q((t_0,T);l))$ for every $t_0 \in (0,T)$, and $l > 0$ on $[0,T]$, where $Q((t_0,T);l) := Q(T;l) \cap \{t > t_0\}$.
(S2): $\{u(t,x), l(t)\}$ solves (1).

Our main results on the unique solvability is summarized as follows [5].

**Theorem 1** *Suppose that $l_0 > 0$, and a non-negative function $u_0 \in C([0,\infty)$ satisfy $u_0 \in C^{1+\beta}([0,l_0])$ for some $\beta \in (0,1)$, $u_0 > 0$ on $(0,l_0)$, $u_0(0) = 0$, $u_0 = 0$ on $[l_0,\infty)$, and $(u_0)_x > 0$, $(u_0)_x(l_0) < 0$. Then there exists only one solution $\{u,l\}$ of (1) on $(0,\infty)$.*

The proof involves a similar technique as in [9] with necessary modifications.

## 3 Asymptotic Behavior

We turn our attention to the asymptotic behavior of solutions for (1) [6].

The analysis should be associated with the one for the usual fixed Dirichlet boundary value problems

$$\begin{cases} v_t = v_{xx} + v^{1-\alpha} & \text{in } Q(T;l_1) \\ v(t,0) = v(t,l_1) = 0 & \text{for } 0 \le t < T \\ v(0,x) = v_0(x) \ge 0 & \text{on } 0 \le x \le l_1, \end{cases} \tag{2}$$

where $Q(T;l_1) := \{(t,x)\,|\,0 < t < T, 0 < x < l_1\}$, a constnat $l_1 > 0$ is fixed, $0 < \alpha < 1$, and $v_0$ is also a given nonzero smooth function.

We introduce a functional $F_1$ associated with (2):

$$F_1[v] := \int_0^{l_1} \left( \frac{1}{2} v_x^2 - \frac{1}{2-\alpha} v^{2-\alpha} \right) dx.$$

**Theorem 2** *Suppose* $0 < \alpha < 1$. *Every solution of (2) converges as $t \to \infty$ to a unique stationary point of $F_1$ defined above. Moreover, $v \equiv 0$ is not a minimizer for $F_1$.*

It is easy to see that a unique positive satationary point of $F_1$ is given by the unique positive solution to the problem

$$\begin{cases} v_{xx} + v^{1-\alpha} = 0, \quad v > 0 \quad \text{in } 0 < x < l_1 \\ v(0) = v(l_1) = 0. \end{cases} \tag{3}$$

The uniqueness property of (3) leads us to the corollary.

**Corollary 1** *Let $v(x;l_1)$ denote the solution of (3). Then there holds*

$$v(x;l) = l^{2/\alpha} v(x/l;1).$$

Now we show our main theorem concerning the asymptotic profile.

**Theorem 3** *Suppose* $0 < \alpha < 1$. *Then any solution* $\{u(t,x), l(t)\}$ *grows up as* $t \to \infty$ *on the whole* $\mathbf{R}^+ = \{x > 0\}$; *that is, there holds*

$$l(t) \to \infty \quad \textit{as } t \to \infty,$$

*and eventually at every* $x \in \mathbf{R}^+$ *one has*

$$u(t,x) \to \infty \quad \textit{as } t \to \infty.$$

In the above, it is certain that $u(t,x)$ is meaningless initially for any $x > l_0$; however, since $l(t) \to \infty$ as $t \to \infty$, $u(t,x)$ is justified for every $x > l_0$ if $t \geq t_x > 0$ where $t_x$ depends on $x$. Our theorem asserts that $u(t,x)$ grows up as $t \to \infty$, which should be understood only after $u(t,x)$ is defined.

We provide a sketch of proof. Let $\{u(t,x), l(t)\}$ be a solution to (1), which exists for any $T > 0$. We introduce a functional $F_2$ associated with (1), which is similar to $F_1$:

$$F_2[u](t) := \int_0^{l(t)} \left( \frac{1}{2} u_x(t,x)^2 - \frac{1}{2-\alpha} u(t,x)^{2-\alpha} \right) dx.$$

$F_2$ serves as a Lyapunov functional.

The proof proceeds along several steps. In the sequel $C(\alpha, u_0)$ stand for constants which may differ from line to line.

Step 1. $\|u(t,\cdot)\|_{L^\infty} \leq C(\alpha, u_0)(1+t)^{1/\alpha}$ if $t > 0$.

Step 2. $\|u(t,\cdot)\|_{L^\infty}$ cannot tend to zero as $t \to \infty$.

Step 3. $l(t) \leq C(\alpha, u_0)(1 + t^{(2+\alpha)/(2\alpha)})$ for all large $t$.

Step 4. $l(t)$ cannot converge to a finite $l_\infty$ as $t \to \infty$.

Step 5. $\|u(t,\cdot)\|_{L^\infty} \to \infty$ as $t \to \infty$.

Step 6. Conclusion with establishing a pointwise estimate.

## 4 Numerical Justifications

We have analysed one-phase Stefan problems for sublinear heat equations with Dirichlet boundary condition. Here we exhibit numerical computation with the hope of strengthening our current exposition. The details of the numerical procedure and experiments for other types of nonlinearity are reported in our previous paper [14].

Figure 1 simulates the free boundary problem (1) with $\alpha = 1/2$. The initial datum are $u_0(x) = Ax^2(1-x)^2$ ($A = 0.0001$ and $A = 1000$) and $l_0 = 1$. In both values of $A$, the growing-up behavior is apparently observed. Figure 2, on the other hand, illustrates the fixed boundary problem (2) with $\alpha = 1/2$ and $l_1 = 1$. The initial data is similar as in Figure 1. Whatever large or small the initial function is, it converges to a nonzero stationary point of $F_1$ defined in §2; this agrees well with analytical results depicted in Theorem 2.

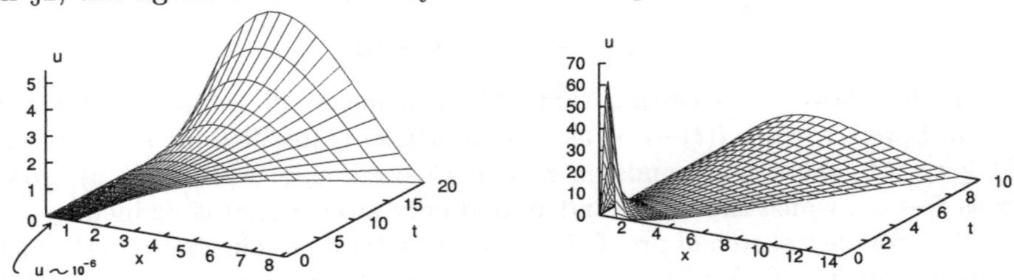


(a) (b)

Figure 1: Free boundary problem with $\alpha = 0.5$.
(a) $A = 0.0001,\ u(t,x) \uparrow +\infty$. (b) $A = 1000,\ u(t,x) \uparrow +\infty$.

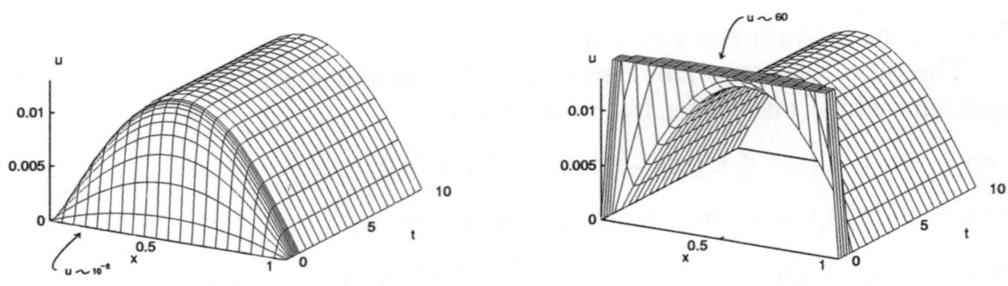


(a) (b)

Figure 2: Fixed boundary problem with $\alpha = 0.5$.
(a) $A = 0.001$. (b) $A = 1000$.

## References


1. J. Aguirre and M. Escobedo, *Ann. Fac. Sci. Toulouse* **8**, 175 (1986-1987).
2. T. Aiki, *Nonl. Anal. T. M. A.* **26**, 707 (1996).
3. T. Aiki and H. Imai in *Free boundary problems, theory and applications*, eds. M. Niezgódka and P. Strelecki, (Pitman Research Notes Math. 363, Longman, Essex, 1996).
4. T. Aiki and H. Imai, *Ann. Mat. Pura Appl. (IV)* **175**, 327 (1998).

5. T. Aiki, H. Imai, N. Ishimura and Y. Yamada, preprint.
6. T. Aiki, H. Imai, N. Ishimura and Y. Yamada, preprint.
7. T. Aiki and N. Kenmochi, *Bull. Fac. Edu., Chiba Univ.* **39**, 15 (1991).
8. T. Cazenave, F. Dickstein and M. Escobedo, *Rend. Mat. Appl. (VII)* **19**, 211 (1999).
9. A. Fasano and M. Primicerio, *J. Math. Anal. Appl.* **72**, 247 (1979).
10. A. Friedman, *Variational Principles and Free-Boundary Problems*, (R.E. Krieger, Florida, 1988).
11. H. Fujita, *J. Fac. Sci. Univ. Tokyo, Sect. I* **13**, 109 (1966).
12. H. Fujita and S. Watanabe, *Comm. Pure Appl. Math.* **21**, 631 (1968).
13. V.A. Galaktionov, J. Hulshof and J.L. Vazquez, *J. Math. Pures Appl.* **76**, 563 (1994).
14. H. Imai, N. Ishimura and T. Aiki in *Free Boundary Problems: Theory and Applications II*, ed. N. Kenmochi, (Gakuto, Tokyo, 2000).

# HOPF BIFURCATION IN A CHEMOSTAT-RELATED MODEL

LORNA S. ALMOCERA
*University of the Philippines in the Visayas, Cebu College*
*Lahug, Cebu City, Philippines*
*E-mail: almocera@cnms.net*

POLLY W. SY
*Department of Mathematics, University of the Philippines, Diliman*
*Quezon City, Philippines*
*E-mail: pweesy@i-manila.com.ph*

In this paper, we investigate an autonomous nonlinear system of ordinary differential equations that model competition in a chemostat with an inhibitor. We will illustrate the simplicity of using Hopf's theorem in establishing the existence of periodic solutions bifurcating from the positive equilibrium point.

## 1 Introduction

The chemostat is a piece of laboratory device frequently used for studying competition between microorganisms. It is a laboratory realization of a simple lake and plays a role in theoretical ecology [7], in wastewater treatment problems[5] and is used as a model for the study of recombinant problems in genetically altered organisms [6]. In its commercial form, the chemostat is used as a model for the manufacture of products by using genetically altered organisms. In this paper, we will investigate the model which is a combination of the model of competition between plasmid-bearing and plasmid-free organisms in a chemostat proposed by Staphanopoulis and Lapidus [6] and the model of a chemostat with an inhibitor investigated by Lenski and Hattingh [3]. The equations of the model take the form

$$\begin{aligned} \dot{x}_1 &= x_1((1-q)f_1(1-x_1-x_2)-1) \\ \dot{x}_2 &= x_2(e^{-\mu p}f_2(1-x_1-x_2)-1)+qx_1f_1(1-x_1-x_2) \\ \dot{p} &= 1-p-\delta\frac{px_1}{K+p} \end{aligned} \tag{1}$$

where $x_i(0)>0$, $\quad p(0)\geq 0$, $\quad f_i(S)=\dfrac{m_iS}{a_i+S}$, $\quad 0<x_1+x_2<1, 0<q<1$. We shall restrict system (1) to a positively invariant region

$$\Gamma=\{(x_1,x_2,p):0<x_1+x_2<1,\quad 0<p<1\}.$$

The variables $x_1$ and $x_2$ denote the concentration of the competing organisms and the variable $p$ represents the concentration of the inhibitor. The definitions of the constants $m_1, m_2, a_1, a_2, \delta, \mu$ and $K$ are given in Hsu *et al*[2].

The equilibrium points of system (1) are obtained by setting the right-hand side of system (1) to zero. A positive equilibrium point $E^* = (x_1^*, x_2^*, p^*)$ exists if

$$(1-q)\frac{m_1}{a_1+1} > 1 \quad \text{and} \quad e^{-\mu}\frac{m_2}{a_2+1} < 1, \quad \text{or}$$
$$(1-q)\frac{m_1}{a_1+1} > 1 \quad \text{and} \quad 1 > \lambda_2^* > \lambda_1^*$$

where the constants $\lambda_1^*$ and $\lambda_2^*$ are defined by the following relations:

$$f_1(\lambda_1^*) = \frac{m_1\lambda_1^*}{a_1+\lambda_1^*} = \frac{1}{1-q} \quad , \quad f_2(\lambda_2^*) = \frac{m_2\lambda_2^*}{a_2+\lambda_2^*} = e^{\mu}.$$

Hsu *et al*[2] studied system (1) and focused on the asymptotic behavior of the solutions to the system as a function of the parameters of the organisms $x_1$ and $x_2$ and the alteration of the parameters by the inhibitor $p$. In this paper we will investigate the qualitative behavior of the system near the positive equilibrium point $E^*$. We will apply Hopf's theorem [4] in establishing the existence of periodic solutions to system (1). This theorem establishes the existence of periodic solutions bifurcating from an equilibrium point $x^*$ for a system such as $\dot{x} = f(x, \alpha)$ if (a) $f(x^*, \alpha_c) = 0$ and (b) the Jacobian matrix of $f$ about $x^*$ has exactly two complex conjugate eigenvalues $u(\alpha) \pm v(\alpha)$ whose real part $u(\alpha)$ vanishes at $\alpha = \alpha_c$ while $\frac{du}{d\alpha} \neq 0$ at $\alpha = \alpha_c$.

This paper is organized as follows. In section 2, we investigate the conditions for the stability of $E^*$. Section 3 discusses the existence of periodic solutions of the Hopf type.

## 2 Stability of the Positive Equilibrium

Let $E^* = (x_1^*, x_2^*, p^*)$ be the positive equilibrium point of system (1). The Jacobian matrix evaluated at this equilibrium point is given by

$$J = \begin{pmatrix} n_{11} & n_{12} & 0 \\ n_{21} & n_{22} & n_{23} \\ n_{31} & 0 & n_{33} \end{pmatrix} \tag{2}$$

where

$$n_{11} = \frac{(1-q)m_1(1-x_1^*-x_2^*)}{a_1+1-x_1^*-x_2^*} - 1 - \frac{(1-q)m_1a_1x_1^*}{(a_1+1-x_1^*-x_2^*)^2}, \quad n_{12} = \frac{-x_1^*(1-q)m_1a_1}{(a_1+1-x_1^*-x_2^*)^2}$$
$$n_{21} = \frac{-m_2a_2e^{-\mu p^*}x_2^*}{(a_2+1-x_1^*-x_2^*)^2} + \frac{qm_1(1-x_1^*-x_2^*)}{a_1+1-x_1^*-x_2^*} - \frac{qm_1a_1x_1^*}{(a_1+1-x_1^*-x_2^*)^2}$$
$$n_{22} = \frac{e^{-\mu p^*}m_2(1-x_1^*-x_2^*)}{a_2+1-x_1^*-x_2^*} - 1 - \frac{x_2^*m_2a_2e^{-\mu p^*}}{(a_2+1-x_1^*-x_2^*)^2} - \frac{qm_1a_1x_1^*}{(a_1+1-x_1^*-x_2^*)^2}$$
$$n_{23} = \frac{-\mu e^{-\mu p^*}m_2(1-x_1^*-x_2^*)}{a_2+1-x_1^*-x_2^*}, \quad n_{31} = \frac{-\delta p^*}{K+p^*}, \quad n_{33} = -1 - \frac{\delta K x_1^*}{(K+p^*)^2}.$$

At the equilibrium point $E^*$ the characteristic equation of the Jacobian matrix $J$ takes the form

$$P(\lambda) = \lambda^3 + C_1\lambda^2 + C_2\lambda + C_3 = 0 \tag{3}$$

where

$$\begin{aligned} C_1 &= -(n_{11} + n_{22} + n_{33}) \\ C_2 &= n_{11}(n_{22} - n_{21}) + n_{33}(n_{11} + n_{22}) \\ C_3 &= -n_{11}n_{33}(n_{22} - 21) - n_{12}n_{23}n_{31}. \end{aligned} \tag{4}$$

With $0 < q < 1$, $0 < x_1^* + x_2^* < 1$ and with the rest of the constants positive, we conclude that $C_1 > 0, C_2 > 0$, and $C_3 > 0$.

**Lemma 1. (Routh-Hurwitz stability criterion)** All roots of (3) have negative real part if and only if $C_1 > 0, C_3 > 0, C_1C_2 - C_3 > 0$. Equation (3) has one negative root and a pair of complex roots with a positive real part if and only if $C_1 > 0, C_3 > 0, C_1C_2 - C_3 < 0$.

**Lemma 2.** The positive equilibrium point $E^*$ is stable if $C_1C_2 - C_3 > 0$ and unstable if $C_1C_2 - C_3 < 0$.

## 3 Existence of Periodic Solutions of the Hopf Type

By substituting the entries of (2) to (4), we can easily see that $C_1, C_2$, and $C_3$ are functions of $K$. After some algebraic computations, $C_1C_2 - C_3 = 0$ reduces to

$$A(K) = A_4K^4 + A_3K^3 + A_2K^2 + A_1K + A_0 = 0 \tag{5}$$

where

$$\begin{aligned} A_4 &= (n_{11} + n_{22})((n_{11} + n_{22}) - n_{11}(n_{22} - n_{21}) - 1), \\ A_3 &= 4p^*A_4 + \delta x_1^* A_{00} - \delta p^* n_{12}n_{23}, \\ A_2 &= 6(p^*)^2A_4 + 2\delta p^* x_1^* A_{00} - \delta^2(x_1^*)^2(n_{11} + n_{22}), \\ &\quad -3\delta(p^*)^2 n_{12}n_{23}, \\ A_1 &= 4(p^*)^3A_4 + \delta x_1^*(p^*)^2A_{00} - 3\delta(p^*)^3 n_{12}n_{23}, \\ A_0 &= (p^*)^4A_4 - \delta(p^*)^4 n_{12}n_{23}, \\ A_{00} &= (n_{11} + n_{22})(n_{11} + n_{22} - 2). \end{aligned}$$

If $A_2 > 0$ and $A_1 < 0$ then A(K)=0 has a unique positive solution. We take $K$ as the bifurcation parameter and leave the rest of the parameters fixed.

**Lemma 3.** Let $K^*$ be the positive root of (5). Then one root of (3) is negative and the other two are complex conjugates $u(K) \pm iv(K)$ such that

$$u(K^*) = 0, v(K^*) \neq 0 \quad \text{and} \quad \frac{du}{dK}|_{K=K^*} \neq 0.$$

**Proof** Substituting $\lambda(K) = u(K) + iv(K)$ into (3) and separating real and imaginary parts, we have

$$u^3 - 3uv^2 + C_1(u^2 - v^2) + C_2u + C_3 = 0 \tag{6}$$

$$3u^2v - v^3 + 2C_1uv + C_2v = 0 \tag{7}$$

Since $v(K) \neq 0$, from (7) we get

$$v^2 = 3u^2 + 2C_1u + C_2. \tag{8}$$

Substituting (8) to (6) we obtain

$$u(8u^2 + 8uC_1 + 2(C_1^2 + C_2)) + C_1C_2 - C_3 = 0. \tag{9}$$

Now, (9) has a root $u(K^*)$ if and only if

$$C_1(u(K^*)) \cdot C_2(u(K^*)) - C_3(u(K^*)) = 0. \tag{10}$$

Furthermore, in this case, $u(K^*) = 0$ is the only real root since the roots of

$$8u^2 + 8uC_1 + 2(C_1^2 + C_2) = 0$$

are imaginary. Hence, at $K = K^*$, the roots of (3) are

$$\pm i(C_2)^{1/2}, \quad -C_1,$$

which means that $E^*$ is a center. We will now show that $\frac{d(\mathrm{Re}\lambda(K))}{dK}|_{K=K^*} = \frac{du}{dK}|_{K=K^*} \neq 0$. Differentiating (9) with respect to $K$ gives

$$24u^2\frac{du}{dK} + 16uC_1\frac{du}{dK} + 8u^2\frac{dC_1}{dK}$$
$$+2\frac{du}{dK}(C_1^2 + C_2) + 2u(2C_1\frac{dC_1}{dK} + \frac{dC_2}{dK}) + \frac{d(C_1C_2 - C_3)}{dK} = 0. \tag{11}$$

Evaluating (11) at $K = K^*$, and using the fact that $u(K^*) = 0$, we obtain

$$\frac{du}{dK}(K^*) = \frac{-\frac{d(C_1C_2 - C_3)}{dK}|_{K=K^*}}{2(C_1^2 + C_2)} \neq 0 \tag{12}$$

since

$$\frac{d}{dK}(C_1C_2 - C_3)\ |_{K=K^*} = \frac{(n_{11} + n_{22})(n_{11} + n_{22} + 2n_{33})\delta x_1^*(p^* - K^*)}{(K^* + p^*)^3} + \frac{n_{12}n_{23}\delta p^*}{(K^* + p^*)^2} \neq 0.$$

Lemma 3 shows that the hypothesis of Hopf's theorem are fully satisfied and we have the following

**Theorem 1.** Let $E^*$ be the positive equilibrium of system (1). If $C_j(K)$ at $K = K^*$ satisfy the following conditions

$$\begin{aligned} C_j(K^*) > 0, &\quad j = 1, 2, 3 \\ C_1C_2 - C_3|_{K=K^*} &\quad = 0 \\ \frac{d(C_1C_2 - C_3)}{dK}|_{K=K^*} &\quad \neq 0 \end{aligned}$$

then Hopf bifurcation from $E^*$ occurs at $K = K^*$. As the value of $K$ passes through $K^*$, the periodic solution of the Hopf type of system (1) appears which surrounds $E^*$. Here, $K^*$ is the Hopf bifurcation value satisfying $A(K^*) = 0$.

The existence of periodic solutions of the Hopf type, although a local property, demostrates the case of oscillatory coexistence of the two organisms $x_1$ and $x_2$ in the presence of an inhibitor $p$.

## References


1. G. D'ans *et al*, *IEEE Trans.* AC **62**, 341 (1971).
2. S.B. Hsu *et al*, *J.Math. Biol.* **34**(2), 225 (1995).
3. R.E. Lenski and S. Hattingh in *J.Theor. Bio.* **122**, 83(1986).
4. J.E. Marsden and M. McCracken in *The Hopf bifurcation and its applications*, (Springer, New York, 1976).
5. E.B. Pike and C.R. Curds in *The microbial ecology of activated sludge process*, eds. G. Skykes and F. A. Skinner (Academic Press, New York, 1971).
6. G. Stephanopoulis and G. Lapidus, *Chem. Engr. Science* **43**, 49(1988).
7. P. Waltman, *Rockey Mt. Jour. of Math.* **20**, 777(1990).
8. P. Waltman in *Competition models in population biology* (Philadelphia: Society for Industrial and Applied Mathematics, 1983).

# MODEL-THEORETIC PROPERTIES OF BANACH SPACES[a]

S. Baratella
*Dipartimento di Matematica - Università di Trento*
*via Sommarive 14, 38050 Povo (TN)*
*Italy*
*E-mail: baratell@science.unitn.it*

S.-A. Ng[b]
*School of Mathematics - University of Natal*
*Pietermaritzburg 3209, South Africa*
*E-mail: ngs@nu.ac.za*

We survey some results on Banach spaces and related structures, viewed as first-order structures in the model-theoretic setting. We study some geometric properties defined by the neocompact formulas: back-and-forth, density, quantifier elimination, winning strategy. Various kinds of saturation (compactness) are considered. Included also are applications to definable mappings, fixed point theorem and the Banach-Mazur problem on characterizing Hilbert spaces.

## 1 Introduction

In recent years, the study of Banach spaces has been notably influenced by methods and techniques from other part of mathematics, such as group theory, combinatorics, set theory and mathematical logic. To name two well-known deep results, Gowers dichotomy theorems (Gowers [14] and also Benyamini and Lindenstrauss [7] ) and the construction of Tsirelson space (see Benyamini and Lindenstrauss [7]) are basically applications of combinatorics and set-theory.

Meanwhile application of mathematical logic is less well-known – except the Banach ultraproduct construction, i.e. nonstandard hull constructions (see Dacunha-Castelle and Krivine[11] and Luxemburg[23]), whose foundation actually lies in the model theory of mathematical logic. In a typical use of Banach ultraproduct, one is concerned with explicit classical properties of the original space and no commitment is made to the investigation of general properties of similar structures. We believe that the latter should not be overlooked.

The present short survey paper is aimed at beginning research students and professional researchers who are interested in learning model-theoretic

---

[a] *Mathematics Subject Classification* Primary: 03C10. Secondary: 03H05 03C65 46B08 46B04.
*Keywords*: Model theory, Banach space structure, quantifier elimination, neocompact, back-and-forth, Banach-Mazur problem.
[b] Supported by South African NRF 2039556

properties of Banach spaces and their applications. This is also a good place for those seeking new research problems. Apparently, the present subject is still in its infancy, only the future will tell whether the direction we take here is the correct one. The choice of topics is neither complete nor balanced, in fact it only reflects those results (both published previously and new ones) that we are interested in. We made little reference to the earlier important investigation by Henson and Iovino on the same subject from a different approach, but we recommend the interested readers to Iovino [21].

In the §2, we first consider Banach spaces as a more general class of structures in an appropriate first order logic. Then we consider properties in the *neocompact* language (Definition 2) which is roughly an approximate higher order language. This language was first introduced by Fajardo and Keisler [12] in abstract settings, but we apply it in formulating concrete geometric properties of Banach spaces. For example, in Theorem 6 we consider the *back-and-forth* property and its relevance to quantifier elimination — a well-known topic in classical model theory. In finite dimensional spaces, back-and-forth is the same as transitivity. But they are different in general spaces. In fact, from model-theoretic point of view, back-and-forth is a more natural notion. Another natural property is *density*. These properties can be used to describe the shape of the unit sphere, and hence the geometry of the Banach space. We prove a fixed point result from it (Proposition 14).

Saturation is a prominent notion in model theory. In our context, there are several natural definitions of saturation (Definition 3). They can be taken as notions of compactness for Banach spaces. Saturation is responsible for the richness of Banach ultraproducts (nonstandard hulls). We give some examples of other usages such as quantifier elimination and back-and-forth.

Hilbert spaces satisfy quantifier elimination (Theorem 10). It is quite possible that it already characterizes Hilbert spaces. (See Question 11.)

In general, neocompact definable mappings may be hard to describe. But under quantifier elimination, they could only be the trivial maps (Theorem 17). This is yet further evidence that quantifier elimination represents the good behaviour of Hilbert spaces.

For the remainder of this paper, we consider game-theoretic type properties such as *winning strategy* (Definition 19) and its relation to the open problem of Banach-Mazur: *Is every separable transitive Banach space a Hilbert space?* Winning strategy is a weaker notion than transitivity (Proposition 20). It is unusual in being a self-dual notion in reflexive spaces (Theorem 23). In Theorem 21, we give the equivalence of reflexivity and Radon-Nikodým property under some weak back-and-forth assumption. We end our paper by a list of questions related to the Banach-Mazur problem.

For all the unexplained notions in functional analysis, see Beauzamy [6], Benyamini and Lindenstrauss [7] or any classical book on functional analysis. For the model theory notions see Chang and Keisler [10] or Hodges [20].

## 2 Preliminaries and general results

From the point of view of first order logic, we can view a Banach space $\mathcal{M}$ over the real numbers as a *structure* of the form

$$\mathcal{M} = \langle M, +, 0, (r_-)_{r\in\mathbf{Q}}, \|\ \|\rangle,$$

where
- $r_-$ is the operation of multiplication by the rational scalar $r$;
- $\langle M, +, 0, (r_-)_{r\in\mathbf{Q}}\rangle$ is the underlying vector space structure over the rational numbers;
- $\|\ \|$ is the norm.

In applications, a Banach space may be equipped with additional functions, for instance a family of seminorms, lattice operations in a Banach lattice, multiplication operation in a Banach algebra, inner product in a Hilbert space... So, in general, we deal with a structure

$$\mathcal{N} = \langle \mathcal{M}, (f_i)_{i\in I}\rangle,$$

where $\mathcal{M}$ is a Banach space and $(f_i)_{i\in I}$ is a countable family of functions from finite powers of $\mathcal{M}$ to $\mathcal{M}$ or to $\mathbf{R}$ that are uniformly continuous on each bounded set.

We will say that $\mathcal{N}$ is a *Banach space structure*. A detailed description of these structures (in a slightly more general setting) can be found in Iovino [21].

The language for $\mathcal{N}$ is the same one adopted by Baratella and Ng [2] (see also Iovino [21]). Here we recall that we work in the language of vector spaces over the rationals, augmented by symbols for the rational bounds of the norm. Furthermore, for each vector-valued $f_i$, we simply add the symbol $f_i$ and, for each real-valued $f_i$, we include function symbols for rational bounds of values of $f_i$.

In the *pure language* (i.e. the language adequate for a structure like $\mathcal{M}$ above), the typical term $t(x_1, \ldots, x_n)$ has the form $\sum_{i=0}^{n} r_i x_i$, where the $x_i$'s are variables and the $r_i$'s are rationals.

Among the simplest formulas that we can write in the above language are those of the form $r \leq \|t(x_1, \ldots, x_n)\| \leq s$, for some rational numbers $r, s$.

For example, a unit vector is expressed by $1 \leq \|x\| \leq 1$. But in order to express irrational norm such as $\sqrt{2}$, we are forced to use an infinitary formula such as $\bigwedge(r \leq \|x\| \leq s)$, where the conjunction is taken over all $r, s \in \mathbf{Q}$, $r < \sqrt{2} < s$.

Consequently, it is natural to work with a family of infinitary expressions called *neocompact formulas* and with its subfamily of finitary *positive bounded formulas* (see Baratella and Ng[2]). *Neocompact formulas* (briefly: *nc*) have been introduced in an abstract framework by Keisler[22]. Positive bounded formulas (briefly: *pb*) are due to Henson[17]. We recall now the recursive definition of *pb* formula. (As a technical convenience, we assume that *in all the formulas all the variables are bound in norm to the closed unit ball.*)

When we refer to the the norm of a tuple, we mean the *supremum norm*.

**Definition 1** *Let $\mathcal{L}$ be the language for some Banach space structure. $\mathcal{L}$-terms are constructed as in the first order case, by using function symbols for vector-valued functions only. We define the class of pb formulas over $\mathcal{L}$ as follows:*

1. *If $t$ is a term and $r$ is a rational number, then*

$$\|t\| \leq r \quad \textit{and} \quad \|t\| \geq r \quad \textit{are pb formulas.}$$

2. *If $f_i$ is an $m$-ary real-valued function, $t_1, \ldots, t_m$ are terms and $r$ is a rational number, then*

$$f_i(t_1, \ldots, t_m) \leq r \quad \textit{and} \quad f_i(t_1, \ldots, t_m) \geq r \quad \textit{are pb formulas.}$$

3. *If $\theta$ and $\psi$ are pb formulas, then $\theta \vee \psi$ and $\theta \wedge \psi$ are pb formulas.*

4. *If $\psi$ is a pb formula and $\theta$ is a quantifier free pb formula, then*

$$\exists \bar{y} \psi(\bar{x}, \bar{y}) \quad \textit{and} \quad \forall \bar{y}(\theta(\bar{y}) \rightarrow \psi(\bar{x}, \bar{y})) \quad \textit{are pb formulas.}$$

The formulas without quantifiers are called *quantifier free pb* (briefly: *qfpb*). We also recall the definition of *nc* formula.

**Definition 2** 1. *Atomic neocompact formulas have the form*

$$\bigwedge_{i \in \omega} \phi_i(\bar{x}),$$

*where each $\phi_i$ is qfpb and $\bar{x}$ is a finite list of variables.*

2. *If $\phi$ and $\psi$ are nc, then $\phi \vee \psi$ is nc.*

3. *If* $\{\phi_i\}_{i<\omega}$ *is a countable family of nc in finitely many variables, then their conjunction* $\bigwedge_{i<\omega} \phi_i$ *is nc.*

4. *If* $\theta$ *is nc, then* $\exists \bar{y}\theta$ *is nc.*

5. *If* $\theta$ *is atomic nc and* $\psi$ *is nc, then* $\forall \bar{y}(\theta(\bar{y}) \to \psi(\bar{y}, \bar{x}))$ *is nc.*

Let $\mathcal{M}$ be a Banach space structure and let $\bar{a} \in \mathcal{M}^n$. The *(qf)pb type* of $\bar{a}$ in $\mathcal{M}$ is the set of (quantifier free) *pb* formulas without parameters that are true of $\bar{a}$ in $\mathcal{M}$. We will use the notation $\mathrm{tp}^{\mathcal{M}}(\bar{a})$ for the *qfpb* type and we will drop reference to the structure, when there is no ambiguity.

An *n-type* is a maximal consistent set of *qfpb* formulas in $n$ free variables.

We call the Banach space structures in the pure language simply *Banach spaces.* A *Hilbert space* is a Banach space whose norm is given by an inner product. Notice that in a Hilbert space, the inner product does not appear explicitly as a function of the structure. Indeed it is a remark in Baratella and Ng [3] that if $\mathcal{M}$ is a Hilbert space in the pure language and $\mathcal{M}^+$ is the expansion of $\mathcal{M}$ to a language with a symbol for the inner product then for any type $p(\bar{x})$ in the pure language, the set $q(\bar{x})$ of *qfpb* formulas in the expanded language that are implied by $p(\bar{x})$ is a type, and the two types define the same set in the corresponding structures.

By $\mathcal{B}_{\mathcal{M}}$ we denote the closed unit ball of $\mathcal{M}$ (we simply write $\mathcal{B}$ when it is safe). We use $\mathcal{S}_{\mathcal{M}}$ (or $\mathcal{S}$) for the unit sphere.

We are interested in Banach space structures being "sufficiently rich" in the sense made precise by the definition below:

**Definition 3** *We say that a Banach space structure* $\mathcal{M}$ *is* pb*-saturated if every finitely satisfiable set in* $\mathcal{M}$ *of* pb *formulas in finitely many free variables and finitely many parameters is satisfiable in* $\mathcal{M}$*. Similarly for* qfpb- *and* nc-*saturation.*

*We say that* $\mathcal{M}$ *is* $\forall$qfpb-*saturated (*$\forall$nc-*saturated) if it is saturated for sets made of* qfpb *and one universal* pb*(*nc*) in finitely many free variables and parameters.*

*When we refer to saturation for sets of formulas without parameters, we make the index* $_0$ *follow the appropriate kind of formulas (so* $\mathrm{pb}_0$ *saturation means saturation for sets of* pb *formulas without parameters).*

It is easy to show that finite dimensional Banach space structure and Hilbert spaces are *nc*-saturated (see Baratella and Ng [2] for the details).

Keeping in mind that in every formula all free variables are bound to the closed unit ball, we formulate the following

**Definition 4** *(Formulas are assumed to be parameter-free.) Let $\mathcal{M}$ be a Banach space structure in the language $\mathcal{L}$. We say that $\mathcal{M}$ has Quantifier Elimination (briefly: QE) if, for every $n$, every subset of $\mathcal{B}^n$ definable in $\mathcal{M}$ by a nc formula in $\mathcal{L}$ is also definable by an atomic nc in $\mathcal{L}$, i.e. it is $\Pi^0_1$ over the qfpb formulas.*

Keisler[22] has proved quantifier elimination theorems in a very general setting. As a consequence we have that, in Banach space structures satisfying certain conditions, every set defined by a *nc* also admits a simpler definition by means of a countable conjunction of *qfpb*. The relevant conditions are singled out in Definition 4.1[2], where equivalent formulations of Keisler's properties in the setting of Banach space structures are provided.

In particular, we recall the following:

**Definition 5** *Let $\mathcal{M}$ be a Banach space structure.*

1. *$\mathcal{M}$ is* dense *if, for each $m$, any two $m$-tuples which satisfy the same* qfpb *formulas in $\mathcal{M}$ also satisfy the same* existential formulas *in $\mathcal{M}$. (Recall that an existential formula is obtained by existential quantification from a* qfpb *formula.)*

2. *$\mathcal{M}$ has the* back-and-forth property *(briefly: BF) if, whenever* $\text{tp}(\bar{a}_1) = \text{tp}(\bar{a}_2)$ *and $\bar{b}_1 \in \mathcal{B}^m$ for some $m$, then there is $\bar{b}_2$ such that* $\text{tp}(\bar{a}_1, \bar{b}_1) = \text{tp}(\bar{a}_2, \bar{b}_2)$.

3. *A sequence* $\text{tp}(\bar{a}_i)$ *of types converges to a type $p$ (notation:* $\text{tp}(\bar{a}_i) \to p$*) if all* qfpb *$\theta$ that appear infinitely often in the sequence are also in $p$. (See the topology defined by Keisler[22] on the set of types and Definition 4.1[2].)*

4. *$\mathcal{M}$ has the* strong open mapping property *(briefly: SOM) if for every $\bar{a} \in \mathcal{M}^n$, for any sequence* $\text{tp}(\bar{a}_i)$ *such that* $\text{tp}(\bar{a}_i) \to \text{tp}(\bar{a})$ *and $\bar{b} \in \mathcal{B}^m$, for some $m$, there is a sequence $\bar{c}_i \in \mathcal{M}^n$, such that* $\text{tp}(\bar{a}_i) = \text{tp}(\bar{c}_i)$ *and* $\text{tp}(\bar{c}_i, \bar{b}) \to \text{tp}(\bar{a}, \bar{b})$.

We are interested in Banach space structures satisfying the above properties because of the following (see Baratella and Ng[2]):

**Theorem 6** *For a* $\forall \text{nc}_0$*-saturated Banach space structure* $\mathcal{M}$ *the following are equivalent:*

1. $\mathcal{M}$ *has BF;*

2. $\mathcal{M}$ *has BF and SOM;*

3. $\mathcal{M}$ *has QE.*

The next result shows how *BF* and *SOM* are related to density and to various degrees of saturation.

**Lemma 7** *Let* $\mathcal{M}$ *be a* $\text{qfpb}_0$*-saturated Banach space structure. Then in* $\mathcal{M}$

1. *BF if and only if* qfpb*-saturation and density;*

2. *QE if and only if* $\forall \text{nc}_0$*-saturation,* qfpb*-saturation and density.*

In the sequel we deal with *nonstandard hulls* of Banach space structures. (Banach ultraproducts form a special subclass of such spaces.) First we briefly recall the nonstandard hull construction. We recall that we always work in nonstandard universes that are at least $\omega_1$-saturated. Now fix a nonstandard universe. Let $\mathcal{M}$ be an internal Banach space structure. Let $\| \ \|$ be the norm and let $\text{fin}(\mathcal{M})$ be the $\mathbf{R}$-vector subspace of norm-finite elements of $\mathcal{M}$. Let $\mathcal{M}_0$ denote the Banach space part of $\mathcal{M}$, i.e. the reduct to the pure language. Recall that the nonstandard hull $\widehat{\mathcal{M}_0}$ of $\mathcal{M}_0$ is, as $\mathbf{R}$-vector space, the quotient of $\text{fin}(\mathcal{M})$ modulo the $\mathbf{R}$-vector subspace of elements having infinitesimal norm. We denote by $[\bar{a}]$ the image of a tuple $\bar{a}$ of elements of $\text{fin}(\mathcal{M})$ under the canonical homomorphism $\pi : \text{fin}(\mathcal{M}) \to \widehat{\mathcal{M}}$ and let $^\circ : \text{fin}({}^*\mathbf{R}) \to \mathbf{R}$ be the *standard part* map. The norm of $[a]$ given by $^\circ\|a\|$ turns $\widehat{\mathcal{M}_0}$ into a Banach space (topological completeness is ensured by saturation, even if we start from an internal normed space only). Furthermore, if the extra functions in $\mathcal{M}$ are $S$-continuous and $S$-bounded on bounded sets, then as in Keisler [22], if we include also the standard part (defined in the natural way) of extra functions from $\mathcal{M}$, then we get a Banach space structure (still denoted by $\widehat{\mathcal{M}}$): the nonstandard hull of the Banach space structure $\mathcal{M}$.

If $\mathcal{M}$ is an ordinary Banach space structure (like those introduced at the beginning of this section), we write $\widehat{\mathcal{M}}$ for $\widehat{{}^*\mathcal{M}}$ and, as customary, we say that $\widehat{\mathcal{M}}$ is the nonstandard hull of the Banach space structure $\mathcal{M}$.

There is a canonical isometric embedding of an ordinary Banach space structure $\mathcal{M}$ into its nonstandard hull $\widehat{\mathcal{M}}$. We always identify elements of $\mathcal{M}$ and their images under such an embedding.

The following is a straightforward corollary of Lemma 7.

**Corollary 8** *The nonstandard hull of an internal Banach space structure has $QE$ if and only if it is dense and* $\forall$nc$_0$*-saturated.*

In Baratella and Ng [2] it is proved that if $\mathcal{M}$ is is an ordinary Banach space structure that is *qfpb*-saturated and if its nonstandard hull has $QE$ then $\mathcal{M}$ also has $QE$. They also proved that $QE$ is invariant with respect to different nonstandard hull constructions.

**Theorem 9** *Let $\widehat{\mathcal{M}}$ and $\tilde{\mathcal{M}}$ be nonstandard hulls of a Banach space structure $\mathcal{M}$. Then*

$$\widehat{\mathcal{M}} \text{ has } QE \;\Leftrightarrow\; \tilde{\mathcal{M}} \text{ has } QE.$$

Furthermore, in Baratella and Ng [2] $QE$ is compared with a notion of quantifier elimination introduced by Henson (briefly: $QE_H$). $QE_H$ is related to the *approximate satisfiability* of Henson [17], which is weaker than first order satisfiability. $Q_E$ and $QE_H$, although different, agree on nonstandard hulls of internal Banach spaces that are $\forall nc_0$-saturated.

In an arbitrary Banach space structure, for any rational $\lambda \geq 1$ we define the relation of $\lambda$-*equivalence* on tuples $\bar{a} = (a_1, \ldots, a_n)$ and $\bar{b} = (b_1, \ldots, b_n)$ as follows:

$$\bar{a} \sim_\lambda \bar{b} \quad \Leftrightarrow \quad \lambda^{-1}\|\sum \alpha_i a_i\| \leq \|\sum \alpha_i b_i\| \leq \lambda \|\sum \alpha_i a_i\| \quad \text{for all } \alpha \in \mathbf{Q}^n.$$

In the pure language it is not difficult to show that the properties of *BF* and *SOM* are equivalent to the following:

*BF*: Given $\|a_i\| = 1$, $i \leq n+1$ and $\|b_i\| = 1$, $i \leq n$, if $(a_1, \ldots, a_n) \sim_1 (b_1, \ldots, b_n)$ then there is $b_{n+1}$ such that $(a_1, \ldots, a_{n+1}) \sim_1 (b_1, \ldots, b_{n+1})$.

*SOM*: For any $n$-tuples $\bar{a}$, $(\bar{a}_k)_{k<\omega}$ and $m$-tuple $\bar{b}$, if $\|\sum_{i=1}^n \alpha_i a_{ki}\| \to \|\sum_{i=1}^n \alpha_i a_i\|$ for any $\alpha \in \mathbf{Q}^n$, then there are $n$-tuples $(\bar{c}_k)_{k<\omega}$ such that $\bar{c}_k \sim_1 \bar{a}_k$ for all $k < \omega$ and $\|\sum_{i=1}^n \alpha_i c_{ki} + \sum_{i=1}^m \beta_i b_i\| \to \|\sum_{i=1}^n \alpha_i a_i + \sum_{i=1}^m \beta_i b_i\|$, for any $\alpha \in \mathbf{Q}^n$ and $\beta \in \mathbf{Q}^m$.

In the sequel we will frequently use the above characterizations.

A natural question concerns the existence of Banach space structures that satisfy $QE$. In Baratella and Ng[3] it is proved that Hilbert spaces have $QE$ :

**Theorem 10** *Hilbert spaces satisfies QE.*

Here is a sketch of the proof: since Hilbert spaces are $\forall nc_0$-saturated, by Theorem 6 it suffices to show that they have $BF$. Note that the condition $(a_1, \ldots, a_n) \sim_1 (b_1, \ldots, b_n)$ in $BF$ is equivalent to saying that the spaces $\text{span}\{a_1, \ldots, a_n\}$ and $\text{span}\{b_1, \ldots, b_n\}$ are linearly isometric. Furthermore, each Hilbert space $\mathcal{H}$ is *transitive* in a very strong sense: any surjective linear isometry between subspaces of $\mathcal{H}$ can be extended to a surjective linear isometry of the full $\mathcal{H}$. $BF$ follows immediately.

Indeed, among finite dimensional Banach spaces, $BF$ characterizes precisely those that are Hilbert. For, in the *finite dimensional* case, the following hold:

1. $BF$ is equivalent to *transitivity*, namely to the property that, for each pair of unit vectors, there is a linear isometric isomorphism of the space that maps one to the other. (The equivalence also holds in the separable case.)

2. Transitivity is equivalent to the space being Hilbert (for this equivalence and more about transitivity, see the survey paper by Cabello Sánchez[8]).

**Question 11** *Does $QE$ characterize Hilbert spaces? A related question is whether there exist* $\forall nc_0$*-saturated Banach spaces (or Banach space structures) that are infinite dimensional and not Hilbert.*

## 3 Consequences of $QE$, $BF$ and related properties

In this section we provide a series of examples showing how the properties introduced in the previous section have a wide range of consequences and applications.

### *3.1 A consequence on the geometry of a Banach space*

From Keisler[22], we have that, for each Banach space structure

$$QE \Rightarrow BF \Rightarrow \text{density}$$

**Remark 12** *The above chain of implications can be used to disprove $QE$. For instance if on the unit sphere of a Banach space $\mathcal{M}$ there are an* extreme *and a* nonextreme *point, $\mathcal{M}$ is not dense. (Note that, in the pure language, two vectors satisfy the same parameter-free* qfpb *formulas if and only if they have the same norm. Moreover being a nonextreme point can be expressed by an existential formula.) Hence $\mathcal{M}$ does not have $QE$.*

Properties like those introduced in Definition 5 have relevant consequences on the geometry of a Banach space structure. In particular the following is proved by Baratella and Ng [3]:

**Proposition 13** *For a* $\text{qfpb}_0$ *saturated Banach space $\mathcal{M}$ that is dense exactly one of the following holds:*

1. *$\mathcal{M}$ is reflexive and all points on the unit sphere are extreme.*
2. *$\mathcal{M}$ is nonreflexive and none of the points on the unit sphere is extreme.*

*Moreover both cases do actually occur.*

The proof of the above dichotomy goes as follows: in the reflexive case it suffices to show that there exists an extreme point on the unit sphere (see Remark 12). This is ensured by the Krein-Milman theorem and the weak compactness of the unit ball. Any Hilbert space is an example of such a structure.

If $\mathcal{M}$ is nonreflexive, then it is not uniformly convex. By refuting uniform convexity one gets a finitely satisfiable set of *qfpb* formulas and an application of $qfpb_0$-saturation yields a nonextreme point. An example of such a structure is the nonstandard hull $\widehat{\mathcal{M}}$ of an atomless $L_1$-space $\mathcal{M}$: $\widehat{\mathcal{M}}$ is an $L_1$ space (see Henson [17]), hence nonreflexive, and it is dense because it has $BF$. The latter claim follows from the results proved in Baratella and Ng [3], building on work of Henson [18].

### *3.2 A consequence on the existence of fixed points*

Fixed points are important for the study of differential equations and also for its own right. The next proposition is just an instance of a well-known theorem in fixed point theory. Nevertheless we state it and provide a sketch of the proof in order to show how it relates to the previous results.

**Proposition 14** *Let $\mathcal{M}$ be a reflexive nonstandard hull that is dense. Then $\mathcal{M}$ has the fixed point property for nonexpansive selfmaps on bounded closed convex sets.*

Here is a sketch of the proof: for reflexive nonstandard hulls, the above fixed point property holds if and only if it holds on the unit sphere (see Theorem 3.2 in Baratella and Ng[4]). If $\mathcal{M}$ does not have the fixed point property then there is a non-singleton bounded closed convex set on the unit sphere. Therefore there is no extreme point on the unit sphere, and hence $\mathcal{M}$ is nonreflexive by Proposition 13.

Let us explain why the previous proposition is just an instance of a well-known theorem. It turns out that a dense nonstandard hull $\mathcal{M}$ is reflexive if and only if it is uniformly convex.

First of all we recall that a nonstandard hull is reflexive if and only if it is superreflexive (see Henson and Moore[19]). It is a deep result of James that superreflexivity is equivalent to renormability with an equivalent uniformly convex norm. So one implication is clear.

In order to prove that reflexivity implies uniform convexity, note that $\mathcal{M}$ is strictly convex, by Proposition 13. Then, by saturation, $\mathcal{M}$ is uniformly convex.

Therefore the previous proposition follows from the result obtained independently by Browder and Göhde that nonexpansive selfmaps on bounded closed convex sets in uniformly convex Banach spaces have fixed points.

### *3.3 Consequences on definability of maps*

We are interested in studying properties of maps that are definable in a Banach space by means of *qfpb* or neocompact formulas. All the results stated in the sequel are proved in Baratella and Ng[3].

For $n$-tuple $\bar{b}$ in a Banach space $\mathcal{M}$, we often write $\bar{b} \in \mathcal{M}$ for $\bar{b} \in \mathcal{M}^n$.

We say that a *qfpb* formula $\theta$ is *qfpb*$^+$ if the scalars used in $\theta$ are nonnegative.

**Definition 15** *Let $\mathcal{M}$ be a Banach space.*

1. *A subset $A$ of $\mathcal{M}^n$ is* qfpb-definable *in $\mathcal{M}$ if there exists a set $\Theta(\bar{x})$ of parameter-free qfpb-formulas in $n$ free variables such that $A = \{\bar{a} \in \mathcal{M}^n : \mathcal{M} \models \bigwedge \Theta(\bar{a})\}$.*

2. *A function $f$ is* qfpb-definable *in $\mathcal{M}$ if its graph is qfpb-definable in $\mathcal{M}$.*

*The notions of qfpb$^+$- and nc-definability are formulated with reference to qfpb$^+$ and nc formulas, respectively.*

Note that *qfpb*-definability is the same as definability by means of an atomic neocompact formula. Moreover, in a Banach space with $QE$, *nc*-definability is the same as *qfpb*-definability.

It can proved that a *qfpb*-definable function $f$ defined on the unit ball $\mathcal{B}$ of a Banach space $\mathcal{M}$ is continuous. Furthermore if $\mathcal{M}$ is *qfpb*$_0$-saturated then $f$ is uniformly continuous.

We let $\pi : \mathcal{M} \setminus \{0\} \to \mathcal{S}$, $x \mapsto x/\|x\|$, i.e. the projection onto the unit sphere.

The following yields an equivalent formulation of strict convexity in terms of $qfpb^+$-definability.

**Proposition 16** *The map $\pi$ is qfpb-definable in all Banach spaces. Furthermore the following are equivalent for a Banach space $\mathcal{M}$:*

1. *$\mathcal{M}$ is strictly convex;*

2. *$\pi$ is qfpb$^+$ definable in $\mathcal{M}$.*

We can prove that, under some conditions, definable functions are continuous radial functions of the form $f(x) = \phi(\|x\|)x$.

**Theorem 17** *The following are equivalent in Banach space $\mathcal{M}$ that has QE:*

1. *The only nc-definable selfmaps on $\mathcal{S}$ are* $\pm$id.

2. *The nc-definable function on $\mathcal{B} \setminus \{0\}$ are exactly the continuous functions of the form $f(x) = \phi(\|x\|)x$, where $\phi : ]0,1] \to \mathbf{R}$ is continuous.*

It is not difficult to show that in a Hilbert space $\mathcal{H}$ the only *qfpb*-definable selfmaps on $\mathcal{S}$ are $\pm$id. Therefore the *nc*-definable on $\mathcal{B} \setminus \{0\}$ are exactly those described in the previous theorem.

We finish this subsection with the following

**Conjecture 18** *In a Banach space $\mathcal{M}$ with QE all qfpb-definable selfmaps on $\mathcal{S}$ are given by* $\pm$id. *Consequently, by Theorem 17, the nc-definable function on $\mathcal{B} \setminus \{0\}$ are exactly the continuous functions of the form $f(x) = \phi(\|x\|)x$, where $\phi : ]0,1] \to \mathbf{R}$ is continuous.*

### *3.4 Consequences on a game-theoretic notion*

In the sequel all isomorphisms are assumed to be *linear isomorphisms.*

We recall from § 2 that a Banach space $\mathcal{M}$ is *transitive* if for each pair of unit vectors $a$ and $b$ there exists an isometric isomorphism $T : \mathcal{M} \to \mathcal{M}$ such that $T(a) = b$.

The following old problem was posed by Banach and Mazur (see Banach[1]):

*Is every separable transitive Banach space necessarily a Hilbert space?*

As far as we know, this problem remains unsolved. (Ref. Guerrero and Palacios[16] for the latest development.) In this section we formulate and study a new property which is equivalent to transitivity in separable spaces but weaker in general. Such property may provide more understanding into the Banach and Mazur problem. We call this property *WS*, for having a winning strategy in the game of extending partial isometries. More specifically, suppose from the previous moves of a play in the game, we have two $n-$tuples $\bar{a}$ and $\bar{b}$ of vectors in a Banach space structure, so that there is an isometric isomorphism $T : \text{span}\{a_1, \ldots, a_n\} \to \text{span}\{b_1, \ldots, b_n\}$ such that $T(a_i) = b_i$, $i = 1, \ldots, n$. Now suppose there is given a new element $a_{n+1}$, we would like to respond by an element $b_{n+1}$ so that $T$ is extendable to an isometric isomorphism that takes $a_{n+1}$ to $b_{n+1}$. The motivation behind the proposing of such property comes from similar consideration in model theory of first order logic, as mentioned in Hodges[20], and hence it is related to the Banach space model theory investigated by Henson[17] and Iovino[21] as well as to the $BF$ property presented in § 2.

For simplicity we work with Banach spaces, but some of the results can be generalized to Banach space structures. In the sequel we prove that $WS$ is weaker than transitivity. We also give a sketch of the proof that $WS$ is a self-dual property for reflexive spaces.

We refer the reader to Baratella and Ng[5] for all the missing proofs.

Note that, given two Banach spaces $\mathcal{M}$ and $\mathcal{N}$ and $n$-tuples $\bar{a}$ and $\bar{b}$ from $\mathcal{M}$ and $\mathcal{N}$ respectively, the following are equivalent:

1. $\text{tp}^{\mathcal{M}}(\bar{a}) = \text{tp}^{\mathcal{N}}(\bar{b})$;
2. there exists an isometric isomorphism

$$T : \text{span}\{a_1, \ldots, a_n\} \to \text{span}\{b_1, \ldots, b_n\}$$

   such that $T(a_i) = b_i$, $i = 1, \ldots, n$.

In § 2 we abbreviated $\mathrm{tp}^{\mathcal{M}}(\bar{a}) = \mathrm{tp}^{\mathcal{N}}(\bar{b})$ by $\bar{a} \sim_1 \bar{b}$, when $\mathcal{M} = \mathcal{N}$.
We now give the main definitions.

**Definition 19** *Let $\mathcal{M}$ and $\mathcal{N}$ be Banach spaces and let $a \in \mathcal{M}$ and $b \in \mathcal{N}$ be unit vectors.*

1. *The* game $\Gamma[(\mathcal{M},a),(\mathcal{N},b)]$ *is a 2-player game. We define a* play *of the game by describing the $n+1$-th move of each of the two players. Suppose after $n$ moves we have $\mathrm{tp}^{\mathcal{M}}(a,a_1,\ldots,a_n) = \mathrm{tp}^{\mathcal{N}}(b,b_1,\ldots,b_n)$, then player $\forall$ chooses some $a_{n+1} \in \mathcal{M}$ and player $\exists$ responses by choosing some $b_{n+1} \in \mathcal{N}$, or $\forall$ chooses some $b_{n+1} \in \mathcal{N}$ and $\exists$ responses by choosing some $a_{n+1} \in \mathcal{M}$.*

2. *Player $\exists$ wins the play at step $n+1$ if*

$$\mathrm{tp}^{\mathcal{M}}(a,a_1,\ldots,a_n,a_{n+1}) = \mathrm{tp}^{\mathcal{N}}(b,b_1,\ldots,b_n,b_{n+1}).$$

3. *Player $\exists$ has a* winning strategy *for $\Gamma[(\mathcal{M},a),(\mathcal{N},b)]$ if, in each play, $\exists$ can win at every finite step $n$, no matter how $\forall$ plays.*

4. *The Banach space $\mathcal{M}$ has* winning strategy *(briefly: $WS$) if, whenever $a, b$ are unit vectors, $\exists$ has a winning strategy for $\Gamma[(\mathcal{M},a),(\mathcal{M},b)]$.*

We say that a Banach space $\mathcal{M}$ has $BF_n$ if, whenever $(a_1,\ldots,a_n) \sim_1 (b_1,\ldots,b_n)$ and $a_{n+1} \in \mathcal{M}$, there is $b_{n+1} \in \mathcal{M}$ such that $(a_1,\ldots,a_{n+1}) \sim_1 (b_1,\ldots,b_{n+1})$. So $\mathcal{M}$ has $BF$ if and only if it has $BF_n$ for all positive $n$.

We now remark some elementary facts.

- Note the main difference between $BF$ and $WS$: in the latter, the partial isometry before the next move in a play is not arbitrary; it must be constructed from previous moves.

- If a Banach space $\mathcal{M}$ has $WS$, then for all unit vectors $a, b$ there exists a separable subspace $\mathcal{M}_0 \subset \mathcal{M}$ and an isometric isomorphism $T : \mathcal{M}_0 \to \mathcal{M}_0$ such that $T(a) = b$. For, from $WS$, we get $\{a_i\}_{i\in\omega}$ and $\{b_i\}_{i\in\omega}$ such that $a_0 = a$, $b_0 = b$, $\{a_i\}_{i\in\omega} = \{b_i\}_{i\in\omega}$ and $(a_0,\ldots,a_n) \sim_1 (b_0,\ldots,b_n)$ for all $n \in \omega$. Then we let $\mathcal{M}_0$ be the closed linear span of $\{a_i\}$, and $T : \mathcal{M}_0 \to \mathcal{M}_0$ given by $T(a_i) = b_i$.

- A Banach space $\mathcal{M}$ has $BF$ if and only if, for all tuples $\bar{a}$ and $\bar{b}$ such that $\bar{a} \sim_1 \bar{b}$, player $\exists$ has a winning strategy in $\Gamma[\mathcal{M},\bar{a}),(\mathcal{M},\bar{b})]$. (Here the reader can easily provide the missing definitions.)

- $BF \Rightarrow WS \Rightarrow BF_1$.
- Transitivity implies $WS$ and the two are equivalent in separable Banach spaces.
- $BF$ and $WS$ are not preserved by equivalent renorming. Yet they are preserved by renorming induced by a linear homeomorphism.

We now prove that $WS$ is weaker than transitivity.

**Proposition 20** *$WS$ does not imply transitivity.*
**Proof** Let

$$X = L_p([0,1]^{\omega_1}) \qquad Y = \oplus_{\alpha<\omega_1} L_p([0,1]^{\omega_2}) \qquad Z = X \oplus Y,$$

where by $\oplus$ we denote $p$-sum and $1 \le p < \infty$, $p \ne 2$. It is proved by Greim *et al.*[15] that $Z$ is not transitive. Nevertheless we claim that it has $WS$.

Note that each element $[\theta] \in L_p([0,1]^\kappa)$, as an equivalence class of measurable functions, is determined by a countable set of coordinates (since $L_p([0,1]^\kappa)$ is generated by characteristic functions of Borel sets). More precisely, for $A \subset \kappa$, let $\pi_A$ denote the projection of $[0,1]^\kappa$ onto $[0,1]^A$ via $(x_i)_{i<\kappa} \mapsto (x_i)_{i\in A}$. Then for any $[\theta] \in L_p([0,1]^\kappa)$, there is a countable $A \subseteq \kappa$ and $[\psi] \in L_p([0,1]^A)$ such that $[\theta] = [\psi \circ \pi_A]$.

Note also that $X \subset L_p([0,1]^{\omega_2})$ as a subspace via $[\theta] \mapsto [\theta \circ \pi_{\omega_1}]$. In particular, as a subspace, $Z \subset Y \simeq L_p([0,1]^{\omega_2}) \oplus Y$.

We denote by $G(Y)$ the group of isometric isomorphisms of $Y$ onto itself. Again by Greim *et al.*[15], we have that $G(Y)$ acts transitively on the unit sphere of $Y$.

In order to prove that $Z$ has $WS$ let $a_1, b_1 \in Z$ (denoting both functions and equivalence classes) with $a_1 \sim_1 b_1$ and let $T_1 \in G(Y)$ be such that $T_1(a_1) = b_1$.

Assume now inductively that $(a_1, \ldots, a_n) \sim_1 (b_1, \ldots, b_n)$ and $T_1, \ldots, T_n$ in $G(Y)$ are constructed with $a_i, b_i \in Z$ and $T_j(a_i) = b_i$, $i \le j \le n$. Write $a_i = a_i^1 \oplus a_i^2$ and $b_i = b_i^1 \oplus b_i^2$, where $a_i^1, b_i^1 \in X \subset L_p([0,1]^{\omega_2})$ and $a_i^2, b_i^2 \in Y$. Now let $a_{n+1} \in Z \subset Y$ be given and let $b = T_n(a_{n+1})$. Regard $b \in L_p([0,1]^{\omega_2}) \oplus Y$, i.e. we can decompose $b = c \oplus d$, where $c \in L_p([0,1]^{\omega_2})$ and $d \in Y$. Let $A \subset \omega_2$ be a countable set of coordinate indices on which $c$ depends on.

Let $B \subset \omega_1$ be a countable set of coordinate indices on which $a_1^1, \ldots, a_n^1$ and $b_1^1, \ldots, b_n^1$ depend on and let $\rho : \omega_2 \to \omega_2$ be a bijection that fixes $B$ pointwise and $\rho(A) \subset \omega_1$. Let $\hat{\rho} : [0,1]^{\omega_2} \to [0,1]^{\omega_2}$ be the corresponding bijection.

Let $T_{n+1} \in G(Y)$ be defined as follows: for all $a \in Y$, regarding $T_n(a)$ as an element of $L_p([0,1]^{\omega_2}) \oplus Y$, with $T_n(a) = r \oplus s$, where $r \in L_p([0,1]^{\omega_2})$ and $s \in Y$, define $T_{n+1}(a) = r \circ \hat{\rho}^{-1} \oplus s$.

Define $b_{n+1} = c \circ \hat{\rho}^{-1} \oplus d$. Since $c \circ \hat{\rho}^{-1}$ depends on $\rho(A) \subset \omega_1$, we have: $b_{n+1} \in Z$, $T_{n+1}(a_i) = b_i$ for $i \leq n+1$, and therefore $(a_1, \ldots, a_{n+1}) \sim_1 (b_1, \ldots, b_{n+1})$. □

Along the lines of the proof of Theorem 2.1 in Cabello Sánchez [9], we have that in a Banach space with $BF_1$ some relevant properties coincide.

**Theorem 21** *For a Banach space $\mathcal{M}$ with $BF_1$ the following are equivalent:*

1. *$\mathcal{M}$ has the Radon-Nikodým property;*
2. *$\mathcal{M}$ is reflexive;*
3. *$\mathcal{M}$ is superreflexive;*
4. *$\mathcal{M}$ is uniformly convex.*

**Proof** By a known result, every reflexive space has Radon-Nikodým property and that every uniformly convex space is superreflexive. Therefore it remains to prove that (1) $\Rightarrow$ (4). For $0 \leq \epsilon \leq 1$ and $x \in \mathcal{M}$ let

$$\Delta(x, \epsilon) = \inf\{1 - \lambda : \text{ there exists } y \in \mathcal{M} \text{ such that } \|\lambda x \pm y\| \leq 1 \text{ and } \|y\| \geq \epsilon\}.$$

It is known that if $\inf\{\Delta(x,\epsilon) : x \in \mathcal{S}_{\mathcal{M}}\} > 0$ for all $0 < \epsilon \leq 1$, then $\mathcal{M}$ is uniformly convex (see Cabello Sánchez [8] and Finet [13]).

By Radon-Nikodým property there exists $x_0 \in \mathcal{S}_{\mathcal{M}}$ such that $\Delta(x_0, \epsilon) > 0$ for all $0 < \epsilon \leq 1$ (see Cabello Sánchez [8]). An easy calculation shows that $BF_1$ implies $\Delta(x_0, \epsilon) = \Delta(x, \epsilon)$ for all $x \in \mathcal{S}_{\mathcal{M}}$ and all $0 < \epsilon \leq 1$. Then $\mathcal{M}$ is uniformly convex. □

The following gives dichotomy results with respect to smoothness and convexity property.

**Theorem 22** *Let $\mathcal{M}$ be a Banach space.*

1. *If $\mathcal{M}$ has $WS$ then either all points on its unit sphere are smooth or none is smooth.*
2. *If $\mathcal{M}$ has $BF_1$ then all points on its unit sphere are extreme or none is extreme.*

**Proof**

1. For sake of contradiction, suppose $u, v$ are unit vectors, such that $u$ is a smooth but $v$ is not. Let $w$ be a unit vector witnessing the nonsmoothness for $v$. (By this we mean that there exist two unit functionals $\phi$ and $\theta$ both norming $v$ but $\phi(w) \neq \theta(w)$). Using $WS$ we can find a closed separable subspace $\mathcal{M}_0$ of $\mathcal{M}$ including elements $u, v, w$, and an isometric isomorphism $T : \mathcal{M}_0 \to \mathcal{M}_0$ such that $T(u) = v$.

   By Hahn-Banach theorem, $u$ is a smooth point in $\mathcal{M}_0$ and, by including the witness $w$, we conclude that $v$ is nonsmooth in $\mathcal{M}_0$. We thus get a contradiction since there can be no isometric isomorphism mapping a smooth point to a nonsmooth point.

2. Note that $BF_1$ is equivalent to saying that whenever two vectors of $\mathcal{M}$ have the same *qfpb* type (i.e. the same norm), they also satisfy the same existential *pb* formulas with one existential quantifier.

   Now we can prove the dichotomy by noticing that, for some $\epsilon > 0$, either all unit vectors satisfy the existential *pb* formula

   $$\exists y(\|x\| = \|y\| = \|2x - y\| = 1 \wedge \|x - y\| \geq \epsilon)$$

   or none does.

□

The following shows that, for reflexive Banach spaces, $WS$ is a self-dual property.

**Theorem 23** *Let $\mathcal{M}$ be a reflexive Banach space. Then $\mathcal{M}$ has $WS$ if and only if the dual space $\mathcal{M}'$ has $WS$.*

**Proof** Suppose $\mathcal{M}$ has $WS$. Then, by Theorem 21, it is strictly convex and superreflexive. It follows from the results in Finet [13] (see Definition 7 and Proposition 9) that there are smooth points on the unit sphere of $\mathcal{M}$ (Fréchet differentiability is stronger than Gâteux differentiability). So, by Theorem 22, $\mathcal{M}$ is smooth.

Recall also that, for a reflexive space, strict convexity is equivalent to smoothness of the dual.

We want to prove that, for all unit vectors $\phi$ and $\theta$ in $\mathcal{M}'$, player $\exists$ has a winning strategy for $\Gamma[(\mathcal{M}', \phi), (\mathcal{M}', \theta)]$. Let $a, b$ be unit vectors in $\mathcal{M}$ norming $\phi, \theta$ respectively. The main idea of the proof is that, during a play of $\Gamma[(\mathcal{M}', \phi), (\mathcal{M}', \theta)]$, we keep records of a simultaneous play of $\Gamma[(\mathcal{M}, a), (\mathcal{M}, b)]$.

Suppose that, for some $\bar{\phi}, \bar{\theta} \in (\mathcal{M}')^{n+1}$ and some $\bar{a}, \bar{b} \in \mathcal{M}^{n+1}$ such that $\phi_0 = \phi, \theta_0 = \theta$ and $a_0 = a, b_0 = b$, we already have the following from the first $n$ moves of a play in the game $\Gamma[(\mathcal{M}', \phi_0), (\mathcal{M}', \theta_0)]$ :

- $(\phi_0, \ldots, \phi_n) \sim_1 (\theta_0, \ldots, \theta_n)$;
- $(a_0, \ldots, a_n) \sim_1 (b_0, \ldots, b_n)$;
- (norming) $\|a_i\| = \|\phi_i\| = \phi_i(a_i) = 1, \quad \|b_i\| = \|\theta_i\| = \theta_i(b_i) = 1, \quad i \leq n.$

Let $\phi_{n+1} \in \mathcal{M}'$ with $\|\phi_{n+1}\| = 1$. We are going to respond by producing $\theta_{n+1} \in \mathcal{M}'$ such that the above properties extend to some $(n+2)$-tuples.

By smoothness and reflexivity of $\mathcal{M}'$, let $a_{n+1} \in \mathcal{M}$ be the unique vector so that $\|a_{n+1}\| = \phi_{n+1}(a_{n+1}) = 1$. By $WS$ in $\mathcal{M}$, choose $b_{n+1} \in \mathcal{M}$ so that

$$(a_0, \ldots, a_{n+1}) \sim_1 (b_0, \ldots, b_{n+1}).$$

Now by smoothness and reflexivity of $\mathcal{M}$, there is $\theta_{n+1} \in \mathcal{M}'$ so that

$$\|b_{n+1}\| = \|\theta_{n+1}\| = \theta_{n+1}(b_{n+1}) = 1.$$

**Claim:** $(\phi_0, \ldots, \phi_{n+1}) \sim_1 (\theta_0, \ldots, \theta_{n+1})$.

In order to prove the claim it suffices to construct a closed separable subspace $\mathcal{M}_0 \subset \mathcal{M}$ and an isometric isomorphism $T : \mathcal{M}_0 \to \mathcal{M}_0$ so that

(i) $T(a_i) = b_i, \quad i \leq n+1$;

(ii) $\phi_i = \theta_i \circ T$ on $\mathcal{M}_0, \quad i \leq n+1$;

(iii) $\| \sum_{i=0}^{n+1} \alpha_i \phi_i \|_{\mathcal{M}_0} = \| \sum_{i=0}^{n+1} \alpha_i \phi_i \|_{\mathcal{M}}$ and $\| \sum_{i=0}^{n+1} \alpha_i \theta_i \|_{\mathcal{M}_0} = \| \sum_{i=0}^{n+1} \alpha_i \theta_i \|$ for all $\alpha \in \mathbf{R}^{n+2}$ (i.e. the norms are the same whether one views the functionals as functionals on $\mathcal{M}$ or as functionals on the subspace $\mathcal{M}_0$).

The details of the construction are provided in Baratella and Ng [5].

The same argument applies to $\mathcal{M}'$ to prove that if $\mathcal{M}'$ has $WS$ then $\mathcal{M}$ has $WS$. $\square$

The following is a consequence of the previous theorem and of the equivalence of $WS$ and transitivity for separable spaces.

**Corollary 24** *Let $\mathcal{M}$ be a reflexive, separable Banach space. Then $\mathcal{M}$ is transitive if and only if $\mathcal{M}'$ is transitive.*

**Proof** We notice that reflexivity and separability of $\mathcal{M}$ imply separability of $\mathcal{M}'$. (Indeed, it is known that a separable Banach space has separable dual if and only if the dual has Radon-Nikodým property, and, as already noticed, reflexivity implies the Radon-Nikodým property.) Therefore one can apply the previous theorem. □

**Remark 25** *A stronger form of the previous corollary is proved by Cabello Sánchez*[8]*: he shows that transitivity is a self-dual property for reflexive spaces.*

Recall that two elements in $L_p[0,1]$ are *disjoint* if they have disjoint supports. For $p \neq 2, \infty$ a characterization of disjointness in $L_p[0,1]$ is the following: elements $f, g$ are disjoint if and only if

$$\|\alpha f + \beta g\|^p = |\alpha|^p \|f\|^p + |\beta|^p \|f\|^p \quad \text{for all } \alpha, \beta \in \mathbf{R}.$$

By using such characterization and the remark that $WS$ implies $BF_1$, one can easily prove that $L_p[0,1]$ does not have $WS$ when $1 \leq p < \infty$, $p \neq 2$.

Similarly, the following do not have $WS$: $\ell_p(X)$, when $X$ is any set with at least two elements and $1 \leq p \leq \infty$, $p \neq 2$; the space $\mathcal{C}(X)$ of bounded real valued continuous functions on a compact Hausdorff topological space $X$, equipped with supremum norm, when $X$ has at least two isolated points.

A Banach space is *almost-transitive* if the orbit of each unit vector under the action of the group of isometric isomorphisms is dense in the unit sphere. It is an easy remark that the nonstandard hull of an almost-transitive Banach space is transitive. Since $L_p[0,1]$, $1 \leq p < \infty$, $p \neq 2$ is almost-transitive but not transitive (see Greim *et al.* [15]), we have an example showing that $WS$ is not a *super-property* and it is not preserved under subspace.

We finish with a list of open problems. The main one, of course, is the Banach and Mazur problem posed over 60 years ago, which we restated equivalently as

(1) Is every separable $WS$ Banach space a Hilbert space?

Here are some weakenings of this problem:

(2) Is a (separable) $WS$ Banach space reflexive?

(3) Is $WS$ a self-dual property for (separable) Banach spaces? I.e. does the statement of Theorem 23 hold without the additional assumption of reflexivity?

(4) Is a (separable) transitive Banach space reflexive?

Finally, it would be of interest to clarify further the properties in this section:

(5) Is $WS$ the same as $BF$?

## References

1. S. Banach, **Théorie des Opérations Linéaires**, Chelsea Publishing Company, New York, 1955.
2. S. Baratella and S.-A. Ng, **Neocompact quantifier elimination in structures based on Banach spaces**, to appear in *Annals of Pure and Applied Logic.*
3. S. Baratella and S.-A. Ng, **Applications of neocompact quantifier elimination**, in preparation.
4. S. Baratella and S.-A. Ng, **Fixed points in the nonstandard hull of a Banach space**,*Nonlinear Analysis: Theory, Methods and Applications* **34**, 299-306 (1998).
5. S. Baratella and S.-A. Ng, **Playing isometry games in banach spaces**, in preparation.
6. B. Beauzamy, **Introduction to Banach spaces and their geometry**, North-Holland, 1982.
7. Y. Benyamini and J. Lindenstrauss, **Geometric nonlinear functional analysis**, Vol. 1, American Mathematical Society colloquium publications, 2000.
8. F. Cabello Sánchez, **Regards sur le problème des rotations de Mazur**, *Extracta Mathematicae* **12**, 97-116 (1997).
9. F. Cabello Sánchez, **Maximal symmetric norms on Banach spaces**, *Math. Proc. R. Irish Academy* **98**A, 121-130 (1998).
10. C.C. Chang and H.J. Keisler, **Model theory**, 3rd edition, North-Holland, 1990.
11. D. Dacunha-Castelle and J.-L. Krivine, **Applications des ultraproduits à l'étude des espaces et des algèbres de Banach**, *Studia Mathematica* **41**, 315-334 (1972).
12. S. Fajardo and H.J. Keisler, **Neometric Spaces**, *Advances in Mathematics* **118**, 134-175 (1996).
13. C. Finet, **Uniform convexity properties of norms on a superreflexive Banach space**, *Israel Journal of Mathematics* **53**, 81-92 (1986).
14. W.T. Gowers, **An infinite Ramsey theory and some Banach-space dichotomies**, preprint.
15. P. Greim, J.E. Jamison and A. Kaminska, **Almost transitivity in some function spaces**, *Mathematical Proceedings of the Cambridge Philosophical Society* **116**, 475-488 (1994).
16. P.B. Guerrero and A.R. Palacios, **Transitivity of the norm on Banach spaces having a Jordan structure**, *Manuscripta Mathematica*

**116**, 475-488 (1994).
17. C.W. Henson, **Nonstandard hulls of Banach spaces**, *Israel Journal of Mathematics* **25**, 108-144 (1976).
18. C.W. Henson, **Model theory for structures based on Banach spaces**, abstract of the talk given at "X Latin American Symposium on Mathematical Logic", in *Bulletin of Symbolic Logic* **2**, 223-224 (1996).
19. C.W. Henson and L.C. Moore Jr., **Nonstandard Analysis and the Theory of Banach spaces**, in "Nonstandard analysis - recent developments" (A.E. Hurd ed.), *Lecture notes in Mathematics* **983**, Springer, Berlin-New York 1983.
20. W. Hodges, **Model theory**, Cambridge University Press, 1993.
21. J. Iovino, **A quick introduction to Banach space model theory**, lecture notes, Carnegie Mellon University 1995 (available at http://sphere.math.utsa.edu/~iovino/).
22. H.J. Keisler, **Quantifier elimination for neocompact sets**, *Journal of Symbolic Logic* **63**, 1442-1472 (1998).
23. W.A.J. Luxemburg, **A general theory of monads**, in "Applications of Model Theory to Algebra, Analysis and Probability" (W.A.J. Luxemburg ed.), Holt, Rinehart and Winston, New York, 1969.

# A VERTEX PARTITIONING OF A GRAPH

SERGIO R. CANOY, JR.
ESPERANZA B. ARUGAY
ESAMEL M. PALUGA
*Department of Mathematics, MSU-IIT*
*9200 Iligan City, Philippines*

**Abstract**

Basically, the tree chromatic number of a graph is the minimum number of subsets the vertex set of a graph can be partitioned so that each subset induces a tree. Among others, this paper gives a characterization of graphs which have tree chromatic number greater than or equal to two. We also give a characterization of graphs which have tree chromatic number exactly equal to two.

## 1. Introduction

There have been a great deal of results on studies regarding the partitioning of a vertex set as well as an edge set of a graph such that each element of a partition induces a certain graph theoretic property. This paper considers a partitioning of the vertex set of a graph. Such a kind of partitioning, which was defined and investigated in [5], gives rise to what we call the tree-chromatic number of a graph.

Henceforth, every graph G is simple, and $V(G)$ denotes the vertex set of G.

## 2. Definitions and Results

The first two definitions have the greatest roles to play.

**Definition 2.1** Let $G$ be a simple graph. A family $\Omega = \{V_1, V_2, ..., V_k\}$ of nonempty subsets of the vertex set $V(G)$ is a vertex-tree partition of $G$ if

(i) $\Omega$ is a partition of $V(G)$, that is, $\Omega$ is a pairwise disjoint family and

(ii) each element $V_i$ of $\Omega$ induces a subgraph of $G$ which is a tree.

**Definition 2.2** Let $G$ be a simple graph and $P(G)$ the set of all vertex-tree partitions of $G$. The tree-chromatic number of $G$, is given by

$$\tau(G) = min\{|\Omega| : \Omega \in P(G)\}.$$

We first give the following remarks which are easy to verify.

**Remark 1** $\tau(G)$ exists for every simple graph $G$ and $1 \leq \tau(G)$.

**Remark 2** If $G$ is a graph of order $|G|$, then $\tau(G) \leq |G|$.

**Remark 3** If $G$ and $H$ are isomorphic graphs, then $\tau(G) = \tau(H)$.

**Lemma 2.3** $G$ is a tree if and only if $\tau(G) = 1$.

*Proof:* Suppose $G$ is a tree. Then $\Omega = \{V(G)\}$ is a vertex-tree partition of $G$. By Definition 2.2, $\tau(G) \leq 1$. Combining this with Remark 1 gives us the desired result. Conversely, suppose that $\tau(G) = 1$ and assume that $G$ is not a tree. Then there exists no partition $\Omega$ of $G$ that contains a single element. This means that $\tau(G) \geq 2$, contrary to our assumption. Thus, $G$ is a tree.

The following remark is immediate from the above lemma.

**Remark 4** For all $n \geq 1$, $\tau(P_n) = 1$ and for all $n \geq 3$, $\tau(K_1, n) = 1$, where $P_n$ is the path of order $n$ and $K_1, n$ is the star graph.

**Lemma 2.4** Let $G$ be a graph with exactly two components $G_1$ and $G_2$. Then $\tau(G) = \tau(G_1) + \tau(G_2)$.

*Proof:* First, suppose that $\tau(G) = k$ and choose a vertex-tree partition $\Omega$ of $G$ consisting of $k$ elements. Then $\Omega = \Omega_1 \cup \Omega_2$, where $\Omega_1 \cap \Omega_2 = \emptyset$ and $\Omega_1$ and $\Omega_2$ are vertex-tree partitions of $G_1$ and $G_2$, respectively. By Definition 2.2, we have

$$\tau(G_1) + \tau(G_2) \leq |\Omega_1| + |\Omega_2| = |\Omega| = \tau(G).$$

On the other hand, if $\Omega_1$ and $\Omega_2$ are vertex-tree partitions of $G_1$ and $G_2$, respectively, with $|\Omega_1| = \tau(G_1)$ and $|\Omega_2| = \tau(G_2)$, then $\Omega = \Omega_1 \cup \Omega_2$ is a vertex-tree partition of $G$. Hence, with Definition 2.2, we have

$$\tau(G_1) + \tau(G_2) = |\Omega_1| + |\Omega_2| = |\Omega| \leq \tau(G).$$

The above two inequalities imply that $\tau(G_1) + \tau(G_2) = \tau(G)$.

By induction on the number of components of a graph, we obtain the following extension of Lemma 2.4.

**Theorem 2.5** If $G_1, G_2, ..., G_k$ are the components of a graph $G$, then

$$\tau(G) = \sum_{i=1}^{k} \tau(G_i).$$

**Remark 5** There exist a graph $G$ and a subgraph $H$ of $G$ such that $\tau(G) \leq \tau(H)$.

To see this, let $G$ be a tree such that $E(G) \neq \emptyset$ and let $H$ be a disconnected subgraph of $G$. By Lemma 2.3, $1 = \tau(G) < \tau(H)$.

**Lemma 2.6** Let $G$ be a connected graph. If $G$ is unicyclic, then $\tau(G) = 2$.

*Proof:* Let $C$ be the cycle in $G$ and $v \in V(C)$. Since $G$ is not a tree, $\tau(G) \geq 2$. Now the set $V(G)\backslash\{v\}$ induces a subgraph $H$ of $G$. Let $G_1, G_2, ..., G_m$ $(m \geq 1)$ be the components of $H$. Then each $G_i$ is a tree. If $G_1$ contains the other vertices of $C$, then the set $X = V(G_2) \cup V(G_3) \cup ... \cup V(G_m) \cup \{v\}$ induces a tree. It follows that $\Omega = \{V(G_1), X\}$ is a vertex-tree partition of $G$. By Definition 2.2, $\tau(G) \leq 2$. Thus, $\tau(G) = 2$.

**Lemma 2.7** Let $G$ be a connected graph with exactly $n$ cycles. If these cycles are disjoint, i.e., any two cycles do not share a common vertex, then $\tau(G) \geq n + 1$.

*Proof:* By Lemma 2.6, the assertion holds for $n = 1$. Suppose that the conclusion holds for $n$. Assume that $G$ has exactly $n + 1$ cycles and that these cycles are all disjoint. Let $C_1, C_2, ..., C_n, C_{n+1}$ be the disjoint cycles of $G$. Let $\Omega$ be a vertex-tree partition of $G$ with $\tau(G) = |\Omega|$. Without loss of generality, we may assume that $C_{n+1}$ is an endcycle, i.e., $C_{n+1}$ has only one vertex, say $v$, that is in any shortest path connecting this cycle to another cycle in $G$. Note that such an endcycle in $G$ exists because the cycles in $G$ are mutually disjoint. Let $W$ be the element of $\Omega$ having the following property: (i) $v \notin W$ and (ii) $W$ contains at least one vertex of $C_{n+1}$. Note that $X = V(G)\backslash W$ induces a subgraph $H$ of $G$ which contains exactly $n$ cycles (the cycles $C_1, C_2, ..., C_n$). Naturally, the family $\Omega^+ = \Omega\backslash\{W\}$ is a vertex-tree partition of $H$. Hence by Definition 2.2 and the inductive hypothesis, $|\Omega^+| \geq \tau(H) \geq n+1$. Therefore, $\tau(G) = |\Omega| \geq n + 2$. This proves the lemma.

We now show that the reverse inequality in Lemma 2.8 is also true. Hence, we have the following result.

**Theorem 2.8** Let $G$ be a connected graph with exactly $n$ cycles. If these cycles are disjoint, then $\tau(G) = n + 1$.

*Proof:* We prove this by induction on the number of cycles in $G$. First, suppose that $n = 2$. Then $\tau(G) \geq 3$ by Lemma 2.7. Let $C_1$ and $C_2$ be the disjoint cycles of $G$. Let $P$ be a shortest path connecting $C_1$ to $C_2$. Then every path that connects $C_1$ to $C_2$ has always the path $P$ as a portion (i.e. $P$ is unique). For if there is a path $P^+$ that connects $C_1$ to $C_2$ and yet does not contain $P$, then $P$ and $P^+$, together with some vertices of $C_1$ and $C_2$, will form a cycle different from $C_1$ and $C_2$. This is contrary to our assumption.

Now let $v \in V(C_1)$ be a vertex of the shortest path connecting $C_1$ and $C_2$. Then $V(G)\backslash\{v\}$ induces a subgraph $H$ of $G$. Let $G_1, G_2, ..., G_m (m \geq 2)$ be the components of $H$. One of these components, say $G_1$, contains the other vertices of $C_1$. Hence, $G_1$ is a tree. Let $X = V(G_2) \cup V(G_3) \cup ... \cup V(G_m) \cup \{v\}$. Then $X$ induces a connected unicyclic subgraph $K$ of $G$. By Lemma 2.6, $\tau(K) = 2$. Let $\Omega = \{V_1, V_2\}$ be a vertex tree-partition of $K$. Then $\Omega^+ = \Omega \cup \{V(G_1)\}$ is vertex-tree partition of $G$. By Definition 2.2, it follows that $\tau(G) \leq 3$. Thus, $\tau(G) = 3$ and the assertion holds for $n = 2$.

Next, suppose that the assertion holds for all $k \leq n$. We show that it also holds for $n+1$. To this end, let $C_1, C_2, ..., C_n, C_{n+1}$ be the disjoint cycles of $G$. Then $\tau(G) \geq n+2$ by Lemma 2.7. Let $v \in V(C_{n+1})$. Then $V(G)\backslash\{v\}$ induces a subgraph $H$ of $G$. Consider the following cases:

*Case 1.* Suppose $H$ is connected. Then it contains exactly $n$ cycles (the cycles $C_1, C_2, ..., C_n$). Hence, by assumption, $\tau(H) = n + 1$. Let $\Omega = \{V_1, V_2, , V_{n+1}\}$ be a vertex-tree partition of $H$. Then $\Omega^+ = \Omega \cup \{\{v\}\}$ is a vertex-tree partition of $G$. Hence, $\tau(G) \leq n + 2$.

*Case 2.* Suppose $H$ has components $G_1, G_2, ..., G_m (m \geq 2)$. Then one of these components, say $G_1$, contains the other vertices of $C_{n+1}$. Suppose first that $G_1$ does not contain all the $n$ cycles $C_1, C_2, ..., C_n$. Without loss of generality, assume that $C_1, C_2, ..., C_k (k < n)$ are the only cycles contained in $G_1$. Let $X = \{v\} \cup V(G_2) \cup ... \cup V(G_m)$. Then $X$ induces a connected subgraph $K$ of $G$. Further, $K$ contains the cycles $C_{k+1}, C_{k+2}, ..., C_n$. By the inductive hypothesis, $\tau(G_1) = k+1$ and $\tau(K) = (n - k) + 1$. Let $\Omega_1 = \{V_1, V_2, , V_{k+1}\}$ and $\Omega_2 = \{W_1, W_2, , W_{(n-k)+1}\}$ be vertex-tree partitions of $G_1$ and $K$, respectively. Then $\Omega = \Omega_1 \cup \Omega_2$ is a vertex-tree partition of $G$. Hence, $\tau(G) \leq |\Omega| = (k + 1) + (n - k) + 1 = n + 2$. If $G_1$ contains all the cycles $C_1, C_2, ..., C_n$, then $\tau(G_1) = n + 1$. In this case, the subgraph $K$ induced by $X$ is a tree and so,$\tau(H) = 1$. Consequently, $\tau(G) \leq n + 2$.

In both cases, we see that $\tau(G) \leq n + 2$. Therefore, $\tau(G) = n + 2$. This proves the theorem.

**Theorem 2.9** Let $G$ be a graph with exactly $k$ components. If $G$ has exactly $n$ cyles and these cycles are disjoint, then $\tau(G) = n + k$.

*Proof:* Let $G_1, G_2, ..., G_k$ be the components of $G$. Then $\tau(G) = \sum_{i=1}^{k} \tau(G_i)$ by Theorem 2.5. For each $i \in \{1, 2, , k\}$, denote by $\eta_i$ the number of cycles in $G_i$. Then, by Theorem 2.8, we have

$$\begin{aligned}\tau(G) &= \sum_{i=1}^{k} \tau(G_i) \\ &= \sum (\eta_i + 1) \\ &= n + k.\end{aligned}$$

The following is a quick consequence of Theorem 2.9.

**Corollary 2.10** Let $G$ be a graph with exactly $k$ components. If $G$ is unicyclic, then $\tau(G) = k + 1$.

*Proof:* Apply Theorem 2.9 with $n = 1$.

**Remark 6** For all $n \geq 3$, the cycle graph $C_n$ is connected and unicyclic. Therefore, from Corollary 2.10, $\tau(C_n) = 2$.

**Theorem 2.11** Let $G$ be a graph. Then $\tau(G) \geq 2$ if and only if $G$ is disconnected or else $G$ has at least one cycle.

*Proof:* Suppose that $\tau(G) \geq 2$. If $G$ were disconnected, then we are done. So, suppose that $G$ is connected. If $G$ does not contain a cycle, then it must be a tree. Thus, by Lemma 2.3, $\tau(G) = 1$. This gives a contradiction. Hence, G has at least one cycle.

Conversely, assume that G is disconnected. Then G is not a tree. Hence, $\tau(G) \geq 2$ by Lemma 2.3. The same argument holds if $G$ has at least one cycle.

**Theorem 2.12** Let $G$ be a graph. Then $\tau(G) = 2$ if and only if G is a forest with exactly two components or else $G$ is connected, has a cycle and a subgraph $H$ which is a tree such that $X \backslash V(H)$, in turn, induces a tree.

*Proof:* Suppose that $\tau(G) = 2$ and suppose that $G$ is not a forest with exactly two components. Observe that if $G$ is disconnected with $\tau(G) = 2$, then $G$ cannot have more than two components. Thus our assumptions force $G$ to be connected. Further, since $\tau(G) = 2$, it follows from Lemma 2.3 that G is not a tree. Consequently, G contains at least one cycle.

Now, let $\Omega = \{V_1, V_2\}$ be a vertex-tree partition of $G$ and $H$ the subgraph of $G$ induced by $V_1$. Then $H$ is a tree by Definition 2.1. Again, by Definition 2.1, $V(G) \backslash V(H) = V_2$ induces a tree.

Conversely, suppose that $G$ is a forest with exactly two components $G_1$ and $G_2$. From Theorem 2.10, we have $\tau(G) \geq 2$. Note that by definition of a forest, $G_1$ and $G_2$ are both trees. Let $V_1 = V(G_1)$ and $V_2 = V(G_2)$. Then $\Omega = \{V_1, V_2\}$ is a vertex-tree partition of $G$. By Definition 2.2, $\tau(G) \leq 2$. Therefore, $\tau(G) = 2$.

Suppose now that $G$ is connected, has a cycle and a subgraph $H$ which is tree such that $X \setminus V(H)$ induces a tree. Put $V_1 = V(H)$ and $V_2 = X \setminus V_1$. Then $\Omega = \{V_1, V_2\}$ is a vertex-tree partition of $G$. By Definition 2.2, $\tau(G) \leq 2$. But since $G$ contains a cycle, it follows from Theorem 2.11 that $\tau(G) \geq 2$. These two inequalities mean that $\tau(G) = 2$. This completes the proof of the theorem.

**Corollary 2.13** Let $G$ be a connected graph with cycles. If the cycles of $G$ have a common non-cutvertex $v$ , then $\tau(G) = 2$.

*Proof:* Let $V_1 = \{v\}$ and $V_2 = V(G) \setminus \{v\}$. Let $H$ be the graph induced by $V_2$. Since $v$ is a non-cutvertex, its removal yields a connected graph. Hence, $H$ is a connected subgraph of $G$. Now, if $H$ has a cycle, say $C$, then $C$ must be a cycle in $G$. Hence, by assumption, $v \in V(C)$. But this is not possible because $v$ is not in $V_2 = V(H)$. Therefore, $H$ is a tree. Since $V_1$ induces the path $P_1$ (which is a tree), the desired result now follows from Theorem 2.12.

# References

[1] Akiyama, J., Era, H., Gervacio, S., and Watanabe, M., Path Chromatic Number of Graphs, Journal of Graph Theory, Vol. 13, No. 13, 1989, pp. 569-575.

[2] Arugay, E. B., On the Path Chromatic Number of Graphs, Ph.D. Thesis, Ateneo de Manila University, Philippines, 1990.

[3] Buenavista, R.N., On the Tree Chromatic Index of a Graph, M.S. Thesis, MSU-Iligan Institute of Technology, Philippines, 1996.

[4] Harrary, F., Graph Theory, Addison-Wesley Reading, MA, 1969.

[5] Manla, E., On the Tree Chromatic Number of Graphs, M.S. Thesis, MSU-Iligan Institute of technology, Philippines, 1997.

[6] Toft, B., Graph Colouring Problems Part I, Universitas Othiniensis, Denmark,1987.

# CONSTRUCTION OF LAGRANGIAN AND NON-LAGRANGIAN IP LOOPS

ALEXANDER S. CARRASCAL
*SciTech R&D Center, Polytechnic University of the Philippines*
*E-mail: ascarrascal@yahoo.com*

This paper introduces several practical methods for constructing Lagrangian and non-Lagrangian IP loops. The construction of Lagrangian IP loops $L$ involves three known factorable groups $G_k$, $k = 1, 2, 3$, each of order $2m$, having a common normal subgroup $H$ of order $m$ such that $G_k/H$ is isomorphic to $C_2$, where $C_2$ is the cyclic group of order 2. On the other hand, let $G_k$, $k = 1, 2, 3$, be groups, each of order $m = rs$, $2r < rs$, with a common subgroup $H$ of order $r$. Then non-Lagrangian IP loops of order $(3s - 2)r$ can be constructed.

## 1 Introduction

The study of quasigroups and loops is gaining much interest because of their many applications in both pure and applied mathematics. However, not much is known about these interesting structures because their theory is a relatively young discipline.

A **quasigroup** $\langle Q; \cdot \rangle$ is a non-empty set $Q$ with a binary operation $\cdot$ such that for any two elements $a, b \in Q$, the linear equations

$$a \cdot x = b, \qquad y \cdot a = b$$

have unique solutions for $x$ and $y$ lying in $Q$. A **loop** $\langle L; \cdot \rangle$ is a quasigroup variety with a unique *identity element* $e$ such that $e \cdot x = x = x \cdot e$ for all $x \in L$.

A quasigroup (loop) $\langle Q; \cdot \rangle$ is said to possess the **left inverse property (LIP)** if there exists a bijection $J_\lambda : a \rightarrow a^\lambda$ on $Q$ such that $a^\lambda(ax) = x$ for every $x \in Q$. Similarly, a quasigroup (loop) $\langle Q; \cdot \rangle$ is said to possess the **right inverse property (RIP)** if there exists a bijection $J_\rho : a \rightarrow a^\rho$ such that $(xa)a^\rho = x$ for every $x \in Q$. If a quasigroup (loop) has both LIP and RIP, it is said to have the **inverse property (IP).**

## 2 Parastrophs and Complements

Let $a$ be a fixed element in $Q$. We define the **left** and **right translation maps** $L(a) : Q \rightarrow Q$ and $R(a) : Q \rightarrow Q$, respectively, by

$$xL(a) = a \cdot x \quad \text{and} \quad xR(a) = x \cdot a$$

for all $x \in Q$. Since the mappings $L(a)$ and $R(a)$ are bijections, they have inverse maps $L(a)^{-1}$ and $R(a)^{-1}$. These inverse maps can be exploited to define two binary operations \ and / for the set $Q$ as follows

$$x\backslash y = yL(x)^{-1} \quad \text{and} \quad x/y = xL(y)^{-1}$$

where $x\backslash y = z$ if and only if $x \cdot z = y$ and $x/y = z$ if and only if $z \cdot y = x$. These operations are called **left division** and **right division**, respectively, and the quasigroups $\langle Q; \backslash \rangle$ and $\langle Q; / \rangle$ are called **conjugates** of $\langle Q; \cdot \rangle$.

Using functional notation, one can write $F(a,b) = c$ instead of $a \cdot b = c$ and substitute symbols $F^{-1}$ and ${}^{-1}F$ for operations / and \, respectively. Going a step further, one can define three other operations on $Q$ associated with $F$, as shown in the table below:

| | |
|---|---|
| (1) $F(x,z) = y$ | (4) ${}^{-1}(F^{-1})(y,x) = z$ |
| (2) $F^{-1}(y,z) = x$ | (5) $({}^{-1}F)^{-1}(z,y) = x$ |
| (3) ${}^{-1}F(x,y) = z$ | (6) $({}^{-1}(F^{-1}))^{-1}(z,x) = y$ |

The six quasigroups associated with each of these operations are also referred to as **parastrophs**[1]. It is easy to show that $(F^{-1})^{-1} = F$, ${}^{-1}({}^{-1}F) = F$ and $({}^{-1}(F^{-1}))^{-1} = {}^{-1}(({}^{-1}F)^{-1})$. Moreover, the following pairs of quasigroups: (1) $\langle Q; F^{-1} \rangle$ and $\langle Q; ({}^{-1}F)^{-1} \rangle$, (2) $\langle Q; {}^{-1}F \rangle$ and $\langle Q; {}^{-1}(F^{-1}) \rangle$, and (3) $\langle Q; F \rangle$ and $\langle Q; ({}^{-1}(F^{-1}))^{-1} \rangle$, are transposes of each other.

We also need the following definitions:

**Definition 1** *Suppose $\langle Q; \cdot \rangle$ is a quasigroup and let $a, x \in Q$, where $a$ is a fixed element in $Q$. We shall call $y \in Q$ as the* ***right*** ***a-complement*** *of $x$ in $Q$ if $x \cdot y = a$. Similarly, we shall call $z \in Q$ as the* ***left*** ***a-complement*** *of $x$ in $Q$ if $z \cdot x = a$. We shall denote the right and left a-complements of $x$ by $x_a^*$ and ${}^*x_a$, respectively. If $x_a^* = {}^*x_a$, we shall refer to $x_a^*$ or ${}^*x_a$ simply as the a-**complement** of $x$ and denote it by $\overline{x}_a$.*

**Definition 2** *We shall call the bijections $\gamma^*(a) : x \to x_a^*$ and ${}^*\gamma(a) : x \to {}^*x_a$ as the* ***right*** *and* ***left*** *a-**complement maps**, respectively. If $\gamma^*(a) = {}^*\gamma(a)$, we shall call it simply as the a-**complement map** and denote it by $\gamma(a)$.*

## 3 Multi-$\phi$ Systems and Quasidirect Products

**Definition 3** *Let $\{\langle Q; \phi_k \rangle \,|\, k = 1, ..., r\}$ be a set of $r$ quasigroups each of order $n$ with binary operations $\phi_k$. The collection of such quasigroups, denoted by $\langle Q; \Phi \rangle$, where $\Phi = \{\phi_k \mid k = 1, ..., r\}$, is called a* ***multi-$\phi$ system***[2] *of order $(n; r)$.*

**Definition 4** *Let $\langle P; \cdot \rangle$ be a quasigroup of order $m$ and let $\langle Q; \Phi \rangle$ be* ***multi-$\phi$ system*** *of order $(n; m^2)$, where $\Phi = \{\phi_{i,j} | i, j \in P\}$. The* ***quasidirect***

***product**** $^{3}$ *of $\langle P;\cdot\rangle$ and $\langle Q;\Phi\rangle$, denoted by $\langle P;\cdot\rangle\otimes\langle Q;\Phi\rangle$ or simply by $P\otimes Q$, is a system $\langle G;\diamond\rangle$ such that $G=\{(p,q)|p\in P,\ q\in Q\}$ with composition*

$$(p_i,q_h)\diamond(p_j,q_k)=(p_i\cdot p_j,q_h\phi_{p_i,p_j}q_k)$$

*where each $\phi_{p_i,p_j}$ defines a local operation on $Q$.*

## 4 Construction of Special Classes of Lagrangian IP Loops

We shall now discuss several methods for constructing special classes of Lagrangian IP loops using *parastrophs* and their *isotopes.*

### *4.1 Construction of Totally Symmetric Loops of the Form $L(V,C_{2^m},Q)$*

We begin by constructing a special class of totally symmetric IP loops.

**Theorem 5** *Suppose $V=\langle P;\cdot\rangle$ is the Klein 4 group and $C_{2^m}=\langle Q;\circ\rangle\cong C_2\times\ldots\times C_2$, up to $m$ factors, is the Klein group of order $2^m$, where $P=\{1,\ldots,4\}$ and $Q=\{1,\ldots,2^m\}$. Let $\langle Q;*\rangle$ be a quasigroup of order $2^m$. Form the quasidirect product*

$$\langle G;\diamond\rangle=V\otimes Q=\langle P;\cdot\rangle\otimes\langle Q;\Phi\rangle$$

*such that $G=\{(p,q)|p\in P,\ q\in Q\}$ and $\Phi=\{\phi_{i,j}|i,j\in P\}$ where*

| | | | |
|---|---|---|---|
| $\phi_{1,1}=\circ$ | $\phi_{1,2}=\circ$ | $\phi_{1,3}=\circ$ | $\phi_{1,4}=\circ$ |
| $\phi_{2,1}=\circ$ | $\phi_{2,2}=\circ$ | $\phi_{2,3}={}^{-1}F$ | $\phi_{2,4}=F=*$ |
| $\phi_{3,1}=\circ$ | $\phi_{3,2}={}^{-1}(F^{-1})$ | $\phi_{3,3}=\circ$ | $\phi_{3,4}=F^{-1}$ |
| $\phi_{4,1}=\circ$ | $\phi_{4,2}=({}^{-1}(F^{-1}))^{-1}$ | $\phi_{4,3}=({}^{-1}F)^{-1}$ | $\phi_{4,4}=\circ$ |

*Then $\langle G;\diamond\rangle$ is a totally symmetric IP loop of order $n=2^{m+2}$ which we shall denote by $L(V,C_{2^m},Q)$.*

**Proof.** Let $i,j,k\in\{2,3,4\}$ and let $x,y,z\in G$ where $x=(i,q_x)$, $y=(j,q_y)$, and $z=(k,q_z)$. It is not difficult to show that the parastroph operations satisfy $\phi_{i,k}\phi_{i,j}=\phi_{i,j}\phi_{i,k}=\iota$ and $\phi_{i,k}\phi_{j,k}=\phi_{j,k}\phi_{i,k}=\iota$. This means that if $\phi_{i,k}(q_x,q_z)=q_y$, then $\phi_{i,,j}(q_x,q_y)=q_z$ and $\phi_{j,k}(q_y,q_z)=q_x$. Since $C_{2^m}$ is totally symmetric and $\phi_{1,1}=\phi_{2,2}=\phi_{3,3}=\phi_{4,4}=\circ$ in $\langle G;\diamond\rangle$, it follows that $x\diamond x=(1,1)$ or $x^\lambda=x^\rho=x^{-1}=x$, for all $x\in G$. Hence, it suffices to show that the identities $x^\lambda(xz)=x(xz)=z$ and $(xz)z^\rho=(xz)z=x$ are satisfied by the local operations $\phi_{r,s}$, $r\neq s\neq 1$, i.e., the parastroph operations. Now, $x^\lambda\diamond(x\diamond z)=x\diamond(x\diamond z)=(i,q_x)\diamond[(i,q_x)\diamond(k,q_z)]=$

* The *quasidirect product* is also called the *block product*[2]. If the quasigroup is a loop, the block product is called a *coset product.*

$(i, q_x) \diamond (i \cdot k,\ q_x\phi_{i,k}q_z) = (i, q_x) \diamond (j, q_y) = (i \cdot j,\ q_x\phi_{i,j}q_y) = (k, q_z) = z$. Similarly, $(x \diamond z) \diamond z^\rho = (x \diamond z) \diamond z = [(i, q_x) \diamond (k, q_z)] \diamond (k, q_z) = (i \cdot k,\ q_x\phi_{i,k}q_z) \diamond (k, q_z) = (j, q_y) \diamond (k, q_z) = (j \cdot k,\ q_y\phi_{j,k}q_z) = (i, q_x) = x$. Hence $\langle G; \diamond\rangle = L(V, V_{2^m}, Q)$ is IP and totally symmetric.

## *4.2 Construction of IP Loops of the Form $L(V, D_m, Q)$*

**Theorem 6** *Suppose $\langle D_m; \times\rangle = \langle C_2; \circ\rangle \otimes \langle Q; \Theta\rangle$ is the dihedral group of order $2m$ expressed as a quasidirect product of the cyclic group $\langle C_2; \circ\rangle$ of order 2, and a multi-$\phi$ system, $\langle Q; \Theta\rangle$, where $C_2 = \{1, 2\}$, $Q = \{1, ..., m\}$ and $\Theta = \{\theta_{i,j} | i, j \in C_2\}$ such that*

| | |
|---|---|
| $\theta_{1,1}(x, y) = (x + y - 2) \bmod m + 1$ | $\theta_{1,2} = \theta_{1,1}$ |
| $\theta_{2,1}(x, y) = (x - y) \bmod m + 1$ | $\theta_{2,2} = \theta_{2,1}$ |

*Moreover, let $V = \langle P; \cdot\rangle$ be the Klein 4 group and $\langle Q; *\rangle$ be a quasigroup of order $m$. Form the quasidirect product*

$$\langle G; \star\rangle = V \otimes Q = \langle P; \cdot\rangle \otimes \langle Q; \Phi\rangle$$

*where $\Phi = \{\phi_{i,j} | i, j \in P\}$ such that*

| | | | |
|---|---|---|---|
| $\phi_{1,1} = \theta_{1,1}$ | $\phi_{1,2} = \theta_{1,2}$ | $\phi_{1,3} = \theta_{1,2}$ | $\phi_{1,4} = \theta_{1,2}$ |
| $\phi_{2,1} = \theta_{2,1}$ | $\phi_{2,2} = \theta_{2,2}$ | $\phi_{2,3} = {}^{-1}F$ | $\phi_{2,4} = F = *$ |
| $\phi_{3,1} = \theta_{2,1}$ | $\phi_{3,2} = {}^{-1}(F^{-1})$ | $\phi_{3,3} = \theta_{2,2}$ | $\phi_{3,4} = F^{-1}$ |
| $\phi_{4,1} = \theta_{2,1}$ | $\phi_{4,2} = ({}^{-1}(F^{-1}))^{-1}$ | $\phi_{4,3} = ({}^{-1}F)^{-1}$ | $\phi_{4,4} = \theta_{2,2}$ |

*Then $\langle G; \star\rangle$ is an IP loop of order $n = 4m$, which we shall denote by $L(V, D_m, Q)$.*

**Proof.** Since $\langle Q; \theta_{1,1}\rangle \cong C_m$, where $C_m$ is the cyclic group of order $m$, the local operation $\phi_{1,1} = \theta_{1,1} = (x + y - 2) \bmod m + 1$ implies that $\langle G; \star\rangle$ has a normal subgroup $N \cong C_m$. The fact that $\phi_{p,p} = \theta_{2,2} = \theta_{2,1} = (x - y) \bmod m + 1$, $p = 2, 3, 4$ implies that all $x = (p, q)$, $p \neq 1$ are self-inverse elements since $x \diamond x = (p \cdot p,\ q\phi_{p,p}q) = (1, [(x - y) \bmod m + 1]) = (1, 1)$. Moreover, whenever $i, j, k \in \{2, 3, 4\}$, the local operations satisfy $\phi_{i,k}\phi_{i,j} = \phi_{i,j}\phi_{i,k} = \iota$ and $\phi_{i,k}\phi_{j,k} = \phi_{j,k}\phi_{i,k} = \iota$. Hence, if $x = (i, q_x)$, $y = (j, q_y)$, and $z = (k, q_z)$ such that $\phi_{i,k}(q_x, q_z) = q_y$, then $\phi_{i,j}(q_x, q_y) = q_z$ and $\phi_{j,k}(q_y, q_z) = q_x$. It follows that $x^\lambda \star (x \star z) = x \star (x \star z) = (i, q_x) \star [(i, q_x) \star (k, q_z)] = (i, q_x) \star (i \cdot k,\ q_x\phi_{i,k}q_z) = (i, q_x) \star (j, q_y) = (i \cdot j,\ q_x\phi_{i,j}q_y) = (k, q_z) = z$. Similarly, $(x \star z) \star z^\rho = (x \star z) \star z = [(i, q_x) \star (k, q_z)] \star (k, q_z) = (i \cdot k,\ q_x\phi_{i,k}q_z) \star (k, q_z) = (j, q_y) \star (k, q_z) = (j \cdot k,\ q_y\phi_{j,k}q_z) = (i, q_x) = x$. Thus, the system $\langle G; \star\rangle = L(V, D_m, Q)$ has inverse property.

### *4.3 Construction of IP Loops of the Form $L(G_1, G_2, G_3)$*

We shall now present a construction method that utilizes isotopes of the parastrophs of a given quasigroup.

**Theorem 7** *Suppose* $\langle G_k; \lhd_k\rangle = \langle C_2; \circ\rangle \otimes \left\langle Q; \Theta^k\right\rangle$, $k = 1,2,3$, *are three groups of the same order* $2m$ *expressed as quasidirect products of the cyclic group* $\langle C_2; \circ\rangle$ *of order 2, and the multi-*$\phi$ *system* $\left\langle Q; \Theta^k\right\rangle$, *where* $C_2 = \{1,2\}$, $Q = \{1,...,m\}$, *and* $\Theta^k = \{\theta^k_{i,j} | i,j \in C_2\}$ *such that* $\theta^1_{1,1} = \theta^2_{1,1} = \theta^3_{1,1} = \theta_{1,1}$. *Define the bijections* $\alpha_k : q \rightarrow \overline{q}_1$, *as* 1-*complement maps in* $\langle G_k; \lhd_k\rangle$ *such that* $\alpha_k(q) = \overline{q}_1$, *whenever* $q(\theta^k_{2,2})\overline{q}_1 = \overline{q}_1(\theta^k_{2,2})q = 1$, *where* $q, \overline{q}_1 \in Q$, *i.e.,* $(c_2, q) \lhd_k (c_2, \overline{q}_1) = (c_2, \overline{q}_1) \lhd_k (c_2, q) = (1,1)$. *Finally, let* $V = \langle P; \cdot\rangle$ *be the Klein 4 group and let* $\langle Q; *\rangle$ *be a quasigroup of order* $m$. *Form the quasidirect product*

$$\langle G; \rhd\rangle = V \otimes Q = \langle P; \cdot\rangle \otimes \langle Q; \Phi\rangle$$

*where* $\Phi = \{\phi_{i,j} | i,j \in P\}$ *such that*

| | | | |
|---|---|---|---|
| $\phi_{1,1} = \theta_{1,1}$ | $\phi_{1,2} = \theta^1_{1,2}$ | $\phi_{1,3} = \theta^2_{1,2}$ | $\phi_{1,4} = \theta^3_{1,2}$ |
| $\phi_{2,1} = \theta^1_{2,1}$ | $\phi_{2,2} = \theta^1_{2,2}$ | $\phi_{2,3} = {}^{-1}f$ | $\phi_{2,4} = f$ |
| $\phi_{3,1} = \theta^2_{2,1}$ | $\phi_{3,2} = {}^{-1}(f^{-1})$ | $\phi_{3,3} = \theta^2_{2,2}$ | $\phi_{3,4} = f^{-1}$ |
| $\phi_{4,1} = \theta^3_{2,1}$ | $\phi_{4,2} = ({}^{-1}(f^{-1}))^{-1}$ | $\phi_{4,3} = ({}^{-1}f)^{-1}$ | $\phi_{4,4} = \theta^3_{2,2}$ |

*and*

| | |
|---|---|
| *(1)* $f(x,z) = x * z = y$ | *(4)* ${}^{-1}(f^{-1})(y\alpha_2, x) = z\alpha_3$ |
| *(2)* $f^{-1}(y, z\alpha_3) = x$ | *(5)* $({}^{-1}f)^{-1}(z, y\alpha_2) = x\alpha_1$ |
| *(3)* ${}^{-1}f(x\alpha_1, y) = z$ | *(6)* $({}^{-1}(f^{-1}))^{-1}(z\alpha_3, x\alpha_1) = y\alpha_2$ |

*Then, the system* $\langle G; \rhd\rangle$ *is an IP loop of order* $n = 4m$, *which we shall denote by* $L(G_1, G_2, G_3)$.

**Proof.** For simplicity we shall denote the 1−complement of $q_r$ by $\overline{q}_r$. Since $q(\theta^k_{2,2})\overline{q}_1 = \overline{q}_1(\theta^k_{2,2})q = 1$, it follows that $\alpha_k^2 = \iota$ or $\alpha_k^{-1} = \alpha_k$, i.e., $q\alpha_k = \overline{q}$ and $\overline{q}\alpha_k = q$, where $q, \overline{q} \in Q$. It also follows that if $a = (p,q) \in G$, then $a^{-1} = (p^{-1}, \overline{q}) = (p, \overline{q})$. Moreover, if $f(x,z) = y$, where $x,y,z \in Q$, then we have the following:

| | | |
|---|---|---|
| $\phi_{2,4}(x,z) = y$ | $\phi_{2,3}(\overline{x}, y) = z$ | $\phi_{3,4}(y, \overline{z}) = x$ |
| $\phi_{4,3}(z, \overline{y}) = \overline{x}$ | $\phi_{3,2}(\overline{y}, x) = \overline{z}$ | $\phi_{4,2}(\overline{y}, \overline{z}) = \overline{x}$ |

It follows from this table that if we let $i,j,k \in \{2,3,4\}$ and $q_x, q_z, q_y \in Q$ such that $\phi_{i,k}(q_x, q_z) = q_y$, then $\phi_{i,j}(\overline{q}_x, q_y) = q_z$ and $\phi_{j,k}(q_y, \overline{q}_z) = q_x$. Now, let $a,b,c \in G$, where $a = (i, q_x)$, $b = (j, q_y)$, and $c = (k, q_z)$ such that $\phi_{i,k}(q_x, q_z) = q_y$. Then $a^\lambda \rhd (a \rhd c) = a^{-1} \rhd (a \rhd c) = (i, \overline{q}_x) \rhd [(i, q_x) \rhd (k, q_z)]$

$= (i,\overline{q}_x) \triangleright (i \cdot k,\ q_x\phi_{i,k}q_z) = (i,\overline{q}_x) \triangleright (j,q_y) = (i \cdot j,\ \overline{q}_x\phi_{i,j}q_y) = (k,q_z) = c$. Similarly, $(a \triangleright c) \triangleright c^{\rho} = (a \triangleright c) \triangleright c^{-1} = [(i,q_x) \triangleright (k,q_z)] \triangleright (k,\overline{q}_z) = (i \cdot k,\ q_x\phi_{i,k}q_z) \triangleright (k,\overline{q}_z) = (j,q_y) \triangleright (k,\overline{q}_z) = (j \cdot k,\ q_y\phi_{j,k}\overline{q}_z) = (i,q_x) = a$. Hence, the system $\langle G;\triangleright\rangle = L(G_1,G_2,G_3)$ is IP.

## 5 Construction of Non-Lagrangian IP Loops

We shall now construct a special class of non-Lagrangian IP loops of order $n = (3s-2)\,r$.

**Theorem 8** *Let $\langle G_k;\diamond_k\rangle$, $k = 1,...,3$, be factorable groups of order $rs$ with a normal subgroup $\langle H;\diamond\rangle$ of order $r$ such that $H = \{1,...,r\}$ and $G_k = H \cup H_k$, where $H_k = \{[(k-1)(s-1)+1]r+1, ..., [k(s-1)+1]r\}$. Divide the Cayley table of $\langle G;\diamond_k\rangle$ as follows:*

| | |
|---|---|
| $[H;\diamond_{11}]$ | $[H_k;\diamond^k_{12}]$ |
| $[H_k;\diamond^k_{21}]$ | $[G_k;\diamond^k_{22}]$ |

*where $[H;\diamond_{11}] \equiv \langle H;\diamond\rangle$. Furthermore, $[H_k;\diamond^k_{12}]$ and $[H_k;\diamond^k_{21}]$ are Latin rectangles of dimensions $r\times(s-1)r$ and $(s-1)r\times r$, respectively, and $[G_k;\diamond^k_{22}]$ is an $(s-1)r\times(s-1)r$ block which is not a Latin square. Define the bijections $\alpha_k : h \to h^{-1}$, as 1−complement maps in $[G_k;\diamond^k_{22}]$ such that $\alpha_k(h) = h^{-1}$, whenever $h(\diamond^k_{2,2})h^{-1} = h^{-1}(\diamond^k_{2,2})h = 1$, where $h, h^{-1} \in H_k$. Moreover, let $\langle Q;\Phi\rangle$ be a multi-$\phi$ system, where $Q = \{1,...,(s-1)r\}$ and $\Phi = \{\phi_i | i = 1,...,6\}$ such that $\phi_1 = f$, $\phi_2 = f^{-1}$, $\phi_3 = {}^{-1}f$, $\phi_4 = {}^{-1}(f^{-1})$, $\phi_5 = ({}^{-1}f)^{-1}$, and $\phi_6 = ({}^{-1}(f^{-1}))^{-1}$, where*

| | |
|---|---|
| (1) $f(x,z) = y$ | (4) ${}^{-1}(f^{-1})(y\alpha_2, x) = z\alpha_3$ |
| (2) $f^{-1}(y, z\alpha_3) = x$ | (5) $({}^{-1}f)^{-1}(z, y\alpha_2) = x\alpha_1$ |
| (3) ${}^{-1}f(x\alpha_1, y) = z$ | (6) $({}^{-1}(f^{-1}))^{-1}(z\alpha_3, x\alpha_1) = y\alpha_2$ |

*Form the system $\langle P;*\rangle$ such that $P = H \cup (\bigcup_k H_k)$ with a Cayley table as follows*

| | | | |
|---|---|---|---|
| $[H;\diamond_{11}]$ | $[H_1;\diamond^1_{12}]$ | $[H_2;\diamond^2_{12}]$ | $[H_3;\diamond^3_{12}]$ |
| $[H_1;\diamond^1_{21}]$ | $[G_1;\diamond^1_{22}]$ | $[H_3;\phi_3]$ | $[H_2;\phi_1]$ |
| $[H_2;\diamond^2_{21}]$ | $[H_3;\phi_4]$ | $[G_2;\diamond^2_{22}]$ | $[H_1;\phi_2]$ |
| $[H_3;\diamond^3_{21}]$ | $[H_2;\phi_6]$ | $[H_1;\phi_5]$ | $[G_3;\diamond^3_{22}]$ |

*where the blocks of the form $[H_k;\phi_i] \cong \langle Q;\phi_i\rangle$ are $(s-1)r\times(s-1)r$ Latin squares. Then $\langle P;*\rangle$ is an IP loop of order $n = (3s-2)\,r$.*

**Proof.** The proof is similar to the preceding theorem.

### 5.1 *Sample Construction of a Lagrangian IP Loop*

We shall now construct an IP loop of the form $L(G_1, G_2, G_3)$. To do this, let $G_1 = C_2 \times C_4$, $G_2 = Q$ (**quaternion** group) and $G_3 = D_4$ (**dihedral** group of order 8) [4]. Let $\langle Q; *\rangle$ be a quasigroup of order 4 with Cayley table shown below.

| * | 1 | 2 | 3 | 4 |
|---|---|---|---|---|
| 1 | 4 | 2 | 3 | 1 |
| 2 | 1 | 3 | 4 | 2 |
| 3 | 3 | 1 | 2 | 4 |
| 4 | 2 | 4 | 1 | 3 |

Quasigroup $\langle Q;*\rangle$ of order 4.

The resulting IP loop of order $n = 16$ is shown below. This Cayley table was obtained by renaming the elements $(p, q)$ according to: $(p, q) \equiv 4(p-1)+q$.

| * | 1 | 2 | 3 | 4 | 5 | 6 | 7 | 8 | 9 | 10 | 11 | 12 | 13 | 14 | 15 | 16 |
|---|---|---|---|---|---|---|---|---|---|---|---|---|---|---|---|---|
| 1 | 1 | 2 | 3 | 4 | 5 | 6 | 7 | 8 | 9 | 10 | 11 | 12 | 13 | 14 | 15 | 16 |
| 2 | 2 | 3 | 4 | 1 | 6 | 7 | 8 | 5 | 10 | 11 | 12 | 9 | 14 | 15 | 16 | 13 |
| 3 | 3 | 4 | 1 | 2 | 7 | 8 | 5 | 6 | 11 | 12 | 9 | 10 | 15 | 16 | 13 | 14 |
| 4 | 4 | 1 | 2 | 3 | 8 | 5 | 6 | 7 | 12 | 9 | 10 | 11 | 16 | 13 | 14 | 15 |
| 5 | 5 | 6 | 7 | 8 | 1 | 2 | 3 | 4 | 16 | 14 | 15 | 13 | 12 | 10 | 11 | 9 |
| 6 | 6 | 7 | 8 | 5 | 2 | 3 | 4 | 1 | 13 | 15 | 16 | 14 | 11 | 9 | 12 | 10 |
| 7 | 7 | 8 | 5 | 6 | 3 | 4 | 1 | 2 | 15 | 13 | 14 | 16 | 10 | 11 | 9 | 12 |
| 8 | 8 | 5 | 6 | 7 | 4 | 1 | 2 | 3 | 14 | 16 | 13 | 15 | 9 | 12 | 10 | 11 |
| 9 | 9 | 12 | 11 | 10 | 14 | 16 | 13 | 15 | 1 | 4 | 3 | 2 | 7 | 5 | 8 | 6 |
| 10 | 10 | 9 | 12 | 11 | 16 | 14 | 15 | 13 | 2 | 1 | 4 | 3 | 8 | 6 | 7 | 5 |
| 11 | 11 | 10 | 9 | 12 | 13 | 15 | 16 | 14 | 3 | 2 | 1 | 4 | 5 | 8 | 6 | 7 |
| 12 | 12 | 11 | 10 | 9 | 15 | 13 | 14 | 16 | 4 | 3 | 2 | 1 | 6 | 7 | 5 | 8 |
| 13 | 13 | 16 | 15 | 14 | 11 | 10 | 9 | 12 | 6 | 7 | 8 | 5 | 3 | 2 | 1 | 4 |
| 14 | 14 | 13 | 16 | 15 | 9 | 11 | 12 | 10 | 8 | 5 | 7 | 6 | 4 | 3 | 2 | 1 |
| 15 | 15 | 14 | 13 | 16 | 12 | 9 | 10 | 11 | 7 | 6 | 5 | 8 | 1 | 4 | 3 | 2 |
| 16 | 16 | 15 | 14 | 13 | 10 | 12 | 11 | 9 | 5 | 8 | 6 | 7 | 2 | 1 | 4 | 3 |

A Lagrangian IP Loop of Order $n = 16$.

### 5.2 *Sample Construction of a Non-Lagrangian IP Loop*

Let us choose $G_1 = G_2 = G_3 = C_3$, the cyclic group of order 3. The only possible values for $r$ and $s$ are $r = 1$ and $s = 3$; hence, $Q = \{1, 2\}$. We define the quasigroup $\langle Q; \phi_1\rangle$ as shown below. The result is a non-Lagrangian IP loop order $n = 7$. Up to isomorphism, this is the only IP Loop of order 7 and the smallest.

| $\phi_1$ | 1 | 2 |
|---|---|---|
| 1 | 2 | 1 |
| 2 | 1 | 2 |

Quasigroup $\langle Q;*\rangle$ of Order 2.

| $\cdot$ | 1 | 2 | 3 | 4 | 5 | 6 | 7 |
|---|---|---|---|---|---|---|---|
| 1 | 1 | 2 | 3 | 4 | 5 | 6 | 7 |
| 2 | 2 | 3 | 1 | 7 | 6 | 4 | 5 |
| 3 | 3 | 1 | 2 | 6 | 7 | 5 | 4 |
| 4 | 4 | 6 | 7 | 5 | 1 | 3 | 2 |
| 5 | 5 | 7 | 6 | 1 | 4 | 2 | 3 |
| 6 | 6 | 5 | 4 | 2 | 3 | 7 | 1 |
| 7 | 7 | 4 | 5 | 3 | 2 | 1 | 6 |

The Smallest and the Only IP loop of order 7.

## 6 Summary

We have shown in this paper the existence and the method of construction of a special class of non-associative Lagrangian IP loops for all orders $n = 4m$, where $m \geq 3$. Moreover, we have also demonstrated the existence and the procedure for the construction of a special class of non-Lagrangian IP loops for all orders $n = (3s - 2)\,r$, where $2r < rs$ and $s \geq 3$.

## References


1. Plugfelder, H., *Quasigroups and Loops: Introduction*, Heldermann Verlag, Berlin (1990).
2. Cawagas, R. E., *Terminal Report: Development of the Theory of Finite Pseudogroups* (1998). Research supported by the National Research Council of the Philippines (1996-1998) under Project B-88 and B-95.
3. Chein, O., et al, *Quasigroups and Loops: Theory and Applications*, Heldermann Verlag, Berlin (1990).
4. Thomas, A. D., *Group Tables*, Shiva Mathematics Series, Shiva Publishing Ltd. (1980).

# ON EXTENSIONS OF DIFFERENTIAL OPERATORS IN $L^2[a,\infty)$

MARITINA T. CASTILLO
*Department of Mathematics*
*University of the Philippines*
*Diliman, Quezon City, Philippines*
*E-mail: tina@math01.cs.upd.edu.ph*

For a differential expression $M$ in $L^2[a,\infty)$, structural and spectral properties of extensions generated by M will be presented. The main result establishes that if the essential spectrum of a differential expression $M$ is empty and $T_{M,\lambda}$ is a maximal extension of $M$ at $\lambda \in \mathbb{C}$, then for $\delta > 0$, there exists a constant $K$ and a finite-dimensional subspace $\mathcal{S}$ of $L^2[a,\infty)$ such that the inequality $\left\|T_{M,\lambda} f\right\| \geq K \|f\|$ holds for every $f$ in the set $\mathcal{D}\left(T_{M,\lambda}\right) \cap \left(L^2[a,\infty) \setminus \mathcal{S}\right)$.

We consider a class of differential operators generated by differential expressions having the form

$$M = a_n D^n + a_{n-1} D^{n-1} + \cdots + a_2 D^2 + a_1 D + a_0 \tag{1}$$

on $J = [a,\infty)$ where $D = d/dt$, $a_k \in C^k(J)$, $0 \leq k \leq n$, $a_n(t) \neq 0$, $t \in J$. For each positive integer $n$, let $\mathcal{A}_n(J)$ denote the set of complex-valued functions $f$ for which $f^{(n-1)} = D^{n-1} f$ exists and is absolutely continuous for every compact sub-interval of $J$. For $n = 0$, $\mathcal{A}_0(J) = C(J)$, the space of continuous functions on $J$. We also denote by $M^+$ the Lagrange adjoint of $M$,

$$M^+ = b_n D^n + b_{n-1} D^{n-1} + \cdots + b_2 D^2 + b_1 D + b_0, \tag{2}$$

$b_i := \sum_{j=i}^{n} (-1)^j \binom{j}{i} \overline{a_j}^{(j-i)}$, $0 \leq i \leq n$. If $f^{(n)}$ exists everywhere, then $M^+ f$ exists for almost all $t \in J$. Furthermore, $M^{++} := (M^+)^+ = M$.

Let $T_{M,max}$ denote the **maximal operator** generated by $M$ where

$$\mathcal{D}(T_{M,max}) = \{ f \mid f \in \mathcal{A}_n(J) \cap L^2(J) \text{ and } Mf \in L^2(J) \}$$

$$T_{M,max} f = Mf = \sum_{k=0}^{n} a_k D^k f$$

and define the **minimal operator** generated by $M$, $T_{M,min}$, as follows:

$$T_{M,min} := \overline{T_{M,max}|_{C_0^\infty\binom{0}{J}}}$$

where $C_0^\infty\binom{0}{J}$ is the set of infinitely differentiable functions $f$ in $L^2(J)$ having compact support in the interior of $J$. Properties of the maximal and minimal operators generated by $M$ have been established in the literature [3]. In particular it has been shown that the maximal operator is closed and $T_{M,min}$ is injective. Moreover, $T_{M,max}$ has a dense range with a finite-dimensional null space.

Of special interest to us are operators generated by $M$ in $L^2(J)$ for which the spectrum is minimal. The restriction that will be imposed on $M$ is that its essential spectrum set in $L^2(J)$ is not the entire complex plane.

Since $T_{M,max}$ is a closed operator in $L^2(J)$, then $\mathcal{D}(T_{M,max})$ is a Hilbert space equipped with the inner product $\langle\ .\ ,\ .\ \rangle_{T_{M,max}}$ and $T_{M,max}$-norm $\|.\|_{T_{M,max}}$ defined, respectively, as follows:

$$\langle\ f\ ,\ g\ \rangle_{T_{M,max}} := \langle\ f\ ,\ g\ \rangle + \langle\ T_{M,max}\ f\ ,\ T_{M,max}\ g\ \rangle$$

$$\|f\|_{T_{M,max}} := \|f\| + \|T_{M,max}\ f\|$$

Furthermore if $T$ is a differential operator corresponding to $M$ in $L^2(J)$, then $\mathcal{D}(T)$ is closed in the $\mathcal{D}(T_{M,max})$-space equipped with the $T_{M,max}$-norm. On the other hand, given a closed subspace $\mathcal{Y}$ of the $\mathcal{D}(T_{M,max})$-space such that $\mathcal{D}(T_{M,min}) \subseteq \mathcal{Y}$, the operator $T$ which is the restriction of $T_{M,max}$ to $\mathcal{Y}$ is closed. This establishes a one-to-one correspondence between the differential operators corresponding to $M$ in $L^2(J)$ and the closed subspaces $\mathcal{Y}$ of $\mathcal{D}(T_{M,max})$ containing $\mathcal{D}(T_{M,min})$.

We now define differential operators generated by $M$ having minimal spectrum.

**Definition 1** *Let $T$ be a closed linear operator such that $T_{M,min} \subseteq T \subseteq T_{M,max}$. We call $T$ a* ***differential operator corresponding to $M$ in*** *$L^2(J)$. A differential operator $T$ corresponding to $M$ in $L^2(J)$ is called a*

***maximal extension*** *of $M$ at $\lambda \in \mathbb{C}$ if and only if $\lambda$ is in the resolvent set of $T$. A maximal extension of $M$ at $\lambda \in \mathbb{C}$ will be denoted by $T_{M,\lambda}$ .*

The definition of a maximal extension implies that the operator $T_{M,\lambda} - \lambda I$ is onto and has a bounded inverse. We recall that the essential spectrum of a differential operator $T$, denoted by $\sigma_e(T)$, is the closed set defined as follows:

$$\sigma_e(T) := \{ \lambda \in \mathbb{C} \mid Ran(T - \lambda I) \text{ is not closed} \}$$

The complement of the essential spectrum set in $\mathbb{C}$ is called the **essential resolvent** of $T$ and is denoted by $\rho_e(T)$ . Note that by definition of a maximal extension of $M$, there is no maximal extension of $M$ at any point on its essential spectrum. This justifies the restriction imposed on $M$ that the essential spectrum of $M$ does not cover the entire complex plane.

The basic spectral and extension theory for operators generated by $M$ was developed by G.C. Rota [4]. Some of the results are presented here translated, for our purposes, into the $L^2(J)$-setting. For a detailed discussion of these results the reader is referred to a related work by the author [1].

(i) *The essential spectrum is invariant under differential operators corresponding to $M$.* It is in this regard that the essential spectrum of the differential expression $M$, $\sigma_e(M)$, and its essential resolvent set, $\rho_e(M)$, are defined to be the essential spectrum and essential resolvent, respectively, of the minimal operator generated by $M$.

(ii) *The dimension of the quotient space $\frac{\mathcal{D}(T_{M,max})}{\mathcal{D}(T_{M,min})}$ is finite.* Consequently, for a differential operator $T$ corresponding to $M$ in $L^2(J)$ we can write $\mathcal{D}(T)$ as the direct sum

$$\mathcal{D}(T) = \mathcal{D}(T_{M,min}) \oplus \mathcal{W}$$

where $\mathcal{W}$ is a finite dimensional subspace of $\mathcal{D}(T_{M,max})$. In the same manner, we have the direct sum

$$\mathcal{D}(T_{M,max}) = \mathcal{D}(T_{M,min}) \oplus \mathcal{N}(T_{M,max}) \oplus \mathcal{V}$$

where $\mathcal{N}(T_{M,max})$ is the null space of $T_{M,max}$ and $\mathcal{V}$ is a finite dimensional subspace of $\mathcal{D}(T_{M,max})$.

(iii) *If* $\lambda \in \rho_e(M)$*, then there exists a maximal extension of* $M$ *at* $\lambda$. This result establishes the existence of a maximal extension of $M$ at every point in its essential resolvent set.

(iv) *Suppose* $\lambda \in \rho_e(M)$*. Let* $v_1, v_2, \cdots, v_m$ *be a basis for* $\mathcal{N}\left(T_{M+-\bar{\lambda},max}\right)$*, and* $y_1, y_2, \cdots, y_m$ *in* $L^2(J)$ *such that for* $1 \le i, j \le m$*,*

$$\int_J v_i \, \overline{y_j} = \delta_{ij}$$

*where* $\delta_{ij}$ *is the Kronecker delta.*

*Choose* $f_1, f_2, \cdots, f_m$ *in* $\mathcal{D}(T_{M-\lambda,max})$ *such that for* $1 \le i \le m$*,*

$$(T_{M-\lambda,max}) \; f_i = y_i$$

*and* $\{ g_1, g_2, \cdots, g_m \}$ *an arbitrary set of elements of* $\mathcal{N}(T_{M-\lambda,max})$*. If we consider the operator* $T_{M,rest}$ *which is a restriction of* $T_{M,max}$ *on the domain*

$$\mathcal{D}(T_{M,rest}) := \mathcal{D}(T_{M,min}) \oplus span\{ f_i + g_i \} ,$$

*then* $T$ *is a maximal extension of* $M$ *at* $\lambda$*.*

*Conversely, every maximal extension* $T$ *of* $M$ *at* $\lambda$ *may be obtained by such an extension of* $\mathcal{D}(T_{M,min})$*.*

This result describes the structure of maximal extensions of $M$ in $L^2(J)$. A similar result was established by Goldberg [3] using the notion of solvable operators in the general $L^p$-setting.

(v) *Let* $\lambda \in \rho_e(M)$ *and* $A$ *the connected component of* $\rho_e(M)$ *to which* $\lambda$ *belongs. If* $T_{M,\lambda}$ *is a maximal extension of* $M$ *at* $\lambda$ *then the spectrum of* $T_{M,\lambda}$ *in* $A$ *consists (at most) of a sequence of eigenvalues with limits in* $\sigma_e(M)$*.*

This result gives a characterization of the spectrum of maximal extensions.

## Main Results

**Theorem 2** *Let* $M = \sum_{k=0}^{n} a_k D^k$ *where* $J = [a, \infty)$, $D = d/dt$, $a_k \in C^k(J)$, $0 \leq k \leq n$, *and* $a_n(t) \neq 0$ *for* $t \in J$. *Let* $\lambda \in \rho_e(M)$ *and* $A$ *the connected component of* $\rho_e(M)$ *to which* $\lambda$ *belongs. Let* $E$ *be the boundary of* $A$. *Suppose a maximal extension of* $M$, $T_{M,\lambda}$, *exists at* $\lambda \in A$. *Then for* $\alpha \in A$ *and a constant* $\delta$, $0 < \delta < d(\alpha, E)$, *there exists a finite dimensional space* $\mathcal{S}$ *and a constant* $K$ *dependent on* $\delta$ *such that*

$$\|(T_{M,\lambda} - \alpha) f\| \geq K \|f\|$$

*for all* $f$ *in* $\mathcal{D}(T_{M,\lambda}) \cap \left(L^2(J) \setminus \mathcal{S}\right)$.

***Proof:*** Let $T_{M,\lambda}$ be a maximal extension of $M$ at $\lambda \in A$. Then the spectrum of $T_{M,\lambda}$ in $A$ consists (at most) of a sequence of eigenvalues which accumulate in $\sigma_e(M)$. Consider the disk

$$D = \overline{B(\alpha, \delta)} = \{ z \in \mathbb{C} \mid | z - \alpha | \leq \delta \}$$

where $\alpha \in A$ and $0 < \delta < d(\alpha, E)$. Then the spectrum of $T_{M,\lambda}$ in $\overline{B(\alpha, \delta)}$ consists only of a finite number of eigenvalues. Let $\gamma_1, \gamma_2, \cdots, \gamma_p$ be these eigenvalues and $\mathcal{S}_1, \mathcal{S}_2, \cdots, \mathcal{S}_p$ be the corresponding subspaces of eigenvectors. Define $\mathcal{S}$ such that

$$\mathcal{S} := \mathcal{S}_1 \oplus \mathcal{S}_2 \oplus \cdots \oplus \mathcal{S}_p .$$

Then $\mathcal{S}$ is finite dimensional and the restriction, $T_{M,rest}$ of $T_{M,\lambda}$ on the set

$$\mathcal{D}(T_{M,rest}) := \mathcal{D}(T_{M,\lambda}) \cap \left(L^2(J) \setminus \mathcal{S}\right)$$

is dense in $L^2(J) \setminus \mathcal{S}$ and for $f \in \mathcal{D}(T_{M,rest})$, $T_{M,rest} f$ is in $L^2(J) \setminus \mathcal{S}$. Furthermore $T_{M,rest}$ has no spectrum in $D$, i.e. $\overline{B(\alpha, \delta)}$ is contained in the resolvent set of $T_{M,rest}$.

In particular, $\alpha \in \rho(T_{M,rest})$ and $(T_{M,rest} - \alpha)^{-1}$ exists and is bounded. More generally, the resolvent function

$$R(., T_{M,rest}) : D \longrightarrow \text{set of bounded operators in } L^2(J) \setminus \mathcal{S}$$

where

$$x \longrightarrow (T_{M,rest} - x)^{-1}$$

is holomorphic.

By the maximum-modulus principle [2], $\|R(.,M_{\mathcal{B}})\|$ does not attain its maximum at the interior point of $D$. Hence the maximum of $\|R(.,T_{M,rest})\|$ must be on the boundary of $D$. That is, there exists $\lambda_0$ in $\{ z \in \mathbb{C} \mid |z-\alpha| = \delta \}$ such that for $\lambda \in D$ we have

$$\left\|(T_{M,rest}-\lambda)^{-1}\right\| \leq \left\|(T_{M,rest}-\lambda_0)^{-1}\right\|.$$

Hence for $f \in \mathcal{D}(T_{M,rest})$ and $\lambda \in D$,

$$\begin{aligned} \|f\| &= \left\|(T_{M,rest}-\lambda)^{-1}(T_{M,rest}-\lambda)f\right\| \\ &\leq \left\|(T_{M,rest}-\lambda)^{-1}\right\| \|(T_{M,rest}-\lambda)f\| \\ &\leq \left\|(T_{M,rest}-\lambda_0)^{-1}\right\| \|(T_{M,rest}-\lambda)f\|. \end{aligned}$$

In particular, for $f \in \mathcal{D}(T_{M,\lambda}) \cap \left(L^2(J) \setminus \mathcal{S}\right)$ we have

$$\|(T_{M,\lambda}-\alpha)f\| \geq K\|f\|$$

where $K = \left\|(T_{M,\lambda}-\lambda_0)^{-1}\right\|^{-1}$.

■

**Corollary 3** *Let* $M = \sum_{k=0}^{n} a_k D^k$ *where* $J = [a,\infty)$, $D = d/dt$, $a_k \in C^k(J)$, $0 \leq k \leq n$, *and* $a_n(t) \neq 0$ *for* $t \in J$. *Suppose* $\sigma_e(M)$ *does not exist. If* $T_{M,\lambda}$ *is a maximal extension of* $M$ *at* $\lambda \in \mathbb{C}$, *then for* $\delta > 0$ *there exists a constant* $K$ *and a finite dimensional space* $\mathcal{S}$ *such that for* $f$ *in* $\mathcal{D}(T_{M,\lambda}) \cap \left(L^2(J) \setminus \mathcal{S}\right)$,

$$\|T_{M,\lambda}f\| \geq K\|f\|.$$

***Proof :*** This follows directly from Theorem 2 by setting $A = \mathbb{C}$ and $\alpha = 0$.

■

## Notes on Related Research

Properties of maximal extensions of a differential expression $M$ may be used in establishing the existence of the essential spectrum for large classes of differential expressions. In a related research done by the author, it has been shown that the essential spectrum exists for a class of differential expressions having normal maximal extensions. Theorem 2, in particular, will be useful in generalizing this result.

**Acknowledgments**

This research was made possible through grants from the Office of the Vice-Chancellor for Research and Development, University of the Philippines-Diliman and the Natural Sciences Research Institute, University of the Philippines-Diliman.

**References**

1. M. Castillo, "Maximal Extensions of Differential Operators in $L^2[a,\infty)$", Proceedings of the Third Sino-Philippine Symposium in Analysis, *Matimyas Matematika*, Special Issue, August 2000, pp. 6-12.
2. N. Dunford, and J. T. Schwartz, "Linear Operators," Part I and Part II, Interscience Publishers, Inc., New York, 1963.
3. S. Goldberg, "Unbounded Linear Operators," McGraw-Hill, New York, 1966.
4. G. C. Rota, Extension theory of differential operators, *Comm. Pure Appl. Math* **11**, pp.23-65.
5. J. Weidmann, "Linear Operators in Hilbert Spaces", Springer-Verlag, New York, 1980.

# GENERATION OF NAFIL LOOPS OF SMALL ORDER

RAOUL E. CAWAGAS
*SciTech RD Center, Polytechnic University of the Philippines, Manila*
*E-mail: raoulec@pacific.net.ph*

NAFILs are non-associative finite invertible loops which form a class that includes IP, Moufang, and Bol loops. This paper deals with the computer generation and determination of all non-isomorphic NAFILs of orders $n = 5, 6, 7$. Results show that the number $I_n$ of non-isomorphic NAFILs of order $n$ are: $I_5 = 1$, $I_6 = 33$, and $I_7 = 2,333$. These loops are now being studied to determine their structural and other properties.

## 1 Introduction

In 1998, we completed a two-year research project* at the Polytechnic University of the Philippines to study *Non-Associative Finite Invertible Loops (NAFIL)*. This is a class of non-associative loops (also called *pseudogroups*) in which every element has a unique two-sided inverse and it includes the familiar IP, Moufang, and Bol loops. In particular, the loops generated by the basis vectors of Cayley-Dickson algebras belong to this class. However, most of these interesting loops have not yet been studied as much as other loops. A survey of the literature shows that there has been no systematic study to determine NAFIL loops of small order.

Part of this project involved the determination of all distinct (non-isomorphic) NAFILs of orders $n = 5, 6$, and 7. For this, we developed a Pascal program called *ICONSTRUCT* to generate and determine the number $I_n$ of distinct (non-isomorphic) finite invertible loops of given orders. For NAFILs, the results of this study are summarized as follows:

| Order $n$ | 5 | 6 | 7 |
|---|---|---|---|
| Number $I_n$ | 1 (N) | 33 (7A + 26N) | 16 (A only) |

For orders 5 and 6, the figures given are complete. For order 7, however, only the abelian (A) NAFILs were determined at that time. The non-abelian (N) ones were not completely determined because the program ICONSTRUCT would have taken several months to check the enormous number of possible NAFIL loops for isomorphism. Because of this, we decided in 1999 to use more efficient programs to determine all non-isomorphic NAFILs of

* Research supported by the National Research Council of the Philippines (NRCP).

order 7. This paper shows the results of this undertaking and how it was accomplished.

## 2 Non-Associative Finite Invertible Loops

A *loop* is a variety of *quasigroup* with a unique identity element. An *invertible loop,* usually denoted by $(\mathcal{L},\star)$, is a variety of loop defined by the identities

$$e\star x = x = x\star e = x \quad \text{and} \quad x\star x^{-1} = e = x^{-1}\star e$$

for all $x \in \mathcal{L}$, where $\star$ is a binary operation over the set $\mathcal{L}$, $e$ is a unique *identity* element, and every element $x \in \mathcal{L}$ has a unique *inverse* $x^{-1} \in \mathcal{L}$. Such an algebraic system may satisfy other conditions like *commutativity* and *associativity*.

In this paper, we shall be primarily concerned with the class of invertible loops $(\mathcal{L},\star)$, where $\mathcal{L}$ is of finite order $n$ and the operation $\star$ is non-associative. This is the type of loop we call the *NAFIL (*an acronym for *Non-Associative Finite Invertible Loop).*

## 3 Computer Generation of NAFILs

Being a form of quasigroup, every loop of order $n$ is defined by a Latin square in standard form over the set $\{1,2,...,n\}$, where $1\equiv e$ is the loop identity. Therefore, the problem of determining the exact number of all non-isomorphic loops of a given order is basically equivalent to the generation of all distinct Latin squares in standard form. It is known, for instance, that the number $l_n$ of all loops (both associative and non-associative) of orders $n = 5, 6, 7$ and 8, are given in the following table:

| Order $n$ | 5 | 6 | 7 | 8 |
|---|---|---|---|---|
| Number $l_n$ | 56 | 9,408 | 16,942,080 | 535,281,401,856 |
| Number $k_n$ | 6 | 109 | 23,750 | not known |

In this table, $k_n$ represents the exact number of non-isomorphic loops[2,5] (or *isomorphy classes*) of order $n$. Except for the $k_5 = 6$ loops (which were completely determined by manual construction by A. A. Albert[1] in 1944) and the $k_6 = 109$ loops (which were computer-generated by B. F. Bryant and H. Schneider[2] in 1966), most of the other figures given above were determined by calculation using combinatorial formulas. What these loops actually are were not known because their Cayley tables were not available for analysis.

Our preliminary studies have indicated that only a small portion of the loops in the above table are distinct NAFILs and their exact numbers $I_n$ have not yet been determined by calculation from the known values of $k_n$. To date, not much work has been done to study NAFILs because the theory of loops is still a fairly young discipline.

Because of the importance of studying NAFILs of small order in the development of *loop theory*, we decided in 1996 to actually generate some of these loops (in the form of Cayley tables) and determine their exact number by developing the program ICONSTRUCT. This computer program enabled us to generate the Cayley tables of all non-isomorphic NAFILs of orders $n = 5$ and 6 but it proved to be not efficient enough to determine those of order $n = 7$ in a reasonable period of time – except for the abelian ones. Using a PC (Personal Computer) with a Pentium processor, it is estimated that it would take several months for ICONSTRUCT to process 16,942,080 loops to determine all distinct NAFILs of order 7. Clearly, this would have been impractical to undertake.

In November 1999, it has come to our attention that there are two softwares, SATO[6] and SEM[7], developed by Prof. Hantao Zhang at the University of Iowa, USA. These are finite model generators that were successfully used to study quasigroups. Given a set of logic formulas and a number $n$, SEM will generate models of size $n$ satisfying the formulas. SATO can do the same; however, its input must be a set of propositional formulas.

So, we posed the NAFIL problem to Zhang and told him about the program ICONSTRUCT[3] and how it works. After studying the matter, Zhang agreed to collaborate with us by using SEM and SATO to determine the number of distinct NAFILs of order $n = 7$.

Both SEM and SATO can easily generate all NAFILs of a given order $n \leq 7$. In fact, SEM generated 29,679 (approximately 30K) NAFILs of order 7 in just a couple of minutes. These generated loops, however, were not all distinct because, unlike ICONSTRUCT, both SEM and SATO were not designed to eliminate all NAFILs isomorphic to a given NAFIL. Hence, they can not be used directly to determine the number $I_n$ of non-isomorphic NAFILs of order $n$.

The bottleneck of the problem was therefore how to identify isomorphic NAFILs. Zhang modified SATO so that it can check two given loops for isomorphism. Such a checking takes almost no time (about 0.01 second). Then we asked him to develop a simple shell script (similar to ICONSTRUCT) that removes, from the list of NAFILs generated by SEM, all loops isomorphic to a given NAFIL. Thus, he formed a SEM-SATO system, where SEM generates the loops while SATO removes all isomorphic duplicates of a given loop.

Taking the approximately 30K NAFILs of order 7 yielded by SEM, the script would make approximately $900\text{K}^2$ isomorphism checks if all of the loops were distinct. The script takes about 0.1 second for each checking/removal of duplicate. So it would take about 90K$^2$ seconds or about 35 months to finish the job. Clearly, as with ICONSTRUCT, this again would have been impractical to undertake using a PC with a single processor.

Fortunately, we knew through limited tests that most of these 30K loops were isomorphic. So, Zhang finally decided to run the SEM-SATO system in a supercomputer (with 48 Pentium II 400 processors) and finished the job in about three days[8]. He also run the program for $n = 5$ and 6 and obtained the same results as those obtained by ICONSTRUCT. Hence we now have, *for the first time*, the following complete figures for all distinct NAFILs of orders $n = 5$, 6, and 7:

| Order $n$ | 5 | 6 | 7 |
|---|---|---|---|
| Number $l_n$ | 56 | 9,408 | 16,942,080 |
| Number $k_n$ | 6 | 109 | 23,750 |
| Number $I_n$ | 1 (N) | 33 (7A+26N) | 2,333 (16A+2,317N) |

We also tried to tackle the difficult case of NAFILs of order 8. However, we have not yet been able to complete the job using the SEM-SATO system because of the extremely large number (535,281,401,856) of loops to be checked for isomorphism. We have so far generated about 2,257 non-isomorphic abelian NAFILs of order 8 but we have not yet determined their exact number. Some rough estimates indicate that there are about 4K$^2$ NAFILs of this order. To settle this problem, we are now developing a new loop generator program called *LOOPGEN* which shows some promise of greatly reducing the computing time.

## 4 Analysis of NAFILs

Our research team at the SciTech R&D Center of the Polytechnic University of the Philippines is now studying these NAFIL loops to determine their structural and other properties. To date, our results have revealed many interesting properties of these loops[3].

Because there is no available commercial software for studying non-associative finite algebraic structures, we found it necessary to develop a special software called *FINITAS*[4] for the analysis and construction of finite mathematical systems like loops and quasigroups. Using *FINITAS*, we have so far completed the analysis of all NAFILs of orders 5 and 6. For order 7 only the abelian ones have been completely analyzed[3]; the analysis of the non-abelian ones is not yet complete.

The analysis is based on the following scheme aimed at classifying NAFILs of small order $n \leq 8$. Each NAFIL belongs to an *isomorphism class,* that is, the class of all NAFILs of order $n$ that have the same abstract properties (isomorphic). Each ***class*** (defined by the Cayley table of a ***class representative***) is identified by its *order* and *class number* and is characterized by its ***type*** according to a set of ***basic characteristics*** which include: *Commutativity*; *Structural form*; *Number of subsystems*; *Number of self-inverse elements*; *Number of generators*; *Weak associative properties*; *Others.*

Two classes are distinct if they differ in at least one of their characteristics. The following Class-Type (**CT**) system of notation[3], based on the basic characteristics given above, was therefore used to designate NAFILs of small orders.

> **Class (C)** [Order (n-number); Class designation number (# numeral)] and **Type (T)** [Commutativity (A-abelian, N-nonabelian); Structural form (S-simple, NS-not simple, SE-simple and elementary), Number of subsystems; Number of self-inverse elements; Number of generators (G-generator, LG-left generator, RG-right generator); Weak associative properties (IP, LIP, RIP, PA); and Others (X-additional property, ∅-not required)].

This is simply expressed in compact form by:

> **C**[order; class designation number]**T**[commutativity; structure; subsystems; self-inverse; generators; associative properties; others]

**Example**: **C[6;#1]T[A;NS;2;2;2G;0;∅]** means: a NAFIL of order 6, class #1, abelian, not simple, with 2 subsystems, 2 self-inverse elements, 2 generators, and no weak associative properties. The symbol ∅ indicates that no other property is required to characterize this loop.

One of the most important results of the analysis of NAFILs is the establishment of their existence. It can be shown that any invertible loop of order $n \leq 4$ is a group and that, if $n = 5$, there exists exactly one (up to isomorphism) non-associative invertible loop $(L_5, *)$ of this order whose Cayley table is that shown in Figure 1. Hence, this loop is the smallest NAFIL. Analysis (using *FINITAS*) shows that this NAFIL is designated by: C[5;#1]T[N;S; 5;5;0;PA;∅]. Thus, it is non-abelian, simple, has 5 subsystems, 5 selfinverse elements, no generators, and is power-associative.

Moreover, we have also shown that if $n \geq 5$, there are two special families of NAFILs whose members cover all ODD and EVEN orders from $n = 5$ to infinity[3]. Thus, we have the following

**Theorem 1** *There exists at least one NAFIL loop of every finite order* $n \geq 5$.

The Cayley tables of the first members of the ODD and EVEN families of NAFILs are shown in Figure 1. The loop $(L_5, *)$ is the smallest member of the ODD family, while $(L_6, \star)$, designated by C[6;#7]T[A;S;6;6;PA;∅], is the smallest member of the EVEN family. Both of these loops are simple and power-associative.

$$\begin{array}{cc} ODD & EVEN \\ \begin{bmatrix} 1\,2\,3\,4\,5 \\ 2\,1\,5\,3\,4 \\ 3\,4\,1\,5\,2 \\ 4\,5\,2\,1\,3 \\ 5\,3\,4\,2\,1 \end{bmatrix} & \begin{bmatrix} 1\,2\,3\,4\,5\,6 \\ 2\,1\,4\,5\,6\,3 \\ 3\,4\,1\,6\,2\,5 \\ 4\,5\,6\,1\,3\,2 \\ 5\,6\,2\,3\,1\,4 \\ 6\,3\,5\,2\,4\,1 \end{bmatrix} \\ (L_5, *) & (L_6, \star) \end{array}$$

Figure 1. The *Cayley* tables of the first members of the ODD and EVEN families of NAFILs (non-associative finite invertible loops).

## 5 Summary

The generation and determination of all non-isomorphic NAFILs of orders $n = 5, 6, 7$ have been successfully accomplished by means of three computer programs: ICONSTRUCT, SEM, and SATO. For $n = 7$, the use of a supercomputer (with 48 Pentium II 400 processors) was required. These programs not only generated the loops but also determined the exact number $I_n$ of non-isomorphic loops of order $n \leq 7$. The generated loops are in the form of Cayley tables which can be studied for their structural properties. These Cayley tables are now being studied and can be made available to interested researchers.

## References

1. Albert, A. A., *Quasigroups II*, Trans. Amer. Math. Soc. 55 (1944)
2. Bryant, B. F., Schneider, H., *Principal loop-isotopes of quasigroups*, Canad. J. Math. 18 (1966), 120-125. MR 32 (1966) #5772.
3. Cawagas, R. E., *Terminal Report: Development of the Theory of Finite Pseudogroups* (1998). Research supported by the National Research Council of the Philippines (1996-1998) under Project B-88 and B-95.

[See also: R. E. Cawagas, *On the Existence of Pseudogroups*, Matimyas Matematika, 5 (1981), 8-13.]

4. Cawagas, R. E., *FINITAS - A Software for the construction and Analysis of Finite Algebraic Structures*, PUP Journal of Research and Exposition, Vol 1, No. 1 (1997)
5. Pflugfelder, H., *Quasigroups and Loops: Introduction*, Helderman Verlag, Berlin (1990), pp. 27-28.
6. Zhang, H., Bonacina, M.P., Hsiang, H., *PSATO: a distributed propositional prover and its application to quasigroup problems.* Journal of Symbolic Computation (1996) 21, 543–560
7. Zhang, J., Zhang, H., *Generating models by SEM*, Proc. of International Conference on Automated Deduction (CADE-96), pp. 308–312, Lecture Notes in Artificial Intelligence 1104, Springer-Verlag.
8. Zhang, H., *Generation of NAFILs of Order 7*, Association of Automated Reasoning Newsletter, No. 46, January 2000.

# Contact topology and $CR$ geometry in three dimensions

Jih-Hsin Cheng
Institute of Mathematics, Academia Sinica
Taipei, R.O.C.(Taiwan)
E-mail: cheng@math.sinica.edu.tw

**Abstract**

We study low-dimensional problems in topology and geometry via a study of contact and Cauchy-Riemann ($CR$) structures. A contact structure is called spherical if it admits a compatible spherical $CR$ structure. We will talk about spherical contact structures and our analytic tool, an evolution equation of $CR$ structures. We argue that solving such an equation for the standard contact 3-sphere is related to the Smale conjecture in 3-topology. Furthermore, we propose a contact analogue of Ray-Singer's analytic torsion. This "contact torsion" is expected to be able to distinguish among "spherical space forms" $\{\Gamma \backslash S^3\}$ as contact manifolds. We also propose the study of a certain kind of monopole equation associated with a contact structure. In view of the recently developed theory of contact homology algebras, we will discuss its overall impact on our study.

## 1 Spherical contact structures

Let $(M^3, \xi)$ denote a contact 3-manifold with the contact structure $\xi$. (assume $M^3$ oriented and $\xi$ cooriented if necessary) We call an almost complex structure $J$ on $\xi$ a $CR$ (stands for Cauchy-Riemann) structure (compatible with $\xi$). That is to say, an endomorphism $J : \xi \to \xi$ such that $J^2 = -Id_\xi$. There are no local invariants for $(M^3, \xi)$ according to a well known theorem of Darboux. Also for closed $M^3$, two nearby contact structures are isotopy-equivalent by a theorem of Gray. ([Gr], [Ham]) Therefore a contact structure on a closed $M^3$ has no continuous moduli. On the other hand, we do have local invariants for a $CR$ 3-manifold $(M^3, \xi, J)$. Namely, we can talk about "curvature". Our strategy of studying 3-topology is via a study of contact topology and $CR$ geometry.

There are distinguished $CR$ structures $J$, called spherical, if $(M^3, \xi, J)$ is locally $CR$ equivalent to the standard 3-sphere $(S^3, \hat{\xi}, \hat{J})$, or equivalently if there are contact coordinate maps into open sets of $(S^3, \hat{\xi})$ so that the transition contact maps can be extended to holomorphic transformations of open sets in $C^2$. In 1930's, Elie Cartan ([Ca], [CL1]) obtained a geometric quantity, denoted as $Q_J$, by solving the local equivalence problem for the $CR$ structure so that the vanishing of $Q_J$ characterizes $J$ to be spherical. We will call $Q_J$ the Cartan (curvature) tensor. A contact structure $\xi$ is called spherical if there is a spherical $CR$ structure compatible with it.

Our main concern is the existence problem of spherical contact structures. For instance, we ask if any homology 3-sphere admits a spherical contact structure, or does there exist a nonspherical contact homology 3-sphere? Notice that a spherical (contact) homotopy 3-sphere is contact-diffeomorphic to $(S^3, \hat{\xi})$.

It has been believed that for closed $M^3$, a spherical (contact) structure is tight. (for open $M^3$, Eliashberg gave counterexamples) We probably can prove this conjecture by showing that the contact homology (recently developed by Eliashberg, Givental and Hofer, [EGH]) of a spherical structure does not vanish. So, most likely, {spherical structures} is a restricted class of tight contact structures, to which we can apply more analytic methods.

## 2 The Cartan flow

The tool we would like to use is the so-called Cartan flow, an evolution equation for $CR$ structures $J_{(t)}$ on $(M^3, \xi)$:

$$\partial_t J_{(t)} = Q_{J_{(t)}}. \tag{2.1}$$

Namely, we deform a $CR$ structure in the direction of its Cartan tensor. And we hope that the limit $CR$ structure has the vanishing Cartan tensor, therefore is spherical.

First note that (2.1) is a system of 4th order nonlinear subparabolic equations (up to an action of contact diffeomorphisms). ([CL1]) Second, we'll mention some topological and geometrical implications of solving such an equation. Before doing that, let us see what have been known.

In the late 1980's, it was observed that (2.1) is a downward (negative) gradient flow. In fact, D. Burns and C. Epstein (also J. Lee and myself) ([BE], [CL1]) found an energy functional $\mu_\xi$ defined on a certain space of $CR$ structures (assuming trivial holomorphic tangent bundle for instance) so that

$$\delta\mu_\xi(J) = -Q_J$$

(meaning $D\mu_\xi(J)(E) = -< Q_J, E >$ for any tangent vector $E$ at $J$, in which $< , >$ is the inner product induced by the Levi form).

The short time solution can be proved by adding a gauge-fixing term to the right-hand side of (2.1). The linearization of the resulting equation is subparabolic with the leading space term of the form $-(const)L_\alpha^* L_\alpha u$. Here $L_\alpha$ is the generized Folland-Stein operator and subelliptic if $\alpha$ is not an odd integer. In our case, $\alpha = 4 + i\sqrt{3}$. ([CL1])

Now we come back to the first potential application in 3-topology by solving (2.1) just for $(M^3, \xi) = (S^3, \hat{\xi})$. This will confirm the so-called Smale conjecture: $Diff(S^3) \approx O(4)$ (" $\approx$ " means "homotopy equivalent") as first pointed out by Eliashberg in the early 1990's. In fact, Hatcher ([Hat]) gave a combinatorial proof in 1983. But people are still seeking for more geometric proofs. We can argue that the solution for $(S^3, \hat{\xi})$ implies the Smale conjecture as follows. Since $\hat{J}$ is the unique spherical $CR$ structure on $(S^3, \hat{\xi})$, any other $J$ will converge to $\hat{J}$ through the Cartan flow. This means a certain marked $CR$ moduli space $\Im'/C'$ is contractible. But $\Im'$, the space of certain marked $CR$ structures, is contractible. So $C'$, the group of certain marked contact diffeomorphisms, is contractible too. It follows that $Diff(S^3) \approx O(4)$ by the relation between $Diff(S^3)$ and $C'$.

We remark that $\hat{J}$ is a strict local minimum for $\mu_{\hat{\xi}}$. ([CL2]) The solution to (2.1) for $(S^3, \hat{\xi})$ will imply that $\hat{J}$ is actually a global minimum. Because there seems to be no suitable maximum principle available for 4th order subelliptic operators, a proof of the solution to (2.1) would probably have to be based on a priori integral estimates in place of the usual pointwise estimates for 2nd order parabolic flows. To learn more analytic techniques, we have been working on a comparatively easier flow. Let us define an energy $e_J$ for a contact form $\theta$ as follows:

$$e_J(\theta) = \int_{M^3} (W_{J,\theta})^2 \theta \wedge d\theta.$$

Here $W_{J,\theta}$ denotes the Tanaka-Webster curvature associated with $(J, \theta)$. ([Ta], [We]) We consider the downward gradient flow of $e_J$. If we write $\theta_{(t)} = e^{2\lambda_{(t)}}\hat{\theta}$ with respect to a fixed background contact form $\hat{\theta}$, then the equation can be expressed in $\lambda_{(t)}$ as

$$\partial_t \lambda_{(t)} = \Delta_b W_{J,\theta_{(t)}}. \tag{2.2}$$

Here $\Delta_b$ denotes the (positive) sublaplacian. (notice the sign difference for $\Delta_b$ in [Lee]) The equation (2.2) is a 4th order subparabolic, but scalar, flow. (while (2.1) is a "vector" flow with two independent real unknowns) It is easy to see that the volume $\int_{M^3} \theta_{(t)} \wedge d\theta_{(t)} = \int_{M^3} e^{4\lambda_{(t)}}\hat{\theta} \wedge d\hat{\theta}$ is preserved under the flow (2.2). Under certain conditions, we can establish the following integral estimate: ([CCg])

$$\partial_t \int_{M^3} e^{5\lambda_{(t)}}\hat{\theta} \wedge d\hat{\theta} \leq C. \tag{2.3}$$

Here the constant $C$ may or may not depend on the maximum time according to the applied situations. The $C^0$ estimate, hence the long time solution and asymptotic convergence, follows from (2.3) and other known techniques. The idea of estimating an integral such as the one in (2.3) comes from the study of a certain metric flow related to general relativity. The involved integral quantity is known as the Bondi mass. We wonder if there are Bondi-mass type estimates for the Cartan flow (2.1).

A $CR$ manifold is embeddable if it can be "realized" as the boundary of a compact complex manifold. (with the $CR$ structure being the one induced from the complex structure) The embeddability is a special property for 3-dimensional $CR$ manifolds since any closed $CR$ manifold of dimension $\geq 5$ is embeddable. ([BdM]) Now it is natural to ask the following question:

**Is the embeddability preserved under the Cartan flow (2.1)?**

By a direct construction of an integrable almost complex structure, we can show that if $J_{(0)}$ is embeddable with the torsion $\equiv L_T J_{(0)} = 0$ and $W_{J_{(0)},\theta} > 0$ (or $< 0$), then $J_{(t)}$ stays embeddable (for a short time). ([Ch2]) Here $T$ denotes the Reeb vector field associated with $\theta$. In fact, the torsion stays zero under the flow. Also the existence of a $CR$ vector field $T$ is sufficient to imply the embeddability of the $CR$ structure as pointed out by László Lempert. ([Lem]) So the condition on the Tanaka-Webster curvature is redundant. In view of Lempert's results in [Lem], it seems reasonable to conjecture that the embeddability is preserved under the Cartan flow without any conditions.

On the other hand, the zero torsion condition reduces the complexity of our flow a lot. It seems to be a good starting point. We are in a situation analogous to Hamilton's Ricci flow. Namely given a closed contact 3-manifold $(M^3, \xi)$. Suppose there is a pseudohermitian structure $(J, \theta)$ with vanishing torsion and positive Tanaka-Webster curvature. Then can we conclude that $\xi$ is spherical? A possible proof is to apply the Cartan flow to show that the limit $CR$ structure (together with the fixed contact form $\theta$) has the positive constant Tanaka-Webster curvature. (recall that the torsion stays zero for all time) Therefore it has the vanishing Cartan tensor. So it is spherical.

# 3 Spherical space forms

Since the linearization $\delta Q_J$ of the Cartan tensor $Q_J$ is subelliptic modulo the action of the contact diffeomorphism group $C_\xi$ ([CL1]), the kernel of $\delta Q_J$ is finite-dimensional modulo the action of $C_\xi$. So the "virtual" dimension of the moduli space of spherical $CR$ structures is finite-dimensional. In this section, we will just consider a class of examples for the 0-dimensional case. Let $\Gamma$ denote a fixed point free finite subgroup of the $CR$ automorphism group of the standard $S^3$(which is isomorphic to $PU(2,1)$). Then the quotient space $\Gamma\backslash S^3$ inherits a (spherical) contact structure from $(S^3, \hat{\xi})$. It's natural to work on the following problem.

**Problem: Classify $\{\Gamma\backslash S^3\}$ as contact manifolds.**

It has been believed that $\Gamma_1\backslash S^3$ is contact-diffeomorphic to $\Gamma_2\backslash S^3$ if and only if they are $CR$-diffeomorphic to each other in analogy with the conformal case. Thus to deal with the above problem, we borrow ideas from quantum physics to find a potential invariant in terms of $CR$ geometry.

If we view $\mu_\xi$ as a Lagrangian (action, more accurately) in 2+1 dimensions, spherical $CR$-structures are just classical fields. Therefore, "quantum fluctuations" should give us refined invariants. In practice, we compute the partition function heuristically:

$$\begin{aligned}\mathcal{Z}_k &= \int_{\Im_\xi/C_\xi} \mathcal{D}[J] e^{ik\mu_\xi([J])} \\ &= k^{-\frac{dim}{2}}(\mathcal{Z}_{sc} + O(k^{-1}))\ (k\ large),\end{aligned}$$

in which $\mathcal{Z}_{sc}$ is called the semi-classical approximation. Note that only classical fields make contributions to $\mathcal{Z}_{sc}$. By imitating the finite dimensional case, we can compute the modulus of $\mathcal{Z}_{sc}$:

$$\begin{aligned}|\mathcal{Z}_{sc}| &= lim_{k\to\infty} k^{\frac{dim}{2}} |\mathcal{Z}_k| \\ &= \Sigma_{J:spherical} \left|\frac{det\Box_J}{det'\delta Q_J}\right|^{\frac{1}{2}},\end{aligned}$$

in which $\Box_J$ is a 4th order subelliptic self-adjoint operator related to the $C_\xi$-action, and $\delta Q_J$, the second variation of $\mu_\xi$, is also a 4th order subelliptic self-adjoint operator modulo the $C_\xi$-action. We can regularize two determinants via zeta functions. (*det'* means taking a regularized determinant under a certain gauge-fixing condition.) (see [Ch1] for more details)

**Conjecture: If $J$ is spherical,**

$$Tor(J) \equiv \left|\frac{det\Box_J}{det'\delta Q_J}\right|^{\frac{1}{2}}$$

**is independent of any choice of contact form, i.e., a $CR$ invariant.**

We expect to use $Tor(J)$ to distinguish among spherical space forms $\{\Gamma\backslash S^3\}$. And we note that $Tor(J)$ is a contact-analogue of Ray-Singer's analytic torsion while no contact-analogue is known for the Reidemeister torsion. Also we speculate that if the contact homology of $\Gamma\backslash S^3$ ([EGH]) can distinguish $\Gamma\backslash S^3$'s, it may be possible to identify $Tor(J)$ with a certain quantity composed of elements in the contact homology of $\Gamma\backslash S^3$.

Let us consider the case that $\Gamma = I^*$, the binary icosahedral group. It is known that $I^*\backslash S^3$ is just the Poincaré homology sphere $P$. Therefore its contact structure is spherical. We know that two spherical $CR$ manifolds can be glued together to form the spherical connected sum by an orientation-preserving gluing map. (the gluing map is given by a $CR$ inversion defined on the Heisenberg group $H$ minus the origin in view of a spherical $CR$ manifold being $CR$ equivalent to $H$ locally. In coordinates $(t, z)$ where $t \in R, z \in C$, the $CR$ inversion $I$ defined by $I(t, z) = (-t/|w|^2, z/w)$ in which $w = t + i|z|^2$ satisfies $I^*\theta_0 = |w|^{-2}\theta_0$ where $\theta_0 = dt + izd\bar{z} - i\bar{z}dz$ is the standard contact form on $H$. It is easy to verify that "$I$" is orientation preserving and interchanges the surfaces defined by $|w| = 2$ and $|w| = 1/2$, respectively). It follows that the connected sum $P\#P$ of $P$ and itself is spherical too. On the other hand, we have the following conjecture

**Conjecture: There does not exist any spherical contact structure on** $P\#\bar{P}$.

Here $\bar{P}$ denotes $P$ with the reverse orientation. In [EH], Etnyre and Honda proved that there does not exist any tight contact structure on $P\#\bar{P}$, either positive or negative. So if a spherical contact structure on a closed 3-manifold is tight (a previous conjecture that we mentioned in the end of section 1), then the above-mentioned conjecture holds in view of Etnyre and Honda's result. If so, we then have a homology 3-sphere that does not admit any spherical contact structure.

# 4 Monopoles and contact structures

Given a contact 3-manifold $(M^3, \xi)$ and a background pseudohermitian structure $(J, \theta)$, we have the so-called canonical *spin*$^c$-structure $c_\xi$ on $\xi^\star$. ([CCu]) With respect to $c_\xi$, we will consider the equations for our "monopole" $\Phi$ coupled to the "gauge field" $A$. Here, $A$, the *spin*$^c$-connection, is required to be compatible with the pseudohermitian connection on $M^3$. The Dirac operator $D_\xi$ relative to $A$ is identified with a certain boundary $\bar{\partial}$-operator $\sqrt{2}(\bar{\partial}_b^a + (\bar{\partial}_b^a)^\star)$. In terms of the components $(\alpha, \beta)$ of $\Phi$, our equations read as

$$\text{(4.1)} \qquad \begin{cases} (\bar{\partial}_b^a + (\bar{\partial}_b^a)^\star)(\alpha + \beta) = 0 \\ (\text{or } \alpha^a_{,\bar{1}} = 0,\ \beta^a_{\bar{1},1} = 0) \\ da(e_1, e_2) - W_{J,\theta} = |\alpha|^2 - |\beta_{\bar{1}}|^2, \end{cases}$$

where $A = A_{can} + iaI$ and $W_{J,\theta}$ denotes the Tanaka-Webster curvature. Our first step in understanding (4.1) is as follows:

*Suppose the torsion $L_T J = 0$ ($T$ is the Reeb vector field). Also, suppose $\xi$ is symplectically semifillable, and that the Euler class $e(\xi)$ is not a torsion class. Then (4.1) has nontrivial solutions (i.e., $\alpha$ and $\beta$ are not identically zero simultaneously). ([CCu])*

On the other hand, the Weitzenbock-type formula gives a nonexistence result for $W_{J,\theta} > 0$. Together with the above existence result, we can conclude the following:

*Suppose the torsion vanishes and the Tanaka-Webster curvature $W_{J,\theta} > 0$. Then, either $\xi$ is not symplectically semifillable, or $e(\xi)$ is a torsion class. ([CCu])*

We remark that Rumin ([Ru]) proved that $M^3$ must be a rational homology sphere under the conditions given above using a different method. Originally we were hoping to define contact invariants from the solution space of (4.1). But since $D_\xi$ (also $da(e_1, e_2)$) is not elliptic (not even subelliptic), the solution space might be infinite dimensional. To distinguish such spaces, it seems that we need to know more structures about the solution space. On the other hand, the contact homology algebras recently developed in [EGH] seem to provide such a structure from the algebraic point of view.

# 5 General discussion

About the Cartan flow (2.1), one would like to know under what conditions the solution to (2.1) exists for all time and converges as $t \to +\infty$ to a spherical $CR$ structure. This will be impossible in general. Even if our manifold is the sphere, if we start with an overtwisted contact structure, the solution to (2.1) can not converge since the limit $CR$ structure would perforce be diffeomorphic to the standard one (which is tight). Hence the solution must blow up at a finite time. We then ask what the shape of the blow-up set is.

In [Go], W. Goldman obtained some topological obstruction to the existence of spherical (contact) structures. In particular, $T^3$ does not admit any spherical structures. We hope to be able to obtain some contact topological obstruction in terms of contact homology algebras. To do this, we have to analyze how the contact homology changes under covering and developing maps associated with a spherical structure. K. Mohnke [Mo] has studied the contact homology of certain coverings. His work should be useful for our study. Having obtained some contact obstruction, we can then answer the nonexistence problem of spherical structures in a more refined way. For instance, we might be able to determine which ones among those known tight contact structures are nonspherical for Brieskorn homology spheres $\sum(2,3,6n-1), n \geq 2$. Also we can then easily confirm the following previously mentioned conjecture by showing that the contact homology of a spherical structure does not vanish.

**Conjecture: A spherical structure on a closed 3-manifold is tight.**

**Acknowledgment.** In preparing this article the author benefited from a number of conversations with Yasha Eliashberg, László Lempert, Kaoru Ono, Klaus Mohnke, and Mei-Lin Yau. Also, the author would like to thank the organizers of both the 3rd Asian Mathematical Conference (Quezon City) and the Contact Geometry Conference (Stanford) for their kind invitations. The main part of this article is based on the author's notes for his talks at the above mentioned two conferences.The research was partially supported by the National Science Council, R.O.C., under grant NSC 90-2115-M-001-003.

# References

[BdM] Boutet de Monvel, L., Integration des equations de Cauchy-Riemann induites formelles, Seminar Goulaouic-Lions-Schwartz (1974-75) IX.1-IX.13.

[BE] Burns, D. and Epstein, C., A global invariant for three dimensional $CR$-manifolds, Invent. Math., 92 (1988) 333-348.

[Ca] Cartan, E., Sur la géometrie pseudo-conforme des hypersurfaces de l'espace de deux variables complexe I, Ann. Mat., 11 (1932a) 17-90; Sur la géometrie pseudo-conforme des hypersurfaces de l'espace de deux variables complexe II, Ann. Sc. Norm. Sup. Pisa, 1 (1932b) 333-354.

[CCg] Chang, S.-C. and Cheng, J.-H., The Calabi flow on $CR$ 3-manifolds, in preparation.

[CCu] Cheng, J.-H. and Chiu, H.-L., Monopoles and contact 3-manifolds, to appear in Quademi di Matematica, topical volume: $CR$ and pseudohermitian geometry. (math.DG/9905066)

[Ch1] Cheng, J. H., The geometry of Cauchy-Riemann manifolds in dimension three, Proceedings of the Second Asian Mathematics Conference 1995 (Nakhon Ratchasima), World Sci. Publishing, River Edge, NJ (1998) 201-208.

[Ch2] ——, The Cartan flow and embeddability of $CR$ structures, in preparation.

[CL1] Cheng, J. H. and Lee, J. M., The Burns-Epstein invariant and deformation of $CR$ structures, Duke Math. J., 60 (1990) 221-254.

[CL2] Cheng, J. H. and Lee, J. M., A local slice theorem for 3-dimensional $CR$ structures, Amer. J. Math., 117 (1995) 1249-1298.

[EGH] Eliashberg, Y., Givental, A. and Hofer, H., Introduction to Symplectic Field Theory, math.SG/0010059.

[EH] Etnyre, J. and Honda, K., On the non-existence of tight contact structures, math.GT/9910115.

[Go] Goldman, W., Conformally flat manifolds with nilpotent holonomy and the uniformization problem for 3-manifolds, Trans. Amer. Math. Soc., 278 (1983) 573-583.

[Gr] Gray, J. W., Some global properties of contact structures, Ann. Math., 69 (1959) 421-450.

[Ham] Hamilton, R. S., The inverse function theorem of Nash and Moser, Bull. Amer. Math. Soc., 7 (1982) 65-222.

[Hat] Hatcher, A. E., A proof of the Smale conjecture, $Diff(S^3) \simeq O(4)$, Ann. Math.,117 (1983) 553-607.

[Lee] Lee, J. M., The Fefferman metric and pseudohermitian invariants, Trans. Amer. Math. Soc., 296 (1986) 411-429.

[Mo] Mohnke, K., Contact homology of coverings, preprint.

[Ru] Rumin, M., Formes différentielles sur les variétés de contact, J. Diff. Geom., 39 (1994) 281-330.

[Ta] Tanaka, N., A Differential Geometric Study on Strongly Pseudo-Convex Manifolds, Kinokuniya Co. Ltd., Tokyo (1975)

[We] Webster, S. M., Pseudohermitian structures on a real hypersurface, J. Diff. Geom., 13 (1978) 25-41.

# Upper and lower non-absolute integrals

CHENG LU PIEN AND CHEW TUAN SENG
Department of Mathematics
National University of Singapore
Singapore 119260
email : matcts@nus.edu.sg

Recently, Lee and Zhao [5] have successfully defined the Henstock integral by using only upper and lower Riemann sums. The advantage of this approach is that the proofs of theorems for the Riemann integral can be easily carried over to the Henstock integral. In this note, we shall use this approach for integrals defined on division spaces, which include the Henstock integral and the approximately continuous integral.

## 1. Upper and lower integrals

In this section, we define upper and lower integrals by using the lower and upper Riemann sums respectively. However, in order that a lower Riemann sum $\sum \inf f(\xi)(q-p)$ and an upper Riemann sum $\sum \sup f(\xi)(q-p)$ are well-defined, i.e. $\inf f(\xi)$ and $\sup f(\xi)$ are finite, we have to identify the associated point $\xi$. Hence, we need to introduce a tag function $v$ which maps a closed subinterval of $[a, b]$ to a subset of $[a, b]$. In the above, infimum and supremum are over $\xi \in v[p, q]$.

A set-valued function $v$ defined on the set of all nondegenerate closed subintervals of $[a, b]$ is called a tag function on $[a, b]$ if it maps a nondegenerate closed interval $I \subseteq [a, b]$ to a subset $v(I)$ of $[a, b]$, possibly empty. Let $a = x_0 < x_1 < \cdots < x_n = b$. Then $D = \{[x_{i-1}, x_i] : i = 1, 2, \cdots n\}$ is called a partition of $[a, b]$. For brevity, a partition $D$ is often written as $D = \{[p, q]\}$, in which $[p, q]$ represents a typical interval in $D$, namely $[x_{i-1}, x_i]$. A partition $D = \{[p, q]\}$ of $[a, b]$ is said to be $v$-fine if $v[p, q] \neq \emptyset$ for every

$[p,q]$ in $D$. A tag function $v$ is said to have the partition property if for any subinterval $[c,d]$ of $[a,b]$, there exists a $v$-fine partition of $[c,d]$. We always assume that $\mathbf{B}$ has the partition property i.e., every $v \in \mathbf{B}$ has the partition property. Let $\mathbf{B}$ be a collection of tag functions on $[a,b]$. Let $v_1, v_2 \in \mathbf{B}$. We say $v_1 \leq v_2$ if $v_2(I) \subseteq v_1(I)$ for any closed subinterval $I$ of $[a,b]$. Then $(\mathbf{B}, \leq)$ is a partially ordered set. $B$ is said to be filtering if $(B, \leq)$ is directed. We always assume that $B$ is filtering. In this note, $R$ denotes the set of all real numbers and $|I|$ the Lebesgue measure of a set I.

Let $f : [a,b] \to \mathcal{R}$ and $v \in \mathbf{B}$. Let $D = \{[p,q]\}$ be a $v$-fine partition of $[a,b]$. We define the following sums.

$$\underline{S}(v,D,f) = (D)\sum \inf_{\xi \in v([p,q])} f(\xi)|q-p|$$

$$\overline{S}(v,D,f) = (D)\sum \sup_{\xi \in v([p,q])} f(\xi)|q-p|$$

$$\overline{S}(v,f) = \sup_D \overline{S}(v,D,f)$$

$$\underline{S}(v,f) = \inf_D \underline{S}(v,D,f)$$

where supremum and infimum are taken over all $v$-fine partition $D$. We define the Lower Riemann $\mathbf{B}$-integral as $(RB)\underline{\int_a^b} f = \sup\{\underline{S}(v,f); v \in B\}$ and the Upper Riemann $\mathbf{B}$-integral as $(RB)\overline{\int_a^b} f = \inf\{\overline{S}(v,f); v \in B\}$. If $(RB)\overline{\int_a^b} f = (RB)\underline{\int_a^b} f \neq \pm\infty$, $f$ is said to be Riemann $\mathbf{B}$-integrable on $[a,b]$. We define $(RB)\underline{\int_a^b} f = (RB)\overline{\int_a^b} f = (RB)\int_a^b f$.

We remark that every $v$-fine partition is $v_0$-fine if $v \geq v_0$. Hence, $\underline{S}(v,f) \geq \underline{S}(v_0,f)$. Thus

$$(RB)\underline{\int_a^b} f = \sup_{v \geq v_0} \{\underline{S}(v,f); v \in B\}$$

and $$(RB)\overline{\int_a^b} f = \inf_{v \geq v_0} \{\overline{S}(v,f); v \in B\}$$

The proofs of theorems for the Riemann integral can he easily carried over to the Riemann B-integral [1]. For examples, additive and homogeneous properties of the integral, and criterion for integrability can be proved similarly. Here we shall list down several theorems and only prove one of them to illustrate the similarity.

**Theorem 1.1. Riemann's Criterion for Integrability**

The function $f$ is Riemann **B**-integrable on $[a, b]$ if and only if for each $\epsilon > 0$, there exists a tag function $v \in \mathbf{B}$ such that $\overline{S}(v, D, f) - \underline{S}(v, D, f) < \epsilon$ for every $v$-fine partition $D$ of $[a, b]$

**Theorem 1.2. Cauchy's Criterion**

$f$ is Riemann **B**-integrable on $[a, b]$ if and only if for every $\epsilon > 0$, there exist $v_0 \in \mathbf{B}$ on $[a, b]$ such that for any two $v_0$-fine partitions $D = \{I\}$, $D' = \{I'\}$ of $[a, b]$, we have

$$|(D) \sum_{\xi \in v_0(I)} f(\xi)|I| - (D') \sum_{\xi' \in v_0(I')} f(\xi')|I'|| < \epsilon$$

whenever $D$ and $D'$ are $v_0$-fine partitions of $[a, b]$.

**Theorem 1.3.** $f$ is Riemann **B**-integrable on $[a, b]$ if and only if there exists a real number $A$ such that for every $\epsilon > 0$, there exists $v \in \mathbf{B}$ such that $\left|(D) \sum_{\xi \in v(I)} f(\xi)|I| - A\right| < \epsilon$ whenever $D = \{I\}$ is $v$-fine partition of $[a, b]$.

**Proof.** Suppose that $f$ is Riemann **B**-integrable on $[a, b]$. By Theorem 1.1, for any given $\epsilon > 0$, there exists a tag function $v \in B$ such that $\overline{S}(v, f) - \underline{S}(v, f) < \epsilon$. Thus, $\overline{S}(v, D, f) - \underline{S}(v, f) < \epsilon$ for any $v$-fine partition $D$. Hence, $\overline{S}(v, D, f) < \epsilon + \underline{S}(v, f) \leq \epsilon + \int_a^b f$. Therefore,

$$(D) \sum_{\xi \in v(I)} f(\xi)|I| - \int_a^b f < \epsilon. \tag{1.3.1}$$

On the other hand, $\overline{S}(v,f)-\epsilon<\underline{S}(v,f)\leq\underline{S}(v,D,f)$ for any $v$-fine partition $D$. Thus,

$$\int_a^b f-\epsilon\leq\overline{S}(v,f)-\epsilon\leq\underline{S}(v,D,f)\leq(D)\sum_{\xi\in v(I)}f(\xi)|I|. \tag{1.3.2}$$

Combining (1.3.1) and (1.3.2), we have, $\left|(D)\sum_{\xi\in v(I)}f(\xi)|I|-\int_a^b f\right|<\epsilon$. Conversely, suppose that for every $\epsilon>0$, there exists $v\in B$ such that $\left|(D)\sum_{\xi\in v(I)}f(\xi)(|I|)-A\right|<\epsilon$ whenever $D$ is $v$-fine partition of $[a,b]$. Hence,

$$|\underline{S}(v,D,f)-A|<\epsilon \tag{1.3.3}$$

$$\text{and}\qquad |\overline{S}(v,D,f)-A|<\epsilon. \tag{1.3.4}$$

Then (1.3.3) and (1.3.4) imply $|\overline{S}(v,D,f)-\underline{S}(v,D,f)|<2\epsilon$. By Theorem 1.1, $f$ is Riemann $\mathbf{B}$-integrable on $[a,b]$.

## 2. Examples

**Example 1.** Let $\delta$ be a positive number. For each closed subinterval $I$ of $[a,b]$, define $v_\delta(I)=I$ if $|I|<\delta$ and $v_\delta(I)=\emptyset$ if otherwise. Let $\mathbf{B}_R=\{v_\delta:\delta>0\}$. Then $\mathbf{B}_R$ has the filtering and partition properties. Let $f$ be a real-valued bounded function defined on $[a,b]$. Then it is easy to check that $f$ is Riemann $\mathbf{B}_R$-integrable on $[a,b]$ if and only if $f$ is Riemann integrable on $[a,b]$.

**Example 2.** Let $\delta(x)>0$ on $[a,b]$ and $I$ be a closed subinterval $I$ of $[a,b]$. Define $v_\delta(I)=\{x\in[a,b]:I\subset(x-\delta(x),x+\delta(x))\}$. Let $B_M=\{v_\delta:\delta>0\}$. Then $\mathbf{B}_M$ has the filtering and partition properties. Using Theorem 1.3, it is easy to check that $f$ is Riemann $\mathbf{B}_M$-integrable on $[a,b]$ if and only if $f$ is McShane integrable on $[a,b]$, see [2,3,4,7].

**Example 3.** Let $\delta(x)>0$ on $[a,b]$. Define $v_\delta(I)=\{x:x\in I\subset(x-\delta(x),x+\delta(x))\}$. Let $\mathbf{B}_H=\{v_\delta:\delta>0\}$. Then $\mathbf{B_H}$ has the filtering and partition properties. Using

Theorem 1.3, it is easy to check that $f$ is Riemann $\mathbf{B_H}$-integrable on $[a,b]$ if and only if $f$ is Henstok integrable on $[a,b]$, see [2,3,4,5,7].

**Example 4.** Let $x \in [a,b]$ and $D_x$ be a measurable subset of $[a,b]$ with $x \in D_x$ and of density 1 at x (i.e. $\lim_{|J|\to 0} \frac{D_x \cap J|}{|J|} = 1$). Let $I = [p,q]$ be a closed subset of $[a,b]$. Let $\Delta = \{D_x : x \in [a,b]\}$. Define $v_\Delta(I) = \{x \in [a,b] : x \in I, p, q \in D_x\}$. Let $\mathbf{B_A}$ be the collection of such $v_\Delta$. Then $\mathbf{B_\Delta}$ is filtering and has the partition property. Using Theorem 1.3, we can prove that $f$ is Riemann $\mathbf{B_\Delta}$-integrable on $[a,b]$ if and only if $f$ is $AP$ integrable on $[a,b]$, see [2,3,7].

## 3. Division spaces

In this section, we shall generalize Riemann **B**-integrals to division spaces [3].

Let $T$ be a given set, and $\mathcal{T}$ be a collection of some nonempty subsets $I$ of $T$. An element $I$ in $\mathcal{T}$ is called an interval. A set-valued function $v$ defined on $\mathcal{T}$ is called a tag function if it maps an interval $I$ to a subset of $T$, possibly empty. Let $E$ be an interval set and $D = \{I\}$ be a partition of $E$. A partition $D = \{I\}$ of $E$ is said to be $v$-fine if $v(I) \neq \emptyset$ for every $I \in D$. A tag function $v$ is said to have the partition property if for any interval $E$ there exists a $v$-fine partition of $E$. Let $B$ be a collection of tag functions. $B$ is said to have partition property if each $v \in B$ has the partition property. Let $v_1, v_2, \in B$. We say $v_1 \leq v_2$ if $v_2(I) \subset v_1(I)$ for any $I \in \mathcal{T}$. We assume that $(B, \leq)$ has the filtering property.

Let $h : \mathcal{T} \times T \to R$ be an interval-point function. Let $D = \{I\}$ be a $v$-fine partition of an interval $E$. We define

$$\underline{S}(v, D, h) = (D) \sum \inf_{\xi \in v(I)} h(I, \xi)$$

$$\underline{S}(v, h) = \inf_D \underline{S}(v, D, h)$$

and
$$(RB)\underline{\int}_E dh = \sup\{\underline{S}(v,h); v \in B\}$$

Similarly, we can define $\overline{S}(v,D,h)$, $\overline{S}(v,h)$ and $(RB)\overline{\int}_E dh$.

If $(RB)\overline{\int}_E dh = (RB)\underline{\int}_E dh \neq \pm\infty$, then $h$ is said to be Riemann $B$-integrable on $E$. We define

$$(RB)\int_E dh = (RB)\overline{\int}_E dh = (RB)\underline{\int}_E dh.$$

Similar results of Section 1 can be proved. Here we only mention one of them.

**Theorem 3.1.** A function $h : \mathcal{T} \times T \to R$ is Riemann $B$-integrable on an interval $E$ if and only if there exists a real number $A$ such that for every $\epsilon > 0$, there exists $v \in B$ such that

$$|(D)\sum_{\xi \in v(I)} h(I,\xi) - A| < \epsilon$$

whenever $D = \{I\}$ is a $v$-fine partition of $E$.

Next we shall show that the Riemann $B$-integral and the Henstock integral are equivalent on division spaces. Now we recall division spaces.

Let $U'$ be a collection of some interval-point pairs $(I,\xi)$, where $I \in \mathcal{T}$, $\xi \in T$, and let $\mathcal{C}$ be a family of some subcollections $U \subseteq U'$. Each interval-point pair $(I,\xi) \in U$, where $U \in \mathcal{C}$, is said to be U-fine. Given an interval $E$, a collection $U \in \mathcal{C}$ divides $E$ if the exists a finite subset $D$ of $U$, called a division of $E$ from $U$ , or a $U$-fine division of $E$, such that $\{I : (I,\xi) \in D\}$ is a partition of $E$, i.e., a collection of mutually disjoint intervals whose union is $E$.

**Definition 3.2.** The triple $(T, \mathcal{T}, \mathcal{C})$ is called a division space if the following condition holds: P1 (Division property). For each interval $E$ in $T$, every $U \in \mathcal{C}$ divides $E$. P2(Filtering property). For every pair $U_1, U_2 \in \mathcal{C}$, there exists $U_3 \in \mathcal{C}$ with $U_3 \subset U_1 \cap U_2$.

**Definition 3.3.** [3] Let $(T, \mathcal{T}, \mathcal{C})$ be a division space and $h : \mathcal{T} \times T \to \mathbb{R}$. The function $h$ is said to be Henstock integrable to $\int_E dh \in R$ on an interval $E$ if given any $\epsilon > 0$, the exists $U \in C$ such that

$$|(D)\sum h(I,\xi) - \int_E dh| < \epsilon$$

whenever $D = \{(I, \xi)\}$ is a $U$-fine division of $E$.

Let $(T, \mathcal{T}, \mathcal{C})$ be a division space and $\mathcal{U} \in \mathcal{C}$. Define $v_u(I) = \{\xi \in T : (I, \xi) \in \mathcal{U}\}$. Let $B_{\mathcal{C}} = \{v_u : u \in \mathcal{C}\}$, Then $B_{\mathcal{C}}$ has the partition and filtering properties.

**Theorem 3.4.** Let $(T, \mathcal{T}, \mathcal{C})$ be a division space and $h : \mathcal{T} \times T \to R$. Then $h$ is Henstock integrable on an interval $E$ if and only if $h$ is Riemann $B_{\mathcal{C}}$-integrable on $E$. Furthermore, $(RB)\int_E dh = \int_E dh$.

**Proof.** It follows from Theorem 3.1 and the fact that $(I, \xi) \in U \in C$ if and only if $\xi \in v_u(I)$.

**Remark 3.5.** The above theorem is a generalization of Examples in section 2. It is known that the $B$-variational integral defined by Thomson [7] is a special case of the Henstock integral on division spaces. Hence the $B$-variational integral can also be defined by upper and lower integrals.

In Theorem 3.4, the division space $(T, \mathcal{T}, \mathcal{C})$ is given. Now we shall only start with $(T, \mathcal{T})$ and a family $B$ of tag functions defined on $\mathcal{T}$, which has the partition and filtering properties, we shall show that $B$ induces a division space $(T, \mathcal{T}, \mathcal{C})$. For each $v \in B$, define $\mathcal{U}_v = \{(I, \xi) : \xi \in v(I), I \in \mathcal{T}\}$. Let $\mathcal{C}_B$ be the family of all such $\mathcal{U}_v$. Note that for each interval $E$ in $T$, every $\mathcal{U}_v \in \mathcal{C}$ divides $E$ since $v$ has the partition property. Furthermore for every $U_{v_1}, U_{v_2} \in \mathcal{C}_B$, there exists $U_{v_3} \in \mathcal{C}_B$ with $U_{v_3} \subset U_{v_1} \cap U_{v_2}$, where $v_3(I) \subseteq v_1(I)$, $v_3(I) \subseteq v_2(I)$ for every $I \in \mathcal{T}$.

**Theorem 3.6.** Let $(T, \mathcal{T})$, $B$ and $\mathcal{C}_B$ be given as above. Suppose $h : \mathcal{T} \times T \to R$. Then $h$ is Riemann $B$-integrable on an interval $E$ if and only if $h$ is Henstock integrable on $E$ with respect to the division space $(T, \mathcal{T}, \mathcal{C}_{\mathcal{B}})$.

**Proof.** It follows from Theorem 3.1 and the definition of $\mathcal{C}_{\mathcal{B}}$.

Finally we remark that as in the $B$-variational integral and division spaces, we can introduce the local character [2, 3, 7] and decomposable [3] properties for the Riemann $B$-integral, see [1].

**References**


1. Cheng Lu Pien, The Riemann and variational B-integrals, M.Sc. thesis, NUS, 1999.
2. Chew Tuan Seng and Lee Peng Yee, Nonabsolute Integration Using Vitali Covers, New Zealand Journal of Mathematics 23(1994).
3. R. Henstock, The general theory of integration, Clarendon Press, Oxford, 1991.
4. Lee Peng Yee and R. Vyborny, The Integral : An easy approach after Kurzweil and Henstock, Cambridge University Press, 2000
5. Lee Peng Yee and Zhao Dongsheng, Upper and Lower Henstock Integrals, Real Analysis Exchange 22 (1996/1997), 734-739.
6. Ma Zheng-Min, Lee Peng Yee and Chew Tuan Seng, Absolute Integration Using Vitali Covers, Real Analysis Exchange 18(1992/93) 409-419.
7. B. S. Thomson, Differentiation bases on the real line, I and II, Real Analysis Exchange 7 (1982/83), 67-208;8(1982/83), 278-442.

# CONSTRUCTING GRAPHS THAT ARE ISOMORPHIC TO THEIR CLIQUE GRAPHS

G.L. CHIA

*Institute of Mathematical Sciences, University of Malaya, 50603 Kuala Lumpur, Malaysia*

The clique graph of a graph $G$ is the graph obtained by taking the cliques of $G$ as vertices, and two vertices are adjacent if and only if the corresponding cliques have non-empty intersection. A graph is *self-clique* if it is isomorphic to its clique graph. In this paper, we show how self-clique graphs could be constructed from those graphs that have the Helly property and the $T_1$ property.

Let $G$ be a finite simple graph. By a *clique* in $G$ we mean a maximal complete subgraph in $G$. Let $\mathcal{K}(G)$ denote the set of all cliques of $G$. The *clique graph* of $G$, denoted $K(G)$, is the intersection graph on $\mathcal{K}(G)$. That is, $K(G)$ is the graph whose vertex set is $\mathcal{K}(G)$ and two vertices are adjacent in $K(G)$ if and only if the corresponding cliques have non-empty intersection. For $n \geq 2$, define $K^n(G) = K(K^{n-1}(G))$. A graph is said to be *self-clique* if it is isomorphic to its clique graph, that is $K(G) \cong G$.

While it is known that graphs $G$ for which $K^2(G) \cong G$ exist in abundance, the constructions for self-clique graphs seem relatively unknown. In fact, in [2], Hedetniemi and Slater have proved that if $H$ is a connected triangle-free graph without vertices of degree 1, then $K^2(H) \cong H$. This result has been generalized to a wider class of graphs in [3].

A graph $G$ is said to have the *Helly property* if for every subset $\{Q_i : i \in I\} \subseteq \mathcal{K}(G)$ with $Q_i \cap Q_j \neq \emptyset$ where $i, j \in I$, we have $\cap_{i \in I} Q_i \neq \emptyset$.

A graph $G$ is said to have the $T_1$ *property* if for every distinct vertices $u, v \in G$ with $d_G(u) \geq 2$ and $d_G(v) \geq 2$, there exist $Q_1, Q_2 \in \mathcal{K}(G)$ such that $u \in Q_1$, $v \notin Q_1$ and $u \notin Q_2$, $v \in Q_2$.

It is not difficult to see that if $G$ is a connected triangle-free graph, then $G$ has the Helly property and the $T_1$ property (see Lemmas 3.2 and 3.3 in [3]). In [3], it was proved that if $G$ has the Helly property and the $T_1$ property, and that if $G$ has no vertices of degree 1, then $K^2(G) \cong G$.

**Example 1:** Let $H$ denote the graph of Fig. 1(a). Since $H$ is triangle-free, it has the Helly property and the $T_1$ property. Hence we have $K^2(H) \cong H$. The clique graph $K(H)$ of $H$ is shown in Fig. 1(b). Note that the cliques of $H$ are denoted by $Q_i$ and they become the vertices of $K(H)$.

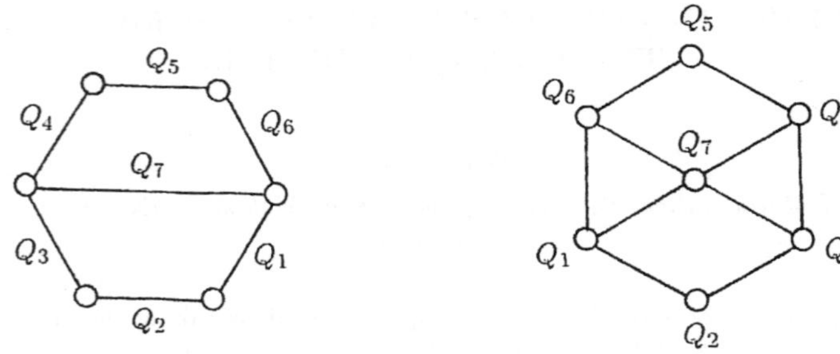


(a) $H$ (b) $K(H)$

**Figure 1. A graph and its clique graph**

In this paper, we wish to present some constructions of self-clique graphs.

(I) Suppose $H$ is a graph such that $K^n(H) \cong H$, where $n \geq 2$. Then it is easy to see that $G = H \cup K(H) \cup \cdots \cup K^{n-1}(H)$ (that is, the disjoint union of $K^i(G)$, $i = 1, 2, \ldots, n$) is a self-clique graph.

The graph in (I) is a disconnected self-clique graph. The following construction yields a connected self-clique graph.

(II) Let $H$ be a connected graph without vertices of degree 1. Suppose $H$ has the Helly property and the $T_1$ property. Then $K^2(H) \cong H$.

For any vertex $x \in H$, let $K(x)$ denote the induced subgraph of $K(H)$ whose vertex set is the set of all cliques of $H$ containing $x$. Since $H$ has the Helly property and the $T_1$ property, $K(x)$ is a clique of $K(H)$ (see Lemma 2.3 of [3]) and hence a vertex of $K^2(H)$.

Let $\varphi$ be an isomorphism from $H$ onto $K^2(H)$. Then we may take $\varphi$ to be of the form

$$\varphi(u) = K(u). \tag{1}$$

Now, take a copy of the graph $H$ and a copy of its clique graph $K(H)$. Let $w$ be a vertex in $H$. Let $G_1(H)$ denote the graph obtained by joining a new vertex $z$ to $w$ and to all vertices of $K(w)$ of $K(H)$. This construction is depicted in Fig. 2. The clique graph of $G_1(H)$ is shown in Fig. 3.

We assert that $G_1(H)$ is self-clique. To see this, we shall establish an

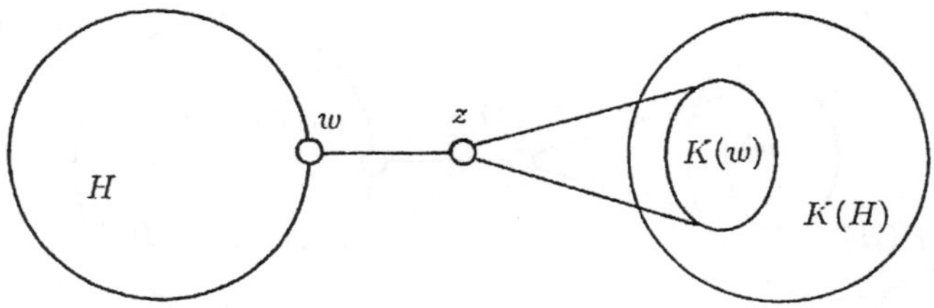


Figure 2. The graph $G_1(H)$

isomorphism from $G_1(H)$ onto $K(G_1(H))$.

First, we observe that the vertex set and the edge set of $G_1(H)$ are

$$V(H) \cup V(K(H)) \cup \{z\}$$

and

$$E(H) \cup E(K(H)) \cup \{(w, z)\} \cup \{(z, v) : v \in K(w)\}$$

respectively while the vertex set and edge set of $K(G_1(H))$ are

$$V(K(H)) \cup V_1 \cup \{\{z, w\}\}$$

and

$$E(K(H)) \cup E(K^2(H)) \cup \{(K(w), \{z, w\})\} \cup \{(\{z, w\}, v) : v \in K(w)\}$$

respectively, where

$$V_1 = \{K(v) : v \in V(H) - w\} \cup \{K(z) \cup \{z\}\}.$$

Next, let $\varphi$ be an isomorphism of $H$ onto $K^2(H)$ as defined in equation (1) and let $\pi_1$ be a mapping from the vertex set $G_1(H)$ to the vertex set of $K(G_1(H))$ defined by

$$\pi_1(v) = \begin{cases} \varphi(v) & if \;\; v \in V(H) - w \\ \varphi(v) \cup \{z\} & if \;\; v = w \\ \{z, w\} & if \;\; v = z \\ v & if \;\; v \in V(K(H)) \end{cases} \tag{2}$$

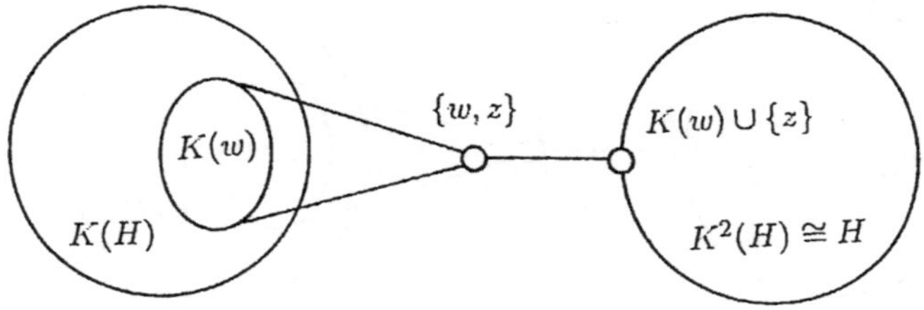


Figure 3. The graph $K(G_1(H))$

Then it is easy to see that $\pi_1$ is one to one and onto. What remains is to show that $\pi_1$ preserves adjacency.

Note that in the process of forming the clique graph $K(G_1(H))$ of $G_1(H)$,

(i) the subgraph $H$ of $G_1(H)$ gives rise to the subgraph $K(H)$ of $K(G_1(H))$,

(ii) the edge $(w, z)$ of $G_1(H)$ gives rise to the vertex $\{w, z\}$ of $K(G_1(H))$ which is adjacent to every vertices of $K(w)$ in $K(H)$, and

(iii) since $z$ is adjacent to every vertices of $K(w)$ in the subgraph $K(H)$ of $G_1(H)$, we see that the clique graph for the subgraph induced by the vertices $K(H) \cup \{z\}$ is isomorphic to the subgraph $K(K(H))$ which is isomorphic to $H$.

(i) and (ii) imply that the subgraph induced by the vertices $K(H) \cup \{z\}$ in $G_1(H)$ is isomorphic to the subgraph induced by the vertices $K(H) \cup \{w, z\}$ in $K(G_1(H))$.

Combining (i), (ii) and (iii) above, we see that $\pi_1$ preserves adjacency.

We shall summarize the above in the following theorem.

**Theorem 1** *Let $H$ be a connected graph without vertices of degree 1. Suppose $H$ has the Helly property and the $T_1$ property. Then $G_1(H)$ is a connected self-clique graph.*

**Example 2:** Apply the construction in (2) to the graph $H$ and its clique graph $K(H)$ of Fig. 1, we obtain the graph $G_1(H)$ in Fig. 4. It is easy to verify that this graph is self-clique.

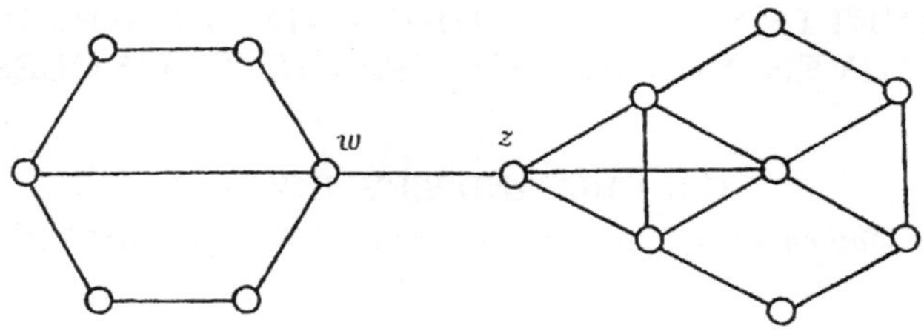


Figure 4. A self-clique graph

**Problem 1** *Suppose $H$ is a connected graph such that $K^n(H) \cong H$ for some $n \geq 3$ and that $K^i(H) \ncong H$ for $1 \leq i < n$. How could one join up the $n$ graphs $H, K(H), \ldots, K^{n-1}(H)$ so as to obtain a connected self-clique graph?*

## References


1. R. Balakrishnan and P. Paulraja, Self-clique graphs and diameters of iterated clique graphs, *Utilitas Math.* **29** (1986) 263-268.
2. S.T. Hedetniemi and P.J. Slater, Line graphs of triangleless graphs and iterated clique graphs, *Graph Theory and Applications*, Lecture Notes in Math., **303** (Springer, Berlin, 1972) 139-147.
3. C.K. Lim, A result on iterated clique graphs, *J. Austral. Math. Soc. (Series A)*, **32** (1982) 289-294.

# ON BARNETTE'S CONJECTURE AND CBP GRAPHS WITH GIVEN NUMBER OF HAMILTON CYCLES

G.L. CHIA AND SIEW-HUI ONG

*Institute of Mathematical Sciences, University of Malaya, 50603 Kuala Lumpur, Malaysia*

Barnette's Conjecture (BC) [1] states that every 3-connected cubic bipartite planar graph admits a Hamilton cycle. It is shown that Barnette's Conjecture is equivalent to the conjecture stated in the paper. As byproduct, by assuming that BC is true, we characterize all those cubic bipartite planar graphs with given cyclic connectivity and whose number of Hamilton cycles is $n$ for $n = 6$ and $n = 12$.

## 1 Introduction

While it is well-known that not all 3-connected cubic planar graphs are hamiltonian (see [10]), it is undecided whether or not the same conclusion still holds when this class of graphs is restricted to those that are bipartite. In [1], Barnette conjectured all 3-connected cubic bipartite planar (CBP) graphs are hamiltonian. This conjecture is now known as Barnette's conjecture (BC) and has been proved to be true by Holton, Manvel and McKay [6] for all such graphs of order up to and including 64. We remark, in passing, that Garey, Johnson and Tarjan [5] have proved that the problem of deciding if a cubic planar graph has a Hamilton cycle is NP-complete. In [9], Thomassen extended the result to CBP graphs. In this paper, while unable to prove that Barnette's conjecture is true, we prove that it is equivalent to Conjecture 1 stated below (Section 3). The proof makes use of the idea of a merger of (2, 3)-regular graphs (discussed in Section 2) and by induction on the number of vertices of these graphs. Earlier, Kelmans [7] has proved that Barnette's conjecture implies Conjecture 1. We refer the readers to the Introduction in [6] for background of Barnette's conjecture.

**Conjecture 1** *Let $G$ be a 3-connected CBP hamiltonian graph. Then, for any two edges $x$ and $y$ on the same face of $G$, there is a Hamilton cycle passing through $x$ and $y$, and another one passing through $x$ but avoiding $y$.*

Let $G$ be a graph. By a *$k$-cut* of $G$ we mean a set $S$ of $k$ edges of $G$ such that $G - S$ is a disconnected graph. If, in addition, $G - S$ consists of two components each of which contains a cycle, then $S$ is called a *cyclic $k$-cut*. If $k$ is the least integer with this property, the $G$ is said to have *cyclic connectivity $k$*. If $G$ is a CBP graph, then the cyclic connectivity of $G$ is no more than 4.

In Section 4 we consider the problem of characterizing CBP graphs with given cyclic connectivities in terms of the number of Hamilton cycles they posses. We attempt to solve this problem based on the assumption that Barnette's conjecture is true.

We will have occasion to use the following result of C.A.B. Smith which appears in [10].

**Theorem 1** *([10])*

*Every edge of a cubic graph lies on an even number of Hamilton cycles. Consequently a cubic hamiltonian graph has at least three Hamilton cycles.*

## 2 (2, 3)-regular graphs and merger of graphs

A graph is (2, 3)-*regular* if it is connected and the degree of each vertex is either 2 or 3. Let $H$ be a (2, 3)-regular graph and let $A$ denote the set of vertices of degree 2 in $H$. For any two vertices $x, y \in A$, let $p_H(x, y)$ or just $p(x, y)$ denote the number of Hamilton paths in $H$ with $x$ and $y$ as end vertices. The set $\{p(x, y) : x, y \in A\}$ is called the *Hamilton path sequence* of $H$ or just *path sequence* of $H$.

Let $H$ be a (2, 3)-regular graph with four vertices $x, y, z$ and $w$ of degree 2. Let $Q_1$ and $Q_2$ be two disjoint paths in $H$ whose end vertices are the vertices of degree 2 in $H$. Then $Q_1$ and $Q_2$ form a 2-*saturated path* (or 2*s*-*path*) if $Q_1 \cup Q_2$ includes all the vertices of $H$. If the end vertices of $Q_1$ are $x$ and $y$, then the 2*s*-path is denoted by $xy \cup zw$ (where $z$ and $w$ are the other two vertices of degree 2 in $H$). There are only three kinds of 2*s*-paths in $H$ depending on the end vertices of $Q_1$ (since there are only three ways to pair up the four end vertices of $H$). Let $q(x, y)$ denote the number of 2*s*-paths of the kind $xy \cup zw$. The set $\{q(x, y), q(x, z), q(x, w)\}$ is called the 2*s*-*path sequence* of $H$.

Let $G_1$ and $G_2$ be two (2, 3)-regular graphs whose sets of vertices of degree 2 are $A_1 = \{x_{1,0}, x_{1,1}, \ldots, x_{1,k-1}\}$ and $A_2 = \{x_{2,0}, x_{2,1}, \ldots, x_{2,k-1}\}$ respectively. By a *merger* of $G_1$ and $G_2$, we mean a graph obtained by joining the vertices $x_{1,i}$ and $x_{2,i}$, $i = 0, 1, \ldots, k - 1$.

The idea of a merger is attributed to A. Kotzig [8]. It was used in [4] to discuss some problems concerning the number of Hamilton cycles in cubic graphs. This was done by confining the attention only to the cases $k = 2$ and $k = 3$. In this paper, we continue the discussion by extending it to the case $k = 4$. For ease of reference, some of the relevant facts concerning the cases $k = 2$ and $k = 3$ are also provided here.

Suppose $k = 2$. Let the path sequences of $G_1$ and $G_2$ be $\{m_1\}$ and $\{m_2\}$

respectively. Then a merger of $G_1$ and $G_2$ will result in a cubic graph (with connectivity 2) whose number of Hamilton cycles is $m_1 m_2$.

Suppose $k = 3$. In this case, the definition of a merger of two graphs is also extended to the case when $G_1$ is a $(2,3)$-regular graph with three vertices of degree 2 and $G_2$ is an isolated vertex. In this case, the *merger* of $G_1$ and $G_2$ is the graph obtained by joining $G_2$ to all the vertices of degree 2 in $G_1$.

For $i = 1, 2$, let $p(x_{i,j}, x_{i,j+1}) = m_{i,j}$, $j = 0, 1, 2$ where the operations on the second subscripts are reduced modulo $k$. Then a merger of $G_1$ and $G_2$ will result in a cubic graph whose number of Hamilton cycles is given by

$$m_{1,0}m_{2,0} + m_{1,1}m_{2,1} + m_{1,2}m_{2,2} \tag{1}$$

In the event that $G_2$ is an isolated vertex, we take $m_{2,0} = 1 = m_{2,1} = m_{2,2}$ so that the number of Hamilton cycles in the merger of $G_1$ and $G_2$ is $m_{1,0} + m_{1,1} + m_{1,2}$.

**Lemma 1** *([4])*

*Suppose $\{m_1, m_2, m_3\}$ is the path sequence of some $(2,3)$-regular graph. Then all the $m_i$'s are of the same parity.*

Suppose $k = 4$. For $i = 1, 2$, let $p(x_{i,j}, x_{i,j+1}) = m_{i,j}$, $j = 0, 1, 2, 3$ and the operations on the second subscripts are reduced modulo $k$. Also, for $i = 1, 2$, let $p(x_{i,0}, x_{i,2}) = m_{i,4}$ and $p(x_{i,1}, x_{i,3}) = m_{i,5}$.

For $i = 1, 2$, let $q(x_{i,0}, x_{i,j}) = n_{i,j}$, $j = 1, 2, 3$.

Let $G$ be a merger of $G_1$ and $G_2$ whose path sequences and $2s$-path sequences are $\{m_{1,0}, m_{1,1}, \ldots, m_{1,5}\}$, $\{n_{1,1}, n_{1,2}, n_{1,3}\}$ and $\{m_{2,0}, m_{2,1}, \ldots, m_{2,5}\}$, $\{n_{2,1}, n_{2,2}, n_{2,3}\}$ respectively. Then the number of Hamilton cycles in $G$ is given by

$$\sum_{i=0}^{5} m_{1,i}m_{2,i} + \sum_{i \neq j} n_{1,i}n_{2,j} \tag{2}$$

We may also extend the definition of a merger to the case when $G_1$ is a $(2,3)$-regular graph with four vertices of degree 2 and $G_2$ is $K_2$. In this case, a merger of $G_1$ and $G_2$ is obtained by joining a vertex of $K_2$ to $x_{1,0}$ and $x_{1,1}$ and the other vertex (of $K_2$) to $x_{1,2}$ and $x_{1,3}$. We may take the path sequence and $2s$-path sequence of $K_2$ to be $\{0, 1, 0, 1, 1, 1\}$ and $\{1, 0, 0\}$ respectively so that the number of Hamilton cycles in a merger of $G_1$ and $G_2$ is $m_{1,1} + m_{1,3} + m_{1,4} + m_{1,5} + n_{1,2} + n_{1,3}$ in this case.

Let $G$ be a 3-connected CBP graph. Let $S$ be a $k$-cut of $G$ for some $k \leq 4$. Suppose $G_1$ and $G_2$ are the components of $G - S$. Then $G_1$ or $G_2$ is a $(2,3)$-regular graph with $k$ vertices of degree 2. We shall term such graph an *s-planar graph*. Notice that an $s$-planar graph can be drawn on the plane in such a way that all the $k$ vertices of degree 2 lie on the same face. An example of a $(2,3)$-regular graph that is not $s$-planar is the bipartite planar graph $K_{2,3}$ whose merger with an isolated vertex yields the non-planar graph $K_{3,3}$.

We are most interested in the case when $G_i$ is an $s$-planar graph with four vertices of degree 2. Here, notice that $n_{i,2} = 0$. In this case, let $G_i$ be drawn on the plane in the form of a square so that the four vertices $x_{i,0}, x_{i,1}, x_{i,2}$ and $x_{i,3}$ of degree 2 are located at the four corners in cyclic order. Throughout this paper, an $s$-planar graph is always drawn on the plane in this manner. Pictorially, a merger of $G_1$ and $G_2$ is represented in Figure 1 where $x_{1,i}$ is joined to $x_{2,i}$, $i = 0, 1, 2, 3$.

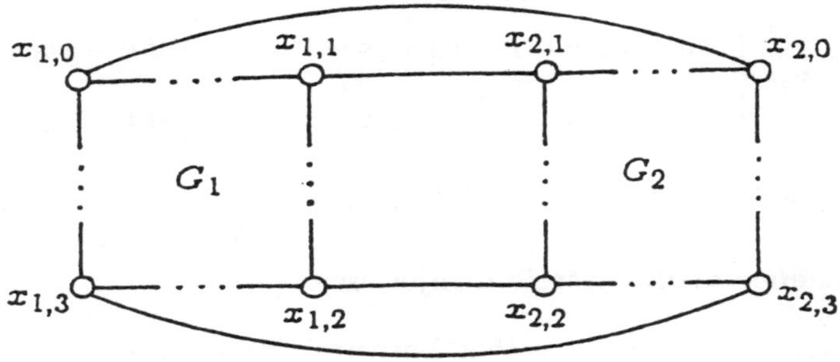


Figure 1. Merger of graphs $G_1$ and $G_2$

Since $G_i$ is bipartite, we may color the vertices of $G_i$ with two colors, say black and white, so that vertices of the same color are not adjacent. Then there are exactly two black vertices of degree 2 and two white vertices of degree 2. (See [6] p. 290 - 291.) In the event that $x_{i,0}$ and $x_{i,2}$ are of the same color, then $G_i$ is said to be of *Type-1*; otherwise it is said to be of *Type-2*. In this paper, whenever is needed, all the vertices in a CBP graph are colored with black and white.

**Example 1** Let $L_k$ denote the ladder graph $P_k \times K_2$, $k \geq 2$. Then $L_k$ is Type-1 if $k$ is even and Type-2 otherwise. Note that the graph $L_{2n}$ has path sequence $\{1,1,1,1,0,0\}$ and $2s$-path sequence $\{1,0,2n-1\}$ and that the graph $L_{2n+1}$ has path sequence $\{0,1,0,1,1,1\}$ and $2s$-path sequence $\{1,0,2n\}$. The graphs $L_2$ and $L_3$ are depicted in Figure 2.

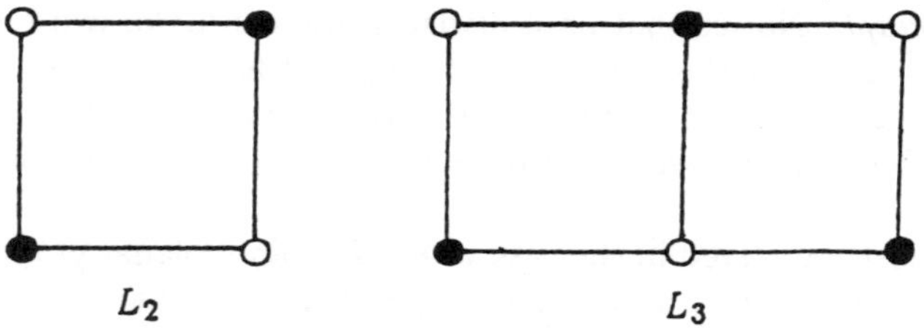

Figure 2. The graphs $L_2$ and $L_3$

Let $h(G)$ denote the number of Hamilton cycles in the graph $G$.

**Lemma 2** *Let $G$ be a merger of $L_k$ and an s-planar graph with path sequence $\{m_0, m_1, \ldots, m_5\}$ and 2s-path sequence $\{n_1, 0, n_3\}$. Then*

$$h(G) = \begin{cases} m_0 + m_1 + m_2 + m_3 + (k-1)n_1 + n_3 & \text{if } k \text{ is even} \\ \\ m_1 + m_3 + m_4 + m_5 + (k-1)n_1 + n_3 & \text{if } k \text{ is odd} \end{cases}$$

**Proof** This follows from Example 1 and Equation (2).

## 3 Equivalence of Barnette's Conjecture

In this section, we shall prove that Barnette's conjecture is equivalent to Conjecture 1 (Theorem 4). We begin by stating a result of Holton, Manvel and McKay [6] and then proving a structural property (Theorem 3) concerning 3-connected CBP graphs.

**Theorem 2** *([6])*
*Let $G$ be a 3-connected CBP graph with $n$ vertices. If $n \leq 64$, then $G$ is hamiltonian.*

Let $H$ be an $s$-planar graph whose vertices of degree 2 are $x_0, x_1, x_2$ and $x_3$ (in cyclic order). Then $H$ is said to have the *$\mathcal{A}$-property* if

(i) $(x_0, x_3), (x_1, x_2) \notin E(H)$ and

(ii) $x_0$ and $x_3$ (and likewise $x_1$ and $x_2$) are of different colors.

An example of an $s$-planar graph that has the $\mathcal{A}$-property is the ladder graph $L_{2n}$, $n \geq 2$. Note that $L_{2n+1}$ is an $s$-planar graph but does not have the $\mathcal{A}$-property.

**Theorem 3** *Let $G$ be a 3-connected CBP graph which is not the cube. Then $G$ is a merger of an $s$-planar graph having the $\mathcal{A}$-property and the ladder graph $L_k$ for some $k \geq 2$.*

**Proof** Let $f_i$ denote the number of $i$-faces in $G$. By using Euler's formula for plane graphs, we have

$$\sum_{i \geq 2} (3-i) f_{2i} = 6$$

from which it follows that $G$ has at least six 4-faces.

If $G$ contains precisely six 4-faces, then $G$ is the cube. Hence $G$ contains a $2i$-face, for some $i \geq 3$. We may then assume that there is a 4-face in $G$ that is adjacent to a $2i$-face $F$ for some $i \geq 3$. Thus $G$ is a merger of a $(2,3)$-regular graph $H$ and the ladder graph $L_k$ for some $k \geq 2$ (see Figure 3). Since $G$ is a 3-connected CBP graph, $H$ is an $s$-planar graph.

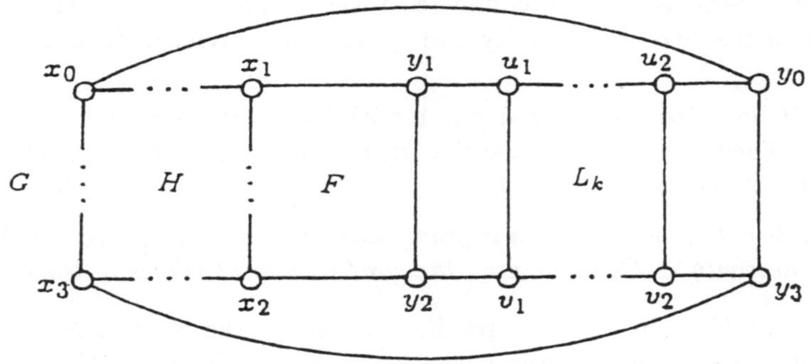


**Figure 3. Merger of $H$ and $L_k$**

Since $F$ is a $2i$-face for some $i \geq 3$, $x_1$ and $x_2$ are not adjacent in $H$. Without loss of generality, we may assume that $x_0$ and $x_3$ are not adjacent in $H$. This is because otherwise $G$ may be taken to be a merger of an $s$-planar graph $H'$ and a ladder graph $L_n$ for some $n > k$.

It remains to see that $x_0$ and $x_3$ are of different colors (and likewise for the pair of vertices $x_1$ and $x_2$). If $x_0$ and $x_3$ are of the same color, then so are the vertices $y_0$ and $y_3$. However this is impossible because $y_0$ and $y_3$ are adjacent. Similar argument applies to the vertices $x_1$ and $x_2$ to conclude that they are of different colors.

This completes the proof.

Let $H$ be an $s$-planar graph with the $\mathcal{A}$-property. Throughout this paper we shall let $H^{\oplus}$ denote the graph obtained from $H$ by joining $x_0$ to $x_3$ and $x_1$ to $x_2$ by two new edges (see Figure 4). Then $H^{\oplus}$ is a CBP graph.

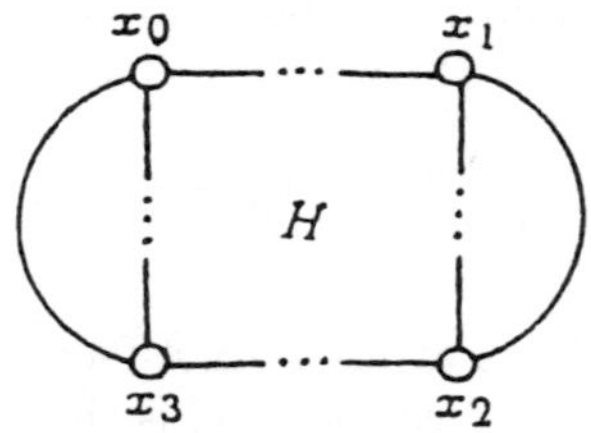


Figure 4. The graph $H^{\oplus}$

Before proving the next two lemmas, we shall make the following remark.

**Remark 1** Suppose $G$ is a connected CBP graph with connectivity $c$. If $c \leq 2$, then the edge connectivity and cyclic connectivity of $G$ are also equal to $c$.

Let $H$ be a Type-1 $s$-planar graph with four vertices $x_0, x_1, x_2$ and $x_3$ of degree 2. Then $H$ is said to have the *strong* $\mathcal{A}$*-property* if $(x_i, x_{i+1}) \notin E(H)$ for $i = 0, 1, 2$ and $(x_3, x_0) \notin E(H)$.

**Lemma 3** *Let $H$ be an $s$-planar graph with the strong $\mathcal{A}$-property. Then $H$ is of connectivity* 2. *Consequently, the connectivity of $H^{\oplus}$ is at least* 2.

**Proof** Since $H$ is an $s$-planar graph, it is one of the components of a disconnected graph obtained from some 3-connected CBP graph $G$ by deleting a 4-cut.

Suppose $H$ is of connectivity 1. Then there exists an edge $e$ in $H$ such that $H - e$ consists of two components $Y_1$ and $Y_2$. Let $A_i$ denote the set of vertices of degree 2 in $Y_i$.

*Case (1):* $|A_1| = 3 = |A_2|$.

Then all the vertices in $A_1$ are white and all the vertices in $A_2$ are black. (See [6] p. 290 - 291.) But this means that $H$ is a Type-2 $s$-planar graph, a contradiction.

*Case (2):* $|A_1| = 2$ and $|A_2| = 4$.

Suppose $A_1 = \{v_1, v_2\}$ where $v_1$ is incident with the edge $e$ and $v_2$ is incident with an edge $f$ which is in $G$ but not in $H$. But this implies that $\{e, f\}$ is a 2-cut in $G$ contradicting the 3-connectedness of $G$.

If $H$ has the strong $\mathcal{A}$-property, then we may obtain another CBP graph $H_{\oplus}$ (other than the graph $H^{\oplus}$) by joining $x_0$ to $x_1$ and $x_2$ to $x_3$ by two new edges.

We shall now describe a special kind of a Type-1 $s$-planar graph which we shall call *generalized 4-gon graph* ($g4g$ graph).

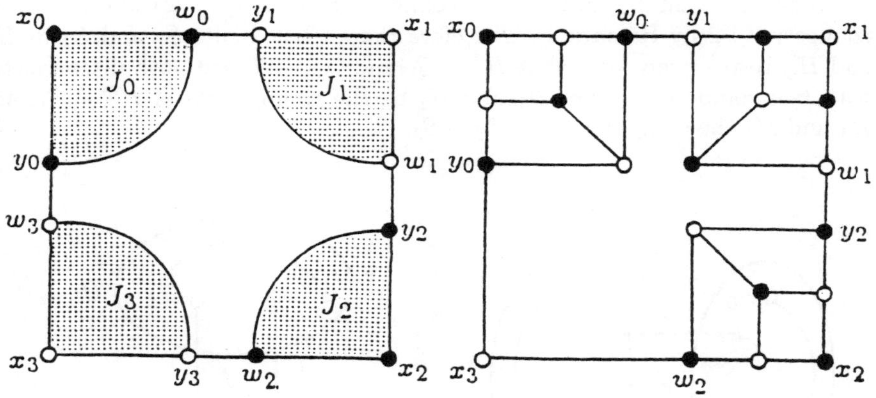


Figure 5. $g4g$ graphs

The vertices of a $g4g$ graph $H$ are defined by $V(H) = \cup_{i=0}^{3} V(J_i)$ where $J_i$ is either an isolated vertex $\{x_i\}$ or else an $s$-planar graph having precisely three vertices of degree 2. If $J_i$ is not an isolated vertex, let $\{x_i, y_i, w_i\}$ denote its vertices of degree 2.

The edges of $H$ are defined by $E(H) = \cup_{i=0}^{3} E(J_i) \cup E_1$ where $E_1 = \{(w_i, y_{i+1}) : i = 0, 1, 2, 3\}$ with the operations on the subscripts reduced modulo 4. In the event that $J_i$ is an isolated vertex, we let $x_i = y_i = w_i$. Further, in this case, we require that $J_{i+1}$ to be a non-isolated vertex. The general situation for a $g4g$ graph is shown in Figure 5(a) and an example is given in Figure 5(b). We may call $J_0, J_1, J_2$ and $J_3$ the *g-vertices* of a $g4g$ graph.

Notice that a $g4g$ graph $H$ has the strong $\mathcal{A}$-property. Further $H^{\oplus}$ and $H_{\oplus}$ are both of connectivity 2. The following lemma shows that $g4g$ graphs are the only Type-1 $s$-planar graphs with the strong $\mathcal{A}$-property such that $H^{\oplus}$ and $H_{\oplus}$ are both of connectivity 2.

**Lemma 4** *Let $H$ be a Type-1 s-planar graph with the strong $\mathcal{A}$-property. Then $H^{\oplus}$ or $H_{\oplus}$ are both of connectivity 2 if and only if $H$ is a g4g graph.*

**Proof** The sufficiency is clear. We prove the necessity.

Since $H$ is a Type-1 $s$-planar graph, it is one of the components of a disconnected graph obtained from some 3-connected CBP graph $G$ by deleting a 4-cut.

Since $H^{\oplus}$ and $H_{\oplus}$ are both of connectivity of 2, they are both of cyclic connectivity 2 by Remark 1. Therefore there exist 2-cuts $S_1$ and $S_2$ in $H^{\oplus}$ and $H_{\oplus}$ respectively such that $H^{\oplus} - S_1$ and $H_{\oplus} - S_2$ are both disconnected with two components. Let $H_1$ and $H_2$ be the components of $H^{\oplus} - S_1$, and $H_3$ and $H_4$ the components of $H_{\oplus} - S_2$.

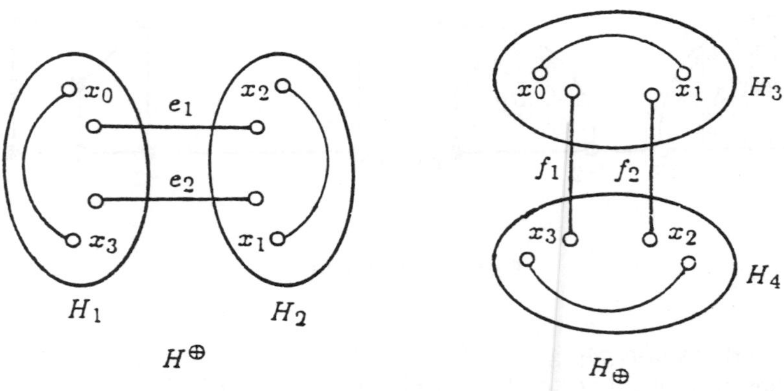


Figure 6. The graphs $H^{\oplus}$ and $H_{\oplus}$

Since $G$ is 3-connected, it must be the case that $x_0, x_3 \in H_1$ and $x_1, x_2 \in H_2$. (See Figure 6.) Similarly we have $x_0, x_1 \in H_3$ and $x_2, x_3 \in H_4$.

Suppose $S_1 = \{e_1, e_2\}$ and $S_2 = \{f_1, f_2\}$.

Let $z_1$ and $z_2$ be two vertices such that $z_1 \in H_1 \cap H_3$ and $z_2 \in H_2 \cap H_3$. Then $z_1$ and $z_2$ belong to different components of $H^{\oplus} - S_1$. Hence any path in $H^{\oplus}$ with $z_1$ and $z_2$ as end vertices must contain some edges in $S_1$. This means that $S_1 \cap E(H_3) \neq \emptyset$. Suppose $e_1 \in E(H_3)$.

In a like manner, by choosing pair of vertices from the subgraphs $H_1 \cap H_4$ and $H_2 \cap H_4$, we see that $S_1 \cap E(H_4) \neq \emptyset$. Consequently, we have $e_2 \in E(H_4)$.

Similarly, we have $S_2 \cap E(H_1) \neq \emptyset$ and $S_2 \cap E(H_2) \neq \emptyset$ and as such we may then assume that $f_1 \in E(H_1)$ and $f_2 \in E(H_2)$.

Let $e_1 = u_{1,3}u_{2,3}$, $e_2 = u_{1,4}u_{2,4}$, $f_1 = v_{1,3}v_{1,4}$ and $f_2 = v_{2,3}v_{2,4}$ where $u_{i,j}, v_{i,j} \in H_i \cap H_j$ and $(i,j) \in \{(1,3),(1,4),(2,3),(2,4)\}$. The situation for the graph $H$ is now depicted in Figure 7.

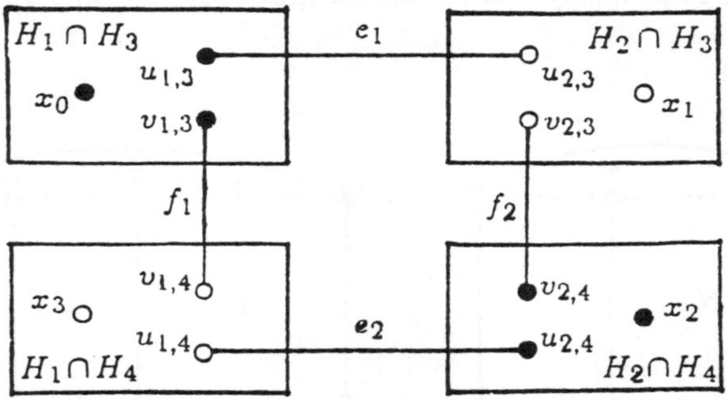


Figure 7. The graph $H$

Note that if $H_i \cap H_j$ is an isolated vertex, then $u_{i,j} = v_{i,j} = x_r$ for some $r$. In this case, it follows that neither $H_i \cap H_l$ nor $H_k \cap H_j$ ($l \neq j$ and $k \neq i$) is an isolated vertex because $H$ has the strong $\mathcal{A}$-property.

If $H_i \cap H_j$ is not an isolated vertex, we observe that vertices in $H_i \cap H_j - \{u_{i,j}, v_{i,j}\}$ are adjacent only to vertices in $H_i \cap H_j$. Clearly, $H_i \cap H_j$ has precisely three vertices of degree 2. It remains to show that $H_i \cap H_j$ is an $s$-planar graph.

Let $X_{i,j}$ denote the graph obtained by joining a new vertex to all the vertices of degree 2 in $H_i \cap H_j$. Then clearly $X_{i,j}$ is a CBP graph. If $X_{i,j}$ is of connectivity $\leq 2$, then so is $G$. Thus $X_{i,j}$ is 3-connected.

**Remark 2** Suppose $H$ is a $g4g$ graph with $g$-vertices $J_0, J_1, J_2$ and $J_3$. Let $M_i$ denote the subgraph of $H$ induced by the vertices of $J_i \cup J_{i+1}$ for some $0 \leq i \leq 3$. (Here $J_4 = J_0$.) Let $M_i^*$ denote the graph obtained from $M_i$ by joining $x_i$ to $x_{i+1}$ and $y_i$ to $w_{i+1}$ by two new edges. Then $M_i^*$ is a 3-connected CBP graph.

**Theorem 4** *Barnette's conjecture is equivalent to Conjecture 1.*

**Proof** As mentioned in Section 1, Kelman [7] has proved that BC implies Conjecture 1. Let $G$ be a 3-connected CBP graph. We shall prove that Conjecture 1 implies BC by induction on the number of vertices in $G$.

The smallest 3-connected CBP graph is the cube which is hamiltonian. In fact, by Theorem 2, any 3-connected CBP graph on no more than 64 vertices is hamiltonian.

Since $G$ is not the cube, by Theorem 3, $G$ is a merger of an $s$-planar graph $H$ having the $\mathcal{A}$-property and the ladder graph $L_k$ for some $k \geq 2$ (see Figure 3).

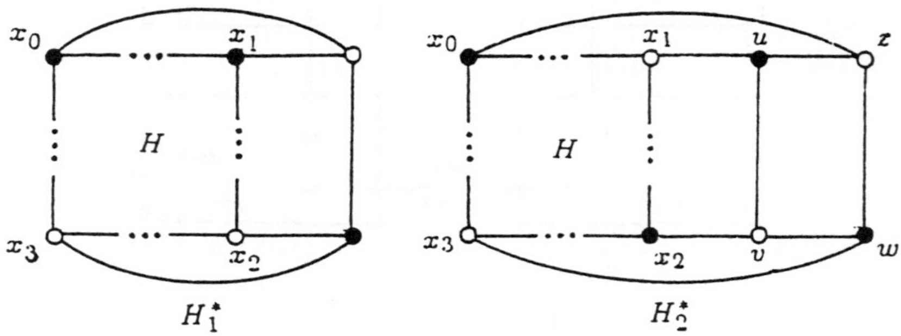


**Figure 8. $H_1^*$ and $H_2^*$**

There are two cases to consider:

*Case (1):* $x_0$ and $x_1$ have the same color.

In this case, $k$ is odd and so $k \geq 3$. Let $H_1^*$ be a merger of $H$ and $K_2 = \{u, v\}$ (see Figure 8). Then $H_1^*$ is a 3-connected CBP graph and $|H_1^*| < |G|$. By the induction hypothesis, $H_1^*$ is hamiltonian. Since we assume that Conjecture 1 is true, there is a Hamilton cycle in $H_1^*$ passing through the edge $x_0u$ but avoiding the edge $uv$. Now this Hamilton cycle must also pass through the edges $x_1u, x_2v$ and $x_3v$. This implies that there is a $2s$-path $x_0x_3 \cup x_1x_2$ in $H$. Combining this $2s$-path with the paths $x_0y_0u_2 \cdots u_1y_1x_1$ and $x_3y_3v_2 \cdots v_1y_2x_2$ (see Figure 3), we have a Hamilton cycle in $G$.

*Case (2):* $x_0$ and $x_1$ have different colors.

In this case, since the vertices $x_2$ and $x_3$ also have different colors, we see that $H$ is a Type-1 $s$-planar graph. Hence $k$ is even and $k \geq 2$.

*Subcase (2.1):* $k \geq 4$.

Let $H_2^*$ be a merger of $H$ and $L_2 = \{u, v, w, z\}$ (see Figure 8). Then $H_2^*$ is a 3-connected CBP graph and $|H_2^*| < |G|$. By the induction hypothesis and by Conjecture 1, there is a Hamilton cycle in $H_2^*$ passing through the edge $uv$ but avoiding the edge $zw$. Now this Hamilton cycle must pass through the edges $x_0z, zu, vw$ and $wx_3$. This implies that there is a Hamilton path in $H$ with $x_0$ and $x_3$ as end vertices. Combining this Hamilton path with the path $x_0y_0u_2 \cdots u_1y_1y_2v_1 \cdots v_2y_3x_3$ (see Figure 3), we have a Hamilton cycle in $G$.

*Subcase (2.2):* $k = 2$.

Since $H$ has the $\mathcal{A}$-property, obtain the new graph $H^{\oplus}$ from $H$ (see Figure 4). There are two subcases to consider.

*Subcase (2.2.1):* $H^{\oplus}$ is 3-connected.

Since $|H^{\oplus}| < |G|$, by the induction hypothesis, $H^{\oplus}$ is hamiltonian.

By Conjecture 1, there is a Hamilton cycle in $H^{\oplus}$ passing through the edge $x_0x_3$ and avoiding the edge $x_1x_2$. This implies that there is a Hamilton path in $H$ with $x_0$ and $x_3$ as end vertices. Combining this path with the path $x_0y_0y_1y_2y_3x_3$ (see Figure 3), we have a Hamilton cycle in $G$.

*Subcase (2.2.2):* $H^{\oplus}$ is not 3-connected.

In this subcase, we may further assume that $(x_0, x_1), (x_2, x_3) \notin E(H)$. This is because if $(x_0, x_1) \in E(H)$ or $(x_2, x_3) \in E(H)$, then $G$ is a merger of an $s$-planar graph $H'$ and a ladder graph $L_m$, for some $m \geq 3$ (see Figure 9). Here $H'$ is such that $(x_0', x_1'), (x_2', x_3') \notin E(H')$ and that $x_0'$ and $x_1'$ have different colors (and likewise for $x_2'$ and $x_3'$). In other words, $H'$ has the $\mathcal{A}$-property. The situation is symmetrically similar to that of either Case (1) or Subcase (2.1). Hence the previous argument applies and we have a Hamilton cycle in $G$.

Therefore $H$ has the strong $\mathcal{A}$-property. Since $H^{\oplus}$ is not 3-connected, it has connectivity 2 by Lemma 3. Consider the graph $H_{\oplus}$. If $H_{\oplus}$ is 3-connected, the situation is symmetrically similar to that of Subcase (2.2.1) and will lead to a Hamilton cycle in $G$.

If $H_{\oplus}$ is not 3-connected, then it has connectivity 2 by Lemma 3. (Note that Lemma 3 applies to $H_{\oplus}$ as well because $H$ has the strong $\mathcal{A}$-property.) By Lemma 4, $H$ is a $g4g$ graph.

By Remark 2, the graph $M_3^*$ and $M_1^*$ are both 3-connected CBP graphs. Since $|M_3^*|, |M_1^*| < |G|$, by the induction hypothesis, $M_3^*$ and $M_1^*$ are hamiltonian. By Conjecture 1, there is a Hamilton cycle $C_3$ in $M_3^*$ (respectively $C_1$ in $M_1^*$) passing through the edges $x_0x_3$ and $w_0y_3$ (respectively $y_1w_2$ and $x_1x_2$). Such Hamilton cycles in $M_3^*$ and $M_1^*$ can be combined in the following

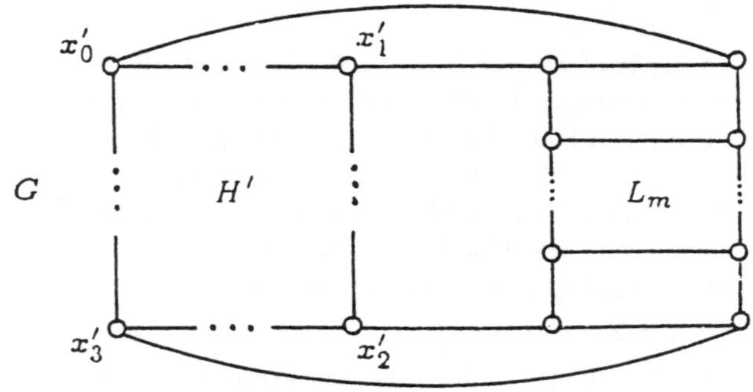


Figure 9. The graph $G$

way to yield a Hamilton cycle $C$ in $H^{\oplus}$. Delete the two edges $w_0y_3$ and $y_1w_2$ in $C_3$ and $C_1$ respectively and replace them by the edges $w_0y_1$ and $y_3w_2$.

Since the Hamilton cycle $C$ in $H^{\oplus}$ passes through the edges $x_0x_3$ and $x_1x_2$, there is a $2s$-path $x_0x_1 \cup x_2x_3$ in $H$. This $2s$-path can be extended easily to a Hamilton cycle in $G$.

In all cases, we have a Hamilton cycle in $G$. By induction, all 3-connected CBP graphs are hamiltonian.

## 4 CBP graphs with given number of Hamilton cycles

In this section, we consider the following problem.

**Problem** For a given integer $n$,. determine all CBP graphs with precisely $n$ Hamilton cycles.

Under the assumption that every 3-connected CBP graph is hamiltonian, we solve this problem for certain values of $n$. In particular, we characterize all CBP graphs with given cyclic connectivity and having precisely $n$ Hamilton cycles for $n = 6$ and $n = 12$.

We shall state a result of Bosák [2] here because it is not only relevant to the above problem (in that we need only consider those $n$ that are even) but it is also needed in the proofs of some of the results in this section.

**Theorem 5** ([2])

*Every cubic bipartite graph has an even number of Hamilton cycles.*

**Lemma 5** *Suppose BC is true. Let H be an s-planar graph with path sequence* $\{m_0, m_1, m_2\}$. *Then* $m_i \geq 2$ *is even for each* $i$.

**Proof** Let $a_0, a_1, a_2$ denote the three vertices of degree 2 in $H$. Let $H^*$ denote the merger of $H$ and an isolated vertex $x$. Then $H^*$ is a 3-connected CBP graph.

By Theorem 4, since $H^*$ is hamiltonian, there is a Hamilton cycle in $H^*$ passing through $xa_i$ and $xa_j$ for some $j \neq i$. This means that $m_r \geq 1$ for each $r$. By Theorem 2 of [3] and Theorem 5, $h(H^*) = m_0 + m_1 + m_2 \geq 6$ is even. Since the $m_i$'s are of the same parity (by Lemma 1), it follows that $m_r \geq 2$ is even for each $r$.

**Remark 3** Suppose BC is true. Let $H$ be a $g4g$ graph with path sequence $\{m_0, m_1, \ldots, m_5\}$ and $2s$-path sequence $\{n_1, 0, n_3\}$. Then $m_i \geq 4$ for $0 \leq i \leq 3$ and $n_1, n_3 \geq 4$.

To see the conclusion of Remark 3, notice that any Hamilton path in $H$ having $x_i$ and $x_{i+1}$ as end vertices must pass through the edges $w_{i+1}y_{i+2}$, $w_{i+2}y_{i+3}$ and $w_{i+3}y_i$ and avoid the edge $w_iy_{i+1}$ (see Figure 5(a)). Moreover it must include the Hamilton path with end vertices $x_i$ and $y_i$ in the subgraph $J_i$, the Hamilton path with end vertices $w_s$ and $y_s$ in the subgraph $J_s$ for $s = i+2, i+3$ and the Hamilton path with end vertices $w_{i+1}$ and $x_{i+1}$ in the subgraph $J_{i+1}$. Thus

$$m_i = p_{J_i}(x_i, y_i)p_{J_{i+3}}(w_{i+3}, y_{i+3})p_{J_{i+2}}(w_{i+2}, y_{i+2})p_{J_{i+1}}(w_{i+1}, x_{i+1}).$$

Since $H$ has at most two isolated $g$-vertices, by Lemma 5, we have $m_i \geq 4$.

Any $2s$-path $x_0x_1 \cup x_2x_3$ must avoid the edges $y_0w_3$ and $w_1y_2$. Moreover the path with end vertices $x_0$ and $x_1$ (respectively $x_2$ and $x_3$) must pass through the edge $w_0y_1$ (respectively $w_2y_3$). Thus

$$n_1 = p_{J_0}(x_0, w_0)p_{J_1}(y_1, x_1)p_{J_2}(x_2, w_2)p_{J_3}(y_3, x_3).$$

Since $H$ has at most two isolated $g$-vertices, by Lemma 5, we have $n_1 \geq 4$.

Similarly, we have $n_3 \geq 4$.

Let $G$ be a graph. Suppose $e$ is an edge in $G$. Let $h_e$ denote the number of Hamilton cycles containing the edge $e$.

**Lemma 6** *Suppose BC is true. Let G be a 3-connected CBP graph. If G is not the cube, then G has at least eight Hamilton cycles.*

**Proof** The proof is by induction on the number of vertices of $G$. It is based on the same idea employed in the proof of Theorem 4. The smallest 3-connected CBP graph which is not the cube is the graph $C_6 \times K_2$. It has precisely eight Hamilton cycles as can be easily verified.

Since $G$ is not the cube, then by Theorem 3, $G$ is a merger of an $s$-planar graph $H$ having the $\mathcal{A}$-propety and the ladder graph $L_k$ for some $k \geq 2$ (see Figure 3). Let $\{m_0, m_1, \ldots, m_5\}$ and $\{n_1, 0, n_3\}$ denote the path-sequence and the $2s$-path sequence of $H$ respectively.

Suppose $k \geq 3$. If $k$ is odd, let $H_1^*$ be a merger of $H$ and $K_2 = \{u, v\}$ (see Figure 8). Then $H_1^*$ is a 3-connected CBP graph and is not the cube because it contains a $2i$-face $F$ for some $i \geq 3$. Since $|H_1^*| < |G|$, by the induction hypothesis, $h(H_1^*) \geq 8$.

By Lemma 2, $h(H_1^*) = m_1 + m_3 + m_4 + m_5 + n_3 \geq 8$. By the same lemma, we have $h(G) = m_1 + m_3 + m_4 + m_5 + (k-1)n_1 + n_3 \geq 8$.

If $k$ is even, let $H_2^*$ be a merger of $H$ and $L_2 = \{u, v, w, z\}$ (see Figure 8). Then as before $H_2^*$ is a 3-connected CBP graph and is not the cube because it contains a $2i$-face $F$ for some $i \geq 3$. Since $|H_2^*| < |G|$, by the induction hypothesis, $h(H_2^*) \geq 8$.

By Lemma 2, $h(H_2^*) = m_0 + m_1 + m_2 + m_3 + n_1 + n_3 \geq 8$. By the same lemma, we have $h(G) = m_0 + m_1 + m_2 + m_3 + (k-1)n_1 + n_3 \geq 8$.

Now suppose $k = 2$. We may assume that $H$ has the strong $\mathcal{A}$-property (because otherwise the situation is symmetrically similar to the previous case already discussed).

*Case (1):* $H^{\oplus}$ or $H_{\oplus}$ is 3-connected

By symmetry, we may just assume that $H^{\oplus}$ is 3-connected.

Now in the graph $H^{\oplus}$, $h_{x_0x_3} = m_3 + n_1$. By Corollary 1 of [3],

$$h_{x_0x_3} = m_3 + n_1 \geq 4 \tag{3}$$

Applying Theorem 4 to the 4-face in $L_2$, we see that, for any edge $e$ on $L_2$, there is a Hamilton cycle in $G$ passing through all the edges but $e$ in $L_2$. This leads eventually to the fact that $m_i \geq 1$ for $0 \leq i \leq 3$. By Lemma 2, $h(G) \geq m_0 + m_1 + m_2 + m_3 + n_1 \geq 3 + 4$. Thus $h(G) \geq 8$ by Theorem 5.

*Case (2):* Neither $H^{\oplus}$ nor $H_{\oplus}$ is 3-connected

By Lemma 3, $H^{\oplus}$ and $H_{\oplus}$ are both of connectivity 2. By Lemma 4, $H$ is a $g4g$ graph. By Remark 3, if the path sequence and $2s$-path of $H$ are $\{m_0, m_1, \ldots, m_5\}$ and $\{n_1, 0, n_3\}$ respectively, then $m_i \geq 4$ for $0 \leq i \leq 3$ and $n_1, n_3 \geq 4$. By Lemma 2, $h(G) = m_0 + m_1 + m_2 + m_3 + n_1 + n_3 \geq 24$.

**Theorem 6** *Suppose BC is true. Let G be a CBP graph. Then G has precisely six Hamilton cycles if and only if G is the cube.*

**Proof** Suppose $G$ has precisely six Hamilton cycles. By Lemma 6, $G$ can only be the cube. It is easy to see that the cube has precisely six Hamilton cycles.

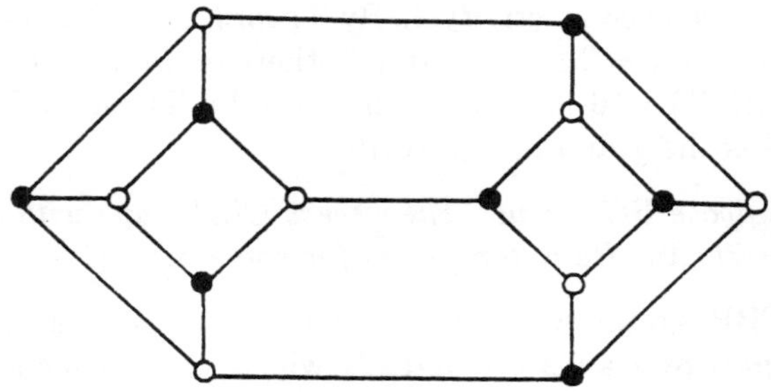

Figure 10. A CBP graph with precisely twelve Hamilton cycles

**Theorem 7** *Suppose BC is true. Let G be a CBP graph with cyclic connectivity* 3*. Then G has precisely twelve Hamilton cylces if and only if G is the graph in Figure 10.*

**Proof** Let $G$ be the graph in Figure 10. Then $G$ is the merger of $H$ and itself where $H$ is the graph obtained from the cube by deleting a vertex. Since $H$ has path sequence $\{2, 2, 2\}$, $G$ has precisely twelve Hamilton cylces by Equation (1).

Conversely, suppose $G$ has precisely twelve Hamilton cycles. Since $G$ is of cyclic connectivity 3, $G$ is a merger of two $(2, 3)$-regular graphs $G_1$ and $G_2$ with path sequences $\{m_{1,0}, m_{1,1}, m_{1,2}\}$ and $\{m_{2,0}, m_{2,1}, m_{2,2}\}$ respectively. Here, neither $G_1$ nor $G_2$ is an isolated vertex. Then from Equation (1),

$$h(G) = m_{1,0}m_{2,0} + m_{1,1}m_{2,1} + m_{1,2}m_{2,2} = 12 \tag{4}$$

Now clearly $G_i$ is an $s$-planar graph whose vertices of degree 2 are either all black or all white (see [6] p. 290 - 291). Let $G_i^*$ be the merger of $G_i$ and an isolated vertex $x_i$. Then $G_i^*$ is a CBP graph whose number of Hamilton cycles is

$$h(G_i^*) = m_{i,0} + m_{i,1} + m_{i,2} \tag{5}$$

Suppose $G_i^*$ is of cyclic connectivity $\leq 3$. Then $h(G_i^*) \geq 12$ by Theorems 3 and 4 of [3]. Since $m_{i,j} \geq 2$ for any $i$ and $j$ (by Lemma 5), there is at least one $i$ and $j$ (say $i = 1$) for which $m_{i,j} \geq 4$. But then we have $h(G) > 2(m_{2,0} + m_{2,1} + m_{2,2}) \geq 12$, a contradiction to Equation (4).

Thus $G_i^*$ is of cyclic connectivity 4. By Theorem 5 of [3], $h(G_i^*) \geq 6$. It has to be the case that $m_{i,j} = 2$ for all $i$ and $j$ otherwise $h(G) > 12$ (as explained before). Hence $h(G_i^*) = 6$ and $G_i^*$ is the cube by Theorem 8, for $i = 1, 2$. Consequently, $G$ is the graph in Figure 10.

**Corollary 1** *Suppose BC is true. Then every CBP graph with cyclic connectivity* 3 *has precisely* $4m$ *Hamilton cycles for some* $m \geq 3$.

**Proof** Every CBP graph with cyclic connectivity 3 is a merger of two (2, 3)-regular bipartite graphs $G_1$ and $G_2$ whose path sequences are $\{m_{1,0}, m_{1,1}, m_{1,2}\}$ and $\{m_{2,0}, m_{2,1}, m_{2,2}\}$ respectively. Since $m_{i,j} \geq 2$ is even for each $i$, (Lemma 5), we see that

$$h(G) = m_{1,0}m_{2,0} + m_{1,1}m_{2,1} + m_{1,2}m_{2,2} \equiv 0 \pmod 4$$

and this finishes the proof.

## References


1. D. Barnette, Conjecture 5, *Recent Progress in Combinatorics*, Academic Press, New York (1969) 343.
2. J. Bosák, Hamiltonian lines in cubic graphs, *Theory of Graphs, International Symposium, Rome, July 1966*, Gordon & Breach, New York (1967) 35-46.
3. G.L. Chia and Siew-Hui Ong, Lower bounds for the number of Hamilton cycles in cubic bipartite hamiltonian graphs, *Algebras and Combinatorics*, An International Congress, ICAC 97, Hong Kong, Springer-Verlag (1999) 113-118.
4. G.L. Chia and Roger Q.L. Yu, On the number of Hamilton cycles in cubic graphs, *Congressus Numerantium* **110** (1995) 13-32.
5. M.R. Garey, D.S. Johnson and R.E. Tarjan, The planar Hamiltonian circuit problem is $NP$-complete, *SIAM J. Comput.* **5** (1976) 704-714.
6. D.A. Holton, B. Manvel and B.D. McKay, Hamiltonian cycles in cubic 3-connected bipartite planar graphs, *J. Combinat. Theory Ser. B* **38** (1985) 279-297.
7. A.K. Kelmans, Constructions of cubic bipartite 3-connected graphs without Hamiltonian cycles, *Amer. Math. Soc. Transl. (2)*, **158** (1994) 127-140.
8. A. Kotzig, Construction of third-order Hamiltonian graphs (Russian, Czech and German summaries), *Časopis. Pěst. Mat.* **87** (1962) 148-168.

9. C. Thomassen, Triangulating a surface with a prescribed graph, *J. Combinat. Theory Ser. B* **57** (1993) 196-206.
10. W.T. Tutte, On Hamiltonian circuits, *J. London Math. Soc.* **21** (1946) 98-101.

# MULTIPLIER PROBLEMS

Christian DATRY & Gilbert MURAZ

**Abstract**

This work gives general properties of the multipliers from $L^1(G)$ into an $L^1(G)$-module and characterizes various spaces of this multipliers. A relation with the problem of difference is proposed. The connection between the compactness of a multiplier and the almost-periodicity is established.

The study of multipliers has an important role in many areas of harmonic analysis as well as in probability, optimization theory, differential equations, theory of signal, acoustic, mathematical finance and economics.

The general framework of this paper is the study of the multiplier from $L^1(G)$ into $E$ (an $L^1(G)$-module). Many authors consider particular cases; for example: general theory, R. Larsen ([La1], [La2]); $M_1$ the Wiener space, J.K. Wang [W], $L^1(G)$ where is $G$ non abelian, T.S. Quek and L.Y.H. Yap ([QY1], [QY2]), homogeneous Banach space, H.G. Feichtinger [F] and Banach valued function spaces, H.C. Lai ([Lai][LaiC]), see also ([Co], [ST]).

This paper has 3 parts. In the first, the framework of $L^1(G)$-module is described.

The second groups together most results from ([DMu], [DMuQY]) and characterizes various spaces of multipliers.

The third part associates the notion of relative completion and the problem of difference.

Finally, the connection between the compactness and the almost periodicity of the range of a multiplier is established.

# 1 NOTATIONS AND GENERAL FRAMEWORK

The terminology and the notations are the same as those in [DMu], [Mu2]. Let $G$ be a locally compact abelian group and $L^p(G)$ $(1 \leq p \leq +\infty)$ the classical Banach spaces defined with the Haar measure $dx$ on $G$.

## 1.1 General framework

A Banach space $E$ is called:

- a *G-module* if there exists a group representation $L$ of $G$ into the group of isometries of $E$ with $L_0 = I$ (identity) and $L_{x+y} = L_x \circ L_y$.

- an $L^1(G)$-*module* if there exists an algebra representation $T$ (continuous) of $L^1(G)$ into $\mathcal{L}(E,E)$ the algebra of bounded operators of $E$ into itself with
  - $\|Tf\| \leq \|f\|_1$.
  - $T_{f*g} = T_f T_g$ where $*$ is the convolution product in $L^1(G)$. For $e$ in $E$, $(T_f)(e)$ is denoted $f * e$.

An $L^1(G)$-module $E$ has a *compatible translation* if $E$ is also a $G$-module with $L_x T_f = T_f L_x = T(f_x)$ where $f_x(y) = f(y-x)$.

An $L^1(G)$-module $E$ is *order free* if for each $e \in E$, $e \neq 0$ there exists $f \in L^1(G)$ such that $f * e \neq 0$ and $E_{\text{deg}} = \{e \in E\ ,\quad f * e = 0\ ,\quad \forall f \in L^1(G)\}$ is called the degenerate part of $E$.

## 1.2 Essential part

For an $L^1(G)$-module $E$ with compatible translation it follows, its *essential part* is defined by:

$$\begin{aligned} E_{\text{ess}} &= \{f * e\ ,\quad f \in L^1(G)\ ,\quad e \in E\} \\ &= \{e \in E\ ,\quad \lim_{x \to 0} L_x e = e\} \\ &= \{e \in E\ ,\quad \lim_{\alpha} \mu_e * e = e\}\ . \end{aligned}$$

where $\{\mu_\alpha\}_\alpha$ is a bounded approximate identity (u.a.b., unit approche borne) of $L^1(G)$.

$E_{\text{ess}}$ is an order free $L^1(G)$-module with a compatible translation.

All the $L^1(G)$-modules considered here are with compatible translation.

The classical Banach spaces of functions or measures defined on $G$ have this property, but they are not essential.

## 1.3 Dual module

The dual $E'$ of a $L^1(G)$-module $E$ is also an $L^1(G)$-module for the action:

$$\langle f * e', e\rangle = \langle e', f * e\rangle,\quad f \in L^1,\quad e' \in E',\quad e \in E\ .$$

It is the same for a $G$-module.

The degenerate part $(E')_{\text{deg}}$ in $E'$ is the orthogonal of $E_{\text{ess}}$ and $E'$ is order free if an only if $E$ is essential.

## 1.4 Spectrum

The spectrum (of Beurling) of an element $e \in E$, where $E$ is an $L^1(G)$-module, is the closed subset of $\Gamma = \widehat{G}$, dual group of $G$ defined by:

$$\text{Sp}(e) = \{\gamma \in \Gamma,\quad \hat{f}(\gamma) = 0,\quad f \in I_e\}$$

where $I_e$ is the closed ideal $I_e = \{f \in L^1(G), \quad f * e = 0\}$ and $\hat{f}$ the Fourier transform of $f$.

$E$ is order free if and only if the space $E_{\text{deg}} = \{e, \ \text{Sp}(e) = \emptyset\}$ of degenerated elements contains only 0. In this case the subspace

$$E_c = \{e \in E, \ \text{Sp}(e) \text{ compact}\}$$

is dense in $E_{\text{ess}}$.

The subspace $E_{\text{fin}}$ of elements of $E$ with finite spectrum has an important role in the study of almost periodicity and ergodicity.

For $E$, order free, it follows: if $E$ is order free, the following equalities hold

$$\begin{aligned} E_\gamma &= \{e \in E, \quad \text{Sp}(e) \subset \gamma\} \\ &= \{e \in E, \quad f * e = \hat{f}(\gamma) \cdot e, \quad \forall f \in L^1(G)\} \\ &= \{e \in E, \quad L_x e = \overline{\langle \gamma, x \rangle} \cdot e, \quad \forall x \in G\} \end{aligned}$$

moreover $E_{\text{fin}} = \bigcup\limits_{F_i \in \mathcal{F}} \sum\limits_{\gamma_i \in F_i} E_{\gamma_i}$, where $\mathcal{F}$ is the family of finite subsets of $\Gamma$.

# 2 GENERAL PROPERTIES

## 2.1 Homomorphism properties

Because of the two structures on the essential part of a Banach $L^1(G)$-module there exist also two homomorphism notions for an operator:

A bounded linear operator $T$ from $L^1(G)$ into $L^1(G)$-module is

- *invariant* if $T$ commutes with the "translation operators", $TL_x = L_x T$, $\forall x \in G$; that is $T \in H_G(L^1(G), E)$.
- *multiplier* if $T$ commutes with the "convolution operators", $T*f = f*T$ $(TT_f = T_f T)$, $\forall f \in L^1(G)$; that is $T \in H_{L^1(G)}(L^1(G), E)$.

The equivalence of this two notions in the general case is a classical problem of harmonic analysis.

**Théorème 2.1** *Let $E$ be an $L^1(G)$-module with $G$ compatible translation, then:*

$$H_{L^1}(L^1, E) \subset H_G(L^1, E)\,.$$

*If Eis order free, then:*
- $H_{L^1}(L^1, E) = H_G(L^1, E)$
- *$T \in H_{L^1}(L^1, E)$ if and only if for any $f$ and $g \in L^1$ such that $f * g = 0$ then $f * T(g) = 0$.*

*Démonstration.* The first part of the proof may be found, for example in [DMu]. The second equivalence is the classical equivalence between multiplier and subordinative operator ([Ma], [Mu1]).

Each $e' \in E'$ defines an operator $T_{e'} : L^1(G) \to L^\infty(G)$, given by

$$T_{e'}(f)(x) = (L_x Tf, e') = (TL_x f, e') \ .$$

It follows that $T_{e'}$ satisfies the hypothesis of the theorem in [Mul], p. 846 and that $T_{e'}$ is a multiplier for all $e' \in E'$ furthermore, since $E$ is order free then $T$ is a multiplier.

Throughout this paper the space of all invariant multiplier is denoted $(L^1, E)$. □

## 2.2 Automatic continuity

The continuity of linear $G$-module on $L^1(G)$-module homomorphisms is studied by many authors: Sinclair [Si], Johnson [J] and Datry-Muraz [DMu].

Let $S$ a linear operator, commuting with the convolution; let $\Lambda(S)$ be the set of points of discontinuity of $S$, that is

$$\Lambda(S) = \{\gamma \in \Gamma,\ \forall f \in L^1(G),\ \hat{f}(\gamma) \neq 0 \text{ implies } f * S \text{ is discontinuous.}\}$$

This definition of $\Lambda(S)$ given in [DMu] corresponds to a natural definition of local continuity.

The theorem below, follows from the results obtained in [DMu] for this case:

**Théorème 2.2** *Let $S$ a linear homomorphism from $L^1(G)$ into $E$. Then $\Lambda(S)$ is finite .*

*If $E$ is order free then $S$ is continuous.*

## 2.3 Module properties

The space $(L^1, E)$ is an $L^1(G)$-module with $G$ compatible translation and since $L^1(G) * L^1(G) = L^1(G)$, $(L^1, E)$ is always order free.

The natural continuous homomorphism $j : E \to (L^1, E)$ with $j(e)(f) = f * e$ is an injection if and only if $E$ is order free. In this case $(L^1, E)$ is said the *completion* of $E$ and $E$ is *complete* if $j(E) = (L^1, E)$ ([CLM], Ch. III.2.9).

In general, $E$ is not complete (for example $\big(L^1(G), L^1(G)\big) \equiv M(G)$) but it follows ([DMuQY]):

**Théorème 2.3** *$j(E)$ is dense in $(L^1, E)$ for the strong topology of operators with $j(E_c) = (L^1, E)_c$.*

*Démonstration.* It is sufficient to prove that each $T \in (L^1, E)_c$ admits the representation $T(f) = f * e$ for $e \in E_c$, because $L^1(G)$ has a u.a.b. $\{\mu_\alpha\}_\alpha$, where $\hat{\mu}_\alpha$ has compact support.

Let $K$ be compact $\subset F$ and $T \in (L^1, E)_c$ such that $\mathrm{Sp}(T) \subset K$ and let $f \in L^1(G)$ such that $\hat{f}(\gamma) = 1, \forall \gamma \in K$. Then, it follows that $T(g) = T(f * g) = f * T(g) = g * f = g * e$ with $e = T(f)$.

For the characterisation of $(L^1, E)$ the following results are useful: □

**Théorème 2.4** *The $L^1(G)$-module isomorphisms follow:*

*i)* $(L^1, E) \cong (L^1, E_{\text{ess}})$,

*ii)* $(L^1, E)_{\text{ess}} \cong E_{\text{ess}}$,

*iii)* $(L^1, (L^1, E)) \cong (L^1, E)$,

*iv)* $(L^1, E') \cong (E_{\text{ess}})'$.

*Démonstration.* *i)* and *ii)* are obvious, because the essential part is the closure of the subspace of elements with compact spectrum.

The equality $L^1(G) * L^1(G) = L^1(G)$ implies *iii)*.

The classical result $H_A(X, Y') = (X \otimes_A Y)'$ (see for example [R]) gives here

$$H_{L^1(K)}(L^1, F') = (L^1 \otimes_{L^1(G)} E)' = (L^1 * E)' = (E_{\text{ess}})' . \qquad \square$$

**Corollaire 2.5** *If $E$ is an essential $L^1(G)$-module then $E'$ is an order free complete $L^1(G)$-module .*

**Corollaire 2.6** *If $E$ is an order free and reflexive $L^1(G)$-module then $E$ is essential and complete.*

*Démonstration.*
1. $(L^1, E') = (E_{\text{ess}})' = E'$.
2. $(L^1, E) = (L^1, (E')') = ((E')_{\text{ess}})' = (E')' = E$. $\square$

## 2.4 Examples

- $\big(L^1(G), M(G)\big) = \big(L^1(G), M(G)_{\text{ess}}\big) = \big(L^1(G), L^1(G)\big) = \big(L^1(G), C_0(G)'\big) = \big(L^1(G) * C_0(G)\big)'$ $= \big(C_0(G)\big)' = M(G)$ , ([La1]).
- $\big(L^1(G), L^p(G)\big) = \big(L^{p'}(G)\big)' = L^p(G)$, $1 < p < +\infty$, $\frac{1}{p} + \frac{1}{p'} = 1$.
- For $p = \infty$, $(L^1, L^\infty) = \big(L^1, L^\infty(G)_{\text{ess}}\big) = \big(L^1, C_{ub}(G)\big) = L^\infty(G)$.
- $\big(L^1, C_0(G)\big) \neq C_0$.
- For the case $G$ non abelian, see [QY2].
- $\big(L^1(G), \ell^p(L^p(G))\big) = \big[\ell^{p'}(L^{q'}(G))\big]' = \ell^p(L^q(G))$, $p > 1$, $q > 1$, $\frac{1}{p} + \frac{1}{p'} = 1$ et $\frac{1}{q} + \frac{1}{p'} = 1$ where $\ell^p(L^q(G))$ is the amalgam space defined on $G$. For the definition of amalgam space, see [BeDD].
- $\big(L^1, \ell^1(L^q)\big) = \big(L^1, (c_0(L^{q'}))'\big) = \ell^1(L^q)$, $q > 1$, $\frac{1}{q} + \frac{1}{q'} = 1$.
- For $q = \infty$, $(L^1, \ell^1(L^\infty)) = \big(L^1, \ell^1(L^\infty)_{\text{ess}})\big) = \big(L^1, \ell^1(C)\big) = \ell^1(L^\infty)$, see [W].

- $(L^1, \ell^p(M)) = (L^1, (\ell^p(M))_{\text{ess}}) = (L^1, \ell^p(L^1)) = (L^1, (\ell^{p'}(C))') = (\ell^{p'}(C)_{\text{ess}})' = \ell^p(M)$, $p > 1$, $\frac{1}{p} + \frac{1}{p'} = 1$.
- For $p = \infty$ and $G$ non abelian see [F].
- A study for $L^p(G, X)$ where $X$ is a Banach space is given in [Lai], [LaiC].

# 3 INCLUSION PROBLEMS

## 3.1 Inclusion property and maximality

Let $E$ be an $L^1(G)$-module contained in an $L^1(G)$-module $X$. Then a multiplier from $L^1(G)$ to $E$ is evidently also a multiplier to $X$ and we have ([DMuQY]):

**Théorème 3.1** *Let $E \subset X$ two $L^1(G)$-modules then $T \in (L^1, X)$ is a multiplier from $L^1(G)$ to $E$ if and only if $T(\mu_\alpha) \in E$ and $\sup_\alpha \|T(\mu_\alpha)\|_E < +\infty$ where $\mu_\alpha$ is a u.a.b. of $L^1(G)$.*

With this result each multiplier in many case has a measure representation ($E \subset M(G)$, Segal algebra [BuG], [F], [QY2]) or function representation ($E \subset L^\infty(G)$, $C_0(G)$, $AP^\infty(G)$, $WAP^\infty(G)$, [CrG], [DMu]).

The inclusion $(L^1, E) \subset (L^1, E'') = ((E')_{\text{ess}})'$ gives a general representation but it is not very useful.

In fact because of the inclusion $E \subset (L^1, E)$, it follows:

**Théorème 3.2** *If $E$ is an essential $L^1(G)$-module then $(L^1, E)$ is the maximal order free $L^1(G)$-module such that $(L^1, E)_{\text{ess}} = E$.*

*Démonstration.* Let $X$ be an $L^1(G)$-module order free which contains $E$ such that $X_{\text{ess}} = E$. Then $L^1(G) * X = X_{\text{ess}} = E$ and if $X$ is order free then there is an continuous injection from $X$ to $(L^1, E)$.

In this sense the maximal order free $L^1(G)$-module $X$ such that $X_{\text{ess}} = E$ is identified with the subspace of the elements $\sigma$ of $(E'_{\text{ess}})'$ such that $\sup_\alpha \|\mu_\alpha * \sigma\|$ is bounded where $\{\mu_\alpha\}_\alpha$ is a a.u.b. of $L^1(G)$. □

## 3.2 Relative completion

Let $F$ be a sub-$L^1(G)$-module of $E$ and $\widetilde{F}$ the *relative completion* of $F$ in $E$, that is $\widetilde{F} = \{e \in E,\ f * e \in F\}$. The definition given by Burnham and Goldberg in [BuG] is not exactly equivalent.

Let $I_F(e)$ be the ideal $I_F(e) = \{f \in L^1(G),\ f * e \in F\}$ and $\sigma_F(e) = \{\gamma \in \Gamma,\ \hat{f}(\gamma) = 0,\ f \in I_F(e)\}$, the relative spectrum of $e$.

If $E$ is order free $\widetilde{F}$ is embedded naturally in $(L^1, E)$ and in this case $\widetilde{F} = (L^1, F) \cap E \subset (L^1, E)$, is a sub-$L^1(G)$-module with $F \subset \widetilde{F} \subset E$ and so:

**Théorème 3.3** *Let $F$ be a sub-$L^1(G)$-module of $E$ and $e \in E$; then*
$e \in \widetilde{F}$ *if and only if* $\{\mu_\alpha * e \ : \ \mu_\alpha \text{ is u.a.b. of } L^1(G)\} \subset F$,
$e \in \widetilde{F}$ *if and only if* $\sigma_F(e) = \emptyset$.
*If* $F = \{0\}$ *then* $\widetilde{F} = E_{\text{deg}}$.
*If $E$ is essential then* $\widetilde{F} = F$.

Below, in the section of "compact multiplier", the result: "let $\varphi \in C_{ub}(G)$ such that $\varphi * f \in AP^\infty(G)$ for $\forall f \in L^1(G)$, then $\varphi \in AP^\infty(G)$", is obvious.

## 3.3 Problem of difference

The preceding section is connected to the classical problem of difference (see for example [Br], [DMu], [GMuS], [K1], [K2], [Lo]) by the following result:

**Théorème 3.4** *Let $F$ and $E$ be two $L^1(G)$-modules where $F \subset E$. The following conditions are equivalent:*
*i)* $\sigma_F(e) \subset \{\gamma_0\}$ ;
*ii)* $I_{\gamma_0} = \{f \in L^1(G), \hat{f}(\gamma_0) = 0\} \subset I_F(e) = \{f \in L^1(G), f * e \in F\}$ ;
*iii)* $\forall f \in L^1(G),\ f * e - \hat{f}(\gamma_0)e \in \widetilde{F}$ ;
*iv)* $\forall f \in L^1(G),\ \hat{f}(\gamma_0) = 1,\ f * e - e \in \widetilde{F}$.
*If $E$ is order free and admits a compatible $G$-translation then:*
*v)* $\forall x \in G,\ L_x e - \overline{\langle x, \gamma_0\rangle} e \in \widetilde{F}$.

*Démonstration.* *i)* is equivalent to: $e$ lives locally in $F$ at each point of $\Gamma$ except perhaps at $\gamma_0$; it is equivalent to *ii)* by definition.

*ii)* $\Rightarrow$ *iii)*. Let $f \in L^1(G)$ and $g \in L^1(G)$ then $f * g - \hat{f}(\gamma_0)g \in I_{\gamma_0}$ and $g * (f * e - \hat{f}(\gamma_0)e) \in F$ that is $f * e - \hat{f}(\gamma_0)e \in \widetilde{F}$.

*iii)* $\Rightarrow$ *iv)* is obvious.

*ii)* $\Rightarrow$ *v)*. Let $f \in L^1(G)$; for each $x \in G$, $(L_x f - \overline{\langle x, \gamma_0\rangle} f) \in I_{\gamma_0}$ and so $(L_x f - \overline{\langle x, \gamma_0\rangle} f) * e = f * (L_x f - \overline{\langle x, \gamma_0\rangle} e) \in F$.

*v)* $\Rightarrow$ *iii)*. Let $g \in L^1(G)$ then $g * e \in E_{\text{ess}}$, that is $x \mapsto L_x(g * e)$ is continuous; then it follows that

$$L_x e - \overline{\langle x, \gamma_0\rangle} e \in F \quad \text{and} \quad g * \left(L_x e - \overline{\langle x, \gamma_0\rangle} e\right) \in F \ ,$$

$$\int \left(L_x(g * e) - \overline{\langle x, \gamma_0\rangle} g * e\right) f(x\, dx = g * (f * e - \hat{(}\gamma_0)e \in F$$

or $f * e - \hat{f}(\gamma_0)e \in \widetilde{F}$. □

**Définition 3.5** *Let $F \subset E$ be a $G$-module. Then $F$ has the property of difference for $E$ at $\gamma_0 \in \Gamma$ if for each $e \in E$ such that $L_x e - \overline{\langle x, \gamma_0\rangle} \in F$, $\forall x \in G$ then $e \in F$.*

But the last result which says that if $E$ is an $L^1(G)$-module with $G$-translation compatible, when $F = \widetilde{F}$, the property of difference is equivalent to:
for each $e \in E$ such that $f * e - \hat{f}(\gamma_0)e \in F$ for all $f \in L^1(G)$ then $e \in F$ ;

for each $e \in E$, s.t. $f * e \in F$ for all $f \in I_{\gamma_0}$ then $e \in F$.
When $E = L^\infty(G)$ and $F = AP^\infty(G)$, the famous result of Loomis ([Lo]) is obtained; for variants or generalizations, see [BE], [Br], [GMuS], [DMu], [H], [HMu]. A proof for $C_{ub}(G)$ or for the essential part $E_{\text{ess}}$ of $E$ is given in [GMuS], [DMu] and in particular it is the Cauchy problem.

The $C_0(G)$ space of functions vanishing at infinity does not have the property of difference.

# 4 COMPACT MULTIPLIERS

The property of compactness for a multiplier is connected with the concept of almost periodicity, see for example [Cr], [CrG], [Mu2], [T].

A main theorem of [DMu] for almost periodic element in an $L^1(G)$-module gives the characterization of the compact multipliers.

**Théorème 4.1** *An element $e$ of an $L^1(G)$-module $E$ is almost periodic if it satisfies the following equivalent conditions:*
*i)* $\{L_x e,\ x \in G\}$ *is relatively compact;*
*ii)* $\{f * e,\ f \in L^1(G), \|f\|_1 \le 1\}$ *is relatively compact ($f \mapsto f * e$ is compact from $L^1(G)$ to $E$);*
*iii) the map $x \in G \mapsto L_x e \in E$ is continuous for the Bohr topology $\overline{G}$, restricted to $G$;*
*iv)* $e \in \overline{E_{\text{fin}}}$.

The identification of $(L^1, E)$ with $(L^1, (L^1, E))$ gives the following corollary, because $(L^1, E)$ is also an $L^1(G)$-module.

**Corollaire 4.2** *Let $E$ be an order free $L^1(G)$-module and $T$ a multiplier from $L^1$ to $E$. Then the following conditions are equivalent*
*i)* $\{L_x T,\ x \in G\}$ *is compact in $(L^1, E)$ ($T$ is almost periodic);*
*ii) $T$ is compact from $L^1$ to $E$;*
*iii)* $T \in AP(E)$.

For the general case of compact multipliers between two $L^1(G)$-modules, see [Mu2].

The weak topology on $E$ is denoted $w^*$ and the strong operator (weak operator) topology on $(L^1, E)$ is $\omega_F$ ($\omega_f$).

**Théorème 4.3** *Let $E$ be an order free $L^1(G)$-module and let* $\Sigma_T = \{T(f),\ f \in L^1(G),\ \|f\|_1 \le 1\} \subset E \subset (L^1, E)$. *Then*
*i)* $\Sigma_T$ *is relatively compact in $E$ for the norm-topology iff* $T \in AP(E)$;
*ii)* $\Sigma_T$ *is relatively compact in $E$ for the $\omega^*$-topology iff* $T \in WAP(E)$;
*iii)* $\Sigma_T$ *is relatively compact in $(L^1, E)$ for the $\omega_F$-topology iff* $T \in (L^1, AP(E))$;
*iv)* $\Sigma_T$ *is relatively compact in $(L^1, E)$ for the $\omega_f$-topology iff* $T \in (L^1, WAP(E))$.

*i)* is exactly the last corollary. The proofs of *ii)*, *iii)* and *iv)* are similar using the density of $\{f * \mu_\alpha,\ \mu_\alpha \text{ a.u.b.}\}$ in $L^1(G)$. The results *iii)* and *iv)* for the case $E = L^\infty(G)$ are the proposition 1, p. 28 of [CrG].

The inclusions $AP(E) \subset (L^1, AP(E))$ and $WAP(E) \subset (L^1, WAP(E))$ are strict in general but the theory of $L^1(G)$-module [DMu] shows:

$\left(L^1, AP(E)\right)_{\text{ess}} = AP(E)$

$\left(L^1, WAP(E)\right)_{\text{ess}} = WAP(E)$.

The next concept of almost periodicity for $f \in L^\infty(G)$ using the $\varepsilon$-period, denoted *u.r. (uniformly recurrent)* [DMu] is equivalent to the compactness of the orbit only if $f$ is continuous.

**Définition 4.4** $f \in (L^\infty(G))_{ur}$ *if for every* $\varepsilon > 0$ *there exists a compact set* $K$ *of* $G$ *such that* $K + P_\varepsilon = G$ *where* $P_\varepsilon = \{x \in G,\ \|L_x f - f\|_\infty < \varepsilon\}$.

$L^\infty(G)_{ur}$ is not a linear space but,

$L^\infty(G)_{ur} \cap C(G) = AP^\infty(G)$

$L^\infty(G)_{ur} \subset (L^1, AP^\infty(G))$ .

This proves the strict inclusions:

$$AP^\infty(G) \subset L^\infty(G)_{ur} \subset \left(L^1, AP^\infty(G)\right) \subset WAP^\infty(G) \subset \left(L^1, WAP^\infty(G)\right) .$$

# Conclusion

This presentation is not exhaustive and you can find many extensions and applications depending on the choice of $E$ and also on the basic algebra of the module.

The references give examples of various possibilities.

The problem of difference is inexhaustive.

The natural action of $\mathbb{Z}$ on the Banach algebra $\mathcal{L}(E, E)$ of bounded operators of $E$, $n \mapsto T^n$ where $T$ is invertible $T \in \mathcal{L}(E, E)$ admits an extension to $\ell^1(\omega)$ with $\omega(n) = \sup(\|T^n\|, \|T^{-n}\|)$.

The problem of invariant subspace is connected with the study of multipliers from $\ell^1(\omega)$ into $\mathcal{L}(E, E)$.

# Acknowledgments

Thanks go to M.P. Navarro for her assistance in the preparation of the english version of this paper.

# References

[BE] B. Basit, M. Emam, *Differences of functions in locally convex spaces and applications to almost periodic and almost automorphic functions*, Ann. Polonici Math. **XLI** (1983), 193–201.

[BeDD] J.P. Bertrandias, C. Datry, C. Dupuis, *Unions et intersections d'espaces $L^p$ invariantes par translation ou convolution*, Ann. Inst. Fourier, **29** (1979), 189–209.

[Br] N.G. Brujin, *Functions whose differences belong to a given class*, Nieuw. Arch. Wisk., **23** (1951), 194–218.

[BuG] J.T. Burnham, R.R. Goldberg, *Multipliers from $L^1(G)$ to a Segal algebra*, Bull. Inst. Math. Acad. Sinica, **2** (1974), 153–154.

[CLM] J. Cigler, V. Losert, P. Michor, *Banach modules and functors on categories of Banach spaces*, Lecture Notes in Pure and Applied Mathematics, M. Dekker, inc., 1979.

[Co] C.V. Comisky, *Multipliers of Banach modules*, Indag. Math., **33** (1971), 32–38.

[Cr] G. Cromez, *Almost periodicity in groups and compactness of operators*, Bull. Soc. Math. de Belgique, 31 (2) B (1979), 197–201.

[CrG] G. Cromez, W. Gowaerts, *Completely continuous multipliers from $L^1(G)$ into $L^\infty(G)$*, Ann. Inst. Fourier (Grenoble), **32**-2 (1984), 137–154.

[DMu] C. Datry, G. Muraz, *Analyse harmonique dans les modules de Banach,*
*I. Proprits gnrales*, Bull. Sci. Math., **119** (1995), 299–337,
*II. Presque-priodicit et ergodicit*, Bull. Sci. Math., **120** (1996), 493–536.

[DMuQY] C. Datry, G. Muraz, T.S. Quek, L.Y.H. Yap, *Homomorphismes de $L^1(G)$-modules*, Prépublication de l'Institut Fourier n° 76,Grenoble, 1987.

[F] H.G. Feichtinger, *Multipliers from $L^1(G)$ to a homogeneous Banach space*, J. Math. Anal. and Appl., **61** (1977), 341–356.

[GMuS] F. Galvin, G. Muraz, P. Szeptycki, *Fonctions aux diffrences continues $f(x) - f(x+a)$ continues*, C. R. Acad. Sci. Paris Sér. I Math., **315** (1992), 397–400.

[H] A.E. Hamza, *Solutions of functional equations*, PhD Thesis, december 93, Cairo, Egypt.

[HMu] A.E. Hamza, G. Muraz, *Spectral criteria of abstract functions, integral and difference problem*, Acta Mathematica Vietnamica, **28**-1 (1998), 171–184.

[J] B.E. Johnson, *Continuity of homomorphisms of Banach G-modules*, Pacific J. of Math., **120**-1 (1985), 111–121.

[K1] T. Keleti, *Periodic $L_p$ functions with $L_q$ difference functions*, Real Analysis Exchange, **23**-2 (1997/8), 431–440.

[K2] T. Keleti, *Difference functions of periodic measurable functions*, PhD Thesis, Budapest 1996.

[Lai] H.C. Lai, *Multipliers of Banach valued function spaces*, Lecture Notes, series **23**, march 1984.

[LaiC] H.C. Lai, T.K. Chang, *Multipliers and translation invariant operators*, Thoku Math. J., **41** (1989), 31–41.

[La1] R. Larsen, *The multiplier problem*, Springer-Verlag, Lecture notes in math., **105** (1969).

[La2] R. Larsen, *An introduction to the theory of multipliers*, Springer-Verlag (1971).

[Lo] L.H. Loomis, *The spectral characterisation of a class of almost periodic functions*, Ann. Math., **72** (1960), 362–368.

[Ma] P. Masani, *The normality of time invariant subordinative operators in a Hilbert space*, Bull. Amer. Math. Soc., **71** (1965), 546–550.

[Mu1] G. Muraz, *Multiplicateurs de $L^p(G)$) dans $L^{p'}(G)$*, C. R. Acad. Sci. Paris Sér. I Math.**285** (1977), 845–847.

[Mu2] G. Muraz, *Multiplicateurs compacts*, Bull. Belg. Math. Soc., **4** (1997), 601–612.

[QY1] T.S. Quek, L.Y.H. Yap, *Multipliers from $L^1(G)$ to a Lipschitz space*, J. Math. Anal. and Appl., **69**-2 (1979), 531–539.

[QY2] T.S. Quek, L.Y.H. Yap, *Measures in $M(G)$ with satisfy certain integrability conditions*, Analysis Mathematica, **23** (1997), 35–44.

[R] M.A. Rieffel, *Multipliers and tensor products on $L^p$-spaces of locally compact groups*, Studia Math., **33** (1969), 71–82.

[ST] S. Saeki, E.L. Thome, *Lorentz spaces as $L^1$-modules and multipliers*, Hokkaido Math. J., **23** (1994), 55-92.

[Si] A.M. Sinclair, *Automatic continuity of linear operators*, Cambridge Univ. Press, 1976.

[T] U.B. Tewari, *Compactness and almost periodicity of multipliers*, (manuscrit).

[W] J.K. Wang, *On multipliers from $L^1$ to $M_1$*, Bull. of the Institute of Mathematics, Academia Sinica, **23**-3, september 1995, 189–195.

Christian DATRY & Gilbert MURAZ
INSTITUT FOURIER
Laboratoire de Mathmatiques
UMR5582 (UJF-CNRS)
BP 74
38402 St MARTIN D'HRES Cedex (France)

Christian.Datry@ujf-grenoble.fr
Gilbert.Muraz@ujf-grenoble.fr

# The Henstock Integral and Measures on Infinite Dimensional Spaces

**Elvira P. de Lara-Tuprio**
Mathematics Department, Ateneo de Manila University
P. O. Box 154, Manila 0917, Philippines

**Chew Tuan Seng**
Department of Mathematics, National University of Singapore
2, Science Drive 2, Singapore 117543
email: matcts@nus.edu.sg

**Abstract**

The classical construction of measures in infinite dimensional spaces not only needs finite dimensional measure spaces but also requires extensions from the finite to the countable and finally to the uncountable infinite dimensional spaces. Establishing the countable additive property is also involved [4]. In this note, we shall construct measures in infinite dimensional spaces from the infinite product of division spaces, which is introduced by Henstock. The countable additive property is a consequence of the monotone convergence theorem, which is proved by using the fully decomposable property.

## 1. Division spaces

Let $T$ be a given set, and $\mathcal{T}$ be a collection of some nonempty subsets $I$ of $T$. An element $I$ in $\mathcal{T}$ is called an interval. Let $U'$ be a collection of some interval-point pairs $(I, t)$, where $I \in \mathcal{T}$, $t \in T$, and let $A$ be a family of some subcollections $U \subseteq U'$. Each interval-point pair $(I, t) \in U$, where $U \in A$, is said to be $U$-fine.

An elementary set $E$ is an interval or a union of a finite number of mutually disjoint intervals. Given an elementary set $E$, a collection $U \in A$ divides $E$ if there exists a finite subset $D$ of $U$, called a division of $E$ from $U$ or a $U$-fine division of $E$, such that $\{I : (I,t) \in D\}$ is a collection of mutually disjoint intervals whose union is $E$.

**Definition 1 [1].** *The triple $(T, \mathcal{T}, A)$ is called a division space if the following conditions hold:*

**P1.** *For each elementary set $E$ in $T$, every $U \in A$ divides $E$.*

**P2.** *For every pair $U_1$, $U_2 \in A$, there is $U_3 \in A$ with $U_3 \subset U_1 \cap U_2$.*

**P3.** *For every $U \in A$ and any elementary set $E$ there is $U_1 \subset U$ such that $U_1 \in A$, and $I \subset E$ for every $(I,t) \in U_1$.*

**P4.** *Given disjoint elementary sets $E_1$ and $E_2$, if $U_1 \in A$ with $I \subset E_1$ for all $(I,t) \in U_1$ and $U_2 \in A$ with $I \subset E_2$ for all $(I,t) \in U_2$, then there is $U_3 \in A$ with $U_3 \subset U_1 \cup U_2$.*

**Definition 2 [1].** A division space $(T, \mathcal{T}, A)$ is said to be fully decomposable if to every family $\mathcal{F}$ of mutually disjoint subsets $X$ of $T$ and every family of $V_X \in A$, $X \in \mathcal{F}$, there exists a $V \in A$ such that

$$V[X] \subset V_X \quad \text{for every} \quad X \in \mathcal{F},$$

where $V[X] = \{(I,x) \in V; x \in X\}$.

Let $(T, \mathcal{T}, A)$ be a given fully decomposable division space. Let $(T_\lambda, \mathcal{T}_\lambda, A_\lambda) = (T, \mathcal{T}, A)$ for $\lambda = 1, 2, \ldots, n$. Let

$$T^n = \prod_{\lambda=1}^{n} T_\lambda = \{(x_1, x_2, \ldots, x_n); \quad x_\lambda \in T \text{ for } \lambda = 1, 2, \ldots, n\}$$

and

$$\mathcal{T}^n = \{I_1 \times I_2 \times \cdots \times I_n; \quad I_\lambda \in \mathcal{T}, \quad \text{for } \lambda = 1, 2, \ldots, n\}.$$

Now we shall construct a family, denoted by $A^n$, of some collections of interval-point pairs $(I,x)$, where $I \in \mathcal{T}^n$ and $x \in T^n$. First for each $x \in T^n$ with $x = (x_1, x_2, \ldots, x_n)$, choose $V_i(x) \in A$, $i = 1, 2, \ldots, n$. Let

$$V = \{(I_1 \times I_2 \times \cdots I_n, \; (x_1, x_2, \ldots, x_n)) : x \in T^n, \; (I_i, x_i) \in V_i(x), \; i = 1, 2, \ldots, n\}$$

and $A^n$ be the family of all such $V$. In other words, if $V \in A^n$, then for each $x \in T^n$, there exist $V_i(x) \in A$, $i = 1, 2, \ldots, n$, such that $(I_1 \times I_2 \times \cdots \times I_n, \; (x_1, x_2, \ldots, x_n)) \in V$

if and only if $(I_i, x_i) \in V_i(x)$. The approach used above is local. We start with $x \in T^n$, then choose $V_i(x)$ and finally construct $V$ from $V_i(x)$.

**Theorem 3 [1, p.149].** *If $(T, \mathcal{T}, A)$ is a fully decomposable division space, then $(T^n, \mathcal{T}^n, A^n)$ is also a fully decomposable division space.*

Now we shall consider the infinite dimensional product space $\prod_{t\in[0,+\infty)} T_t$, where $T_t = T$ for each $t$. That is,

$$\prod_{t\in[0,+\infty)} T_t = \{x;\ x : [0,+\infty) \to T\}.$$

Hence, $\prod_{t\in[0,+\infty)} T_t$ is the space of all functions defined on $[0,+\infty)$ with values in $T$. We shall denote $\prod_{t\in[0,+\infty)} T_t$ by $T^{[0,+\infty)}$.

Let $N = \{t_1, t_2, \ldots, t_n\}$, where $t_i \in [0,+\infty)$ and $t_1 < t_2 < \cdots < t_n$. For $x \in T^{[0,+\infty)}$, let $x_i = x(t_i)$ and $x(N) = (x_1, x_2, \ldots, x_n)$.

An interval $I$ in $T^{[0,+\infty)}$ is

$$I = I[N] = \prod \{I(t);\ t \in N\} \times \prod \{T_t;\ t \in [0,+\infty)\backslash N\}$$

where $I(t)$ is an interval in $T$ and $N$ is a finite subset of $[0,+\infty)$. $N$ is said to be the dimension set of $I$. We denote $\prod \{I(t) : t \in N\}$ by $I(N)$. Let $\mathcal{T}^{[0,+\infty)}$ be the collection of all intervals in $T^{[0,+\infty)}$.

An elementary set in $T^{[0,+\infty)}$ is an interval or a finite union of mutually disjoint intervals in $T^{[0,+\infty)}$. We shall assume that $T^{[0,+\infty)}$ is an elementary set.

Assume that the length $\ell$ of an interval $I$ is given i.e., $\ell : \mathcal{T}^{[0,+\infty)} \to (0,+\infty)$ is a finite additive function. The length of an interval $I = I[N]$, denoted by $|I|$, may depend on $N$. For example if $|I|$ is the classical Wiener measure of $I$, then $|I|$ depends on $N$, see [2, p.36].

Now we shall construct $A^{[0,+\infty)}$. Again, the construction is local, as in $A^n$. For each $x \in T^{[0,+\infty)}$ and each finite subset $N = \{t_1, t_2, \ldots, t_n\}$ of $[0,+\infty)$, choose a finite subset $L(x)$ of $[0,+\infty)$ and $U_1(x, N) \in A^N$, where $A^N = A^n$. Let

$$U = \{(I[N], x) : (I(N), x(N)) \in U_1(x, N), N \supset L(x), x \in T^{[0,+\infty)}\} \text{ and } A^{[0,+\infty)}$$

be the family of all such $U$. In other words, if $U \in A^{[0,+\infty)}$, then for each $x \in T^{[0,+\infty)}$ and each finite subset $N = \{t_1, t_2, \ldots, t_n\}$ of $[0,+\infty)$, there exist a finite subset $L(x)$

of $[0,+\infty)$ and $U_1(x,N) \in A^N$, such that $(I[N],x) \in U$ if and only if $(I(N),x(N)) \in U_1(x,N)$ and $N \supset L(x)$.

From now onwards, we alway assume that $(T^{[0,+\infty)}, \mathcal{T}^{[0,+\infty)}, A^{[0,+\infty)})$ has P1 property, since it may not have this property in general. For sufficient conditions that P1 property holds, see [1, p. 160].

**Theorem 4 [1, pp. 159-161].** *If $(T,\mathcal{T},A)$ is a fully decomposable division space, then so is $(T^{[0,+\infty)}, \mathcal{T}^{[0,+\infty)}, A^{[0,+\infty)})$.*

For convenience, denote $\mathcal{T}^{[0,+\infty)}$ by $\mathcal{I}$ and $A^{[0,+\infty)}$ by $\mathcal{C}$.

## 2. The Henstock Integral

Let $(T^{[0,+\infty)}, \mathcal{I}, \mathcal{C})$ be a division space induced by $(T,\mathcal{T},A)$. We are going to define the Henstock integral of a function $f : E \to \mathbb{R}$, where $E$ is an elementary set in $T^{[0,+\infty)}$ and $\mathbb{R} = (-\infty,\infty)$, and state its properties.

**Definition 5.** *Let $E$ be an elementary set in $T^{[0,+\infty)}$ and $f : E \to \mathbb{R}$. The function $f$ is said to be* **Henstock integrable** *to $K \in \mathbb{R}$ if given $\epsilon > 0$, there exists $V \in \mathcal{C}$ such that for any $V$-fine division $D = \{(I,x)\}$ of $E$, we have*

$$\left|(D)\sum f(x)|I| - K\right| < \epsilon.$$

*The number $K$ is called the* **Henstock integral of** *$f$* **on** *$E$ and will be denoted by $\int_E f$.*

**Theorem 6 [1, p. 171]** *(Henstock's Lemma). Let $f$ be Henstock integrable on an elementary set $E$. Then given $\epsilon > 0$, there exists $V \in \mathcal{C}$ such that whenever $D = \{(I,x)\}$ is a partial $V$-fine division of $E$, we have*

$$(D)\sum \left| f(x)|I| - \int_I f \right| \leq \epsilon. \tag{1}$$

Since $(T^{[0,+\infty)}, \mathcal{I}, \mathcal{C})$ is fully decomposable, (for the role of decomposable property, see remark by Henstock in [1, p. 126]) we have

**Theorem 7 [1, p.126]** *(Monotone Convergence Theorem). Let $\{f_n\}$ be a sequence of functions defined on an elementary set $E$ such that*

1. *each* $f_n$ *is Henstock integrable on* $E$,
2. *for each* $x \in E$, $f_n(x) \to f(x)$ *as* $n \to +\infty$,
3. $f_1(x) \leq f_2(x) \leq \ldots \quad \forall x \in E$, *and*
4. $\{\int_E f_n\}$ *is bounded.*

*Then* $f$ *is Henstock integrable on* $E$ *and*

$$\lim_{n\to\infty} \int_E f_n = \int_E f.$$

**Theorem 8 [1, p. 96].** *If* $f_1$ *and* $f_2$ *are Henstock integrable on an elementary set* $E$, $g(x) \leq f_i(x) \leq h(x)$, $\forall x \in E$, $i = 1, 2$, *where* $g$ *and* $h$ *are Henstock integrable functions on* $E$, *then* $\max\{f_1, f_2\}$ *and* $\min\{f_1, f_2\}$ *are Henstock integrable on* $E$.

**Theorem 9 [1, p. 136]** *(Dominated Convergence Theorem). Let* $\{f_n\}$ *be a sequence of functions defined on an elementary set* $E$ *such that*

1. *each* $f_n$ *is Henstock integrable on* $E$,
2. *for each* $x \in E$, $f_n(x) \to f(x)$ *as* $n \to +\infty$, *and*
3. $g(x) \leq f_n(x) \leq h(x)$ $\forall x \in E$, $\forall n$, *where* $g$ *and* $h$ *are both Henstock integrable on* $E$.

*Then* $f$ *is Henstock integrable on* $E$ *and* $\lim\limits_{n\to\infty} \int_E f_n = \int_E f$.

The proofs of the theorems 6 – 9 in $\mathbb{R}$, see [3], can be easily carried over to the division spaces.

## 3. Measures on $T^{[0,+\infty)}$

Recall that we assume that $T^{[0,+\infty)}$ is an elementary set and that the length of $T^{[0,+\infty)}$ is finite. For any subset $E$ of $T^{[0,+\infty)}$, let $\chi_E$ denote the characteristic function of $E$. In order to construct a measure space on $T^{[0,+\infty)}$, we will consider the Henstock integral of $\chi_E$ on the elementary set $T^{[0,+\infty)}$.

**Definition 10.** *Let* $\mathcal{B}$ *be the collection of all subsets* $E$ *of* $T^{[0,+\infty)}$ *such that* $\chi_E$ *is Henstock integrable on* $T^{[0,+\infty)}$. *If* $E \in \mathcal{B}$, *we say that* $E$ *is an integrable set.*

**Theorem 11.** $(T^{[0,+\infty)}, \mathcal{B})$ *is a measurable space.*

**Proof.** We have to show that $\mathcal{B}$ is a $\sigma$-algebra. Clearly, $\emptyset \in \mathcal{B}$. Let $E \in \mathcal{B}$. Then $\chi_E$ is Henstock integrable. Since

$$\chi_{E^c} = \chi_{T^{[0,+\infty)}} - \chi_E ,$$

then $\chi_{E^c}$ is also Henstock integrable. Hence, $E^c \in \mathcal{B}$.

If $A$, $B \in \mathcal{B}$, then $A \cup B \in \mathcal{B}$ since $\chi_{A\cup B} = \max\{\chi_A, \chi_B\}$ is Henstock integrable. It follows that finite union of integrable sets is also integrable.

Finally, let $\{E_i\}_i$ be a sequence of sets in $\mathcal{B}$. Define

$$F_1 = E_1, \quad F_2 = E_1 \cup E_2, \ldots, F_n = \bigcup_{i=1}^{n} E_i, \ldots.$$

Then

(1) each $\chi_{F_n}$ is Henstock integrable on $T^{[0,+\infty)}$.

(2) $\chi_\emptyset \le \chi_{F_n} \le \chi_{T^{[0,+\infty)}}$ $\forall n$ and both $\chi_\emptyset$ and $\chi_{T^{[0,+\infty)}}$ are Henstock integrable on $T^{[0,+\infty)}$.

(3) $\chi_{F_n} \to \chi_{\bigcup_{i=1}^{+\infty} E_i}$ as $n \to +\infty$.

By Dominated Convergence Theorem, $\chi_{\bigcup_{i=1}^{+\infty} E_i}$ is Henstock integrable on $T^{[0,+\infty)}$. Thus $\bigcup_{i=1}^{+\infty} E_i \in \mathcal{B}$.

**Definition 12.** *For $E \in \mathcal{B}$, define $mE = \int_{T^{[0,+\infty)}} \chi_E$.*

**Theorem 13.** *$(T^{[0,+\infty)}, \mathcal{B}, m)$ is a complete measure space.*

**Proof.** Let $\{E_i\}$ be a sequence of pairwise disjoint integrable sets. Since

$$\chi_{\bigcup_{i=1}^{+\infty} E_i} = \lim_{n\to+\infty} \chi_{\bigcup_{i=1}^{n} E_i} = \lim_{n\to+\infty} \sum_{i=1}^{n} \chi_{E_i},$$

then

$$m\left(\bigcup_{i=1}^{+\infty} E_i\right) = \int_{T^{[0,+\infty)}} \chi_{\bigcup\limits_{i=1}^{+\infty} E_i}$$
$$= \int_{T^{[0,+\infty)}} \lim_{n\to+\infty} \sum_{i=1}^{n} \chi_{E_i}$$
$$= \lim_{n\to+\infty} \sum_{i=1}^{n} \int_{T^{[0,+\infty)}} \chi_{E_i}$$
$$= \sum_{i=1}^{+\infty} \int_{T^{[0,+\infty)}} \chi_{E_i}$$
$$= \sum_{i=1}^{+\infty} m(E_i).$$

Hence, $m$ is a measure on $(T^{[0,+\infty)}, \mathcal{B})$. Now we shall prove that the measure space is complete. Let $E \in \mathcal{B}$ with $mE = 0$. Let $F \subset E$. Observe that

$$(D) \sum \chi_F(x)|I| \leq (D) \sum \chi_E(x)|I|.$$

Hence $\chi_F$ is Henstock integrable on $T^{[0,+\infty)}$ with $mF = 0$. Thus $F \in \mathcal{B}$.

The following theorem, which is a consequence of Henstock's Lemma, shows that a measurable set is very nearly a nice set, which is similar to the structure of a Lebesgue measurable set.

**Theorem 14.** *Let $E \in \mathcal{B}$. Then for each $\epsilon > 0$, there exists a finite union of disjoint intervals $U$ such that $m(U \triangle E) < 2\epsilon$, where $U \triangle E = (U \backslash E) \cup (E \backslash U)$.*

**Proof.** Let $\epsilon > 0$. By Henstock's Lemma there exists $V \in \mathcal{C}$ such that

$$(\mathcal{D}) \sum \left| \chi_E(x)|I| - \int_I \chi_E \right| < \epsilon$$

whenever $\mathcal{D} = \{(I, x)\}$ is a $V$-fine division of $T^{[0,+\infty)}$. Hence

$$(\mathcal{D}) \sum |\chi_E(x)|I| - m(E \cap I)| < \epsilon.$$

Write

$$\mathcal{P}' = \{I : x \in E, (I, x) \in D\}$$

and

$$\mathcal{P}'' = \{I : x \notin E, (I, x) \in D\}.$$

Then $|m(\cup_{I\in\mathcal{P}'}I) - m(E\cap(\cup_{I\in\mathcal{P}'}I))| < \epsilon$ and $m(E\cap(\cup_{I\in\mathcal{P}''}I)) < \epsilon$. Thus $m(U\triangle E) < 2\epsilon$ where $U = \cup_{I\in\mathcal{P}'}I$.

It is shown in [5] that the Classical Wiener measure can be constructed by the above approach.

From the proofs, we notice that a measure can be constructed on the infinite product of fully decomposable division spaces because the infinite product is fully decomposable and every fully decomposable division space can induce a measure.

## References

[1] R. Henstock, The general theory of integration, *Clarendon Press, Oxford*, 1991.

[2] Hui-Hsiung Kuo, Gaussian Measures in Banach spaces, Lecture Notes in Maths 463, *Springer-Verlag*, 1975.

[3] Lee Peng Yee and R. Vyborny, The integral: an easy approach after Kurzweil and Henstock, *Cambridge*, 2000.

[4] Y. Yamasaki, Measures on infinite dimension spaces, *World Scientific*, 1985.

[5] Yang Chien Hwa and Chew Tuan Seng, On McShane-Wiener integral, *Proceedings of the International Mathematics Conference*, 1998, Manila, Philippines, pp. 39-46, 1999.

# STABILITY OF A QUADRATURE METHOD FOR THE BIHARMONIC DIRICHLET PROBLEM


V. D. DIDENKO

*Universiti Brunei Darussalam, Faculty of Science, Tungku BE 1410, Brunei*

*E-mail: victor@fos.ubd.edu.bn*



A quadrature method based on locally non-equidistant meshes is proposed for the approximate solution of the biharmonic Dirichlet problem on a non-smooth domain. Necessary and sufficient conditions for stability are established.


## 1 Introduction

The aim of this paper is to study an approximation method for the biharmonic Dirichlet problem for the function $u = u(x, y)$, viz.

$$\begin{array}{lcll} \Delta^2 u(x,y) & = & 0, & (x,y) \in D \\ u(x,y) & = & f_1(x,y), & (x,y) \in \Gamma \\ \dfrac{\partial u}{\partial \mathbf{n}}(x,y) & = & f_2(x,y), & (x,y) \in \Gamma \end{array} \tag{1}$$

where $\Delta^2$ denotes the biharmonic operator

$$\Delta^2 = \left( \frac{\partial^2}{\partial x^2} + \frac{\partial^2}{\partial y^2} \right)^2,$$

$\Gamma$ is a simple piece-wise smooth contour, $D$ is the interior domain bounded by $\Gamma$, and $f_1$ and $f_2$ are prescribed functions. It is also assumed that $0 \in D$.

The biharmonic problem (1) arises in various areas of applied mathematics, including the theory of 'slow' viscous flows [10] and elasticity theory [11, 12, 13]. A popular solution method is to reduce (1) to a boundary integral equation [1, 3, 7, 8, 9], which can be solved approximately. Although a variety of such boundary integral equations are known, not all have been used to find approximate solutions of (1). Most approaches are based on the integral formulations proposed in [2] (see also [1]). However, as far as we know the stability of the corresponding approximation methods have been proven for smooth boundaries only. In the present paper, the construction of an approximate solution of (1) via the Muskhelishvili equation is considered, which is applicable to piece-wise smooth domains.

## 2 Meshes and local operators

Let $\gamma = \gamma(s), s \in \mathbb{R}$ be a 1-periodic parametrization of $\Gamma$, and let $0 \leq s_1 < s_2 < \ldots < s_{k_0} < 1$ denote the values of $s$ which correspond to the corner points of $\Gamma$. We set $s_{k_0+1} := s_1 + 1$ and assume that $\gamma$ is twice continuously differentiable in each of the intervals $(s_j, s_{j+1})$ and that there exist finite one-sided limits $\gamma'(s_j \pm 0), \gamma''(s_j \pm 0), j = 1, 2, \ldots, k_0$. Using the parametrization $\gamma$ one can generate a sequence of meshes on $\Gamma$. In particular, let us consider the natural numbers $n$ which have the form $n = lk_0, l = 1, 2, \ldots$, and for any such number introduce the mesh

$$\mathcal{P}_n := \{\gamma(k/n),\ k = 0, 1, \ldots, n-1\}.$$

Note that if

$$|\gamma'(s_j + 0)| = |\gamma'(s_j - 0)|, \tag{2}$$

then the corresponding mesh is called locally equidistant at the point $t_j = \gamma(s_j)$. The present paper discusses an approximation method constructed on meshes which can be locally non-equidistant at some corner points $t_j$, i.e. instead of (2) the parametrization $\gamma$ satisfies the equations

$$|\gamma'(s_j + 0)| = p_j |\gamma'(s_j - 0)|,\ j = 1, 2, \ldots, k_0 \tag{3}$$

with some (or all) of $p_j$ being different from 1.

The stability of approximation methods using locally non-equidistant meshes has been presented for the first time in [4, 5], where such methods were respectively applied to the double layer potential equation and to the singular integral equations of the Cauchy type. Here we consider an analogous problem for the Muskhelishvili equation:

$$M\varphi(t) \equiv -\overline{\varphi(t)} - \frac{1}{2\pi i}\int_\Gamma \overline{\varphi(\tau)} d\log\frac{\overline{\tau} - \overline{t}}{\tau - t} - \frac{1}{2\pi i}\int_\Gamma \varphi(\tau) d\frac{\overline{\tau} - \overline{t}}{\tau - t} = f_0(t). \tag{4}$$

Let $\alpha$ denote the angle between the outward normal $\mathbf{n}$ to $\Gamma$ and the real axis, $\nu$ the unit vector which makes up the angle $\pi/2 - \alpha$ with the real axis. It is known [11, 12] that the biharmonic Dirichlet problem (1) the corresponding Muskhelishvili equation has the right-hand side

$$f_0(t) = -\frac{1}{2}f(t) + \frac{1}{2\pi i}\int_\Gamma \frac{f(\tau)d\tau}{\tau - t},\ t = x + iy \in \Gamma \tag{5}$$

where

$$f(t) = f_2(t) + i\frac{\partial f_1}{\partial \nu}(t). \tag{6}$$

If $\varphi$ is a solution of (4) with the right-hand side (5), and if the function $\psi$ is obtained from the equation

$$\psi'(t) = f(t) - \overline{\varphi(t)} - \bar{t}\varphi'(t), \, t \in \Gamma, \tag{7}$$

then the solution of (1) is given by the formulae

$$U(x, y) = \operatorname{Re}\left[\bar{z}\varphi(z) + \psi(z)\right], \, z = x + iy \tag{8}$$

with

$$\psi(z) = \frac{1}{2\pi i}\int_\Gamma \frac{\psi(t)dt}{t - z}, \quad z \in D$$

and

$$\varphi(z) = \frac{1}{2\pi i}\int_\Gamma \frac{\varphi(t)dt}{t - z}, \quad z \in D \; .$$

Thus, the main problem is to approximate the solution of (4).

We start by investigating a quadrature method for a special curve. Let $\beta \in [0, 2\pi), \omega \in (0, 2\pi)$ be real numbers, and let $\Gamma_{\beta,\omega}$ refer to the angle

$$\Gamma_{\beta,\omega} := \{e^{i(\beta+\omega)}\mathbb{R}^+\} \cup \{e^{i\beta}\mathbb{R}^+\}$$

assuming the ray $e^{i(\beta+\omega)}\mathbb{R}^+$ is directed to 0 whereas the ray $e^{i\beta}\mathbb{R}^+$ is directed away from 0. We also choose real numbers $p > 0$ and $\delta, \varepsilon \in (0, 1), \delta \neq \varepsilon$, fix $n \in \mathbb{N}$ and introduce the points $t_k^{(n)}, \tau_k^{(n)}, k \in \mathbb{Z}$ as follows

$$t_k^{(n)} := \begin{cases} p\dfrac{k+\delta}{n}e^{i\beta} & \text{if} \quad k \geq 0 \\ -\dfrac{k+\delta}{n}e^{i(\beta+\omega)} & \text{otherwise} \end{cases}, \tag{9}$$

$$\tau_k^{(n)} := \begin{cases} p\dfrac{k+\varepsilon}{n}e^{i\beta} & \text{if} \quad k \geq 0 \\ -\dfrac{k+\varepsilon}{n}e^{i(\beta+\omega)} & \text{otherwise} \end{cases}. \tag{10}$$

The points (9) and (10) correspond to a mesh on $\Gamma_{\beta,\omega}$ which is locally non-equidistant at the point 0.

In the space $L_2(\Gamma_{\beta,\omega})$, consider the corresponding Muskhelishvili operator (4) for the contour $\Gamma_{\beta,\omega}$ and the integral equation

$$M_{\Gamma_{\beta,\omega}}\phi = y \tag{11}$$

with the right-hand side $y$ belonging to appropriate class $R_2(\Gamma_{\beta,\omega})$ of the Riemann integrable functions on $\Gamma_{\beta,\omega}$. Approximate values $\xi_j^{(n)}, j \in \mathbb{Z}$ for the exact solution $\phi$ of the equation (11) at the points $t_j^{(n)}$ are found from the system of algebraic equations

$$-\xi_k^{(n)} - \frac{1}{2\pi i}\sum_{j\in\mathbb{Z}}\left[\frac{\Delta\tau_j^{(n)}}{\tau_j^{(n)} - t_k^{(n)}} - \frac{\overline{\Delta\tau_j^{(n)}}}{\overline{\tau_j^{(n)}} - \overline{t_k^{(n)}}}\right]\overline{\xi_j^{(n)}} -$$

$$\frac{1}{2\pi i}\sum_{j\in\mathbb{Z}}\left[\frac{\overline{(\tau_j^{(n)} - t_k^{(n)})}\Delta\tau_j^{(n)}}{(\tau_j^{(n)} - t_k^{(n)})^2} - \frac{\overline{\Delta\tau_j^{(n)}}}{\tau_j^{(n)} - t_k^{(n)}}\right]\xi_j^{(n)} = y(t_k^{(n)}),\ k \in \mathbb{Z}, \tag{12}$$

where $\Delta\tau_j^{(n)}, j \in \mathbb{Z}$ are defined by

$$\Delta\tau_j^{(n)} := \begin{cases} \frac{p}{n}e^{i\beta} & \text{if} \quad j \geq 0 \\ -\frac{1}{n}e^{i(\beta+\omega)} & \text{otherwise} \end{cases}. \tag{13}$$

After the coefficients $\xi_j^{(n)}, j \in \mathbb{Z}$ have been determined, an approximate solution $\phi_n$ of equation (11) can be written in the form

$$\phi_n(t) = \sum_{k\in\mathbb{Z}} \xi_k^{(n)}\phi_k^{(n)}(t), \quad t \in \Gamma_{\beta,\omega} \tag{14}$$

where

$$\phi_k^{(n)}(t) := \begin{cases} \begin{cases} 1 & \text{if} \quad t \in pe^{i\beta}\left[\frac{k}{n}, \frac{k+1}{n}\right) \\ 0 & \quad\text{otherwise} \end{cases} & k \geq 0 \\ \\ \begin{cases} 1 & \text{if} \quad t \in e^{i(\beta+\omega)}\left(\frac{-k+1}{n}, \frac{-k}{n}\right] \\ 0 & \quad\text{otherwise} \end{cases} & k < 0 \end{cases} \tag{15}$$

Let $\widetilde{L}_n$ denote the orthogonal projection onto the smallest closed subspace of $L_2(\Gamma_{\beta,\omega})$ which contains all functions (15). Then one can rewrite the approximation method (12)-(15) as the operator equation

$$\widetilde{A}_n^{\omega,p}\phi_n = f_n, \tag{16}$$

with the operator $\widetilde{A}_n^{\omega,p}$ acting on the space $\operatorname{im}\widetilde{L}_n$. The task now is to study properties of the operators $\widetilde{A}_n^{\omega,p}, n \in \mathbb{N}$. To this end, consider operators on the Hilbert space $l_2$ of all sequences $\{\eta_j\}_{j=0}^{\infty}$ of complex numbers $\eta_j$ satisfying the condition

$$\sum_{j=0}^{\infty} |\eta_j|^2 < \infty.$$

Let $l_2^2$ refer to the Cartesian product of two spaces $l^2$. The following result can be proved by immediate calculation.

**Proposition 1** *The operator* $\widetilde{A}_n^{\omega,p} : \operatorname{im}\widetilde{L}_n \to \operatorname{im}\widetilde{L}_n$ *is isometrically isomorphic to the operator*

$$A_{\omega,p} = A^1_{\omega,p} + A^2_{\omega,p}V : l_2^2 \to l_2^2, \tag{17}$$

*where*

$$V(\{\eta_j\} \times \{\zeta_k\}) = \{\overline{\eta_j}\} \times \{\overline{\zeta_k}\}$$

$$A^s_{\omega,p} = \begin{pmatrix} A^s_{11} & A^s_{12} \\ A^s_{21} & B^s_{22} \end{pmatrix}, \quad s = 1, 2$$

*and*

$$A^1_{11} = A^1_{22} = 0$$

$$A^1_{12} = \left( \frac{e^{-2i\beta}\sin\omega}{\pi} \frac{p^{-1}(k+1-\delta)}{((j+\varepsilon) - p^{-1}(k+1-\delta)e^{i\omega})^2} \right)_{k,j=0}^{+\infty}$$

$$A^1_{21} = \left( \frac{e^{-i2(\beta+\omega)}\sin\omega}{\pi} \frac{p(k+\delta)}{((j+1-\varepsilon) - p(k+\delta)e^{-i\omega})^2} \right)_{k,j=0}^{+\infty}$$

$$A^2_{11} = A^2_{22} = -I$$

$$A^2_{12} = \left( \frac{1}{2\pi i} \left[ \frac{1}{(j+\varepsilon) - p^{-1}(k+1-\delta)e^{i\omega}} - \frac{1}{(j+\varepsilon) - p^{-1}(k+1-\delta)e^{-i\omega}} \right] \right)_{k,j=0}^{+\infty}$$

$$A^2_{21} = \left( \frac{1}{2\pi i} \left[ \frac{1}{(j+1-\varepsilon) - p(k+\delta)e^{i\omega}} - \frac{1}{(j+1-\varepsilon) - p(k+\delta)e^{-i\omega}} \right] \right)_{k,j=0}^{+\infty}$$

An immediate consequence of Proposition 1 is the stability condition for the quadrature method (12)-(15). Let us recall that a sequence $\{\widetilde{A}_n\}$, $\widetilde{A}_n$ : $\operatorname{im}\widetilde{L}_n \to \operatorname{im}\widetilde{L}_n$ is called stable if the operators $\widetilde{A}_n$ are invertible for all $n \geq n_0$ and

$$\sup_{n\geq n_0} ||\widetilde{A}_n^{-1}\widetilde{L}_n|| < \infty.$$

**Corollary 2** *The sequence $\{\widetilde{A}_n^{\omega,p}\}$ is stable if and only if the operator $A_{\omega,p}$ is invertible.*

Thus, the stability of the quadrature method (12)-(15) depends on the invertibility of the operator $A_{\omega,p}$ in the space $l_2^2$. Actually, at present there are no tools to study this problem completely but a weaker invertibility (Fredholmness) can be established.

**Proposition 3** *1. The operator $A_{\omega,p} : l_2^2 \to l_2^2$ belongs to the smallest closed subalgebra of $\mathcal{L}_{add}(l_2^2)$ which contains the operator $V$ and all Toeplitz operators with piece-wise constant pre-symbols.*

*2. For all $p > 0$ and for all $\omega \in (0,\pi)\cup(\pi,2\pi)$ the operator $A_{\omega,p}$ is Fredholm and its index vanishes.*

The proof of this result is rather lengthy and is omitted here. (It follows the proof of the Lemma 3.8 and Proposition 3.9 of [6], and it requires statements like the Propositions 2.5 and 2.8 from same paper.)

## 3 Quadrature method for the Muskhelishvili equation

Let us now consider the approximate solution of the Muskhelishvili equation on the $\Gamma$. Assume the parametrization $\gamma$ satisfies the requirements of the previous section. As before $n = k_0 l, l \in \mathbb{N}$. By $[\gamma(k/n), \gamma((k+1)/n)), k = 1, 2, \ldots, n-1$ we denote the half-open arc of $\Gamma$ with the starting point $\gamma(k/n)$ and with the end point $\gamma((k+1)/n)$ and consider the functions

$$\varphi_j^{(n)}(t) = \begin{cases} 1 & \text{if} \quad t \in [\gamma(j/n), \gamma((j+1)/n)) \\ 0 & \text{otherwise} \end{cases}.$$

The approximate solution of the Muskhelishvili equation (4) is sought in the form

$$x_n(t) = \sum_{j=1}^{n-1} \xi_j^{(n)} \varphi_j^{(n)}(t). \tag{18}$$

Let $\varepsilon, \delta$ be some real numbers in $(0,1)$ such that $\varepsilon \neq \delta$. Then the unknown coefficients $\xi_j^{(n)}$ of the approximate solution (18) of the Muskhelishvili equation (4) are defined from the system of algebraic equations

$$-\xi_k^{(n)} - \frac{1}{2\pi i}\sum_{j=0}^{n-1}\left[\frac{\Delta\tau_j^{(n)}}{\tau_j^{(n)} - t_k^{(n)}} - \frac{\overline{\Delta\tau_j^{(n)}}}{\overline{\tau_j^{(n)}} - \overline{t_k^{(n)}}}\right]\overline{\xi_j^{(n)}} -$$

$$\frac{1}{2\pi i}\sum_{j=0}^{n-1}\left[\frac{(\overline{\tau_j^{(n)}} - \overline{t_k^{(n)}})\Delta\tau_j^{(n)}}{(\tau_j^{(n)} - t_k^{(n)})^2} - \frac{\overline{\Delta\tau_j^{(n)}}}{\tau_j^{(n)} - t_k^{(n)}}\right]\xi_j^{(n)} + (T_n x_n)(t_k^{(n)}) = f_0(t_k^{(n)}),$$

$$k = 0, 1, \ldots, n-1 \qquad (19)$$

where $\Delta\tau_j^{(n)} = \gamma((j+1)/n) - \gamma(j/n)$, and $T$ is the following compact operator

$$(T\varphi)(t) = \frac{1}{2\pi i}\int_\Gamma \frac{\varphi(\tau)d\tau}{\tau} + \frac{1}{t}\frac{1}{2\pi i}\int_\Gamma\left(\frac{\varphi(\tau)}{\tau^2}d\tau + \frac{\overline{\varphi(\tau)d\tau}}{\overline{\tau}^2}\right).$$

The stability conditions of the approximation method (18)-(19) can be expressed in terms of invertibility of model operators. Let $\gamma(s_j), j = 1, 2, \ldots, k_0$ be the corner points of $\Gamma$, $p_j$ be defined by (3), $\omega_j$ be the angle between the corresponding semi-tangents at the point $\gamma(s_j)$ and $\beta_j$ be the angle between the right semi-tangent at $\gamma(s_j)$ to $\Gamma$ and the real axis. By $A_{\omega_j,p_j}$ we denote the operator (17) in which $\omega, p$ and $\beta$ are replaced by $\omega_j, p_j$ and $\beta_j$, respectively.

**Theorem 4** 1. *The quadrature method (18)-(19) is stable if and only if the operators $A_{\omega_j,p_j}, j = 1, 2, \ldots, k_0$ are invertible.*

2. *If the quadrature method (18)-(19) is stable and if the function $f$ of (6) belongs to the Sobolev space $H^{3/2}(\Gamma)$, then the sequence of approximate solutions (18) converges to the solution of the Muskhelishvili equation (4) in $L_2(\Gamma)$.*

The proof of this result follows the proof of the Theorem 1 of [5]. However, we have to take into account the invertibility of the operator $M_\Gamma + T$ in the space $L_2(\Gamma)$.

Returning to the biharmonic Dirichlet problem, an approximate solution can be obtained using the formula (8), where the function $\varphi$ is approximated by (18) and approximations of $\psi$ are found from (7). The corresponding sequence of approximate solutions converges to the exact solution of (1) in the norm $L_2(\Gamma)$.

# 4 Acknowledgements

This work was supported in part by University Brunei Darussalam under the Grant UBD/PNC2/2/RG/1(12).

## References

[1] S. Christiansen, *J. Comput. Appl. Math.* 91 (1998).

[2] S. Christiansen and P. Hougaard, *J. Inst. Mat. Appl.* 22 (1978).

[3] C. Constanda, *Mathem. Met. Appl. Sc.* 20 (1997).

[4] V.D. Didenko and B. Silbermann, *IMACS J. of Appl. Numer. Math.* **26,** 1-2 (1998).

[5] V.D. Didenko and B. Silbermann, *J. of Integ. Equat. Appl.* **11,** 3 (1999).

[6] V.D. Didenko and B. Silbermann, *Preprint MATHS 2000/05/006,* (University Brunei Darussalam, 2000).

[7] D.B. Ingham and M.A. Kelmanson, *Boundary Integral Equations Analysis of Singular, Potential and Biharmonic Problems. Lecture Notes in Eng.* (Springer, Berlin, 1984).

[8] M.A. Jaswon and G.T. Symm, *Integral Equation Method in Potential Theory and Electrostatics,* (Academic Press, London, 1977).

[9] A.I. Kalandia, *Mathematical Methods of Two-dimensional Elasticity,* (Mir Publishers, Moscow, 1975).

[10] V.V. Meleshko, *Appl. Sci. Res.* 1 - 4 (1998).

[11] N.I. Muskhelishvili, *Fundamental Problems in the Theory of Elasticity,* (Nauka, Moscow, 1966).

[12] V.Z. Parton and P.I. Perlin, *Integral equations of elasticity theory,* (Nauka, Moscow, 1977).

[13] S. Timoshenko and S. Woinowsky-Krieger, *Theory of Plates and Shells,* (McGraw-Hill, New-York, 1959).

# SHADOWS OF CODES AND LATTICES

STEVEN T. DOUGHERTY AND PATRICK SOLÉ

ABSTRACT. We give the major developments of shadows of self-dual codes and unimodular lattices and generalize the given constructions for using shadows to build new codes and lattices.

## 1. INTRODUCTION

In this note we shall present the main theorems about shadows of codes and lattices, generalizing the constructions of self-dual codes and unimodular lattices using shadows. Shadow codes and shadow lattices have been important in both proving the non-existence of codes and lattices for specific parameters and for constructing new codes and lattices. We shall restrict our study of lattices to unimodular lattices, avoiding modular lattices. The theory of shadows has been developed in a sequence of papers by the authors and others. The notion of a shadow of a binary singly-even self-dual code (Type I code)[6] and of an odd unimodular lattice (Type I lattice)[7] was introduced by Conway and Sloane. However, the object first appeared early in the study of self-dual codes in Ward's paper.[24] This notion was extended to Type I codes over the rings $\mathbb{Z}_{2k}$ and $\mathbb{F}_2 + u\mathbb{F}_2$ in Bannai *et al.*,[2] Dougherty *et al.*[12] and Dougherty *et al.*[17]

Shadows play an important role in the study of Type I codes and Type I lattices. For example, shadows were used to determine the highest possible minimum norms of Type I lattices [9] and the highest possible minimum weights of Type I binary[6] and $\mathbb{Z}_4$-codes.[17] Several constructions based on shadows of unimodular lattices and self-dual codes over $\mathbb{F}_2$,[5] $\mathbb{F}_q$,[11] and $\mathbb{Z}_{2k}$[18] are also known.

In this paper we shall present the major results of shadows and lattices and show their connection. In Section 4 of this note, we give general constructions of unimodular lattices and self-dual codes using shadows which generalize known constructions, together with a number of examples. These generalized constructions give a complete framework for constructions using shadows. We shall show that all known shadow constructions can be viewed from a single unified perspective. This investigation was motivated by a comment by an anonymous referee to Dougherty *et al.*[18]

## 2. UNIMODULAR LATTICES AND SELF-DUAL CODES

We begin by giving the standard definitions of lattices and codes.

2.1. **Unimodular Lattices.** Let $\mathbb{R}^n$ be $n$-dimensional Euclidean space with inner-product $x \cdot y = x_1y_1 + x_2y_2 + \cdots + x_ny_n$ for $x = (x_1, x_2, \ldots, x_n)$ and $y = (y_1, y_2, \ldots, y_n)$. An $n$-dimensional lattice $\Lambda$ in $\mathbb{R}^n$ is a free $\mathbb{Z}$-module spanned by $n$ linearly independent vectors $v_1, \ldots, v_n$. An $n$ by $n$ matrix whose rows generate $\Lambda$ is called a generator matrix $G$ of $\Lambda$. Two lattices $\Lambda$ and $\Lambda'$ are *equivalent* if there is an orthogonal matrix $M$ with $\Lambda' = \Lambda M = \{xM \mid x \in \Lambda\}$. The norm of a vector

$x \in \Lambda$ is $x \cdot x$. The minimum norm of $\Lambda$ is the smallest norm among all nonzero vectors of $\Lambda$. The theta series $\theta_\Lambda(q)$ of a lattice $\Lambda$ is the formal power series in the indeterminate $q$,

$$\theta_\Lambda(q) = \sum_{x \in \Lambda} q^{x \cdot x}$$

The first non-trivial coefficient of the theta series is the kissing number of $\Lambda$, which is equal to the number of the minimum norm vectors in $\Lambda$.

The *dual* lattice $\Lambda^*$ is given by $\Lambda^* = \{x \in \mathbb{R}^n \mid x \cdot a \in \mathbb{Z} \text{ for all } a \in \Lambda\}$. The Poisson-Jacobi formula [10, (19) Chap. 4] relates the theta series of a lattice and its dual. Let $q = e^{\pi i z}$ then

$$\theta_{L^*}(-1/z) = \sqrt{|L|}(\frac{i}{z})^{n/2}\theta_L(z).$$

$$\theta_{L^*}(z) = (\det L)^{1/2}(\frac{i}{z})^{n/2}\theta_L(\frac{-1}{z}).$$

A lattice $\Lambda$ is *integral* if $\Lambda \subseteq \Lambda^*$. An integral lattice with $\Lambda = \Lambda^*$ is called *unimodular*. If the norm $x \cdot x$ is an even integer for all $x \in \Lambda$, then $\Lambda$ is said to be *even*. An even unimodular lattice $\Lambda$ is called *Type II* and a unimodular lattice which is not Type II is called *Type I*. It is known that $n$-dimensional Type II lattices exist if and only if $n \equiv 0 \pmod 8$.

2.2. **Self-Dual Codes over $\mathbb{Z}_{2k}$.** Here we give definitions for self-dual codes over $\mathbb{Z}_{2k}$ (see Bannai *et al.*[2] for a complete description). A code $C$ of length $n$ over $\mathbb{Z}_{2k}$ (or a $\mathbb{Z}_{2k}$-code of length $n$) is a $\mathbb{Z}_{2k}$-submodule of $\mathbb{Z}_{2k}^n$ where $\mathbb{Z}_{2k}$ is the ring of the integers modulo $2k$. An element of $C$ is called a codeword of $C$. A *generator* matrix of $C$ is a matrix whose rows generate $C$. The *Hamming weight* $wt_H(x)$ of a vector $x$ in $\mathbb{Z}_{2k}^n$ is the number of non-zero components. The *Euclidean weight* $wt_E(x)$ of a vector $x = (x_1, x_2, \ldots, x_n)$ is $\sum_{i=1}^n \min\{x_i^2, (2k - x_i)^2\}$. The *Lee weight* $wt_L(x)$ of a vector $x$ is $\sum_{i=1}^n \min\{|x_i|, |2k - x_i|\}$. The minimum Hamming, Lee and Euclidean weights, $d_H, d_L$ and $d_E$, of $C$ are the smallest Hamming, Lee and Euclidean weights among all non-zero codewords of $C$, respectively. The *symmetrized weight enumerator* of $C$ is defined as

$$swe_C(x_0, x_1, \ldots, x_k) = \sum_{c \in C} x_0^{n_0(c)} x_1^{n_1(c)} \cdots x_{k-1}^{n_{k-1}(c)} x_k^{n_k(c)},$$

where $n_0(x), n_1(x), \ldots, n_{k-1}(c), n_k(c)$ are the numbers of $0, \pm 1, \ldots, \pm(k-1), k$ components of $c$, respectively.

For $x = (x_1, \ldots, x_n)$ and $y = (y_1, \ldots, y_n)$, we define the inner-product of $x$ and $y$ in $\mathbb{Z}_{2k}^n$ by $x \cdot y = x_1 y_1 + \cdots + x_n y_n$. The dual code $C^\perp$ of $C$ is defined as $C^\perp = \{x \in \mathbb{Z}_{2k}^n \mid x \cdot y = 0 \text{ for all } y \in C\}$. $C$ is *self-dual* if $C = C^\perp$. Self-dual $\mathbb{Z}_{2k}$-codes with the property that all Euclidean weights are divisible by $4k$ are called *Type II*.[2] A self-dual code which is not Type II is called *Type I*.

2.3. **Connections.** Now we give a construction of lattices from self-dual $\mathbb{Z}_{2k}$-codes, which is called *Construction $A_{2k}$*. Define the reduction modulo $2k$

$$\rho : \mathbb{Z}^n \to \mathbb{Z}_{2k}^n$$

by $\rho(x_1, \ldots, x_n) = (x_1 \pmod{2k}, \ldots, x_n \pmod{2k})$.

**Theorem 1** (Bannai *et al.*[2]). *If $C$ is a self-dual code of length $n$ over $\mathbb{Z}_{2k}$, then the lattice*

$$A_{2k}(C) = \frac{1}{\sqrt{2k}}\{x \in \mathbb{Z}^n \mid \rho(x) \in C\}$$

*is an $n$-dimensional unimodular lattice. The minimum norm is* $\min\{2k, d_E/2k\}$ *where $d_E$ is the minimum Euclidean weight of $C$. Moreover if $C$ is Type II then the lattice $A_{2k}(C)$ is a Type II lattice.*

This connection is very important because results in one space can be transferred across this connection, for example orthogonality relations and existence results.

**Theorem 2** (Dougherty *et al.*[18])). *Let $C$ be a Type I code over $\mathbb{Z}_{2k}$ and $A_{2k}(C)$ the Type I lattice constructed by Theorem 1. Then*

$$\begin{array}{ll}(A_{2k}(C))_j = A_{2k}(C_j) & \textit{for } j = 0, 2, \textit{ and,}\\ (A_{2k}(C))_j = A_{2k}(C_j) & \textit{for } j = 1, 3\end{array}$$

*up to labeling.*

## 3. Shadows of Type I Lattices and Codes

We shall begin with the definitions of shadows in both spaces and then describe the major results.

3.1. **Shadows of Lattices.** Let $L$ be a Type I lattice and let $L_0$ denote its subset of even norms vectors. The set $L_0$ is a sublattice of $L$ of index 2 in $L$. Let $L_2$ be that unique nontrivial coset of $L_0$ into $L$. Then $L_0^*$ can be written as a union of cosets of $L_0$: $L_0^* = L_0 \cup L_2 \cup L_1 \cup L_3$. The *shadow* of $L$ is defined to be $S = L_1 \cup L_3$. Throughout this paper, $L_i$ $(i = 0, 1, 2, 3)$ denotes a coset of $L_0$.

**Lemma 3** (Dougherty *et al.*[18]). *The glue group $L_0^*/L_0$ of $L_0$ is isomorphic to the Klein 4-group if $n$ is even and to the cyclic 4-group if $n$ is odd.*

**Lemma 4** (Dougherty *et al.*[18]). *Suppose that $C$ is a Type I lattice in dimension $n$.*

(1) *Suppose that $n$ is even. Let $a$ be a binary element such that $a \equiv \frac{n}{2}$ (mod 2), then Table 1 holds where $(x \cdot y) \in \mathbb{Z} + \alpha$ and $\alpha$ is the value in position $(i, j)$ for any vector $x \in L_i$ and any vector $y \in L_j$.*
(2) *Suppose that $n$ is odd. Let $a$ be $1$ or $-1$ such that $n \equiv a$ (mod 4) then Table 2 holds.*

Table 1. Orthogonality Relations for Even $n$

| | $L_0$ | $L_1$ | $L_2$ | $L_3$ |
|---|---|---|---|---|
| $L_0$ | 0 | 0 | 0 | 0 |
| $L_1$ | 0 | $\frac{a}{2}$ | $\frac{1}{2}$ | $\frac{1-a}{2}$ |
| $L_2$ | 0 | $\frac{1}{2}$ | 0 | $\frac{1}{2}$ |
| $L_3$ | 0 | $\frac{1-a}{2}$ | $\frac{1}{2}$ | $\frac{a}{2}$ |

Recall the three Jacobi theta functions:

$$\begin{aligned}\theta_3(z) &:= \theta_{\mathbb{Z}}(z),\\ \theta_4(z) &:= \theta_3(z+1),\\ \theta_2(z) &:= \theta_{\mathbb{Z}+\frac{1}{2}}(z).\end{aligned}$$

TABLE 2. Orthogonality Relations for Odd $n$

| | $L_0$ | $L_1$ | $L_2$ | $L_3$ |
|---|---|---|---|---|
| $L_0$ | 0 | 0 | 0 | 0 |
| $L_1$ | 0 | $\frac{2-a}{4}$ | $\frac{1}{2}$ | $\frac{-a}{4}$ |
| $L_2$ | 0 | $\frac{1}{2}$ | 0 | $\frac{1}{2}$ |
| $L_3$ | 0 | $\frac{-a}{4}$ | $\frac{1}{2}$ | $\frac{2-a}{4}$ |

**Theorem 5** (Conway and Sloane[9]). *Let $L$ be a Type I lattice. If*

$$\theta_L = \sum_{j=0}^{\lfloor n/8 \rfloor} a_j \theta_3(q)^{n-8j} \Delta_8(q)^j$$

*then*

$$\theta_S = \sum_{j=0}^{\lfloor n/8 \rfloor} \frac{(-1)^j}{16^j} a_j \theta_2(q)^{n-8j} \theta_4(q^2)^{8j},$$

*where*

$$\Delta_8(q) = q \prod_{m=1}^{\infty} (1-q^{2m-1})^8 (1-q^{4m})^8$$

*and $\theta_2, \theta_3$ and $\theta_4$ are the Jacobi theta series. The norms of vectors of $S$ are congruent to $n/4$* (mod 2).

3.2. **Shadows of Codes.** Let $C$ be a Type I $\mathbb{Z}_{2k}$-code of length $n$. Let $C_0$ be the subcode of codewords whose Euclidean weights are a multiple of $4k$. $C_0$ is a subcode of index 2 in $C$. Let $C_2$ be that unique nontrivial coset of $C_0$ into $C$. Then $C_0^\perp$ can be written as a union of cosets of $C_0$: $C_0^\perp = C_0 \cup C_2 \cup C_1 \cup C_3$. The *shadow* of $C$ is defined to be $S = C_1 \cup C_3$ (see Bannai *et al.*[2] and Dougherty *et al.*[18] for the details). Throughout this paper, $C_i$ $(i = 0, 1, 2, 3)$ denotes a coset of $C_0$.

**Theorem 6.** **(Bannai** *et al.*[2]) *Let $C$ be a Type I code over $\mathbb{Z}_{2k}$ and let $S$ be its shadow. Then the cwe and swe of $S$ is related to the cwe and swe of $C$ by the relation*

$$cwe_S(x_0, x_1, \ldots, x_{2k-1}) = cwe_C(A(x_0, x_1, \ldots, x_{2k-1}))$$
$$swe_S(x_0, x_1, \ldots, x_k) = swe_C(B(x_0, x_1, \ldots, x_k))$$

*where $A = (a_{ij})$ is the $2k$ by $2k$ matrix with $a_{ij} = \frac{1}{\sqrt{2k}} \eta_{4k}^{i^2+2ij}$, and $B = (b_{ij})$ is the $(k+1)$ by $(k+1)$ matrix with $b_{ij} = \sum_{i' \equiv i} a_{i'j}$ where $i' \equiv i$ if $i' = i$ or $i' = -i$.*

**Theorem 7** (Dougherty *et al.*[18]). *Suppose that $C$ is a self-dual code over $\mathbb{Z}_{2k}$ of length $n$.*

(1) *If $n \equiv 0$ and 2* (mod 4) *then Tables 3 and 4 hold, respectively, where the symbol $\perp$ in position $(i, j)$ means that $x \cdot y \equiv 0$* (mod $2k$) *for any vector $x \in C_i$ and any vector $y \in C_j$, and the symbol $\not\perp$ means that $x \cdot y \equiv k$* (mod $2k$) *for any vector $x \in C_i$ and any vector $y \in C_j$.*
(2) *If $n \equiv 1$* (mod 2) *then Table 5 holds where the symbol $\perp$ in position $(i, j)$ means that $x \cdot y \equiv 0$* (mod $2k$) *for any vector $x \in C_i$ and any vector $y \in C_j$, and the symbol $\not\perp$ means that $x \cdot y \not\equiv 0$* (mod $2k$) *for any vector $x \in C_i$ and any vector $y \in C_j$.*

**Remark.** For the case $k = 1$ and 2, the above theorem was proved in Brualdi and Pless[5] and Dougherty *et al.*,[17] respectively.

TABLE 3. Orthogonality Relations for $n \equiv 2 \pmod 4$ (The Klein Case)

| | $C_0$ | $C_1$ | $C_2$ | $C_3$ |
|---|---|---|---|---|
| $C_0$ | $\perp$ | $\perp$ | $\perp$ | $\perp$ |
| $C_1$ | $\perp$ | $\not\perp$ | $\not\perp$ | $\perp$ |
| $C_2$ | $\perp$ | $\not\perp$ | $\perp$ | $\not\perp$ |
| $C_3$ | $\perp$ | $\perp$ | $\not\perp$ | $\not\perp$ |

TABLE 4. Orthogonality Relations for $n \equiv 0 \pmod 4$

| | $C_0$ | $C_1$ | $C_2$ | $C_3$ |
|---|---|---|---|---|
| $C_0$ | $\perp$ | $\perp$ | $\perp$ | $\perp$ |
| $C_1$ | $\perp$ | $\perp$ | $\not\perp$ | $\not\perp$ |
| $C_2$ | $\perp$ | $\not\perp$ | $\perp$ | $\not\perp$ |
| $C_3$ | $\perp$ | $\not\perp$ | $\not\perp$ | $\perp$ |

TABLE 5. Orthogonality Relations for $n \equiv 1 \pmod 2$

| | $C_0$ | $C_1$ | $C_2$ | $C_3$ |
|---|---|---|---|---|
| $C_0$ | $\perp$ | $\perp$ | $\perp$ | $\perp$ |
| $C_1$ | $\perp$ | $\not\perp$ | $\not\perp$ | $\not\perp$ |
| $C_2$ | $\perp$ | $\not\perp$ | $\perp$ | $\not\perp$ |
| $C_3$ | $\perp$ | $\not\perp$ | $\not\perp$ | $\not\perp$ |

## 4. CONSTRUCTIONS OF UNIMODULAR LATTICES AND SELF-DUAL CODES

In this section, we give constructions of unimodular lattices using shadows, which are a generalization of the construction given in Dougherty *et al.*[18] As a corollary, we also give a construction of self-dual codes over $\mathbb{Z}_{2k}$ which includes known constructions of binary self-dual codes and self-dual $\mathbb{Z}_4$-codes using shadows.

**Theorem 8.** *Let $L$ and $L'$ be Type I lattices in dimensions $n$ and $m$, respectively with the four cosets $L_i$ and $L'_i$ $(i = 0, 1, 2, 3)$. Let*

$$M = (L_0, L'_0) \cup (L_1, L'_1) \cup (L_2, L'_2) \cup (L_3, L'_3),$$
$$M' = (L_0, L'_0) \cup (L_1, L'_3) \cup (L_2, L'_2) \cup (L_3, L'_1),$$
$$N = (L_0, L'_0) \cup (L_1, L'_2) \cup (L_2, L'_3) \cup (L_3, L'_1),$$

*where $(L_i, L'_i) = \{(l, l') \mid l \in L_i, l' \in L'_i\}$.*

(1) *Suppose that $n \equiv m \equiv 0 \pmod 4$. Then $M, M'$ and $N$ are unimodular lattices in dimensions $n+m$. Moreover $M$ and $M'$ are Type II if and only if $n+m \equiv 0 \pmod 8$ and $N$ is a Type II if and only if $n \equiv m \equiv 4 \pmod 8$.*

(2) *Suppose that $n \equiv m \equiv 2 \pmod 4$. Then $M$ and $M'$ are unimodular lattices in dimension $n+m$. Moreover $M$ and $M'$ are Type II if and only if $n+m \equiv 0 \pmod 8$.*

(3) *Suppose that $n$ and $m$ are odd such that $n+m \equiv 0 \pmod 4$. Then $M$ is a unimodular lattice in dimension $n+m$. Moreover $M$ is Type II if and only if $n+m \equiv 0 \pmod 8$.*

*Proof.* By Lemma 3, it is easy to see that $M, M'$ and $N$ are $\mathbb{Z}$-modules, that is, lattices. $M, M'$ and $N$ consist of the 4 cosets of the lattice $(L_0, L_0')$ of determinant 4. Thus these are lattices of determinant 1. It follows from Tables 1 and 2 that $M, M'$ and $N$ are integral under each assumption. Hence $M, M'$ and $N$ are unimodular. Also Tables 1 and 2 give the norm $\pmod{\mathbb{Z}}$ of a vector in each coset $L_i$ and $L_i'$. This determines when $M, M'$ and $N$ are Type II. □

Constructions of unimodular lattices using shadows are given in Dougherty *et al.*[18] By putting $m = 1,2,3,4$ and $L' = \mathbb{Z}^m$ in the above theorem, the construction is obtained.

In order to demonstrate that not every $M$ is the same as direct sum $(L, L')$, we give a trivial example. Consider $\mathbb{Z}^n$ and $\mathbb{Z}^{8-n}$ as $L$ and $L'$ in the above theorem, then $M$ is a Type II lattice in dimension 8 which is the $E_8$-lattice. Of course, the direct sum of $\mathbb{Z}^n$ and $\mathbb{Z}^{8-n}$ is $\mathbb{Z}^8$.

We now consider constructions by shadows for self-dual $\mathbb{Z}_{2k}$-codes.

Lemma 2, together with the previous theorem, gives essentially the following constructions of self-dual $\mathbb{Z}_{2k}$-codes.

**Corollary 9.** *Let $C$ and $D$ be Type I $\mathbb{Z}_{2k}$-codes of lengths $n$ and $m$, respectively. Let*

$$E = (C_0, D_0) \cup (C_1, D_1) \cup (C_2, D_2) \cup (C_3, D_3),$$
$$F = (C_0, D_0) \cup (C_1, D_3) \cup (C_2, D_2) \cup (C_3, D_1),$$
$$G = (C_0, D_0) \cup (C_1, D_2) \cup (C_2, D_3) \cup (C_3, D_1),$$

*where $(C_i, D_i) = \{(c,d) \mid c \in C_i, d \in D_i\}$.*

(1) *Suppose that $n \equiv m \equiv 0 \pmod 4$. Then $E$ and $F$ are self-dual codes of length $n+m$. Moreover $E$ and $F$ are Type II if and only if $n+m \equiv 0 \pmod 8$. $G$ is a Type II code if and only if $n \equiv m \equiv 4 \pmod 8$.*
(2) *Suppose that $n \equiv m \equiv 2 \pmod 4$. Then $E$ and $F$ are self-dual codes of length $n+m$. Moreover $E$ and $F$ are Type II if and only if $n+m \equiv 0 \pmod 8$.*
(3) *Suppose that $n$ and $m$ are odd such that $n+m \equiv 0 \pmod 4$. Then $E$ is a self-dual code of length $n+m$. Moreover $E$ is Type II if and only if $n+m \equiv 0 \pmod 8$.*

*Proof.* The linearity follows from Lemmas 3 and 2 and the fact that the code is self-orthogonal follows from Lemmas 4 and 2. Now

$$|E| = |F| = |G| = 4(|C_i||D_i|) = 4(\frac{|C|}{2}\frac{|D|}{2}) = |C||D| = (4k)^{\frac{n+m}{2}},$$

giving that each code is self-dual under each assumption. The Euclidean weight of any vector in the shadow of a Type I $\mathbb{Z}_{2k}$-code of length $n$ is congruent to $\frac{kn}{2} \pmod{4k}$. This determines the condition that codes are Type II.[18] □

*Remark.* It is worthwhile noting that

$$\cup_{i=0}^{3}(A_{2k}(C_i), A_{2k}(D_i)) = \cup_{i=0}^{3}(A_{2k}(C)_i, A_{2k}(D)_i).$$

The symmetrized weight enumerators of these codes are obtained from symmetrized weight enumerators of $C_i$ and $D_i$, for example, $E$ has symmetrized weight enumerator

$$swe_E(x_0, x_1, \ldots, x_k) = \sum_{i=0}^{3} swe_{C_i}(x_0, x_1, \ldots, x_k) \times swe_{D_i}(x_0, x_1, \ldots, x_k).$$

Now we consider the case $k = 1$ and 2, i.e., binary and $\mathbb{Z}_4$-codes. Let $C$ be the binary self-dual code $\{(0,0),(1,1)\}$ of length 2 in the above theorem. Then the above theorem is the same as Theorem 1 in Brualdi and Pless.[5] Moreover let $C$ be the self-dual code

$$\{(0,0,0,0),(1,1,0,0),(0,0,1,1),(1,1,1,1)\}$$

of length 4 then the above theorem is the same as Theorem 2 in Brualdi and Pless.[5] Thus the above theorem is a generalization of Theorems 1 and 2 in this paper.[5] Moreover constructions of self-dual $\mathbb{Z}_4$-codes using the shadows were also given in Dougherty *et al.*,[17] which are generalizations of Theorems 1 and 2 in Brualdi and Pless.[5] The constructions given in Dougherty *et al.*[17] are obtained from the above theorem considering $C$ as a Type I code of length $1, 2, 3$ or 4.

There are exactly four inequivalent Type II codes of length 8, namely $K_8, K_8', O_8$ and $Q_8$ in Conway and Sloane.[8] The four codes are distinguished by comparing the coefficients of $a^6c^2$ in the symmetrized weight enumerators $swe_C(a,b,c)$. Also there are exactly four inequivalent Type II codes of length 7, namely $A^7, D_4^{\oplus}A^3, D_6^{\oplus}A$ and $E_7^+$ where $A$ denotes the unique self-dual code $\{0,2\}$ of length 1. We consider a Type II code $E$ constructed from $A$ and a self-dual code $B$ of length 7 by Corollary 9. Let $N(B)$ be the number of codewords corresponding to $a^6c^2$ in $swe_E(a,b,c)$. Since $A_0 = \{0\}$, $A_2 = \{2\}$, $A_1 = \{1\}$ and $A_3 = \{3\}$, $N(B)$ is the sum of the numbers of codewords corresponding to $a^5c^2$ and $a^6c$ in $swe_B(a,b,c)$. Hence

$$N(A^7) = 28, N(D_4^{\oplus}A^3) = 12, N(D_6^{\oplus}A) = 4 \text{ and } N(E_7^+) = 0,$$

where the symmetrized weight enumerator of $B$ is listed in Conway and Sloane.[8] Therefore we have the following:

**Proposition 10.** *Type II $\mathbb{Z}_4$-codes of length 8 are in one-to-one correspondence with Type I codes of length 7 by Corollary 9.*

There are exactly three Type I codes of length 6, and there is a unique Type I code of length 2. Thus not every Type II code is constructed from Type I codes of lengths 2 and 6 by Corollary 9.

## 5. Applications to Unimodular Lattices and Self-Dual Codes

5.1. **Unimodular Lattices.** First let us look at a 12-dimensional lattice. Consider the six-dimensional lattice $\mathbb{Z}^6$ as $L$ and $L'$. Then both the minimum norms of $(L_0, L_0')$ and $(L_2, L_2')$ are 2 since $L_0$ and $L_0'$ consist of vectors of even norms and $L_2$ and $L_2'$ have norms $\equiv 1 \pmod 2$. Since the norms of vectors of $L_1$ and $L_3$ are congruent to $n/4 \pmod 2$, both the minimum norms of $(L_1, L_1')$ and $(L_3, L_3')$ are 3. Therefore $M$ is the unique 12-dimensional unimodular lattice $D_{12}^+$ with minimum norm 2, while $(L, L')$ is the lattice $\mathbb{Z}^{12}$. Similarly, some Type II lattices in dimension 16 and some Type I lattices with minimum norm 2, which is the highest minimum norm among all unimodular lattices in that dimension, are constructed.

**Lemma 11.** *All Type II $\mathbb{Z}_4$-codes of length $8n$ are constructed from Type I codes of lengths 1 and $8n-1$ by Corollary 9.*

*Proof.* Let $A$ be the unique Type I $\mathbb{Z}_4$-code of length 1 where $A_0 = \{0\}$, $A_1 = \{1\}$, $A_2 = \{2\}$ and $A_3 = \{3\}$. Since a Type II $\mathbb{Z}_4$-code $C$ contains a vector whose elements are 1 or 3,(see Harada *et al.*[20]) every coordinate of $C$ consists of $0, 1, 2$, and 3. □

*Remark.* The converse assertion of the above lemma is Lemma 1 in Fields et al.[19]

**Theorem 12.** *All unimodular lattices in dimensions $n \leq 17$ and $n = 23$ are constructed from some $\mathbb{Z}_4$-codes by Construction $A_4$.*

*Proof.* Our proof is divided into the following cases.

- $n = 7, 15, 23$: $E_8$, the unique Type II lattice in dimension 8, is constructed from any Type II $\mathbb{Z}_4$-code of length 8 by Construction $\mathrm{A}_4$.[3] Similarly, all Type II lattices in dimensions 16 and 24 are constructed from Type II $\mathbb{Z}_4$-codes.[4] If a Type II $\mathbb{Z}_{2k}$-code $E$ of length $n+m$ is constructed from some Type I codes $C$ and $D$ of lengths $n$ and $m$ by Corollary 9 then the lattice $A_{2k}(E)$ is constructed from $A_{2k}(C)$ and $A_{2k}(D)$ by Theorem 8. Thus, by Lemmas 2 and 11, all Type I lattices in dimensions $7, 15$ and 23 are constructed from $\mathbb{Z}_4$-codes.
- $1 \leq n \leq 13$: In this case, we give a $\mathbb{Z}_4$-code $C$ for each lattice $L$ in Table 6 where $C_8$ is a Type II code of length 8.

TABLE 6. $\mathbb{Z}_4$-code construction of unimodular lattices

| Lattices | $\mathbb{Z}^n$ $(1 \leq n \leq 13)$ | $E_8$ | $E_8 \oplus \mathbb{Z}^{n-8}$ $(9 \leq n \leq 13)$ | $D_{12}^+$ | $D_{12}^+ \oplus \mathbb{Z}$ |
|---|---|---|---|---|---|
| Codes | $A^n$ $(1 \leq n \leq 13)$ | $C_8$ | $C_8 \oplus A^{n-8}$ $(9 \leq n \leq 13)$ | $K_{12}$ | $K_{12} \oplus A$ |

- $n = 14$: There is a unique extremal unimodular lattice, that is, its minimum norm is 2. A corresponding $\mathbb{Z}_4$-code is known.[4] A Type I lattice with minimum norm 1 is constructed as $\mathbb{Z} \oplus L_{13}$ where $L_{13}$ is a Type I lattice in dimension 13. For every Type I lattice in dimension $n \leq 13$, a corresponding $\mathbb{Z}_4$-code is listed in Table 6. Thus all Type I lattices in dimension 14 are constructed from $\mathbb{Z}_4$-codes.
- $n = 16$: Since a Type I lattice with minimum norm 1 is constructed as $\mathbb{Z} \oplus L_{15}$ where $L_{15}$ is a Type I lattice in dimension 15, such a Type I lattice is constructed from some $\mathbb{Z}_4$-code. There is a unique Type I lattice with minimum norm 2. A Type I $\mathbb{Z}_4$-code with minimum Euclidean weight 8 is known.[17]
- $n = 17$: There is a unique Type I lattice with minimum norm 2. A Type I $\mathbb{Z}_4$-code with minimum Euclidean weight 8 is known.[17] Since all 16-dimensional unimodular lattices are constructed from some $\mathbb{Z}_4$-codes, all Type I lattices are constructed by Construction $\mathrm{A}_4$. □

5.2. **Self-Dual Codes over $\mathbb{Z}_4$.** All self-dual codes of length up to 15 and Type II codes of length 16 over $\mathbb{Z}_4$ are known, Conway and Sloane,(see [8] Fields *et al.*[19] and Pless *et al.*[22])

We construct Euclidean-optimal self-dual codes of length 20 and minimum Euclidean weight 8, that is, self-dual codes having the highest minimum Euclidean weight among all self-dual codes of that length, by Corollary 9.

- $(n,m) = (4,16), (5,15), (6,14), (7,13), (8,12)$: First recall that all Euclidean weights of vectors in the shadow of a self-dual $\mathbb{Z}_4$-code of length $n$ is congruent to $n \pmod 8$. Thus in this case, $E$ is a Euclidean-optimal self-dual code of length 20.
- $(n,m) = (10,10)$: It is easy to see that only two codes [10,3]-d4⊕d6 and [10,4]-d10a in Fields *et al.*[19] have shadows of minimum Euclidean weight 10. Thus if at least one of $C$ and $D$ is [10,3]-d4⊕d6 or [10,4]-d10a then the obtained code $E$ is Euclidean-optimal.
- $(n,m) = (9,11)$: There is no indecomposable Type I code of length 9 and every code is the direct sum of the self-dual code of length 1 and a self-dual code of length 8 (see Conway and Sloane[8]). Every Type I code obtained from a Type I code of length 8 has a shadow which has minimum Euclidean weight 9, while every code obtained from a Type II code has a shadow with minimum Euclidean weight 1. Thus if $C$ is a Type I code of length 9 obtained from a Type I code of length 8 and $D$ is any Type I code of length 11 then $E$ is Euclidean-optimal.

  For length 11, only [11,4]-d4⊕e7 in Fields *et al.*[19] has a shadow with minimum Euclidean weight 11. Thus if $C$ is a Type I code of length 9 and $D$ is [11,4]-d4⊕e7 then $E$ is Euclidean-optimal.
- $(n,m) = (1,19), (2,18), (3,17)$: If there is a Type I code of lengths $17, 18$ and $19$ whose shadows of minimum Euclidean weights $\geq 11, \geq 10$ and $\geq 9$, respectively, then the obtained code $E$ is Euclidean-optimal.

5.3. **Binary Self-Dual Codes.** For binary codes Corollary 9 gives that $E$ and $F$ are self-dual codes for all possible cases since self-dual codes exist only for even lengths. In this section, we construct an extremal self-dual $[44,22,8]$ code. Let $C$ and $D$ both be the unique self-dual $[22,11,6]$ code $G_{22}$ which is called the shorter Golay code.

$$W_C(1,y) = 1 + 77y^6 + 330y^8 + 616y^{10} + \cdots,$$
$$W_{C_1}(1,y) = W_{C_3}(1,y) = 176y^7 + 672y^{11} + \cdots.$$

By Corollary 9, $E$ and $F$ are extremal self-dual $[44,22,8]$ codes. The numbers of codewords of weights 8 and 10 are 660 and 0, respectively. Thus these codes correspond to the second weight enumerator with $\beta = 154$ in Conway and Sloane.[6]

It is well known that $C$ is constructed from the Golay $[24,12,8]$ code $G$ by subtracting. Thus let $G_{ij} = \{(x_3,\ldots,x_{24}) \mid (x_1,\ldots,x_{24}) \in G, x_1 = i, x_2 = j\}$. Then $G_{00} = C_0$, $G_{10} = C_1$, $G_{11} = C_2$ and $G_{01} = C_3$. Since the automorphism group of $G$ is the Mathieu group $M_{24}$ which acts 5-fold transitively on the coordinates of $G$, $C_1 = C_3$. Hence our two codes $E$ and $F$ are equivalent.

We now compare our code with the code constructed in Yorgov and Ruseva.[25] It was shown[25] that the following matrix

$$\begin{pmatrix} \mathbf{1} & & \mathbf{1} & \\ & \mathbf{1} & & \mathbf{1} \\ M & N & & \\ & & M & N \end{pmatrix},$$

generates a self-dual $[44, 22, 8]$ code corresponding to the second weight enumerator with $\beta = 154$, where $M$ and $N$ are $10 \times 11$ circulant matrices with first rows (01111111111) and (11011100010), respectively, and $\mathbf{1}$ is the all-one vector of length 11. It is easy to see that the matrix

$$\begin{pmatrix} \mathbf{1} & \mathbf{1} \\ M & N \end{pmatrix},$$

generates $G_{22}$ where $(\ M\ N\ )$ generates the doubly-even subcode $G_0$ of even weight in $G_{22}$. Moreover from the definition of the shadow we have that $G_0 + (\ \mathbf{1}\ ,\ \mathbf{0}\ ) = C_1$ and $G_0 + (\ \mathbf{0}\ ,\ \mathbf{1}\ ) = C_3$, up to indexes. Therefore our code is equivalent to the code in Yorgov and Ruseva[25] and the construction there is a special case of Corollary 9.

## 6. Other Fields and Rings

In this section, we give modifications of Corollary 9 to self-dual codes over finite fields $\mathbb{F}_q$ where $q$ is a prime power, and to Type I codes over $\mathbb{F}_2 + u\mathbb{F}_2$.

6.1. **Self-Dual Codes over $\mathbb{F}_q$.** Let $C$ be a self-dual code over $\mathbb{F}_q$ of length $n$, and let $C_0$ be any subcode of codimension 1. Let $s$ and $t$ be the vectors such that $C = \langle C_0, t\rangle$ and $C_0^\perp = \langle C, s\rangle$. Of course, $t$ can be chosen so that $s \cdot t = 1$. Define the cosets $C_{i,j} = C_0 + it + js$ $(i, j \in \mathbb{F}_q)$.

Let $D$ be a self-dual code over $\mathbb{F}_q$ of length $m$. Let $s'$ be a vector not in $D$ and let $D_0$ be the subcode of vectors of $D$ orthogonal to $s'$. We shall describe how $s'$ is chosen shortly. Let $t'$ and $s'$ be the vectors such that $D = \langle D_0, t'\rangle$ and $D_0^\perp = \langle D, s'\rangle$. Here we chose $t'$ so that $s' \cdot t' = -1$. If $s'$ can be chosen such that $s' \cdot s' = -s \cdot s$ then $x_1 \cdot x_2 = -y_1 \cdot y_2$ for $x_1 \in D_{i,j}, x_2 \in D_{l,m}$ and $y_1 \in C_{i,j}, y_2 \in C_{l,m}$.

**Proposition 13.** *Let $C$ and $D$ be self-dual codes over $\mathbb{F}_q$ of lengths $n$ and $m$, respectively. Let $C_{i,j}$ and $D_{i,j}$ be as above. Then the following*

$$E = \bigcup_{i,j \in \mathbb{F}_q} (C_{i,j}, D_{i,j})$$

*is a self-dual code of length $n + m$.*

*Proof.* The self-orthogonality follows from that $x_1 \cdot x_2 = -y_1 \cdot y_2$ for $x_1, x_2 \in D_{i,j}$ and $y_1, y_2 \in C_{i,j}$. $|E| = q^2(|C|/q)(|D|/q) = |C||D| = q^{\frac{n+m}{2}}$. Thus $E$ is self-dual. □

The above proposition is a generalization of Dougherty.[11] As an example, we consider the case that $q = 5, n = 2$ and $m = 4$. Let $C$ be the self-dual code over $\mathbb{F}_5$ of length 2 and let $D$ be the self-dual code of length 4 which is the direct sum of $C$. Take $s = (1, 1)$, $t = (2, 4)$ then $C_0 = \{(0, 0)\}$. Also take $s' = (1, 1, 1, 0)$ and $t' = (1, 2, 1, 2)$ then $D_0 =< (1, 2, 2, 4) >$. Indeed, $s \cdot t = 1$, $s' \cdot t' = -1$, and $s \cdot s = -s' \cdot s'$. By the above proposition, $E$ is a self-dual code of length 6. It can be easily seen that $E$ has generator matrix of the form

$$\begin{pmatrix} 1\,0\,0\,3\,1\,3 \\ 0\,1\,0\,1\,3\,3 \\ 0\,0\,1\,2\,2\,4 \end{pmatrix}.$$

Thus $E$ is equivalent to $F_6$ in Leon *et al.*[21] which is the indecomposable self-dual code of length 6 with minimum weight 4.

6.2. **Type I Codes over $\mathbb{F}_2 + u\mathbb{F}_2$.** Another important ring of order 4 besides $\mathbb{Z}_4$ in self-dual codes is $\mathbb{F}_2 + u\mathbb{F}_2$ introduced in Bachoc[1] to construct lattices (cf. e.g. Bachoc[1] and Dougherty *et al.*[12]) We give necessary definitions for self-dual codes over $\mathbb{F}_2 + u\mathbb{F}_2$. The ring $\mathbb{F}_2 + u\mathbb{F}_2$ is commutative with elements $0, 1, u, u+1$ with $u^2 = 0$. A linear code $C$ in this space of length $n$ is an $(\mathbb{F}_2 + u\mathbb{F}_2)$-submodule of $(\mathbb{F}_2 + u\mathbb{F}_2)^n$. The *Hamming weight* of a codeword is the number of non-zero components. The *Lee weight* $w_L(x)$ of $x = (x_1, x_2, \dots, x_n)$ is defined as $n_1(x) + 2n_2(x)$ where $n_0(x)$ is the number of $x_i = 0$, $n_2(x)$ the number of $x_i = u$ and $n_1(x) = n - n_0(x) - n_2(x)$.

The dual code $C^\perp$ of $C$ is defined as $\{x \in (\mathbb{F}_2 + u\mathbb{F}_2)^n \mid x \cdot y = 0 \text{ for all } y \in C\}$ where $x \cdot y = x_1y_1 + x_2y_2 + \cdots + x_ny_n$ for $x = (x_1, x_2, \dots, x_n)$ and $y = (y_1, y_2, \dots, y_n)$. As usual, a code is *self-dual* if $C = C^\perp$. A self-dual code $C$ is *Type II* if the Lee weights of all codewords of $C$ are divisible by four and $C$ is *Type I* otherwise.[12] It is known that a Type II code exists if and only if the length is divisible by four. We define the doubly-even Lee weight subcode $C_0$ of a Type I code $C$ as the subcode consisting of all codewords of $C$ with Lee weights a multiple of 4. The subcode $C_0$ is of index 2 in $C$ and $C$ of index 2 in $C_0^\perp$. Naturally we denote $C_0^\perp = C_0 \bigcup C_1 \bigcup C_2 \bigcup C_3$, and $C = C_0 \cup C_2$; defining the *shadow* $S$ of $C$ as $S = C_1 \cup C_3$.

**Proposition 14.** *(**Dougherty** et al.[12]) If $C$ is Type I then $\phi(C_j) = \phi(C)_j$ for $j = 0, 1, 2, 3$, that is,*

$$\begin{array}{ccc} C & \xrightarrow{\;\;\phi\;\;} & \phi(C) \\ \text{coset}\big\downarrow & & \big\downarrow\text{coset} \\ C_j & \xrightarrow[\;\;\phi\;\;]{} & \phi(C_j) = \phi(C)_j \end{array}$$

*Moreover $\phi(S)$ is the shadow of $\phi(C)$. Hence if the shadow of $C$ has minimum Lee weight $d_S$ then $\phi(C)$ is a binary $[2n, n]$ code whose shadow has minimum Hamming weight $d_S$.*

**Theorem 15.** *(**Dougherty** et al.[12]) If $C$ is a Type I code, then the swe of $C_0$ is*

$$swe_{C_0}(a, b, c) = \frac{1}{2}(swe_C(a, b, c) + swe_C(a, ib, -c)).$$

*The swe of $S$ is related to the swe of $C$ by the relation*

$$swe_S(a, b, c) = swe_C(b + (a+c)/2, i(a-c)/2, b - (a+c)/2)$$

*If the swe of a Type I code $C$ can be expressed as*

$$\sum_{j,k} \alpha_{jk}(a+c)^{n-2j-4k}(ac - b^2)^j(b^2(a-c)^2)^k,$$

*then the swe of its shadow is*

$$\sum_{j,k} \alpha_{jk}(-1)^k 2^{n-2j-6k}(b)^{n-2j-4k}(b^2 - ac)^j(a^2 - c^2)^{2k}.$$

*In particular the Lee weight enumerator $W_L(y)$ of the shadow is*

$$\sum_k \alpha_{0k}(-1)^k 2^{n-6k} y^{n-4k}(1 - y^4)^{2k}.$$

*The Lee weights of codewords of $S$ are congruent to $n$ modulo 4.*

**Theorem 16** (Dougherty *et al.*[12]). *The cosets for a Type I code over $\mathbb{F}_2 + u\mathbb{F}_2$ have the orthogonality relations given in Table 7 if n is odd and Table 8 if n is even, with the value in position $(i, j)$ is $x \cdot y$ for any vector $x \in C_i$ and any vector $y \in C_j$. Moreover the glue group $C_0^{\perp}/C_0$ of a Type I code $C$ over $\mathbb{F}_2 + u\mathbb{F}_2$ is isomorphic to the Klein 4-group for all lengths $n$.*

TABLE 7. Orthogonality Relations for $n$ odd

| | $C_0$ | $C_1$ | $C_2$ | $C_3$ |
|---|---|---|---|---|
| $C_0$ | 0 | 0 | 0 | 0 |
| $C_1$ | 0 | 1 | $u$ | $1+u$ |
| $C_2$ | 0 | $u$ | 0 | $u$ |
| $C_3$ | 0 | $1+u$ | $u$ | 1 |

TABLE 8. Orthogonality Relations for $n$ even

| | $C_0$ | $C_1$ | $C_2$ | $C_3$ |
|---|---|---|---|---|
| $C_0$ | 0 | 0 | 0 | 0 |
| $C_1$ | 0 | 0 | $u$ | $u$ |
| $C_2$ | 0 | $u$ | 0 | $u$ |
| $C_3$ | 0 | $u$ | $u$ | 0 |

**Proposition 17.** *Let $C$ and $D$ be Type I codes over $\mathbb{F}_2 + u\mathbb{F}_2$ of lengths $n$ and $m$, respectively. Let*

$$E = (C_0, D_0) \cup (C_1, D_1) \cup (C_2, D_2) \cup (C_3, D_3),$$
$$F = (C_0, D_0) \cup (C_1, D_3) \cup (C_2, D_2) \cup (C_3, D_1),$$
$$G = (C_0, D_0) \cup (C_1, D_2) \cup (C_2, D_3) \cup (C_3, D_1),$$

*where $(C_i, D_i) = \{(c, d) \mid c \in C_i, d \in D_i\}$. Then if $n \equiv m \pmod 2$ then $E$ and $F$ are self-dual codes. Moreover if both $n$ and $m$ are even then $G$ is a self-dual code.*

*Proof.* The linearity follows from Lemma 16 and the fact that the code is self-orthogonal follows from Lemma 16. Now

$$|E| = |F| = |G| = 4(|C_i||D_i|) = 4(\frac{|C|}{2}\frac{|D|}{2}) = |C||D| = 4^{\frac{n+m}{2}},$$

giving that each code is self-dual under each assumption. □

*Remark.* This proposition can be also proved by Proposition 3.8 in Dougherty *et al.*[12] and Corollary 9.

Theorems 10.6 and 10.7 in Dougherty *et al.*[12] are specific cases of the above proposition. Since self-dual codes of length up to 8 are classified,[12] we construct new optimal self-dual codes of length 10 as an example. $\mathcal{K}_1 \oplus \mathcal{K}_4$ and $\mathcal{K}_1^5$ are self-dual codes[12] of length 5. We define the Lee weight enumerator $W_C(z)$ of $C$ as

$\sum_{c\in C} z^{w_L(c)}$. It is not difficult to see that

$$W_{\mathcal{K}_1\oplus\mathcal{K}_4}(z) = 1 + z^2 + 14z^4 + 14z^6 + z^8 + z^{10}$$
$$W_{(\mathcal{K}_1\oplus\mathcal{K}_4)_1}(z) = W_{(\mathcal{K}_1\oplus\mathcal{K}_4)_3}(z) = z + 14z^5 + z^9,$$
$$W_{\mathcal{K}_1^5}(z) = 1 + 5z^2 + 10z^4 + 10z^6 + 5z^8 + z^{10}$$
$$W_{(\mathcal{K}_1^5)_1}(z) = W_{(\mathcal{K}_1^5)_3}(z) = 16z^5.$$

Only $\mathcal{K}_{10}$ is a known optimal self-dual code[12] of length 10, that is, it has minimum Lee weight 4. Let $E_1$ and $E_2$ be the self-dual codes constructed from the above proposition when $(C, D) = (\mathcal{K}_1 \oplus \mathcal{K}_4, \mathcal{K}_1^5)$ and $(\mathcal{K}_1^5, \mathcal{K}_1^5)$, respectively. It turns out that

$$W_{E_1}(z) = 1 + 29z^4 + 32z^6 + \cdots,$$
$$W_{E_2}(z) = 1 + 25z^4 + 40z^6 + \cdots,$$
$$W_{\mathcal{K}_{10}}(z) = 1 + 45z^4 + 210z^8 + \cdots.$$

Therefore we have the following:

**Proposition 18.** *There are at least three inequivalent optimal Type I codes over* $\mathbb{F}_2 + u\mathbb{F}_2$ *of length* 10.

Similarly to $\mathbb{Z}_4$-codes, we have the following:

**Corollary 19.** *All Type II codes over* $\mathbb{F}_2 + u\mathbb{F}_2$ *of length* $4n$ *are constructed from Type I codes of lengths* 1 *and* $4n - 1$ *by Proposition 17.*

*Proof.* Let $A$ be the unique Type I code $\{(0), (u)\}$ of length 1 where $A_0 = \{0\}$, $A_1 = \{1\}$, $A_2 = \{u\}$ and $A_3 = \{1 + u\}$. Since a Type II code $C$ contains a vector whose elements are 1 or $1+u$,[12] every coordinate of $C$ consists of $0, 1, u$ and $1+u$. □

ACKNOWLEDGMENTS

The authors are grateful to Masaaki Harada for helpful discussions.

## REFERENCES

REFERENCES

[1] Bachoc,C.: Application of coding theory to the construction of modular lattices, *J. Combin. Theory Ser. A* **78**, 92–119, (1997).
[2] Bannai, E., Dougherty, S.T., Harada, M., Oura, M.: Type II codes, even unimodular lattices, invariant rings, *IEEE Trans. Inform. Theory*, **45**, 1194–1205, (1999).
[3] Bonnecaze, A., Solé, P., Calderbank, A.R.: Quaternary quadratic residue codes and unimodular lattices, *IEEE Trans. Inform. Theory* **41**, 366–377, (1995).
[4] Bonnecaze, A., Gaborit, P., Harada, M., Kitazume, M. and Solé, P.: Niemeier lattices and Type II codes over $\mathbb{Z}_4$, *Discrete Math.*, (to appear).
[5] Brualdi, R.A., Pless, V.S.: Weight enumerators of self-dual codes, *IEEE Trans. Inform. Theory* **37**, 1222–1225, (1991).
[6] Conway, J.H., Sloane, N.J.A.: A new upper bound on the minimal distance of self-dual codes, *IEEE Trans. Inform. Theory* **36**, 1319–1333, (1990).
[7] Conway, J.H., Sloane, N.J.A.: A new upper bound for the minimum of an integral lattice of determinant 1, *Bull. of the Ameri. Math. Soc.* **23**, 383–387, (1990).
[8] Conway, J.H., Sloane, N.J.A.: Self-dual codes over the integers modulo 4, *J. Combin. Theory Ser. A* **62**, 30–45, (1993).

[9] Conway, J.H., Sloane, N.J.A.: A note on optimal unimodular lattices, *J. Number Theory* **72**, 357–362, (1998).

[10] Conway, J.H., Sloane, N.J.A.: "Sphere Packing, Lattices and Groups (3rd ed.)," New York: Springer-Verlag, (1999).

[11] Dougherty, S.T.: Shadow codes and weight enumerators, *IEEE Trans. Inform. Theory* **41**, 762–768, (1995).

[12] Dougherty, S.T., Gaborit, P., Harada, M., Solé, P.: Type II codes over $\mathbb{F}_2 + u\mathbb{F}_2$, *IEEE Trans. Inform. Theory* **45**, 32–45, (1999).

[13] Dougherty, S.T., T.A. Gulliver, and M. Harada, Extremal binary self-dual codes, *IEEE Trans. Inform. Theory*, 2036-2047, November, (1997).

[14] Dougherty, S.T., T.A. Gulliver, and M. Harada, Type II Codes, Even Unimodular Lattices, and Invariant Rings, Journal of Algebraic Combinatorics, **9**, 3, 233–250, (1999).

[15] Dougherty, S.T., Harada, M.: Shadow optimal self-dual codes, Kyushu Journal of Matheamtics, **53**, No 2. 223–237, (1999).

[16] Dougherty, S.T., Harada, M.: New extremal self-dual codes of length 68, *IEEE Trans. Inform. Theory*, **46**, No. 6, 2133–2136, September (1999).

[17] Dougherty, S.T., Harada, M., Solé, P.: Shadow codes over $\mathbb{Z}_4$, *Finite Fields and Their Appl.*, to appear.

[18] Dougherty, S.T., Harada, M., Solé, P.: Shadow lattices and shadow codes, *Discrete Math.*, **219**, 49–64, (2000).

[19] Fields, J., Gaborit, P., Leon, J., , Pless, V.S.: All self-dual $\mathbb{Z}_4$ codes of length 15 or less are known, *IEEE Trans. Inform. Theory* **44**, pp. 311–322, (1998).

[20] Harada, M., Solé, P., Gaborit, P.: Self-dual codes over $\mathbb{Z}_4$ and unimodular lattices: a survey, in the Proc. of ICAC (International Congress in Algebras and Combinatorics 1997 August, Hong Kong), Springer-Verlag, 255–275, (1999). (to appear).

[21] Leon, J.S., Pless, V.S., Sloane, N.J.A.: Self-dual codes over $GF(5)$, *J. Combin. Theory Ser. A*, **32**, pp. 178–194, (1982).

[22] Pless, V.S., Leon, J., Fields, J.: All $\mathbb{Z}_4$ codes of Type II and length 16 are known, *J. Combin. Theory Ser. A* **78**, pp. 32–50, (1997).

[23] Quebbemann, H.-G.: A shadow identity and an application to isoduality, preprint.

[24] H.N. Ward, A restriction on the weight enumerator of a self-dual code, *J. Combin. Theory Ser. A* **21**, 253–255, (1976).

[25] Yorgov, V., Ruseva, R.: Two extremal codes of length 42 and 44, (in Russian) *Probl. Perda. Inform.* **29**, pp. 99–103, (1993). English translation in *Probl. Inform. Transm.* **29**, pp. 385–388, (1994).

Department of Mathematics, University of Scranton, Scranton, PA 18510,, USA

CNRS, I3S, ESSI, BP 145, Route des Colles,, 06 903 Sophia Antipolis, France

# NON-LINEAR $N$-TERM APPROXIMATIONS OF SMOOTH FUNCTIONS USING WAVELET DECOMPOSITIONS

DINH DUNG

*Institute of Information Technology*
*Nghia Do, Cau Giay, Hanoi, Vietnam*
*E-mail: ddung@ioit.ncst.ac.vn*

In the present paper, a central question to be considered is to compute asymptotic orders of non-linear $n$-term approximations of mixed smoothness classes of functions. We study non-linear $n$-term approximations with regard to the wavelet family $\mathbf{V}$ formed from the integer translates of the mixed dyadic scales of the tensor product multivariate de la Vallée Poussin kernel, and optimal continuous algorithms of $n$-term approximation in terms of $n$-widths, for functions from mixed smoothness Besov classes $\mathbf{SB}^{\Omega}_{p,\theta}$. Their mixed smoothness is defined via mixed modulus of smoothness dominated by a function $\Omega$. For a given $\Omega$ of a special form, we give the asymptotic order of these quantities.

## 1 Introduction

In an approximation problem, we resolve a complicated so-called target function by simplier and easier to compute approximants. In linear approximation, the approximants constitute a linear manifold. In non-linear approximation, they are a non-linear set. In the present paper, we will not discuss the advantages of non-linear approximations over linear ones. The interested reader can consult [3], [10] for details on this question. We will restrict ourselve to investigate asymptotic orders of non-linear $n$-term approximations for smoothness classes of target functions. In non-linear $n$-approximations with regard to a family $\Phi$, a target function is approximated by linear combinations of $n$ free terms from $\Phi$, which constitute a non-linear set. As in linear approximation in non-linear approximation, the smoothness of target functions is more conveniently and, maybe more naturally given by boundedness of Besov quasi-norm. We study $n$-term approximations with regard to the wavelet family $\mathbf{V}$ formed from the integer translates of the mixed dyadic scales of the tensor product multivariate de la Vallée Poussin kernel, and optimal continuous algorithms of $n$-term approximation in terms of non-linear $n$-widths, for mixed smoothness Besov classes $\mathbf{SB}^{\Omega}_{p,\theta}$ of functions on the $d$-torus $\mathbf{T}^d := [-\pi, \pi]^d$.

Let us first introduce mixed smoothness Besov spaces $\mathbf{B}^{\Omega}_{p,\theta}$. For a positive integer $l$, the univariate symmetric difference operator $\Delta^l_h, h \in \mathbf{T}$, is defined inductively by $\Delta^l_h := \Delta^1_h \Delta^{l-1}_h$, starting from $\Delta^1_h f := f(\cdot + h/2) - f(\cdot - h/2)$.

Further, for $l \in \mathbf{N}^d$, we let the multivariate mixed $l$th difference operator $\Delta_h^l, h \in \mathbf{T}^d$, be defined by

$$\Delta_h^l f := \Delta_{h_1}^{l_1} \Delta_{h_2}^{l_2} \cdots \Delta_{h_d}^{l_d} f$$

where the univariate operator $\Delta_{h_j}^{l_j}$ is applied to the variable $x_j$, and for $x \in \mathbf{R}^d$, we denote $x := (x_1, ..., x_d)$. Let

$$\Omega_l(f,t)_p := \sup_{|h_j|<t_j, j=1,\dots,d} \|\Delta_h^l f\|_p, \quad t \in \mathbf{R}_+^d,$$

is the $l$th mixed modulus of smoothness of $f$.

We now define the class $\mathbf{MS}_l, l \in \mathbf{N}^d$, of functions of mixed modulus of smoothness type. It consists of all non-negative functions $\Omega$ on $\mathbf{R}_+^d$ such that:

**(i)** $\Omega(t) = 0$ if $t_1 t_2 \cdots t_d = 0$,

**(ii)** $\Omega(t) \le \Omega(t')$ if $t \le t'$,

**(iii)** $\Omega(k_1 t_1, ..., k_d t_d) \le (k_1 k_2 \cdots k_d)^l \Omega(t)$, for any $k \in \mathbf{N}^d$,

**(iv)** $\Omega$ satisfies Condition $\mathrm{Z}_l$, i.e., there exist positive numbers $a_j < l_j$, $j = 1, 2, ..., d$, and positive constant $C_l$ such that for any $h > 0$

$$\int_h^\infty \Omega(t) t_j^{-a_j} \frac{dt_j}{t_j} \le C_l t^{-a_j} \Omega(t_1, ..., t_{j-1}, h, t_{j+1}, ..., t_d), \ j = 1, ..., d,$$

**(v)** $\Omega$ satisfies Condition BS, i.e., there exist a positive number $b$ and positive constant $C$ such that

$$\Omega(t)(t_1 t_2 \cdots t_d)^{-b} \le C\Omega(t')(t_1' t_2' \cdots t_d')^{-b}, \ t \le t', \ t, t' \in [0,1]^d. \quad (1)$$

Notice that $\mathbf{MS}_l \subset \mathbf{MS}_{l'}$ if $l < l'$. For $\Omega \in \mathbf{MS}_l$ and $1 \le p \le \infty$, $0 < \theta \le \infty$, let $\mathbf{B}_{p,\theta}^\Omega$ denote the Besov space of all functions on $\mathbf{T}^d$, for which the quasi-norm

$$\|f\|_{\mathbf{B}_{p,\theta}^\Omega} := \|f\|_p + |f|_{\mathbf{B}_{p,\theta}^\Omega} \quad (2)$$

is finite, where $\|\cdot\|_p$ is the usual $p$-integral norm in $L_p := L_p(\mathbf{T}^d)$ and

$$|f|_{\mathbf{B}_{p,\theta}^\Omega} := \left( \int_{\mathbf{R}_+^d} \{\Omega_l(f,t)_p / \Omega(t)\}^\theta \prod_{j=1}^d t_j^{-1} dt \right)^{1/\theta}, \theta < \infty, \quad (3)$$

(the integral changed to the supremum for $\theta = \infty$). The definition of $\mathbf{B}_{p,\theta}^\Omega$ does not depend on $l$, i. e., for a given $\Omega$, (2)–(3) determine equivalent quasi-norms for all $l$ such that $\Omega \in \mathbf{MS}_l$ (see Lemma 1).

Let $A$ is a given compact subset of $\mathbf{R}^d_+$ such that

$$\rho(A) := \min\{\max\{a_j : \ a_j e^j \in A\} : \ j = 1, ..., d\} > 0,$$

where $\{e^j\}_{j=1}^d$ the canonical basis in $\mathbf{R}^d$. Define the function $\Omega_A$ on $\mathbf{R}^d_+$ by

$$\Omega_A(t) := \inf_{\alpha \in A} t_1^{\alpha_1} t_2^{\alpha_2} \cdots t_d^{\alpha_d}.$$

If

$$l_j > \max\{\alpha_j : \alpha \in A\}, \ j = 1, 2, ..., d, \tag{4}$$

for some $l \in \mathbf{N}^d$, then it is easy to check that $\Omega_A \in \mathbf{MS}_l$. Therefore, (2)–(3) define the Besov space $\mathbf{B}^A_{p,\theta} := \mathbf{B}^{\Omega_A}_{p,\theta}$ for any $l$ satisfying the condition (4). We say that the space $\mathbf{B}^A_{p,\theta}$ consists of the functions with the common mixed Besov smoothness $A$. An important case of the Besov space $\mathbf{B}^A_{p,\theta}$ is that when $A$ is finite. In the last case, there holds the following quasi-norms equivalence

$$C\|f\|_{\mathbf{B}^A_{p,\theta}} \ \le \ \|f\|_p \ + \ \sum_{\alpha \in A} |f|_{\mathbf{B}^{\{\alpha\}}_{p,\theta}}. \ \le \ C'\|f\|_{\mathbf{B}^A_{p,\theta}}$$

Notice that the classical isotropic and anisotropic Besov spaces are special case of $\mathbf{B}^A_{p,\theta}$. For $r > 0$, let $A_r := \{r(e) : e \subset \{1, 2, ..., d\}\}$ where $r(e)$ denotes the element of $\mathbf{R}^d_+$ such that $r(e)_j = r$ for $j \in e$, and $r(e)_j = 0$ for $j \notin e$. We use the abbreviation: $\mathbf{B}^r_{p,\theta} := \mathbf{B}^{A_r}_{p,\theta}$.

Let $X$ be a quasi-normed linear space and $\Phi = \{\varphi_k\}_{k=1}^\infty$ a family of elements in $X$. Denote by $\mathbf{M_n}(\mathbf{\Phi})$ the non-linear set of all linear combinations of $n$ free terms from $\Phi$ of the form

$$\varphi = \sum_{k \in Q} a_k \varphi_k,$$

where $Q$ is a set of natural numbers with $|Q| = n$. Here and later $|Q|$ denotes the cardinality of $Q$. Let $W$ be a subset in $X$. Consider the $n$-term approximation to elements $f \in W$ with regard to the family $\Phi$, i.e., the approximation to $f$ by elements from $\mathbf{M}_n(\Phi)$. Then the worst case error of the $n$-term approximation to elements from $W$ with regard to the family $\Phi$ is given by

$$\sigma_n(W, \Phi, X) \ := \ \sup_{f \in W} \sigma_n(f, \Phi, X). \ := \ \sup_{f \in W} \inf_{\varphi \in \mathbf{M}_n(\Phi)} \|f - \varphi\|.$$

An other way to deal with $n$-term approximation by $\mathbf{M}_n(\Phi)$ is to impose the continuity assumption on algorithms of $n$-term approximation. An (continuous) algorithm of $n$-term approximation to elements from $W$ with regard to

$\Phi$, is represented as a (continuous) mapping $S$ from $W$ into $\mathbf{M_n}(\mathbf{\Phi})$. The continuity assumption is quite natural: the closer objects are the closer their reconstructions should be. On the one hand, for many interesting approximation systems $\Phi$ and classes of functions with common smoothness $W$, the asymptotic order of $\sigma_n(W, \Phi, X)$ is achieved by a continuous algorithm $S$ from $W$ into $\mathbf{M}_n(\Phi^*)$, where $\Phi^*$ is a finite subset of $\Phi$. (see [8], [9]). This is shown again in this paper for the family $\mathbf{V}$ and the class of functions with mixed Besov smoothness. On the other hand, the continuity assumption on approximating algorithms tends to guarantee a lower bound for $n$-term approximation. The continuity assumption goes back to the classical Alexandroff non-linear $n$-width $a_n(W, X)$ which is defined by

$$a_n(W, X) := \inf_{S,K} \sup_{f \in W} \|f - S(f)\|,$$

where the infimum is taken over all complexes $K \subset X$ of dimensions $\leq n$ and all continuous mappings $S$ from $W$ into $K$ (for definition see, e. g., [8], [12]).

Denote by $\mathcal{F}(X)$ the set of all bounded $\Phi$ whose intersection $\Phi \cap L$ with any finite-dimensional subspace $L$ in $X$, is a finite set. The non-linear $n$-width $\tau_n(W, X)$ [9] is defined by

$$\tau_n(W, X) := \inf_{\Phi,S} \sup_{f \in W} \|f - S(f)\|,$$

where the infimum is taken over all continuous mappings $S$ from $W$ into $\mathbf{M}_n(\Phi)$ and all families $\Phi \in \mathcal{F}(X)$. Observe that the condition $\Phi \in \mathcal{F}(X)$ is quite mild and natural for well-known approximation systems. Similar to $\tau_n(W, X)$, the non-linear $n$-width $\tau'_n(W, X)$ [9] which is defined in the same way as $\tau_n(W, X)$ but the infimum is taken over all continuous mappings $S$ from $W$ into a finite-dimensional subset of $\mathbf{M}_n(\Phi)$, or equivalently, over all continuous mappings $S$ from $W$ into $\mathbf{M}_n(\Phi)$ and all finite families $\Phi$ in $X$.

Let $l_\infty$ be the normed linear space of all bounded sequences of numbers $x = \{x_k\}_{k=1}^\infty$, equipped with the norm $\|x\|_\infty := \sup_{1 \leq k < \infty} |x_k|$, and $\mathbf{M}_n$ the subset in $l_\infty$ of all $x \in l_\infty$ for which $x_k = 0, k \notin Q$, for some set of natural numbers $Q$ with $|Q| = n$. Consider the mapping $R_\Phi$ from the metric space $\mathbf{M}_n$ into $X$ defined by

$$R_\Phi(x) := \sum_{k \in Q} x_k \varphi_k,$$

if $x = \{x_k\}_{k=1}^\infty$ and $x_k = 0, k \notin Q$, for some $Q$ with $|Q| = n$. From the definitions we can easily see that if the family $\Phi$ is bounded, then $R_\Phi$ is a continuous mapping from $\mathbf{M}_n$ into $X$ and moreover, $\mathbf{M}_n(\Phi) = R_\Phi(\mathbf{M}_n)$. On the other hand, any algorithm $S$ of $n$-term approximation with regard to $\Phi$

can be represented as a composition $S = R_\Phi \circ G$ for some mapping $G$ from $W$ into $\mathbf{M}_n$. Therefore, if $G$ is required to be continuous, then the algorithm $S$ will also be continuous. Although the class of continuous algorithms of the form $S = R_\Phi \circ G$, does not exhaust the continuous algorithms of $n$-term approximation with regard to $\Phi$, in many cases one can construct such a continuous algorithm $S$ which is asymptotically optimal even for $\sigma_n(W, \Phi, X)$, i.e., there holds the following inequanlity

$$\sup_{f \in W} \|f - S(f)\| \leq C\sigma_n(W, \Phi, X),$$

with some absolute constant $C$, where $S = R_\Phi \circ G$ for some mapping $G$ from $W$ into $\mathbf{M}_n$. The continuity of algorithms of the form $S = R_\Phi \circ G$ are easier verified, they are easier to contruct and to be applied in practical problems. Finally, as mentioned above, the boundedness assumption on $\Phi$ does not lose generality. These preliminary remarks are a basis for the notion of the non-linear $n$-width $\alpha_n(W, X)$ [8], [9] which is given by

$$\alpha_n(W, X) := \inf_{\Phi, G} \sup_{f \in W} \|f - R_\Phi(G(f))\|,$$

where the infimum is taken over all continuous mappings $G$ from $W$ into $\mathbf{M}_n$ and all bounded families $\Phi$ in $X$. In what follows, the families $\Phi$ are conveniently represented in the form $\Phi = \{\varphi_k\}_{k \in Q}$ where $Q$ is an at most countable set of indices.

There are other notions of non-linear $n$-widths which are based on continuous algorithms of non-linear approximations different from $n$-term approximation, and related to problems discussed in the present paper. They are the non-linear manifold $n$-width $\delta_n(W, X)$ introduced in [4], [14], and the non-linear $n$-width $\beta_n(W, X)$ introduced in [7].

Wavelets were discovered and rapidly developed in the 1980's. The most important discoveries in Wavelet Theory are the multiresolution analysis of Mallat and Meyer (see [13]) and the discovery of Daubechies [1] of compactly supported orthogonal wavelets with arbitrary smoothness. The interested reader can consult [2], [15] for basic ideas and knowlegde on wavelets. Wavelets are powerful tools for approximations and numerical computations. They provide simple and useful wavelet decompositions of functions on $\mathbf{R}^d$ into a wavelet series

$$f = \sum_{k,s} \lambda_{k,s} \phi_{k,s},$$

of the wavelets which are the integer translates of dyadic dilates

$$\phi_{k,s}(x) := \phi(2^k x - s), \quad s \in \mathbf{Z}^d, \ k \in \mathbf{Z}$$

of a scaling function $\phi$. Periodic wavelet decompositions of functions defined on the $d$-torus $\mathbf{T}^d$, are constituted from the periodic wavelets which are the integer translates

$$\phi_{k,s}(x) := \phi_k(x - 2\pi s/2^k), \quad s \in \mathbf{Z}_+^d, s_j = 0, ..., 2^k - 1, \ k \in \mathbf{Z}$$

of different scaling functions $\phi_k$for every dyadic scale. The coefficients $\lambda_{k,s} := \lambda_{k,s}(f)$ are certain functionals of $f$.

Wavelet decompositions are quite appropriate to non-linear approximations, in particular, $n$-term approximation because of their good approximation properties. Firstly, they provide a simultaneous time and frequency localization. This allows us to select different numbers of terms $\phi_{k,s}$ at each $k$th dyadic scale for $n$-term approximation, depending on a given target function. Secondly, they give discrete descriptions of equivalent norms and semi-norms for Sobolev and Besov spaces in terms of coefficients functionals $\lambda_{k,s}(f)$. Using these discrete characterizations, we can process a quantization or discretization of our approximation problems. In the discrete form, they are more convenient for study and numerical computation. In non-linear $n$-approximations with ragard to a wavelet family $\Phi = \{\phi_{k,s}\}$, a target function is approximated by linear combinations of $n$ free terms from different dyadic levels depending on its behavior. The asymptotic orders of $\sigma_n(SB^r_{p,\theta}, \Phi, L_q)$ of $n$-term approximations with regard to B-spline wavelets and periodic wavelets generated from de la Vallée Pussin kernels for classical Besov and Sobolev classes, and of their $n$-widths based on continuous algorithms of $n$-term approximation were obtained in [7]. The reader can also consult [3] for a survey on $n$-term approximation using wavelet decompositions.

Our purpose is to invetigate the $n$-term approximation to functions from of the Besov class $\mathbf{SB}^A_{p,\theta}$ with regard to multivariate wavelets generated from de la Vallée Poussin kernels, as well as optimal continuous algorithms of $n$-term approximation to them in terms of the non-linear $n$-widths which have been introduced. The Besov class of the mixed smoothness $A$

$$\mathbf{SB}^A_{p,\theta} := \ \{f \in \mathbf{B}^A_{p,\theta} : \ \|f\|_{\mathbf{B}^A_{p,\theta}} \leq 1\}$$

is defined as the unit ball of the space $\mathbf{B}^A_{p,\theta}$. Periodic wavelets and corresponding wavelet decompositions mentioned above, do not appropriate to the mixed structure of smoothness $A$. For $n$-term approximations of functions from $\mathbf{SB}^A_{p,\theta}$, we have to construct so-called mixed wavelets and mixed wavelet decompositions. Let

$$V_m(t) := \frac{1}{3m^2} \sum_{k=m}^{2m-1} D_k(t) \ = \ \frac{\sin(mt/2)\sin(3mt/2)}{3m^2 \sin^2(t/2)}$$

be the de la Vallée Poussin kernel of order $m$, where $D_m(t) := \sum_{|k|\le m} e^{ikt}$ is the univariate Dirichlet kernel of order $m$. Next we define

$$\varphi_0 := 1, \quad \varphi_k := V_{2^{k-1}}, \ k = 1, 2, ...$$

For $k \in \mathbf{Z}^d_+ = \{s \in \mathbf{Z}^d : s_j \ge 0, j = 1, ..., d\}$, we let the mixed dyadic scaling functions

$$\varphi_k(x) := \varphi_{k_1}(x_1)\varphi_{k_2}(x_2)\cdots\varphi_{k_d}(x_d)$$

be defined as the tensor product of the univariate scaling functions $\varphi_{k_j}(x_j)$, and the mixed wavelets

$$\varphi_{k,s} := \varphi_k(\cdot - 2\pi s/2^k), \ s \in Q_k, \tag{5}$$

be defined as the integer translates of $\varphi_k$, where $Q_k := \{s \in \mathbf{Z}^d : \ 0 \le s_j < 2^{k_j}, \ j = 1, ..., d\}$ and $2\pi s/2^k := 2\pi(s_1/2^{k_1}, s_2/2^{k_2}, ..., s_d/2^{k_d})$. From well-known propeties of de la Vallée Poussin kernels one can easily prove that a function $f \in L_q(\mathbf{T}^d)$, $1 \le q \le \infty$, is decomposed into a series of mixed wavelets $\varphi_{k,s}$, converging in $L_q(\mathbf{T}^d)$. We will give in Section 2 a mixed wavelet decomposition into $\varphi_{k,s}$ for functions from $\mathbf{B}^\Omega_{p,\theta}$ and $\mathbf{B}^A_{p,\theta}$.

For $n$-term approximation of functions from $\mathbf{SB}^A_{p,\theta}$, we take the family of mixed wavelets:

$$\mathbf{V} := \{\varphi_{k,s} : k \in \mathbf{Z}^d_+; \ s \in Q_k\}.$$

We use the notation $F \asymp F'$ if $F \ll F'$ and $F' \ll F$, and $F \ll F'$ if $F \le CF'$ with $C$ an absolute constant. Denote by $\gamma_n$ any one of $\alpha_n, \tau_n, \tau'_n, \beta_n, a_n$ and $\delta_n$.

The main results of the present paper are the asymptotic order of $\sigma_n(\mathbf{SB}^A_{p,\theta}, \mathbf{V}, L_q)$ and $\gamma_n(\mathbf{SB}^A_{p,\theta}, L_q)$. It turns out that they are closely related to the convex problem in $\mathbf{R}^d$

$$(\mathbf{1}, x) \ \to \ \sup, \ x \in A^o_+, \tag{6}$$

where $A^o_+ := \{x \in \mathbf{R}^d : (\alpha, x) \le 1, \ \alpha \in A, \quad x_j \ge 0, \ j = 1, ..., d\}$, $\mathbf{1} := (1, 1, \dots, 1) \in \mathbf{R}^d$. Let $1/r = 1/r(A)$ be the optimal value of of this problem, i. e.,

$$1/r := \sup\{(\mathbf{1}, x) : \ x \in A^o_+\}.$$

Further, let $\nu = \nu(A)$ be the linear dimensions of the set of solutions of (6). i. e.,

$$\nu := \dim\{x \in A^o_+ : \ (\mathbf{1}, x) = 1/r\},$$

$\mu = \mu(A) = d - 1 - \nu$ and

$$v(h) := \mathrm{Vol}_{d-1}\{x \in A^o_+ : \ (\mathbf{1}, x) = 1/r - h\},$$

where $\mathrm{Vol}_m G$ denotes the $m$-dimensional volume of $G \subset \mathbf{R}^d$. In [6]we explicitly constructed from the set $A$ a function $w = w(A, \cdot)$ on $[0, \infty)$ so that $w$ is a modulus of continuity if $\nu < d-1$, and $w = 1$ if $\nu = d-1$ and

$$v(h) \asymp w^\mu(h) \text{ as } h \to 0.$$

Notice that $r(A) = \min\{t > 0 : t\mathbf{1} \in \mathrm{co}A\}$ and $\mu$ is the linear dimension of the minimal extreme subset of $\mathrm{co}A$ containing the point $r\mathbf{1}$, where $\mathrm{co}A$ denotes the convex hull of $A$. The quantity $r(A)$ is, in some sense, a characterization of "average smoothness" of $\mathbf{B}^A_{p,\theta}$. If the set $A$ is finite, then we have [6]

$$w(h) = h \text{ if } \nu < d-1.$$

Put $a_+ := \max\{a, 0\}$. We proved the following

**Theorem 1** *Let $1 < p, q, \theta < \infty$. Assume that either $\rho(A) > 1/p$ and $\theta \ge p$ or $\rho(A) > (1/p - 1/q)_+$ and $\theta \ge \min\{q, 2\}$. Then we have*

$$\begin{aligned}\gamma_n(\mathbf{SB}^A_{p,\theta}, L_q) &\asymp \sigma_n(\mathbf{SB}^A_{p,\theta}, \mathbf{V}, L_q)\\ &\asymp n^{-r}(w^\mu(1/\log n)\log^{d-1} n)^{r+1/2-1/\theta},\end{aligned}$$

*where $r = r(A)$, $\mu = \mu(A)$ and $w = w(A, \cdot)$. In addition, we can explicitly construct a finite subset $\mathbf{V}^*$ of $\mathbf{V}$ and an asymptotically optimal (continuous) algorithm $S = R_{\mathbf{V}^*} \circ G^*$ of $n$-term approximation with regard to $\mathbf{V}^*$, so that*

$$\sup_{f \in \mathbf{SB}^A_{p,\theta}} \|f - S(f)\|_q \ll n^{-r}(w^\mu(1/\log n)\log^{d-1} n)^{r+1/2-1/\theta},$$

*where $G^*$ is a mapping from $\mathbf{SB}^A_{p,\theta}$ into $\mathbf{M}_n$.*

If in Theorem 1 $A$ is finite, then we have

$$\gamma_n(\mathbf{SB}^A_{p,\theta}, L_q) \asymp \sigma_n(\mathbf{SB}^A_{p,\theta}, \mathbf{V}, L_q) \asymp n^{-r}(\log n)^{\nu(r+1/2-1/\theta)}.$$

Special cases of the last results were proved in [8]. Theorem 1 was proved in [9] for the Besov class $\mathbf{SB}^r_{p,\theta}$. For this class, under the assumptions of Theorem 1 we have

$$\gamma_n(\mathbf{SB}^r_{p,\theta}, L_q) \asymp \sigma_n(\mathbf{SB}^r_{p,\theta}, \mathbf{V}, L_q) \asymp n^{-r}(\log n)^{(d-1)(r+1/2-1/\theta)}.$$

For $n$-term approximation with regard to the family $U^d$ formed from the integer translates of the mixed dyadic scales of the tensor product multivariate Dirichlet kernel, the asymptotic order of $\sigma_n(\mathbf{SB}^r_{p,\infty}, U^d, L_q)$ has been obtained in [18].

If the Besov scale $\theta$ are given, then the asymptotic order of $\sigma_n(\mathbf{SB}^A_{p,\theta}, \mathbf{V}, L_q)$ and $\gamma_n(\mathbf{SB}^A_{p,\theta}, L_q)$ in Theorem 1 is completely determined by the "average smoothness" $r(A)$, the quantity $\mu(A)$ and the concave modulus of continuity $w(A, \cdot)$ which are explicitly constructed from the smoothness $A$, and does not depend on $p, q$. Moreover, $r = r(A)$ defines the main term $n^{-r}$ of this asymptotic order. The second term $(w^{\mu}(1/\log n)\log^{d-1} n)^{r+1/2-1/\theta}$ appears only in the multivariate case $(d > 1)$. In the univariate approximation this term is absent.

A proof of Theorem 1 is given in Section 3. In the proof we employ mixed wavelet deompositions of functions $f \in \mathbf{SB}^A_{p,\theta}$ into the wavelets $\varphi_{k,s}$, and discrete equivalent quasi-norms of $\mathbf{B}^A_{p,\theta}$ in terms of the coefficient functionals in wavelet decompositions. Using wavelet decompositions we preliminarly decompose $f$ into a special series of functions $f_\eta, \eta = \xi, \xi + 1, ....,$ with integer parameter $\xi$ tending to infinity. The component functions $f_\eta$ are constructed according to the smoothness properties of $\mathbf{SB}^A_{p,\theta}$. Then using discrete equivalent quasi-norms we carry out a discretization for $n$-term approximation to each $f_\eta$. This allows us to reduce the proof to a problem of $n$-term approximation in finite-dimensional spaces equipped with mixed quasi-norms. Auxiliary results are given and proved in Section 2.

## 2 Wavelet decompositions of Besov spaces and other auxiliaries

As usual, $\hat{f}(k)$ denotes the $k$th Fourier coefficient of $f \in L_p$ $(1 \le p \le \infty)$. We define the operator $\delta_k, k \in \mathbf{Z}^d_+$, in $f \in L_p$ by

$$\delta_k(f) \ := \ \sum_{s \in P_k} \hat{f}(s) e^{i(s,\cdot)},$$

where $P_k := \{s \in \mathbf{Z}^d : \ [2^{k_j - 1}] \le |s_j| < 2^{k_j}, \ j = 1, ..., d\}$ and $[a]$ denotes the integer part of $a$. The Littlewood-Paley Theorem (see, e.g., [16]) states that for $1 < p < \infty$, there holds the following norms equivalence

$$\|f\|_p \ \asymp \ \|(\sum_{k \in \mathbf{Z}^d_+} |\delta_k(f)|^2)^{1/2}\|_p. \tag{7}$$

For $m \in \mathbf{Z}^d_+$, we let the tensor product de la Vallée Poussin kernel $V_m$ of order $m$ be defined by

$$V_m(x) := V_{m_1}(x_1) V_{m_2}(x_2) \cdots V_{m_d}(x_d),$$

and $f \in L_p(\mathbf{T}^d)$, the function $V_m(f)$ by

$$V_m(f) = \sum_{k \in K_m} f(hk) V_m(\cdot - hk), \tag{8}$$

where $K_m := \{k \in \mathbf{Z}^d : 0 \leq k_j < 4m_j, \ j = 1, ..., d\}$ and $h := (\pi/2)(m_1^{-1}, ..., m_d^{-1})$. Notice that for $m$ with $m_j > 0, j = 1, ..., d$, and $V_m(f)$ is a trigonometric polynomial of order $\leq 2m_j - 1$ in the variable $x_j$. For the following equivalence see [8].

$$\|V_m(f)\|_p \asymp M^{-1/p}\|\{f(hk)\}\|_{l_p^M}, \ 1 \leq p \leq \infty, \tag{9}$$

where $M := |K_m| = 4^d \prod_{j=1}^d m_j$. Denote by $\mathcal{T}_m$ the space of all trigonometric polynomials of order $\leq m_j$ in the variable $x_j, \ j = 1, ..., d$. It is easy to check that for every $f \in \mathcal{T}_m$

$$V_m(f) = f. \tag{10}$$

Next for univariate functions $f \in L_p(\mathbf{T})$, we define

$$u_0(f) := U_0(f); \ u_1(f) := U_1(f) - U_0(f);$$

$$u_k(f) := U_{2^{k-1}}(f) - U_{2^{k-2}}(f), \ k = 2, 3, ...$$

where $U_m(f) = f * U_m$ as the convolution of $f$ and $U_m := 3mV_m$. For multivariate functions $f \in L_p(\mathbf{T}^d)$, the mixed operators $u_k, k \in \mathbf{Z}_+^d$, are defined by

$$u_k := u_{k_1} u_{k_2} \cdots u_{k_d},$$

where the univariate operator $u_{k_j}$ is applied to the variable $x_j$. Note that $u_k(f)$ are a trigonometric polynomial of order $\leq 2^{k_j} - 1$ in the variable $x_j, j = 1, ..., d$. Put $2^{-k} := (2^{-k_1}, ..., 2^{-k_d})$ for $k \in \mathbf{Z}_+^d$. For any function $f \in L_p$, we have

$$\|u_k(f)\|_p \ll \Omega_l(f, 2^{-k})_p \ (1 \leq p \leq \infty), \tag{11}$$

and

$$\|\delta_k(f)\|_p \ll \Omega_l(f, 2^{-k})_p \ (1 < p < \infty). \tag{12}$$

These inequalities are easily deduced from the following so-called Jackson-Stechkin type's inequality.

For $m \in \mathbf{N}^d$, denote by $T(m)$ the set of functions $f$ of the form

$$f(x) = \sum_{j=1}^d \sum_{|s| \leq m_j} e^{isx_j} g(x_1, ..., x_{j-1}, x_{j+1}, ..., x_d).$$

Let $1 \leq p \leq \infty$. Then for every function $f \in L_p$, we can construct a function $g \in T(m)$ such that there hold the Jackson-Stechkin type's inequality

$$\|f - g\|_p \ll \Omega_l(f, 1/m)_p, \tag{13}$$

where $1/m := (1/m_1, ..., 1/m_d)$. This inequality proved by Stechkin for the univariate case (see, e.g., Theorem 2.3 in Chapter 7 of [5]). In the general case it can be proved by a standard method, which is similar to the proof of Lemma 1.1 in Chapter II of [17]. For completeness we give a sketch of proof. Let $K_{i,j}$ be the univariate generalized Jackson kernels (see, e.g., p. 203 in [5]). For a given $f \in L_p$, the function

$$g_1(x) := (2\pi)^{-1} \int_{\mathbf{T}} (f(x) - \Delta_{h_1}^{l_1} f(x)) K_{m_1,l_1}(x) dh_1$$

belongs to $T(m)$. Define $f_1(x) := f(x) - g_1(x)$. Further, we construct the sequenses $g_j \in T(m)$ and $f_j = f_{j-1} - g_j$, $j = 1, 2, ..., d$ by putting

$$g_j(x) := (2\pi)^{-1} \int_{\mathbf{T}} (f_{j-1}(x) - \Delta_{h_j}^{l_j} f_{j-1}(x)) K_{m_j,l_j}(x_j) dh_j.$$

Using properties of $K_{i,j}$ and of mixed modulus of smoothness $\Omega_l(f,\cdot)_p$ and Minkowski's inequality we prove that

$$\|\Delta_{1/m_d}^{l_d} ... \Delta_{1/m_{j+1}}^{l_{j+1}} f_j\|_p \ll \Omega_l(f, 1/m)_p, \; j = 1, ..., d.$$

Defining $g := \sum_{j=1}^d g_j$, we obtain $g \in T(m)$, $f - g = f_d$ and, therefore, the inequality (13).

Assume that the function $f \in L_p, 1 \leq p \leq \infty$, can be represented as the series

$$f = \sum_{s \in \mathbf{Z}_+^d} f_s, \; f_s \in \mathcal{T}_{2^s}, \tag{14}$$

converging in $L_p$-norm. Then from the well-known Bernstein inequality and the inequality $\|\Delta_h^l g\|_p \leq |h|^l \|g^{(l)}\|_p$ for univariate functions $g$, we can prove the following estimate

$$\Omega_l(f, 2^{-k})_p \ll \sum_{s \in \mathbf{Z}_+^d} \|f_s\|_p \prod_{j=1}^d \min\{1, 2^{l_j(s_j - k_j)}\}. \tag{15}$$

Let $\phi$ be a positive function on $(0,\infty)^d$ and $0 < q \leq \infty$. We define the quasi-norm $\|a\|_{q,\phi}$ for a sequence $a = \{a_k\}_{k \in \mathbf{Z}^d}$ as follows

$$\|a\|_{q,\phi} := (\sum_{k \in \mathbf{Z}^d} \{|a_k|/\phi(2^{-k}\}^q)^{1/q}, \; (q < \infty)$$

(the sum is changed to sup when $q = \infty$). The following inequality is a mixed type generalization of the discrete Hardy inequality (see, e.g., p. 27 in [5]).

Let $0 < q \leq \infty$ and $\Omega \in \mathbf{MS}_l$. If the sequences $a = \{a_k\}_{k \in \mathbf{Z}^d}$ and $b = \{b_k\}_{k \in \mathbf{Z}^d}$ with $a_k, b_k \geq 0$, satisfy the condition

$$b_k \leq M \sum_{s \in \mathbf{Z}^d_+} a_s \prod_{j=1}^{d} \min\{1, 2^{l_j(s_j - k_j)}\},$$

then

$$\|b\|_{q,\Omega} \leq CM\|a\|_{q,\Omega}, \tag{16}$$

with $C$ depending only on $q, \Omega$. The proof of this inequality is similar to that for the one-dimensional case in [11].

Using the discrete Hardy inequality (16), from (11), (12) and (14)–(15) we prove the following

**Lemma 1** *Let $1 \leq p \leq \infty$, $0 < \theta \leq \infty$. Then there holds the following quasi-norms equivalence*

$$\|f\|_{\mathbf{B}^{\Omega}_{p,\theta}} \asymp \Big(\sum_{k \in \mathbf{Z}^d_+} \{\|u_k(f)\|_p / \Omega(2^{-k})\}^{\theta}\Big)^{1/\theta} \ (\theta < \infty). \tag{17}$$

*Moreover, a function $f$ belongs to $\mathbf{B}^{\Omega}_{p,\theta}$ iff $f$ can be represented as a series*

$$f = \sum_{s \in \mathbf{Z}^d_+} f_s, \ f_s \in \mathcal{T}_{2^s},$$

*converging in $L_p$-norm, and the quasi-norm*

$$\Big(\sum_{k \in \mathbf{Z}^d_+} \{\|f_k\|_p / \Omega(2^{-k})\}^{\theta}\Big)^{1/\theta} < \infty \ \ (\theta < \infty) \tag{18}$$

*is finite. In addition, the quasi-norm (18) is equivalent to $\|f\|_{\mathbf{B}^{\Omega}_{p,\theta}}$. (The sum in (17) and (18) is changed to the supremum for $\theta = \infty$).*

Similarly, there holds the following quasi-norms equivalence for $1 < p < \infty$

$$\|f\|_{\mathbf{B}^{\Omega}_{p,\theta}} \asymp \Big(\sum_{k \in \mathbf{Z}^d_+} \{\|\delta_k(f)\|_p / \Omega(2^{-k})\}^{\theta}\Big)^{1/\theta} \ (\theta < \infty), \tag{19}$$

Let us describe a mixed wavelet decomposition and equivalent discrete quasi-norm for the Besov space $\mathbf{B}^{\Omega}_{p,\theta}$. From Lemma 1 and (10) it follows that a function $f \in L_p$ belongs to the Besov space $B^{\Omega}_{p,\theta}$ iff $f$ has a wavelet decomposition

$$f = \sum_{k \in \mathbf{Z}^d_+} \sum_{s \in Q_k} \lambda_{k,s} \varphi_{k,s} \tag{20}$$

with the convergence in the space $\mathbf{B}^{\Omega}_{p,\theta}$, where the wavelets $\varphi_{k,s}$ are defined in (5) and $\lambda_{k,s} = \lambda_{k,s}(f)$ are linear coefficients functionals of $f$ . Moreover, there holds the following quasi-norms equivalence

$$\|f\|_{\mathbf{B}^{\Omega}_{p,\theta}} \asymp (\sum_{k=0}^{\infty}(\|\{\lambda_{k,s}\}\|_p/2^{|k|/p}\Omega(2^{-k})^{\theta})^{1/\theta}. \tag{21}$$

For $\Omega = \Omega_A$, this quasi-norms equivalence is of the form:

$$\|f\|_{\mathbf{B}^{A}_{p,\theta}} \asymp (\sum_{k=0}^{\infty}(2^{S(A,k)-|k|/p}\|\{\lambda_{k,s}\}\|_p)^{\theta})^{1/\theta}, \tag{22}$$

where $S(A,x) := \sup\{(\alpha,x) : \alpha \in A\}$ is the support function of $A$.

For $0 < p, \theta \le \infty$ and $\Omega$ a postive function on $(0,\infty)^d$, let $\mathbf{B}^{\Omega}_{p,\theta}$ be the quasi-normed linear space of all functions $f$ on $\mathbf{T}^d$ such that, $f$ are represented by the wavelet series (20) for which the discrete quasi-norm defined in the right-hand side of (21) is finite. We also use the notation $\mathbf{B}_{p,\theta} = \mathbf{B}^{\Omega}_{p,\theta}$ for $\Omega \equiv 1$. From Lemma 1 we can see that if $1 \le p \le \infty$ and $\Omega \in \mathbf{MS}_l$, this definition coincides with the definition of $\mathbf{B}^{\Omega}_{p,\theta}$ given by (2)–(3). Because asymptotic ordes of approximation are invariant to different equivalent quasi-norms, in what follows, for $0 < p, \theta \le \infty$, we denote again by $\|\cdot\|_{\mathbf{B}^{\Omega}_{p,\theta}}$ and $\|\cdot\|_{\mathbf{B}^{A}_{p,\theta}}$ the discrete quasi-norms in the right-hand side of (21) and (22), respectively.

We say that $\Omega$ satisfies Condition $\mathbf{R}$(p,q) if $\Omega(t)(t_1t_2\cdots t_d)^{-(1/p-1/q)_+}$ satisfies Condition BS (see (1)). For the case $\Omega = \Omega_A$ Condition $\mathbf{R}$(p,q) is equivalent to the inequality $\rho(A) > (1/p-1/q)_+$. If $1 \le q \le \infty$ and $\Omega$ satisfies Condition $\mathbf{R}$(p,q), then the space $\mathbf{B}^{\Omega}_{p,\theta}$ is compactly embedded into the space $L_q$ and and the coefficient functionals $\lambda_{k,s}$ in the wavelet decomposition (20) can be chosen to depend continuously on $f \in \mathbf{B}^{\Omega}_{p,\theta}$ in the norm of $L_q$ and $\mathbf{B}_{q,\tau}$ $(0 < \tau \le \infty)$. Such a possible choice of $\lambda_{k,s}$ are given by Lemma 1 and (10). This fact in the case $\Omega = \Omega_A$ with $A$ a finite set follows from Lemma 9 and its proof in [8]. The general case can be proved similarly.

For $0 < p \le \infty$, denote by $l^m_p$ the space of all sequences $x = \{x_k\}^m_{k=1}$ of (complex) numbers, equipped with the quasi-norm

$$\|\{x_k\}\|_{l^m_p} = \|x\|_{l^m_p} := (\sum_{k=1}^{m}|x_k|^p)^{1/p}$$

with the change to the max norm when $p = \infty$. Denote by $B^m_p$ the unit ball in $l^m_p$. Let $\mathcal{E} = \{e_k\}^m_{k=1}$ be the canonical basis in $l^m_q$, i. e., $x = \sum_{k=1}^m x_k e_k$.

Let $0 < q \leq \infty$ and $L_s$ be a $s$-dimensional linear subspace in $l_q^m$ $(s \leq m)$. Then we have for any possitive integer $n < s - 1$ [10]

$$\sigma_n(B_\infty^m \cap L_s, \mathcal{E}, l_q^m) \;\geq\; (m - n - 1)^{1/q}. \tag{23}$$

Let $0 < p, \theta \leq \infty$, $N = \{N_k\}_{k\in Q}$ be a sequence of natural numbers with $Q$ a finite set of indices. Denote by $\mathbf{b}_{p,\theta}^N$ the space of all such sequences $x = \{x^k\}_{k\in Q} = \{\{x_j^k\}_{j=1}^{N_k}\}_{k\in Q}$, for which the mixed quasi-norm

$$\|\{\{x_j^k\}\}\|_{\mathbf{b}_{p,\theta}^N} \;=\; \|x\|_{\mathbf{b}_{p,\theta}^N} := \;(\sum_{k\in Q} \|x^k\|_{X^k}^\theta)^{1/\theta},\; \theta < \infty,$$

is finite (the sum is changed to supremum for $\theta = \infty$), where $X^k := l_p^{N_k}$. Let $\mathbf{S}_{p,\theta}^N$ be the unit ball in $\mathbf{b}_{p,\theta}^N$ and $\mathcal{E} = \{\mathcal{E}^k\}_{k\in Q}$ be the canonical basis in $\mathbf{b}_{p,\theta}^N$, where $\mathcal{E}^k = \{e_j^k\}_{j=1}^{N_k}$ is the canonical basis in $l_p^{N_k}$.

We define the *greedy algorithm* $G_n$ and the continuous algorithm $G_n^C$ for the $n$-term approximation with regard to $\mathcal{E}$ of the unit ball $\mathbf{S}_{p,\theta}^N$ in the space $\mathbf{b}_{q,\tau}^N$ as follows. For $x = \{x^k\}_{k\in Q} \in \mathbf{b}_{p,\theta}^N$, we let the set $\{(k,j) : k \in Q, 1 \leq j \leq N_k\}$ be ordered as a sequense $\{(k_s, j_s)\}_{s=1}^m$ so that

$$|x_{j_1}^{k_1}| \geq |x_{j_2}^{k_2}| \geq \cdots |x_{j_s}^{k_s}| \geq \cdots |x_{j_m}^{k_m}|.$$

Then, we define

$$G_n(x) := \sum_{s=1}^n x_{j_s}^{k_s} e_{j_s}^{k_s}, \tag{24}$$

and

$$G_n^C(x) := \sum_{s=1}^n (x_{j_s}^{k_s} - |x_{j_{n+1}}^{k_{n+1}}| \mathrm{sign} x_{j_s}^{k_s}) e_{j_s}^{k_s}, \tag{25}$$

Let $0 < p, \theta, \tau \leq \infty$ and $n < m = \sum_{k\in Q} N_k$. Then we have for $\theta \geq p$ [9]

$$\begin{aligned} \sup_{x\in\mathbf{S}_{p,\theta}^N} \|x - G_n(x)\|_{\mathbf{b}_{\infty,\tau}^N} \;&\leq\; \sup_{x\in\mathbf{S}_{p,\theta}^N} \|x - G_n^C(x)\|_{\mathbf{b}_{\infty,\tau}^N} \\ &\leq\; n^{-1/p}|Q|^{1/p+1/\tau-1/\theta}. \end{aligned} \tag{26}$$

If in addition $p \leq q \leq \infty$ and $N_k \leq N_0$ for all $k \in Q$, then we have for $p \leq q$ and $\theta \geq \tau$ [9]

$$\begin{aligned} \sup_{x\in\mathbf{S}_{p,\theta}^N} \|x - S_n^G(x)\|_{\mathbf{b}_{q,\tau}^N} \;&\leq\; \sup_{x\in\mathbf{S}_{p,\theta}^N} \|x - S_n^C(x)\|_{\mathbf{b}_{q,\tau}^N} \\ &\leq\; C(p) n^{1/q-1/p}|Q|^{1/p-1/q+1/\tau-1/\theta}. \end{aligned} \tag{27}$$

Let $\varepsilon \geq 0$ and $r = r(A) > \varepsilon$. We define

$$B := \{x \in \mathbf{R}^d : \ S(A,x) - (\varepsilon \mathbf{1}, x) \leq r - \varepsilon\}, \ B_+ := B \cap \mathbf{R}^d_+. \tag{28}$$

Notice that $B_+$ is a compact convex set and

$$\sup\{(\mathbf{1}, x) : x \in B\} = \sup\{(\mathbf{1}, x) : x \in B_+\} = 1.$$

Let $E = \{v^j\}_{j=1}^d$ be an orthnormal basis in $\mathbf{R}^d$ such that $v^d = -d^{-1/2}\mathbf{1}$, and $(y,t) = (y_1, ..., y_{d-1}, t)$ the new coordinates of $x \in \mathbf{R}^d$ with regard to $E$, i. e., $x = y_1 v^1 + \cdots + y_{d-1} v^{d-1} + t v^d$. Represent $\mathbf{R}^d$ as $\mathbf{R}_E^{d-1} \times \mathbf{R}_E$, where $\mathbf{R}_E^{d-1} = \text{span}\{v^1, v^2, ..., v^{d-1}\}$ and $\mathbf{R}_E = \text{span}\{v^d\}$. Let the function $f$ on $\mathbf{R}_E^{d-1}$ be defined by $f(y) := \ \inf\{t : (y,t) \in B\}$. Then it is easy to check that

$$B = \{(y,t) \in \mathbf{R}_E^{d-1} \times \mathbf{R}_E : f(y) \leq t\}$$

Notice that $f(y) \geq -1$. Also, for simplicity we assume that $f(0) = -1$, i. e., 0 belongs to the non-empty compact set $\{y \in \mathbf{R}_E^{d-1} : f(y) = -1\}$.

Let $\lambda > 1$ be a fixed number. Since $f$ is convex and $f(0) = -1$ we have $f(y/2\lambda) \leq f(y)/2\lambda - (1 - 1/2\lambda)$. Hence, $h(y) \leq f(y)$, $y \in \mathbf{R}_E^{d-1}$ and, consequently, $B \subset B'$, where $h(y) = 2\lambda f(y/2\lambda) + 2\lambda - 1$ and

$$B' := \{x = (y,t) : h(y) \leq t\}. \tag{29}$$

Define

$$D(\xi, \eta) := \ \{(\xi + \eta)B_+ \setminus \xi B'\} \cap \{(\xi + \eta)B_+ \setminus (\xi + \eta - 1)B_+\}. \tag{30}$$

**Lemma 2** *Let $\varepsilon \geq 0$ and $\lambda > 1$ be fixed. Then for any pair $\xi, \eta$ satisfying the condition $\eta \leq (2\lambda - 1)\xi$, we have*

$$\text{Vol}_d D(\xi, \eta) \ \leq \ C\xi^{d-1} w^\mu (1/\xi) \eta^\mu,$$

*where $C$ is a constant depending on $\varepsilon, \lambda$ and $A$ only.*

*Proof.* Observe that for any $\xi > 0$, $\xi B = \{(y,t) : \xi f(y/\xi) \leq t\}$, $\xi B' = \{(y,t) : \xi h(y/\xi) \leq t\}$. Hence,

$$\text{Vol}_d D(\xi, \eta) \ \leq \ \int_G F(y) dy, \tag{31}$$

where $G := \{y \in B_+^E; \ F(y) \geq 0\}$, $B_+^E$ is the orthgonal projection of $B_+$ onto $\mathbf{R}_E^{d-1}$ and

$$F(y) := \min\{\xi h(y/\xi), (\xi + \eta - 1) f(y/(\xi + \eta - 1))\} - (\xi + \eta) f(y/(\xi + \eta)).$$

Further, putting $g(y) := f(y) + 1$, by the equality $h(y) := \lambda g(y/2\lambda) - 1$ we have

$$\begin{aligned} F(y) &\le \xi h(y/\xi) - (\xi+\eta) f(y/(\xi+\eta)) \\ &= \lambda\xi g(y/2\lambda\xi) - (\xi+\eta) g(y/(\xi+\eta) + \eta. \end{aligned}$$

Since $g$ is convex and $g(0) = 0$, there holds the inequality

$$\lambda\xi g(y/2\lambda\xi) \le (\xi+\eta) g(y/(\xi+\eta),$$

and consequently,

$$F(y) \ \le \ \eta - \lambda\xi g(y/2\lambda\xi). \tag{32}$$

On the other hand, for any $y \in B_+^E$

$$\begin{aligned} F(y) &\le (\xi+\eta-1) f(y/(\xi+\eta-1)) - (\xi+\eta) f(y/(\xi+\eta)) \\ &= (\xi+\eta-1) g(y/(\xi+\eta-1)) - (\xi+\eta) g(y/(\xi+\eta)) + 1. \end{aligned}$$

Using the convexity of $g$ and the relations $g(y/(\xi+\eta)) \ge 0$ and $g(0) = 0$, we obtain $(\xi+\eta-1) g(y/(\xi+\eta-1)) \le g(y)$. Therefore,

$$F(y) \le g(y) \le \max\{g(z) : z \in B_+^E\} + 1 = \text{const}.$$

By (30)–(31) and the inequality $\mathrm{Vol}_{d-1}\{g(y) \le t\} \ll w^\mu(t)$, we have

$$\begin{aligned} \mathrm{Vol}_d D(\xi,\eta) &\ll \int_H dy = (2\lambda\xi)^{d-1} \mathrm{Vol}_{d-1}\{g(y) \le \eta/\lambda\xi\} \\ &\ll (2\lambda\xi)^{d-1} w^\mu(\eta/\lambda\xi) \ll \xi^{d-1} w^\mu(1/\xi)\eta^\mu, \end{aligned}$$

where $H := \{y \in \mathbf{R}_E^{d-1} : g(y/2\lambda\xi) \le \eta/\lambda\xi\}$. This proves the lemma because the relation $\ll$ in our estimations always means the inequality $\le$ with some constants depending on $\varepsilon, \lambda$ and $A$ only. $\square$

## 3 Proof of Theorem 1

We prove Theorem 1 from the following

**Theorem 2** *Let* $0 < p, q, \theta \le \infty$, $1 \le \tau \le \infty$. *Assume that either* $\rho(A) > 1/p$ *and* $\theta \ge p$ *or* $\rho(A) > (1/p - 1/q)_+$ *and* $\theta \ge \tau$. *Then we have*

$$\gamma_n(\mathbf{SB}_{p,\theta}^A, \mathbf{B}_{q,\tau}) \ \asymp \ \sigma_n(\mathbf{SB}_{p,\theta}^A, \mathbf{V}, \mathbf{B}_{q,\tau}) \ \asymp \ E_{\theta,\tau}(n), \tag{33}$$

*where*

$$E_{\theta,\tau}(n) := \ n^{-r}(w^\mu(1/\log n)\log^{d-1} n)^{r+1/\tau-1/\theta},$$

*and* $r = r(A)$, $\mu = \mu(A)$ *and* $w = w(A, \cdot)$. *In addition, we can explicitly construct a finite subset* $\mathbf{V}^*$ *of* $\mathbf{V}$ *and an asymptotically optimal (continuous) algorithm* $S = R_{\mathbf{V}^*} \circ G^*$ *of n-term approximation with regard to* $\mathbf{V}^*$, *so that*

$$\sup_{f \in \mathbf{SB}^A_{p,\theta}} \|f - S(f)\|_q \ll E_{\theta,\tau}(n), \tag{34}$$

*where* $G^*$ *is a mapping from* $\mathbf{SB}^A_{p,\theta}$ *into* $\mathbf{M}_n$.

*Proof of Theorem 2.* We first construct a positive homogeneous (continuous) mapping $G^* : \mathbf{B}^A_{p,\theta} \longrightarrow \mathbf{M}_n$ satisfying (34), and therefore, prove the upper bound of (33).

For a positive integer $\xi$, we put $\Delta_\xi := \{k \in \mathbf{Z}^d_+ : k \in \xi B'\}$ and $\Delta_\eta := \{k \in \mathbf{Z}^d_+ : k \in D(\xi, \eta - \xi)\}$ for any integer $\eta > \xi$, where the sets $B'$ and $D(\xi, \eta - \xi)$ are defined in (29), (28) and (30) with parameters $\varepsilon$ and $\lambda$. Observe that $r \geq \rho(A)$. We put $\varepsilon := 1/p$ for the case $\rho(A) > 1/p$, and $\varepsilon := (1/p - 1/q)_+$ for the case $\rho(A) > (1/p - 1/q)_+$ and $\theta \geq \tau$. An appropriate value of the other parameter $\lambda$ will be choosen later. By the remarks in Section 2, the asumption $\rho(A) > (1/p - 1/q)_+$ and the wavelet decomposition (22) yield that every $f \in \mathbf{B}^A_{p,\theta}$ is represented as the series

$$f = \sum_{\eta=\xi}^{\infty} f_\eta, \tag{35}$$

converging in the quasi-norm of $\mathbf{B}_{\infty,\tau}$, where

$$f_\eta = \sum_{k \in \Delta_\eta} \sum_{s \in Q_k} \lambda_{k,s} \varphi_{k,s}, \tag{36}$$

with the coefficient functionals $\lambda_{k,s}$ continuously depending on $f \in \mathbf{B}^A_{p,\theta}$ in the space $\mathbf{B}_{\infty,\tau}$. Let $D_\eta$ be the positive homogeneous continuous mapping from $\mathbf{B}^A_{p,\theta}$ into $\mathbf{b}^{N^\eta}_{\infty,\tau}$, given by $D_\eta(f) := \{\lambda_{k,s}\}_{s \in Q_k, k \in \Delta_\eta}$, where $N^\eta := \{N^\eta_k\}_{k \in \Delta_\eta}$, $N^\eta_k = |Q_k| = 3^d 2^{|k|}$. By (20) and (22) we have

$$\|f_\eta\|_{\mathbf{B}^A_{p,\theta}} = 2^{(r-1/p)\eta} \|D_\eta(f)\|_{\mathbf{b}^{N^\eta}_{p,\theta}}, \quad \|f_\eta\|_{\mathbf{B}_{\infty,\tau}} = \|D_\eta(f)\|_{\mathbf{b}^{N^\eta}_{\infty,\tau}}. \tag{37}$$

In what follows, we use the abbreviation $R_\xi := \xi^{d-1} w^\mu(1/\xi)$. Given a possitive integer $n$, we take a function $\xi = \xi(n)$ satisfying the condition

$$C 2^\xi R_\xi \leq n \asymp 2^\xi R_\xi, \tag{38}$$

where $C$ is an absolute constant whose value will be chosen below.

We first consider the case $\rho(A) > 1/p$. By the inequality $\|\cdot\|_{\mathbf{B}_{q,\tau}} \leq \|\cdot\|_{\mathbf{B}_{\infty,\tau}}$ and the inclusion $\mathbf{SB}^A_{p,\theta} \subset \mathbf{SB}^A_{p,\max\{p,\theta\}}$, it is enough to prove the upper bound for the case $p \leq \theta$ and $q = \infty$.

Fix $0 < \delta < p(r - 1/p)$ and let the sequence $\{n_\eta\}_{\eta=0}^\infty$ be given by $n_\eta := \sum_{k\in\Delta_\xi} |Q_k|$ if $\eta = \xi$, and $:= \left[2^{-\delta(\eta-\xi)}2^\xi R_\xi\right]$ if $\eta > \xi$. We have

$$\sum_{\eta=\xi}^{\infty} n_\eta \le \sum_{k\in\Delta_\xi} |Q_k| + \sum_{\eta>\xi} 2^{-\delta(\eta-\xi)}2^\xi R_\xi. \tag{39}$$

From Theorem 1.3 of [6] we derive that $\sum_{k\in\Delta_\xi} |Q_k| \le C'2^\xi R_\xi$. Combining the last estimate and (39) gives

$$\sum_{\eta=\xi}^{\infty} n_\eta \;\le\; (C' + \sum_{\eta>\xi} 2^{-\delta(\eta-\xi)})2^\xi R_\xi,$$

where $C'$ is an absolute constant. Defining the absolute constant in the right-hand side of the last inequality as $C$, we have

$$\sum_{\eta=\xi}^{\infty} n_\eta \le n. \tag{40}$$

It is easy to verify that $n_\eta > 0$ for any $\eta \le \xi(1+\delta)/\delta + (d-1)\log\xi$ and $n_\eta = 0$ for any $\eta > \xi(1+\delta)/\delta + (d-1)\log\xi$. Notice that $(1+\delta)/\delta > r/(r-1/p)$. Now in the defintion of $\Delta_\eta$ we take a fixed value of the parameter $\lambda > 1$ so that $(1+\delta)/\delta > 2\lambda - 1 > r/(r-1/p) > 1$. Put $\eta^* = [(2\lambda-1)\xi]$. Then for $\xi$ large enough, $n_\eta > 0$ for any $\eta \le \eta^*$.

Let $\xi < \eta \le \eta^*$. Denote by $G'_\eta$ either the greedy algorithm $G_{n_\eta}$ or the continuous algorithm $G^C_{n_\eta}$ defined in (24)–(25) for the $n_\eta$-term approximation with regard to the canonical basis $\mathcal{E}_\eta$ in $\mathbf{b}^{N^\eta}_{q,\tau}$ of $\mathbf{S}^{N^\eta}_{\infty,\tau}$. By (26) we have

$$\|x - G'_\eta(x))\|_{\mathbf{b}^{N^\eta}_{\infty,\tau}} \;\le\; n_\eta^{-1/p}|\Delta_\eta|^{1/p+1/\tau-1/\theta}\|x\|_{\mathbf{b}^{N^\eta}_{p,\theta}}. \tag{41}$$

We then define the algorithm of $n_\eta$-term approximation with regrad to $\mathbf{V}$ $S_\eta$ from $\mathbf{B}^A_{p,\theta}$ into $\mathbf{M}_{n_\eta}(\mathbf{V})$ by

$$S_\eta(f) := \sum_{k\in\Delta_\eta}\sum_{s\in Q_k} \lambda^*_{k,s}\varphi_{k,s},$$

where $\{\{\lambda^*_{k,s}\}_{s\in Q_k}\}_{k\in\Delta_\eta}\} := G'_\eta(D_\eta(f))$. Hence, by (41) and (37) it is easy to check that for any $\eta \le \eta^*$

$$\|f_\eta - S_\eta(f))\|_{\mathbf{B}_{\infty,\tau}} \;\ll\; n_\eta^{-1/p}|\Delta_\eta|^{1/p+1/\tau-1/\theta}2^{-(r-1/p)\eta}\|f_\eta\|_{\mathbf{B}^A_{p,\theta}}.$$

Lemma 2 gives $|\Delta_\eta| \;\ll\; \mathrm{Vol}_d D(\xi, \eta-\xi) \;\ll\; R_\xi(\eta-\xi)^\mu$. From the last relation it follows that

$$\|f_\eta - S_\eta(f))\|_{\mathbf{B}_{\infty,\tau}} \;\ll\; A(\eta)\|f_\eta\|_{\mathbf{B}^A_{p,\theta}}, \tag{42}$$

where

$$A(\eta) \;=\; 2^{-r\xi} R_\xi^{1/\tau-1/\theta} 2^{-\beta(\eta-\xi)} (\eta-\xi)^{\mu(1/p+1/\tau-1/\theta)}$$

with $\beta = (r - 1/p - \delta/p) > 0$.

For $\eta > \eta^*$, we define the trivial mapping $S_\eta$ from $\mathbf{B}^A_{p,\theta}$ into $\mathbf{M}_{n_\eta}(\mathbf{V})$ by $S_\eta(f) = 0$ for any $f \in \mathbf{B}^A_{p,\theta}$. Using (37), the Hölder inequality, we obtain for any $\eta > \eta^*$

$$\|f_\eta - S_\eta(f))\|_{\mathbf{B}_{\infty,\tau}} \;\ll\; 2^{-(r-1/p)\eta} |\Delta_\eta|^\alpha \|f_\eta\|_{\mathbf{B}^A_{p,\theta}}, \tag{43}$$

where $\alpha = \max\{0, 1/\tau - 1/\theta\}$.

Finally, we define

$$S_\xi(f) \;=\; \sum_{k\in\Delta_\xi} \sum_{s\in Q_k} \lambda_{k,s}\varphi_{k,s}.$$

For approximation of functions $f \in \mathbf{SB}^A_{p,\theta}$ we take the following algorithm

$$S := \sum_{\eta=\xi}^{\infty} S_\eta.$$

From (35)–(36) we have

$$f - S(f) = \sum_{\eta>\xi}^{\infty} (f_\eta - S_\eta(f)).$$

By (42)–(43), the inequlities $2\lambda - 1 > r/(r-1/p)$ and $\|f_\eta\|_{\mathbf{B}^A_{p,\theta}} \le \|f\|_{\mathbf{B}^A_{p,\theta}}$, and the relation $\eta^* = [(2\lambda-1)\xi]$, we get the following estimates for any $f \in \mathbf{SB}^A_{p,\theta}$

$$\begin{aligned}\|f - S(f))\|_{\mathbf{B}_{\infty,\tau}} &\le \sum_{\eta>\xi} \|f_\eta - S_\eta(f))\|_{\mathbf{B}_{\infty,\tau}} \\ &\ll \sum_{\xi<\eta\le\eta^*} A(\eta) \;+\; \sum_{\eta>\eta^*} 2^{-(r-1/p)\eta} |\Delta(\eta)|^\alpha.\end{aligned}$$

Observe that

$$|\Delta_\eta| \ll \mathrm{Vol}_d\{\eta B_+ \setminus (\eta-1)B_+\} \asymp \eta^{d-1}.$$

Hence, by (38)

$$\begin{aligned}\|f - S(f))\|_{\mathbf{B}_{\infty,\tau}} &\ll 2^{-r\xi} R_\xi^{1/\tau-1/\theta} \sum_{\eta>\xi} 2^{-\beta(\eta-\xi)} (\eta-\xi)^{\mu(\alpha+1/p)} \\ &\quad + 2^{-(r-1/p)\eta^*} \sum_{\eta>\eta^*} 2^{-(r-1/p)(\eta-\eta^*)} (\eta-\eta^*)^{(d-1)\alpha} \\ &\ll 2^{-r\xi} R_\xi^{1/\tau-1/\theta} \asymp E_{\theta,\tau}(n).\end{aligned}$$

Let $\mathbf{V}_\eta^*$ be the family of the $\varphi_{k,s}, s \in Q_k, k \in \Delta_\eta$, and $\mathbf{V}^* = \mathbf{V}_\xi^* \cup \cdots \cup \mathbf{V}_{\eta^*}^*$. Clearly, $\mathbf{V}^*$ is a finite subset of $\mathbf{V}$. Consider $X^* := \mathbf{b}_{\infty,\tau}^{N^\xi} \oplus \cdots \oplus \mathbf{b}_{\infty,\tau}^{N^{\eta^*}}$ as a subset of $l_\infty$. Then $S = R_{\mathbf{V}^*} \circ G^*$ where

$$G^*(f) := \sum_{\xi \leq \eta \leq \eta^*} G'_\eta(D_\eta(f)).$$

is a mapping from $\mathbf{SB}_{p,\theta}^A$ into $\mathbf{M}_n \cap X^*$ by (40). Moreover, $S$ is an asymptotically optimal algorithm of $n$-term approximation with regard to $\mathbf{V}^*$ satisfying (34). If $G'_\eta = G_{n_\eta}^C$ in the definition of $S_\eta$, then $S$ is continuous. By the way the upper bound of (33) is also proved.

The case $\rho(A) > (1/p - 1/q)_+$ and $\theta \geq \tau$ can be proved similarly, but the estimate (26) is replaced the estimate (27).

Let us prove the lower bound of (33). We prove it first for $\sigma_n$. Because of the inclusion $\mathbf{SB}_{\infty,\theta}^A \subset \mathbf{SB}_{p,\theta}^A$, it is sufficient to treat the case $p = \infty$. For a positive integer $\xi$, we define $m(\xi) := \max\{(\mathbf{1}, k) : k \in \xi B_+ \cap \mathbf{Z}_+^d\}$, and take the set $M(\xi) := \{k \in \xi B_+ \cap \mathbf{Z}_+^d : (\mathbf{1}, k) = m(\xi) - d\}$. Denote by $B(\xi)$ the space of all trigonometric polynomials $f$ of the form

$$f = \sum_{k \in M(\xi)} \sum_{s \in Q_k} \lambda_{k,s}\varphi_{k,s},$$

and for $0 < a, b \leq \infty$ denote by $B(\xi)_{a,b}$ the subspace in $\mathbf{B}_{a,b}$ which consists of all $f \in B(\xi)$. By (20)–(22) we have

$$\|f\|_{\mathbf{B}_{p,\theta}^A} = 2^{r\xi}\|f\|_{\mathbf{B}_{\infty,\theta}} \leq 2^{r\xi}|M(\xi)|^{1/\theta}\|f\|_{\mathbf{B}_{\infty,\infty}}$$

for any $f \in B(\xi)_{\infty,\theta}$. This implies $2^{-r\xi}|M(\xi)|^{-1/\theta} SB(\xi)_{\infty,\infty} \subset \mathbf{SB}_{\infty,\theta}^A$, where $SB(\xi)_{\infty,\infty}$ is the unit ball in $B(\xi)_{\infty,\infty}$. Therefore,

$$\sigma_n(\mathbf{SB}_{\infty,\theta}^A, \mathbf{V}, \mathbf{B}_{q,\tau}) \gg 2^{-r\xi}|M(\xi)|^{-1/\theta}\sigma_n(SB(\xi)_{\infty,\infty}, \mathbf{V}, \mathbf{B}_{q,\tau}).$$

Let the linear projection $P_\xi$ from $\mathbf{B}_{q,\tau}$ onto $B(\xi)_{q,\tau}$ be defined by

$$P_\xi(f) = \sum_{k \in M(\xi)} \sum_{s \in Q_k} \lambda_{k,s}\varphi_{k,s}.$$

Then we have $\sigma_n(SB(\xi)_{\infty,\infty}, \mathbf{V}, \mathbf{B}_{q,\tau}) \geq \sigma_n(SB(\xi)_{\infty,\infty}, \mathbf{V}', B(\xi)_{q,\tau})$, where $\mathbf{V}' = P_\xi(\mathbf{V})$, and consequently,

$$\sigma_n(\mathbf{SB}_{\infty,\theta}^r, \mathbf{V}, \mathbf{B}_{q,\tau}) \gg 2^{-r\xi}|M(\xi)|^{-1/\theta}\sigma_n(SB(\xi)_{\infty,\infty}, \mathbf{V}', B(\xi)_{q,\tau}). \quad (44)$$

Let us now give a lower bound for $\sigma_n(SB(\xi)_{\infty,\infty}, \mathbf{V}', B(\xi)_{q,\tau})$. Define $\xi = \xi(n)$ from the condition

$$n \asymp |M(\xi)|2^\xi \asymp \dim B(\xi) > 2n. \quad (45)$$

We have by (20)–(22)

$$\|f\|_{B(\xi)_{\infty,\infty}} = \|D'(f)\|_{\mathbf{b}^{N'}_{\infty,\infty}}, \ \|f\|_{B(\xi)_{q,\tau}} = 2^{-\xi/q}\|D'(f)\|_{\mathbf{b}^{N'}_{q,\tau}}, \tag{46}$$

where $N' := \{N_k\}_{k\in M(\xi)}, N_k = |Q_k|$ and $D'$ is the positive homogeneous continuous mapping from $B(\xi)_{q,\tau}$ into $\mathbf{b}^{N'}_{\infty,\infty}$, given by $D'(f) := \{\lambda_{k,s}\}_{s\in Q_k, k\in M(\xi)}$. Clearly, $D'(\mathbf{V}') = \mathcal{E}'$ and $D'(B(\xi)_{q,\tau}) = \mathbf{b}^{N'}_{q,\tau}$, where $\mathcal{E}'$ is the canonical basis in $\mathbf{b}^{N'}_{q,\tau}$. Also, if $S$ is an algorithm of $n$-term approximation with regard to $\mathbf{V}'$ in $B(\xi)_{q,\tau}$, then $D'\circ S$ will be an algorithm of $n$-term approximation with regard to $\mathcal{E}'$ in $\mathbf{b}^{N'}_{q,\tau}$. Hence, by (46) we obtain

$$\sigma_n(SB(\xi)_{\infty,\infty}, \mathbf{V}', B(\xi)_{q,\tau}) = 2^{-\xi/q}\sigma_n(\mathbf{S}^{N'}_{\infty,\infty}, \mathcal{E}', \mathbf{b}^{N'}_{q,\tau}). \tag{47}$$

Corollary 3.1 of [6] yields $|M(\xi)| \asymp R_\xi$. On the other hand, it is easy to check that

$$\|\cdot\|_{\mathbf{b}^{N'}_{q,\tau}} \gg R_\xi^{1/\tau-1/\rho}2^{\xi(1/q-1/\rho)}\|\cdot\|_{\mathbf{b}^{N'}_{\rho,\rho}},$$

for some $0 < \rho \le \min\{q,\tau\}$. Therefore, by (23)

$$\begin{aligned}\sigma_n(\mathbf{S}^{N'}_{\infty,\infty}, \mathcal{E}', \mathbf{b}^{N'}_{q,\tau}) &\gg R_\xi^{1/\tau-1/\rho}2^{\xi(1/q-1/\rho)}\sigma_n(\mathbf{S}^{N'}_{\infty,\infty}, \mathcal{E}', \mathbf{b}^{N'}_{\rho,\rho}) \\ &\gg R_\xi^{1/\tau-1/\rho}2^{\xi(1/q-1/\rho)}(m-n-1)^{1/\rho} \gg 2^{\xi/q}R_\xi^{1/\tau},\end{aligned}$$

where $m := \dim\mathbf{b}^{N'}_{\rho,\rho} = \dim B(\xi)$. Combining the last estimate, (44), (45) and (47) gives

$$\sigma_n(\mathbf{SB}^A_{\infty,\theta}, \mathbf{V}, \mathbf{B}_{q,\tau}) \gg 2^{-r\xi}R_\xi^{1/\theta-1/\tau} \asymp E_{\theta,\tau}(n). \tag{48}$$

Thus, we have proved lower bound of (33) for $\sigma_n(\mathbf{SB}^A_{\infty,\theta}, \mathbf{V}, \mathbf{B}_{q,\tau})$, and therefore, the asymptotical optimality of of the algorithm $S$.

We next prove the lower bound of (33) for $\gamma_n(\mathbf{SB}^A_{\infty,\theta}, \mathbf{B}_{q,\tau})$. Because of the inequalities between $\alpha_n, \tau_n, \tau'_n, \beta_n, a_n$ and $\delta_n$ (see [8], [12]), it is enough to prove the last lower bound for any of them, for example, for $\alpha_n$. It can be proved in the same way as the lower bound (48), but by using Lemma 3 of [9] and the inequality $\alpha_n(W,X) \ge a_n(W,X)$ (see [8]). □

*Proof of Theorem 1.* By (7) and (19) it is easy to verify the inequalities

$$\|\cdot\|_{\mathbf{B}_{\min\{q,2\},2}} \ll \|\cdot\|_q \ll \|\cdot\|_{\mathbf{B}_{q,\min\{q,2\}}}$$

for $1 < q < \infty$. Hence it follows that

$$\gamma_n(\mathbf{SB}^A_{p,\theta}, \mathbf{B}_{\min\{q,2\},2}) \ll \gamma_n(\mathbf{SB}^A_{p,\theta}, L_q) \ll \gamma_n(\mathbf{SB}^A_{p,\theta}, \mathbf{B}_{q,\min\{q,2\}}).$$

The same inequalities are true for $\sigma_n$. This and Theorem 2 imply Theorem 1. □

**References**


[1] I. Daubechies, *Orthonormal bases of compactly supported wavelets*, Comm. Pure and Appl. Math. **41**(1988), 909–996.

[2] I. Daubechies, *Ten Lectures on Wavelets*, CBMS-NSF Regional Conference Series in Applied Mathematics, **61**, SIAM, Philadelphia, 1992.

[3] R. DeVore, *Nonlinear approximation*, Acta Numerica **7**(1998), 51–150.

[4] R. DeVore, R. Howard and C. Micchelli, *Optimal non-linear approximation*, Manuscripta Math. **63**(1989), 469–478.

[5] R. DeVore and G. Lorentz, *Constructive approximation*, Springer-Verlag, 1993.

[6] Dinh Dung (Din' Zung), *Approximation by trigonometric polynomials of functions of several variables on the torus*, Mat. Sb. **131(173)**(1986), 251–267. [English transl. in Math. USSR Sb. **59**(1988), 247–267]

[7] Dinh Dung, *On non-linear $n$-widths and $n$-term approximations*, Vietnam J. Math. **26**(1998), 165–176.

[8] Dinh Dung, *Continuous algorithms in $n$-term approximation and non-linear $n$-widths*, J. Approx. Theory **102**(2000), 217–242.

[9] Dinh Dung, *Asymptotic orders of optimal non-linear approximations*, East J. Approx. **7**(2001).

[10] Dinh Dung, *non-linear approximations using wavelet decompositions*, Vietnam J. Math **29**(2001).

[11] Dinh Dung, *Best multivariate approximations by trigonometric polynomials with frequencies from hyperbolic crosses*, J. Approx. Theory **91**(1997), 205–225.

[12] Dinh Dung and Vu Quoc Thanh, *On non-linear $n$-widths*, Proc. Amer. Math. Soc. **124**(1996), 2757–2765.

[13] S. Mallat, *Multiresolution approximation and wavelets*, TAMS **315**(1989), 69–88.

[14] P. Mathé, *s-Number in information-based complexity*, J. Complexity **6**(1990), 41–66.

[15] Y. Meyer, Ondelettes et Opérateurs, Vol. **1** and **2**, Hermann, Paris, 1990.

[16] S. Nikol'skii, *Approximation of functions of several variables and embedding theorems*, Springer-Verlag, Berlin, 1975.

[17] V. Temlyakov, *Approximation of functions with mixed derivative*, Trudy Mat. Inst. Steklov. **178)**(1986), pp. 1–115. [English transl. in Proc. Steklov Inst. Math. **N**o 1(1989), pp. 1–121]

[18] V. Temlyakov, *Greedy algorithms with regard to the multivariate systems with a special structure*, Constr. Approx. **16**(2000), 399–425.

# Double series expansions and their convergence domains

Keiko FUJITA
Saga University, Saga 840-8502, Japan
E-mail: keiko@cc.saga-u.ac.jp

Abstract

A holomorphic function $f$ in a neighborhood in $\mathbf{C}^{n+1}$ can be expanded into the double series: $f(z) = \sum_{k=0}^{\infty} \sum_{l=0}^{[k/2]} (z^2)^l f_{k,k-2l}(z)$, where $f_{k,k-2l}$ is a homogeneous harmonic polynomial of degree $k-2l$ and $z^2 = z_1^2 + \cdots + z_{n+1}^2$. In case of $\mathbf{C}^{n+1}, n \geq 2$, we characterized holomorphic functions and analytic functionals on an $N_p$-ball which lies between the Lie ball and the dual Lie ball by the growth behavior of homogeneous harmonic polynomials. In this note, we study the Taylor expansion of holomorphic functions on the $L_p$-ball which is a ball with respect to the $L_p$-norm in $\mathbf{C}^{n+1}$, then we consider a relation between the double series expansion on the $N_p$-ball in $\mathbf{C}^2$ and the Taylor expansion on the $L_p$-ball.

## 1 $N_p$-norm and $N_p$-ball

Let

$$L(z) = \sqrt{\|z\|^2 + \sqrt{\|z\|^4 - |z^2|^2}}$$

be the Lie norm on $\tilde{\mathbf{E}} = \mathbf{C}^{n+1}$, where $z \cdot w = z_1 w_1 + z_2 w_2 + \cdots + z_{n+1} w_{n+1}$, $z^2 = z \cdot z$, and $\|z\|^2 = z \cdot \overline{z}$. Put

$$M(z) = |z^2|/L(z) = \sqrt{\|z\|^2 - \sqrt{\|z\|^4 - |z^2|^2}}$$

and consider the function

$$N_p(z) = \left(\frac{1}{2}\left(L(z)^p + M(z)^p\right)\right)^{1/p}, \quad p \geq 1.$$

Then $N_p(z)$ is a norm on $\tilde{\mathbf{E}}$ (see [2] or [3]). Note that $N_2(z) = \|z\|$ is the complex Euclidean norm and

$$N_1(z) = \sqrt{(\|z\|^2 + |z^2|)/2} = L^*(z)$$

is the dual Lie norm.

It is easy to see that $N_p(z) \le L(z)$. Moreover, $N_p(z)$ is monotone increasing in $p$ and $\lim_{p\to\infty} N_p(z) = L(z)$. We interpret $N_\infty(z) = L(z)$ (see [3]).

Put

$$\tilde{B}_p(r) = \left\{z \in \tilde{\mathbf{E}}; N_p(z) < r\right\}$$

and call it the open $N_p$-ball of radius $r > 0$. We denote by $\mathcal{O}(\tilde{B}_p(r))$ the space of holomorphic functions on $\tilde{B}_p(r)$ equipped with the topology of uniform convergence on compact sets. Especially for $p = 1$ and $p = \infty$, we call $\tilde{B}_1(r) = \tilde{B}^*(r)$ the dual Lie ball and $\tilde{B}_\infty(r) = \tilde{B}(r)$ the Lie ball.

## 2 Double series expansion

We denote by $\mathcal{P}^k_\Delta(\tilde{\mathbf{E}})$ the space of homogeneous harmonic polynomials of degree $k$ on $\tilde{\mathbf{E}}$; that is, if $f \in \mathcal{P}^k_\Delta(\tilde{\mathbf{E}})$, then $f$ is a homogeneous polynomial of degree $k$ and satisfies $\Delta_z f(z) \equiv (\partial^2/\partial z_1^2 + \partial^2/\partial z_2^2 + \cdots + \partial^2/\partial z_{n+1}^2)(z) = 0$. For the Legendre polynomial $P_{k,n}(t)$ of degree $k$ and of dimension $n+1$, define

$$\tilde{P}_{k,n}(z,w) = (\sqrt{z^2})^k(\sqrt{w^2})^k P_{k,n}\left(\frac{z}{\sqrt{z^2}} \cdot \frac{w}{\sqrt{w^2}}\right).$$

Then $\tilde{P}_{k,n}(z,w) = \tilde{P}_{k,n}(w,z)$ and $\tilde{P}_{k,n}(\,\cdot\,,w) \in \mathcal{P}^k_\Delta(\tilde{\mathbf{E}})$.

A holomorphic function $f$ on a neighborhood of 0 in $\tilde{\mathbf{E}}$ can be expanded locally into homogeneous polynomials $f_k$:

$$f(z) = \sum_{k=0}^{\infty} f_k(z), \quad f_k(z) = \frac{1}{2\pi i}\int_{|t|=\rho} \frac{f(tz)}{t^{k+1}}dt,$$

where $\rho > 0$ is sufficiently small. For a homogeneous polynomial $f_k$ of degree $k$, define

$$f_{k,k-2l}(z) = N(k-2l,n)\int_S f_k(\tau)\tilde{P}_{k-2l,n}(z,\tau)\dot{d\tau},$$

where $\dot{d\tau}$ is the normalized invariant measure on the real unit sphere $S$ and $N(k,n)$ is the dimension of $\mathcal{P}^k_\Delta(\tilde{\mathbf{E}})$. Then $f_{k,k-2l} \in \mathcal{P}^{k-2l}_\Delta(\tilde{\mathbf{E}})$ and we have $f_k(z) = \sum_{l=0}^{[k/2]}(z^2)^l f_{k,k-2l}(z)$. Thus $f$ can be expanded by means of homogeneous harmonic polynomials as follows:

$$f(z) = \sum_{k=0}^{\infty}\sum_{l=0}^{[k/2]}(z^2)^l f_{k,k-2l}(z). \tag{1}$$

We call the right-hand side in (1) the double series expansion of $f$ and $f_{k,k-2l}$ the $(k, k-2l)$-harmonic component of $f$.

For $n \ge 2$, we can characterize holomorphic functions $f$ on $\mathcal{O}(\tilde{B}_p(r))$ by the growth behavior of their harmonic components $f_{k,k-2l}$:

THEOREM **2.1** *Let $f \in \mathcal{O}(\{0\})$ be expanded into* (1), *and let $r > 0$, $1 \leq p \leq \infty$. If $f \in \mathcal{O}(\tilde{B}_p(r))$, then the sequence $\{f_{k,k-2l}\}$ of $(k, k-2l)$-harmonic components of $f$ satisfies*

$$\limsup_{k\to\infty}\left(\left(\frac{2^k l!(k-l)!}{k!}\right)^{\frac{1}{p}}\|f_{k,k-2l}\|_{C(S)}\right)^{\frac{1}{k}} \leq 1/r, \tag{2}$$

*where $\|f\|_{C(S)}$ is the supremun norm of $f$ on the unit sphere $S$.*

*Conversely, if a sequence $\{f_{k,k-2l}\}$ of homogeneous harmonic polynomials satisfies* (2), *then the double series*

$$\sum_{k=0}^{\infty}\sum_{l=0}^{[k/2]}(z^2)^l f_{k,k-2l}(z)$$

*converges to a holomorphic function $f$ in $\mathcal{O}(\tilde{B}_p(r))$ and the $(k, k-2l)$-harmonic component of $f$ is equal to the given $f_{k,k-2l}$.*

For a proof and some references, see [1].

# 3 $L_p$-norm and $L_p$-ball

For $w = (w_1, w_2, \cdots, w_{n+1}) \in \tilde{\mathbf{E}}$, put

$$\|w\|_p = (|w_1|^p + |w_2|^p + \cdots + |w_{n+1}|^p)^{1/p}, \quad p \geq 1,$$
$$\|w\|_\infty = \sup\{|w_j|; j = 1, \cdots, n+1\}.$$

$\|w\|_p$ is called the $L_p$-norm on $\tilde{\mathbf{E}}$. We denote by

$$B_p(r) = \{w \in \tilde{\mathbf{E}}; \|w\|_p < r\}$$

the open ball of radius $r > 0$ with respect to the $L_p$-norm.

Let $f \in \mathcal{O}(\{0\})$ and denote the Taylor expansion of $f$ as follows:

$$f(w) = \sum_\alpha a_\alpha w^\alpha, \tag{3}$$

where $\alpha = (\alpha_1, \alpha_2, \cdots, \alpha_{n+1})$ is a multi-index and $w^\alpha = w_1^{\alpha_1} w_2^{\alpha_2} \cdots w_{n+1}^{\alpha_{n+1}}$. Note that $a_\alpha$ is given by

$$a_\alpha = \left(\frac{1}{2\pi i}\right)^{n+1}\int_{|w_1|=s_1}\cdots\int_{|w_{n+1}|=s_{n+1}}\frac{f(w_1, \cdots, w_{n+1})}{w_1^{\alpha_1+1}\cdots w_{n+1}^{\alpha_{n+1}+1}}dw_1\cdots dw_{n+1}$$

for sufficiently small $s_1, \cdots, s_{n+1}$.

**Theorem 3.1** *Let $f \in \mathcal{O}(\{0\})$ be expanded into (3), and let $r > 0$, $1 \le p \le \infty$. If $f \in \mathcal{O}(B_p(r))$, then the sequence $\{a_\alpha\}$ satisfies*

$$\limsup_{|\alpha|\to\infty} \sqrt[|\alpha|]{|a_\alpha| \left(\frac{\alpha!}{|\alpha|!}\right)^{\frac{1}{p}}} \le \frac{1}{r}, \tag{4}$$

*where $\alpha! = \alpha_1!\alpha_2!\cdots\alpha_{n+1}!$ and $|\alpha| = \alpha_1 + \alpha_2 + \cdots + \alpha_{n+1}$.*

*Conversely, if a sequence $\{a_\alpha\}$ satisfies (4), then*

$$\sum_\alpha a_\alpha w^\alpha$$

*converges to a holomorphic function $f$ in $\mathcal{O}(B_p(r))$.*

Note that (4) is equivalent to

$$\limsup_{|\alpha|\to\infty} \sqrt[|\alpha|]{|a_\alpha| \left(\frac{\alpha^\alpha}{|\alpha|^{|\alpha|}}\right)^{\frac{1}{p}}} \le \frac{1}{r} \tag{5}$$

by the Stirling formula.

Proof. For simplicity of notation, we will give a proof for $n = 1$. Because a proof for $n \ge 2$ is given by the same way for $n = 1$. If $p = \infty$, then the condition (5) reads

$$\limsup_{|\alpha|\to\infty} \sqrt[|\alpha|]{|a_\alpha|} \le \frac{1}{r}$$

and the theorem is the well-known Cauchy-Hadamard theorem. We shall assume $1 \le p < \infty$.

Let $f(w_1, w_2) = \sum_{\alpha_1,\alpha_2=0}^{\infty} a_{\alpha_1\alpha_2} w_1^{\alpha_1} w_2^{\alpha_2}$ be the Taylor expansion of $f \in \mathcal{O}(B_p(r))$. Then for any $s_1 > 0, s_2 > 0$ with $s_1^p + s_2^p < r^p$, we have

$$(\frac{1}{2\pi i})^2 \int_{|w_1|=s_1} \int_{|w_2|=s_2} \frac{f(w_1, w_2)}{w_1^{\alpha_1+1} w_2^{\alpha_2+1}} dw_1 dw_2 = a_{\alpha_1\alpha_2}.$$

Thus putting

$$M_\epsilon = \sup\left\{|f(w_1, w_2)|; |w_1|^p + |w_2|^p \le (\frac{r}{1+\epsilon})^p\right\}, \quad \epsilon > 0,$$

for $s_1^p + s_2^p = (\frac{r}{1+\epsilon})^p$ we have the Cauchy inequality

$$|a_{\alpha_1\alpha_2}| \le \frac{M_\epsilon}{s_1^{\alpha_1} s_2^{\alpha_2}}.$$

Since

$$\max_{s_1^p+s_2^p=(\frac{r}{1+\epsilon})^p}\{s_1^{\alpha_1}s_2^{\alpha_2}\} = \left((\frac{\alpha_1}{\alpha_1+\alpha_2})^{1/p}\frac{r}{1+\epsilon}\right)^{\alpha_1}\left((\frac{\alpha_2}{\alpha_1+\alpha_2})^{1/p}\frac{r}{1+\epsilon}\right)^{\alpha_2}, \quad (6)$$

we obtain

$$|a_{\alpha_1\alpha_2}| \leq M_\epsilon \left(\frac{(\alpha_1+\alpha_2)^{\alpha_1+\alpha_2}}{\alpha_1^{\alpha_1}\alpha_2^{\alpha_2}}\right)^{1/p}(\frac{r}{1+\epsilon})^{\alpha_1+\alpha_2},$$

and we have

$$\limsup_{\alpha_1+\alpha_2\to\infty} \sqrt[\alpha_1+\alpha_2]{|a_{\alpha_1\alpha_2}|\left(\frac{\alpha_1^{\alpha_1}\alpha_2^{\alpha_2}}{(\alpha_1+\alpha_2)^{\alpha_1+\alpha_2}}\right)^{\frac{1}{p}}} \leq \frac{1+\epsilon}{r}.$$

Since $\epsilon > 0$ is arbitrary, we have (5) or (4).

Conversely, assume (4) or (5). If $s_1^p + s_2^p < r^p$, then

$$(s_1^p)^{\alpha_1}(s_2^p)^{\alpha_2} \leq r^{p(\alpha_1+\alpha_2)}\frac{\alpha_1^{\alpha_1}}{(\alpha_1+\alpha_2)^{\alpha_1}}\frac{\alpha_2^{\alpha_2}}{(\alpha_1+\alpha_2)^{\alpha_2}}$$

by (6), and hence

$$\limsup_{\alpha_1+\alpha_2\to\infty} \sqrt[\alpha_1+\alpha_2]{|a_{\alpha_1\alpha_2}|s_1^{\alpha_1}s_2^{\alpha_2}} \leq r \limsup_{\alpha_1+\alpha_2\to\infty} \sqrt[\alpha_1+\alpha_2]{|a_{\alpha_1\alpha_2}|\left(\frac{\alpha_1^{\alpha_1}\alpha_2^{\alpha_2}}{(\alpha_1+\alpha_2)^{\alpha_1+\alpha_2}}\right)^{\frac{1}{p}}} \leq 1.$$

Therefore, by the Cauchy-Hadamard theorem, $\sum a_{\alpha_1\alpha_2}w_1^{\alpha_1}w_2^{\alpha_2}$ converges uniformly on any compact sets in $\{(w_1,w_2); |w_1| < s_1, |w_2| < s_2, s_1^p + s_2^p \leq r^p\}$. Thus $f(w_1,w_2) = \sum a_{\alpha_1\alpha_2}w_1^{\alpha_1}w_2^{\alpha_2}$ is holomorphic on

$$\bigcup_{s_1^p+s_2^p\leq r^p}\{(w_1,w_2); |w_1| < s_1, |w_2| < s_2\} = \{(w_1,w_2); |w_1|^p + |w_2|^p < r^p\} = B_p(r).$$

q.e.d.

# 4 2-dimensional case

In the 2-dimensional case, $L(z)$ and $M(z)$ are given by

$$L(z) = \max\{|z_1 \pm iz_2|\}, \quad M(z) = \min\{|z_1 \pm iz_2|\},$$

and we have

$$N_p(z) = \left(\frac{|z_1+iz_2|^p + |z_1-iz_2|^p}{2}\right)^{1/p},$$

$$\begin{aligned}\tilde{B}_p(r) &= \left\{(z_1,z_2);\left(\frac{|z_1+iz_2|^p+|z_1-iz_2|^p}{2}\right)^{\frac{1}{p}}<r\right\}\\ &= \{(z_1,z_2);|z_1+iz_2|^p+|z_1-iz_2|^p<2r^p\}.\end{aligned}$$

Therefore , by the change of variables

$$w_1 = z_1 + iz_2, \quad w_2 = z_1 - iz_2$$

the $L_p$-ball $B_p(2^{1/p}r)$ corresponds with the $N_p$-ball $\tilde{B}_p(r)$.

Consider the relation between the Taylor expansion in $(z_1+iz_2, z_1-iz_2)$ and the double series expansion in $(z_1, z_2)$:

$$\begin{aligned}f(z_1,z_2) &= \sum_{m,n=0}^{\infty} a_{mn}(z_1+iz_2)^m(z_1-iz_2)^n \qquad (7)\\ &= \sum_{k=0}^{\infty}\sum_{l=0}^{[k/2]}(z^2)^l f_{k,k-2l}(z). \qquad (8)\end{aligned}$$

Comparing the terms in (7) and (8), we have

$$f_{k,k-2l}(z) = \begin{cases} a_{k-l,l}(z_1+iz_2)^{k-l} + a_{l,k-l}(z_1-iz_2)^{k-l}, & k>2l,\\ f_{k,0}(z) = a_{k/2,k/2}, & k=2l.\end{cases}$$

Therefore,

$$\|f_{k,k-2l}\|_{C(S)} = \begin{cases} |a_{k-l,l}| + |a_{l,k-l}|, & k>2l,\\ |a_{k/2,k/2}|, & k=2l,\end{cases}$$

which implies

$$\limsup_{k\to\infty}\sqrt[k]{|a_{k-l,l}|\left(\frac{(k-l)!l!}{k!}\right)^{\frac{1}{p}}} = \limsup_{k\to\infty}\sqrt[k]{\|f_{k,k-2l}\|_{C(S)}\left(\frac{(k-l)!l!}{k!}\right)^{\frac{1}{p}}}.$$

This proves that Theorem 2.1 is valid for $n=1$ (2-dimensional case).

# References


[1] K.Fujita and M.Morimoto, On the double series expansion of holomorphic functions, submitted.

[2] T.Kimura, Norm related to the Lie norm, to appear in Tokyo J. Math.

[3] M.Morimoto and K.Fujita, Between Lie norm and dual Lie norm, to appear in Tokyo J. Math..

# EMPIRICAL PSEUDORANDOM NUMBER GENERATORS

KIM GARGAR
*National Institute of Physics, College of Science, University of the Philippines-Diliman, Quezon City 1101 Philippines*
*E-mail: kim@nip.upd.edu.ph*

The most common pseudorandom number generator or PRNG, the linear congruential generator or LCG, belongs to a whole class of rational congruential generators. These generators work by multiplicative congruential method for integers, which implements a *"grow-and-cut procedure"*. We extend this concept to real numbers and call this the real congruence, which produces another class of random number generators called real congruential generators or RCG. The method in RCG inherits the procedure in LCG. Let $m$ be a positive integer and **I** the interval (0,1). Consider a mapping $f_m$:**I** $\rightarrow$ [-$m$, $m$] and generate a sequence $(x_n)$ as follows: (1) Let $x_0 \epsilon \mathbf{I}$ be the seed; (2) For every positive integer $n$, $x_{n+1} = [[f_m(x_n)]]$, where [[x]] = x - [x], [ ] is the greatest integer function. In this paper, we investigate the sequence generated by $f_m(x) = m\sin(\pi x)$. It turns out that the finite nature of computer numbers contributes to the "randomness" of the sequence. Simple applications and empirical testing were implemented.

## 1 Introduction

"A reliable RNG is a basic building block on which all stochastic simulations depend; without it, everything else collapses."[6] This statement tells us how important RNGs are in simulations of random processes, which include cosmic ray flux measurements, laser ablation, traffic flows, radioactive decay, etc.. To simulate or mimic these processes in a computer, we need to know the distribution function involved and generate numbers according to this distribution, and here enters RNGs. Our main objective in this paper is to produce (pseudo)random numbers that are uniformly distributed in the unit interval **I**=(0,1). For an introduction to random number generation, the reader is referred to Anderson et al.[1] and Hellekalek[4] give an enough introduction to the subject of random number generation.

## 2 The grow-and-cut procedure

Multiplicative congruential generators(LCG) generate sequence of integers, which can be easily converted to real numbers in **I** by dividing them by the modulus. A congruential generator often creates a randomly-looking[a] sequence of integers. In generating the sequence, a number is multiplied (*grow*) by the multiplier, the result of which is then compared to the modulus. When the product exceeds the modulus, it will then be reduced (*cut*) to an equivalent number that is less than the modulus through linear congruential technique. We call this process the grow-and-cut process–the numbers are grown and then cut to fit the required bounds determined from the modulus. It turned out that the randomness of the resulting sequence produced by applying this process to integers has some disadvantages.[5]

### *2.1 The real congruential generators*

We can do the same to real numbers. Applying the grow-and-cut process to real numbers is illustrated in Fig. 1. The generator is described as follows: Let $m$ be a positive integer and $\mathbf{I}$ the interval (0,1). Consider a mapping $f_m$:$\mathbf{I}\rightarrow(-m, m)$ and generate a sequence $(x_n)$ as follows: Let $x_0 \epsilon \mathbf{I}$ be the seed; For every positive integer $n$,

$$x_{n+1} = [[\mathbf{f_m}(x_n)]], \tag{1}$$

where $[[x]] = x$ - $[x]$, $[x]$ is the greatest integer function of $x$. In short, we are taking only the decimal part of the value of the function $f_m(x_n)$. In this generator, the "modulus" is equal to 1; i.e., the numbers in the sequence are inside $\mathbf{I}$. The function $f_m$ is so chosen as to make the sequence *nonlinear* while the function [[ ]] makes the sequence uniformly scattered in the square $\mathbf{I}$x$\mathbf{I}$ for some number $m$. In the present study, the following parameters were found to give a random number generator:

$$\mathbf{f_m}(x) = msin(\pi x),$$

$$\pi = 3.141592,$$

$$m = 1x10^5.$$

## 3 Results and discussion

The testing of the random number generator is patterned from Bacala,[2] which uses scatter plot techniques and distribution analysis with N=10000 as the length of the sequence. The method employed in this paper to test the PRNG

[a]We recognize "randomness" here as the inability of the user to predict the succeeding events in the process or, in short, the unpredictability of the process.

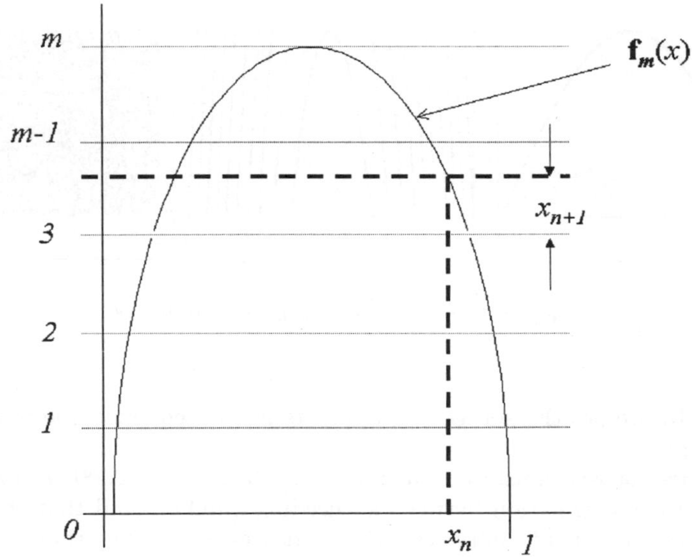


Figure 1. The grow-and-cut process: the next number ($x_{n+1}$) *is obtained by taking only the decimal part of* $f_m(x_n)$.

is purely empirical. As discussed by L'ecuyer,[6] there is still a lot of unsolved problems in the generation of random numbers. The hardest (forever unsolvable perhaps) is the issue of randomness. In this paper, to suggest possible source of randomness (unpredictability) a simple computational errors calculation is conducted for the specific example $m$=1. A plot of $\chi^2$ versus N is shown. Finally, simple application is also presented.

### *3.1 Scatter plot*

Fig. 2 shows the scatter plots for m = 1, 10, 100000. A scatter plot shows the relationship between adjacent numbers in the sequence. So it is obtained by plotting ($x_n$,$x_{n+1}$). It becomes obvious by looking at these scatter plots how the sequence behaves. However, given a number in the sequence, one may not

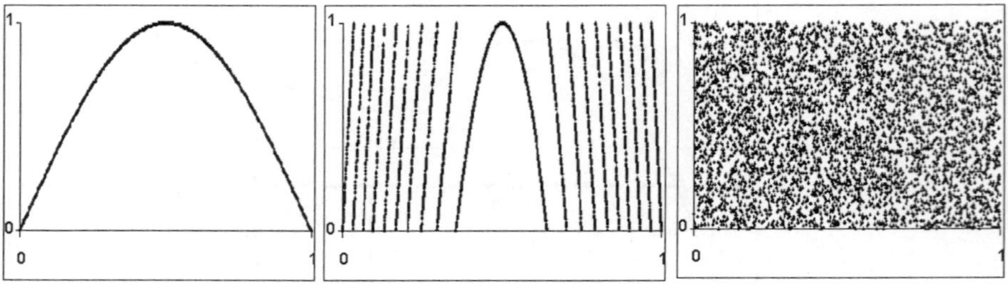


Figure 2. Scatter plots for $m = 1$, 10, 100000.

be able to predict the succeeding numbers in the sequence due to its chaotic behavior.[3,7]

Notice the relationship of the numbers when $m = 100000$. It appears that there is weak correlation between succeeding numbers and that the numbers are "scattered". Uniformity test (See next section) reveals that the numbers are "uniformly scattered" in the area.

### *3.2 Distribution analysis and uniformity test*

Distribution of the generated numbers are displayed in Fig. 3. The distribution for the simplest case, when $m = 1$, can be derived using one-dimensional mapping analysis.[b] If we denote the distribution of the $n$th number by $P_n(x)$, then the next number is distributed as $P_{n+1}(x)$ given by the equation

$$P_{n+1}(x) = \frac{P_n(\frac{sin^{-1}x}{\pi}) + P_n(1 - \frac{sin^{-1}x}{\pi})}{\pi\sqrt{1-x^2}} \tag{2}$$

where for the special case $P_0$=1(uniform distribution) the first number has the elementary distribution

$$P_1(x) = \frac{2}{\pi\sqrt{1-x^2}}. \tag{3}$$

[b]The analysis stands on the assumption that the mapped portion is infinitesimal and continuous whereas, actually, in computer, only discrete (that is, only one point) portion is mapped. The question of whether the one-dimensional mapping analysis is valid or not for this problem was not attended to.

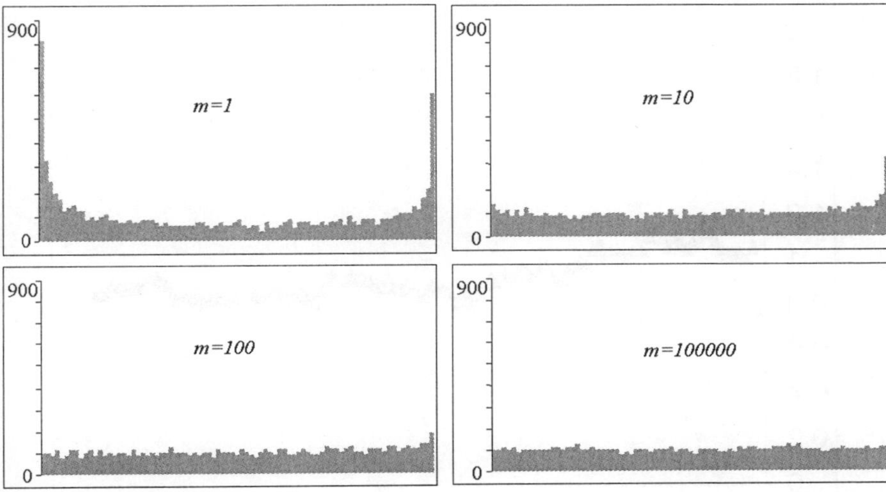


Figure 3. Empirical invariant measures (N=10000) of the pseudorandom numbers using the RCG.

Cahoy[3] used the first numbers as the generated random numbers using LCG as the seed generator.[c]

For $m$ =1, the invariant probability measure must satisfy the equation

$$P_\infty(x) = \frac{P_\infty(\frac{sin^{-1}x}{\pi}) + P_\infty(1 - \frac{sin^{-1}x}{\pi})}{\pi\sqrt{1-x^2}} \tag{4}$$

We are really more interested in the invariant measure of the sequence for large $m(\to \infty)$ since the scatter plot for large $m$ shows a uniform random number generator scatter plot. Unfortunately, as of this writing, the exact form of the distribution cannot be written in terms of an elementary function but as seen from the empirical result (See Fig. 4), the distribution is appreciably uniform with a $\chi^2$ close to and less than one as the number of generated numbers is increased.

[c]In his paper, as well as his colleague at the Mindanao Polytechnic State College, Cahoy employed the logistic map to generate a chaotic sequence. The method, however, used in the current paper was not initially inspired by chaotic generators. The chaotic behavior in the sequence produced by the sine map was discovered only later during the search for explanation of the inherently long sequence.

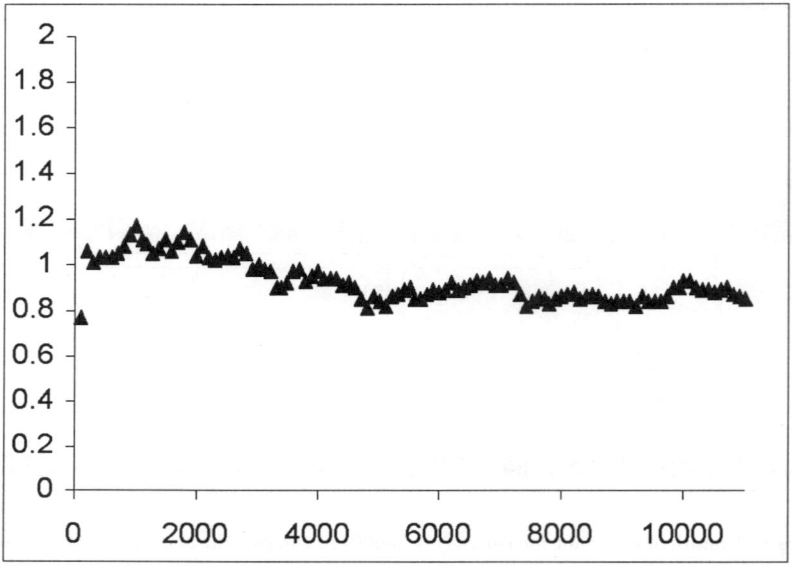


Figure 4. Plot of $\chi^2$ vs N.

### *3.3 Computational errors*

Since the actual value ($\pi^*$) of $\pi$ is not equal to but less than its real value, which is irrational, we can write

$$\pi^* = \pi - \delta\pi$$

The actual value of $x_1$ can then be approximated by $x_1^* = x_1 - \delta x_1$ where

$$\delta x_1 = -\delta\pi x_0 cos(\pi x_0).$$

Then the succeeding actual values of the numbers in the sequence can be approximated by $x_{n+1}^* = x_{n+1} + \delta x_{n+1}$, where

$$\delta x_{n+1} = \pi\delta x_n cos(\pi x_n)$$

This computational error may add to the unpredictability of the generated sequence. To observe this phenomenon of computational error greatly affecting the sequence, a computation of the sequence was done using *float* and *double* precision numbers. While almost preserved in the latter, the chaotic behavior of the sequence was clearly lost in the former, which showed a very short periodic sequence (result not shown).

| N | result | | | absolute error | |
|---|---|---|---|---|---|
| | theoretical | RCG | LCG | RCG | LCG |
| 1000 | 3.141592 | 3.140000 | 3.084000 | -0.001592 | -0.057592 |
| 2000 | 3.141592 | 3.124000 | 3.082000 | -0.017592 | -0.059592 |
| 3000 | 3.141592 | 3.148000 | 3.069333 | 0.006408 | -0.072259 |
| 4000 | 3.141592 | 3.166000 | 3.104000 | 0.024408 | -0.037592 |
| 5000 | 3.141592 | 3.164000 | 3.128000 | 0.022408 | -0.013592 |
| 6000 | 3.141592 | 3.159333 | 3.134000 | 0.017741 | -0.007592 |
| 7000 | 3.141592 | 3.159429 | 3.125714 | 0.017837 | -0.015878 |
| 8000 | 3.141592 | 3.154000 | 3.137500 | 0.012408 | -0.004092 |
| 9000 | 3.141592 | 3.152889 | 3.142667 | 0.011297 | 0.001075 |
| 10000 | 3.141592 | 3.147600 | 3.144000 | 0.006008 | 0.002408 |

Figure 5. Evaluation of $\int_0^1 4\sqrt{1-x^2}dx$ using monte carlo integration.

### *3.4 Monte carlo application*

The most common test for random number generators is that of evaluating definite integrals using monte carlo techniques. We will compare the result using the said integration technique with the actual value of the integral

$$\int_0^1 4\sqrt{1-x^2}dx$$

which is equal to $\pi$=3.141592.... Comparison is also made in terms of the performance of real congruential generator (RCG) and of linear congruential generator (LCG). The result is summarized in the table. (See Fig. 5).

## 4 Conclusion

The generation of pseudorandom numbers was carried on using the so-called real congruential generators which does not implement the original congruential techniques for integers. This implementation suggests a more easy handling of computer arithmetic by avoiding the very large numbers (e.g. $2^{31}-1$)

although another problem arises on computational errors. The use of the concept of chaotic maps was necessary to be able to produce unpredictability and "scatteredness" of the sequence. Although the distribution of the generated numbers are quite hard to calculate, the use of the grow-and-cut procedure for large $m$ eliminated this difficulty if the target distribution is a uniform one.

Of course, other mappings aside from the one used here maybe investigated for their credits.

## Acknowledgments

Thanks to Perry Esguerra of Theory Group at the National Institute of Physics for a valuable discussion on one-dimensional mapping. This work started when the author was still an undergraduate (1999) at Mindanao State University-Iligan Institute of Technology.

## References

1. J A Anderson and J M Bell, *Number Theory With Applications*, Prentice Hall, USA, pp. 90 and 250 (1997).
2. A M Bacala, Random number generators, *The Mindanao Forum*, MSU-IIT, Philippines, December 1995.
3. D O Cahoy, Generation of random numbers from chaotic dynamical systems, and D A Tarepe, Comparison of the efficiency of congruential generators and chaotic dynamical systems in generating random numbers, *Mindanao Higher Education Journal*, CHED-Mindanao, Philippines, Vol. 1, No. 1 (1998).
4. P Hellekalek, *Study of algorithms for primitive polynomials*, Research Institute of Software Technology, University of Salzburg, April 1994.
5. G Marsaglia, Random numbers fall mainly in the planes, *Proc. Nat. Acad. Sci.*, 61: 25-28 (1968).
6. P L'ecuyer, Random numbers and statistical aspects, *Proc. Winter Simulation Conf.* (1999)
7. A Mees and C Sparrow, Some Tools for Analyzing Chaos, *Proc. IEEE*, 75(8), August (1987). D W Jordan and P Smith, *Nonlinear Ordinary Differential Equations (An Introduction to Dynamical Systems)*, 3rd edition, Oxford University Press, England (1999). K T Alligood, T D Sauer, and J A Yorke, *Chaos (An Introduction to Dynamical Systems)*, Springer-Verlag, USA, (1996).

# PARAMETER ESTIMATION ON A SMART MATERIAL STRUCTURE: NUMERICAL COMPUTATIONS WITH IMPLICIT FILTERING OPTIMIZATION

N.C. GOMINTONG AND R.C.H. DEL ROSARIO
*Department of Mathematics, University of the Philippines*
*Diliman, Quezon City 1101, Philippines*

We perform numerical experiments in estimating the effective parameters of a smart material structure. The structure we consider consists of a thin cylindrical shell bonded with piezoceramic patches as actuators. The model-based system identification problem in the time domain is then formulated as a least-square cost functional representing the difference between the predicted response and a numerically simulated data. The very nature of the objective function prompts for an optimizer that can hurdle many local minima. Hence, we test the performance of the implicit filtering algorithm in generating the set of acceptable parameters. Numerical results indicating the feasibility of using this technique in smart material applications are presented.

## 1 Introduction

In this paper, we propose the use of the Implicit Filtering For Constrained Optimization (IFFCO) in estimating the effective parameters of a smart material structure. IFFCO has produced excellent results from a vast array of applications. Notable ones include optimization of gas transmission pipeline networks across the United States by Carter *et al*[3], optimal design of high speed mechanical systems (Choi *et al*[4], David *et al*[5]), and control of groundwater temperature (Battermann *et al*[2]).

We believe this is the first time IFFCO will be used for parameter estimation in smart material structures. Parameter estimation is a very important step in the design of optimal control strategies for smart material structures. An incorrect parameter used in the control design could yield to instabilities when the control is applied to the real physical system. Although we chose a cylindrical shell mounted with piezoceramic patches as our smart material structure, the techniques we present can be readily extended to other types of smart material structures.

The model describing the dynamics of the shell is PDE-based, hence the resulting cost function requires intensive computation. Each time the optimizing algorithm changes the parameters, a system of three linear PDEs in two dimensions would have to be integrated from a starting time $t_0$ up to a certain final time $t_f$. The numerical algorithm we use in approximating the solution to the PDE is the Galerkin method. This has been implemented by del Rosario and Smith [7,8,9] and we will modify the FORTRAN code used in those papers to accommodate parameter estimation. In modifying the code, extra effort has been exerted to put only the essential computation inside the cost function.

In Section 2, we briefly present the modeling equations of the thin cylindrical shell structure. The numerical method employed in discretizing the PDE model is presented in Section 3, then the computation of the cost function is described in Section 4. The optimization algorithm IFFCO is described in Section 5 and finally, numerical results are presented in Section 6.

## 2 PDE The Model

The smart material structure we consider consists of a homogeneous thin cylindrical shell with uniform thickness. The shell is bonded in pairs with piezoceramic patches to the inner and outer surfaces. For our purposes, we will take the Donnell-Mushtari form of the shell equations throughout the entire treatise. Thus to keep the model reasonably accurate, as discussed in Banks *et al*[1] and Kraus[12], we have to presume that the shell is relatively short with respect to its radius. Specifically, we take the ratio of the length $\ell$ with the radius $R$ of the shell to be less than 4. We also have to assume that the patches and the shell have the same curvature so as to remove the pre-stresses on the structure that would otherwise lead to nonlinear dynamics. As shown in Figure 1, points on the shell have coordinates $(x, \theta)$, where the $x$-axis is aligned with the longitudinal axis of the shell and the $\theta$-axis is oriented along the circumference. The radius of the shell is denoted by $R$, the thickness by $h$, the center of the $i^{th}$ patch pair by $(\bar{x}_i, \bar{\theta}_i)$, and the longitudinal, circumferential and tangential displacements of the middle surface by $u$, $v$ and $w$, respectively. Furthermore, we denote the density of the shell by $\rho$, Young's modulus by $E_{pe}$, Kelvin-Voigt damping parameters by $c_D$ and Poisson ratio by $\nu$. Corresponding measures for the patches, assumed uniform for all the $s$ pairs, are denoted with the subscript *pe* and are given by $h_{pe}, \rho_{pe}, E_{pe}, c_{D_{pe}}$ and $\nu_{pe}$. Edge coordinates of the $i^{th}$ patch pair are given by $x_{1i}, x_{2i}, \theta_{1i}, \theta_{2i}$. The actuating capabilities of the piezoceramic patches come from material deformations which occur in response to applied voltages, and the proportionality constant $d_{31}$ relates the generated strain to the voltage input. In addition, passive patch contribution due to the bonding layer are discarded. For systems incorporating passive patch contributions, see Banks *et al*[1] and del Rosario and Smith*et al*[8], and discussions regarding incorporation of bonding layers in the model could be found in Banks *et al.* [1]

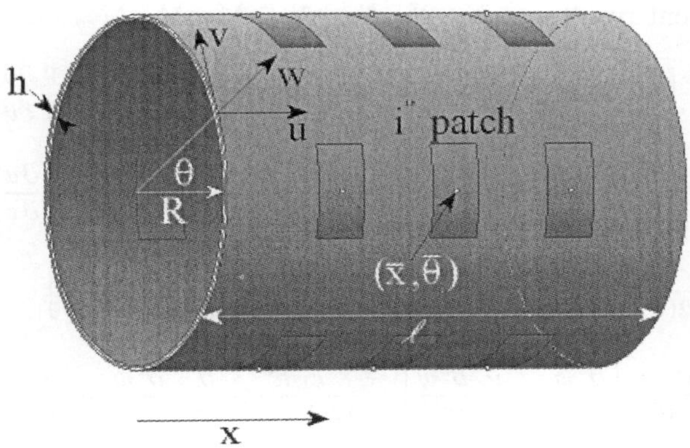


Figure 1. Thin cylindrical shell with surface mounted piezoceramic patches.

The shell model is described using the modified Donnell-Mushtari equations

$$
\begin{aligned}
&R\rho h\frac{\partial^2 u}{\partial t^2} - R\frac{\partial N_x}{\partial x} - \frac{\partial N_{\theta x}}{\partial \theta} = R\hat{q}_x - R\sum_{i=1}^{s}\frac{\partial (N_x)_{pe_i}}{\partial x}\mathcal{S}_{pe_i}(x,\theta) \\
&R\rho h\frac{\partial^2 v}{\partial t^2} - \frac{\partial N_\theta}{\partial \theta} - R\frac{N_{x\theta}}{\partial x} = R\hat{q}_\theta - \sum_{i=1}^{s}\frac{\partial (N_\theta)_{pe_i}}{\partial \theta}\mathcal{S}_{pe_i}(x,\theta) \\
&R\rho h\frac{\partial^2 w}{\partial t^2} - R\frac{\partial^2 M_x}{\partial x^2} - \frac{1}{R}\frac{\partial^2 M_\theta}{\partial \theta^2} - 2\frac{\partial^2 M_{x\theta}}{\partial x \partial \theta} + N_\theta \\
&\qquad = R\hat{q}_n - \sum_{i=1}^{s}\left[R\frac{\partial^2 (M_x)_{pe_i}}{\partial x^2} + \frac{1}{R}\frac{\partial^2 (M_\theta)_{pe_i}}{\partial \theta^2}\right].
\end{aligned}
\tag{1}
$$

These equations are derived in Banks *et al*[1] and Leissa [13] by considering strain-displacement relations, stress-strain relations, internal force and moment resultants, and equations of motion. External surface forces in the longitudinal, circumferential and transverse directions are modeled by the functions $\hat{q}_x, \hat{q}_\theta$ and $\hat{q}_n$ respectively. As detailed in Banks *et al*[1], the internal force and moment resultants $N_x, N_\theta, N_{x\theta}, N_{\theta x}, M_x, M_\theta, M_{x\theta}$ have the form

$$
\begin{aligned}
N_x &= \frac{Eh}{(1-\nu^2)}\left[\frac{\partial u}{\partial x} + \nu\left(\frac{1}{R}\frac{\partial v}{\partial \theta} + \frac{w}{R}\right)\right] + \frac{c_D h}{(1-\nu^2)}\frac{\partial}{\partial t}\left[\frac{\partial u}{\partial x} + \nu\left(\frac{1}{R}\frac{\partial v}{\partial \theta} + \frac{w}{R}\right)\right] \\
N_\theta &= \frac{Eh}{(1-\nu^2)}\left[\frac{1}{R}\frac{\partial v}{\partial \theta} + \frac{w}{R} + \nu\frac{\partial u}{\partial x}\right] + \frac{c_D h}{(1-\nu^2)}\frac{\partial}{\partial t}\left[\frac{1}{R}\frac{\partial v}{\partial \theta} + \frac{w}{R} + \nu\frac{\partial u}{\partial x}\right] \\
N_{x\theta} = N_{\theta x} &= \frac{Eh}{2(1+\nu)}\left[\frac{\partial v}{\partial x} + \frac{1}{R}\frac{\partial u}{\partial \theta}\right] + \frac{c_D h}{2(1+\nu)}\frac{\partial}{\partial t}\left[\frac{\partial v}{\partial x} + \frac{1}{R}\frac{\partial u}{\partial \theta}\right] \\
M_x &= -\frac{Eh^3}{12(1-\nu^2)}\left[\frac{\partial^2 w}{\partial x^2} + \frac{\nu}{R^2}\frac{\partial^2 w}{\partial \theta^2}\right] - \frac{c_D h^3}{12(1-\nu^2)}\frac{\partial}{\partial t}\left[\frac{\partial^2 w}{\partial x^2} + \frac{\nu}{R^2}\frac{\partial^2 w}{\partial \theta^2}\right] \\
M_\theta &= -\frac{Eh^3}{12(1-\nu^2)}\left[\frac{1}{R^2}\frac{\partial^2 w}{\partial \theta^2} + \nu\frac{\partial^2 w}{\partial x^2}\right] - \frac{c_D h^3}{12(1-\nu^2)}\frac{\partial}{\partial t}\left[\frac{1}{R^2}\frac{\partial^2 w}{\partial \theta^2} + \nu\frac{\partial^2 w}{\partial x^2}\right] \\
M_{x\theta} = M_{\theta x} &= -\frac{Eh^3}{12R(1+\nu)}\frac{\partial^2 w}{\partial x \partial \theta} - \frac{c_D h^3}{12R(1+\nu)}\frac{\partial}{\partial t}\left[\frac{\partial^2 w}{\partial x \partial \theta}\right] .
\end{aligned}
\tag{2}
$$

The patch-induced external resultants $(N_x)_{pe_i}$ and $(N_\theta)_{pe_i}$ produce forces which are opposite in direction at points symmetric to the center $(\bar{x}_i, \bar{\theta}_i)$ of the $i^{th}$ patch. This is modeled by the the indicator function $\mathcal{S}_{pe_i}(x,\theta) = S_{1,2}(x)\hat{S}_{1,2}(\theta)$, where

$$
S_{1,2}(x) = \begin{cases} 1 & , x < (x_{1i}+x_{2i})/2 \\ 0 & , x = (x_{1i}+x_{2i})/2 \\ -1 & , x > (x_{1i}+x_{2i})/2 \end{cases} , \quad \hat{S}_{1,2}(\theta) = \begin{cases} 1 & , \theta < (\theta_{1i}+\theta_{2i})/2 \\ 0 & , \theta = (\theta_{1i}+\theta_{2i})/2 \\ -1 & , \theta > (\theta_{1i}+\theta_{2i})/2 \end{cases} . \tag{3}
$$

Denoting the outer and inner input voltages by $V_{1i}(t)$ and $V_{2i}(t)$, respectively, the external force and moment resultants, as derived in Banks *et al*[1], are given by

$$
\begin{aligned}
(M_x)_{pe_i} &= \frac{-E_{pe}d_{31}}{(1-\nu_{pe})h_{pe}}\left[\left(\frac{a_2}{2} + \frac{a_3}{3R}\right)V_{1i} - \left(\frac{a_2}{2} - \frac{a_3}{3R}\right)V_{2i}\right]\chi_{pe_i} \\
(M_\theta)_{pe_i} &= \frac{-E_{pe}d_{31}a_2}{2(1-\nu_{pe})h_{pe}}\left[V_{1i} - V_{2i}\right]\chi_{pe_i} \\
(N_x)_{pe_i} &= \frac{-E_{pe}d_{31}}{(1-\nu_{pe})h_{pe}}\left[\left(h_{pe} + \frac{a_2}{2R}\right)V_{1i} + \left(h_{pe} - \frac{a_2}{2R}\right)V_{2i}\right]\chi_{pe_i}\mathcal{S}_{pe_i} \\
(N_\theta)_{pe_i} &= \frac{-E_{pe}}{1-\nu_{pe}}d_{31}[V_{1i} - V_{2i}]\chi_{pe_i}\mathcal{S}_{pe_i} ,
\end{aligned}
\tag{4}
$$

where the constants $a_2 = (h/2 + h_{pe})^2 - h^2/4$ and $a_3 = (h/2 + h_{pe})^3 - h^3/8$ arise from integrating through the thickness of the patch. The characteristic function $\chi_{pe_i}(x, \theta)$ with value 1 in the region covered by the $i^{th}$ patch and zero elsewhere indicates the patch contributions to regions of the shell covered by the patches. Note that Equations (2), (3) and (4) describe completely the terms in the PDE model (1).

We will assume that the shell is perfectly clamped on both edges hence we use the clamped edge boundary conditions expressed as

$$\begin{aligned} u(t,0,\theta) = u(t,\ell,\theta) = v(t,0,\theta) = v(t,\ell,\theta) &= 0 \\ w(t,0,\theta) = w(t,\ell,\theta) = \frac{\partial w(t,0,\theta)}{\partial x} = \frac{\partial w(t,\ell,\theta)}{\partial x} &= 0\,. \end{aligned} \tag{5}$$

For approximation and theoretical analysis (see del Rosario and Smith[7,9]) we consider the weak form of the equations

$$\begin{aligned} &\int_{\Gamma_0} \left\{ R\rho h \frac{\partial^2 u}{\partial t^2}\eta_1 + RN_x \frac{\partial \eta_1}{\partial x} + N_{\theta x}\frac{\partial \eta_1}{\partial \theta} - R\hat{q}_x\eta_1 - R\sum_{i=1}^{s}(N_x)_{pe_i}\frac{\partial \eta_1}{\partial x} \right\} d\gamma = 0 \\ &\int_{\Gamma_0} \left\{ R\rho h \frac{\partial^2 v}{\partial t^2}\eta_2 + N_\theta \frac{\partial \eta_2}{\partial \theta} + RN_{x\theta}\frac{\partial \eta_2}{\partial x} - R\hat{q}_\theta\eta_2 - \sum_{i=1}^{s}(N_\theta)_{pe_i}\frac{\partial \eta_2}{\partial \theta} \right\} d\gamma = 0 \\ &\int_{\Gamma_0} \left\{ R\rho h \frac{\partial^2 w}{\partial t^2}\eta_3 + N_\theta\eta_3 - RM_x \frac{\partial^2 \eta_3}{\partial x^2} - \frac{1}{R}M_\theta \frac{\partial^2 \eta_3}{\partial \theta^2} - 2M_{x\theta}\frac{\partial^2 \eta_3}{\partial x \partial \theta} \right. \\ &\qquad \left. - R\hat{q}_n\eta_3 + \sum_{i=1}^{s}\left[ R(M_x)_{pe_i}\frac{\partial^2 \eta_3}{\partial x^2} + \frac{1}{R}(M_\theta)_{pe_i}\frac{\partial^2 \eta_3}{\partial \theta^2} \right] \right\} d\gamma = 0 \end{aligned} \tag{6}$$

for all test functions $\Psi = (\eta_1, \eta_2, \eta_3)$ (see Banks *et al*[1] for details in the weak form derivation). Here $\Gamma_0$ denotes the region occupied by the middle surface of the shell, i.e., $\Gamma_0 = \{(x,\theta) : 0 \le x \le \ell, 0 \le \theta \le 2\pi\}$. The variables $y = (u, v, w)$ are taken from the state space

$$\Phi = L^2(0, T; H)$$

where $H = L^2(\Gamma_0) \times L^2(\Gamma_0) \times L^2(\Gamma_0)$. In order to impose the boundary conditions (5) on the solution $y$, we choose test functions $\Psi \in V$ where $V =$

$H_0^1(\Gamma_0) \times H_0^1(\Gamma_0) \times H_0^2(\Gamma_0)$ and

$$H_0^1(\Gamma_0) = \{\eta \in H^1(\Gamma_0) : \eta(0,\cdot) = \eta(\ell,\cdot) = 0\}$$
$$H_0^2(\Gamma_0) = \left\{\eta \in H^2(\Gamma_0) : \eta(0,\cdot) = \frac{\partial \eta(0,\cdot)}{\partial x} = \eta(\ell,\cdot) = \frac{\partial \eta(\ell,\cdot)}{\partial x} = 0\right\}.$$

Note that $V \subset H$.

## 3 Galerkin Approximation

The first step in finite-dimensional approximation is to choose basis functions $\{\mathcal{B}_{u_i}\}_{i=1}^{\mathcal{N}_u}$, $\{\mathcal{B}_{v_i}\}_{i=1}^{\mathcal{N}_v}$ and $\{\mathcal{B}_{w_i}\}_{i=1}^{\mathcal{N}_w}$ to form the $\mathcal{N}$-dimensional subspace $H^{\mathcal{N}} \subset V \subset H$

$$H^{\mathcal{N}} = \text{span}\{\mathcal{B}_{u_i}\}_{i=1}^{\mathcal{N}_u} \times \text{span}\{\mathcal{B}_{v_i}\}_{i=1}^{\mathcal{N}_v} \times \text{span}\{\mathcal{B}_{w_i}\}_{i=1}^{\mathcal{N}_w} \,. \tag{7}$$

The numerical solution $y^{\mathcal{N}}$ is then obtained by setting the Galerkin expansions

$$y^{\mathcal{N}}(t,x,\theta) = \begin{bmatrix} u^{\mathcal{N}}(t,x,\theta) \\ v^{\mathcal{N}}(t,x,\theta) \\ w^{\mathcal{N}}(t,x,\theta) \end{bmatrix} = \begin{bmatrix} \sum_{i=1}^{\mathcal{N}_u} u_i(t)\mathcal{B}_{u_i}(x,\theta) \\ \sum_{i=1}^{\mathcal{N}_v} v_i(t)\mathcal{B}_{v_i}(x,\theta) \\ \sum_{i=1}^{\mathcal{N}_w} w_i(t)\mathcal{B}_{w_i}(x,\theta) \end{bmatrix}, \tag{8}$$

and then solving the weak form (6) for $y^{\mathcal{N}}$ where the test function space is restricted to $H^{\mathcal{N}}$. By choosing the appropriate finite dimensional approximating subspaces $H^{\mathcal{N}}$ which satisfy the conditions

**(C1)** $H^{\mathcal{N}} \subset V$ and if $P^N : H \longrightarrow H^{\mathcal{N}}$ is the orthogonal projection of $H$ onto $H^{\mathcal{N}}$, then for each $\psi \in V$, $|\psi - P^N\psi|v \longrightarrow 0$ as $N \longrightarrow 0$, and

**(C2)** The monotonicity condition $H^{\mathcal{N}} \subset H^{\mathcal{N}+1}$

then we can guarantee that the convergence of the finite dimensional approximation $y^{\mathcal{N}} \longrightarrow y$ (see Banks *et al*[1] for details in the convergence analysis).

The convergence, accuracy and efficiency of the methods depend heavily on the type of basis elements used in creating the finite dimensional system. We choose cubic splines in the axial direction and trigonometric polynomials in the circumferential direction as inspired by the study made by del Rosario and Smith. [7,8,9] Thus the Galerkin expansions are of the form

$$y^{\mathcal{N}}(t,x,\theta) = \begin{bmatrix} u^{\mathcal{N}}(t,x,\theta) \\ v^{\mathcal{N}}(t,x,\theta) \\ w^{\mathcal{N}}(t,x,\theta) \end{bmatrix} = \begin{bmatrix} \sum_{m=-M_u}^{M_u} \sum_{n=1}^{\hat{N}_u} u_{mn}(t) e^{im\theta} B_{u_n}(x) \\ \sum_{m=-M_v}^{M_v} \sum_{n=1}^{\hat{N}_v} v_{mn}(t) e^{im\theta} B_{v_n}(x) \\ \sum_{m=-M_w}^{M_w} \sum_{n=1}^{\hat{N}_w} w_{mn}(t) e^{im\theta} B_{w_n}(x) \end{bmatrix} . \tag{9}$$

Here, $B_{u_n}(x), B_{v_n}(x)$ and $B_{w_n}(x)$ denote the spline components in the longitudinal, circumferential and transverse directions, respectively. $M_u$, $M_v$ and $M_w$ give the Fourier limits whereas $\hat{N}_u$, $\hat{N}_v$ and $\hat{N}_w$ denote the number of splines used in each expansion. Hence, the total number of basis is given by $\mathcal{N} = \mathcal{N}_u + \mathcal{N}_v + \mathcal{N}_w = \hat{N}_u \cdot (2M_u + 1) + \hat{N}_v \cdot (2M_v + 1) + \hat{N}_w \cdot (2M_w + 1)$.

The details in the creation of the system matrices can be found in detail in del Rosario[6], and also in del Rosario and Smith. [7,8,9]. We just point out that by consolidating the generalized coefficients $u_i(t), v_i(t)$ and $w_i(t)$ into one vector

$$\vartheta^{\mathcal{N}}(t) = [u_1(t), \ldots, u_{\mathcal{N}_u}(t), v_1(t), \ldots, v_{\mathcal{N}_v}(t), w_1(t), \ldots, w_{\mathcal{N}_w}(t)]^T \in \mathbb{R}^{\mathcal{N}} ,$$

then the resulting matrix system becomes

$$\begin{aligned} \begin{bmatrix} K_E^{\mathcal{N}} & 0 \\ 0 & M^{\mathcal{N}} \end{bmatrix} \begin{bmatrix} \dot{\vartheta}^{\mathcal{N}}(t) \\ \ddot{\vartheta}^{\mathcal{N}}(t) \end{bmatrix} &= \begin{bmatrix} 0 & K_E^{\mathcal{N}} \\ -K_E^{\mathcal{N}} & -K_{C_D}^{\mathcal{N}} \end{bmatrix} \begin{bmatrix} \vartheta^{\mathcal{N}}(t) \\ \dot{\vartheta}^{\mathcal{N}}(t) \end{bmatrix} + \begin{bmatrix} 0 \\ \hat{B}^{\mathcal{N}} \end{bmatrix} U(t) + \begin{bmatrix} 0 \\ \hat{F}^{\mathcal{N}}(t) \end{bmatrix} \\ \begin{bmatrix} K_E^{\mathcal{N}} & 0 \\ 0 & M^{\mathcal{N}} \end{bmatrix} \begin{bmatrix} \vartheta^{\mathcal{N}}(0) \\ \dot{\vartheta}^{\mathcal{N}}(0) \end{bmatrix} &= \begin{bmatrix} y_1^{\mathcal{N}} \\ y_2^{\mathcal{N}} \end{bmatrix} . \end{aligned} \tag{10}$$

Here, the matrices $M^{\mathcal{N}}, K_E^{\mathcal{N}}, K_{C_D}^{\mathcal{N}}$ and $\hat{F}^{\mathcal{N}}$ represent the mass, stiffness, damping, and the force vector respectively.

As detailed in Section 4.4 of del Rosario and Smith[8] the submatrices of $K_E^{\mathcal{N}}, M^{\mathcal{N}}, K_{C_D}^{\mathcal{N}}, \hat{B}^{\mathcal{N}}$ and force vector $\hat{F}^{\mathcal{N}}$ contain individual components arising from orthogonalizing the residual with respect to the test functions from $H^{\mathcal{N}}$. To write the system in a form suitable for simulations, parameter estimation and control applications, we invert the first-order system mass matrix to yield a first-order Cauchy equation of the form

$$\begin{aligned} \dot{z}^{2\mathcal{N}}(t) &= A^{2\mathcal{N}} z^{2\mathcal{N}}(t) + B^{2\mathcal{N}} U(t) + G^{2\mathcal{N}}(t) \\ z^{2\mathcal{N}}(0) &= z_0^{2\mathcal{N}} , \end{aligned} \tag{11}$$

where $z^{2\mathcal{N}}(t) = [\vartheta^{\mathcal{N}}(t), \dot{\vartheta}^{\mathcal{N}}(t)] \in \mathbb{R}^{2\mathcal{N}}$.

## 4 Cost Function and the Inverse Problem

In this section, the formulation of the cost functional as a scalar least-square function will be presented. This function evaluates the difference between the predicted response and the numerically simulated data. The parameters, which are the arguments of the cost function, are given by the vector $q = (q_1, q_2, q_3, q_4)$, where $q_1 = \rho h R$, $q_2 = Eh$, $q_3 = c_D h$, $q_4 = d_{31} E_{pe}$. Instead of estimating directly the parameters $\rho, E, c_D$ and $d_{31}$, we multiply them by constants in order to normalize their magnitudes. The parameter $E$ typically has exponent $10^{10}$ while $d_{31}$ typically has exponent $10^{-12}$ and the huge difference in magnitudes could lead to difficulties in solving the optimal value for the cost function. Thus, by considering $q_1, q_2, q_3$ and $q_4$ as the parameters, we are able to limit their magnitudes between $10^1$ and $10^7$.

It should be noted that the patch parameters $\rho_{pe}, E_{pe}$ and $c_{D_{pe}}$ are treated as constants in this paper. Although in actual applications these parameters need to be estimated, we decided to limit the parameter estimation problem to four unknowns due to limitations in computing power.

Due to unavailability of experimental setups, numerically simulated data were used as the set of observation denoted by $\{y_i\}_{i=1}^{N_t}$. This set consists of displacements at different points on the shell at different time instances. We measure the displacements at four points on the shell (we give the actual locations Section 6) and record 1000 time instances at each point. Thus the data set is actually given by $\{y(t_i, x_i, \theta_i)\}_{i=1}^{N_t} = \{y(t_j, x_k, \theta_k, j = 1, \ldots, 1000, k = 1, \ldots 4\}$. We use the simpler notation $\{y_i\}$ in order to simplify our exposition.

The data $\{y_i\}$ was obtained by numerically solving the system (6) based on some predefined values of the parameters. We will refer to these predefined values as the "true" values of the parameters. In real situations, the "true" values are the physical values of the parameters.

Letting $Q$ be the set of admissible parameters, then the least square inverse problem can be formulated as

$$\min_{q \in Q} J(q) = \min_{q \in Q} \sum_{i=1}^{N_t} \|C_2 \{C_1 (y(t_i, x_i, \theta_i; q))\} - \{y_i\}\|^2 \ , \tag{12}$$

where $\{y(t_i, x_i, \theta_i; q)\}_{i=1}^{N_t}$ are the parameter dependent solutions of the model (i.e., for a certain set of parameters $q$, $y(t_i, x_i, \theta_i; q)$ is the solution to (6) at time $t = t_i$).

The observation operators $C_1$ and $C_2$ depend on the sensor being used in the experiment. If a proximity probe sensor is used in collecting the data, the actual displacement of the shell is collected hence the cost functional to be minimized is of the form

$$J(q) = \sum_{i=1}^{N_t} \|y(t_i, x_i, \theta_i; q) - \{y_i\}\|^2 \ . \tag{13}$$

In this paper, we assume the cost function to be of the form (13).

In the process of solving the optimal value $q^*$, the optimizer will have to evaluate the cost function (13) for different values of $q$. Each function evaluation involves the solution of a time dependent system of three PDES in two (space) dimensions hence is very computationally intensive. Therefore, we need to use an optimization algorithm which can hopefully attain the minimum $q^*$ in a few number of steps. We propose the use of the implicit filtering algorithm which we present in the next section.

## 5 Implicit Filtering

IFFCO (Implicit Filtering For Constrained Optimization) is a projected quasi-Newton algorithm with decreasing sequence of finite difference steps. This reduction scheme has the effect of filtering out unwanted harmonics at the start (when the current solution is presumed to be far from the optimum) and allowing them near the end (when the current solution is presumed to be near the optimum). Such scheme is found to be effective, in practice, in dealing with noisy type of problems.

Quantitatively, we consider an objective function $f$ defined on $R^n$ that is a perturbation of a smooth function $f_s$ by a small function $\phi$, i.e.,

$$f(q) = f_s + \phi(q). \tag{14}$$

For our problem, this $f$ would be the cost function $J$ described in the previous section (see Equation (13)). It should be noted that computation of $f_s$ and $\phi$ is not a requirement for IFFCO. It is only necessary that $f$ behaves like a function that satisfies (14).

What follows is a basic outline of the algorithm used in the program IFFCO. For a more complete discussion of the program see Gilmore and Kelley. [11,10]. The goal of IFFCO is to solve the constrained optimization problem of the form

$$\min_{q \in \Omega} f(q) \tag{15}$$

where $q$ is the vector of unknown parameters to be estimated.

The set of feasible solutions $\Omega$ is defined by simple bounds

$$\Omega = \{q | l_i \leq q_i \leq u_i\}. \tag{16}$$

For $q \in R^n$, the $\ell^2$ projection of $P(q)$ onto $\Omega$ is

$$P_i(q) = \begin{cases} l_i & , q_i \leq l_i \\ u_i & , q_i \geq u_i \\ q_i & , l_i < q_i < u_i \end{cases} \tag{17}$$

where $l_i$ and $u_i$ represent the lower and upper bounds, respectively. The (simplified) IFFCO algorithm is given by the following steps.

- Scale the hyper-box $\Omega$ to the unit cube. ($l_i = 0$ and $u_i = 1$ for all $i$)
- Pick an initial $q$ and $h$. Find $f(q)$ and the difference gradient $\nabla_h f(q)$.
- Initialize the model Hessian $H$ to identity matrix.
- While $h$ and $\nabla_h f(q)$ satisfy conditions do
  - Use $\nabla_h f(q)$ and $H$ to calculate a descent direction $d$. This is a quasi-Newton step. Here, $d = -H^{-1}\nabla_h f(q)$.
  - Perform a line search in the direction of $d$, and signal success if some criteria are met.
  - If line search is successful then "Accept new point and project into the box" $\Omega$, else reduce $h$ using $h = h/2$.
  - Calculate $\nabla_h f(q)$.
  - Update $H$ with either rank-one SR1 (or rank-two BFGS).
- end while

The gradients are approximated by finite differences, and in our computations we used centered differences given by

$$\nabla f \approx (\nabla_h f(q))_i = \frac{f(q + he_i) - f(q - he_i)}{2h} \tag{18}$$

where $e_i$ is the $i^{th}$ canonical basis vector and $h$ is the current scale.

The value of $q$ is updated using the gradient projection iteration given by

$$q_+ = P(q_c - \alpha \nabla f(q_c)) \tag{19}$$

where $\alpha$ is a step length parameter that would make $f(q_+)$ smaller than $f(q_c)$.

The Hessian matrix, on the other hand is updated using either a rank-one SR1 (or rank-two BFGS). For our purposes, and as the author of IFFCO suggests, we will use SR1-update (which is proven to be good in handling constrained optimization). The SR1 update is given by

$$H_+ = H_c + \frac{(y - H_c s)(y - H_c s)^T}{(y - H_c s)^T s}, \tag{20}$$

whereas BFGS update is done using:

$$H_+ = H_c + \frac{yy^T}{y^T s} - \frac{(H_c s)(H_c s)^T}{s^T H_c s}. \tag{21}$$

In both methods, $s = q_+ - q_c$ and $y = \nabla_h f(q_+) - \nabla_h f(q_c)$.

The iteration is terminated when

$$||q_+ - q_c|| \leq \tau h, \tag{22}$$

where $\tau$ is a prescribed tolerance.

## 6 Numerical Results

We assume that the shell has dimensions and true parameters given in Table 1. Four pairs of patches were assumed to be bonded to the inner and outer surfaces of the shell: two pairs centered at $\theta = 0$ and the other two pairs centered at $\theta = \pi$. The location, dimensions and the (fixed) parameters of the four patch pairs are indicated in Table 2.

The numerically simulated data was obtained by exciting the shell with a voltage spike onto the four pairs of patches. The voltage spike was triangular in shape with a duration of 0.001 seconds and with a maximum magnitude of 50 volts at 0.0005 seconds. The voltages sent to the inner and outer patch pairs were negatives of each other in order to excite transverse displacements. The shell was allowed to vibrate up to 0.05 seconds and 1000 measurements were obtained at the four sensor points. The four measurement points are located at the center of the four patch pairs hence their locations are also given by the second column in Table 2.

In actual applications, engineers and experimentalists are very careful in choosing data to be used in parameter estimation. The data should contain the sufficient amount of frequencies, hence the vibrations should not die down easily. The numerically simulated data produced by the set-up described above provided a rich amount of frequencies. The vibrations show a large number of oscillations together with exponential decay.

Table 1. Shell dimensions and parameters.

| Dimensions | Parameters |
|---|---|
| $h = 0.00127m$ | $\rho = 2700.0kg/m^3$ |
| $R = 0.4m$ | $E = 7.1 \times 10^{10}N/m^2$ |
| $l = 1.0m$ | $c_D = 1.47004 \times 10^5 Nms$ |
| $\nu = 0.33$ | $d_{31} = 190 \times 10^{-12}m/V$ (from patch) |

Table 2. Dimensions, (fixed) parameters and center location of the patches.

| Dimensions | Centers: $(\hat{x}, \hat{\theta})$ |
|---|---|
| $h_{pe} = .0001778m$ | $(.25, 0)$ |
| $\rho_{pe} = 7600.0kg/m^3$ | $(.75, 0)$ |
| $E_{pe} = 6.3 \times 10^{10}N/m^2$ | $(.25, \pi)$ |
| $cD_{pe} = 1.7 \times 10^5 Nms$ | $(.75, \pi)$ |
| $\nu_{pe} = .31$ | $size : .15m \times \pi/3$ |

The first-order ODE system given by (11) was numerically solved using VODE, a publicly available variable order stiff ODE solver. We would like to point out that most of the computation time involved the integration of the ODE and hence we need to keep the number of function evaluations at a minimum.

The program IFFCO has several inputs, but in our case the only ones we provided were:

- name of objective function,
- initial guess of the solution,
- the bounds $\{l_i\}$ and $\{u_i\}$,
- estimate $f_{scale}$ for the maximum absolute value of $f$,
- termination criteria constant $termTol$,
- smallest finite difference step $minh$,
- largest finite difference step $maxh$.

The parameter estimation program was ran with different initial guesses and error on the numerically simulated data. For each set of error configuration, we tuned the IFFCO inputs above in order for the optimizer to attain the optimal parameter. Our experience indicated that a good value for $f_{scale}$ is $10^{-5}$. This is about 5 to 10 times the initial cost. The termination criteria constant $termTol$ we used was from $10^{-4}$ to $10^{-6}$ in magnitude. Another crucial parameter for IFFCO in our problem was $minh$, where we used a value from $2^{-8}$ to $2^{-12}$. However, we just allowed $maxh$ take a value of either 0.1 or 0.2.

From here on, we will refer to "initial guess error" as the maximum difference of the initial estimate values (provided by the user) and the true values of the parameters. On the other hand, "data error" denotes the maximum difference between the true values of the data and the actual data used in the cost function. The introduction of "data error" allows us to simulate scenarios where the measuring devices introduce errors on the observed data. Furthermore, in actual applications, the initial guesses for the parameters come from book values provided by the manufacturers. These book values are claimed by the manufacturers to be correct within a certain percentage hence "initial guess error" allows us to simulate this possibility.

First, we tried using different degrees of errors in our initial guess (no error on the data yet) using $q_0 = q_t \pm deltaq \cdot q_t$, where

- $deltaq$ = 5%, 10% and 15%,
- $q_t$ contains the true values of the parameters,
- and $q_0$ is the initial guess.

Then we introduced randomly generated errors in the numerically simulated data giving a maximum effect of 1, 5, and 10 percent differences from the "true" data. Specifically, we let $y_{data} = y_{truedata} \pm deltay \cdot y_{truedata}$ where

- $deltay$ = 1%, 5% and 10%,
- $y_{truedata}$ contains the true data,
- and $y_{data}$ is the data containing errors.

This yielded a total of 12 error configurations: 3 (possible initial guess errors) $\times$ 4 (possible data error, including no error in data) = 12. Plots of displacements in $u$, $v$ and $w$ directions with 10% error in the initial guess and no data error are given in Figure 2. Similar plots were obtained with the other 11 configurations of initial estimate and data errors, hence we only present one plot (Figure 2). It can be seen from Figure 2 that model displacements using the initial guess as parameters are distinguishable from numerical data,

Table 3. Comparison of initial and final cost value for each setup of initial guess error and data error.

| **Initial Guess Error** (%) | **Data Error** (%) | **Initial Cost** | **Final Cost** |
|---|---|---|---|
| 5 | 0 | $3.01749906 \times 10^{-6}$ | $5.49663197 \times 10^{-9}$ |
| 5 | 1 | $3.02395692 \times 10^{-6}$ | $6.91562264 \times 10^{-9}$ |
| 5 | 5 | $3.0500731 \times 10^{-6}$ | $2.92823392 \times 10^{-8}$ |
| 5 | 10 | $3.08334552 \times 10^{-6}$ | $5.74815095 \times 10^{-8}$ |
| 10 | 0 | $3.2634917 \times 10^{-6}$ | $3.65370777 \times 10^{-9}$ |
| 10 | 1 | $3.26975749 \times 10^{-6}$ | $6.07738407 \times 10^{-9}$ |
| 10 | 5 | $3.29509164 \times 10^{-6}$ | $2.23204041 \times 10^{-8}$ |
| 10 | 10 | $3.32735743 \times 10^{-6}$ | $5.70045293 \times 10^{-8}$ |
| 15 | 0 | $3.45318516 \times 10^{-6}$ | $2.58587164 \times 10^{-9}$ |
| 15 | 1 | $3.45908048 \times 10^{-6}$ | $6.44424054 \times 10^{-9}$ |
| 15 | 5 | $3.48293108 \times 10^{-6}$ | $2.91326044 \times 10^{-8}$ |
| 15 | 10 | $3.51333998 \times 10^{-6}$ | $5.82026292 \times 10^{-8}$ |

Table 4. Estimated parameters with 5% error on initial guess and 0%, 1%, 5% and 10% error in data.

| **True Parameters** | **Estimated Values** | | | |
|---|---|---|---|---|
| | 0% **Data** | 1% **Data** | 5% **Data** | 10% **Data** |
| $(q_1)$ $1.3716 \times 10^{0}$ | $1.3773 \times 10^{0}$ | $1.3902 \times 10^{0}$ | $1.3932 \times 10^{0}$ | $1.3945 \times 10^{0}$ |
| $(q_2)$ $9.0170 \times 10^{7}$ | $9.0541 \times 10^{7}$ | $9.1390 \times 10^{7}$ | $9.1337 \times 10^{7}$ | $9.1668 \times 10^{7}$ |
| $(q_3)$ $1.8670 \times 10^{2}$ | $1.8877 \times 10^{2}$ | $1.9025 \times 10^{2}$ | $1.9183 \times 10^{2}$ | $1.9298 \times 10^{2}$ |
| $(q_4)$ $1.1970 \times 10^{1}$ | $1.2045 \times 10^{1}$ | $1.2212 \times 10^{1}$ | $1.2460 \times 10^{1}$ | $1.2843 \times 10^{1}$ |

while model displacements using the estimated parameters are indistinguishable from the data. This is true for all 12 runs.

In Table 3 we give a summary of the initial cost and the final cost for each combination of the initial guess error and data error. Notice that the final cost is always smaller than the initial cost, at least 100 times smaller. We also present the optimal parameter values for the 12 runs in Tables 4 through 6.

Next, we tried to test how IFFCO would perform when the initial estimate is close to the optimal point (using a 1% error on initial estimate and no error in data) and when the initial estimate is far from the optimal point (using a 20% error on initial estimate and no error in data). We obtained

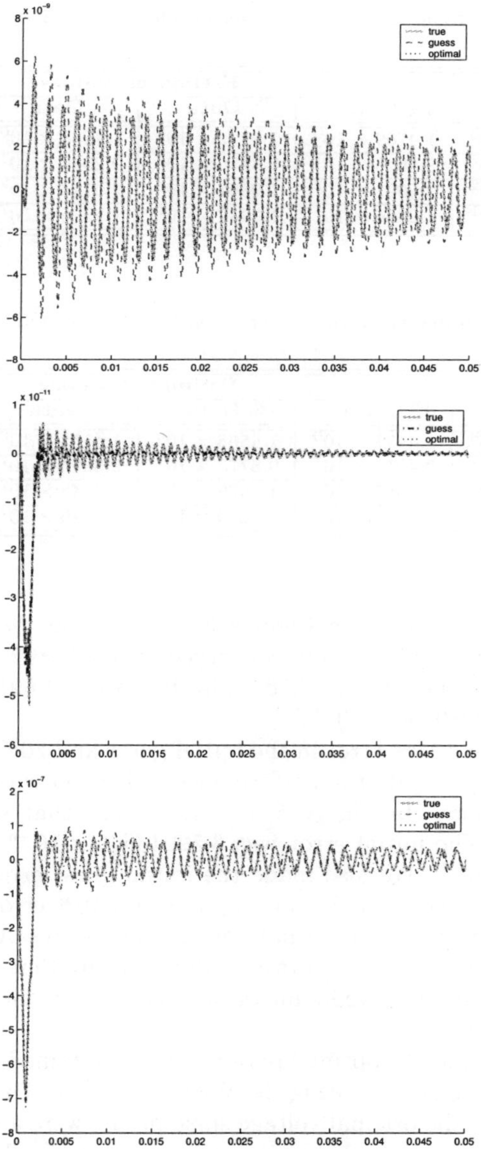


Figure 2. Displacements in $u$ (top), $v$ (middle) and $w$ (bottom) directions with 10% initial guess error and no data error. Note that simulations with estimated parameters are indistinguishable from the data while simulations with initial guess parameters are distinguishable.

Table 5. Estimated parameters with 10% error on initial guess and 0%, 1%, 5% and 10% error in data.

| True Parameters | Estimated Values | | | |
|---|---|---|---|---|
| | **0% Data** | **1% Data** | **5% Data** | **10% Data** |
| ($q_1$) $1.3716 \times 10^0$ | $1.4709 \times 10^0$ | $1.4172 \times 10^0$ | $1.4088 \times 10^0$ | $1.4422 \times 10^0$ |
| ($q_2$) $9.0170 \times 10^7$ | $9.6696 \times 10^7$ | $9.3170 \times 10^7$ | $9.2570 \times 10^7$ | $9.4811 \times 10^7$ |
| ($q_3$) $1.8670 \times 10^2$ | $2.0021 \times 10^2$ | $1.9301 \times 10^2$ | $2.0113 \times 10^2$ | $1.9740 \times 10^2$ |
| ($q_4$) $1.1970 \times 10^1$ | $1.2836 \times 10^1$ | $1.2447 \times 10^1$ | $1.2875 \times 10^7$ | $1.3219 \times 10^1$ |

Table 6. Estimated parameters with 15% error on initial guess and 0%, 1%, 5% and 10% error in data.

| True Parameters | Estimated Values | | | |
|---|---|---|---|---|
| | **0% Data** | **1% Data** | **5% Data** | **10% Data** |
| ($q_1$) $1.3716 \times 10^0$ | $1.4437 \times 10^0$ | $1.4568 \times 10^0$ | $1.4155 \times 10^0$ | $1.4238 \times 10^0$ |
| ($q_2$) $9.0170 \times 10^7$ | $9.4911 \times 10^7$ | $9.5773 \times 10^7$ | $9.3054 \times 10^7$ | $9.3604 \times 10^7$ |
| ($q_3$) $1.8670 \times 10^2$ | $1.9586 \times 10^2$ | $1.9529 \times 10^2$ | $1.9548 \times 10^2$ | $1.9917 \times 10^2$ |
| ($q_4$) $1.1970 \times 10^1$ | $1.2586 \times 10^1$ | $1.3289 \times 10^1$ | $1.2719 \times 10^7$ | $1.3131 \times 10^1$ |

acceptable results as shown in Figures 3 and 4. The 1% run started with $1.69117929 \times 10^{-6}$ initial cost and converged at $3.5360899 \times 10^{-10}$ final cost. On the other hand, the 20% run had an initial cost of $3.69189481 \times 10^{-6}$ and a final cost of $8.32089994 \times 10^{-9}$.

By introducing a 10% error in the initial guess and 20% error in the data, we were able to investigate the performance of IFFCO when the observation data has huge errors. In Figure 5, it can be seen that we still obtain acceptable results. The initial error was $3.52916645 \times 10^{-6}$, with a final cost of $4.79138711 \times 10^{-8}$. We also made the initial guess error bigger than the data error, i.e., 15% error in the initial guess and 10% error in the data. We were able to obtain the optimal parameter and the results are not plotted anymore. These two error configurations demonstrate that there is an inverse relationship between the maximum initial guess error and data error (that yield acceptable results).

We now show that the optimal parameters we obtained will perform effectively when the manner of exciting the shell is changed. The optimal parameters obtained using the original voltage spike profile were applied on a system with different magnitude and duration of the voltage spike. The results of simulations with 95 volts magnitude and durations 0.001 and 0.01 seconds are shown in Figures 6 and 7, respectively. Note that the simulations using

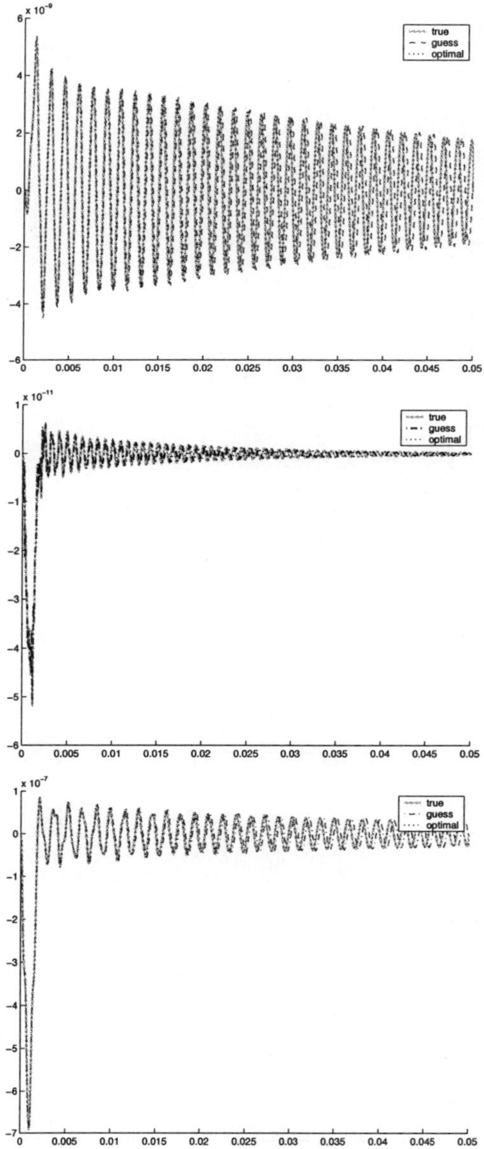


Figure 3. Displacements in $u$ (top), $v$ (middle) and $w$ (bottom) directions with 1% initial guess error and no data error. Note that simulations with estimated parameters are indistinguishable from the data.

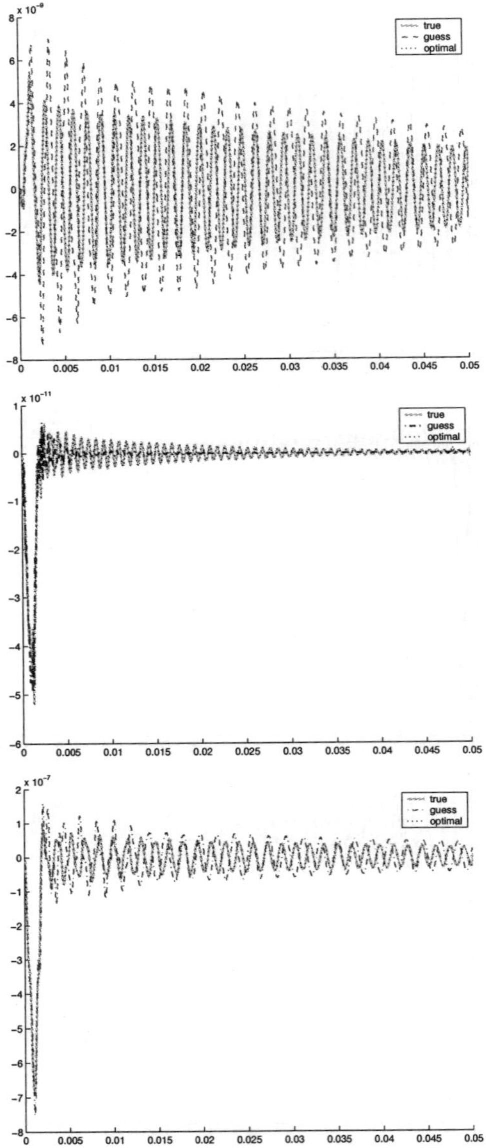


Figure 4. Displacements in $u$, $v$ and $w$ direction with 20% initial guess error and no data error. Note that simulations with estimated parameters are indistinguishable from the data while simulations using initial guess parameters are distinguishable.

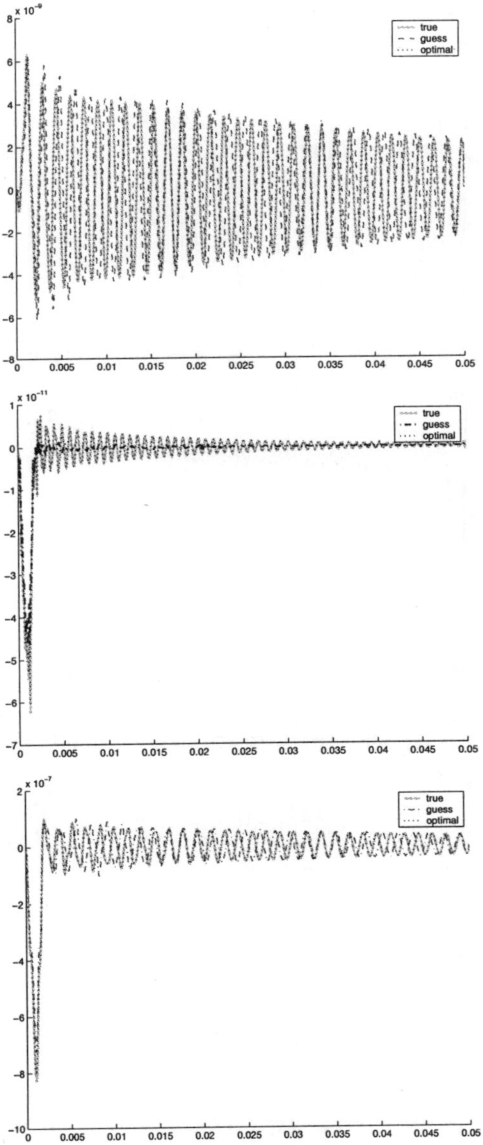


Figure 5. Displacements in $u$, $v$ and $w$ direction with 10% initial guess error and 20% data error. Note that simulations with estimated parameters are indistinguishable from the data while simulations using initial guess parameters are distinguishable.

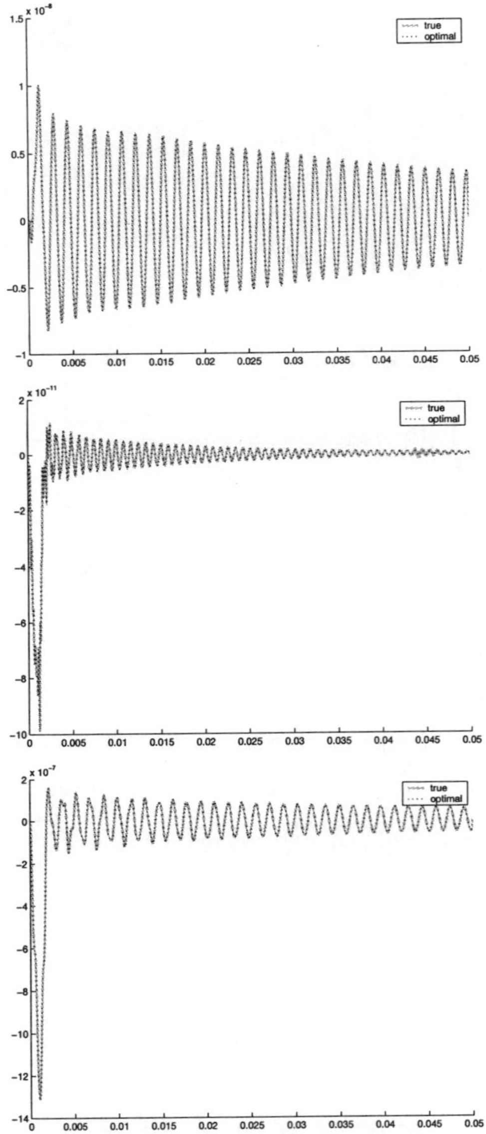


Figure 6. Performance of optimal values when excitation is changed. The estimated parameters were obtained with 20% initial guess error, no data error and original patch excitation. The simulations presented above were done using a voltage spike of with 95 volts magnitude and 0.001 second duration. Note that simulations using optimal values are indistinguishable from the true displacements.

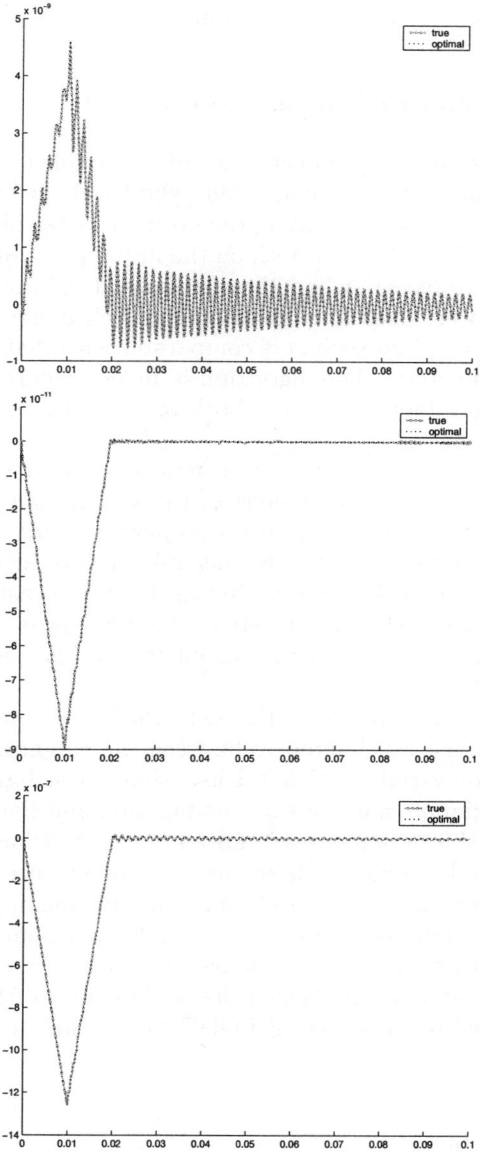


Figure 7. Performance of optimal values when excitation is changed. The estimated parameters were obtained with 20% initial guess error, no data error and original patch excitation. The simulations presented above were done using a voltage spike with 95 volts magnitude and 0.01 second duration. Note that simulations using optimal values are indistinguishable from the true displacements.

optimal values were indistinguishable from the (true-)system performance.

## 7 Conclusions and Recommendations

The results indicated the promising use of the implicit filtering algorithm in parameter estimation on a smart thin cylindrical shell structure. IFFCO performed an excellent job in finding the optimal values of the parameters in different configurations of errors both on the initial guess and the numerically simulated data. Moreover, when these estimated values of the parameters found by IFFCO were used in simulating the shell dynamics using different voltages and duration, the results are comparable with that of using the actual values of the parameters. The algorithm is indeed robust in the sense that it can be tuned to adopt to changes both on the data and the initial guess estimates of the user.

In addition, it is very important to note that the parameters being estimated are the "effective" parameters of the system and not the individual values of the smart material structure's properties, since the usual situation in many smart material structures is that different products are combined in a composite material configuration, altering their inherent properties. Thus we use combinations of the parameters as our unknowns. It also has the effect of normalizing the values of the parameters which greatly helped in the convergence of IFFCO.

Our experience indicated that IFFCO might fail to obtain the optimal parameters if there are a big differences between the magnitudes of the components of the decision variable. Also, the user should also have some knowledge regarding the maximum value of the cost function and bounds on the search region. Although this is not possible for all kinds of estimation problems, in our case it is possible to meet this requirement due to the availability "book values" for the parameters. We also point out that the algorithm will fail to converge if the initial guess error is more than 30% off from the optimal value.

Each optimization run took 12 hours and most of the computation was done in the ODE integration. Hence the performance of this technique can be greatly improved by using parallel ODE solvers and also in using parallel IFFCO.

## Acknowledgments

The authors would like to acknowledge Prof. C.T. Kelley of North Carolina State University for helpful input and insight regarding IFFCO.

**References**


1. H. T. BANKS, R. C. SMITH, AND Y. WANG, *Smart Material Structures: Modeling Estimation and Control*, Masson/John Wiley, Paris/Chichester, 1996.
2. A. BATTERMANN, J. M. GABLONSKY, A. PATRICK, C. T. KELLEY, K. R. KAVANAGH, T. COFFEY, AND C. T. MILLER, *Solution of a groundwater control problem with implicit filtering*, Tech. Report CRSC-TR00-30, Center for Research in Scientific Computation, NCSU, Raleigh, 2000.
3. R. G. CARTER, J. M. GABLONSKY, A. PATRICK, C. T. KELLEY, AND O. J. ESLINGER, *Algorithms for noisy problems in gas transmission pipeline optimization*, Tech. Report CRSC-TR00-10, Center for Research in Scientific Computation, NCSU, Raleigh, 2000.
4. T. D. CHOI, O. J. ELSINGER, C. T. KELLEY, J. W. DAVID, AND M. ETHERIDGE, *Optimization of automotive valve train components with implicit filtering*, Tech. Report CRSC-TR98-44, Center for Research in Scientific Computation, NCSU, Raleigh, 1998.
5. J. W. DAVID, C. Y. CHENG, T. D. CHOI, C. T. KELLEY, AND J. GABLONSKY, *Optimal design of high speed mechanical systems*, Tech. Report CRSC-TR97-18, Center for Research in Scientific Computation, NCSU, Raleigh, 1997.
6. R. C. H. DEL ROSARIO, *Computational Methods for Feedback Control in Structural Systems*, PhD thesis, N. C. State University, 1998.
7. R. C. H. DEL ROSARIO AND R. C. SMITH, *Spline approximation of thin shell dynamics: Numerical examples*, Tech. Report CRSC-TR96-5, Center for Research in Scientific Computation, NCSU, Raleigh, 1996.
8. ——, *Spline approximation of thin shell dynamics*, International Journal for Numerical Methods in Engineering, 40 (1997), pp. 2807–2840.
9. ——, *LQR conrol of shell vibrations via piezoceramic actuators*, International Series in Numerical Mathematics, 126 (1998), pp. 247–265.
10. P. GILMORE AND C. T. KELLY, *An implicit filtering algorithm for optimization of functions with many local minima*, Tech. Report CRSC-TR94-23, Center for Research in Scientific Computation, NCSU, Raleigh, 1994.
11. ——, *An implicit filtering algorithm for optimization of functions with many local minima*, SIAM Journal of Optimization, 5 (1995).
12. H. KRAUS, *Thin Elastic Shells: An Introduction to the Theoretical Foundations and The Analysis of Their Static and Dynamic Behavior*, John Wiley & Sons, Inc., New York, 1967.
13. A. W. LEISSA, *Vibrations of shells.* Reprinted by the Acoustical Society of America through the American Institute of Physics, 1973.

# On a Parametrization of Skew-Symmetrically Perturbed Gaussian Diffusions

AUGUSTO Y. HERMOSILLA
*Department of Mathematics, University of the Philippines*
*Diliman, Quezon City 1101, Philippines*
*Email: auggie@math01.cs.upd.edu.ph*

**Abstract**

In this paper, some of the important results of Hwang, Hwang-Ma, and Sheu [HHS, 1993] are modified, alternatively restated and proved, regardless of how $C$ is skew-symmetrically constructed (see [H3,5]). Specifically, parametrizing the *perturbing drift* $c(x) = Cx$ into $c_\alpha(x) = \alpha Cx$ gives the perturbed Gaussian drift

$$b_\alpha(x) = -\nabla U(x) + \alpha c(x) = Dx + \alpha Cx = (D + \alpha C)x = B_\alpha x .$$

For an underlying Gaussian diffusion, by the stability of the spectrum of matrices consisting only of eigenvalues, the convergences of the mean and covariance reduces to the study of the stable spectrum of the matrices involved in the approximating Ornstein-Uhlenbeck process.

## 1 On General Diffusions

Consider the random variable $x : \Omega \longrightarrow \mathbb{R}^n$ with $\pi : \mathbb{R}^n \longrightarrow [0,1]$ as its fixed *probability density* given by:

$$\pi(x) = \frac{1}{Z}\exp(-U(x)) \tag{1.1}$$

with *norming constant* $Z$, *potential function* $U(x)$ (usually given with nice and desirable properties; e.g., differentiable or compactly supported), mean vector $\mu_x = E\{x\} = \int_{\mathbb{R}^n} x\pi(dx)$ , and the symmetric positive definite covariance matrix

$$\sigma_x = \text{Cov}\{x\} = E\{ (x-\mu_x)(x-\mu_x)^T \} .$$

In practice, sampling from the underlying distribution given by (1.1) is prohibitively expensive or infeasible. To deal with this problem, (1.1) can be simply approximated instead by sampling from a homogeneous diffusion process $\{ \mathbf{X}(t) \}_{t\geq 0}$ described by the *stochastic differential equation* (to be denoted by $SDE$)[1]:

$$d\mathbf{X}(t) = -\nabla U(\mathbf{X}(t))\,dt + \sqrt{2}\,d\mathbf{W}(t) , \quad t>0 , \quad \mathbf{X}(0) = x_0 . \tag{1.2}$$

[1]Here, $\{ \mathbf{W}(t) \}_{t\geq 0}$ is the *standard Brownian Motion* (BM) in $\mathbb{R}^n$ with mean $E\{ \mathbf{W}(t) \} = \mathbf{0}$ and covariance $\text{Cov}\{ \mathbf{W}(t) \} = 2I_n$ .

Let $T_t = e^{tL}$ be the corresponding semigroup with infinitesimal generator $L$ of the diffusion process described by $SDE(1.2)$.

With only the potential function $U(x)$ known to have some desirable properties, the *gradient drift* $-\nabla U(x)$ is usually the *local* ( *global,* for convex programs) *optimal* direction in optimization[2]. However, the gradient drift will now be affected by the random perturbation involving the $BM$ here, making it no longer the best direction[3] for optimization. Thus, to improve from the starting gradient drift $-\nabla U(x)$ direction in accelerating the convergence towards the underlying density $\pi(x)$ at equilibrium, [HHS] considered the following approach to this problem by:

1. studying a *family of approximating diffusions modelled by a* $SDE$ with $\pi(x)$ still remaining as its underlying distribution at equilibrium ; and
2. constructing new *perturbed drifts* $b(x)$ from the *unperturbed* $-\nabla U(x)$ by adding a *perturbing drift* $c(x)$ , which will still be desirable, say, with smoothness and/or with compact supports, such that the corresponding approximating diffusions would accelerate convergence to $\pi(x)$ at equilibrium.

To their objective, [HHS] used the $SDE$

$$d\mathbf{X}(t) = b(\mathbf{X}(t))\,dt + \sqrt{2}\,d\mathbf{W}(t), \quad t > 0\,, \quad \mathbf{X}(0) = x_0\,, \tag{1.3}$$

where $b(\mathbf{X}(t))$ should be an improvement from the starting drift $-\nabla U(\mathbf{X}(t))$, and $\sigma(\mathbf{X}(t)) = \sqrt{2}I_n$ . The coefficient $a(\mathbf{X}(t)) = \sigma\sigma^T(\mathbf{X}(t)) = (\sqrt{2}I_n)(\sqrt{2}I_n) = 2I_n$ is nicely regularly conditioned (e.g., Lipschitz continuous). In particular, $SDE(1.2)$ is a specific case of $SDE(1.3)$ with $b(\mathbf{X}(t)) = -\nabla U(\mathbf{X}(t))$ .

With appropriate regularity condition on the drift vector $b(x)$ (e.g., Lipschitz continuous), stochastic integration applied to the $SDE(1.3)$ yields a general solution

$$\mathbf{X}(t) = x_0 + \int_0^t b(\mathbf{X}(t))\,dt + \sqrt{2}\int_0^t d\mathbf{W}(t)$$

with corresponding approximating probability density of the form

$$p(\mathbf{X}(t)) = \frac{1}{R}\exp\left\{ -\frac{1}{2}\left\langle \sigma_X^{-1}\,(\,\mathbf{X}(t) - \mu_X\,)\,,\ \mathbf{X}(t) - \mu_X \right\rangle \right\},$$

mean vector $\mu_X = E\{\,\mathbf{X}(t)\,\} = \int_{\mathbb{R}^n} \mathbf{X}(t)dp(\mathbf{X}(t))$ , covariance matrix

$$\sigma_X = \mathbf{Cov}\{\,\mathbf{X}(t)\,\} = E\left\{\,(\,\mathbf{X}(t) - \mu_X\,)(\,\mathbf{X}(t) - \mu_X\,)^T\,\right\},$$

[2]This gives the physical-geometric direction towards the *optimum* of $U(x)$ (global optimum, when desirable properties such as *convexity or differentiability* are present).

[3]For example, in simulated annealing (also in quantum and statistical mechanics – see [BF], [CHS], [GH], and the references therein), the gradient drift may no longer give the fastest convergence of an approximating diffusion process towards the underlying distribution at equilibrium (when time will become infinitely large), especially if chaotic instabilities are possible in the process.

and norming constant $R = [\,(2\pi)^n \det(\sigma_X)\,]^{1/2}$ . For this process, let $T_t^b = e^{tL_b}$ be the semigroup with infinitesimal generator $L_b$ given by [HHS] as

$$L_b\, f = \frac{1}{2}\sum_{i=1}^{n} 2\frac{\partial^2 f}{\partial x_i^2} + \sum_{i=1}^{n} b_i(x)\frac{\partial f}{\partial x_i} = \Delta f + \langle\, b(x)\,,\, \nabla f\,\rangle \tag{1.4}$$

and formal adjoint given by

$$L_b^*\, f = \Delta f - \operatorname{div}(fb)\,. \tag{1.5}$$

In particular, for $SDE(1.2)$ with $b(x) = -\nabla U(x)$ , [HHS] denoted its formal *unperturbed infinitesimal generator* operator by

$$L = \Delta + \langle\, -\nabla U(x)\,,\, \nabla\,\rangle\,. \tag{1.6}$$

This paper's results are built upon the following:

**1.1 Theorem** (Varadhan(1980)). *If* $L_b^* \exp(-U(x)) = 0$ *(and there is no explosion, i.e. the stopping time of the process at* $\partial$ *is not in finite time), then* $\pi(x) = \frac{1}{Z}\exp(-U(x))$ *is the equilibrium distribution of* $SDE(1.3)$ *. Conversely, if* $\pi(x)$ *is the equilibrium distribution and the coefficients are smooth enough, then* $L_b^* \exp(-U(x)) = 0$.

**1.2 Proposition** (Proposition 2.1 of [HHS]). *For* $SDE(1.3)$ *with smooth drift* $b(x)$, *if the diffusion is reversible, then* $b(x) = -\nabla U(x)$.

**1.3 Proposition** (Proposition 2.2 of [HHS]). $L_b^* \exp(-U(x)) = 0$ **if and only if** *the drift can be written as* $b(x) = -\nabla U(x) + (\exp(U(x))\mathbf{g}(x)$ , *where* $\mathbf{g}(x) = [\, g_1(x), \ldots, g_n(x)\,]^T$ , $\operatorname{div}\mathbf{g} = 0$ , *and there exist smooth functions* $f_{ij}\,, 1 \le i < j \le n$, *such that* $g_i = (-1)^{i-1}\left\{ \sum_{j<i}(-1)^{j-1}\frac{\partial f_{ij}}{\partial x_j} + \sum_{i<j}(-1)^{j-2}\frac{\partial f_{ij}}{\partial x_j} \right\}$ .

**1.4 Corollary** (Corollary 2.1 of [HHS]). *Let* $c(x)$ *be chosen such that* $\operatorname{div} c = 0$ and $\langle\, c\,,\, \nabla U\,\rangle = 0$ , $\forall\ x \in \mathbb{R}^n$. *Then* $b(x) = -\nabla U(x) + c(x)$ *satisfies* $L_b^* \exp(-U(x)) = 0$ .

Further extending these results are the following:

**1.5 Theorem** (Theorem 5.1 of [H1]; Theorem 2.2.8 of [H3] ).
*Let* $c(x) = e^U\mathbf{g}(x)$ *satisfy Proposition* 2.1.3. *Then*

$$L_b^*\, e^{-U} = -\operatorname{div}\mathbf{g} = -\, e^{-U}[\,\operatorname{div} c + \langle\, c\,,\, -\nabla U\,\rangle\,] = 0\,, \quad \forall\ x \in \mathbb{R}^n\,.$$

**1.6 Theorem** (Theorem 5.2 of [H1]; Theorem 2.2.9 of [H3] ).
$L_b^*\, e^{-U} = 0 \iff$ *we can write the perturbed drift as:* $b(x) = -\nabla U(x) + c(x)$ , *for some perturbing drift* $c(x)$ *satisfying:* $\operatorname{div} c + \langle\, c\,,\, -\nabla U\,\rangle = 0\,, \ \forall\, x \in \mathbb{R}^n$ .

We define the *kernels* of $L$ and $L_b^*$ respectively as follows:

$$\begin{aligned} \ker(L) &= \{\, f \in \operatorname{dom}(L)\ :\ Lf = 0\,\} \\ \ker(L_b^*) &= \{\, f \in \operatorname{dom}(L_b^*)\ :\ L_b^* f = 0\,\}\,. \end{aligned}$$

Consequently, a generalized construction of the perturbing drift using functionals from the kernels of the infinitesimal generators can give an alternative perspective to [HHS]'s construction $c(x) = e^U \mathbf{g}(x)$ given in Proposition 1.3 here, as stated in:

**1.7 Theorem** (Theorem 2.4.1 of [H3] ). *Let* $c(x) = \nabla\phi(x)$ *, for some* $\phi \in \mathrm{dom}(L)$*. Then* $\phi \in \ker(L) \iff e^{-U} \in \ker(L_b^*)$ *. Consequently, density* (1.1) *is the equilibrium distribution of* $SDE(1.3)$*.*

These results help in deriving the important results particularly associated with Gaussian diffusions.

# 2 On Gaussian Diffusions

Let $x \in \mathbb{R}^n$ be an underlying *Gaussian random process* with pdf given by:

$$\pi(x) = \frac{1}{Z}\exp(-U(x)) = \frac{1}{Z}\exp\left( -\frac{1}{2}\langle -Dx, x\rangle \right) \tag{2.1}$$

*mean vector* $\mu_x = \mathbf{0}$ , and *covariance matrix* $\sigma_x = -D^{-1}$ , where $D$ is a *negative definite and symmetric* matrix. Approximating by sampling from a *diffusion* modeled by the *Ornstein-Uhlenbeck (O-U) process* is given by the *SDE* involving a *stability matrix* $B$ as follows:

$$d\mathbf{X}(t) = B\mathbf{X}(t)dt + \sqrt{2}\,d\mathbf{W}(t), \quad t > 0, \quad \mathbf{X}(0) = x_0 \ . \tag{2.2}$$

The relationship[4] between $B$ and $D$ is revealed by:

**2.1 Theorem** (Theorem 3.1 of [HHS]). *The Ornstein-Uhlenbeck process described by* $SDE(2.2)$ *has a Gaussian underlying equilibrium with density* (2.1) *if and only if* $B = D + C$*, where* $C$ *satisfies* $\langle\, Cx \,,\, Dx \,\rangle = 0$ *, for all* $x \in \mathbb{R}^n$*. In fact,* $C = SD$*, where* $S$ *is any skew-symmetric real matrix.*

So, we can now explicitly consider from (2.2) the *SDE*

$$d\mathbf{X}(t) = (D + C)\mathbf{X}(t)dt + \sqrt{2}\,d\mathbf{W}(t), \quad t > 0, \ \mathbf{X}(0) = x_0$$

as a perturbation of the *SDE*

$$d\mathbf{X}(t) = D\mathbf{X}(t)dt + \sqrt{2}\,d\mathbf{W}(t), \quad t > 0, \quad \mathbf{X}(0) = x_0 \ ,$$

where $-\nabla U(x) = Dx$ is the *steepest descent direction* in minimizing the *Gaussian potential* $U(x) = \frac{1}{2}\langle -Dx, x\rangle$ . Among the negative real parts of the crucial eigenvalues of $B$ , we denote the **largest in magnitude** by $\lambda_M(B)$ , while the **least in magnitude**

[4]Essentially, by the results of Varadhan ([V2]) and [HHS] applied to the Gaussian case, the starting drift $-\nabla U(x) = Dx$ is being added by the perturbing drift $c(x) = Cx$ to yield the perturbed drift

$$b(x) = -\nabla U(x) + c(x) = Dx + Cx = (D + C)x = Bx \ .$$

That is, we simply define $B = D + C$ , which should remain as stability matrix.

by $\lambda_m(B)$. Here, only $U(x)$ is known, since $D$ is known to be a symmetric and strictly negative definite matrix ( so, all its eigenvalues are negative).

Using mainly matrix analysis (see [HJ1, 2]) with methods different from those used by **[HHS]**, specifically using here only the *Jordan Canonical Form Decompositions* (JCFD) and spectral mapping, gives the following:

**2.2 Theorem.** (Theorem 3.4.1 of [H3]; Theorem 4.1 of [H7])
$E\{\mathbf{X}(t)\} = e^{tB}x_0 \overset{t\to\infty}{\longrightarrow} \mathbf{0}$ *at the rate of the spectral radius* $e^{t\lambda_M(B)} \overset{t\to\infty}{\longrightarrow} 0$ of $e^{tB}$.

**2.3 Theorem.** (Theorem 3.4.3 of [H3]; Theorem 4.2 of [H7])
$\mathbf{Cov}\{\ \mathbf{X}(t)\ \} \overset{t\to\infty}{\longrightarrow} -D^{-1} \iff 2\int_t^\infty \exp(sB)\exp(sB^T)ds \overset{t\to\infty}{\longrightarrow} \mathbf{O}_n$
*at the rate of the spectral radius* $e^{2t\lambda_M(B)} \overset{t\to\infty}{\longrightarrow} 0$ of the integrand $e^{tB}e^{tB^T}$ .

Some important results from [HHS] to be generalized in their parametrized versions in the main results in this paper are as follows:

A comparison among the eigenvalues of $B$ and $D$ is given in

**2.4 Theorem** (Theorem 3.3 of [HHS]). *Let* $B = D + C$ *with* $\langle\ Cx\ ,\ Dx\ \rangle = 0\ ,\ \forall\, x \in \mathbb{R}^n$ . *Then*

$$\lambda_m(D) \le \lambda_m(B) \le \lambda_M(B) \le \lambda_M(D)\ ,$$

*and* $\lambda_M(B) = \lambda_M(D)$ *if and only if there exist nonzero* $u, v \in \mathbb{R}^n$ , *and* $\rho \in \mathbb{R}$ *such that* $u + iv$ *is an eigenvector of* $B$ *with eigenvalue* $\lambda_M(D) + i\rho$, *and* $u, v$ *are eigenvectors of* $D$ *with eigenvalue* $\lambda_M(D)$. *Moreover,* $u + iv$ *is an eigenvector of* $C$ *with eigenvalue* $i\rho$ , *and* $Cu = -\rho v$ , $Cv = \rho u$. *If* $\rho \neq 0$ , *then* $u \perp v$.

**2.5 Theorem** (Theorem 3.4 of [HHS]). $\exists\, C \in M_n(\mathbb{R})$ *such that* $\langle Cx, Dx\rangle = 0\ ,\ \forall x \in \mathbb{R}^n$ *; and* $\lambda_M(D + C) < \lambda_M(D) \iff$ *the eigenvalues of* $D$ *are not all identical.*

These theorems show the improvement of the acceleration of convergence via the perturbation of the diffusion approximation.

# 3 Skew-Symmetric Perturbations of Gaussian Diffusions

Alternatively constructed in [H7] to [HHS]'s skew-symmetric construction $C = SD$ for the perturbing matrix $C$ is basically by applying the criterion

$$\operatorname{div} c + \langle\ c\ ,\ -\nabla U\ \rangle = 0\ ,\quad \forall\, x \in \mathbb{R}^n\ ,$$

for desirable perturbing drifts as given in Theorem 2.6 here but now applied to the Gaussian case. This leads to the very important results:

**3.1 Theorem** (Theorem 2.1 of [H7]; Theorem 3.7.1 of [H3]).
*The following are equivalent:*

i.) $\operatorname{tr} C + \langle\ Cx\ ,\ Dx\ \rangle = 0\ ,\quad \forall\, x \in \mathbb{R}^n\ ,\quad \forall\, \alpha \in \mathbb{R}$

ii.) $\operatorname{tr} C = 0$ *and* $\langle\ Cx\ ,\ Dx\ \rangle = 0\ ,\quad \forall\, x \in \mathbb{R}^n$ .

The following constructions also include [HHS]'s skew-symmetric construction $C = SD$ as a special case:

**3.2 Theorem** (Theorem 2.2 of [H7]; Theorem 3.7.2 of [H3]). *The constructions* $K = C^T D$ *and* $K^T = DC = DC$ *are skew-symmetric matrices satisfying the results of Theorem 3.1. Equivalently,* $C = D^{-1}K^T$.

It can be easily shown that $\mathcal{S} = \left\{ S \in M_n \;\; S^T = -S \right\}$ is a subspace of $M_n$, the vector space of $n \times n$ matrices. Defining the family of constructions:

$$\begin{aligned} \mathcal{C} &= \{ C \in M_n : \text{tr } C + \langle Cx , Dx \rangle = 0 , \;\; \forall x \in \mathbb{R}^n \} \\ &= \left\{ C \in M_n : C^T D \text{ is skew} - \text{symmetric} \right\} \\ \mathcal{C}_D &= \{ C = SD : S \in \mathcal{S} \} \qquad \text{and} \\ \mathcal{C}_{D^{-1}} &= \left\{ C = D^{-1} S : S \in \mathcal{S} \right\} , \end{aligned}$$

[HHS]'s and our alternative skew-symmetric constructions yield the very interesting

**3.3 Theorem.** (Theorem 3.1 of [H7]; Theorem 3.7.6 of [H3])
$\mathcal{S}$ , $\mathcal{C}_D$ , $\mathcal{C}_{D^{-1}}$ , *and* $\mathcal{C}$ *are isomorphic subspaces of* $M_n$ .

# 4 Main Results

The main objective in the parametrization of the construction of the perturbing drift $c(x) = Cx$ to satisfy the alternatives in the previous sections is on the preservation of the stability matrix $B$ after the perturbation of $D$ by the parametrized perturbing matrix $\alpha C$ , where $\alpha$ is a scaling parameter that can be made sufficiently large. Even if $\alpha \to \pm\infty$ , interesting results can be obtained by the simple linearity of the operators given by the associated matrices in this study.

For the generalization in the parametrized context, and by the linearity of the div, tr, and $\langle \cdot , \cdot \rangle$ operators, the requirement of $C^T D$ being skew-symmetric, and equivalent to $c_\alpha(x) = \alpha C x$ being divergence-free (i.e., $\text{tr } \alpha C = 0$ ), and orthogonality relative to $-\nabla U(x) = Dx$ (i.e. $\langle \alpha Cx , Dx \rangle = 0 , \; \forall x \in \mathbb{R}^n$ ) are invariant[5] under the parametrization of the perturbing drift into $c_\alpha(x) = \alpha Cx$ .

Actually, without focusing on the divergence-free and orthogonality requirements, the crucial equation

$$\text{tr } c_\alpha + \langle c_\alpha , Dx \rangle = 0 , \quad \forall x \in \mathbb{R}^n ,$$

[5]These are given by:

$$\langle c_\alpha(x) , Dx \rangle = \langle \alpha Cx , Dx \rangle = \alpha \underbrace{\langle Cx , Dx \rangle}_{=0} = \alpha \cdot 0 = 0 , \quad \text{and}$$

$$\text{tr}(c_\alpha(x)) = \text{tr}(\alpha Cx) = \alpha \underbrace{\text{tr}(Cx)}_{=0} = \alpha \cdot 0 = 0 .$$

is also invariant under the parametrization, since

$$\begin{aligned}\text{tr } c_\alpha + \langle c_\alpha , Dx \rangle &= \text{tr}(\alpha Cx) + \langle \alpha Cx , Dx \rangle \\ &= \alpha \text{tr } C + \alpha \langle Cx , Dx \rangle \\ &= \alpha \underbrace{[ \text{ tr } C + \langle Cx , Dx \rangle ]}_{=0} = \alpha \cdot 0 = 0 .\end{aligned}$$

The significant condition to be imposed to get all the results here is the following: for all values of the parameter $\alpha$, even as $\alpha \longrightarrow \pm \infty$,

$$\text{tr } C_\alpha = \alpha \text{tr } C = \alpha \cdot 0 = 0 .$$

Basically, taking the *parametrically perturbed Ornstein-Uhlenbeck process*

$$d\mathbf{X}_\alpha(t) = (D + \alpha C)\mathbf{X}_\alpha(t)dt + \sqrt{2}\, d\mathbf{W}(t), \quad t > 0, \ \mathbf{X}_\alpha(0) = x_0 \tag{4.1}$$

as a parametrization of the perturbed $SDE$[6]

$$d\mathbf{X}(t) = (D + C)\mathbf{X}(t)dt + \sqrt{2}\, d\mathbf{W}(t), \quad t > 0, \ \mathbf{X}(0) = x_0 . \tag{4.2}$$

gives the parametrized version of Varadhan's and [HHS]'s results as motivated by Theorems 1.1 and 1.7 here. This is by considering first the functional construction to get:

$$\phi_\alpha(x) = \frac{1}{2} \langle \alpha Cx , x \rangle + k \ \in \ker(L) \quad \text{and} \quad k \in \mathbb{R} \qquad \text{such that}$$

$$\begin{aligned} c_\alpha(x) &= \nabla \phi_\alpha(x) = \alpha Cx , \quad \forall\, x \in \mathbb{R}^n , \qquad \text{and} \\ \Delta \phi_\alpha(x) &= \text{div}(\nabla \phi_\alpha(x)) = \text{div}(\alpha Cx) = \text{tr } \alpha C = \alpha \text{tr } C .\end{aligned}$$

Hence, letting $-\nabla U(x) = Dx \iff U(x) = \frac{1}{2} \langle -Dx , x \rangle$ alternatively restates Theorem 2.1 in

**4.1 Theorem.** *The parametrically perturbed Ornstein-Uhlenbeck process given in* $SDE(4.1)$ *has a Gaussian equilibrium with density (2.1) if and only if* $B_\alpha = D + \alpha C$ *such that* $C^T D$ *is skew-symmetric.*

**Proof:**

( $\Longrightarrow$:) By converse of Theorem 1.1, if (2.1) is the equilibrium distribution of $SDE(4.1)$ with the functional construction given by Theorem 1.7 here, then

$$L^*_{b_\alpha} e^{-U(x)} = 0 \iff L\phi_\alpha = 0$$

$$\iff 0 = L\phi_\alpha = \Delta \phi_\alpha + \langle \nabla \phi_\alpha , \nabla U \rangle = \text{tr } \alpha C + \langle \alpha Cx , Dx \rangle , \ \forall\, x \in \mathbb{R}^n$$

$$\begin{aligned} &\iff b_\alpha(x) = -\nabla U(x) + c_\alpha(x) = -\nabla U(x) + \nabla \phi_\alpha(x) = \nabla(- U + \phi_\alpha)(x) \\ &\iff b_\alpha(x) = Dx + \alpha Cx = (D + \alpha C)x = B_\alpha x \\ &\iff B_\alpha = D + \alpha C . \end{aligned}$$

[6]This is as a special case when $\alpha = 1$ as given in all earlier results of the related papers by this author, the parametrized perturbing $\alpha C = \alpha D^{-1} K$ with $K$ being any skew-symmetric matrix (including [HHS]'s $C = SD$ , where $S$ is also skew-symmetric).

So, by Theorem 3.2 here, $\alpha C^T D$ is skew-symmetric. By linearity, we equivalently get $C^T D$ being skew-symmetric.

($\Longleftarrow$:) Let $B_\alpha = D + \alpha C$ such that $C^T D$ is skew-symmetric. By Theorem 3.2 applied to the functional construction above,

$$0 = \text{tr}\, \alpha C + \langle \alpha Cx , Dx \rangle = L\phi_\alpha , \quad \forall\, x \in \mathbb{R}^n ,$$

gives the perturbed drift

$$b_\alpha(x) = -\nabla U(x) + c_\alpha(x) = Dx + \alpha Cx = (D + \alpha C)x = B_\alpha x .$$

By Theorem 1.7 ,

$$L^*_{B_\alpha x}\, e^{-\frac{1}{2}\langle -Dx , x \rangle} = L^*_{b_\alpha}\, e^{-U} = -e^{-U} L\phi_\alpha = 0 , \quad \forall\, x \in \mathbb{R}^n .$$

Thus, by Theorem 1.1, (2.1) is the equilibrium distribution of the $SDE(4.1)$ . ■

By linearity of all operators and matrices involved in the crucial conditions, the parametrized versions of the earlier results useful to prove our main results in this paper uses the following lemmas:

**4.2 Lemma** (Theorem 4.3.2.2 of [H3]). *Let $B_\alpha = D + \alpha C$ with eigenvalue $\lambda_\alpha + i\mu$ and corresponding eigenvector $x + iy$. Let $\lambda_1, \ldots, \lambda_n$ be the negative eigenvalues (not necessarily distinct) of $D$ with $\{ e_k : De_k = \lambda_k e_k \}_{k=1}^n$ be a CONBE of $\mathbb{R}^n$. Then, for at least one $k$ , $\lambda_\alpha = \lambda_k + \alpha \langle e_k , Ce_k \rangle$ .*

**4.3 Lemma** (Theorem 4.3.2.3 of [H3]). *Let $B_\alpha = D + \alpha C$ and $D$ have the eigenvalue-eigenvector system given in the assumptions of Lemma 4.2. Suppose that*

$$\text{tr}\, C = - \langle Cz , Dz \rangle \quad , \quad \forall\, z \neq \mathbf{0} .$$

*Then* $\lambda_m(D) - \dfrac{\alpha \text{tr}\, C}{\lambda_m(D)} \leq \lambda(B_\alpha) \leq \lambda_M(B_\alpha) \leq \lambda_M(D) - \dfrac{\alpha \text{tr}\, C}{\lambda_m(D)}$ ; *and* $\lambda_M(B_\alpha) = \lambda_M(D) - \dfrac{\alpha \text{tr}\, C}{\lambda_m(D)}$ *if and only if there exist nonzero* $u, v \in \mathbb{R}^n$ , *and* $\rho \in \mathbb{R}$ *such that* $u + iv$ *is an eigenvector of* $B_\alpha$ *with eigenvalue* $\lambda_M(B_\alpha) + i\rho$ .
*Moreover, the following constructions are possible:*

1. *$u, v$ are eigenvectors of $D$ with eigenvalue $\lambda_M(D)$ , and*
   *$u + iv$ is an eigenvector of $\alpha C$ with eigenvalue* $-\dfrac{\alpha \text{tr}\, C}{\lambda_m(D)} + i\rho$ *such that*

$$\alpha C u = -\frac{\alpha \text{tr}\, C}{\lambda_m(D)} u - \rho v \qquad \textit{and} \qquad \alpha C v = -\frac{\alpha \text{tr}\, C}{\lambda_m(D)} v + \rho u .$$

   *If $\rho \neq 0$ , then*

$$\langle u , v \rangle = \frac{\alpha \text{tr}\, C}{\rho \lambda_m(D)} \left[ \frac{\lambda_m(D)}{\lambda_M(D)} - \|u\|^2 \right] = -\frac{\alpha \text{tr}\, C}{\rho \lambda_m(D)} \left[ \frac{\lambda_m(D)}{\lambda_M(D)} - \|v\|^2 \right] .$$

2. *For* $\mathrm{tr}\ C \le 0$ *and* $\alpha > 0$ , $\lambda_M(B_\alpha) = \lambda_M(D) - \dfrac{\alpha \mathrm{tr}\ C}{\lambda_m(D)} \le \lambda_M(D)$.

   *For* $\mathrm{tr}\ C \le 0$ *and* $\alpha < 0$ , $\lambda_M(B_\alpha) = \lambda_M(D) - \dfrac{\alpha \mathrm{tr}\ C}{\lambda_m(D)} \ge \lambda_M(D)$.

   *For* $\mathrm{tr}\ C \ge 0$ *and* $\alpha > 0$ , $\lambda_M(B_\alpha) = \lambda_M(D) - \dfrac{\alpha \mathrm{tr}\ C}{\lambda_m(D)} \ge \lambda_M(D)$.

   *For* $\mathrm{tr}\ C \ge 0$ *and* $\alpha < 0$ , $\lambda_M(B_\alpha) = \lambda_M(D) - \dfrac{\alpha \mathrm{tr}\ C}{\lambda_m(D)} \le \lambda_M(D)$.

   *And,* $u, v$ *are eigenvectors of* $D$ *with eigenvalue* $\lambda_M(B_\alpha)$ , *and* $u + iv$ *is an eigenvector of* $\alpha C$ *with eigenvalue* $i\rho$ *such that*

$$\alpha C u = -\rho v \qquad \text{and} \qquad \alpha C v = \rho u \ .$$

   *If* $\rho \neq 0$, *then* $u \perp v$ .

**4.4 Lemma** (Theorem 4.3.2.4 of [H3]). *Let* $B_\alpha = D + \alpha C$ *and* $D$ *have the eigenvalue-eigenvector system given in the assumptions of Lemma 4.2. Suppose that*

$$\mathrm{tr}\ C = -\langle\ Cz\ ,\ Dz\ \rangle\ , \quad \forall\ z \neq \mathbf{0}\ .$$

*Then* $\lambda_M(D + \alpha C) < \lambda_M(D) - \dfrac{\alpha \mathrm{tr}\ C}{\lambda_m(D)} \iff$ *the eigenvalues of* $D$ *are not identical.*

Basically, in this paper, $C$ will just be replaced here by the parametrized $C_\alpha = \alpha C$, and then just impose the condition $\mathrm{tr}\ C_\alpha = 0$ on Lemmas 4.2, 4.3, and 4.4 to yield

**4.5 Theorem.** *Let* $B_\alpha = D + \alpha C$ *such that* $\alpha C^T D$ *is skew-symmetric. Then* $\lambda_m(D) \le \lambda(B_\alpha) \le \lambda_M(B_\alpha) \le \lambda_M(D)$ ; *and* $\lambda_M(B_\alpha) = \lambda_M(D)$ *if and only if there exist nonzero* $u, v \in \mathbb{R}^n$ , *and* $\rho \in \mathbb{R}$ *such that* $u + iv$ *is an eigenvector of* $B_\alpha$ *with eigenvalue* $\lambda_M(D) + i\rho$ , *and* $u, v$ *are eigenvectors of* $D$ *with eigenvalue* $\lambda_M(D)$. *Moreover,* $u + iv$ *is an eigenvector of* $\alpha C$ *with eigenvalue* $i\rho$ , *and* $\alpha C u = -\rho v$ , $\alpha C v = \rho u$. *If* $\rho$ *is nonzero, then* $u \perp v$.

**Proof:** Let $B_\alpha = D + \alpha C$ such that $C^T D$ is skew-symmetric . Hence, by Theorems 3.1 and 3.2, we get $\alpha \mathrm{tr}\ C = 0$ and $\langle\ \alpha C x\ ,\ D x\ \rangle = 0\ ,\ \forall\ x \in \mathbb{R}^n$ . But the assumptions here also satisfy all the assumptions in Lemma 4.2, as a rather extended case, particularly satisfying $\alpha \mathrm{tr}\ C = -\langle\ \alpha C z\ ,\ D z\ \rangle\ ,\ \forall\ z \neq 0$ . So, the results of Lemma 4.3 follow. But imposing $\alpha \mathrm{tr}\ C = 0$ implies

$$\lambda_m(D) \le \lambda_m(B_\alpha) \le \lambda_M(B_\alpha) \le \lambda_M(D)\ .$$

Futhermore, imposing $\alpha \mathrm{tr}\ C = 0$ on the other results of Lemma 4.3 yields:

$\lambda_M(B_\alpha) = \lambda_M(D) - \dfrac{\alpha \mathrm{tr}\ C}{\lambda_m(D)} = \lambda_M(D)$ if and only if there exist nonzero $u, v \in \mathbb{R}^n$, and $\rho \in \mathbb{R}$ such that $u + iv$ is an eigenvector of $B_\alpha$ with eigenvalue $\lambda_M(B_\alpha) + i\rho$ .

For $\alpha \mathrm{tr}\ C \le 0$ , $\lambda_M(B_\alpha) = \lambda_M(D) - \dfrac{\alpha \mathrm{tr}\ C}{\lambda_m(D)} \le \lambda_M(D)$ .

For $\alpha \mathrm{tr}\ C \ge 0$ , $\lambda_M(B_\alpha) = \lambda_M(D) - \dfrac{\alpha \mathrm{tr}\ C}{\lambda_m(D)} \ge \lambda_M(D)$ .

Thus, imposing $\alpha \text{tr } C = 0$ yields $\lambda_M(B_\alpha) = \lambda_M(D)$ .

Moreover, further imposing $\alpha \text{tr } C = 0$ , the following constructions are possible:

1. $u + iv$ is an eigenvector of $\alpha C$ with eigenvalue $-\dfrac{\alpha \text{tr } C}{\lambda_m(D)} + i\rho = i\rho$ , and $u, v$ are eigenvectors of $D$ with eigenvalue $\lambda_M(D)$ such that

$$\alpha C u = -\frac{\alpha \text{tr } C}{\lambda_m(D)} u - \rho v = -\rho v \quad \text{and} \quad \alpha C v = -\frac{\alpha \text{tr } C}{\lambda_m(D)} v + \rho u = \rho u .$$

   If $\rho \neq 0$ , then $u \perp v$ , since

$$\langle u , v \rangle = \frac{\alpha \text{tr } C}{\rho \lambda_m(D)} \left[ \frac{\lambda_m(D)}{\lambda_M(D)} - \|u\|^2 \right] = -\frac{\alpha \text{tr } C}{\rho \lambda_m(D)} \left[ \frac{\lambda_m(D)}{\lambda_M(D)} - \|v\|^2 \right] = 0 .$$

2. $u + iv$ is an eigenvector of $\alpha C$ with eigenvalue $i\rho$ , and $u, v$ are eigenvectors of $D$ with eigenvalue $\lambda_M(B_\alpha)$ such that

$$\alpha C u = -\rho v \qquad \text{and} \qquad \alpha C v = \rho u .$$

   If $\rho \neq 0$, then $u \perp v$ .

In other words, imposing $\alpha \text{tr } C = 0$ makes the two constructions above coincide from the results of Lemma 4.3 to give the following conclusions:

$$\lambda_m(D) \le \lambda_m(B_\alpha) \le \lambda_M(B_\alpha) \le \lambda_M(D) .$$

And $\lambda_M(B_\alpha) = \lambda_M(D)$ if and only if there exist nonzero $u, v \in \mathbb{R}^n$ , and $\rho \in \mathbb{R}$ such that $u + iv$ is an eigenvector of $B_\alpha$ with eigenvalue $\lambda_M(D) + i\rho$ , and $u, v$ are eigenvectors of $D$ with eigenvalue $\lambda_M(D)$. Moreover, $u + iv$ is an eigenvector of $\alpha C$ with eigenvalue $i\rho$ , and $\alpha C u = -\rho v$ , $\alpha C v = \rho u$. If $\rho \neq 0$ , then $u \perp v$. ■

Eigenvector-eigenvalue relations involved here come in conjugate pairs:
$D$ with eigenvectors $u$ and $v$ of eigenvalue $\lambda_M(B_\alpha) = \lambda_M(D) - \dfrac{\alpha \text{tr } C}{\lambda_m(D)} = \lambda_M(D)$,
$B_\alpha$ with eigenvector $u \pm iv$ of eigenvalue $\lambda_M(B_\alpha) \pm i\rho = \lambda_M(D) \pm i\rho$ ,
$\alpha C$ with eigenvector $u \pm iv$ of eigenvalue $-\dfrac{\alpha \text{tr } C}{\lambda_m(D)} \pm i\rho = \pm i\rho$ .
Here, $\alpha \text{tr } C$ is like a measure of orthogonality and divergence-free for the perturbing drift $c_\alpha(x) = \alpha C x$ . In fact, if $\theta$ is the angle between the vectors $\alpha C x$ and $Dx$ , then

$$-\alpha \text{tr } C = \langle \alpha C x , Dx \rangle = \|\alpha C x\| \, \|Dx\| \cos\theta ,$$

giving the relationship

$$\cos\theta = \frac{-\text{tr } C}{\|Cx\| \, \|Dx\|} \iff \theta = \text{Arccos}\left( \frac{-\alpha \text{tr } C}{\|\alpha C x\| \, \|Dx\|} \right) .$$

With $\alpha \mathrm{tr}\, C = -\langle \alpha Cx , Dx \rangle = 0$ , surely $\cos\theta = 0 \iff \theta = \frac{\pi}{2}$ , as expected of the constructions of $\alpha C$.

Similarly, imposing the requirement $\alpha \mathrm{tr}\, C = 0$ on Lemma 4.4 yields the alternative restatement of Theorem 2.4 in

**4.6 Theorem.** *Let* $B_\alpha = D + \alpha C$ *such that* $C^T D$ *is skew-symmetric.*
*Then* $\lambda_M(D + \alpha C) < \lambda_M(D) \iff$ *the eigenvalues of* $D$ *are not identical.*
**Proof:** Let $B_\alpha = D + \alpha C$ such that $\alpha C^T D$ is skew-symmetric . Again, by Theorems 3.1 and 3.2, we get $\alpha \mathrm{tr}\, C = 0$ and $\langle \alpha Cx , Dx \rangle = 0 , \ \forall\, x \in \mathbb{R}^n$ . But the assumptions here also satisfy all the assumptions in Lemma 4.4, as a rather extended case also, particularly satisfying

$$\alpha \mathrm{tr}\, C = -\langle \alpha Cz , Dz \rangle , \quad \forall\, z \neq 0 .$$

So, the results of Lemma 4.4 follow. Hence, imposing $\alpha \mathrm{tr}\, C = 0$ on the results of Lemma 4.4 yields the conclusion:

Then $\lambda_M(D + \alpha C) < \lambda_M(D) - \dfrac{\alpha \mathrm{tr}\, C}{\lambda_m(D)} = \lambda_M(D) \iff$ the eigenvalues of $D$ are not identical. ∎


## Acknowledgement

The author would like to thank his hospitable host **Dr. Chii-Ruey Hwang** of the *Institute of Mathematics, Academia Sinica, Taipei, Taiwan*, for the basic motivations of this study and the use of their excellent library facilities.


## References


[CHS] Chiang, T.S., Hwang, C.R. & Sheu, S.J. (1987): *Diffusion for Global Optimization*, SIAM J. Control Optim., 25:735-753

[CZ] Chung, K.L. & Zhao, Z.X.(1982): *Lectures from Markov Processes to Brownian Motion*, Springer-Verlag, New York

[F] Friedrichs, K. (1965): *Perturbation of Spectra in Hilbert Space*, American Mathematical Society (AMS), Academic Press, New York

[GH] Geman, S. & Hwang, C.R. (1986): Diffusion for Global Optimization, *SIAM J. Control Optim.*, 24:1031-1043

[H1] Hermosilla, A.Y. (1997): On Approximation Using General Diffusions, *Matimyas Matematika*, Vol.#20, No. 2, pp. 26-36, May 1997

[H2] Hermosilla, A.Y. (1997): On the Convergence Rate of the Covariance of Some Special Ornstein-Uhlenbeck Processes, *Proc. 1997 Symposium on Mathematical Aspects of Computer Science*, U.P. College Baguio, pp. 77-84, May 28 - June 1, 1997

[H3] Hermosilla, A.Y. (1998): On the Approximation and Acceleration of Diffusion Processes, *Unpublished Ph.D. Dissertation*, University of the Philippines-Diliman, January 1998

[H4] Hermosilla, A.Y. (1998): Non-Skewsymmetric Perturbation of General Diffusions, *Proc. Int.'l Conf. on Inverse Problems and Applications*, Univ. of the Philippines-Diliman, pp. 154-165, February 23-27, 1998

[H5] Hermosilla, A.Y. (1999): Convergence of Skew-Symmetrically Perturbed Gaussian Diffusions, *Proc. Int.'l Mathematics Conf. (in honorof Prof. Lee Peng Yee on his* $60^{th}$ *Birthday and Applications*, Ateneo de Manila, pp. 47-65, December 16-18, 1999

[H6] Hermosilla, A.Y. (1999): Perturbing Diffusions Using the Kernel of its Infinitesimal Generator, *The Philippine Statistician, 1999*, Vol.#48, No. 1-4, pp. 49-64,

[H7] Hermosilla, A.Y. (2000): Skew-Symmetric Gaussian Diffusions, *Proc.* $3^{rd}$ *Sino-Philippine Symposium in Analysis*, Univ. of the Philippines-Diliman, pp. 47-59, March 21-25, 2000

[HJ1] Horn, R.A. & Johnson, C.R. (1990): *Matrix Analysis*, Cambridge University Press, Cambridge

[HJ2] Horn, R.A. & Johnson, C.R. (1991): *Topics in Matrix Analysis*, Cambridge University Press, Cambridge

[HHS] Hwang, C.R., Hwang-Ma, S.Y., & Sheu, S.J. (1993): Accelerating Gaussian Diffusion, *The Annals of Applied Probability*, Vol.#3, No. 3, pp. 897-913

[HS] Hwang, C.R. & Sheu, S.J. (1990): Large Time Behavior of Perturbed Diffusion Markov Processes with Applications to the Second Eigenvalue Problem of Fokker-Planck Operators and Simulated Annealing, *Acta Applicande Mathematicae*, 19:253-295, Kluwer Academic Publishers, Dordrecht, Netherlands

[K] Kato, T. (1966): *Perturbation of Linear Operators*, Springer-Verlag, New York

[SV1] Stroock, D. & Varadhan, S.R.S. (1969): *Multidimensional Diffusion Processes*, Springer-Verlag, New York

[SV2] Stroock, D. & Varadhan, S.R.S. (1982): *Lectures on Topics in Stochastic Differential Equations*, Springer-Verlag, New York

[T] Taira, K. (1988): *Diffusion Processes and Partial Differential Equations*, Academic Press, San Diego

[V1] Varadhan, S.R.S. (1968) : *Stochastic Processes*, Courant Institute of Mathematical Sciences, New York

[V2] Varadhan, S.R.S. (1980) : Lectures on Diffusion Processes and Partial Differential Equations, *Tata Institute Lecture Series*, Springer-Verlag, New York

# ON NON-AMBIGUOUS BÜCHI $V$- AUTOMATA

PHAN TRUNG HUY

*Departement of Applied Mathematics, Hanoi Institute of Technology*
*E-mail: phanhuy@hn.vnn.vn*

DO LONG VAN

*Hanoi Institute of Mathematics, P.O.Box 631,Boho,10000 Hanoi,Vietnam*
*E-mail: dlvan@thevinh.ncst.ac.vn*

The notion of V-automata on infinite words, where V is a variety of finite monoids, was introduced and considered in an earlier paper of the authors [VH]. In the present paper we show that, if V is closed under Schützenberger product, then for every Büchi V-automaton there exists an equivalent non-ambiguous Büchi V-automaton in the sense that they both recognize the same $\omega$-language. When V is the M-variety of all finite monoids this is nothing but the well-known result due to A. Arnold [Ar] saying that every regular $\omega$-language can be recognized by a non-ambiguous Büchi automaton.

## 1 Preliminaries

Let us recall some notions and results necessary in the sequel. Throughout the paper all alphabets and all monoids, except for free monoids, are supposed to be finite, and for morphisms we mean monoid morphisms. A monoid $N$ is a *divisor* of a monoid $M$, denoted by $N \prec M$, if $N$ is a homomorphic image of a submonoid of $M$. A family $\underline{V}$ of finite monoids is called an *M-variety* if, for any monoids $P, Q, R$, the conditions $P \prec Q \times R$ & and $Q, R \in \underline{V}$ imply $P \in \underline{V}$. If $\underline{V}$ contains the monoid $U_1 = \{0,1\}$ with $01 = 10 = 00 = 0$ and $11 = 1$, we call $\underline{V}$ a *$U_1$-variety.* Let $A$ be an alphabet. As usual we denote by $A^*$ the free monoid generated by $A$ whose elements are called *finite words.* The *empty word* is denoted by $\epsilon$ and $A^+ = A^* - \{\epsilon\}$. The set of all *infinite words* over $A$ is denoted by $A^N$. A subset of $A^*$ is called a *language* while a subset of $A^N$ is an *$\omega$-language.* For any language $X \subseteq A^*$, $X^*$ is the submonoid of $A^*$ generated by $X$, and $X^\omega$, $\vec{X}$ the $\omega$-languages

$X^\omega = \{w \in A^N | w = u_1u_2u_3... \textit{ with } u_i \in X\}$.

$\vec{X} = \{w \in A^N | w = u_1u_2u_3... \textit{ with } u_1u_2..u_i \in X\}$.

The families of regular languages and of regular $\omega$-languages on the alphabet $A$ are denoted by $ARec^*$ and $ARec^N$ respectively.

Let $\varphi : A^* \to M$ be a morphism from $A^*$ into a monoid $M$. Let $L$ be a language and $W$ an $\omega$-language on $A$. We say that $\varphi$ *saturates* $L$ if, for every $e \in M, \varphi^{-1}(e) \cap L \neq \emptyset$ implies $\varphi^{-1}(e) \subseteq L$. Similarly, $\varphi$ *saturates* the $\omega$-language $W$ if, for every $e, f \in M, \varphi^{-1}(e)[\varphi^{-1}(f)]^\omega \cap W \neq \emptyset$ implies $\varphi^{-1}(e)[\varphi^{-1}(f)]^\omega \subseteq W$. In that case we also say that $L$ (resp., $W$) is *saturated by the kernel congruence* $\underline{\varphi}$ induced by $\varphi$ or $L$ (resp., $W$) is *saturated by the monoid* $M$. The largest congruence on $A^*$ saturating $L$ (resp., $W$) is called the *syntactic congruence* of $L$ (resp., of $W$), denoted by $\sim_L$ (resp., by $\approx_W$). The quotient monoid $M_L = A^*/\sim_L$ (resp., $I_W = A^*/\approx_W$) is called *syntactic monoid* of $L$ (resp., of $W$). For more details we refer to [HV], [HLV], [Ei].

Let $\underline{V}$ be an $M$- variety. For every alphabet $A$ we put

$$A\mathcal{V}^* = \{L \in ARec^* | M_L \in \underline{V}\},$$

and

$$A\mathcal{V}^N = \{W \in ARec^\omega | I_W \in \underline{V}\}.$$

The family $\mathcal{V}^* = \{A\mathcal{V}^*\}$ of all $A\mathcal{V}^*$ is called the $*$-*variety corresponding to* $\underline{V}$. If $\underline{V}$ is a $U_1$-variety then the family $\mathcal{V}^N = \{A\mathcal{V}^N\}$ of all $A\mathcal{V}^N$ is called the $N$-*variety corresponding to* $\underline{V}$.

In the sequel, for every $*$-variety $\mathcal{V}^*$ we consider also the families $\mathcal{V}^\omega = \{A\mathcal{V}^\omega\}$ and $\vec{\mathcal{V}} = \{A\vec{\mathcal{V}}\}$ (see [Pe]) with

$$A\mathcal{V}^\omega = \{XY^\omega : \ X, Y^* \in A\mathcal{V}^*\}^{\{\cup\}}$$

and

$$A\vec{\mathcal{V}} = \{\vec{X} : \ X \in A\mathcal{V}^*\}^{\mathcal{B}}$$

where $\mathcal{C}^{\mathcal{B}}$ and $\mathcal{C}^{\{\cup\}}$ denote respectively the Boolean closure and the *union* closure of $\mathcal{C}$.

The following result is due to Reutenauer [Re]

**Proposition 1.1** *Let* $\underline{V}$ *be a* $M$*-variety and* $\mathcal{V}^*$ *the corresponding* $*$*-variety. Then* $\underline{V}$ *is closed under Schützenberger product if and only if* $\mathcal{V}^*$ *is closed under concatenation* .

An *automaton* over an alphabet $A$ is given by a finite set $Q$ of *states*, a subset $E$ of $Q \times A \times Q$ whose elements are called *arrows*, and two subsets $I$ and $T$ of $Q$ which are called respectively the sets of *initial* and *terminal* states. We write $\mathcal{A} = (Q, E, I, T)$. A *path* in $\mathcal{A}$ is a finite sequence $P$ of consecutive arrows, $p = \{(p_i, a_i, p_{i+1}), 0 \leq i \leq n\}$. The word $a_0a_1..a_n$ is the *label* of the path. The path $p$ is said to be *successful* if $p_0 \in I$ and $p_{n+1} \in T$. The set of labels of all successful paths in $\mathcal{A}$ is the *language recognized by* $\mathcal{A}$, denoted by $|\mathcal{A}|$. An infinite sequence $p$ of consecutive arrows, $p = \{(p_k, a_k, p_{k+1}), \ k \geq 0\}$, is an $\omega$-*path* in $\mathcal{A}$ with the *label* $a_0a_1..a_n...$ The $\omega$-path $p$ is *successful* if $p_0 \in I$ and $p_k \in T$ for infinitely many $k$. The set of labels of all successful $\omega$-paths in $\mathcal{A}$ is the *Büchi behaviour* of $\mathcal{A}$ or the *$\omega$-language recognized by* $\mathcal{A}$, denoted by $|\mathcal{A}|_B$. The automaton $\mathcal{A}$ itself is called a *Büchi automaton.* If in $\mathcal{A}$ the set $T$ is replaced by a family $\mathcal{T}$ of subsets of $Q$ we obtain a *Müller automton* $\mathcal{A} = (Q, E, I, \mathcal{T})$. In that case an $\omega$-path $p = \{(p_k, a_k, p_{k+1}), k \geq 0\}$ is *successful* if $p_0 \in I$ and the set of all states appearing infinitely in $p$ concides with some subset $T$ in $\mathcal{T}$. The set of labels of all successful paths in $\mathcal{A}$ is the *Müller behaviour* of $\mathcal{A}$, denoted by $|\mathcal{A}|_M$. Whenever it is not necessary to specify the sets of initial and terminal states, instead of $\mathcal{A} = (Q, E, I, T)$ or $\mathcal{A} = (Q, E, I, \mathcal{T})$ we write simply $\mathcal{A} = (Q, E)$. The automaton $\mathcal{A}$ is *deterministic* if for any two arrows $(p, a, q)$, $(p', a, q')$ with the same label $a$, $p = p'$ implies $q = q'$.

Let $\mathcal{A} = (Q, E)$ be an automaton on the alphabet $A$. With every $a \in A$ we associate a binary relation $r(a)$ on $Q$, $r(a) = \{(p, q)|(p, a.q) \in E\}$. The submonoid $M_\mathcal{A}$ generated by the set $\{r(a)|\ a \in A\}$ in the monoid of all binary relations on $Q$ is called the *relation monoid* of $\mathcal{A}$.

As remarked in [VH] (page 324) we have

**Proposition 1.2** *For any language $X$ in $ARec^*$ there exists a deterministic automaton $\mathcal{A} = (Q, E, i, T)$ over $A$ with a unique initial state $i$ such that $M_\mathcal{A} \cong M_X$*

Every subset $P$ of $Q \times Q$ defines a *congruence* $\equiv_P$ on $M_\mathcal{A}$ given by

$$r \equiv_P r' \iff \forall s, s' \in M_\mathcal{A}, srs' \cap P \neq \emptyset \Leftrightarrow sr's' \cap P \neq \emptyset.$$

The quotient monoid $M_\mathcal{A}/ \equiv_P$ is denoted by $M_\mathcal{A}(P)$. In particular, if $P = I \times T$ with $I, T \subseteq Q$, we use the notations $\equiv_{I,T}$ and $M_\mathcal{A}(I, T)$ instead of $\equiv_{I \times T}$ and $M_\mathcal{A}(I \times T)$ respectively.

**Proposition 1.3** ([VH], Proposition 3.1) *If $\mathcal{A} = (Q, E, I, T)$ is an automaton and $X = |\mathcal{A}|$ then*

$$M_X \cong M_\mathcal{A}(I, T)$$

Let $\underline{V}$ be an $M$- variety. An automaton $\mathcal{A} = (Q, E)$ is a *$V$-automaton* if $M_\mathcal{A} \in \underline{V}$. A Büchi automaton $\mathcal{A} = (Q, E, I, T)$ is a *Büchi $V$-automaton* if $M_\mathcal{A} \in \underline{V}$.

A Müller automaton $\mathcal{A} = (Q, E, I, \mathcal{T})$ is a *Müller $V$-automaton* if $M_\mathcal{A}$, $M_{\mathcal{A}_T} \in \underline{V}$ for all $T \in \mathcal{T}$, or equivalently (see [VH], Remark 2.2) if $M_\mathcal{A}$, $M_{\mathcal{A}_T}(t, t') \in \underline{V}$ for all $T \in \mathcal{T}$ and all $t, t' \in T$, where $\mathcal{A}_T$ denotes the restriction of $\mathcal{A}$ on $T$, namely

$$\mathcal{A}_T = (T, E_T) \textit{ with } E_T = E \cap (T \times A \times T).$$

If $\mathcal{A}$ is deterministic then $\mathcal{A}$ is called a *deterministic Müller automaton.*

By virtue of Proposition 3.6 in [VH] and Propositon 1.1 we see that if $\underline{V}$ is a $M$-variety closed under Schützenberger product then a deterministic Müller automaton $\mathcal{A}$ is a Müller $V$-automaton if and only if $M_{\mathcal{A}} \in \underline{V}$ .

The following proposition is a part of Theorem 2.3 in [VH]

**Proposition 1.4** *Let $\underline{V}$ be a non-trivial $M$-variety closed under Schützenberger product and $\mathcal{V}^N = \{A\mathcal{V}^N\}$ be the corresponding $N$-variety. The following conditions are equivalent*

*(i) $W \in A\mathcal{V}^N$ ;*

*(ii) $W$ is recognized by a Büchi $V$-automaton on the alphabet $A$;*

*(iii) $W$ is recognized by a deterministic Müller $V$-automaton over the alphabet $A$.*

The main result of this paper is Theorem 3.3 saying that if $\underline{V}$ is closed under Schüzenberger product then for every Büchi $V$-automaton there exists a non-ambiguous Büchi $V$ automaton recognizing the same $\omega$-language as the former. When $\underline{V}$ is the variety of all finite monoids this is the well-known result due to Arnold [Ar]: every regular $\omega$-language can be recognized by a non-ambiguous Büchi automaton. The proof of Theorem 3.3 is based on *non-ambiguous representations* of $\omega$-languages presented in section 3. Thus our approach here, from language to automaton, is in some sense inverse to the Arnold's approach in [Ar], from automaton to language. For this we used deterministic Müller automata as an intermediate tool and combinatorial computations on languages were done.

## 2 Basic lemmas

In this section we shall formulate two lemmas which will be used in proving the main results in Section 3.

**Lemma 2.1** *Let $\mathcal{A} = (Q, E)$ be an automaton. Let $\underline{V}$ be an $M$- variety and $\mathcal{V}^*$ the corresponding $*$-variety. The following conditions are equivalent*

*(i) $M_{\mathcal{A}} \in \underline{V}$*

*(ii) $\forall p, q \in Q : |(Q, p, q)| \in A\mathcal{V}^*$*

*(iii) $\forall p, q \in Q : M_{\mathcal{A}}(p, q) \in \underline{V}$.*

*Proof.* By Proposition 1.3 we have $|(Q,p,q)| \in A\mathcal{V}^* \Leftrightarrow M_{\mathcal{A}}(p,q) \in \underline{V}$. This implies the equivalence $(ii) \Leftrightarrow (iii)$. Denote by $\rho$ the intersection of all congruences $\equiv_{p,q}$ for $p, q \in Q$, $\rho = \cap_{p,q \in Q} \equiv_{p,q}$. Put $M_\rho = M_{\mathcal{A}}/\rho$. Obviously

$$M_{\mathcal{A}}(p,q) \prec M_{\mathcal{A}} \text{ and } M_\rho \prec \prod_{i,t \in Q} M_{\mathcal{A}}(i,t).$$

The congruence $\rho$ is easily verified to be the identity relation on $M_{\mathcal{A}}$. Indeed, let $r$ and $r'$ be two different elements of $M_{\mathcal{A}}$. There exists then a couple $(p, q)$ belonging to one of the relations $r$ and $r'$ not being in the other. By definition of the $\equiv_{p,q}$ this means $r$ and $r'$ are not in $\equiv_{p,q}$ and therefore not in the congruence $\rho$. Thus $\rho$ is the identity relation on $M_{\mathcal{A}}$. It follows that $M_{\mathcal{A}} \cong M_\rho$. Therefore

$$M_{\mathcal{A}}(p,q) \prec M_{\mathcal{A}} \prec \prod_{i,t \in Q} M_{\mathcal{A}}(i,t)$$

for all $p, q \in Q$. This implies the equivalence $(i) \Leftrightarrow (iii)$ which completes the proof .||

**Lemma 2.2** *Let $\mathcal{A} = (T, E, t, \{T\})$ be a deterministic Müller automaton over $A$ with the only initial state $t$, the only set $T$ of terminal states which coincides with the set of all states of the automaton. Let $L = |\mathcal{A}|_{\mathcal{M}}$. Then*

*$(i)$ There exist prefix regular languages $U, V$ such that*

$$U \subseteq V^+ \ , \quad V^+U = (V^+U)^+ \qquad (1)$$

$$L = (VU)^\omega = (V^+U)^\omega = \overrightarrow{(V^+U)} \quad (2)$$

*$(ii)$ There exists a deterministic Büchi automaton $\mathcal{B}$ recognizing $L$, $L = |\mathcal{B}|_B$.*

*$(iii)$ If the automaton $\mathcal{A}$ mentioned in $(i)$ is a $\underline{V}$-automaton, where $\underline{V}$ is a non-trivial $M$- variety closed under Schützenberger product, and $\mathcal{V}^*, \mathcal{V}^N$ are the corresponding $*$-variety and $N$-variety respectively, then*

$$U, V, V^+U \in A\mathcal{V}^*, L \in A\mathcal{V}^N$$

*and the automaton $\mathcal{B}$ in $(ii)$ may be chosen to be a $\underline{V}$-automaton.*

*Proof.* If $T$ consists of only one state $t$, then obviously $L = A_1^\omega$, where
$A_1 = \{a \in A |\ (t, a, t) \in E\}$.

By chosing $U = V = A_1$, it is easy to check that all the assertions of the lemma are true trvially.

Without loss of generality we may assume $T = \{t_1, .., t_n\}$ with $n \geq 2$ and $t = t_1$. Put

$L_{ij} = |(T, E, t_i, t_j)|$,
$U_1 = \cap_{j \neq 1} L_{1j} L_{j1}$ if $n > 1$ or $U_1 = L_{11} \cap A^+$ if $n = 1$,
$U = U_1 - U_1 A^+$, $V = (L_{11} - L_{11}A^+) \cap A^+$.

(i) Evidently we have facts

$(f_1)$ $U$ and $V$ are regular prefix ;

$(f_2)$ $L_{11} = V^*$.

Since $\mathcal{A}$ is deterministic, $U_1$ consists of all the words which are labels of the paths in $\mathcal{A}$ going from $t_1$ to $t_1$ and passing all states in $T$. The following facts can be easily verified consecutively

$(f_3)$ $U_1^\omega = L$, $U_1^+ = U_1$.

$(f_4)$ $U_1 \subseteq L_{11}$, $U \subseteq U^+ \subseteq U_1$, $U_1 = UL_{11}$

$(f_5)$ $L_{11} \cap A^+ = V^+$; $U \subseteq V^+$

$(f_6)$ $(V^+U)^*(V^+) \subseteq V^+$;

Since $(V^+U)^+ \subseteq ((V^+U)^*(V^+))U \subseteq V^+U$, it follows

$(f_7)$ $V^+U = (V^+U)^+$.

Using $(f_1)$, $(f_5)$ and $(f_6)$ it is not difficul to show

$(f_8)$ $(\overrightarrow{V^+U})^+ = (V^+U)^\omega = (\overrightarrow{V^+U})$

The obvious equality $L = VUL$ implies $L = (VU)^\omega \subseteq (V^+U)^\omega$. But $(V^+U)^\omega \subseteq L$, therefore we have

$(f_9)$ $L = (VU)^\omega = (V^+U)^\omega$.

By combining $(f_1)$,$(f_5)$, $(f_7)$, $(f_8)$ and $(f_9)$ we obtain

$U \subseteq V^+$, $V^+U = (V^+U)^+$

$L = (VU)^\omega = (V^+U)^\omega = \overrightarrow{V^+U}$

where $U$, $V$ are regular prefix sets.

(ii) Since $V^+U$ is regular and $L = \overrightarrow{V^+U}$, as a folklore there exists a Büchi deterministic automaton $\mathcal{B}$ such that

$|\mathcal{B}| = V^+U$ and $|\mathcal{B}|_B = L$.

(iii) By Proposition 1.1, $V^*$ is closed under concatenation. Because $\underline{V}$ is non-trivial, it is a $U_1$-variety. Therefore $\{\epsilon\}$, $A^*$, $A^+ \in A\mathcal{V}^*$. The hypothesis that $\mathcal{A}$ is a Müller V-automaton implies $M_{\mathcal{A}} \in \underline{V}$. By Lemma 2.1, $L_{ij} = |(T, E, t_i, t_j) \in A\mathcal{V}^*$. Because $A\mathcal{V}^*$ is closed under concatenation and Boolean operations it follows $U, V, VU, V^+, V^+U \in A\mathcal{V}^*$. Therefore $L = \overrightarrow{V^+U} \in A\overrightarrow{\mathcal{V}}$. But $A\overrightarrow{\mathcal{V}} = A\mathcal{V}^N$ (see [Pe]), hence $L \in A\mathcal{V}^N$. Since $V^+U \in A\mathcal{V}^*$, there exists a deterministic $V$-automaton such that $|\mathcal{B}| = V^+U$ and $|\mathcal{B}|_B = L$ (see [VH], Section 3) .||

## 3 Main results

This section is devoted to prove the recognizing equivalence of Büchi $V$-automata and non-ambiguous ones (Theorem 3.3). For this we introduce the notion of a limit-nonambiguous representation of an $\omega$-language (Definition 3.1) and prove a result ( Theorem 3.2) which is meaningful on its own right.

Let $(X_i, Y_i)_{i\in I}$ be a family of couples of subsets of $A^*$. This famly is said to be *limit-non-ambiguous* if every word $w$ in $A^N$ admits at most one factorization of the form $w = xu$ with $x \in X_i$, $u \in \overrightarrow{Y_i}$ and $i \in I$.

**Definition 3.1** Let $\underline{V}$ be an $M$- variety and $\mathcal{V}^*$ be the corresponding $*$-variety. Let $W$ be an $\omega$-language on the alphaber $A$, $W \subseteq A^N$. We say that $W$ has a *limit-non-ambiguous representation* (*LNA-representation*, for short) *in* $\mathcal{V}^*$ if it can be represented in the form of a finite union

$$W = \bigcup_{i=1}^{n} X_i \; \overrightarrow{Y_i} \, ,$$

where $X_i$ and $Y_i$ are in $A\mathcal{V}^*$, $Y_i = Y_i^+$, $\overrightarrow{Y_i} = Y_i^\omega$ and the family $(X_i, Y_i)$ is limit-non-ambiguous.

**Theorem 3.2** *Let $\underline{V}$ be a nontrivial $M$- variety closed under Schützenberger product. Let $\mathcal{V}^*$ and $\mathcal{V}^N$ be respectively the $*$-variety and $N$-variety corresponding to $\underline{V}$. For any $W \subseteq A^N$, $W \in A\mathcal{V}^N$ if and only if it has an LNA-representation in $\mathcal{V}^*$*

*Proof.* The "if" part is derived directly from the fact in [Pe] that if $X, Y \in A\mathcal{V}^*$ and $Y = Y^+$ then $XY^\omega \in A\mathcal{V}^N$ and that $A\mathcal{V}^N$ is closed under the union (see [HV]). We now prove the "only if" part. Let $W \in A\mathcal{V}^N$. By Proposition 1.4 there exists a deterministic Müller $V$-automaton $\mathcal{A} = (Q, E, q_0, \mathcal{T})$ recognizing $W$. Evidently,

$$W = \bigcup_{T\in\mathcal{T}} |(Q, E, q_0, \{T\})|_M.$$

Because this is a disjoint union, without loss of generality we may assume that the family $\mathcal{T}$ consists of only one set $T$, that is $W = |(Q, E, q_0, \{T\})|_M$. Let $T = \{t_1, t_2, \cdots, t_n\}$ with $n \geq 1$. Set $E_T = E \cap (T \times A \times T)$, and for every $i$, $1 \leq i \leq n$, consider the deterministic Müller automaton $\mathcal{A}_i = (T, E_T, t_i, \{T\})$, the restriction of $\mathcal{A}$ on $T$ with the initial state $t_i$. After that we put

$N_i = |\mathcal{A}_i|_M$, the $\omega$-languages recognized by $\mathcal{A}_i$;

$L_q = |(Q, E, q_0, q)|$ for every $q \in Q - T$;

$L_{qi} = |(Q, E, q, t_i)| \cap A$ for $q \in Q, t_i \in T$;

$X_i = \bigcup_{q\in Q-T} L_q.L_{qi}$, $i = 1, .., n$; $\;W_i = X_i.N_i$.

We now distinguish two cases.

*Case 1*: $q_0 \notin T$ : One can verify that each $\omega$-path $\alpha$ is devided uniquely into two paths $\alpha = \alpha_1\alpha_2$ : the first one $\alpha_1$ is of the form

$$\alpha_1 : \quad q_0 - - \to q \xrightarrow{a} t_i, \; a \in A, \; t_i \in T \;\; for some \;\; i \leq n, \; q \notin T$$

and the second path $\alpha_2$ is an $\omega$-path running forever in the deterministic Müller automaton $\mathcal{A}_i$. That is for each $\omega$-word $w$ in $W$, there exists exactly one $i, 1 \leq i \leq n$ such that $w$ is devided uniquely into two words: $w = uv$ where $u, v$ are two labels of two paths $\alpha_1, \alpha_2$ of the forms as above. Since $u \in X_i$, $v \in N_i$, applying Lemma 2.2 to $N_i$ and the automata $\mathcal{A}_i$ we can see that $N_i = (V_iU_i)^\omega$ for some prefix $U_i, V_i$. Hence, $W$ admits a unique decomposition of the form

$$W = \bigcup_{i \leq n} X_i N_i = \bigcup_{i \leq n} X_i (V_i U_i)^\omega,$$

where $(V_iU_i)^\omega = \overrightarrow{(V_i^+U_i)} = (V_i^+U_i)^\omega$, $(V_i^+U_i) = (V_i^+U_i)^+$. Hence, putting $Y_i = V_i^+U_i$ we have

$$W = \bigcup_{i \leq n} X_i \overrightarrow{Y_i}$$

which is a $LNA$-representation in $\mathcal{V}^*$.
*Case 2*: $q_0 \in T$. Suppose $q_0 = t_k$ for some $1 \leq k \leq n$. By the same argument as above, in this case we have to examine one more kind of successful $\omega$-paths in $\mathcal{A}$, namely the $\omega$-path in $\mathcal{A}_k$ with initial state $q_0 = t_k$. Hence one can check easily that $W$ has the following form

$$W = \bigcup_{i \leq n} X_i N_i \bigcup N_k = \bigcup_{i \leq n} X_i (V_i U_i)^\omega \bigcup (V_k U_k)^\omega$$

which is $LNA$-representation in $\mathcal{V}^*$ when putting $Y_i = V_i^+U_i, i = 1, ..., n$ and applying Lemma 2.2. This completes the proof.||

Now we shall prove the main result of this note which is formulated as follows

**Theorem 3.3** *Let $\underline{V}$ be a nontrivial $M$-variety closed under Schützenberger product. Then for every Büchi $V$-automaton there exists a non-ambiguous Büchi $V$-automaton equivalent to the former in the sense that they both recognize the same $\omega$-language.*

*Proof.* Given a Büchi $V$-automaton $\mathcal{A}$ over an alphabet $A$ with $W = |\mathcal{A}|_B$. Let $\mathcal{V}^*$ and $\mathcal{V}^N$ be respectively the $*$-variety and $N$-variety corresponding to

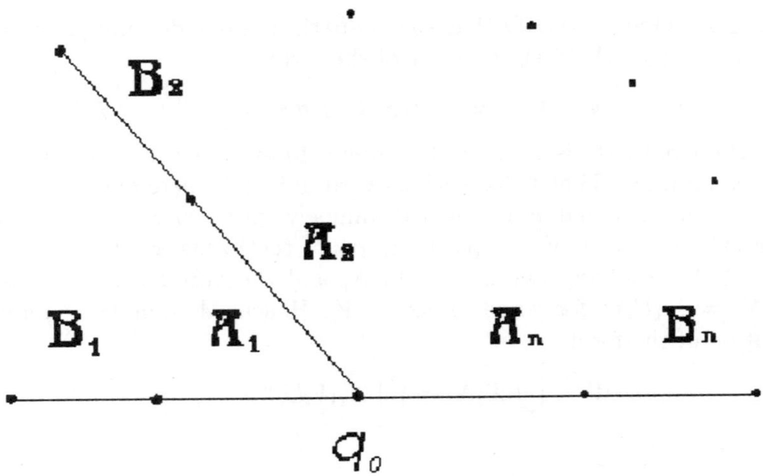


Figure 1. The automaton A'.

$\underline{V}$. We have $W \in A\mathcal{V}^N$ by Proposition 1.4. Then, by Theorem 3.2, $W$ has a LNA-representation in $V^*$,

$$W = \bigcup_{i=1}^{n} X_i \overrightarrow{Y_i},$$

Since $X_i, Y_i \in A\mathcal{V}^*$, by Proposition 1.2 there exist deterministic $V$-automata $\mathcal{A}_i$ and $\mathcal{B}_i$, having each a unique initial state, such that $|\mathcal{A}_i| = X_i$, $|\mathcal{B}_i| = Y_i$. Then as a folklore we have $|\mathcal{B}_i|_B = \overrightarrow{Y_i}$. Denote by $\mathcal{A}'$ the automaton obtained by connecting in a standard way the automata $\mathcal{A}_i$ and $\mathcal{B}_i$ as depicted in the Figure 1.

Evidently $|\mathcal{A}'|_B = W$. The fact that $\mathcal{A}'$ is a $V$-automaton is easily seen by applying Lemma 2.1 with the notice that $V$ is closed under Schützenberger product and therefore, by Proposition 1.1, $\mathcal{V}^*$ is closed under concatenation. Finally, it follows from the non-ambiguity of the above limit-representation of $W$ that the automaton $\mathcal{A}'$ is non-ambiguous. The proof is completed .||

## Acknowledgments

This work was supported in part by the National Program for Basic Research in Natural Sciences of Vietnam. The first of the authors would like to thank the Hanoi Institute of Mathematics for the support and excellent conditions during his two month stay at the institute where the paper was completed.


## References


[Ar] A. Arnold, *Deterministic and non-ambiguous rational $\omega$-languages*, Lecture Notes in Computer Science **192** (1985), 138–146.

[Ei] S. Eilenberg, *Automata, languages and machines*, Acad. Press, New York, Vol. **A**, 1974; Vol. **B**, 1976 .

[HLV] P.T. Huy, I. Litovsky, D.L. Van, *Which finite monoids are syntactic monoids of rational $\omega$-languages*, Inform. Proc. Lett., No.3, **42** (1992) 127-132, North-Holland.

[HV] P.T. Huy, D.L. Van, *Syntactic monoids of $\omega$-languages and Eilenberg Theorem for $\omega$-languages*, A Supplement to Proceedings of the 17th Symposium on Semigroup, Languages and their related fields, CHUO University, Tokyo, 1993.

[Pe] D. Perrin, *Variété de langages et mots infinis*, C.R.Acad.Sci.Paris **295** (1982) 595-598.

[Re] Ch. Reutenauer, *Sur les varietes de languages et de monoides*, 4th. GI. Conference, Lecture Notes in Computer Science, No. **67** Springer (1979) 260-265.

[VH] D.L. Van, P.T. Huy, *Varieties of finite monoids and Büchi-McNaughton Theorem*, Theoretical Computer Science, **98** (1992) 321-337.

# The Consequence of Controlled Densed Theorem of Henstock-Kurzweil Integral in n-Dimensional Euclidean Space

CHRISTIANA RINI INDRATI
Mathematics Department Gadjah Mada University
Yogyakarta, Indonesia
e-mail:chrini.indrati@lycos.com

SOEPARNA DARMAWIJAYA
Mathematics Department Gadjah Mada University
Yogyakarta, Indonesia
e-mail:soeparna@astaga.com

**Abstract**

Controlled densed theorem of Henstock-Kurzweil integral has been generalized in n-dimensional Euclidean space. This theorem gives the characterization of Henstock-Kurzweil integral in Lebesgue integral. By that theorem it will be charachterized the Henstock-Kurzweil integral in Lebesgue integral in term of FSRS (*Functionally Small Riemann Sum*). Then, from FSRS the characterization will be continued in ESRS (*Essentially Small Riemann Sum*).

# 1 Introduction

Recently, many researches focus on Henstock-Kurzweil integral. This integral is special because it solves more problems than Lebesgue integral does. For example, we have function $f(x)$,

$$f(x) = \begin{cases} x \; sin \; \frac{1}{x} & , x \in (0,1] \\ 0 & , \mathrm{x} = 0, \end{cases}$$

is not Lebesgue integrable on [0, 1], but Henstock-Kurzweil integrable on [0,1] to sin 1.

Although Henstock-Kurzweil integral is more general than Lebesgue integral, the Lebesgue integral is a powerful integral. Untill now, Lebesgue integral is still used by mathematics itself and by others which need integral. So, it is important to give more attention on that integral. In one dimensional space, based on controlled densed theorem for Henstock-Kurzweil integral, Darmawijaya charachterized the Henstock-Kurzweil integral in Lebesgue integral in term of small Riemann sums. His important result is that Lebesgue integrable functions space is dense in Henstock-Kurzweil integrable functions space (Darmawijaya, 1995). So, every Hentosck-Kurzweil integrable function can be approached by Lebesgue integrable functions.

According to the generalization of Henstock-Kurzweil integral in n-dimensional space, every Lebesgue integrable function is Henstock-Kurzweil integrable on the same space. And, Lebesgue integral still equivalent with absolute Henstock-Kurzweil integral (Pfeffer, 1993 and Lee, 1996-1997). This paper gives the charachterization of Henstock-Kurzweil integral in Lebesgue integral in n-dimensional space in term of small Riemann sums. To understand the paper, we give some concepts which will be used.

As usual we have $\mathcal{R}^n$ is an n-dimensional Euclidean space. For $\mathbf{x} \in \mathcal{R}^n$ we mean $\mathbf{x} = (x_1, x_2, ..., x_n)$ with $x_i \in \mathcal{R}, i = 1, 2, ..., n$. In this paper, $E$ is stands for non degenerate interval or cell, i.e., $E = [\mathbf{a}, \mathbf{b}] = [a_1, b_1] \times [a_2, b_2] \times \ldots \times [a_n, b_n]$ with $a_i < b_i$ for $i = 1, 2, ..n$. A union of finite cells is called **an elementary set** (Lee P.Y.,1989).

Let $E$ be a cell in $\mathcal{R}^n$, and $\delta$ be a positive function on $E$. An open ball center at $\mathbf{x}$ with radius $\delta(\mathbf{x})$ is $B(\mathbf{x}, \delta(\mathbf{x}))$ with

$$B(\mathbf{x}, \delta(\mathbf{x})) = \{\mathbf{y}; d(\mathbf{x}, \mathbf{y}) < \delta(\mathbf{x})\}$$

where d($\mathbf{x}$, $\mathbf{y}$) = max $\{|x_i - y_i|$, i = 1, 2, ..., n$\}$.
A collection of cells, $\{I\}$, is called to be non overlapping, if $I_i^o \cap I_j^o = \emptyset$ for every i $\neq$ j. A partition of $E$ is a collection of finite non overlapping cells $\mathcal{D} = \{I\}$, with $\cup_{I \in \mathcal{D}} D = E$. A collection of cell-point $\mathcal{D} = \{(I, \mathbf{x})\} = \{(I_1, \mathbf{x}_1), (I_2, \mathbf{x}_2), ..., (I_p, \mathbf{x}_p)\}$ is called **$\delta$-fine partition of** $E$, if $E = \cup_{i=1}^p I_i$, $\mathbf{x}_i \in I_i \subseteq B(\mathbf{x}_i, \delta(\mathbf{x}_i))$, and $I_i^o \cap I_j^o = \emptyset$ for every i $\neq$ j. If the condition $\mathbf{x}_i \in I_i$ is violated, the partition is called **$\delta$-fine McShane partition or McShane partition**. Hence, every $\delta$-fine parti-

tion of $E$ is McShane partition of $E$. If $\cup_{i=1}^{p} I_i \subset E$ then the partition is called **$\delta$-fine partial partition in $E$**. For every positive function $\delta$ defined on a cell $E$ there exists a $\delta$-fine partition of $E$.

A real valued function $f$ defined on $E$ is said to be Henstock-Kurzweil integrable ($R^*$-integrable) on $E$, written $f \in R^*(E)$, if there is a number A such that for any $\epsilon > 0$ there exists a positive function $\delta$ on $E$ such that for any $\delta$-fine partition $\mathcal{D} = \{(\mathrm{I}, \mathbf{x})\} = \{(I_1, \mathbf{x}_1), (I_2, \mathbf{x}_2), ..., (I_p, \mathbf{x}_p)\}$ on $E$ implies

$$|(\mathcal{D}) \sum f(\mathbf{x})|I| - A| < \epsilon,$$

with $(\mathcal{D}) \sum f(\mathbf{x})|I| = \sum_{i=1}^{p} f(\mathbf{x}_i)|I_i|$.
If the $\delta$-fine partition is replaced by McShane partition, the integral is called McShane Integral.
If the function $f$ is $R^*$-integrable on $E$ then the number A is unique. Futhermore, the number A is called the value of the $R^*$-integral of $f$ on $E$. And it will be written

$$A = (R^*) \int_E f.$$

It is easy to see that $R^*$(E), the set of all $R^*$-integrable functions on $E$, is a linear space.

From the definition of Henstock-Kurzweil integral, we have Cauchy's criteria to prove the Theorem 2.

**Theorem 1** *A real valued function $f$ defined on $E$ is said to be Henstock-Kurzweil integrable ($R^*$-integrable) on $E$ if and only if for any $\epsilon > 0$ there exists a positive function $\delta$ on $E$ such that for any two $\delta$-fine partitions of $E$, $\mathcal{D}_1$ and $\mathcal{D}_2$, implies*

$$|(\mathcal{D}_1) \sum f(\mathbf{x})|I| - (\mathcal{D}_2) \sum f(\mathbf{x})|I|| < \epsilon.$$

By the Cauchy's criteria it can be proved that every Henstock-Kurzweil integrable function on a cell $E$ is Henstock-Kurzweil integrable on every its sub-cell as we see from this following theorem.

**Theorem 2** *If $f$ is Henstock-Kurzweil integrable on a cell $E$ and $B$ sub-cell of $E$ then $f$ is also Henstock-Kurzweil integrable on $B$.*

Therefore, it is easy to see that if a real valued function $f$ defined on a cell $E$ then $f$ is Henstock-Kurzweil integrable on every elementary set of $E$.
Moreover, by the Theorem 2, we define the Henstock-Kurzweil primitive as the following.
If $F$ is an interval function on $\mathcal{E}$, the collection of all cell subset $E$, and for every cell $I \subseteq E$, $F(I)$ is the value of the Henstock-Kurzweil integral of $f$ on $I$, then $F$ is called an Henstock-Kurzweil primitive function ($R^*$-primitive) of $f$.

Obviously, the R*-primitive $F$ is additif, i.e., $F(A \cup B) = F(A) + F(B)$ for any non overlapping cells, $A$ and $B$, i.e., $A^o \cap B^o = \emptyset$ (Henstock, 1988).
According to the $R^*$-primitive definition, we rewrite a function to be Henstock-Kurzweil integrable as follow.

A real valued function f defined on E is said to be Henstock-Kurzweil integrable on E if and only if given $\epsilon > 0$ there exist a positive function $\delta$ and a function F on E such that for any $\delta$-fine partition $\mathcal{D}$ of E implies

$$|(\mathcal{D})\sum(f(\mathbf{x})|I| - F(I))| < \epsilon.$$

Sometimes, we just need the partial sum of our $\delta$-fine partition we have. To overcome this problem, Henstock gives his lemma as follow.

**Theorem 3 (Henstock's Lemma)**

*If $f$ is Henstock-Kurzweil integrable on E with $R^*$-primitive function F, i.e., for any $\epsilon > 0$ there is a positive function $\delta$ on E such that for any $\delta$-fine partitions $\mathcal{D}$ of E implies*

$$|(\mathcal{D})\sum(f(\mathbf{x})|I| - F(I))| < \epsilon,$$

*then for any partial $\delta$-fine partition $\mathcal{D}_1$ of $\mathcal{D}$ implies*

$$|(\mathcal{D}_1)\sum(f(\mathbf{x})|I| - F(I))| < 2\epsilon.$$

Since every $\delta$-fine partition of $E$ is McShane partition of $E$, then we have every McShane integrable function is Henstock-Kurzweil integrable. Pfeffer (1993) said that Lebesgue integral is equivalent with McShane integral. We have McShane integral is an absolute integral, i.e., if $f$ is McShane integrable then $|f|$ is integrable too. On the other hand, Henstock-Kurzweil integral is not an absolute integral. However, if a function is absolutely Henstock-Kurzweil integrable on $E$ then it is McShane integrable. One of the sufficience condition of a function to be absolutely Henstock-Kurzweil integrable on a cell $E$ is that it is a bounded function. The following theorems talk about absolutely Henstock-Kurzweil integral.

**Theorem 4** *(Indrati, 1998)*
*If $f$ is a bounded Henstock-Kurzweil integrable function on a cell $E$ then $f$ is absolutely Henstock-Kurzweil integrable on $E$.*

As the corollary, the truncated function defined bellow will be an absolutely Henstock-Kurzweil integrable on a cell $E$ if the original function is Henstock-Kurzweil integrable on $E$.

**Theorem 5** *Let a function $f$ is defined on $E$ to $R$.*
*We define the truncated function, $f_n$, with*

$$f_n(\mathbf{x}) = \begin{cases} f(\mathbf{x}) & , \textit{if } |f(\mathbf{x})| \leq n \\ 0 & , \textit{otherwise.} \end{cases}$$

*If $f$ is Henstock-Kurzweil integrable on $E$, then $f_n$ is absolutely Henstock-Kurzweil integrable on $E$.*

For any set $X$ subset of $E$, we define the restriction function, $f_X$, as follows.

$$f_X(\mathbf{x}) = \begin{cases} f(\mathbf{x}) & , \text{if } \mathbf{x} \in X \\ \\ 0 & , \text{if } \mathbf{x} \notin X \end{cases}$$

Futhermore, if a function is Lebesgue integrable then it is measurable. This is still true for Henstock-Kurzweil integrable function as stated in Theorem 6.

**Theorem 6** *If $f$ is Henstock-Kurzweil integrable on $E$ then $f$ is measurable on $E$.*

# 2 Main Results

Based on the fact that every Henstock-Kurzweil integrable function is measurable, we characterize the Henstock-Kurzweil integral in terms of FSRS and ESRS properties. The FSRS property is developed based on controlled densed theorem. Then, FSRS property implies ESRS property.

Before we start talking the small Riemann sums, we remind the definition of Lebesgue integral and the controlled densed theorems. Let $E$ be a cell and $f : E \to \bar{\mathcal{R}}$ be a function. Function $f$ is Lebesgue integrable on $E$ if and only if there is an absolutely continuous function $F$ on $E$ and $F'(\mathbf{x}) = f(\mathbf{x})$ almost everywhere in $E$ (Gordon, 1994). According to regular derivative in n-dimensional space we have a characterization that if $F$ is the primitive of Henstock-Kurzweil integrable function $f$ on a cell $E$ then $F'(\mathbf{x}) = f(\mathbf{x})$ almost everywhere in $E$. From that characterization we have controlled densed theorem given in Theorem 7(Indrati, 2000). That theorem help us to solve the Henstock-Kurzweil integral in Lebesgue integral.

**Theorem 7** *If $f$ is Henstock-Kurzweil integrable on $E$ then there exists a sequence of closed set, $\{E_n\}$, $E_n \subseteq E_{n+1}$, $E = \cup E_n$ and $f$ is Lebesgue integrable on $E_n$ for every $n$.*

From Theorem 7 we have that

$$\lim_{n\to\infty} \int_{E_n} f = \int_E f.$$

Therefore we have revised theorem as follows.

**Theorem 8** *Controlled Densed Theorem I*

*If $f$ is Henstock-Kurzweil integrable on $E$ then there exists a sequence of closed set, $\{E_n\}$, $E_n \subseteq E_{n+1}$, $E = \cup E_n$, $f$ is Lebesgue integrable on $E_n$ for every $n$ and*

$$\lim_{n\to\infty} \int_{E_n} f = \int_E f.$$

A measurable function $f$ is said to have **FSRS** (Functionally Small Riemann Sum) property on $E$ if for every number $\epsilon > 0$ there are a nonnegative Lebesgue function g and a positive function $\delta$ on $E$ such that for any $\mathcal{D} = \{(I, \mathbf{x})\}$ $\delta$-fine partition of $E$ we have

$$|(\mathcal{D}) \sum_{|f(\mathbf{x})|>g(\mathbf{x})} f(\mathbf{x})|I|| < \epsilon.$$

Based on the Controlled Dense Theorem of Henstock-Kurzweil integral in Euclidean space, we have the charachteristic of this integral in term of FSRS property.

**Theorem 9** *A measurable function f has FSRS property on E if and only if f is Henstock-Kurzweil integrable on E.*

**Proof :** Let $\epsilon > 0$ be given.
(**Necessity**) Since $f$ has FSRS property on $E$ there are a nonnegative Lebesgue function $g$ and a positive function $\delta_1$ on $E$ such that for every $\mathcal{D}$, $\delta_1$-fine partition of $E$ we have

$$|(\mathcal{D}) \sum_{|f(\mathbf{x})|>g(\mathbf{x})} f(\mathbf{x})|I|| < \frac{\epsilon}{4}. \tag{2.1}$$

Define h($\mathbf{x}$) as follows

$$h(\mathbf{x}) = \begin{cases} f(\mathbf{x}) & , \text{ if } |f(\mathbf{x})| \leq g(\mathbf{x}) \\ 0 & , \text{ otherwise.} \end{cases}$$

We have h is measurable and dominated by g.
Since $g$ is Lebesgue integrable on $E$ then h is Lebesgue integrable on $E$. Therefore, $h$ is absolutely Henstock-Kurzweil integrable on $E$. There is a positive $\delta_2$ on $E$ such that for any two $\delta_2$-fine partitions of $E$, $\mathcal{D}_1$ and $\mathcal{D}_2$, we have

$$|(\mathcal{D}_1) \sum h(\mathbf{x})|I| - (\mathcal{D}_2) \sum h(\mathbf{x})|I|| < \frac{\epsilon}{4}. \tag{2.2}$$

Take $\delta(\mathbf{x}) = \min\{\delta_1(\mathbf{x}), \delta_2(\mathbf{x})\}$. Then for every $\mathcal{D} = \{(\mathrm{I}, \mathbf{x})\}$ $\delta$-fine satisfies (2.1) and (2.2). Therefore, for any two $\delta$-fine partitions of E, $\mathcal{D}^1$ and $\mathcal{D}^2$, we have

$$|(\mathcal{D}^1) \sum f(\mathbf{x})|I| - (\mathcal{D}^2) \sum f(\mathbf{x})|I||$$

$$
\begin{aligned}
&\le |(\mathcal{D}^1) \sum_{|f(\mathbf{x})|>g(\mathbf{x})} f(\mathbf{x})|I|| + |(\mathcal{D}^2) \sum_{|f(\mathbf{x})|>g(\mathbf{x})} f(\mathbf{x})|I|| \\
&+ |(\mathcal{D}^1) \sum_{|f(\mathbf{x})|\le g(\mathbf{x})} f(\mathbf{x})|I|| - (\mathcal{D}^1) \sum_{|f(\mathbf{x})|\le g(\mathbf{x})} f(\mathbf{x})|I||| \\
&< \frac{\epsilon}{4} + \frac{\epsilon}{4} + |(\mathcal{D}^1)\sum h(\mathbf{x})|I| - (\mathcal{D}^2)\sum h(\mathbf{x})|I|| \\
&< \frac{\epsilon}{4} + \frac{\epsilon}{4} + \frac{\epsilon}{4} + \frac{\epsilon}{4} = \epsilon.
\end{aligned}
$$

(**Sufficiency**) Since $f$ is Henstock-Kurzweil integrable on $E$ then $f$ is measurable. By the controlled dense convergence theorem there is a sequence of closed sets $\{X_n\}$, $X_n \subseteq X_{n+1}$ and $f$ is Lebesgue integrable on $E_n$ for every n. Define $f_n(\mathbf{x})$ = f($\mathbf{x}$), if $\mathbf{x} \in X_n$ and $f_n(\mathbf{x}) = 0$, otherwise. For every n, $f_n$ is Henstock-Kurzweil integrable on E. And, there exists a positive integer $n_o$ such that for every n $\ge n_o$

$$|(R^*)\int_E f - (R^*)\int_{X_n} f| < \epsilon.$$

For every n, there exists a positive function $\delta_n$ on E such that for any $\mathcal{D}$ $\delta_n$-fine of E we have

$$|(\mathcal{D})\sum f_n(\mathbf{x})|I| - (R^*)\int_E f_n| < \epsilon.$$

There is a positive function $\delta_o$ on E such that for any $\mathcal{D}$ $\delta_o$-fine of E we have

$$|(\mathcal{D})\sum f(\mathbf{x})|I| - (R^*)\int_E f| < \epsilon.$$

Put g($\mathbf{x}$) = $|f_{n_o}(\mathbf{x})|$. We have g is nonnegative and measurable. So, g is Lebesgue integrable on E. And take $\delta(\mathbf{x})$ = min $\{\delta_o(\mathbf{x}), \delta_{n_o}(\mathbf{x})\}$. Then for every $\mathcal{D}$ $\delta$-fine partition of E, we have

$$
\begin{aligned}
|(\mathcal{D}) \sum_{|f(\mathbf{x})|>g(\mathbf{x})} f(\mathbf{x})|I|| &= |(\mathcal{D})\sum f(\mathbf{x})|I| - (\mathcal{D}) \sum_{|f(\mathbf{x})|\le g(\mathbf{x})} f(\mathbf{x})|I|| \\
&\le |(\mathcal{D})\sum f(\mathbf{x})|I| - (R^*)\int_E f| + |(R^*)\int_E f - (R^*)\int_E f_{n_o}| \\
&+ (R^*)\int_E f_{n_o} - (\mathcal{D}) \sum_{\mathbf{x}\in X_{n_o}} f(\mathbf{x})|I|| \\
&< \epsilon + \epsilon + |(R^*)\int_E g - \sum g(\mathbf{x})|I|| \\
&< 2\epsilon + \epsilon = 3\epsilon.
\end{aligned}
$$

Finally, it will be given the ESRS as the other type of small Riemann Sum of Henstock-Kurzweil integral in n-dimensional Euclidean space.

A measurable function $f$ is said to have **ESRS** (Essentially Small Riemann Sum) property on $E$ if for every number $\epsilon > 0$ there are a Lebesgue function g and a positive function $\delta$ on $E$ such that for any $\mathcal{D} = \{(I, \mathbf{x})\}$ $\delta$-fine partition of $E$ we have

$$|(\mathcal{D}) \sum_{f(\mathbf{x}) \neq g(\mathbf{x})} f(\mathbf{x})|I|| < \epsilon,$$

with the sum is taken over all $f(\mathbf{x}) \neq g(\mathbf{x})$.

**Theorem 10** *A measurable function $f$ has ESRS property on $E$ if and only if $f$ is Henstock-Kurzweil integrable on $E$.*

**Proof :(Necessity)** Let $\epsilon > 0$ be given.
Since $f$ has ESRS property on $E$, there are a Lebesgue integrable function $g$ and a positive function $\delta^*$ on $E$ such that for any $\delta$-fine partition of $E$, $\mathcal{D}_* = \{(I, \mathbf{x})\}$, we have

$$|(\mathcal{D}_*) \sum_{f(\mathbf{x}) \neq g(\mathbf{x})} f(\mathbf{x})|I|| < \epsilon,$$

Since $g$ is Lebesgue integrable on $E$, then it is Henstock-Kurzweil integrable on $E$. So, there exists a positive function $\delta^*$ on $E$ such that for any two $\delta^*$-fine partitions of $E$, $\mathcal{D}_1^*$ and $\mathcal{D}_2^*$, we have

$$|(\mathcal{D}_1^*) \sum g(\mathbf{x})|I| - (\mathcal{D}_2^*) \sum g(\mathbf{x})|I|| < \epsilon,$$

Take a positive function $\delta$ with $\delta(\mathbf{x}) = \min\{\delta_*(\mathbf{x}), \delta^*(\mathbf{x})\}$. Then for any two $\delta^*$-fine partitions of $E$, $\mathcal{D}_1$ dan $\mathcal{D}_2$, we have

$$\begin{aligned}
|(\mathcal{D}_1) \sum f(\mathbf{x})|I| - (\mathcal{D}_2) \sum f(\mathbf{x})|I|| &< |(\mathcal{D}_1) \sum_{f(\mathbf{x}) \neq g(\mathbf{x})} f(\mathbf{x})|I|| + |(\mathcal{D}_2) \sum_{f(\mathbf{x}) \neq g(\mathbf{x})} f(\mathbf{x})|I|| \\
&+ |(\mathcal{D}_1) \sum_{f(\mathbf{x}) = g(\mathbf{x})} f(\mathbf{x})|I| - (\mathcal{D}_2) \sum_{f(\mathbf{x}) = g(\mathbf{x})} f(\mathbf{x})|I|| \\
&\leq \epsilon + \epsilon + |(\mathcal{D}_1) \sum g(\mathbf{x})|I| - (\mathcal{D}_2) \sum g(\mathbf{x})|I|| \leq 4\epsilon.
\end{aligned}$$

So, $f$ is Henstock-Kurzweil integrable on $E$.

(**Sufficiency**) Let $\epsilon > 0$ be given.

Since $f$ is Henstock-Kurzweil integrable on $E$, then $f$ has FSRS property. Therefore, there exist a non negative Lebesgue integrable function $h$ and a positive function $\delta$ on $E$ such that for every $\mathcal{D}$, $\delta$-fine partition of $E$, we have

$$|(\mathcal{D}) \sum_{|f(\mathbf{x})|>h(\mathbf{x})} f(\mathbf{x})|I|| < \epsilon.$$

Defined a function $g$ as follows,

$$g(\mathbf{x}) = \begin{cases} f(\mathbf{x}) & , \text{ if } |f(\mathbf{x})| \leq h(\mathbf{x}) \\ 0 & , \text{ if } |f(\mathbf{x})| > h(\mathbf{x}). \end{cases}$$

Then $g$ is Lebesgue integrable on $E$. Therefore, for any $\delta$-fine partition of $E$, $\mathcal{D}$, we have

$$|(\mathcal{D}) \sum_{f(\mathbf{x})\neq g(\mathbf{x})} f(\mathbf{x})|I|| = |(\mathcal{D}) \sum_{|f(\mathbf{x})|>h(\mathbf{x})} f(\mathbf{x})|I|| < \epsilon.$$

So, $f$ has ESRS property. □

## References

Darmawijaya, S., 1995, Posisi Integral Lebesgue di dalam Integral Henstock, *Laporan Penelitian*, FMIPA-UGM, Yogyakarta.

Gordon, B.C., 1994, The Integral of Lebesgue, Denjoy, Perron & Henstock, *American Mathematical Society* Vol. 4.

Henstock,R.,1988, *Lectures on the Theory of Integration*, World Scientific.

Indrati, Ch.R., and Darmawijaya, S., 1998, Syarat Cukup Fungsi Terintegral Henstock-Kurzweil Mutlak dalam Ruang Euclide Berdimensi-n, *Berkala Ilmiah MIPA* No. 1 Th VIII, 1998.

Indrati, Ch.R., and Darmawijaya, S., 2000, Teorema Rapat Terkendali untuk Fungsi Terintegral Henstock-Kurzweil di dalam Ruang Euclide Berdimensi-n, *Journal of Indonesian Mathematical Society, MIHMI* Vol.6 N0. 3, 134-138

Lee P.Y.,1989, *Lanzhou Lectures on Henstock Integration*, World Scientific.

Lee T.Y., Chew T.S., Lee P.Y., 1996-1997, On Henstock Integrability in Euclidean Space, *Real Analysis Exchange* Vol. 22 N0.1, 382-389.

Pfeffer, W.F., 1993, *The Riemann Approach to Integration*, Cambridge University Press, New-York, USA.

# LOCAL STABILITY ANALYSIS FOR A MODEL OF AIDS TRANSMISSION

MARIA EVELYN V. JACINTO

*Department of Mathematics, University of the Philippines, Diliman, Quezon City, Philippines*

In this paper, we investigated a model of AIDS transmission. The conditions for the existence of equilibrum points and their local stability are given.

## 1 A Model for AIDS Transmission

This paper presents an analysis of the following transmission model for acquired immune deficiency syndrome (AIDS).

$$\begin{cases} \dot{X}_0 = \alpha - (\gamma+\mu)X_0 + 2(\sigma+\mu)X_2 + (\delta+\sigma)X_3 - 2\phi(X_0,X_0) - \phi(X_0,X_1) \\ \dot{X}_1 = \gamma X_0 - \delta X_1 + (\sigma+\mu)X_3 + 2(\delta+\sigma)X_4 - \phi(X_0,X_1) - 2\phi(X_1,X_1) \\ \dot{X}_2 = \phi(X_0,X_0) - (\gamma+\sigma+2\mu)X_2 \\ \dot{X}_3 = \phi(X_0,X_1) - (\mu+\sigma+\delta+\beta)X_3 + \gamma X_2 \\ \dot{X}_4 = \phi(X_1,X_1) - (2\delta+\sigma)X_4 + \beta X_3 \end{cases} \tag{1}$$

where

$X_0$ = number of ***non-infected*** individuals without sexual partners
$X_1$ = number of ***infected*** individuals without sexual partners
$X_2$ = number of pairs with both parties not infected
$X_3$ = number of pairs with only one of the partners infected
$X_4$ = number of pairs with both parties infected
$\alpha$ = entry rate into the $X_0$ category
$\sigma$ = rate of separation of pairs (whether $X_2, X_3$ or $X_4$)
$\mu$ = natural mortality rate of non-infected individuals
$\delta$ = mortality rate of infected individuals
$\beta$ = transmission probability per partner
$\gamma$ = rate at which non-infected individuals contract the HIV virus thru bl contamination (e.g., blood transfusion, needle accidents)

$$\phi(X_i, X_j) = \frac{\rho_{ij} X_i X_j}{X_0 + X_1} \qquad , i, j = 0, 1$$

where $\rho_{ij}$ = rate at which $X_i$ individuals form pairs w/ $X_j$ individuals

We make the assumption that all parameters are non-negative.

This model (1) is a revised version of the one analyzed in a paper by Jing Zhujun and Liu Zhengrong [1]. It incorporates an additional parameter $\gamma$ that represents transmission of the disease thru a-sexual means. The original model, which considers transmission solely thru a sexual partner, looks like this:

$$\begin{cases} \dot{X}_0 = \alpha - \mu X_0 + 2(\sigma+\mu)X_2 + (\delta+\sigma)X_3 - 2\phi(X_0,X_0) - \phi(X_0,X_1) \\ \dot{X}_1 = -\delta X_1 + (\sigma+\mu)X_3 + 2(\delta+\sigma)X_4 - \phi(X_0,X_1) - 2\phi(X_1,X_1) \\ \dot{X}_2 = \phi(X_0,X_0) - (\sigma+2\mu)X_2 \\ \dot{X}_3 = \phi(X_0,X_1) - (\mu+\sigma+\delta+\beta)X_3 \\ \dot{X}_4 = \phi(X_1,X_1) - (2\delta+\sigma)X_4 + \beta X_3 \end{cases} \tag{2}$$

Note that (1) is reduced to this model when $\gamma = 0$.

## 2 Equilibrium Points

We now solve for the equilibrium points of (1), and determine the conditions for their existence. The equilibrium points are the solutions $(X_0, X_1, X_2, X_3, X_4) \in \mathbb{R}^5_+$ [a] that satisfy the following equations.

$$\alpha - (\gamma+\mu)X_0 + 2(\sigma+\mu)X_2 + (\delta+\sigma)X_3 - \frac{2\rho_{00}X_0^2}{X_0+X_1} - \frac{\rho_{01}X_0X_1}{X_0+X_1} = 0 \tag{3}$$

$$\gamma X_0 - \delta X_1 + (\sigma+\mu)X_3 + 2(\delta+\sigma)X_4 - \frac{\rho_{01}X_0X_1}{X_0+X_1} - \frac{2\rho_{11}X_1^2}{X_0+X_1} = 0 \tag{4}$$

$$\frac{\rho_{00}X_0^2}{X_0+X_1} - (\gamma+\sigma+2\mu)X_2 = 0 \tag{5}$$

$$\frac{\rho_{01}X_0X_1}{X_0+X_1} - (\mu+\sigma+\delta+\beta)X_3 + \gamma X_2 = 0 \tag{6}$$

$$\frac{\rho_{11}X_1^2}{X_0+X_1} - (2\delta+\sigma)X_4 + \beta X_3 = 0 \tag{7}$$

[a] $\mathbb{R}^5_+ = \{(X_0,X_1,X_2,X_3,X_4) : X_j \geq 0, j = 0,\ldots,4\}$

From equation (5), we obtain

$$X_2 = \frac{\rho_{00} X_0^2}{(\gamma + \sigma + 2\mu)(X_0 + X_1)} \tag{8}$$

Substituting (8) into (6), and defining $\mu_1 = \mu + \sigma + \delta + \beta$

$$\begin{aligned} X_3 &= \frac{1}{\mu_1}\left[\frac{\rho_{01} X_0 X_1}{X_0 + X_1} + \frac{\gamma \rho_{00} X_0^2}{(\gamma + \sigma + 2\mu)(X_0 + X_1)}\right] \\ &= \frac{(\gamma + \sigma + 2\mu)\rho_{01} X_0 X_1 + \gamma \rho_{00} X_0^2}{\mu_1 (\gamma + \sigma + 2\mu)(X_0 + X_1)} \end{aligned} \tag{9}$$

Substituting (9) into (7), we get

$$\begin{aligned} X_4 &= \frac{1}{2\delta + \sigma}\left[\frac{\rho_{11} X_1^2}{X_0 + X_1} + \frac{(\gamma + \sigma + 2\mu)\beta\rho_{01} X_0 X_1 + \gamma\beta\rho_{00} X_0^2}{\mu_1 (\gamma + \sigma + 2\mu)(X_0 + X_1)}\right] \\ &= \frac{\mu_1 \rho_{11}(\gamma + \sigma + 2\mu) X_1^2 + (\gamma + \sigma + 2\mu)\beta\rho_{01} X_0 X_1 + \gamma\beta\rho_{00} X_0^2}{\mu_1 (2\delta + \sigma)(\gamma + \sigma + 2\mu)(X_0 + X_1)} \end{aligned} \tag{10}$$

Substituting (9) and (10) into (4), we get

$$\gamma X_0 - \delta X_1 + (\sigma + \mu)\left[\frac{(\gamma + \sigma + 2\mu)\rho_{01} X_0 X_1 + \gamma \rho_{00} X_0^2}{\mu_1 (\gamma + \sigma + 2\mu)(X_0 + X_1)}\right] +$$

$$2(\delta + \sigma)\left[\frac{\mu_1 \rho_{11}(\gamma + \sigma + 2\mu) X_1^2 + (\gamma + \sigma + 2\mu)\beta\rho_{01} X_0 X_1 + \gamma\beta\rho_{00} X_0^2}{\mu_1 (2\delta + \sigma)(\gamma + \sigma + 2\mu)(X_0 + X_1)}\right]$$

$$-\frac{\rho_{01} X_0 X_1 + 2\rho_{11} X_1^2}{X_0 + X_1} = 0$$

Combining fractions, and equating the numerator to zero, we have:

$$k_1 X_1^2 + k_2 X_0 X_1 + k_3 X_0^2 = 0$$

where

$$k_1 = -\delta\mu_1(\gamma+\sigma+2\mu)(2\delta+\sigma+2\rho_{11}) \tag{11}$$

$$k_2 = (2\delta+\sigma)(\gamma+\sigma+2\mu)(\mu_1\gamma-\mu_1\delta-\rho_{01}\delta)+\sigma\rho_{01}\beta(\gamma+\sigma+2\mu) \tag{12}$$

$$k_3 = \gamma\mu_1(2\delta+\sigma)(\gamma+\sigma+2\mu)+\gamma\rho_{00}(\sigma+\mu)(2\delta+\sigma)+ 2\gamma\beta\rho_{00}(\delta+\sigma) \tag{13}$$

From which we obtain:

$$X_1 = \frac{-k_2 \pm \sqrt{k_2^2 - 4k_1k_3}}{2k_1}X_0 = lX_0 \tag{14}$$

If $\gamma = 0$ then $k_3 = 0$. In this case,

$$l = \frac{-k_2 \pm |\,k_2|}{2k_1} = \begin{cases} \frac{-k_2+k_2}{2k_1} = 0 \\ \frac{-k_2-k_2}{2k_1} = -\frac{k_2}{k_1} \end{cases}$$

If $k_2 > 0$, $-\frac{k_2}{k_1} > 0$ since $k_1 < 0$; and we get two non-negative values for $l$ yielding two distinct equilibrium points $D_0$ with $X_1 = 0$ and $D_1$ with $X_1 \neq 0$.

If $k_2 < 0$, $-\frac{k_2}{k_1} < 0$ since $k_1 < 0$; and we get only one non-negative value for $l$ yielding only one equilibrium point $D_0$.

These are the same existence conditions and equilibrium points as presented in [1].

Next, we consider the case when $\gamma \neq 0$. Then $k_3 > 0$ and

$$|\,k_2\,| \quad < \quad \left|\sqrt{k_2^2 - 4k_1k_3}\,\right|$$

Thus, taking the + sign in (14) will yield a negative value for $X_1$, while taking the $-$ sign will yield a positive value for $X_1$. We therefore get only one equilibrium point $D_1$, where

$$\begin{aligned} X_1 &= \frac{-k_2 - \sqrt{k_2^2 - 4k_1k_3}}{2k_1}X_0 \qquad , k_1 \neq 0 \\ &= lX_0 \end{aligned} \tag{15}$$

In order to ensure that $k_1 \neq 0$ we make the assumptions that $\delta \neq 0$ and $\sigma \neq 0$. Furthermore, since $k_2^2 - 4k_1k_3 \geq 0$ and $l \geq 0$, $X_1$ will always be defined for any given positive $X_0$.

Finally, substituting (8), (9), and (15) into (3), and defining

$$n = \gamma + \sigma + 2\mu \tag{16}$$

we get

$$X_0 = \frac{\alpha\mu_1 n(1+l)}{\mu_1 n(\gamma+\mu)(1+l) + \gamma\rho_{00}(\delta+\sigma+2\mu+2\beta) + 2\rho_{00}\mu + n\rho_{01}l(\mu+\beta)} \tag{17}$$

These computations are summarized in the following theorem.

**Theorem 1** *If $\gamma = 0$, then the system (1) is reduced to the model (2). Equilibrium points and the conditions for their existence are as presented in [1]. If $\gamma > 0$, $\delta > 0$ and $\sigma > 0$, then system (1) has a single equilibrium point $D_1 = (X_0, X_1, X_2, X_3, X_4) \in \mathbb{R}^5_+$ whose coordinates can be computed using equations (17), (15), (8), (9), and (10).*

## 3 Stability of Equilibrium Points

We now derive the conditions for local asymptotic stability for $D_1$ when $\gamma \neq 0$. The following lemma will be applied.

**Lemma 1** *(Routh-Hurwitz stability condition)[2] Let the characteristic polynomial of the matrix $|A - \lambda I|$ be of the form*

$$f(\lambda) = \lambda^5 + C_4\lambda^4 + C_3\lambda^3 + C_2\lambda^2 + C_1\lambda + C_0 \tag{18}$$

*Then all roots of (18) have negative real parts if the following conditions are satisfied:*

$$\begin{cases} C_j > 0, \qquad j = 0,..,4 \\ C_4C_3C_2 - C_2^2 - C_4^2C_1 > 0 \\ (C_2C_1 - C_3C_0)(C_4C_3 - C_2) - (C_4C_1 - C_0)^2 > 0 \end{cases} \tag{19}$$

The Jacobian at $D_1$ is given by:

$$J_1 = \begin{bmatrix} a_{00} & a_{01} & a_{02} & a_{03} & 0 \\ a_{10} & a_{11} & 0 & a_{13} & a_{14} \\ a_{20} & a_{21} & a_{22} & 0 & 0 \\ a_{30} & a_{31} & a_{32} & a_{33} & 0 \\ a_{40} & a_{41} & 0 & a_{43} & a_{44} \end{bmatrix} \tag{20}$$

where

$$a_{00} = \frac{-(\gamma+\mu)(1+l)^2 - 2\rho_{00}(1+2l) - \rho_{01}l^2}{(1+l)^2}, \qquad a_{01} = \frac{2\rho_{00} - \rho_{01}}{(1+l)^2},$$

$$a_{02} = 2(\sigma+\mu), \qquad a_{03} = \delta+\sigma, \qquad a_{10} = \gamma + \frac{-\rho_{01}l^2 + 2\rho_{11}l^2}{(1+l)^2},$$

$$a_{11} = -\delta - \frac{\rho_{01} + 2\rho_{11}l(2+l)}{(1+l)^2}, \qquad a_{13} = \sigma+\mu, \qquad a_{14} = 2(\delta+\sigma),$$

$$a_{20} = \frac{\rho_{00}(2l+1)}{(1+l)^2}, \qquad a_{21} = \frac{-\rho_{00}}{(1+l)^2}, \qquad a_{22} = -(\gamma+\sigma+2\mu), \quad (21)$$

$$a_{30} = \frac{\rho_{01}l^2}{(1+l)^2}, \qquad a_{31} = \frac{\rho_{01}}{(1+l)^2}, \qquad a_{32} = c, \qquad a_{33} = -\mu_1,$$

$$a_{40} = \frac{-\rho_{11}l^2}{(1+l)^2}, \qquad a_{41} = \frac{\rho_{11}l(2+l)}{(1+l)^2}, \qquad a_{43} = \beta, \qquad a_{44} = -(2\delta+\sigma)$$

Next, we compute for the characteristic polynomial $P(\lambda) = det(\lambda I - J_1)$. Letting $\rho_{00} = \rho_{01} = \rho_{11} = \rho$, we get the following:

$$P(\lambda) = \lambda^5 + C_4\lambda^4 + C_3\lambda^3 + C_2\lambda^2 + C_1\lambda + C_0 = 0 \qquad (22)$$

where

$$C_4(\beta) = \beta + L_4 \qquad (23)$$
$$C_3(\beta) = \beta M_3 + L_3 \qquad (24)$$
$$C_2(\beta) = \beta M_2 + L_2 \qquad (25)$$
$$C_1(\beta) = \beta M_1 + L_1 \qquad (26)$$
$$C_0(\beta) = \beta M_0 + L_0 \qquad (27)$$

where

$$L_4 = 2\gamma + 4\mu + 4\delta + 3\sigma + 3\rho + 2\frac{\rho l}{(1+l)^2} \qquad (28)$$

$$M_3 = 2\gamma + 3\mu + 3\delta + 2\sigma + 3\rho + 2\frac{\rho l}{(1+l)^2} \qquad (29)$$

$$\begin{aligned} L_3 = {} & 8\gamma\delta + 5\delta^2 + 3\sigma^2 + 9\mu\sigma + 7\rho\delta + 15\mu\delta + 5\gamma\sigma + 5\mu^2 + 5\gamma\mu \\ & +2\rho^2 + 5\gamma\rho + 9\delta\sigma + \gamma^2 + 11\rho\mu + 6\rho\sigma + \frac{1}{(1+l)^2}\{-4\rho\mu \\ & +4\rho\delta - 2\gamma\rho + (4\rho\sigma + 2\rho\mu + 10\rho\delta + 4\rho^2 + 2\gamma\rho)\,l\} \end{aligned} \qquad (30)$$

$$M_2 = \gamma^2 + M_{21}\gamma + M_{20} \tag{31}$$

$$M_{21} = 3\,\mu + 3\,\sigma + 6\,\delta + 5\,\rho + \frac{-2\,\rho + 2\,\rho\, l}{(1+l)^2} \tag{32}$$

$$\begin{aligned} M_{20} &= \sigma^2 + 2\,\delta^2 + 2\,\mu^2 + 9\,\mu\,\delta + 2\,\rho^2 + 4\,\mu\,\sigma + 5\,\rho\,\delta + 4\,\rho\,\sigma \\ &\quad +4\,\delta\,\sigma + 8\,\rho\,\mu + \frac{\rho\,\delta - 2\,\rho\,\sigma - 3\,\rho\,\mu + (6\,\rho\,\delta + 4\,\rho^2)\,l}{(1+l)^2} \end{aligned} \tag{33}$$

$$\begin{aligned} L_2 = {}& 2\,\delta^3 + 2\,\gamma^2\,\rho + \sigma^3 + 2\,\rho^2\,\delta + 4\,\rho\,\delta^2 + 17\,\mu\,\delta^2 + 6\,\delta\,\sigma^2 + 8\,\delta^2\,\sigma \\ & +10\,\rho\,\delta\,\sigma + 23\,\rho\,\delta\,\mu + 23\,\mu\,\delta\,\sigma + 3\,\rho\sigma^2 + 17\,\mu^2\,\delta + 12\,\rho\,\mu^2 + 2\,\rho^2\,\sigma \\ & +6\,\rho^2\,\mu + 14\,\rho\,\mu\,\sigma + 6\,\mu\,\sigma^2 + 8\,\mu^2\,\sigma + 2\,\mu^3 + 7\,\gamma\,\rho\,\sigma + 11\,\gamma\,\rho\,\mu \\ & +9\,\gamma\,\sigma\,\mu + 11\,\gamma\,\rho\,\delta + 4\,\gamma\,\sigma^2 + 3\,\gamma\,\mu^2 + \gamma^2\,\mu + 2\,\gamma^2\,\sigma + 18\,\gamma\,\delta\,\mu + \\ & 14\,\gamma\,\delta\,\sigma + 10\,\gamma\,\delta^2 + 4\,\gamma^2\,\delta + 2\,\rho^2\,\gamma + \frac{1}{(1+l)^2}\{-2\,\gamma^2\,\rho + 4\,\rho^2\,\delta \\ & +8\,\rho\,\delta^2 + 4\,\rho\,\delta\,\sigma - 8\,\rho\,\mu^2 - 4\,\rho^2\,\mu - 4\,\rho\,\mu\,\sigma - 6\,\gamma\,\rho\,\mu - 2\,\gamma\,\rho\,\delta \\ & -2\,\gamma\,\rho\,\sigma + (12\,\rho^2\,\delta + 4\,\rho^2\,\gamma + 16\,\rho\,\delta^2 + 12\,\rho\,\delta\,\sigma + 10\,\rho\,\delta\,\mu + 2\,\rho\sigma^2 \\ & +4\,\rho^2\,\sigma + 4\,\rho^2\,\mu + 4\,\rho\,\mu\,\sigma + 8\,\gamma\,\rho\,\delta + 4\,\gamma\,\rho\,\sigma + 2\,\gamma\,\rho\,\mu)\,l\} \end{aligned} \tag{34}$$

$$M_1 = M_{12}\gamma^2 + M_{11}\gamma + M_{10} \tag{35}$$

$$M_{12} = 3\,\delta + \sigma + 2\,\rho - 2\,\frac{\rho}{(1+l)^2} \tag{36}$$

$$\begin{aligned} M_{11} &= 3\,\rho\sigma + 2\,\rho^2 + \sigma^2 + 5\,\delta\,\sigma + 3\,\mu\,\sigma + 9\,\mu\,\delta + 4\,\delta^2 + 6\,\rho\,\mu \\ &\quad +7\,\rho\,\delta + \frac{(2\,\rho\,\sigma + 6\,\rho\,\delta + 4\,\rho^2)\,l - 3\,\rho\,\mu - \rho\,\delta}{(1+l)^2} \end{aligned} \tag{37}$$

$$\begin{aligned} M_{10} &= 6\,\mu\,\delta^2 + \delta\,\sigma^2 + 2\,\delta^2\,\sigma + 6\,\rho\,\delta\,\sigma + 12\,\rho\,\delta\,\mu + 6\,\mu\,\delta\,\sigma + \rho\,\sigma^2 \\ &\quad +6\,\mu^2\,\delta + 4\,\rho\,\mu^2 + 2\,\rho^2\,\sigma + 4\,\rho^2\,\mu + 2\,\rho^2\,\delta + 2\,\rho\,\delta^2 + \mu\,\sigma^2 \\ &\quad +2\,\mu^2\,\sigma + 4\,\rho\,\mu\,\sigma + \frac{1}{(1+l)^2}\{(4\,\rho\,\delta^2 - 2\,\rho\,\sigma^2 + 4\,\rho^2\,\delta - 4\,\rho\,\delta\,\sigma \\ &\quad -4\,\rho^2\,\sigma)\,l - 4\,\rho^2\,\sigma - 2\,\rho^2\,\delta + 2\,\rho\,\delta^2 - 2\,\rho\,\mu^2 - 4\,\rho\,\mu\,\sigma \\ &\quad -2\,\rho^2\,\mu - 2\,\rho\,\sigma^2 - 6\,\rho\,\delta\,\mu - 4\,\rho\,\delta\,\sigma\} \end{aligned} \tag{38}$$

$$\begin{aligned} L_1 = {}& 3\,\gamma^2\,\delta\,\mu + 5\,\gamma^2\,\delta\,\sigma + 11\,\gamma\,\delta^2\,\sigma + 6\,\gamma\,\rho\,\mu^2 + 9\,\gamma\,\rho\,\delta\,\sigma + 7\,\gamma\,\delta\,\sigma^2 \\ & +6\,\gamma\,\rho\,\delta^2 + 9\,\gamma\,\delta\,\mu^2 + 19\,\gamma\,\delta^2\,\mu + 19\,\rho\,\gamma\,\delta\,\mu + 20\,\gamma\,\delta\,\mu\,\sigma + \end{aligned}$$

$$
\begin{aligned}
&8\,\rho\,\mu^2\,\sigma + 2\,\gamma\,\rho^2\,\delta + 5\,\gamma^2\,\delta^2 + 2\,\gamma\,\rho^2\,\mu + \gamma^2\,\sigma^2 + 3\,\sigma\,\gamma\,\mu^2 \\
&+\sigma\,\gamma^2\,\mu + 4\,\gamma\,\delta^3 + 3\,\rho\,\delta\,\sigma^2 + 2\,\delta^3\,\sigma + 4\,\rho\,\delta^2\,\sigma + 6\,\mu\,\delta^3 + 3\,\delta^2\,\sigma^2 \\
&+2\,\sigma\,\rho^2\,\delta + 2\,\sigma\,\mu^3 + \mu\,\sigma^3 + 3\,\mu^2\,\sigma^2 + \delta\,\sigma^3 + \gamma\,\sigma^3 + 4\,\gamma\,\sigma^2\,\mu \\
&+9\,\rho\,\gamma\,\sigma\,\mu + 4\,\gamma^2\,\rho\,\delta + 2\,\gamma^2\,\rho\sigma + 2\,\rho\,\gamma\,\sigma^2 + 2\,\gamma^2\,\rho\,\mu + 2\,\rho^2\,\mu\,\sigma \\
&+3\,\rho\,\mu\,\sigma^2 + 4\,\rho^2\,\mu^2 + 4\,\rho\,\mu^3 + 20\,\rho\,\delta\,\mu^2 + 12\,\rho\,\delta^2\,\mu + 16\,\mu^2\,\delta\,\sigma \\
&+9\,\mu\,\delta\,\sigma^2 + 16\,\mu\,\delta^2\,\sigma + 6\,\rho^2\,\mu\,\delta + 6\,\mu^3\,\delta + 16\,\mu^2\,\delta^2 + 18\,\rho\,\delta\,\sigma\,\mu \\
&+\frac{1}{(1+l)^2}\{(4\,\sigma\,\rho^2\,\delta + 10\,\gamma\,\rho\,\delta\,\sigma + 8\,\rho\,\delta^3 + 6\,\rho\,\gamma\,\delta\,\mu + 4\,\gamma\,\rho^2\,\sigma \\
&+4\,\gamma\,\rho^2\,\mu + 10\,\gamma\,\rho\,\delta^2 + 8\,\gamma\,\rho^2\,\delta + 8\,\rho\,\delta^2\,\sigma + 2\,\rho\,\gamma\,\sigma^2 + 2\,\rho\,\gamma\,\sigma\,\mu \\
&+2\,\rho\,\delta\,\sigma^2 + 2\,\rho\,\mu\,\sigma^2 + 16\,\rho\,\delta^2\,\mu + 8\,\rho^2\,\delta^2 + 4\,\rho^2\,\mu\,\sigma + 12\,\rho\,\delta\,\sigma\,\mu \\
&+12\,\rho^2\,\mu\,\delta)l + 2\,\gamma\,\rho^2\,\sigma - 4\,\rho\,\mu^2\,\sigma - 2\,\gamma^2\,\rho\sigma - 4\,\gamma^2\,\rho\,\delta + 4\,\rho^2\,\delta^2 \\
&-2\,\gamma^2\,\rho\,\mu - \gamma\,\rho^2\,\mu + \gamma\,\rho\,\delta^2 - 6\,\rho\,\gamma\,\delta\,\mu - 5\,\gamma\,\rho\,\mu^2 - 4\,\rho\,\gamma\,\sigma\,\mu \\
&+4\,\rho\,\delta^2\,\sigma + 3\,\gamma\,\rho^2\,\delta - 4\,\rho^2\,\mu^2 + 4\,\rho\,\delta^3 - 8\,\rho\,\delta\,\mu^2 + 8\,\rho\,\delta^2\,\mu \\
&-4\,\rho\,\mu^3\} \qquad (39)
\end{aligned}
$$

$$M_0 = M_{02}\gamma^2 + M_{01}\gamma + M_{00} \qquad (40)$$

$$M_{02} = -2\,\frac{\rho\,\delta}{(1+l)^2} + 2\,\delta^2 + 2\,\rho\,\delta + \delta\,\sigma \qquad (41)$$

$$
\begin{aligned}
M_{01} &= 3\,\mu\,\delta\,\sigma + 2\,\delta^2\,\sigma + 2\,\rho\,\delta^2 + 6\,\rho\,\delta\,\mu + 6\,\mu\,\delta^2 + \delta\,\sigma^2 + 2\,\rho^2\,\delta + 3\,\rho\,\delta\,\sigma \\
&+\frac{-\rho\,\mu\,\sigma + \rho\,\delta\,\sigma - 4\,\rho\,\delta\,\mu + 2\,\rho\,\delta^2 + (4\,\rho^2\,\delta + 4\,\rho\,\delta^2 + 2\,\rho\,\delta\,\sigma)\,l}{(1+l)^2} \qquad (42)
\end{aligned}
$$

$$
\begin{aligned}
M_{00} &= \frac{1}{(1+l)^2}\{-2\,\rho^2\,\mu\,\sigma - 4\,\rho\,\mu^2\,\delta - \rho\,\mu\,\sigma^2 - 2\,\rho\,\mu^2\,\sigma - 4\,\rho^2\,\mu\,\delta - 2\,\sigma\,\rho^2\,\delta \\
&-\rho\,\delta\,\sigma^2 - 2\,\rho\,\delta^2\,\sigma - 2\,\rho\,\mu\,\sigma\,\delta + (-2\,\rho\,\delta\,\sigma^2 - 4\,\rho\,\delta^2\,\sigma - 4\,\sigma\,\rho^2\,\delta)\,l\} \\
&+2\,\mu\,\delta^2\,\sigma + 4\,\rho\,\mu^2\,\delta + \mu\,\delta\,\sigma^2 + 4\,\rho\,\mu\,\sigma\,\delta + 4\,\rho\,\delta^2\,\mu \\
&+\rho\,\delta\,\sigma^2 + 2\,\rho\,\delta^2\,\sigma + 2\,\mu^2\,\delta\,\sigma + 4\,\rho^2\,\mu\,\delta + 2\,\sigma\,\rho^2\,\delta + 4\,\mu^2\,\delta^2 \qquad (43)
\end{aligned}
$$

$$
\begin{aligned}
L_0 &= 3\,\mu^2\,\delta\,\sigma^2 + \mu\,\delta\,\sigma^3 + 2\,\sigma\,\mu^3\,\delta + 4\,\mu^3\,\delta^2 + 4\,\mu^2\,\delta^3 + 3\,\mu\,\delta^2\,\sigma^2 \\
&+2\,\mu\,\delta^3\,\sigma + 8\,\rho\,\mu^2\,\delta^2 + 4\,\rho\,\mu\,\sigma\,\delta^2 + 8\,\mu^2\,\delta^2\,\sigma + 2\,\gamma^2\,\delta^3 + \gamma\,\delta\,\sigma^3 \\
&+8\,\sigma\,\rho\,\mu^2\,\delta + 2\,\sigma\,\rho^2\,\mu\,\delta + 3\,\rho\,\mu\,\sigma^2\,\delta + 11\,\gamma\,\delta^2\,\mu\,\sigma + 9\,\sigma\,\rho\,\gamma\,\delta\,\mu \\
&+6\,\delta\,\gamma\,\rho\,\mu^2 + 6\,\gamma\,\delta^2\,\mu^2 + 2\,\gamma^2\,\delta^2\,\mu + 3\,\gamma^2\,\delta^2\,\sigma + \gamma^2\,\delta\,\sigma^2 + 6\,\gamma\,\delta^3\,\mu \\
&+2\,\gamma\,\delta^3\,\sigma + 3\,\gamma\,\delta^2\,\sigma^2 + 8\,\rho\,\gamma\,\delta^2\,\mu + 3\,\sigma\,\gamma\,\delta\,\mu^2 + \sigma\,\gamma^2\,\delta\,\mu + 2\,\delta\,\gamma\,\rho^2\,\mu
\end{aligned}
$$

$$+2\,\gamma\,\rho\,\delta\,\sigma^2+2\,\gamma\,\rho\,\delta^2\,\sigma+4\,\gamma\,\delta\,\mu\,\sigma^2+4\,\rho\,\mu^3\,\delta+4\,\rho^2\,\delta\,\mu^2+2\,\rho\,\gamma^2\,\delta^2$$
$$+2\,\rho\,\gamma^2\,\delta\,\mu+2\,\rho\,\gamma^2\,\delta\,\sigma+\frac{1}{(1+l)^2}\{3\,\gamma\,\rho\,\delta^2\,\sigma+4\,\rho^2\,\mu\,\delta^2+4\,\rho\,\mu\,\sigma\,\delta^2$$
$$+4\,\rho\,\delta^3\,\mu-2\,\rho\,\gamma\,\delta^2\,\mu+2\,\gamma\,\rho\,\delta^3+2\,\gamma\,\rho^2\,\delta^2-2\,\rho\,\gamma^2\,\delta^2-6\,\sigma\,\rho\,\gamma\,\delta\,\mu$$
$$-4\,\delta\,\gamma\,\rho\,\mu^2-4\,\rho\,\mu^3\,\delta-2\,\rho\,\gamma^2\,\delta\,\sigma-4\,\rho^2\,\delta\,\mu^2-4\,\sigma\,\rho\,\mu^2\,\delta-2\,\rho\,\gamma^2\,\delta\,\mu$$
$$+\sigma\,\gamma\,\rho^2\,\delta+\sigma\,\gamma\,\rho\,\mu^2+\sigma\,\gamma\,\rho^2\,\mu+(8\,\rho^2\,\mu\,\delta^2+8\,\rho\,\mu\,\sigma\,\delta^2+8\,\rho\,\delta^3\,\mu$$
$$+4\,\sigma\,\rho^2\,\mu\,\delta+2\,\rho\,\mu\,\sigma^2\,\delta+2\,\sigma\,\rho\,\gamma\,\delta\,\mu+4\,\rho\,\gamma\,\delta^2\,\mu+4\,\gamma\,\rho\,\delta^3$$
$$+4\,\gamma\,\rho^2\,\delta^2+4\,\delta\,\gamma\,\rho^2\,\mu+2\,\gamma\,\rho\,\delta\,\sigma^2+6\,\gamma\,\rho\,\delta^2\,\sigma+4\,\sigma\,\gamma\,\rho^2\,\delta)\,l\,\} \qquad (44)$$

We define the following variables:

$$\beta_0=-\frac{L_0}{M_0};\qquad \gamma_0=maximum\ root\ of\ \ M_0(\gamma)=0 \qquad (45)$$
$$\beta_1=-\frac{L_1}{M_1};\qquad \gamma_1=maximum\ root\ of\ \ M_1(\gamma)=0 \qquad (46)$$
$$\beta_2=-\frac{L_2}{M_2};\qquad \gamma_2=maximum\ root\ of\ \ M_2(\gamma)=0 \qquad (47)$$

$$G_0(\beta)=C_4C_3C_2-C_2^2-C_4^2C_1 \qquad (48)$$
$$=G_{03}\beta^3+G_{02}\beta^2+G_{01}\beta+G_{00} \qquad (49)$$
$$G_{03}=M_2M_3-M_1 \qquad (50)$$
$$G_{02}=L_2M_3+M_2M_3L_4+M_2L_3-M_2^2-L_1-2M_1L_4 \qquad (51)$$
$$G_{01}=M_3L_2L_4+L_2L_3-2L_2M_2+M_2L_3L_4-L_4^2M_1-2L_4L_1 \qquad (52)$$
$$G_{00}=L_2L_3L_4-L_2^2-L_4^2L_1 \qquad (53)$$

$$G_1(\beta)=(C_2C_1-C_3C_0)(C_4C_3-C_2)-(C_4C_1-C_0)^2 \qquad (54)$$
$$=G_{14}\beta^4+G_{13}\beta^3+G_{12}\beta^2+G_{11}\beta+G_{10} \qquad (55)$$
$$G_{14}=M_1M_2M_3-M_0M_3^2-M_1^2 \qquad (56)$$
$$G_{13}=M_1M_2M_3L_4-M_0M_3^2L_4+M_1M_2L_3-M_0M_3L_3-M_1M_2^2$$
$$+M_0M_2M_3-2M_1^2L_4+2M_1L_1-2M_0M_1 \qquad (57)$$
$$G_{12}=M_3L_1L_2-M_3L_0L_3+M_1M_2L_3L_4-M_0M_3L_3L_4-M_1M_2L_2$$
$$+M_0M_3L_2+M_1M_3L_2L_4+M_2M_3L_1L_4-M_0M_3L_3L_4$$
$$-M_3^2L_0L_4+M_1L_2L_3+M_2L_1L_3-M_0L_3^2-M_3L_3L_0$$
$$-M_1M_2L_2-M_2^2L_1+M_0M_2L_3+M_2M_3L_0-4M_1L_1L_4$$
$$+2M_1L_0-M_1^2L_4^2+2M_0M_1L_4-L_2^2+2M_0L_1-M_0^2 \qquad (58)$$

$$G_{11} = M_1L_2L_3L_4 + M_2L_1L_3L_4 - M_0L_3^2L_4 - M_3L_0L_3L_4 - M_1L_2^2$$
$$-M_2L_1L_2 + M_0L_2L_3 + M_3L_0L_2 + M_3L_1L_2L_4 + L_1L_2L_3$$
$$-M_2L_1L_2 - M_3L_0L_3L_4 - L_0L_3^2 + M_2L_0L_3 - 2M_1L_1L_4^2$$
$$+2M_1L_0L_4 - 2L_1^2L_4 + 2L_0L_1 + 2M_0L_1L_4 - 2M_0L_0 \quad (59)$$
$$G_{10} = L_1L_2L_3L_4 - L_1L_2^2 - L_0L_3^2L_4 + L_0L_2L_3 - L_1^2L_4^2 + 2L_0L_1L_4$$
$$-L_0^2 \quad (60)$$

Next, we define

$$\beta_6 = maximum\ root\ of\ \ G_0(\beta) = 0 \quad (61)$$
$$\beta_7 = maximum\ root\ of\ \ G_1(\beta) = 0 \quad (62)$$

**Theorem 2** *Suppose the following conditions are satified:*
$\beta > max(\beta_0, \beta_1, \beta_2, \beta_6, \beta_7)$ *(see (45), (46), (47), (61), (62))*
$\gamma > max(\gamma_0, \gamma_1, \gamma_2)$ *(see (45), (46), (47))*
$G_{03} > 0$ *(see (50))*
$G_{14} > 0$ *(see (56))*
*Then $D_1$ is locally asymptotically stable.*

**Proof.** We show that the conditions of Lemma 1 are satisfied under the conditions of this theorem. $C_4$ and $C_3$ are always non-negative. Whereas, if $\beta > -\frac{L_2}{M_2} = \beta_2$ and $M_2 > 0$, then $C_2 > 0$. We get $M_2 > 0$ if $\gamma > \gamma_2$. Similarly, if $\beta > -\frac{L_1}{M_1} = \beta_1$ and $M_1 > 0$, then $C_1 > 0$. We get $M_1 > 0$ if $\gamma > \gamma_1$. Finally, if $\beta > -\frac{L_0}{M_0} = \beta_0$ and $M_0 > 0$, then $C_0 > 0$. We get $M_0 > 0$ if $\gamma > \gamma_0$. The remaining two conditions of Lemma 1 are satisfied if $G_0(\beta) > 0$ and $G_1(\beta) > 0$. If we choose $\beta > \beta_6$ and $\beta > \beta_7$ when $G_{03} > 0$ and $G_{14} > 0$, respectively, then these two conditions will be satisfied. ■

## References

[1] Jing Zhujun and Liu Zhengrong, Qualitative Analysis for a Mathematical Model of AIDS, Acta Mathematicae Applicatae Sinica, 9(4), pp 302 –316 (1993).

[2] Lancaster, P., Theory of Matrices, Academic Press, New York, 1969.

# AN ANALYSIS OF CONJUNCTIVE MODEL REDUCTION ALGORITHMS FOR SOLVING LINEAR PROGRAMMING PROBLEMS

**V.C. Jacob, T. R. Natesan, V. Rhymend Uthariaraj and P. Narayanasamy**

Ramanujan Computing Centre, Anna University, Chennai-600 025, India.
E-mail: trn@annauniv.edu

---

Mathematical models are an integral part of everyday life and is concerned with optimal decision making and modelling deterministic and probabilistic systems that originate from real life. These models form a bridge between powerful mathematical techniques and computers. Presence of redundancies in Linear Programming models is hardly disputed and therefore they form an integral part of an originally constructed model. Redundancy may occur in the formulation phase because of bad source data or not to run the risk of omitting some relevant constraints while modelling. These redundant constraints / variables when present in the model will bring down the computational efficiency and can play havoc with LP solution procedures. Very few researchers have attempted to identify and remove redundancy apriori. In this paper some of the earlier developed algorithms for identification of redundancies apriori are discussed and an improved version has been proposed which make use of the gradient matrix of the constraints and the Matrix of Intercepts Algorithms. This approach is a conjunctive use of different algorithms and is able to detect redundancies, which is 20% more than the earlier researcher algorithms. A detailed study of the behaviour of the model reduction algorithms has been made by mixing the lower bound and equality constraints, and mixed set of constraints for minimization objective and upper bound and equality constraints and mixed set of constraints for maximization objective respectively. The proposed algorithm gives substantial improvement in the computational results with respect to time and identification of redundancies for both minimization and maximization objective functions when compared to the earlier researcher's approaches for many LP problems that were randomly generated and down loaded from the net library apart from those chosen from textbooks, journals and Ph.D. dissertations.

## 1. INTRODUCTION

The formulation of a linear programming model involves a detailed study of the system, data collection, identification of decision variables, construction of the objective function and system constraints. In the course of formulating a model, model builders often tend to include either inadvertently or innocently, constraints and variables that may not play a role at all in defining the feasible set. The presence of such constraints / variables is hardly disputed and can play havoc with Linear Programming solution procedures and greatly hamper the solution effort. It is well known that every additional constraint / variable in a Linear Programming Problem demands more computational effort and computer memory. The identification of such embedded non-role play constraints / variables in the model without affecting the character of the system is as difficult as solving the Linear Programming Problem itself.

It is very hard to determine how much redundancy is present in an LP model unless it is solved. Thus it is always a mute question whether it is worth to spend time to identify redundancy when there are sophisticated computers to solve large scale LP problems in a less time? The existence of redundancies in a LP model can not only play havoc in solution procedures but also mislead good decisions, which can indirectly affect the organization. It is therefore worth spending some effort on eliminating redundancy in a model instead of simply ignoring it.

The cost to eliminate redundancy can be calculated by three options *viz.* a) Ignore b) identify redundancy c) Identify and remove redundancy. This paved a way to many researchers to develop algorithms for the identification of redundancies and remove them to get a reduced model.. There always had been a need for developing simple algorithms to detect apriori redundant constraints at minimal computational effort. The main objective of these algorithms had been to remove all redundancies in a LP model, which is not always possible in practice. This has lead to many improvements in developing algorithms to identify as many redundancies as possible in a given LPP.

The earlier researchers have proposed several methods [1-10] for minimizing the computational effort by identifying redundant constraints and variables. But most of these methods proposed for the identification of redundancies are computationally expensive. However, the heuristic algorithms proposed by recent researchers [11-14] have shown significant improvement in computational efficiency. But still these algorithms did not identify all the redundancies leaving scope for further improvement. This has motivated to think and develop an integrated algorithm, which will further reduce the size of the original model and speed up the process of computation saving memory too.

Consider the Linear Programming model with m constraints and n variables.

Extremize Z = CX

Subject to AX $\{\leq , = , \geq\}$ b; X >= 0 with $A \varepsilon R^{mxn}$, $b\varepsilon R^{m}$, $C \varepsilon R^{n}$ and $X \varepsilon R^{n}$

The proposed heuristic model reduction algorithm is applied to the original LP model to identify apparent redundancies and the reduced model is solved for optimal solution using the simplex algorithm. The apparent redundant constraints are validated by substituting the optimal solution obtained from the reduced model and necessary amendments are made to incorporate the faulting constraints into the reduced model and re-seek the optimal solution with minimal computational effort by using the primal-dual properties and post-optimality analysis.

This paper is divided into four sections. The first section gives a brief introduction, section 2 describes the already existing heuristic model reduction algorithm for apriori identification of redundancies in a LP model and section 3 describes the conjunctive model algorithm and its analysis. Lastly section 4 deals with computational results and conclusions are drawn.

## 2. HEURISTIC MODEL REDUCTION ALGORITHMS

The model reduction is based on the following theorem:

**Theorem:** One or more constraint(s) of $\sum_{j=1}^{n} a_{ij}x_j \ \{\leq, =, \geq\}\ b_i$ for $i=1,2....m$; $x_j \geq 0$ is redundant for the system of constraints $AX\ \{\leq, =, \geq\}\ b$, if and only if its associated surplus or artificial variable(s) $s_i$ / $r_i$ are in the basis of every primary sub-system of the LP problem. In other words if a surplus/ artificial variable stays in the basis in all the iterations with a value ( $\geq 0$) then the associated constraint is either strongly (if value>0) or weakly (if value=0) redundant for the LP model.

The step by step procedure of the already developed heuristic algorithms for apriori identification of redundancies using Constraint Gradient matrix algorithm and Matrix of Intercepts algorithm is given below.

### 2.1 CONSTRAINT GRADIENT MATRIX ALGORITHM: The following are the steps :

***Step 1*** : A gradient matrix is constructed by taking the transpose of the constraint matrix $A_{mxn}$. That is Gradient Matrix $(g_{ij}) = A^t$.

***Step 2*** : Identify the largest element in each row, $\beta_j = \max_i \{ g_{ij} \}$ for all j

***Step 3*** : Score out the rows and columns corresponding to the entering and leaving variables. If a column has more than one largest element score out those rows also.

***Step 4*** : The constraints associated with the surplus variables in the unscored columns, if any, abinitio are assumed and predicted to be redundant.

***Step 5*** : Remove these redundant constraints tentatively from the original model.

The application of the above steps results in a reduced model.

**2.2 MATRIX OF INTERCEPTS ALGORITHM:** The following are the steps:

***Step 1***: A matrix of intercepts is constructed with the decision and surplus variables as rows and columns respectively. This matrix is of dimension n x m.

$$\theta_{ji} = b_i / a_{ij} \quad ; \qquad a_{ij} > 0 \text{ for all } i=1,2 \ldots\ldots m \text{ and } j=1,2 \ldots\ldots n$$

***Step 2***: Identify the pivot element in each row

$\beta_j = \min\{\theta_{ji}\}$ for all j while the objective is maximization and constraints are upper bound

$\beta_j = \max\{\theta_{ji}\}$ for all j while the objective is minimization and constraints are lower bound

***Step 3***: Score out the rows and columns corresponding to the entering and leaving variables. If a column has more than one pivot element score out those rows also.

***Step 4:*** The constraints associated with the surplus variables in the unscored columns, if any are assumed and predicted to be redundant.

***Step 5***: Remove these redundant constraints tentatively from the original model.

The application of the above steps to the original LP model results in an apparent reduced primal model. The matrix of intercepts algorithm for maximization objective and the same for minimization objective are referred to as Theta Min. and Theta Max. algorithm respectively.

## 3. CONJUNCTIVE MODEL REDUCTION ALGORITHM AND ITS ANALYSIS

The already developed heuristic algorithms did not find all the redundant constraints in a LP model, thus leaving a scope for further improvement. These algorithms were applied to linear inequalities and some of the algorithms were based on the objective function. Moreover the size of the problem and the computational time increase when more number of equality constraints, which are written as two simultaneous inequalities, are present in the model.

This motivated the authors to think of some alternate algorithms which will make use of the already developed matrix of intercepts and constraint gradient matrix algorithms to identify more number of redundant constraints. The union of the algorithms showed substantial improvements in the identification of redundancies. The advantage of the proposed algorithm has been it is simple in construction and easy to implement without any extra computational effort. The computational effort required to identify the apriori redundant constraints in the matrix of Intercepts algorithms depends on the size of the $\theta$ matrix since mxn number of divisions have to be performed at start in the worst case. The gradient matrix algorithm comparatively requires no extra computational effort at start.

The hybrid algorithm [15] for minimization and maximization objective have shown a larger increase in the identification of redundancies than the individual algorithms. However these algorithms could not identify all the redundancies and was based on the objective function. The quest for identification of more redundancies and a general algorithm for identifying redundancies for a LP model has lead to the development of the Conjunctive algorithm.

The Conjunctive algorithm which is the union of all the three algorithms *viz*– the Gradient, Theta-Max. and Theta-Min., identified more number of redundancies which otherwise escaped detection by the individual algorithms. The performance of the individual, hybrid and proposed conjunctive algorithms were tested on a variety of problems for each of the objective function partly downloaded from the net-library and partly from other sources. It was observed that the conjunctive algorithm identifies more number of redundancies than the individual irrespective of the objective function. Thus the proposed algorithm could be adopted for any general LP model. Figure-I gives the schematic of the model reduction algorithm.

A detailed analysis of the algorithms was done based on a set of 102 LP Problems containing mixed set of constraints for each of the minimization and maximization objectives respectively. The number of constraints ranges between 10 and 70, and the number of variables on an average is three times the number of constraints. The gradient algorithm (Method A / Method R), Theta Max. (Method B) and the hybrid algorithm for minimization objective, which is the union of Method A and B ( Method C) is compared with conjunctive algorithm (Method D / Method U) for a minimization objective. Similarly the gradient algorithm (Method R), Theta Min. (Method S) and the hybrid algorithm for maximization objective, which is the union of method R and S (Method T) is compared with the conjunctive algorithm (Method D / Method U) for a maximization objective

This study has proved that the proposed conjunctive model reduction algorithm is the best in identifying the maximum number of redundant constraints among the various methods for each of the objective function. Some of the factors that were considered for the study for each of the algorithm are initial and final identified redundant constraints, number of iterations, number of multiplication/division operations and time in milli-seconds. The computational results of the 102 problems that were considered for each of the objective function are highlighted in the next section. Figure II gives the average computational time taken by the original and reduced model solved by the simplex algorithm (204 problems were solved). The conjunctive algorithm identifies 20% more redundancies and takes less computational time to solve the problem than the earlier algorithms.

## 4. COMPUTATIONAL RESULTS

To evaluate the performance of the proposed algorithm, it is first tested whether the mean vector – the number of iterations, the number of multiplication and division operations and the time taken to obtain the optimal solution using the usual revised simplex method and the proposed algorithm for each of the objective function are equal. Lastly the performance of the conjunctive algorithm is tested for any objective function by combining the results in minimization and maximization objective. That is to test the hypothesis that $Ho: \mu^{RS} = \mu^{PA}$ where $\mu^{RS}$ and $\mu^{PA}$ contain the above mean vector for the revised simplex method and the proposed algorithm, against the alternative hypothesis ($H_1$) that the mean vectors are not equal. The redundant constraint is not included as one of the factors in the above study because the revised simplex algorithm does not identify redundant constraints whereas the proposed algorithm does. The main interest in the above analysis has been the saving of computational time. That is to know whether the proposed algorithm is computationally efficient when compared to solving LPP by the revised simplex method. The statistical software-SPSS is used to test the above hypothesis using one-way classification in Multivariate Analysis Of Variance (MANOVA). In both the cases it was found that the conjunctive algorithm takes less computational time to solve the LP model than the original model using simplex algorithm. Table-I gives the results of MANOVA for any objective function.

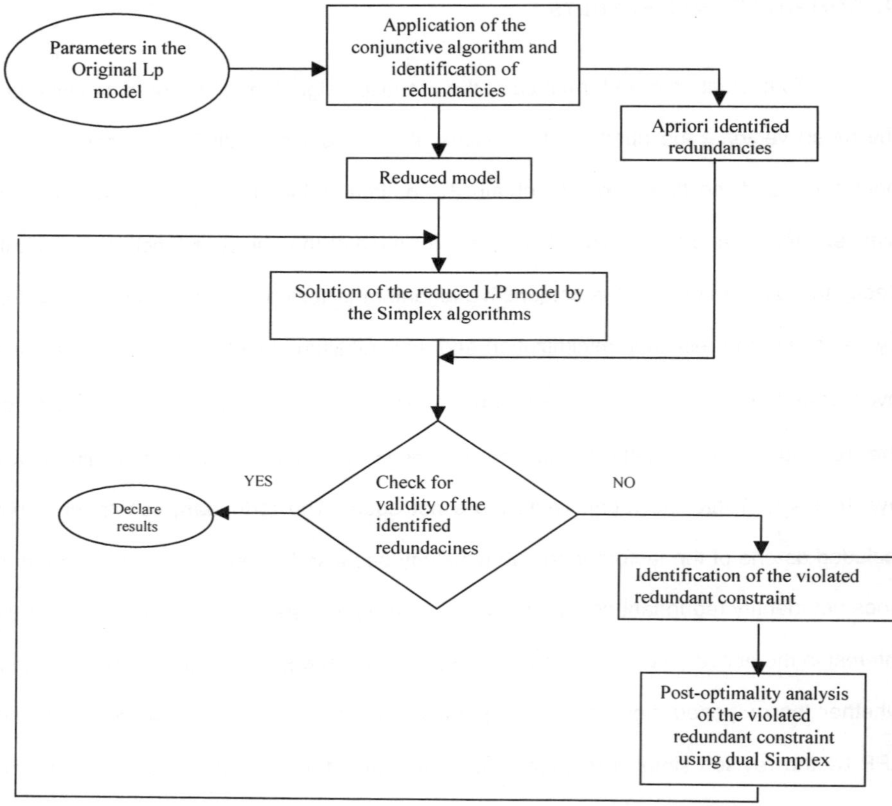


**Figure-I Schematic of the model reduction algorithm**

Figure-III and IV gives the average number of redundancies identified by various methods in percentages for minimization and maximization objective respectively. A comparison of the computational time in milli-seconds with simplex algorithm and various model reduction algorithms for each of the objective function is shown in Figure-V and VI. Thus it may be seen that the conjunctive algorithm identifies maximum number for redundancies for each of the objective function and takes less computational time.

**Table I – Results of F-statistic value using MANOVA**

| Variable | F- calc | Sig. |
|---|---|---|
| Iterations | 0.650 | 0.42 |
| Mult_division | 80.309 | 0.00 |
| Time | 75.475 | 0.00 |

**Figure II- Average computational time taken by the Original and Reduced model using Simplex Algorithm**

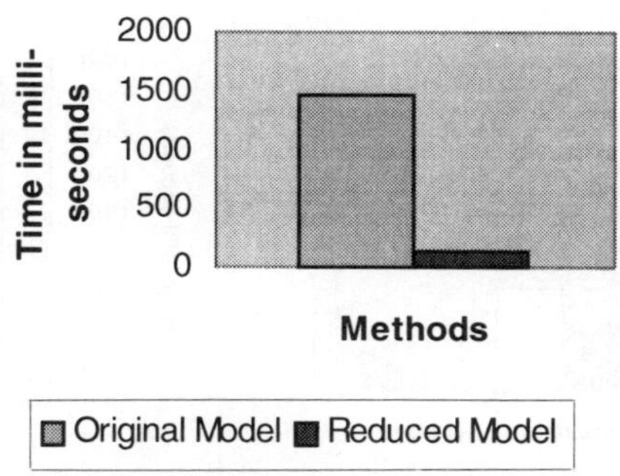

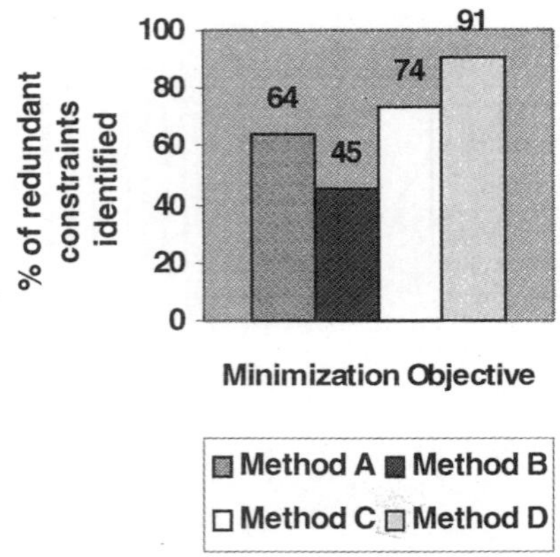


**Figure III - Average percentage of redundancies identified by model reduction algorithms with minimization objective**

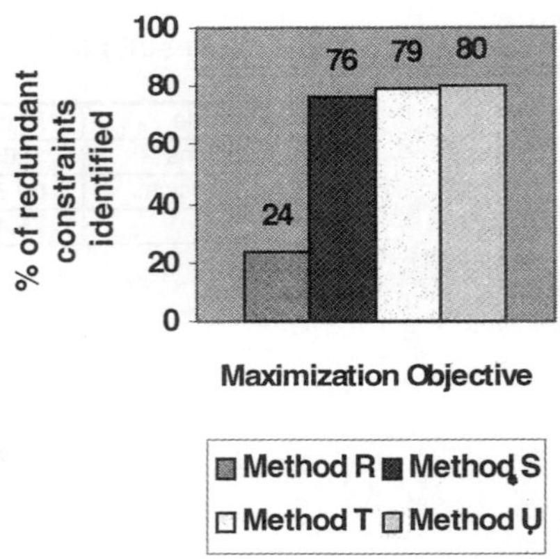


**Figure IV - Average percentage of redundancies identified by model reduction algorithms with maximization objective**

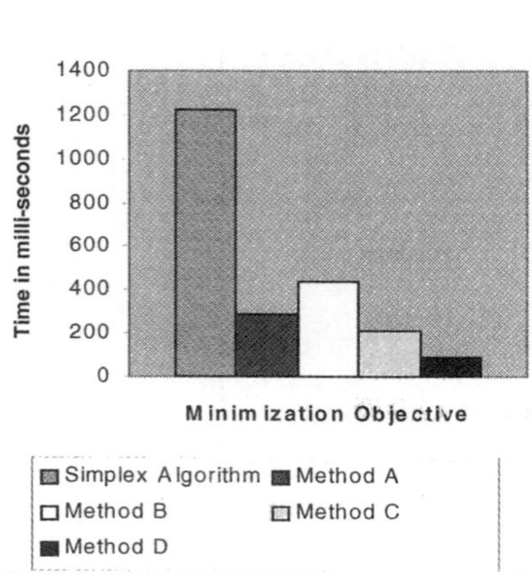


**Figure V - Average computational time taken by Original and reduced models by the simplex algorithm for Minimization Objective**

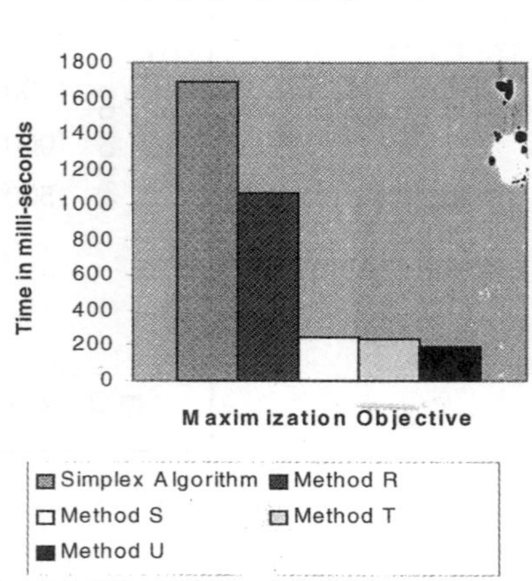


**Figure VI - Average computational time taken by Original and reduced models by the simplex algorithm for Maximization Objective**

## 5. CONCLUSIONS

Thus the conjunctive algorithm can be used as a general algorithm for any objective function and it yields better results than any other algorithms discussed above. The proposed algorithm identifies 20% more redundancies than the earlier algorithms and takes less computational time. Moreover it can be inferred from the above computational results that the conjunctive algorithm takes less time to solve the LP model and is able to detect more number of redundant constraints than other model reduction algorithms for each of the objective function. These algorithms are simple and can be adopted very easily by the existing packages. Some more improvement in the identification of redundancies can be obtained by improving the post-optimality analysis routine. Also the analysis shows that there is some need for improvement in the existing gradient matrix algorithm for the identification of redundancies for maximization objective.

**ACKNOWLEDGEMENT**

Prof. T.R. Natesan , Professor Emeritus , Anna University records his sincere thanks and gratitude to the All India Council for Technical Education, New Delhi for the financial support extended to him to carry out this work.

## References

1. G. B. Dantzig, "Upper Bounds, Secondary Constraints and Block-triangularity" , Econometrica, Vol. 23, No. 2, pp 174-183. (1955)
2. M.L. Balinsky, "An Algorithm for finding all Vertices of Convex Polyhedral Sets", Journal of the Society for Industrial and Applied Mathematics, Vol. 9, No. 1, pp. 72-78,(1961).
3. A. Shefi, "Reduction of Linear Inequality Constraints and Determination of All possible Extreme points", Ph.D. Dissertation, Stanford University,(1969).
4. T.H. Matthesis, "An Algorithm for Determining Irrelevant Constraints and All vertices in Systems of Linear Inequalities", Operations Research, Vol. 21, No. 1, pp. 247-260 (1973)
5. M.H. Karwan, Vahid Lotfi, Jan Telgen and Stanley Zionts, Redundancy in Mathematical Programming, Lecture notes in Economics and Mathematical Systems, Springer –verlag(1982)
6. R. Bixby and D. Wagner, " A Note on Detecting Simple Redundancies in Linear System" , Operations Research Letters 6(1), pp. 15-17, (1987)
7. C. Danny, Myers and Wei Shih, " A Constraint Selection Technique for a class of Linear Programs", Operations Research Letters,7(4), pp. 191-195(1988).
8. J.L. Imbert, Pascal Van Hentenryck," A Note on Redundant Linear Constraints", Technical Report No. CS-92-11, Department of Computer Science, Brown University (1992).
9. J.L. Lassez, T. Huynh and K. McAloon," Simplification and Elimination of Redundant Linear Arithmetic Constraints", Constraint Logic Programming: Selected Research, pp. 73-87, MIT Press, USA (1993).
10. J.L. Imbert, Pascal Van Hentenryck," Redundant Elimination with a Lexicographic Solved Form", Technical Report No. CS-95-02, Department of Computer Science, Brown University (1995).
11. T.R. Natesan, S. Paulraj and C. Chellappan," A New Algorithm for Determining Redundant Constraints from Linear Programming Problems", IASTED- International conference on applied Modeling and Simulation, Havaii, U.S.A. (1998)
12. S. Paulraj, " Studies of the Computational Aspects of Linear Programming Problems", Ph.D. thesis, Anna University(1998).
13. V. Rhymend Uthariaraj, T.R. Natesan and V. Sankaranarayanan, " Algorithms for Identification of Redundancies in Linear Programming Problems", in the 19th IFIP international conference on Systems Modeling and Optimization, Cambridge , London (1999)
14. V. Rhymend Uthariaraj, " Algorithms for Model Reduction of Linear Programming Problems and Computational Efficiency " Ph.D. thesis (1999).
15. V.C. Jacob, T.R. Natesan, V. Rhymend Uthariaraj and P. Narayanasamy," Analysis of model reduction algorithms for solving Linear Programming Problems", in the X congreso Latino Ibero Americano De Investigacion De Operaciones conference , Mexico city (2000).
16. H.A. Taha, Operations Research an Introduction, fifth edition, Prentice Hall International Inc.

# WEAK CONVERGENCE OF INJECTIVE TENSOR PRODUCTS OF VECTOR MEASURES

JUN KAWABE

*Shinshu University, 4-17-1 Wakasato, Nagano 380-8553, JAPAN*
*E-mail: jkawabe@gipwc.shinshu-u.ac.jp*

The purpose of this paper is to explain some results concerning the joint continuity of the injective tensor product of two vector measures with respect to the weak convergence. The joint continuity is studied in two cases that the vector measures take values in certain nuclear spaces and in the positive cone of Banach lattices.

## 1 Introduction

The notion of the injective tensor product of vector measures was introduced by Duchon and Kluvánek[7] in 1967: Let $(\Omega, \mathcal{A})$ and $(\Gamma, \mathcal{B})$ be measurable spaces. Let $X$ and $Y$ be real locally convex spaces, and $X\widehat{\otimes}Y$ the injective tensor product of $X$ and $Y$. Let $\mu : \mathcal{A} \to X$ and $\nu : \mathcal{B} \to Y$ be vector measures. Then the injective tensor product of $\mu$ and $\nu$, which is denoted by $\mu\widehat{\otimes}\nu$, is a vector measure defined on the product $\sigma$-field $\mathcal{A} \times \mathcal{B}$ with values in $X\widehat{\otimes}Y$ such that $\mu\widehat{\otimes}\nu(A \times B) = \mu(A) \otimes \nu(B)$ for all $A \in \mathcal{A}$ and $B \in \mathcal{B}$.

On the other hand, Dekiert[5] recently introduced the notion of weak convergence of vector measures, which is a natural generalization of the weak convergence of probability measures, and März and Shortt[20] gave a sufficient condition for a set of vector measures with values in a Banach space to be sequentially compact with respect to this weak convergence. This compactness criterion has been extended to vector measures which are defined on a completely regular space whose compact subsets are all metrizable, and take values in certain locally convex spaces (see Kawabe[13,16]).

The purpose of this paper is to explain some results concerning the joint continuity of the injective tensor product of two vector measures with respect to the weak convergence in the following two cases: One is the case that the vector measures take values in some nuclear spaces such as the space $\mathscr{S}$ of all rapidly decreasing, infinitely differentiable functions, the space $\mathscr{D}$ of all test functions, and the strong duals of those spaces. The other is the case that they take values in the positive cone of Banach lattices.

For any two probability measures, the injective tensor product of them is just the usual product measure, and so its joint continuity is well known in the case that the underlying topological spaces, on which measures are defined, are separable metric spaces (see Billingsley[2]), and more generally completely regular spaces (see Vakhania *ea al.*[25]). As to the results in the similar direction, it was shown in Csiszár[4] that the convolution of two probability measures on an arbitrary topological group is jointly continuous. These results are important in probability theory, and are unified within the framework of the weak convergence of compound probability measures (see Kawabe[10]).

All the topological spaces and uniform spaces in this paper are Hausdorff, and the scalar fields of topological vector spaces are taken to be the field $\mathbb{R}$ of all real numbers.

## 2 Definitions and Notation

Let $\langle E, F\rangle$ be a dual system over $\mathbb{R}$ with the canonical bilinear form $(x,y) \in E \times F \to \langle x, y\rangle$, and $\tau$ a locally convex topology on $E$ which is consistent with this duality, that is, the dual of $(E,\tau)$ is identical with $F$. The weak (resp. strong) topology on $E$ means the $\sigma(E,F)$ (resp. $\beta(E,F)$)-topology on $E$, and $E_\sigma$ (resp. $E_\beta$) denotes the space $E$ with the topologie $\sigma(E,F)$ (resp. $\beta(E,F)$).

We first recall some definitions and basic facts of vector measures with values in the locally convex space $E$. Let $(\Omega, \mathcal{A})$ be a measurable space and $\mu : \mathcal{A} \to E$ be a finitely additive set function. We say that $\mu$ is a *vector measure* if it is countably additive, i.e., for any disjoint sequence $\{A_k\}_{k=1}^\infty$ in $\mathcal{A}$ with $A = \bigcup_{k=1}^\infty A_k$, we have $\sum_{k=1}^\infty \mu(A_k) = \mu(A)$. Denote by $\mathcal{M}(\Omega; E)$ the set of all vector measures $\mu : \mathcal{A} \to E$. When $E = \mathbb{R}$, we write $\mathcal{M}(\Omega)$ instead of $\mathcal{M}(\Omega;\mathbb{R})$. If $\mu$ is a vector measure, then $(y\mu)(A) \equiv \langle \mu(A), y\rangle$, $A \in \mathcal{A}$, is a real measure for each $y \in F$. Conversely, a theorem of Orlicz and Pettis ensures that a finitely additive set function $\mu : \mathcal{A} \to E$ is countably additive if so is $y\mu$ for each $y \in F$.

Let $\mu : \mathcal{A} \to E$ be a finitely additive set function and $p$ be a seminorm on $E$. Then the *p-semivariation* of $\mu$ is the set function $\|\mu\|_p : \mathcal{A} \to [0,\infty]$ defined by $\|\mu\|_p(A) = \sup_{y\le p} |y\mu|(A)$, $A \in \mathcal{A}$, where $|y\mu|(\cdot)$ is the total variation of the real measure $y\mu$ and $y \le p$ means $\langle x, y\rangle \le p(x)$ for all $x \in E$. See Diestel and Uhr[6], Lewis[19], and Kluvánek and Knowles[18] for some properties of semivariations.

Let $\mathcal{V}$ be a subset of $\mathcal{M}(\Omega; E)$. We say that $\mathcal{V}$ is *uniformly bounded* if $\sup_{\mu\in\mathcal{V}} \|\mu\|_p(\Omega) < \infty$ for each $\tau$-continuous seminorm $p$ on $E$. Then, $\mathcal{V}$ is uniformly bounded if and only if $\sup_{\mu\in\mathcal{V}} |y\mu|(\Omega) < \infty$ for every $y \in F$.

Next we recall the definition and basic facts of weak convergence of vector measures. Let $X$ be a sequentially complete locally convex space. Let $S$ be a completely regular space and $\mathcal{B}(S)$ the $\sigma$-field of all Borel subsets of $S$. Denote by $C(S)$ the Banach space of all bounded, continuous real functions on $S$ with the norm $\|f\|_\infty \equiv \sup_{s\in S}|f(s)|$. Let $\{\mu_\alpha\}$ be a net in $\mathcal{M}(S;X)$ and $\mu \in \mathcal{M}(S;X)$. The following definition is a natural generalization of the weak convergence of real measures, and it is reduced to that of Varadarajan[26] in the case that $X = \mathbb{R}$ and is reduced to that of Dekiert[5] in the case that $X$ is a Banach space. We say that $\{\mu_\alpha\}$ *converges weakly* to $\mu$ for the topology $\tau$ on $X$ if, for every $f \in C(S)$, we have

$$\lim_\alpha \int_S f d\mu_\alpha = \int_S f d\mu \quad \text{for the topology } \tau \text{ on } X.$$

See Lewis[19], and Kluvánek and Knowles[18] for the definition and basic properties of the integral above. We only notice that every bounded, measurable, real function on an arbitrary measurable space is integrable with respect to any vector measure with values in a sequentially complete locally convex space.

## 3 Injective Tensor Product of Two Vector Measures

Let $X$ and $Y$ be locally convex spaces. Denote by $X^*$ and $Y^*$ the topological duals of $X$ and $Y$, respectively. Let $(\Omega, \mathcal{A})$ and $(\Gamma, \mathcal{B})$ be measurable spaces. Denote by $X\widehat{\otimes}Y$ and $X\widehat{\otimes}_\pi Y$ the injective and projective tensor products of $X$ and $Y$, respectively. Let $\mu \in \mathcal{M}(\Omega; X)$ and $\nu \in \mathcal{M}(\Gamma; Y)$. If a set $C$ is of the form

$$C = \bigcup_{k=1}^{n}(A_k \times B_k), \tag{1}$$

where the union is disjoint and $A_k \in \mathcal{A}$, $B_k \in \mathcal{B}$, then the set function

$$\lambda(C) = \sum_{k=1}^{n} \mu(A_k) \otimes \nu(B_k) \tag{2}$$

is unambiguously defined on the field of all sets of the form (1), and is finitely additive. Then, Duchon and Kluvánek[7] proved that $\lambda$ is countably additive and can be uniquely extended to a countably additive set function, which is denoted by $\mu\widehat{\otimes}\nu$, on the $\sigma$-field $\mathcal{A}\times\mathcal{B}$ generated by all sets of the form (1) with values in $X\widehat{\otimes}Y$. This vector measure is called the *injective tensor product* of $\mu$ and $\nu$. This fact is not true in the case of the projective tensor product of $X$ and $Y$, as it was shown in Kluvánek[17]. However, if $X$ is nuclear, then the projective tensor product $X\widehat{\otimes}_\pi Y$ coincides with the injective tensor product $X\widehat{\otimes}Y$, and hence there exists a unique vector measure on $\mathcal{A}\times\mathcal{B}$ with values in $X\widehat{\otimes}_\pi Y$ satisfying condition (2). This vector measure is also denoted by $\mu\widehat{\otimes}\nu$ without any notational confusion.

## 4 Joint Continuity Problem – Nuclear Space Valued Case

In this section we consider the joint continuity problem of vector measures with values in certain nuclear spaces. Let $(\Omega, \mathcal{A})$ and $(\Gamma, \mathcal{B})$ be measurable spaces. Throughout this section, let $X$ be a strict inductive limit of an increasing sequence $\{X_n\}$ of nuclear Fréchet spaces and $Y$ a strict inductive limit of an increasing sequence $\{Y_n\}$ of Fréchet spaces. Denote by $X\overline{\otimes}Y$ a strict inductive limit of the increasing sequence $\{X_n\widehat{\otimes}_\pi Y_n\}$ of the projective tensor products of $X_n$ and $Y_n$. In this case, for $\mu \in \mathcal{M}(\Omega; X)$ and $\nu \in \mathcal{M}(\Gamma; Y)$ there exists a unique vector measure $\mu\overline{\otimes}\nu : \mathcal{A} \times \mathcal{B} \to X\overline{\otimes}Y$ such that

$$\mu\overline{\otimes}\nu(A \times B) = \mu(A) \otimes \nu(B) \tag{3}$$

for every $A \in \mathcal{A}$ and $B \in \mathcal{B}$. For, since $X$ and $Y$ are strict inductive limits of increasing sequences $\{X_n\}$ and $\{Y_n\}$ respectively, we can find a natural number $n_0$ such that $\mu \in \mathcal{M}(\Omega; X_{n_0})$ and $\nu \in \mathcal{M}(\Gamma; Y_{n_0})$. Since $X_{n_0}$ is nuclear, the projective tensor product $X_{n_0}\widehat{\otimes}_\pi Y_{n_0}$ coincides with the injective tensor product $X_{n_0}\widehat{\otimes}Y_{n_0}$, and hence there exists a vector measure $\mu\widehat{\otimes}\nu$ : $\mathcal{A} \times \mathcal{B} \to X_{n_0}\widehat{\otimes}_\pi Y_{n_0}$ satisfying condition (3). Then, it is obvious that $\mu\widehat{\otimes}\nu$ can be viewed as a vector measure with values in $X\overline{\otimes}Y$, and we call it the *inductive tensor product* of $\mu$ and $\nu$ and detnote it by $\mu\overline{\otimes}\nu$.

We also obtain an inductive tensor product of two vector measures with values in dual spaces. Since $X^*_\beta$ is nuclear, for any $\mu \in \mathcal{M}(\Omega; X^*_\beta)$ and $\nu \in \mathcal{M}(\Gamma; Y^*_\beta)$, there exists a unique vector measure $\mu\widehat{\otimes}\nu \in \mathcal{M}(\Omega \times \Gamma; X^*_\beta\widehat{\otimes}_\pi Y^*_\beta)$ satisfying condition (3). Since $(X^*_\beta\widehat{\otimes}_\pi Y^*_\beta)^* = (X\overline{\otimes}Y)^*_\beta$, we may view the measure as a vector measure with values in $(X\overline{\otimes}Y)^*$. We call it the inductive tensor product of $\mu$ and $\nu$, and denote it $\mu\overline{\otimes}\nu$ again.

**Example 1.** (1) Let $\mathscr{S}(\mathbb{R}^m)$ and $\mathscr{S}(\mathbb{R}^n)$ be the spaces of all rapidly decreasing, infinitely differentiable functions on Euclidean spaces $\mathbb{R}^m$ and $\mathbb{R}^n$ respectively. Then, these are examples of nuclear Fréchet spaces. The strong dual spaces $\mathscr{S}^*(\mathbb{R}^m)$ and $\mathscr{S}^*(\mathbb{R}^n)$ are called the spaces of all *slowly increasing distributions.* Then, we have the canonical isomorphisms (see Grothendieck[9]):

$$\mathscr{S}(\mathbb{R}^m)\widehat{\otimes}_\pi\mathscr{S}(\mathbb{R}^n) = \mathscr{S}(\mathbb{R}^{m+n}) \quad \text{and} \quad \mathscr{S}^*(\mathbb{R}^m)\widehat{\otimes}_\pi\mathscr{S}^*(\mathbb{R}^n) = \mathscr{S}^*(\mathbb{R}^{m+n}).$$

Consequently, for $\mu \in \mathcal{M}(\Omega; \mathscr{S}(\mathbb{R}^m))$ and $\nu \in \mathcal{M}(\Gamma; \mathscr{S}(\mathbb{R}^n))$, the inductive tensor product $\mu\overline{\otimes}\nu$ exists and takes values in $\mathscr{S}(\mathbb{R}^{m+n})$. When $\mu \in \mathcal{M}(\Omega; \mathscr{S}^*(\mathbb{R}^m))$ and $\nu \in \mathcal{M}(\Gamma; \mathscr{S}^*(\mathbb{R}^n))$, then $\mu\overline{\otimes}\nu$ also exists and takes values in $\mathscr{S}^*(\mathbb{R}^{m+n})$.
(2) Let $U \subset \mathbb{R}^m$ and $V \subset \mathbb{R}^n$ be open sets. Denote by $\mathscr{D}(U)$, $\mathscr{D}(V)$ and $\mathscr{D}(U \times V)$ the spaces of all test functions on $U$, $V$ and $U \times V$, respectively.

Then, these are examples of locally convex spaces whose type is a strict inductive limit of an increasing sequence of nuclear Fréchet spaces. The strong dual spaces $\mathscr{D}^*(U)$, $\mathscr{D}^*(V)$, and $\mathscr{D}^*(U\times V)$ are called the spaces of all *distributions.* Then, we have the canonical isomorphisms (see Grothendieck[9]):

$$\mathscr{D}(U)\overline{\otimes}\mathscr{D}(V) = \mathscr{D}(U \times V) \quad \text{and} \quad \mathscr{D}^*(U \times V) = \mathscr{D}^*(U)\widehat{\otimes}_\pi \mathscr{D}^*(V).$$

Consequently, for $\mu \in \mathcal{M}(\Omega; \mathscr{D}(U))$ and $\nu \in \mathcal{M}(\Gamma; \mathscr{D}(V))$, the inductive tensor product $\mu\overline{\otimes}\nu$ exists and takes values in $\mathscr{D}(U \times V)$. When $\mu \in \mathcal{M}(\Omega; \mathscr{D}^*(U))$ and $\nu \in \mathcal{M}(\Gamma; \mathscr{D}^*(V))$, then $\mu\overline{\otimes}\nu$ also exists and takes values in $\mathscr{D}^*(U \times V)$.

In the following, let $S$ and $T$ be completely regular spaces which satisfy $\mathcal{B}(S \times T) = \mathcal{B}(S) \times \mathcal{B}(T)$ (this condition is satisfied, for instance, both $S$ and $T$ are separable metric spaces, and more generally have countable bases for open sets or are Suslin spaces; see Schwartz[22]). Then, we have an affirmative answer for the problem of joint continuity of the inductive tensor product of two vector measures with values in nuclear spaces above. The following two theorems insist that the weak convergence of a net of tensor products of uniformly bounded vector measures follows from that of the corresponding net of real product measures. We recall that for $\mu \in \mathcal{M}(S; X)$ and $\nu \in \mathcal{M}(T; Y)$, the tensor product $\mu\overline{\otimes}\nu$ exists and takes values in $Z \equiv X\overline{\otimes}Y$, and $Z^*_\beta$ can be identified with $X^*_\beta\widehat{\otimes}_\pi Y^*_\beta$ as a topological linear space. For proofs of Theorems 2 and 3 see Kawabe[11].

**Theorem 2.** *Let $\{\mu_\alpha\} \subset \mathcal{M}(S; X)$ and $\{\nu_\alpha\} \subset \mathcal{M}(T; Y)$ be uniformly bounded nets. Let $\mu \in \mathcal{M}(S; X)$ and $\nu \in \mathcal{M}(T; Y)$. Assume that, for each $x^* \in X^*$ and $y^* \in Y^*$, the net $\{x^*\mu_\alpha \times y^*\nu_\alpha\}$ of real product measures converges weakly to the real product measure $x^*\mu \times y^*\nu$. Then $\{\mu_\alpha\overline{\otimes}\nu_\alpha\} \subset \mathcal{M}(S \times T; Z)$ converges weakly to $\mu\overline{\otimes}\nu \in \mathcal{M}(S \times T; Z)$ for $\sigma(Z, Z^*)$. Further, if $Y$ is nuclear, it also converges weakly for the inductive limit topology on $Z$.*

In the case of vector measures with values in dual spaces, we have

**Theorem 3.** *Let $\{\mu_\alpha\} \subset \mathcal{M}(S; X^*_\beta)$ and $\{\nu_\alpha\} \subset \mathcal{M}(T; Y^*_\beta)$ be uniformly bounded nets. Let $\mu \in \mathcal{M}(S; X^*_\beta)$ and $\nu \in \mathcal{M}(T; Y^*_\beta)$. Assume that, for each $x \in X$ and $y \in Y$, $\{x\mu_\alpha \times y\nu_\alpha\}$ converges weakly to $x\mu \times y\nu$. Then $\{\mu_\alpha\overline{\otimes}\nu_\alpha\} \subset \mathcal{M}(S \times T; Z^*_\beta)$ converges weakly to $\mu\overline{\otimes}\nu \in \mathcal{M}(S \times T; Z^*_\beta)$ for $\sigma(Z^*, Z)$. Further, if $Y$ is nuclear, it also converges weakly for $\beta(Z^*, Z)$.*

## 5 Banach Lattice Valued Measures

In this section we recall definitions and some basic facts of vector measures with values in a Banach space, and introduce the notion of positivity of vector measures when the Banach space is equipped with the additional structure of a Banach lattice.

Let $(\Omega, \mathcal{A})$ be a measurable space, $X$ a Banach space, and $X^*$ the topological dual of $X$. Let $\mu : \mathcal{A} \to X$ be a finitely additive set function. As is defined in Section 2, we say that $\mu$ is a *vector measure* if it is countably additive, i.e., for any disjoint sequence $\{A_k\}_{k=1}^{\infty} \subset \mathcal{A}$ with $A = \bigcup_{k=1}^{\infty} A_k$, we have $\lim_{n\to\infty} \| \sum_{k=1}^{n} \mu(A_k) - \mu(A)\| = 0$. The *semivariation* of $\mu$ is the set function $\|\mu\|(A) = \sup\{|x^*\mu|(A) : \|x^*\| \leq 1\}$. We say that $\mu$ has *bounded semivariation* if $\|\mu\|(\Omega) < \infty$. Countably additive set functions have bounded semivariation. A control measure for $\mu$ is a finite positive measure $\sigma$ on $\mathcal{A}$ such that $\sigma(A) \to 0$ if and only if $\|\mu\|(A) \to 0$. A theorem of Bartle, Dunford and Schwartz[3] states that such a measure always exists if $\mu$ is countably additive. Moreover, this measure can be chosen as $\sigma = |x^*\mu|$, for certain $x^* \in X^*$, by a result of Rybakov (see, e.g., Diestel and Uhr[6]).

When the Banach space $X$ is equipped with the additional structure of a Banach lattice, we may introduce the notion of positivity for vector measures. We say that a vector measure $\mu : \mathcal{A} \to X$ is *positive* if $\mu(A) \geq 0$ for every $A \in \mathcal{A}$. By a result of Shortt[23], for every positive vector measure $\mu$, we have $\|\mu\|(A) = \|\mu(A)\|$ for all $A \in \mathcal{A}$. Further, it is easy to verify that for any $\mu$-integrable, $\mathcal{A}$-measurable real functions $f$ and $g$ with $|f| \leq g$ almost everywhere, we have

$$\left|\int_{\Omega} f d\mu\right| \leq \int_{\Omega} |f| d\mu \leq \int_{\Omega} g d\mu \quad \text{and} \quad \left\|\int_{\Omega} f d\mu\right\| \leq \left\|\int_{\Omega} g d\mu\right\|.$$

These facts greatly facilitate the analysis of positive vector measures. For further properties of positive vector measures on metric spaces, see März and Shortt[20], and Shortt[23]. For other applications of positivity of vector measures, see Kawabe[12]. We refer the reader to the book of Schaefer[21] for the basic theory of Banach lattices.

We define several notions of regularity for vector measures on a topological space. Let $S$ be a topological space and $\mathcal{B}(S)$ the $\sigma$-field of all Borel subsets of $S$. Let $\mu : \mathcal{B}(S) \to X$ be a vector measure. We say that $\mu$ is *Radon* if for every $\varepsilon > 0$ and $A \in \mathcal{B}(S)$ there exists a compact subset $K$ of $A$ such that $\|\mu\|(A - K) < \varepsilon$, and it is *tight* if this condition is satisfied for $A = S$. We say that $\mu$ is *$\tau$-smooth* if, for any increasing net $\{G_\alpha\}$ of open subsets of $S$ with $G = \bigcup_\alpha G_\alpha$, we have $\lim_\alpha \|\mu\|(G - G_\alpha) = 0$. It follows from the existence of a control measure for $\mu$ and a result of Rybakov cited above that $\mu$ is Radon (resp. tight, $\tau$-smooth) if and only if, for each $x^* \in X^*$, the total variation of $x^*\mu$ is Radon (resp. tight, $\tau$-smooth). Consequently, all of the regularity properties which are valid for finite positive measures remain true for vector measures. For instance, every vector measure on a topological space with a countable basis (in particular, on a separable metric space) is $\tau$-smooth. Further, every vector measure on a complete separable metric space is Radon, and hence it is $\tau$-smooth and tight; see Proposition I.3.1 and Theorem I.3.1 of Vakhania *et al.*[25]

Let $\{\mu_\alpha\}$ be a net in $\mathcal{M}(S;X)$ and $\mu \in \mathcal{M}(S;X)$. We say that $\{\mu_\alpha\}$ *converges weakly* to $\mu$, and we write $\mu_\alpha \xrightarrow{w} \mu$, if for each $f \in C(S)$, we have

$$\lim_\alpha \int_S f d\mu_\alpha = \int_S f d\mu \quad \text{in the norm of } X \tag{4}$$

(see Shortt[23]). The following proposition asserts that in the case of *positive* vector measures, weak convergence follows form the validity of (4) for only bounded *uniformly* continuous functions $f$ on $S$; cf. Theorem 8.1 (Portmanteau Theorem) of Topsøe[24]. For a proof of Proposition 4 see Kawabe[15].

**Proposition 4.** *Let $S$ be a uniform space and $X$ a Banach lattice. Let $\{\mu_\alpha\}$ be a net in $\mathcal{M}^+(S;X)$ and $\mu \in \mathcal{M}^+(S;X)$. Assume that $\mu$ is tight. Then the following two conditions* (i) *and* (ii) *are equivalent*:

(i) *For every $f \in U(S)$, we have*

$$\int_S f d\mu_\alpha \to \int_S f d\mu.$$

(ii) *For every $f \in C(S)$, we have*

$$\int_S f d\mu_\alpha \to \int_S f d\mu.$$

## 6 Injective Tensor Integral

In this section, we first recall the so-called Bartle bilinear integration which was developed by Bartle[1] in 1956: Let $X$, $Y$, and $Z$ be Banach spaces and $(\Omega, \mathcal{A})$ be a measurable space. Let $b : X \times Y \to Z$ be a continuous bilinear map, $\varphi : \Omega \to X$ a vector function, and $\nu : \mathcal{A} \to Y$ a vector measure. Denote by $\chi_A$ the indicator function of a set $A$. The *Bartle semivariation* is the set function $\|\nu\|_b(A) = \sup \| \sum_{k=1}^m b(x_k, \nu(A_k))\|$, where the supremum is taken over all $\{x_k\}_{k=1}^m \subset X$ with $\|x_k\| \leq 1$ and all measurable partitions $\{A_k\}_{k=1}^m$ of $A$. We say that $\varphi$ is *$\nu$-measurable* if it is the limit in measure of a sequence of simple functions. The function $\varphi$ is said to be *$\nu$-integrable* in the sense of Bartle with respect $b$ if (a) there exists a sequence $\{\varphi_n\}$ of simple functions that converges in measure to $\varphi$, and (b) for each $A \in \mathcal{A}$, the sequence $\{\int_A b(\varphi_n, d\nu)\}$ of indefinite integrals converges in the norm of $Z$, where $\int_A b(\varphi_n, d\nu) = \sum_{k=1}^m b(x_k, \nu(A_k \cap A)) \in Z$, if $\varphi_n = \sum_{k=1}^m x_k \chi_{A_k}$. The limit is usually denoted by $\int_A b(\varphi, d\nu)$. It is easy to verify that the limit of the sequence $\{\int_A b(\varphi_n, d\nu)\}$ exists in the norm of $Z$ uniformly for $A \in \mathcal{A}$, and so the indefinite integral $\int_A b(\varphi, d\nu)$ is a countably additive set function on $\mathcal{A}$.

We assume that the Bartle semivariation of $\nu$ with respect to $b$ is dominated by a finite positive measure $\sigma$ on $\mathcal{A}$, that is, $\sigma(A) \to 0$ if and only if $\|\nu\|_b(A) \to 0$ (this is called *-property in the original paper of Bartle). In this case, we may replace the requirement of convergence in measure in the definitions above by almost everywhere convergence. For precise statements of definitions and results concerning Bartle bilinear integration see the original paper of Bartle[1].

In the succeeding sections, we are mainly interested in Bartle bilinear integration with respect to the injective tensor product, that is, in the case that $Z = X\widehat{\otimes}Y$ and the bilinear map $b_\otimes : X \times Y \to Z$ is given by $b_\otimes(x, y) = x \otimes y$. We recall that the injective tensor product $X\widehat{\otimes}Y$ is the completed space of $X \otimes Y$, the space of weak*-to-weak continuous finite rank operators from $X^*$ into $Y$, endowed with the uniform norm of operators. In this case it is easy to see that the Bartle semivariation of $\nu$ with respect to $b_\otimes$ coincides with the semivariation of $\nu$, so it is dominated by a control measure of $\nu$. For simplicity, we say that the $\varphi$ is *$\nu$-integrable* if it is $\nu$-integrable in the sense of Bartle with respect to $b_\otimes$. Denote its indefinite integral $\int_A b_\otimes(\varphi, d\nu)$ by $\int_A \varphi\widehat{\otimes}d\nu$, and call it the *injective tensor integral of $\varphi$ with respect to $\nu$*. See a recent paper of Freniche and García-Vázquez[8] for further properties of injective tensor integrals such as some characterizations of integrable functions and a general Fubini theorem in this setting.

Let $T$ be a topological space and $\mathcal{B}(T)$ the $\sigma$-field of all Borel subsets of $T$. Here and in what follows, $C(T; X)$ denotes the Banach space of all bounded continuous functions $\varphi : T \to X$ with the norm $\|\varphi\|_\infty = \sup_{t\in T} \|\varphi(t)\|$. When $X = \mathbb{R}$, we write $C(T)$ instead of $C(T; \mathbb{R})$. Before ending this section, we remark that every $\varphi \in C(T; X)$ is integrable with respect to any *tight* vector measure $\nu : \mathcal{B}(T) \to Y$; cf. White[27]. For a proof of Proposition 5 see Kawabe[15].

**Proposition 5.** *Let $T$ be a topological space, and $X$ and $Y$ be Banach spaces. Let $\nu : \mathcal{B}(T) \to Y$ be a tight vector measure and $\varphi \in C(T; X)$. Then, $\varphi$ is $\nu$-integrable, and for each $A \in \mathcal{B}(T)$ we have*

$$\left\| \int_A \varphi\widehat{\otimes}d\nu \right\| \le \sup_{t\in A} \|\varphi(t)\| \cdot \|\nu\|(A).$$

## 7 An Diagonal Convergence Theorem

Let $T$ be a uniform space and $X$ a Banach space. Denote by $U(T;X)$ the Banach space of all bounded uniformly continuous functions $\varphi : T \to X$ with the norm $\|\varphi\|_\infty = \sup_{t\in T} \|\varphi(t)\|$. When $X = \mathbb{R}$, we write $U(T)$ instead of $U(T;\mathbb{R})$. Let $Y$ be a Banach lattice. Denote by $\mathcal{M}^+(T;Y)$ the set of all positive vector measures $\nu : \mathcal{B}(T) \to Y$, and denote by $\mathcal{M}_t^+(T;Y)$ the set of all $\nu \in \mathcal{M}^+(T;Y)$ which are tight.

In this section, we give a kind of diagonal convergence theorem for injective tensor integrals with respect to positive vector measures. The following theorem is not only crucial to prove our main Theorems 7 and 8, but seems to be of some interest.

**Theorem 6.** *Let $T$ be a uniform space, $X$ a Banach space, and $Y$ a Banach lattice. Let $\{\nu_\alpha\}$ be a net in $\mathcal{M}_t^+(T;Y)$. Assume that $\nu \in \mathcal{M}_t^+(T;Y)$ is $\tau$-smooth. We also assume that a net $\{\varphi_\alpha\} \subset U(T;X)$ and $\varphi \in U(T;X)$ satisfy the following three conditions* (i), (ii), *and* (iii):

(i) *$\varphi_\alpha(t) \to \varphi(t)$ for every $t \in T$.*

(ii) *$\{\varphi_\alpha\}$ is uniformly bounded, i.e., $\sup_\alpha \|\varphi_\alpha\|_\infty < \infty$.*

(iii) *$\{\varphi_\alpha\}$ is uniformly equicontinuous on $T$, i.e., for any $\varepsilon > 0$, we can find a uniformity $V$ on $T$ such that $(t,t') \in V$ implies $\sup_\alpha \|\varphi_\alpha(t) - \varphi_\alpha(t')\| < \varepsilon$.*

*If for every $g \in U(T)$,*

$$\lim_\alpha \int_T g d\nu_\alpha = \int_T g d\nu,$$

*then we have*

$$\lim_\alpha \int_T \varphi_\alpha \widehat{\otimes} d\nu_\alpha = \int_T \varphi \widehat{\otimes} d\nu.$$

A proof of this theorem, together with some auxiliary results, is given in Kawabe[14].

## 8 Joint Continuity Problem – Banach Lattice Valued Case

In Section 4, we treated joint continuity problem for vector measures with values in certain nuclear spaces. The way of proving the joint continuity is essentially based on a finite dimensional character of nuclear spaces (i.e., the weak topology coincides with the original topology on every bounded subset of any barreled, quasi-complete nuclear space). Therefore, the same method may not apply to the case of vector measures with values in Banach spaces, which enjoy some infinite dimensional characters in the true sense of the world. In this section, we explains that this kind of joint continuity of tensor product measures remains true for the injective tensor products of positive vector measures in certain Banach lattices. Our approach to this problem is based on Bartle bilinear vector integration[1].

Let $S$ and $T$ be uniform spaces, and $X$ and $Y$ be Banach lattices. Then, by a result of Duchon and Kluvánek[7], for any vector measures $\mu \in \mathcal{M}(S;X)$ and $\nu \in \mathcal{M}(T;Y)$, there exists an injective tensor product $\mu\widehat{\otimes}\nu : \mathcal{B}(S) \times \mathcal{B}(T) \to X\widehat{\otimes}Y$ such that

$$\mu\widehat{\otimes}\nu(A \times B) = \mu(A) \otimes \nu(B) \tag{5}$$

for all $A \in \mathcal{B}(S)$ and $B \in \mathcal{B}(T)$. In this section, as an application of the diagonal convergence theorem in Section 7, we consider a problem of joint continuity of the injective tensor product of positive vector measures in certain Banach lattices.

In the rest of this paper, we assume that $S$ and $T$ satisfy

$$\mathcal{B}(S \times T) = \mathcal{B}(S) \times \mathcal{B}(T). \tag{6}$$

The restriction (6), however, may be dropped if, for instance, both $\mu$ and $\nu$ are $\tau$-smooth, positive vector measures, and either of the ranges of $\mu$ and $\nu$ is separable, since in this case the injective tensor product measure can be uniquely extended to a $\tau$-smooth positive vector measure on $\mathcal{B}(S \times T)$, which contains $\mathcal{B}(S) \times \mathcal{B}(T)$ in general (see Kawabe[14]). We can also obtain the same form of a general Fubini theorem as Theorem 13 of Freniche and Caría-Vázquez[8] for this extended injective tensor product measure.

Anyway, under assumption (6), we can view the injective tensor product $\mu\widehat{\otimes}\nu$ as a vector measure defined on $\mathcal{B}(S \times T)$, and integrate every (uniformly) continuous real function with respect to $\mu\widehat{\otimes}\nu$.

As an application of the diagonal convergence theorem in Section 7, we have the following result which seems to be of some interest. For a proof of Theorem 7 see Kawabe[15].

**Theorem 7.** *Let $X$ and $Y$ be Banach lattices. Let $\{\mu_\alpha\}$ be a net in $\mathcal{M}^+(S;X)$ and $\mu \in \mathcal{M}^+(S;X)$. Let $\{\nu_\alpha\}$ be a net in $\mathcal{M}_t^+(T;Y)$ and $\nu \in \mathcal{M}_t^+(T;Y)$. Assume that $\nu$ is $\tau$-smooth. If, for every $f \in U(S)$ and $g \in U(T)$, we have*

$$\int_S f d\mu_\alpha \to \int_S f d\mu \quad \textit{and} \quad \int_T g d\nu_\alpha \to \int_T g d\nu,$$

*then for every* $h \in U(S \times T)$ *we have*

$$\int_{S\times T} hd(\mu_\alpha \widehat{\otimes} \nu_\alpha) \to \int_{S\times T} hd(\mu \widehat{\otimes} \nu).$$

Let $X$ and $Y$ be Banach lattices. Then, in general, the injective tensor product $X\widehat{\otimes}Y$ as well as the projective tensor product is not a vector lattice (and much less a Banach lattice) for the natural ordering. The injective tensor products of some important examples of Banach lattices, however, are also Banach lattices as are illustrated after the following theorem.

Let $X$ and $Y$ be Banach lattices such that the injective tensor product $X\widehat{\otimes}Y$ is also a Banach lattice satisfying the condition that $x\otimes y \geq 0$ for every $x \geq 0$ and $y \geq 0$. Let $(\Omega, \mathcal{A})$ and $(\Gamma, \mathcal{B})$ be measurable spaces, and $\mu : \mathcal{A} \to X$ and $\nu : \mathcal{B} \to Y$ be vector measures. Then it is easy to verify that if $\mu$ and $\nu$ are positive, so is the injective tensor product $\mu\widehat{\otimes}\nu$. In this case, we have an affirmative answer for the problem of joint continuity of the injective tensor products with respect to the weak convergence of vector measures. For a proof of Theorem 8 see Kawabe[15].

**Theorem 8.** *Let* $X$ *and* $Y$ *be Banach lattices such that the injective tensor product* $X\widehat{\otimes}Y$ *is also a Banach lattice satisfying the condition that* $x\otimes y \geq 0$ *for every* $x \geq 0$ *and* $y \geq 0$*. Let* $\{\mu_\alpha\}$ *be a net in* $\mathcal{M}^+(S;X)$ *and* $\mu \in \mathcal{M}^+(S;X)$*. Let* $\{\nu_\alpha\}$ *be a net in* $\mathcal{M}_t^+(T;Y)$ *and* $\nu \in \mathcal{M}_t^+(T;Y)$*. Assume that* $\mu$ *is tight and* $\nu$ *is* $\tau$*-smooth. If* $\mu_\alpha \xrightarrow{w} \mu$ *and* $\nu_\alpha \xrightarrow{w} \nu$ *then* $\mu_\alpha\widehat{\otimes}\nu_\alpha \xrightarrow{w} \mu\widehat{\otimes}\nu$*.*

In the special case that $X = Y = \mathbb{R}$, the proof of Theorem 8 is executed by a well known criterion that one can prove weak convergence of $\mu_\alpha$ to $\mu$ by showing that $\mu_\alpha(A) \to \mu(A)$ for some special class of sets $A$; see, e.g., Corollary 1 of Theorem I.3.5 and Proposition I.4.1 of Vakhania *et al.*[25] However, it seems that the usual proof of the criterion above does not work well for positive vector measures, since the notions of limit infimum and limit supremum can not be well defined in an arbitrary Banach lattice.

We end this paper giving examples of Banach lattices $X$ and $Y$ such that the injective tensor product $X\widehat{\otimes}Y$ is also a Banach lattice satisfying the condition that $x\otimes y \geq 0$ for every $x \geq 0$ and $y \geq 0$; see examples in Schaefer[21] and Grothendieck[9].

**Example 9.** (1) If $K$ is a compact space and $Y$ be any Banach lattice, then $C(K)\widehat{\otimes}Y$ is isometrically lattice isomorphic to the Banach lattice $C(K;Y)$. Especially, when $Y = C(L)$ for some compact space $L$, $C(K)\widehat{\otimes}C(L)$ is isometrically lattice isomorphic to $C(K \times L)$.

(2) Let $P$ be a locally compact space and $Y$ be any Banach lattice. Denote by $C_0(P;Y)$ the Banach lattice with its canonical ordering of all continuous functions $\varphi : P \to Y$ such that for every $\varepsilon > 0$, the set $\{s \in P : \|\varphi(s)\| \geq \varepsilon\}$ is compact. We write $C_0(P)$ instead of $C_0(P;\mathbb{R})$. Then $C_0(P)\widehat{\otimes}Y$ is isometrically lattice isomorphic to $C_0(P;Y)$. Especially, when $Y = C_0(Q)$ for some locally compact space $Q$, $C_0(P)\widehat{\otimes}C_0(Q)$ is isometrically lattice isomorphic to $C_0(P \times Q)$.

(3) Let $(\Omega, \mathcal{A}, \sigma)$ be a measure space and $Y$ be any Banach lattice. Denote by $L^\infty(\Omega;Y)$ the Banach lattice of all (equivalence classes of) $\sigma$-essentially bounded measurable functions $\varphi : \Omega \to Y$ with its canonical ordering. We write $L^\infty(\Omega)$ instead of $L^\infty(\Omega;\mathbb{R})$. Then, $L^\infty(\Omega)\widehat{\otimes}Y$ is a Banach lattice. However, in general, $L^\infty(\Omega)\widehat{\otimes}Y$ is a proper closed subset of $L^\infty(\Omega;Y)$.

(4) Denote by $c_0(Y)$ the Banach lattice of all sequences in a Banach lattice $Y$ converging to 0 with its canonical ordering. We write $c_0$ instead of $c_0(\mathbb{R})$. Then $c_0\widehat{\otimes}Y$ is isometrically lattice isomorphic to $c_0(Y)$.

## Acknowledgments

This work has been supported by Grant-in-Aid for General Scientific Research No. 11640160, the Ministry of Education, Science, Sports and Culture, Japan.

## References

1. R. G. Bartle, *A general bilinear vector integral*, Studia Math. **15** (1956), 337-352.
2. P. Billingsley, *Convergence of Probability Measures*, John Wiley & Sons, New York, 1968.
3. R. G. Bartle, N. Dunford and J. Schwartz, *Weak compactness and vector measures*, Canad. J. Math. **7** (1955), 289-305.
4. I. Csiszár, *On the weak* continuity of convolution in a convolution algebra over an arbitrary topological group*, Studia Sci. Math. Hungar. **6** (1971), 27-40.
5. M. Dekiert, *Kompaktheit, Fortsetzbarkeit und Konvergenz von Vectormaßen*, Dissertation, University of Essen, 1991.
6. J. Diestel and J. J. Uhr, Jr., *Vector Measures*, Amer. Math. Soc. Surveys **15**, Providence, R.I., 1977.
7. M. Duchon and I. Kluvánek, *Inductive tensor product of vector-valued measures*, Mat. Časopis Sloven. Akad. Vied. **17** (1967), 108-112.
8. F. J. Freniche and J. C. García-Vázquez, *The Bartle bilinear integration and Carleman operators*, J. Math. Anal. Appl. **240** (1999), 324-339.

9. A. Grothendieck, *Produits Tensoriels Topologiques et Espaces Nucléaires*, Mem. Amer. Math. Soc. **16**, Providence, R.I., 1955.
10. J. Kawabe, *Weak convergence of compound probability measures on uniform spaces*, Tamkang J. Math. **30** (1999), 269-281.
11. J. Kawabe, *Weak convergence of tensor products of vector measures with values in nuclear spaces*, Bull. Austral. Math. Soc. **59** (1999), 451-460.
12. J. Kawabe, *A type of Strassen's theorem for positive vector measures with values in dual spaces*, Proc. Amer. Math. Soc. **128** (2000), 3291-3300.
13. J. Kawabe, *Sequential compactness for the weak topology of vector measures in certain nuclear spaces*, to appear in Infinite-Dimensional Probability and Related Problems of Analysis.
14. J. Kawabe, *Borel extension of injective tensor products of $\tau$-smooth positive vector measures*, submitted for publication.
15. J. Kawabe, *Joint continuity of injective tensor products of vector measures in certain Banach lattices*, submitted for publication.
16. J. Kawabe, *Compactness criteria for the weak topology of vector measures in certain locally convex spaces*, in preparation.
17. I. Kluvánek, *An example concerning the projective tensor product of vector measures, Mat. Časopis Sloven. Akad. Vied.* **20** (1970), 81-83.
18. I. Kluvánek and G. Knowles, *Vector Measures and Control Systems*, North-Holland, Amsterdam, 1975.
19. D. R. Lewis, *Integration with respect to vector measures*, Pacific J. Math. **33** (1970), 157-165.
20. M. März and R. M. Shortt, *Weak convergence of vector measures*, Publ. Math. Debrecen **45** (1994), 71-92.
21. H. H. Schaefer, *Banach Lattices and Positive Operators*, Springer-Verlag, Berlin, 1974.
22. L. Schwartz, *Radon Measures on Arbitrary Topological Spaces and Cylindrical Measures*, Tata Institute of Fundamental Research, Oxford University Press, 1973.
23. R. M. Shortt, *Strassen's theorem for vector measures*, Proc. Amer. Math. Soc. **122** (1994), 811-820.
24. F. Topsøe, *Topology and Measure*, Lecture Notes in Math. 133, Springer-Verlag, New York, 1970.
25. N. N. Vakhania, V. I. Tarieladze and S. A. Chobanyan, *Probability Distributions on Banach Spaces*, D. Reidel Publishing Company, 1987.
26. V. S. Varadarajan, *Measures on topological spaces*, Amer. Math. Soc. Transl. Ser. II **48** (1965), 161-228.
27. A. J. White, *Convolution of vector measures*, Proc. Roy. Soc. Edinburgh **73A** (1974/75), 117-135.

# LARGE GRADIENTS OF SOLUTIONS TO NONLINEAR PARABOLIC PROBLEMS

M.M.LAVRENTIEV, Jr.
*Sobolev Institute for Mathematics Siberian Division of the Russian Academy of Sciences, pr.Ak.Koptyuga, 4, Novosibirsk 630090, Russia*
*E-mail: mmlavr@nsu.ru*

In a number of dynamical models of Mathematical Physics the *Euler–Lagrange Equation* for a proper integrand stands for the right hand side of the governing evolution equation. Such structure arises, in particular, by minimizing some integrals, for instance, the energy integral. Therefore, solutions to corresponding *Variational Problems* determine steady states for dynamical processes. As was recovered recently, regular variational problems may possess *non smooth minimizers.* "Non–smoothness" of Absolutely Continuous functions (minimizers) necessarily mean, that derivative of such function is unbounded. This paper is devoted to nonlinear parabolic equations of *"Variational Type"*, right hand sides of which are nothing but Euler-Lagrange Equation for some integrand (these integrands are supposed to satisfy classical solvability conditions in Calculus of Variations). Such problems do not satisfy famous "quadratic growth conditions" in gradient terms and, therefore, we can not apply for general solvability theorems for nonlinear parabolic equations. Qualitatively different examples of such parabolic problems have been brought together. Global in time solvability and stabilization of solution for large time for some of the aforementioned problems have been also established.

## 1 Statement of the problem. Factors which causes phenomenon

We shall speak about autonomous quasilinear parabolic equations with a single space variable. Examples and results are clearer in this situation. Moreover, some results, which have been proved for one space dimension, are not valid in general, multidimensional case. For example, Zelenyak [1] (see also [2]) has established *stabilization theorem* (uniformly, in time, bounded in the proper Banach space solution to parabolic problem possesses a limit as time growth to infinity)in case of one space variable. This property (that solution to dynamical problem does possess the limit) is not true even for semilinear equation $u_t = u_{xx} + u_{yy} + f(x, y, u)$, considered on the unit circle $x^2 + y^2 \leq 1$ [3].

In rectangle $(x,t) \in Q_T = (0,1) \times (0,T]$ (or *semistrip* in case $T = \infty$) consider the following problem

$$u_t = a(x, u, u_x)\, u_{xx} + b(x, u, u_x); \tag{1.1}$$

$$\alpha_i\, u_x - \psi_i(u)|_{x=i} = 0 \quad (i = 0,1); \qquad u(x,0) = u_0(x) \tag{1.2}$$

with $\alpha_i = \text{const}$ for $i = 0,1$.

From now on we suppose that $a(x,\xi,\eta) \geq a_0 > 0$ (*condition of parabolicity*), $\alpha_i^2 + [\psi_i'(\xi)]^2 \neq 0$ for $i = 0,1$ (*boundary conditions are nontrivial*), all functions $a$, $b$, $\psi_i$, and $u_0$ are sufficiently smooth (say, from $C^4$ in all variables for simplicity), and, moreover, initial function satisfies compatibility conditions with boundary data (see [4], e.g.). Positive number $T$ (as was afore mentioned) may takes an infinite value $T = \infty$, in which case rectangle $Q_T$ is nothing but half of a strip, which we will denote as $Q_\infty = (0,1) \times R^+$.

Modern "solvability theory" for such problems have been established in the 1960's (see [4, 5, 6, 7] and the recent surveys [8, 9], e.g.). General "existence theorem" for nonlinear problems of the type (1.1)–(1.2) contains the following basic parts:

1. "smoothness conditions" — requirement on the problem data, i.e. functions $a$, $b$, $\psi_i$, $u_0$, to belong to the proper functional space (in case of "smooth solutions" the so–called compatibility conditions of initial function with the boundary data are also requested);

2. two type of restrictions on the nature of dependence of non–linear terms $a$ and $b$ on its arguments $u$ and $u_x$.

Precise formulation of typical existence theorem from [4] takes more then a complete page and, therefore, is here omitted. We would like just to give an idea of the "restrictions of the type 2". First, we should be able to establish that solution $u(x,t)$ itself remains bounded in the uniform norm

$$\sup_{x,t} |u(x,t)| < M. \tag{1.3}$$

It could be the maximum principle conditions, energy–like estimates, and so on.

Second, we should be able to prove an estimate for the first derivative

$$\sup_{x,t} |u_x(x,t)| < M_1, \tag{1.4}$$

provided that the estimate (1.3) is already known.

As it has been mentioned, having known both estimates (1.3) and (1.4) (recall that all coefficients are supposed to be sufficiently smooth), we can prove the existence of classical solution to the above mentioned quasilinear equation (1.1) with no additional restrictions on nonlinear terms.

**Remark 1.1** *In case of more general parabolic equation*

$$u_t = c(x,u,u_x,u_{xx}), \qquad \frac{\partial c}{\partial u_{xx}} \geq \delta > 0 \quad \text{(condition of parabolicity)},$$

*it is necessary also to prove that the senior derivative is bounded $|u_{xx}(x,t)| \leq M_2$ (see [6, 7, 10]).*

The second type of conditions, which provide estimate (1.4), are stated in terms of restrictions on the growth of the quotient $b/a(x,u,u_x)$ in the variable $u_x$. In our case under study, this restriction is nothing but to the only condition

$$\left|\frac{b(x,u,u_x)}{a(x,u,u_x)}\right| \leq K\ \varphi(|u_x|), \qquad \text{with} \qquad \int_0^\infty \frac{\tau\, d\tau}{\varphi(\tau)} = \infty \tag{1.5}$$

for $x \in [0,1]$, $u \in [-M, M]$, see, for instance, [2, 5, 7, 8, 11].

Roughly speaking, function $\varphi(s)$ must possesses not more then quadratic growth, $\varphi(s) \leq 1 + s^2$, as $s \to \infty$. Although the growth rates like $\varphi(s) \sim s^2 \mathrm{ln} s$, $\varphi(s) \sim s^2 \mathrm{ln} s \mathrm{ln}(\mathrm{ln} s)$, etc., are also permitted.

As was shown in a number of examples, violation of the above growth condition (say, **"superquadratic growth"** of the quotient $b/a(x,u,u_x)$ in $u_x$) may cause infinite growth of derivative, $u_x(x,t)$, within finite time.

# 2 Examples of the Derivative Blowing–Up Solutions

In case $b/a(x,u,u_x)$ (see (1.5)) has the growth of the order at least $2+\varepsilon$ in $u_x$, examples (perhaps the first ones) of parabolic problems, which have *bounded* solutions with *unbounded* derivatives, have been given in [12] (see also [4]).

**Example 2.1** *(see [4]). For* $(x,t) \in Q_1 = (0,1) \times (0,1)$ *consider the following problem*

$$u_t = u_{xx} + u\,|u_x|^{\frac{2}{1-\alpha}}, \qquad \alpha \in (0,1);$$

$$u\big|_{x=0} = 0, \qquad u\big|_{x=1} = 1 + \frac{1}{\alpha} + t\left(1 + \frac{1}{\alpha}\right)^{1+\frac{2}{1-\alpha}};$$

$$u\big|_{t=0} = \left(1 + \frac{1}{\alpha}\right) x.$$

*As has been proved in [12], bounded solution* $u(x,t)$ *does not possess bounded derivative. The idea of the proof is to compare solution* $u(x,t)$ *with the barrier function*

$$z(x,t) = \begin{cases} \frac{x}{1-t}, & \text{for } 0 \le x < 1-t \\ \frac{1}{\alpha}(x+t-1)^\alpha + 1, & \text{for } 1-t \le x \le 1 \end{cases};$$

*which has been chosen such that*

$$z(x,t) \le u(x,t) \qquad \textit{and}$$
$$\lim_{x \to +0} z(x,1) = 1.$$

Now such solutions are known as "Derivative Blowing–Up" (DBU) solutions, see [13], e.g. However, since the first example [12], for years the above phenomenon appears not to be under investigation.

Later on it was re–emphasized that these blowing–up solutions (not necessarily the DBU ones) are important from the viewpoint of applications [14]. At the same time the major part of publications on this topic is devoted to unbounded solutions, which growth to infinity in the uniform norm: $\sup|u(x,t)| \to \infty$.

Only recently the new examples of DBU solutions to parabolic equations became available [13, 15, 16], although "the hyperbolic phenomenon" (DBU in our terminology [17]) has been already already observed as the ones, possessing a definite meaning in applications [18, 19].

Here below we are going to mention *qualitatively different* examples of DBU solutions. In particular, as was shown in [15], increase in the length of variation of $x$ may cause appearance of the above DBU solutions. This example, given below, is also very interesting because of a concept of extension of solution beyond the "blowing–up" moment of time.

**Example 2.2** *(see [15]). For* $x \in (0,a)$ *and* $t > 0$ *consider equation*

$$u_t = u_{xx} + f(u_x),$$

*where nonlinearity is such that function* $f(s)$*:*

- *is smooth* ($f \in C^2$)*, positive, and monotone increasing function,*
- *possesses a bounded ratio* $f'(s)/f(s)$ *as* $s \to \infty$*,*
- *possesses "super quadratic growth" at infinity:*

$$\int_0^\infty \frac{s\,ds}{f(s)} < \infty.$$

*Initial and boundary conditions are*

$$u(0,t) = 0, \quad u(a,t) = 0,$$
$$u(x,0) = u_0(x) \quad (u_0(0) = u_0(a) = 0).$$

*As has been demonstrated, there exists a number $A$, such that:*

1. *if $a < A$ solution $u(x,t)$ remains smooth for all $t > 0$;*
2. *if $a = A$ the derivative $u_x$ became unbounded as $t \to \infty$;*
3. *if $a > A$, for any $u_0$, the derivative $u_x$ of solution to the above problem is unbounded*

$$\sup_x |u_x(x,t)| \to \infty \qquad \text{as} \quad t \to T$$

*for proper $T \in (0, \infty)$.*

**Remark 2.1** *The above example has been constructed after studying equation*

$$u_t = u_{xx} + \exp(u_x),$$

*in which case all the steady states could be calculated explicitly.*

*Blowing–up occurs at the end point $x = 0$ of the space interval. Solution, extended beyond the "blowing–up moment", does not satisfy the boundary condition at the left edge $x = 0$ in classical sense.*

In both Examples 2.1 and 2.2 gradient of solution growth to $\infty$ at the end point of the space interval. The next example, proposed in [16], shows an alternative option.

**Example 2.3** *(see [16]). A class of semilinear parabolic equations with bounded solutions, who's derivative blows–up in finite time in the interior of the interval $(-1,1)$, was introduced:*

$$u_t = u_{xx} + f(u)|u_x|^{m-1} u_x,$$

$$u(x,0) = u_0(x), \quad u(\pm 1, t) = A_\pm$$

*with $f \in C^1(R)$, $m > 2$. Note that the maximum principle holds for such equation. It was observed that solution to problem above become singular only at points where $f(u)$ changes the sign. Therefore, if $f(0) = 0$, $f'(0) \neq 0$, and $f(u)$ is nontrivial for $u \neq 0$, under additional assumption that $A_+ \cdot A_- < 0$ we have* interior DBU.

Moreover, the "Derivative Blowing-Up" effect could depend not only on the equation itself, but also on the boundary conditions. Corresponding example, which is simple enough, has been proposed in [20] (see also [2]).

**Example 2.4** *(see [20]). For $(x,t) \in Q_\infty = [0,1] \times [0,\infty)$ let us consider equation*

$$u_t = u_{xx} - x u_x^3$$

*with the initial function*

$$u(x,0) = u_0(x).$$

*Suppose that $u_0(x)$ is smooth and satisfies compatibility conditions for all boundary problems, given below.*

*It has been proved that:*

1. *For the boundary conditions*

$$u(0,t) = 0, \quad u(1,t) = \pi/2$$

*solution* $u(x,t)$ *itself remains bounded, while it exists, but it can not possesses bounded derivative (it means that* $\sup|u_x(x,t)| \to \infty$ *as* $t$ *goes to a certain value* $T$*, finite or infinite – the upper bound of the interval along which classical solution exists). Note that the only steady state with such a boundary condition is* $v(x) = \arcsin x$.

2. *For the boundary conditions*

$$u(0,t) = 0, \quad u_x(1,t) - u(1,t) = 0$$

*classical solution,* $u(x,t)$*, exists for all* $t > 0$ *upon the proper smoothness of initial function,* $u(x,0) = u_0(x)$*. Moreover, this solution remains uniformly bounded with the derivatives and goes to zero as* $t \to \infty$.

# 3 Some applied reasons to study Blowing–Up solutions

Concerning nonlinear parabolic problems, violation of the above mentioned conditions on nonlinear terms (namely, maximum principle or it's analog and/or restrictions in (1.5)) causes the "blowing–up solutions", norm of which growth to infinity within finite time. Blowing-Up Solutions to nonlinear parabolic problems are currently generating much attention (see, e.g. [14, 21] and the references wherein) in view both of various applications and arising theoretical problems. These solutions are smooth for $0 \leq t < t^*$, but $\|u(x,t)\|_B \to \infty$ as $t \to t^* - 0$ for a proper Banach space $B$.

Most of investigations have been focused on *"uniform norm blowing–up"*, in which case the above Banach space is nothing but the space of continuous functions, $B = C$. From the point of view of qualitative theory of parabolic problems such Blowing–Up effect means that the maximum principle fails. At the same time, a number of parabolic models in Mathematical Physics have an Euler–Lagrange Equation (EL–E) for a proper integrand as the right–hand side. Under classical "regularity conditions from the calculus of variations" on these integrands, the maximum principle holds for corresponding parabolic problem. In such cases "Blowing–Up" means that solution itself remains bounded while its space derivatives are blowing–up.

As was mentioned in [14], the fact that Blowing–Up solutions possess a definite physical meaning have been observed even in 60–s (e.g. thermal explosion problem and shock waves cumulation process).

Based on the Introduction to monograph [14], we are going to give some remarks.

1) Within the last 25 years Blowing–Up (BU) solutions have been studied intensively because of different applications. Such areas as self–focussing of light rays in nonlinear media, nonstationary structures in magneto – hydrodynamics ($T$–layers), unshock compression in gas dynamics, and other could be mentioned.

2) Currently general theory of BU solutions leaves much to be investigated. These solutions possess a number of interesting (if not surprising) properties in view of common knowledge about dynamics to dissipative processes.

3) There are no general understanding even of the basic problems, concerning BU solutions, among professionals.

For more details we address the reader to [14]. Research results, which have been obtained during the last years in Keldysh Institute for Applied Mathematics (Russian Academy of Sciences, Moscow), are brought together in this book. We also address the reader to survey [22], where history, references, and *surprising*, qualitatively different results have been exposed.

Here we would like to mention briefly one possible application of Blowing–Up solutions. Original results from the book of Samarskii and Mikhailov [21] will be used (for details see § 2 Chapter *VI* in [21] and the references wherein).

One of the prospective approaches for *"Controlled Laser Fusion"* is based on lasers impulses to heat and to compress the target. Lasers with requested parameters (irradiation of $10^8$–$10^9$ $J$ within $10^{-8}$–$10^{-9}$ *sec*) are not available now, as well as, perhaps, in the nearest future.

Besides, it is possible to achieve an essential reduce of energy barrier by using nonlinear effects of laser plasma, namely, phenomena of *non–shock compression.* To overstep the difficulties, arising along this way, it is necessary to study **Derivative BU** solutions. Indeed, shock waves appears during the target compressions, according to classical linear theory. Such shock waves affect dramatically on the entire process by taking valuable part of the process energy. Shock waves development and propagation could be described just in terms of Derivative BU solutions. Based on analytic representations of DBU solutions, it is possible, for special nonlinear models, to predict conditions, upon which shock waves do not appear.

The above mentioned *"laser fusion models"* have been studied numerically. Reasonable correspondence (both qualitative and quantitative) with physical experiments have been achieved [21]. This improves that "Blowing–Up" effects are reflect correctly the nature of process under study.

In similar models for "dense laser plasma" the "localized burning structures" have been constructed analytically and observed numerically [21]. Process is described by the following model equation

$$\frac{\partial T}{\partial t} = \frac{\partial}{\partial x}\left(k_0 T^{\sigma}\,\frac{\partial T}{\partial x}\right) + \frac{q_0\,T^{\beta}}{1+BT^{b}}. \tag{3.1}$$

For "physical" values of the model parameters $k_0$, $q_0$, $B$, and exponents $\sigma = 2,5$, $\beta = 5,2$, and $b = 3,6$ the so–called localized $LS$– and $S$–regimes (see [21]) have been observed. It means that the solution $T$ growth to infinity only in the bounded area (near central part of the target), while $T$ remains bounded near edges. Such localization could be useful in applications [21].

So, DBU solutions reflect a nature of some physical processes. Here we refer also to papers [18, 19], in which "temperature stratification phenomenon" of ocean water has been studied in the framework of degenerate parabolic models.

As was observed, temperature of the ocean water could be stratified, constituting several layers, stable in time, with more or less constant values of temperature. Such structures are impossible in view of "regular theory for thermal conductivity propagation". So, effective diffusivity for heat and mass transfer in stably stratified turbulent shear flow is expected to be degenerate and, moreover, to be of the changing sign [18, 19]. As such forward–backward diffusion problems may be ill–posed, Barenblatt et al. [18, 19] propose a proper modification

$$T_t = \frac{\partial}{\partial x}[\psi(T_x)] + \tau\,\frac{\partial^2}{\partial x \partial t}\rho(T_x), \tag{3.2}$$

with $\tau > 0$ and a suitable function $\rho$. The point is that nonlinear term $\psi(s)$ is supposed to be non monotone, i.e. derivative $\psi'(T_x)$ takes negative values for bounded interval of variation of the gradient $T_x$. In view of classical models the energy flow from low temperature domain to the higher temperature ones occurs in such intervals of variation of $T_x$.

As was proved, equation (3.2) possesses a solution, but smooth initial conditions may, in finite time, evolve to discontinuous layers. Before solution became discontinuous, it's gradient growth to $\infty$, i.e. Derivative Blowing–Up occurs.

# 4 Degenerate equations.

Here we shall speak about mathematical results concerning *degenerate equations* of the type (1.1). Recently "hyperbolic phenomena" (derivative blowing–up solutions in our terminology) have been established for so–called strongly degenerate parabolic equations [17, 23, 24]. To be precise let us formulate some results from [17].

**Example 4.1** *(see [17]). For* $(x,t) \in R^2_+ = \{(x,t) : x \in R,\, t > 0\}$ *consider the problem*

$$u_t = \frac{\partial}{\partial x}(\varphi(u)\psi(u_x)), \tag{4.1}$$

$$u(x,0) = u_0(x);$$

*where* $\varphi$ *is smooth and strictly positive, and* $\psi$ *is smooth, odd, strictly increasing* ($\psi'(s) > 0$), *and bounded function. Last condition means that equation (4.1) is degenerate, as its leading coefficient,* $\varphi(u)\psi(u_x)$, *tends to zero as the gradient of solution goes to* $\infty$:

$$\lim_{|u_x|\to\infty} \inf \psi'(u_x) = 0. \tag{4.2}$$

*The term "strongly degenerate parabolic equation" has been introduced by the authors of [17] because*

$$\lim_{s\to\infty} \psi(s) = \psi_\infty < \infty. \tag{4.3}$$

*It is important to notice that in this case an estimate of "material flux"*

$$\sup_{(x,t)\in R^2_+} |\psi(u_x(x,t))| \le K$$

*does not imply that the gradient itself is bounded* $|u_x(x,t)| \le K_1$.

*The following result has been established in [17].*

*If* $u_0(x)$ *is strictly increasing, then*

**(1)** *there exists a solution* $u(x,t)$ *to the problem (4.1);*

**(2)** *this solution* $u(x,t)$ *is not necessarily continuous, although "the flux"* $\psi(u_x)$ *is continuous function of* $x$ *(under the convention that* $\psi(\infty) = \lim_{s\to\infty}\psi(s) = \psi_\infty$*);*

**(3)** *solution* $u(x,t)$ *satisfies an* entropy type condition*: at the break point*

$$u^+ = u(x+,t) > u(x-,t) = u^-,$$

*the following inequality holds*

$$\varphi(s) \le \frac{\varphi(u^+) - \varphi(u^-)}{u^+ - u^-}(s - u^-) + \varphi(u^-) \quad \text{for} \quad s \in [u^-, u^+];$$

**(4)** *the above entropy condition is necessary for uniqueness of solution, i.e. in certain cases there exist solutions which do not satisfy the entropy condition;*

**(5)** *if* $\varphi(u)$ *is strictly concave in* $[c_1, c_2]$ *and* $c_1 \le u(x,t) \le c_2$ *for* $(x,t) \in D = (x_1, x_2) \times (t_1, t_2)$ *then* $u_x(x,t) \in L^\infty_{loc}(D)$.

**Remark 4.1** *Mostly monotone initial data* $u_0(x)$ *are considered (because of usage of coordinate transformation* $y = u(x,t)$*) although the authors mentioned that results remain true for a more general class of initial profiles* $u_0$.

This result shows us that "gradient blowing–up phenomena" could be caused by degeneration of the leading coefficient. Similar effect have been observed in [25].

**Example 4.2** *(see [25]). Consider the following parabolic problem*

$$u_t = a(u_x)u_{xx} + f(u)g(u_x), \qquad (4.4)$$
$$u|_{x=0} = u|_{x=1} = 0, \qquad u(x,0) = u_0(x)$$

*setting for simplicity*

$$a(p) = 1/(1+p^2)^\alpha, \quad g(p) = (1+p^2)^\theta, \qquad \alpha \geq 0, \quad \theta \geq 0. \qquad (4.5)$$

*Under a number of assumptions on the problem data (including the ones on initial function and multiplier $f(u)$),* "inside the space interval" derivative blowing–up phenomena *has been demonstrated.*

**Remark 4.2** *Latest Example* 4.2 *essentially differs from Example* 2.3 *by condition, that the function $f(u)$ is nonnegative.*

**Remark 4.3** *Both of the above Examples* 4.1 *and* 4.2 *do not satisfy the growth condition in (1.5). Condition (4.5) shows it clearly in case of Example* 4.2. *As to equation (4.1), it has the form*

$$u_t = \varphi(u)\,\psi'(u_x)\,u_{xx} + \varphi'(u)\,u_x\,\psi(u_x).$$

*Therefore, in terms of condition (1.5), the quotient $b/a(x,u,u_x)$ is nothing but*

$$\frac{b(x,u,u_x)}{a(x,u,u_x)} = \frac{\varphi'(u)\,u_x\,\psi(u_x)}{\varphi(u)\,\psi'(u_x)}.$$

*So, the integral (1.5) converges:*

$$\int_0^\infty \frac{s\,\psi'(s)\,ds}{s\,\psi(s)} = \int_0^\infty d\,ln\psi(s) < \infty.$$

Note, that degenerate equation may possesses even more curious properties. Consider the example, which has been proposed in [26].

**Example 4.3** *(see [26]). The authors propose an exact closed–form solution of "divergent degenerate diffusion equation" such that a discontinuous step closing in a finite time. Problem is as follows:*

$$y_t = D(y_x)\,y_{xx}, \qquad (4.6)$$

$$y(x,0) = \begin{cases} 1, & \text{for } \ x > 0; \\ 1/2, & \text{for } \ x = 0; \\ 0, & \text{for } \ x < 0. \end{cases}$$

*with $D(\theta) = \frac{\alpha^2}{2}\,(e^{-\beta\theta} - e^{-2\beta\theta})$, $\beta > 0$. Explicit solution is discontinuous until given moment $t = t_1 = \beta^2/4\alpha^2$. In spite of the discontinuity of $y(x,t)$ with respect to $x$, its derivative $y_x$ could be regarded as a continuous function on the closure of the real axis, and the "flux" remains continuous at the point of singularity.*

# 5 Lyapunov Functionals and related questions from the Calculus of Variations

In spite of the examples, given in previous Sections, some solvability results are available with no any restrictions on the rate of growth of the quotient $b/a(x, u, u_x)$ (see condition (1.5)). Until now, most of such solvability results have been proved by using the so-called *Lyapunov Functionals* technique.

By definition of Zelenyak [1], we use the concept of *"Lyapunov Functionals"* (LF) (in complete analogy with *Lyapunov Functions* in the theory of *Ordinary Differential Equations*) for integrals in space variable(s) from special nonlinear function of solution and its space derivative(s), which possess the property of *"qualified monotonisity"*, in time.

**Definition 5.1** *We shall say that* **a pair of functions** $\rho$, $\Phi(x, \xi, \eta) \in C^2$ **generates the Lyapunov Functional (LF) along solutions to (1.1)–(1.2)** *if for any solution* $u(x,t) \in C^{2,1}(Q_T)$ *the following identities hold:*

$$\frac{d}{dt} J(t) = \frac{d}{dt} \int_0^1 \Phi(x, u, u_x)\, dx = - \int_0^1 \rho(x, u, u_x)\, u_t^2\, dx \tag{5.1}$$

*with* $\rho(x, \xi, \eta) \geq 0$.

In a general sense, *conservation laws* for dynamical problems have the form of such functionals, $J(t) = \Phi(x, u, u_x)$, for a special case $\rho \equiv 0$. For *parabolic* problems it is more natural to speak about *"energy dissipation inequalities"* (5.1), with $\rho \geq 0$.

For the problems (1.1)–(1.2) under investigation family of LF have been proposed and constructed by Zelenyak [1] (see also [2, 11, 27]). In fact, to construct LF along solution of parabolic problem is nothing but to find out **"variational multipliers"** $\rho(x, u, u_x)$ (which are away from zero $\rho \geq \delta > 0$), such that after multiplication by them the right–hand side of (1.1) becomes an Euler–Lagrange Equation (E–LE) for the corresponding integrand $\Phi(x, u, u_x)$. In other words equation (1.1) could be presented in "variational form":

$$\rho(x, u, u_x) \cdot u_t = \frac{d}{dx} \frac{\partial \Phi(x, u, u_x)}{\partial u_x} - \frac{\partial \Phi}{\partial u}, \qquad \frac{\partial^2 \Phi}{\partial u_x^2} \geq \delta > 0.$$

Keeping analogies with the theory ODE and Calculus of Variations we note that restrictions (1.5) with $\varphi(u_x) = 1 + u_x^2$ (which, remind, make it possible to prove that the gradient of a bounded solution remains bounded) are known in the *Calculus of Variations* and theory of *Ordinary Differential Equations* since the beginning of the century [28, 29, 30]. Thus, Tonelli [28] has proved that "quadratic growth restriction" for Euler–Lagrange Equation is a sufficient condition for absolutely continuous solution to regular variational problem to be smooth (and, therefore, to satisfy in classical sense the above EL–E). Studying second order ODE Nagumo [29] observed that under "quadratic growth condition" a certain class of boundary value problems are solvable (unless solution became unbounded, for details about such additional requirement see [2]). Both "variational" and "differential" applications of the above condition has been studied by Bernstein [30].

In particular, it was proved, that growth conditions (1.5) make it possible to extend bounded solutions to all the Cauchy problems

$$y'' = -\frac{b(x, y, y')}{a(x, y, y')}, \quad y(x_0) = y_0, \quad y'(x_0) = y_1 \tag{5.2}$$

onto the entire interval of the variation of $x$ for arbitrary data $x_0$, $y_0$, and $y_1$. Again, for details of the term "bounded solutions" see [2].

The extendibility of solutions to Cauchy problem (5.2) appears to be exactly existence condition for families of Lyapunov Functionals (LF) along solutions to parabolic problem (1.1)–(1.2) [1, 2, 11, 27].

Originally, Zelenyak [1] has proposed LF to prove stabilization theorem for solution to nonlinear parabolic problem (any solution, which is uniformly in time bounded in the proper functional space, possesses a limit as $t \to \infty$). Later on it was observed that families of LF make it possible also to prove estimates of derivatives, like (1.4), see [2, 11, 27, 35], e.g.

It is interesting that more or less simultaneously with the Derivative Blowing–Up solutions to parabolic problems, investigations of the so–called *"nonsmooth minimizers"* to regular variational problems has been started (see, for example, [31, 32, 33] and the references wherein). Consider the first of such examples.

**Example 5.1** *(see [31]). For analytic and strictly convex in gradient integrand*

$$L(x, u, u_x) = r(k)u_x^2 + (u^3 - x^2)^2\, u_x^{14} \tag{5.3}$$

*we minimize the integral*

$$\int_0^1 L(x, u, u_x)\, dx \quad \to \quad \min. \tag{5.4}$$

*Function $u(x)$ is supposed to be absolutely continuous with the boundary conditions*

$$u(0) = 0, \qquad u(1) = k, \tag{5.5}$$

*where $r(k) = (2k/3)^{12}\,(1 - k^3)\,(13k^3 - 7)$. In other words, $r(k)$ is the positive constant for given parameter $k^3 \in (7/13, 1)$ and, therefore, integrand $L(x, u, u_x)$ is strictly convex in gradient*

$$\frac{\partial^2 L(x, u, u_x)}{(\partial u_x)^2} \geq 2r(k) > 0.$$

*This problem satisfies conditions of classical solvability theorems. As was proved in [31], for $k$ sufficiently close to 1 the exact minimizer is nothing but $v(x) = kx^{2/3}$. Such function is smooth for $x \in (0, 1)$, takes the boundary conditions (3.6), and its derivative is unbounded as $v'(x) \to \infty$ for $x \to 0$.*

**Remark 5.1** *Let us write down the Euler–Lagrange Equation (E–LE) for the above example and set it as the right–hand side for parabolic equation. It is not difficult to verify that this E–LE does satisfy restriction in (1.5) for any fixed pair $x$ and $u$. However, the constant $K$ cannot be chosen as a uniform one, as it tends to infinity when a pair $(x, u)$ approaches the curve $u = x^{2/3}$.*

*Therefore, growth condition (1.5) is violated in a weak sense. Nevertheless, we can not apply general solvability theorems [4] to such parabolic problems.*

In Example 5.1 solution $v(x)$ to variational problem could be regarded as generalized solution to corresponding E–LE, because the minimizer $v(x) = x^{2/3}$ is smooth for $x \in (0, 1)$ and takes the requested boundary values. Such non–smooth minimizer looks exactly the same as the first examples of Derivative Blowing–Up solutions to parabolic problems (see Examples 2.1 and 2.2), which derivatives growth to $\infty$ at the edge point of the space interval. Keeping the analogy between *Parabolic Problems* and *Calculus of Variations*, the other examples appear.

**Example 5.2** *(see [32]). Consider the following problem of the type (5.4)–(5.5):*

$$\int_0^1 (r(k)u_x^2 + (u^5 - x^3)^4\, u_x^{32})\, dx \quad \to \quad \min \tag{5.6}$$

$$u(\alpha) = k\,\alpha^{3/5}, \quad u(\beta) = k\,\beta^{3/5}, \tag{5.7}$$

*where $\alpha < 0 < \beta$. It was proved in [32], that the function $v(x) = k\,x^{3/5}$ is the unique solution to (5.6)–(5.7) for $r(k) = (3k/5)^{30}(1-k^5)(31k^5-16)$ ($r > 0$ for $k \in ((16/31)^{1/5}, 1)$) and $k$ close enough to 1. This problem has no solution among smooth functions and the minimizer $v(x)$ is not a solution to Euler–Lagrange Equation, even in the* generalized *sense, that is defined through a standard integral identity.*

**Remark 5.2** *As was proved in [34], for any closed subset $S \subset [0,1]$ of a zero measure there exists such smooth and strictly convex integrand $F_S(x,u,u_x)$ (perhaps not analytic, but arbitrarily smooth), such that the minimizer for corresponding variational problem*

$$\int_0^1 F_S(x,u,u_x)\,dx \longrightarrow \min$$

*attain "infinite values of derivative" exactly on the set $S$.*

# 6 Some solvability results for degenerate equations and for "super–quadratic growth" equations

Now let us come back to parabolic problems. First of all we would like to show how the family of LF makes it possible to prove estimate of the gradient of the type (1.5).

Following [35], for $(x,t) \in Q_\infty = [0,1] \times R^+$ consider the problem

$$u_t = \frac{\partial}{\partial x}\,\psi(u_x), \tag{6.1}$$

$$u(0,t) = A, \quad u(1,t) = B, \qquad u(x,0) = u_0(x) \tag{6.2}$$

with the same monotone and bounded function $\psi(s)$ as in the Example 4.1 (remind that the function $\psi(s)$ is supposed to be smooth, say, $\psi \to C^4$; odd; strictly increasing $\psi'(s) > 0$; and *bounded* $-\psi_\infty < \psi(s) < \psi_\infty$), see also [17, 35]. Here $A$ and $B$ are constants, compatibility conditions and a proper smoothness conditions on the problem data are supposed to be fulfilled.

**Remark 6.1** *Equation of the type (6.1) could be regarded as "parabolic regularization of the prescribed mean curvature equation" in particular situation $\psi(s) = s/\sqrt{1+s^2}$ (see [36, 37]). For such nonhomogeneous problems with the additional term $f(x)$ in the right side*

$$u_t = \frac{\partial}{\partial x}\,\frac{u_x}{\sqrt{1+u_x^2}} + f(x),$$

*blowing–up solutions have been observed for $|f(x)|$ "large enough" [37]. Note, that in nonhomogeneous case growth conditions (1.5) fails, in spite of homogeneous case.*

**Theorem 6.1** *(see [35]). There exists a classical solution to the problem (6.1)–(6.2) in half of a strip $(x,t) \in (0,1) \times R^+$. Moreover, this solution possesses a limit as $t \to \infty$.*

The key point of the proof is to establish that the gradient, $u_x(x,t)$, is bounded (1.4). In the case under study integrals $I_n(t) = \int_0^1 \int_0^{u_x} \psi^{2n+1}(s)\,ds dx$ serve as the Lyapunov Functionals (see Definition

5.1). Indeed,

$$\frac{d}{dt}\int_0^1\int_0^{u_x}\psi^{2n+1}(s)\,ds\;dx=\int_0^1\psi^{2n+1}(u_x)\,u_{xt}\;dx$$

$$=-(2n+1)\int_0^1\psi^{2n}(u_x)\frac{\partial}{\partial x}\;\psi(u_x)\,u_t\;dx \tag{6.3}$$

$$=-(2n+1)\int_0^1\psi^{2n}(u_x)\,u_t^2\;dx.$$

**Remark 6.2** *Note that "variational multipliers", $\rho_n(u_x)=(2n+1)\,\psi^{2n}(u_x)$, are not away from zero (as $\rho_n(0)=0$, e.g.), which is important for using families of LF.*

As the property $\psi(0)=0$ was assumed (see Example 4.1 and the original papers [17, 35]), the following relations hold:

$$\int_0^1\psi^{2n+2}(u_x)\;dx=\int_0^1\int_0^{u_x}\frac{d}{ds}\psi^{2n+2}(s)\,ds\;dx$$

$$=\int_0^1\int_0^{u_x}(2n+2)\psi^{2n+1}(s)\psi'(s)\,ds\;dx$$

$$\leq(2n+2)\max\psi'(s)\;\int_0^1\int_0^{u_x}\psi^{2n+1}(s)\,ds\;dx.$$

Latter inequality, in view of (6.3), implies that

$$\int_0^1\psi^{2n+2}(u_x)\;dx\leq(2n+2)\max\psi'(s)\;\int_0^1\int_0^{u_0'}\psi^{2n+1}(s)\,ds\;dx$$

$$\leq\max\psi'(s)\int_0^1\int_0^{u_0'}(2n+2)\psi^{2n+1}(s)\;\frac{\psi'(s)}{\min\limits_{|s|<\max|u_0'|}\psi'(s)}\;ds\;dx$$

$$=\frac{\max\psi'(s)}{\min\limits_{|s|<\max|u_0'|}\psi'(s)}\int_0^1\psi^{2n+1}(u_x)\;dx. \tag{6.4}$$

As $\min\psi'(u_0')\geq\delta>0$, then extracting the $2n+2$ root from both sides of (6.4) and then passing to the limit for $n\to\infty$, we obviously prove the inequality

$$\sup_{x,t}|\psi(u_x(x,t))|\leq\sup_{x\in[0,1]}|\psi(u_0'(x))|. \tag{6.5}$$

Note, that $\delta=\min\psi'(u_0')$ does not go to inequality (6.5). That is why (6.4) really gives (6.5).

Desired estimate of the type (1.5) follows (6.5) in view of assumed properties of the function $\psi(s)$.

**Remark 6.3** *Some statement about more general case of equation (4.1) with non constant multiplier $\varphi(u)$ has been also proved in [35].*

Now let us consider the parabolic problem with Euler–Lagrange Equation for the integrand $L(x,u,u_x) = r(k)\,u_x^2 + (u^3 - x^2)^2 u_x^{14}$ (see (5.3)) as the right–hand side:

$$u_t = \frac{d}{dx}\frac{\partial L(x,u,u_x)}{\partial u_x} - \frac{\partial L(x,u,u_x)}{\partial u}, \tag{6.6}$$

with the initial–boundary conditions

$$u(0,t) = 0, \quad u(1,t) = k, \qquad u(x,0) = u_0(x), \tag{6.7}$$

given in the Example 5.1.

**Remark 6.4** *Note that the strong maximum principle (i.e. the statement that the maximum of solution inside the space interval $x \in (0,1)$, for all $t > 0$ does not exceed its maximum on the "parabolic boundary" $\{x = 0\} \cup \{t = 0\} \cup \{x = 1\}$) holds for this problem.*

Let us rewrite equation (6.6) in the form (1.1) with

$$a(x,u,u_x) = \frac{\partial^2 L}{(\partial u_x)^2}, \qquad b(x,u,u_x) = \frac{\partial^2 L}{\partial u \partial u_x}\,u_x + \frac{\partial^2 L}{\partial x \partial u_x} - \frac{\partial L}{\partial u}. \tag{6.8}$$

We observe that the above growth condition (1.5) fails in the "weak sense", described in Remark 5.1 (see also the Example 5.1).

So we can not refer to the global in time existence theorems for (6.6), (6.7), although its classical solution exists for small enough $t$ under proper initial function $u_0(x)$ [4], which is claimed to be smooth and to satisfy compatibility conditions.

Using the Lyapunov Functionals technique, it is possible to prove (see [38]):

**Theorem 6.2** *There exists a classical solution $u(x,t)$, $x \in [0,1]$, $t > 0$ to problem (6.6), (6.7) for the integrand (5.3). Moreover, there exists a smooth steady state $w(x)$, such that*

$$\lim_{t\to\infty} \|u(x,t) - w(x)\|_{C^{2+\alpha}[0,1]} = 0.$$

Note, that all available examples of regular problems with non smooth minimizers in the calculus of variations possess the following property. Presenting the E–LE as

$$a(x,u,u_x)\,u_{xx} + b(x,u,u_x) = 0 \tag{6.9}$$

(see (6.8)) we can see that the function $b$ satisfies the identity

$$b(x,u,0) = 0. \tag{6.10}$$

**Remark 6.5** *From the viewpoint of general theory to parabolic problems the above identity (6.10) means that the strong maximum principle ($\sup\limits_{x,t}|u(x,t)| \le \sup\limits_{x}|u_0(x)|$) holds.*

Let us come back to quasilinear equations of the type (1.1). Consider the following problem

$$u_t = a(x,u,u_x)\,u_{xx} + b(x,u,u_x) = 0; \tag{6.11}$$

$$u(0,t) = A, \quad u(1,t) = B, \qquad u(x,0) = u_0(x) \tag{6.12}$$

under additional assumption (6.10), with constants $A$ and $B$, assuming also the proper smoothness of all the functions $a$, $b$, and $u_0$ and the compatibility conditions between $u_0$ and the boundary data.

In case the nonlinear terms $a$ and $b$ do not satisfy growth condition in (1.5), "extendibility condition" (5.2) fails as well. That means: there exists a "singular solution" to ODE in (5.2) (such solution is bounded in view of (6.10)), which derivative tends to $\infty$ for some $x$. Therefore, it is impossible to construct families of Lyapunov Functinals $\Phi(x, \xi, \eta)$, $\rho(x, \xi, \eta)$ in (5.1) (see Definition 5.1) in the whole layer $(x, \xi, \eta) \in [0, 1] \times R^2$. Besides, as was shown in [39], under condition (6.10), it is possible to construct *"Local Family of Lyapunov Functionals"*, which is determined in the neighborhood of dynamical trajectory $u(\cdot, t)$. In other words, in particular situation of (6.10), a set of initial functions $u_0(x)$ have been described, which generate a correct dynamical problem with no growth conditions.

**Remark 6.6** *Note that an alternative possibility to describe such initial functions $u_0(x)$, which generate correct dynamical problem of the type (6.11) in case of* "superquadratic growth" *of the quotient $b(x, u, u_x)/a(x, u, u_x)$, has been proposed [40, 41].*

To give a precise statement from [39] we need to introduce two definitions.

**Definition 6.1** *We denote by $\mathcal{Z}$ a set of* **singular solutions** *to ODE (5.2), which are bounded and, at the same time, can not be extended regularly onto the entire interval of variation of $x \in [0, 1]$.*

**Definition 6.2** *We shall say, that a smooth function $f(x)$* **possesses a positive generalized distance** *with the above set $\mathcal{Z}$ if*

$$\inf_{z \in \mathcal{Z}} \inf_{x \in [0,1]} \inf_{\eta \in I(x)} (|z(x) - f(x)| + |z'(x) - \eta|) > 0, \tag{6.13}$$

*where $I(x)$ stands for a segment with the endpoints 0 and $f'(x)$.*

**Theorem 6.3** *(see [39]). Suppose that*

1. *the initial function, $u_0(x)$, possesses a positive generalized distance (see Definition 6.2) with the set $\mathcal{Z}$ of singular solutions to ODE (5.2);*
2. *condition (6.10) holds;*
3. *all the functions $a(x, \xi, \eta)$, $b(x, \xi, \eta)$, and $u_0(x)$ are smooth;*
4. *compatibility conditions hold.*

*When:*

**(1)** *there exists a family of Lyapunov Functionals (5.1) along solution, $u(x, t)$ (which, as has been already mentioned, exists at least for small time);*

**(2)** *gradient of solution, $u_x(x, t)$, is uniformly bounded while solution exists (direct corollary of existence of LF family);*

**(3)** *there exists global in time solution, $u(x, t)$, to problem (6.11), (6.12) under study;*

**(4)** *global solution, $u(x, t)$, possesses a limit, $w(x)$, as $t \to \infty$*

$$\lim_{t \to \infty} \|u(x, t) - w(x)\|_{C^{2+\alpha}} = 0, \tag{6.14}$$

*which is nothing but the steady state for the problem (6.11), (6.12).*

Finally, we would like to mention one result concerning "singularly perturbation problems". These are characterized by the fact that the *formal* limiting equation, obtained by setting $\varepsilon = 0$, cannot satisfy the full original set of boundary data. It is a general result that "boundary-layers" are found, that is subdomains of the given space domain whose size is of order of $\varepsilon^\alpha$, for some $\alpha > 0$, located near the boundaries (or even inside the original space domain), where the solution behaves in a dramatically different way. Indeed, the solution exhibits a rather nonuniform character on the whole domain, possessing a multiscale behavior.

Below we give an existence result for nonlinear integroparabolic equation, which has been proved through strong estimates of solution to parabolic problem with the small parameter, in which case derivative blowing up was expected as $\varepsilon \to 0$. Besides, as was shown by the author in cooperation with R.Spigler (University of "Roma Tre", Rome, Italy) and D.Akhmetov (Sobolev Institute for Mathematics, Novosibirsk, Russia) no derivative blowing up occur.

Numerous phenomena, pertaining to Physics, Biology, Medicine, Neural Networks, are reasonably described in terms of large populations of nonlinearly coupled, often noisy, oscillators. A mathematical model is then represented by a large system of possibly stochastic nonlinear ordinary differential equations. In the limit case of infinitely many oscillators, when the interaction is of the so–called "mean–field" type, a single nonlinear parabolic model equation, containing an integral term, was derived by Kuramoto [44]. However, improving the finite–dimensional model to take into account certain observed features, lead to the introduction of second derivatives on the left–hand side of the systems above. This suggested a more general nonlinear partial differential equation, obtained by a procedure similar to that above, [45]. Such a new model equation is a Fokker–Planck type equation. The solution to the equation plus suitable initial data, is required to be positive, $2\pi$–periodic in angle variable $\theta$, and normalized. In this report, we address the problem of existence of such solutions.

The problem above leads to the following statement for partial integrodifferential equation. For $(\theta, \omega, t, \Omega) \in Q_T = [0, 2\pi] \times \mathbf{R} \times [0, T] \times [-G, G]$, find a function $\rho(\theta, \omega, t, \Omega)$ which solves the problem:

$$\frac{\partial \rho}{\partial t} = \frac{\partial^2 \rho}{\partial \omega^2} - \omega \frac{\partial \rho}{\partial \theta} + \frac{\partial}{\partial \omega}(\omega \rho) - \Omega \frac{\partial \rho}{\partial \omega} - \mathcal{K}_s(\theta, t) \frac{\partial \rho}{\partial \omega}, \tag{6.15}$$

$$\rho|_{\theta=0} = \rho|_{\theta=2\pi}, \tag{6.16}$$

$$\rho|_{t=0} = \rho_0(\theta, \omega, \Omega), \tag{6.17}$$

where

$$\mathcal{K}_s(\theta, t) := K \int_{-\infty}^{+\infty} \int_{-\infty}^{+\infty} \int_0^{2\pi} g(\Omega) \sin(\varphi - \theta) \, \rho(\varphi, \omega, t, \Omega) \, d\varphi \, d\omega \, d\Omega. \tag{6.18}$$

The problem (6.15)–(6.18) has been studied under the following assumptions:

**A)** The initial profile, $\rho_0(\theta, \omega, \Omega)$, is supposed to be: $(\mathbf{a_1})$ a continuous function in all variables $(\theta, \omega, \Omega) \in Q = \mathbf{R} \times \mathbf{R} \times [-G, G]$, belonging to the Hölder space $C^{2+\alpha_0}(Q)$, $\alpha_0 \in (0, 1)$ being a real constant; $(\mathbf{a_2})$ $2\pi$–periodic in angle, $\theta$; $(\mathbf{a_3})$ positive; $(\mathbf{a_4})$ normalized for all $\Omega \in [-G, G]$,

$$\int_0^{2\pi} \int_{-\infty}^{+\infty} \rho_0(\theta, \omega, \Omega) \, d\omega \, d\theta = 1;$$

and $(\mathbf{a_5})$ possesses an exponential decay property, in $\omega$, at infinity, namely:

$$\sup_{\theta \in \mathbf{R}, \, \Omega \in [-G, G]} \left| D^{l_1, l_2, l_3}_{\theta, \omega, \Omega} \rho_0(\theta, \omega, \Omega) \right| \leq C_0 \, \mathrm{e}^{-M_0 \omega^2}$$

for $\omega \in \mathbf{R}$ and $l_1 + l_2 + l_3 \leq 2$, where $C_0, M_0 > 0$ are constants and $l_i \geq 0$, $i = 1, 2, 3$, are integers.

**B)** The frequency distribution density, $g(\Omega)$, is assumed to be: $(\mathbf{b_1})$ belonging to the space $L^1(\mathbf{R})$; $(\mathbf{b_2})$ compactly supported on $[-G, G]$; and $(\mathbf{b_3})$ bounded.

**C)** The coupling strength, $K$, is a constant.

The problem should be considered nonstandard by several reasons, namely:

1) The governing equation in (6.15) is of the *second order* with respect to $\omega$, but only of the *first order* with respect to $\theta$ and $t$. Therefore, even besides of the integral term, equation is neither of the parabolic nor of the hyperbolic type.

2) The governing equation in (6.15) is considered in the *slab* $Q_T$, that is in a domain *unbounded* in $\omega$, which serves as a coefficient of the equation. This fact gives rise to *singularity* phenomena typical for equations with *unbounded coefficients.*

3) The governing equation in (6.15) contains an integral term taking on *unbounded* domain.

4) There is an additional variable, the natural frequency of the oscillators, $\Omega$, with respect to which no derivatives appear, but the integral is also made with respect to it.

5) We are interested *only* in solutions, *periodic* in $\theta$, while the governing equation contains the first time–derivative with respect to $\theta$ (cf. "time–periodic solutions to parabolic equations").

Therefore, results available in the literature concerning nonlinear parabolic or even integroparabolic equations cannot be applied.

In order to study the original problem we introduce [46] the *parabolic regularization* in the governing equation. The idea is to "regularize" equation (1) adding diffusion term $\varepsilon\rho_{\theta\theta}$, since equation (6.15) should be considered fully degenerate with respect to $\theta$. Moreover, *to avoid* the *unbounded coefficients* of equation, $\omega$, we also replace $\omega$ by bounding function $F_N(\omega)$, which is supposed to be smooth, bounded, and $F_N = \omega$ for $|\omega| < N$. So, instead of equation (6.15), we first study its *parabolic regularization*, that is a family of equations satisfied by $\rho^{\varepsilon,N}(\theta,\omega,t,\Omega)$,

$$\frac{\partial \rho^{\varepsilon,N}}{\partial t} = \frac{\partial^2 \rho^{\varepsilon,N}}{\partial \omega^2} + \varepsilon\,\frac{\partial^2 \rho^{\varepsilon,N}}{\partial \theta^2} - F_N\,\frac{\partial \rho^{\varepsilon,N}}{\partial \theta} + \frac{\partial}{\partial \omega}\left(F_N \rho^{\varepsilon,N}\right) - \Omega\,\frac{\partial \rho^{\varepsilon,N}}{\partial \omega} - \mathcal{K}_s^{\varepsilon,N}\,\frac{\partial \rho^{\varepsilon,N}}{\partial \omega} \tag{6.19}$$

in $Q_T \cap \{t > 0\}$, for any given $\varepsilon > 0$ and $N > 0$. Here $\varepsilon > 0$ and $N > 0$ are regularization parameters. The term $\mathcal{K}_s^{\varepsilon,N}$ is defined by formula (6.18) with the function $\rho^{\varepsilon,N}$ replacing $\rho$. Adding the second derivative term in $\theta$, we modify the periodic boundary conditions (6.16) to

$$(\rho, \rho_\theta)|_{\theta=0} = (\rho, \rho_\theta)|_{\theta=2\pi} \tag{6.20}$$

for $\omega \in \mathbf{R}$, $t \in (0, T]$, and $\Omega \in [-G, G]$.

After investigation of qualitative properties of convolutions of functions with fundamental solutions to linear parabolic equations, the existence theorem of classical solutions have been proved for the regularized problem (6.17), (6.19), (6.20). Then uniform (with respect to regularization parameters) estimates of these classical solutions have been established. Finally, after passing to the limit, the existence theorem of strong solutions have been proved for original problem (6.15)–(6.17). Here a *strong solution* to the problem (6.15)–(6.17) is a continuous and bounded function in $Q_T$ belonging to the corresponding Sobolev space and satisfying equation (6.17) almost everywhere in $Q_T$ and corresponding initial and boundary data as a continuous function.

**Theorem.** *Suppose that data of the problem (6.15)–(6.17) satisfy all assumptions* **A**, **B**, *and* **C**. *Then there exists a strong solution, $\rho(\theta,\omega,t,\Omega)$, to the problem (6.15)–(6.17). Moreover, this solution $\rho$ is nonnegative in $Q_T$ and is normalized,*

$$\int_0^{2\pi}\int_{-\infty}^{+\infty} \rho(\theta,\omega,t,\Omega)\,d\omega\,d\theta = 1$$

*for all $t \in [0, T]$ and $\Omega \in [-G, G]$.*

# 7 New approach to study "Derivative Blowing–Up Solutions"

Here we describe an alternative concept of *how it is possible to define* solutions to parabolic problems, proposed in [42, 43].

In the rectangle $(x,t) \in Q_T = (0,1) \times (0,T]$ consider the Dirichlet problem for the heat equation

$$u_t(x,t) = u_{xx}(x,t), \tag{7.1}$$

$$u(0,t) = u(1,t) = 0, \qquad u(x,0) = u_0(x). \tag{7.2}$$

Assume that the initial function is smooth enough, say $u_0(x) \in C^2[0,1]$, and satisfies the compatibility conditions $u_0(0) = u_0(1) = 0$. Moreover, suppose that $u_0$ is either strictly convex, or strictly concave

$$u_0''(x) \neq 0, \quad x \in (0,1).$$

It is of common knowledge [4], that under the above assumptions there exists a classical solution, $u(x,t)$, to problem (7.1)–(7.2). Everywhere in $Q_T$ this solution is respectively either convex, or concave with respect to $x$:

$$u_{xx}(x,t) \neq 0, \tag{7.3}$$

and monotone with respect to $t$:

$$u_t(x,t) \neq 0. \tag{7.4}$$

We can easily see that for our heat equation any smooth functions like

$$\rho = \rho(\eta), \qquad \Phi(\eta) = \int_0^\eta (\eta - \tau)\rho(\eta)\, d\tau, \tag{7.5}$$

with $\rho \geq \rho_0 > 0$, generate a *Lyapunov Functional* along solution to (7.1)–(7.2) (see the Definition 5.1):

$$\frac{d}{dt}\int_0^1 \Phi(u_x)\, dx = -\int_0^1 \rho(u_x)\, u_t^2\, dx. \tag{7.6}$$

In fact, for any smooth function $u(x,t) \in C^{2,1}(Q_T)$ (not necessarily a solution to (7.1)–(7.2)) the following identities (see (7.5) and (7.6))

$$\frac{d}{dt}\int_0^1 \Phi(u_x)\, dx = \int_0^1\int_0^{u_x} \rho(\eta)\, d\eta\, u_{xt}\, dx = -\int_0^1 \rho(u_x)\, u_{xx} u_t\, dx \tag{7.7}$$

hold.

Consider any smooth function $u(x,t) \in C^{2.1}(Q_T)$, which satisfy restrictions (7.3), (7.4), and identities (7.6) for all smooth and nonnegative functions $\rho(s)$. Fix an arbitrary value of the time variable, $t^* \in (0,T]$. Due to "convexity" in (7.3), for the given $t = t^*$ there exists a smooth and strictly monotone one–to–one correspondence, $\varphi^*(\xi)$, which maps values of the *derivative trace* $u_x(x,t^*)$ of our given function to interval $[0,1]$ according to the formula

$$\varphi^*(u_x(x,t^*)) = x. \tag{7.8}$$

For each $x^* \in (0,1)$ let us choose a sequence of functions, $\psi_n^*(\tau)$, which converges, as $n \to \infty$, to the Dirac's $\delta$, concentrated at $x = x^*$:

$$\psi_n^*(\tau) = \frac{n}{\pi^{1/2}} \exp\{-n^2(\tau - x^*)^2\}. \tag{7.9}$$

Consider a sequence of pairs $\rho_n$, $\Phi_n$ (7.5) with

$$\rho_n(\eta) = \psi_n^*(\varphi^*(\eta)), \tag{7.10}$$

where the *variational multipliers* $\rho_n$ are not away from zero.

According to our choice of $\rho_n$ ( see (7.10), (7.8), (7.9)) for the given $t = t^*$ on the trace of function $u(x, t^*)$ a sequence $\rho_n(u_x(x, t^*))$ converges to the Dirac's $\delta$, concentrated at $x = x^*$:

$$\rho_n(u_x(x, t^*)) \to \delta(x^*) \qquad \text{as} \quad n \to \infty. \tag{7.11}$$

Having known (7.6), from (7.7) we conclude, that identities

$$\int_0^1 \rho(u_x)\, u_{xx} u_t \, dx = \int_0^1 \rho(u_x)\, u_t^2 \, dx$$

hold with any smooth and nonnegative $\rho(s)$.

Putting here $\rho = \rho_n$ (see (7.10)), for any $t \in (0, T]$ we get

$$\int_0^1 \rho_n(u_x)\, u_{xx} u_t \, dx = \int_0^1 \rho_n(u_x)\, u_t^2 \, dx. \tag{7.12}$$

Consider a sequence of identities (7.12) for $t = t^*$ and pass to the limit as $n \to \infty$. Due to (7.11) we will have

$$u_{xx}(x^*, t^*) u_t(x^*, t^*) = u_t^2(x^*, t^*).$$

From this, because of monotonicity in (7.4), we can deduce that

$$u_t(x^*, t^*) = u_{xx}(x^*, t^*). \tag{7.13}$$

Since the point $(x^*,\, t^*) \in (0, 1) \times (0, T]$ has been chosen as an arbitrary one, the identity in (7.13) means that the equation (7.1) holds in $Q_T$.

Hence we obtain the following:

**Theorem 7.1** *Suppose that a smooth function $u(x, t) \in C^{2,1}(Q_T)$ is strictly convex (or concave) with respect to $x$ for any $t \in (0, T]$ (see (7.3)) and is strictly monotone with respect to $t$ for any $x \in (0, 1)$ (see (7.4)). Moreover, suppose that for all $\rho$, $\Phi$ of the form (7.5) identities (7.6) (see also Definition 5.1) hold.*

*Then the function $u(x, t)$ solves equation (7.1) for $(x, t) \in (0, 1) \times (0, T]$.*

**Remark 7.1** *If under the conditions of Theorem 7.1 function $u(x, t)$ satisfies initial and boundary conditions (7.2), this function solves (7.1)–(7.2).*

Similar theorem related to quasilinear equation (1.1) has been also proved in [42, 43].

# 8 Some problems to be investigated

Here we formulate several questions to be studied in future.

**1.** To study the long time behavior of solutions to parabolic problems in *"variational form"*

$$u_t = \frac{d}{dx} \frac{\partial L(x, u, u_x)}{\partial u_x} - \frac{\partial L(x, u, u_x)}{\partial u} \tag{8.1}$$

in case then the corresponding variational problem

$$\int_0^1 L(x, u, u_x)\, dx \;\to\; \min \tag{8.2}$$

possesses non smooth minimizer and, therefore, "growth conditions"

$$\left|\frac{b(x, u, u_x)}{a(x, u, u_x)}\right| \le K\ \varphi(|u_x|), \qquad \int_0^\infty \tau \frac{d\tau}{\varphi(\tau)} = \infty \tag{8.3}$$

fail.

**2.** To describe better a set of initial functions, which generate correct dynamical problems

$$u_t = a(x, u, u_x)\, u_{xx} + b(x, u, u_x); \tag{8.4}$$

$$\alpha_i\, u_x - \psi_i(u)|_{x=i} = 0 \quad (i = 0, 1); \qquad u(x, 0) = u_0(x). \tag{8.5}$$

in case then condition (8.3) fails.

**3.** To identify necessary and sufficient conditions for derivative blowing–up phenomenon to occur.

**4.** To verify hypothesis: in case initial function $u_0(x)$ possesses positive $C^1$–distance with the set of singular solutions to equation

$$y'' = -\frac{b(x, y, y')}{a(x, y, y')} \tag{8.6}$$

(see Definition 6.1) no derivative blowing–up occur.

**5.** To verify hypothesis that conditions of *"convexity"* in space and *"monotonysity"* in time are not necessary in Theorem 7.1.

Consider a family of variational problems (8.2) with similar boundary conditions and such that Euler–Lagrange Equations are equivalent for all integrands $L(x, u, u_x)$.

**6.** Do all the problems (8.2) possess the same minimizers?

**7.** Is the following statement correct: there exists a sequence of functions, which serve as "monotone in values of integrals" minimizing sequence for all the integrands simultaneously?

Now imagine a set of "level manifolds" for one of the integrals (8.2). Consider two of such manifolds with "close levels". Choose a certain point $u_0(x)$ on the "higher level manifold".

**8.** Is it possible to introduce *"time variable"* in such way, that solution to parabolic problem (8.1) with initial function $u_0(x)$ carries out in some sense *"optimal connection"* between two of above manifolds.

**9.** What does it mean *"optimal"* in the above problem?

Note, that in view of corresponding parabolic problems

$$\rho(x, u, u_x)\, u_t = \frac{d}{dx}\frac{\partial L(x, u, u_x)}{\partial u_x} - \frac{\partial L(x, u, u_x)}{\partial u} \tag{8.7}$$

it does not matter which integrand from the above family of variational problems we choose. Parabolic equations (8.7) for different "integrands" $L(x, u, u_x)$ are equivalent in case $\rho(x, u, u_x) \ge \delta > 0$.

**10.** What is the infuence of $\rho(x, u, u_x)$ on the "connection curve" for "level manifolds" (see question 8)?

**11.** Is that possible to use existence theorems from calculus of variations to prove existence theorems for corresponding parabolic problems and vice versa?

# Acknowledgements

This work have been mostly done in Italy, where the author stay through the CARIPLO Fellowship Program in the fall of 1998, and when finished under support by ARC Small Grant in the School of Mathematics and Applied Statistics, University of Wollongong, Australia.

# References

[1] T.I. Zelenyak, *Differents. Uravneniya*, **4**, 4 (1968).

[2] T.I. Zelenyak, M.M. Lavrentiev, Jr., and M.P. Vishnevskii, *Qualitative Theory of Parabolic Equations. Part I.* (VSP, Utrecht, The Netherlands, 1997).

[3] P. Polacik and K. Rubakovskii, *Preprint* (Universita degli Studi di Trieste, 1994).

[4] O.A. Ladyžhenskaya, V.A. Solonnikov, and N.N. Ural'tseva *Linear and Quasilinear Equations of Parabolic Type* (American Math.Soc., Providence, RI, 1968).

[5] O.A. Oleinik and S.N. Kruzhkov, *Uspekhi Mat. Nauk*, **16**, 5 (1961).

[6] S.N. Kruzhkov, *Trudy Moskovskogo Mat. Obsch.*, **16** (1967).

[7] S.N. Kruzhkov in *Trudy Seminara I.G.Petrovskogo*, **5** (Nauka, Moscow, 1979).

[8] N.Y. Krylov, *Nonlinear Elliptic and Parabolic Equations of Second Order* (D.Reidel, Dordrecht NL, 1987).

[9] O.A. Ladyžhenskaja and N.N.Ural'tseva, *Uspekhi Mat. Nauk*, **41**, 5 (1986).

[10] M.M. Lavrentiev, Jr., *Sib. Math. Zhurn.*, **32**, 1 (1991).

[11] V.S. Belonosov and T.I.Zelenyak, *Nonlocal problems in the theory of quasilinear parabolic equations*, Novosibirsk: Eds. Novosibirsk State Univ. (1975).

[12] A.F. Filippov, *Dokl. Akad. Nauk. SSSR*, **141**, 3 (1961).

[13] T. Dlotko, *J. Math. Anal. Appl.*, **154** (1991).

[14] A.A. Samarskii *et al*, *Blow–Up in Quasilinear Parabolic Equations*, (de Gruyter Exposition in Mathematics, **19**, Berlin, 1995).

[15] M. Fila and G. Lieberman, *Diff. and Integral Eq.*, **7**, 3/4 (1994).

[16] S. Angenent and M. Fila, *Diff and Int. Eq.*, **9** (1996).

[17] M. Bertsch and R. Dal Passo, *Arch. Rat. Mech. An.*, **117** (1992).

[18] G.I. Barenblatt *et al*, *SIAM J. Math. Anal.*, **24**, 6 (1993).

[19] G.I. Barenblatt *et al*, *J. Fluid. Mech.*, **253** (1993).

[20] M.P. Vishnevskii, T.I. Zelenjak, and M.M. Lavrentiev, Jr., *Sib. Mat. Zhurn.*, **36**, 3 (1995).

[21] A.A. Samarskii and A.P. Mikhailov *Mathematical modelling. Ideas. Methodics. Examples.* (Nauka, Moscow, 1997).

[22] V.A. Galaktionov and J.L. Vazquez, *Comm. Pure Appl. Math.*, **L** (1997).

[23] M. Bertsch and R. Dal Passo in *Nonlinear Diffusion Equations and Their Equlibrium States* 3, Eds. N.G.Lloyd et al. (Birkhauser Boston Basel Berlin, 1992).

[24] M. Bertsch *et al, Math. Ann.*, **294** (1992).

[25] Y. Giga, *Discrete and Continuous Dynamical Systems*, **1**, 3 (1995).

[26] P. Broadbridge, J.M. Goard, and M. Lavrentiev, Jr., *Studies in Applied Mathematics*, **99** (1997).

[27] T.I. Zelenyak, *Mat. Sbornik*, **104**, 3 (1977).

[28] L. Tonelli *Fondamenti di calcolo della variazioni* (Eds. Zanichelli, Bologna, **2**, 1923).

[29] M. Nagumo, *Japan J. Math.*, **6** (1929).

[30] S.N. Bernstein in *S.N.Bernstein, Collected Works, Vol.3: Differential Equations, Calculus of Variations and Geometryu* (Izdat. Akademii Nauk SSSR, eds. 1960).

[31] J. Ball and V. Mizel, *Bull. Amer. Math. Soc. (N.S.)*, **11** (1984).

[32] M.A. Sychev, *Mat. Sbornik*, **183**, 4 (1992).

[33] M.A. Sychev, *Prepritn no.11* (Inst. for Math., Siberian Branch of the Russian Acad. Sci., Novosibirsk, 1994).

[34] A.M. Davie, *Arch. Rational Mech. Anal.*, **101** (1988).

[35] M.M. Lavrentiev, Jr., P. Broadbridge, and V. Belov, *Comm. in PDE*, **22**, 1–2 (1997).

[36] P. Marcellini and K. Miller, *Preprint no.53* (Pubblicazioni Dell'Istituto di Analisi Globale e Applicazioni, C.N.R., Firenze, 1996).

[37] P. Marcellini and K. Miller, *J. of Diff. Eq.*, **137** (1997).

[38] M.M. Lavrentiev, Jr., *IMA Preprint Series, no. 1385, March 1996* (University of Minnesota, Minneapolis, USA, 1996).

[39] M.M. Lavrentiev, Jr., *Sib. Mat. Zhurn*, **34**, 6 (1993).

[40] V.L. Kamynin, *Differential Equations*, **19**, 5 (1983).

[41] O.N. Barsov, *Moscow University Math. Bull.*, **49**, 6 (1994).

[42] M.M. Lavrentiev, Jr., *Preprint no. 18/95* (Dpt. of Math. Univ. of Wollongong, Wollongong, Australia, 1995).

[43] M.M. Lavrentiev, Jr., in *Prague Math. Conf. 1996. Proceedings* (Prague, Czech Rep., 1996).

[44] Y. Kuramoto, *Lecture Notes in Physics*, **39** (1975).

[45] J.A. Acebrón and R. Spigler, *Phys. Rev. Lett.*, **81**, 11 (1998).

[46] M.M. Lavrentiev, Jr. and R. Spigler, *Differential Integral Equations*, **13**, 4-6 (2000).

# Timetabling Highly Constrained System via Genetic Algorithms

HO SUNG C. LEE
*Department of Management Engineering*
*Ateneo de Manila University*
*Loyola Heights, Quezon City*
*Philippines*
*hslee@admu.edu.ph*

and

AUGUSTO Y. HERMOSILLA
*Department of Mathematics*
*University of the Philippines*
*Diliman, Quezon City*
*Philippines*
*auggie@math01.cs.upd.edu.ph*

**Abstract**

This paper intends to address the university lecture timetabling problem that are highly constrained via customized Genetic Algorithm. Empirical study is applied with the background of School of Management and the School of Science and Engineering, Ateneo de Manila University, but its method should be easy to extend to other schools and universities as educational institutions often have similar structure or similar types of constraints. This paper also intends to enhance its practical usability by presenting a user-friendly interface to the actual personnel, who schedules classes but may not be operations research or computer science specialist. Attempts to further enhance the heuristic's efficiency through other modifications such as explicit parallelism are discussed.

## 1 Introduction

As educational institutions are challenged to grow in number and complexity with often no corresponding increase in resources due to usual budget constraints, their operations need to be streamlined more and more to maximize efficiency while minimizing resource usage. One of the areas of possible improvement is the task of class timetabling, or assigning the faculty and students that belong to a class to specific time and room without conflicts. University timetabling problem is a very common and seemingly simple, but yet a very difficult problem to solve in practice. While solution definitely exists (evidenced by the fact that we do hold classes!), but may not be satisficing or desirable for the ones concerned, an automated optimal schedule is very difficult to derive at present.

There had been many attempts to address this problem using classical methods, such as integer programming and graph coloring without much success. There were also more successful attempts to address the problem using several heuristics search methods. However, to our present date, university timetabling is still largely done by hand, because a typical university setting requires numerous customized complicating constraints that are difficult to model or difficult to automate. In addition, there is a problem of certain constraints being inviolable, while others are merely desirable.

The objective of this study is to look into some heuristics to solve an intrinsically complicated university timetabling problem with inviolable constraints. There are many inviolable constraints in university scheduling, especially given the situation in the Philippines, in general, and in Ateneo, in particular. For instance, even full-time faculty members, who are valuable resources to industries as well, have to spend considerable amount of time outside the university during normal office hours due to numerous other activities he/she may be involved in or committed to.

Since heuristic methods are often not exact but more of approximate to give at best near-optimal solutions within reasonable time in its search process, it is normally difficult to enforce inviolable constraints. Consider, for example, constraint A, which must be absolutely, non-negotiably, and uncompromisingly kept at all times, and condition B which is merely (but not absolutely) desirable. We would normally assign a big penalty to A, while assigning a relatively small penalty to B. Supposedly, this approach should work, but it is still possible that numerous quantities of condition B might outweigh condition A by its sheer number into a possible explosive geometric order. If we try to address the problem by making the penalty of A even bigger, there might be computational problems with respect to speed and size of the problem, and some of the smaller penalties might lose their relevance altogether, even making a feasible solution meaningless or useless.

Because school timetabling problem is a known *NP-Complete* problem, we consider some heuristic approach to solving it. Specifically, we consider in this paper the *Genetic Algorithms*(GAs), also known as *Evolutionary Algorithms.* We formulate the chromosome such that we can accommodate inviolable constraints without serious degradation in its over-all performance. For this paper, the particular university setting uses the School of Management and School of Science and Engineering, Ateneo de Manila University. The structure should be easily extendable to other university settings, where inviolable constraints are involved.

# 2 Genetic Algorithm Implementation

This section discusses the implementation of the modified GAs used to solve the models for our highly constrained timetabling system.

## 2.1 Input from User

Input from user can be divided into two parts: *Time Slots requirement* and *Room requirement.* Time slot requirement refers to time specific constraints and desired conditions, such as a teacher cannot teach two classes simultaneously or a teacher does not wish to teach for more than 3 consecutive hours. The biggest time slot requirement is that faculty cannot be assigned a class when they are not available. In other words, faculty will specify when he/she is available, and the model will optimize with that given. Of course, in practice, this powerful option must be used with moderation else, it will be impossible to avoid conflicts.

Room requirements refer to conditions, such as a certain class must be scheduled in a certain room, and faculty desiring to teach in a proximate building from his/her office. As a kind of *"divide-and-conquer"* method to enhance the performance of a heuristic, it can be shown that the problem can be solved in two phases by first assigning the time slots, and then later assigning the rooms. This is made possible by embedding additional room related constraints into time slot constraints. Any room requirements such as the maximum rooms available or a class needing a special room can be translated to corresponding time slot requirement, such that after assigning the time slots, the rooms can be assigned without conflict.

### 2.1.1 Time Slots Requirements

The first step in the implementation is to get the classes to be assigned, identify faculties and determine the various requirements to each class and/or faculty. Information gathered for 244 classes and 78 faculties consists of class data of SY 2000-2001, 2nd Semester from School of Management, Computer Science Department and Mathematics Department. Each faculty is given a customized alphabet set for Phase I for their time availability. Time slots 1 through 7 refer to TTh time slots of 1:30, each beginning at 7:30 AM and ending at 6:00 PM. Time slots 8 through 16 refer to MWF time slots of 1:00, each beginning at 7:30 AM and ending at 4:30 PM. On TTh days, 4:30 - 6:00 is an acceptable regular class hour, but on MWF, 4:30 - 6:00 is designated activity hours for various co-curricular and extra-curricular organizations. Class hours starting after 6:00 PM are considered night classes, which our model does not schedule. This is because most of the faculties teaching at night are part-time faculties with really no negotiable time requirements, and the university has relatively few night classes to schedule as demanded by faculties without producing conflicts.

While some full-time faculties do not mind any particular schedule, some part-time faculties are available only at certain exact hours. At this point, we do not consider room assignment yet, since it will be handled by Phase II, but it suffices to say that treatment is identical except that it will be sorted by then assigned time slots instead of faculty.

### 2.1.2 Room Requirements

There are basically two inviolable constraints and one desired condition in room assignment. First and most obvious is that there exist a finite maximum number or rooms to be used. Although in the Ateneo setting, this number is not specified per school basis. Schools simply request for rooms to central admin and the administration tries to give as much as possible. This is often not concluded in one time negotiation and months are spent in trying to iron out the different issues. Therefore, the model will assign the different classes in minimum number of rooms possible. If this number happens to be greater than given number of rooms, then the problem is infeasible to begin with. This constraint is transformed into time slot constraint of no more than $R$ simultaneous classes allowed, where R is the number or rooms available. Second inviolable room requirement is that certain classes must be taught in certain rooms. This can also be translated into time slot requirements. It is translated as special classes requiring the use of $X$ type of room can be held at most $X_n$ at a time, where $X_n$ is the number of type $X$ room. For example, if 80 student-classes can be held in rooms 116, 117 and 118 only, then the simultaneous occurrence of 80 student classes can be up to 3 only. Then, in Phase II, the alphabets of these classes will be modified to allow only rooms 116, 117 and 118 only.

The desired, but not required, condition in room assignment is that teachers desire to teach in a building close to their office. This is addressed in Phase II by penalizing a class held in a room by a teacher whose department is distant. In our model, we consider 4 buildings, where 3 buildings house 3 different departments and one building is approximately in the middle of three buildings. Therefore, if a faculty teach in his/her building or in the middle building, no penalty is given. However, if a faculty teaches in any other building, penalty is assigned.

A sample raw data to be worked on can be sorted by time slots. Since the structure of the problem is exactly the same as Phase I, except that it is sorted by time slots instead of teachers, and we have to make sure no two classes are held in the same room (at the same time) instead of making sure no two classes are held in the same time (by the same faculty). The logic is the same and minimal adjustment is made in evaluation, mutation and initialization setting to make the Phase I GA work here.

Important to note, however, is that the Phase II problem is almost trivial. Since we have N rooms, and we already made sure that no more than N classes happen in each time slot, it is just a matter of distributing the N rooms to N classes in that time slot. Since we want to honor the teacher's preference of having the class near their offices, we assign the classes in each building first to the classes held by the faculty of that department. Then, we assign whatever else we have by first utilizing the common building, and later by using whatever other rooms we have. The problem is also linear and it can be computed efficiently by LP. Therefore, the real important goal of the entire algorithm lies in solving Phase I, which satisfies all time slot requirements and also satisfies room requirements translated to time requirements.

Also, notable is that Phase II can be solved per time slot, because each time slot does not affect the other time slots. In fact, that is how the GA is applied in this case, and GA usually finds the optimal solution to Phase II per time slot within 10 seconds, except for two time slots where 1 faculty has to teach in a non-favorite building. (However, since Ateneo is small, the building is still within 5-minute walking distance.)

Since Phase II is so trivial in solving, most of the experiments and comparisons are based on Phase I computation alone.

## 2.2 Fitness Evaluation

Fitness, or desirability of a timetable, must be computed for GA to move on. Each chromosome is searched for some undesirable condition and is penalized according to the gravity of the condition. The evaluation is a rather easy part of this application of GA, since the alphabet sets are actual integers representing time slots. This is a desired condition for GA to be effective. The solution space can be combinatorially complex but evaluation must be rather quick. The conditions given penalty are outlined according to their relative importance. Since it is easier to model a maximizing GA, actual fitness to be used is:

$$FITNESS = \frac{1}{1 + Condition_1 \times Penalty_1 + \cdots + Condition_n \times Penalty_n} .$$

### 2.2.1 Time Slot Conflict

The first and most important violation would be assigning two or more classes of the same faculty into the same time slot. Actually, since the time slots are already hard coded depending on the availability of faculty, this would be the only real critical violation to the problem. After this constraint had been satisfied, the resulting schedule is already conflict-free and satisfies the requests of the teachers.

### 2.2.2 Occurrence of More Than 3 Straight Lecture Hours

It is quite difficult for a professor to deliver quality lecture for 4 or more consecutively straight hours. Therefore, each faculty must not be given more than 3 straight hours of lecture on any given day. This would be interpreted as up to 2 straight classes on TTh days and up to 3 straight classes on MWF days.

### 2.2.3 Possibility of Student Conflict

We can largely ignore the possibility of student conflicts in a schedule because most courses are offered in several sections. However, there exist in Ateneo some very small major programs where the entire batch is only 1 class big, meaning less than 30 students. The major subjects of these students must not be scheduled at the same time. This is, however, not as big a problem as it might sound, because these specialized groups of students are only few. In the above example of 244 classes, only 8 sets of classes must not be in conflict due to student constraints. So, it naturally does not tend to occur. Also, since those are strictly major subjects of each departments, the department can easily make certain that those classes are not in conflict. Nonetheless, however rare this occurrence is, when it does occur, it is penalized heavily since it is a definitely unacceptable schedule.

### 2.2.4 Minimize the Room Usage

In the Ateneo setting, different schools in the university share classrooms. Therefore, the number of classrooms available to a certain school is not strictly determined. However, it is desirable to use as small number of rooms as possible to give more freedom for other schools to schedule their classes. Also, this means that the classes of the school can be more compactly arranged geographically, as a way to minimize the distances or diameter of the utilized area. This is a desired result of each department. To implement this, each room used adds little penalty points to the fitness function. If in the future, the maximum allowable classrooms for a school is more rigidly defined, the penalty can be big for any number greater than that and only a small increment for numbers less than that.

## 2.3 Selection Process

One of the biggest problems GA researchers face is the possibility of having a super-individual early on the iterations. Such super-individual given proportionally equal chance of reproduction will totally eliminate all other individuals by cloning itself in succeeding generations and make the heuristic converge prematurely. But we also cannot refuse to give stronger individuals more chance of reproduction, since doing so will reduce Genetic Algorithm to pure random search. There are various methods to solve this problem and two often-used procedures are discussed here.

### 2.3.1 Fitness Scaling

Before selection process, the fitness values can be remapped into numbers of narrower range to cause the relative differences to become smaller while maintaining the ranks. The result is that super-individuals will have more chance than other individuals, but will not be as dominating as had there been no fitness scaling. We do not just arbitrarily scale the fitness values. We would like to at least make sure that an average individual prior to scaling remains to be an average individual after scaling. A simple but useful fitness scaling is the *Linear Fitness Scaling*:

$$f^* = a + bf$$

where $f^*$ is the scaled fitness value with $a$ and $b$ as scaling operators. However, if a population contain an individual far too low fitness compared to the average, then some of the mapped fitness values can go negative and this is not allowable. So, in these instances, fitness scaling cannot be used and becomes ineffective. In general, fitness scaling helps but not greatly [Goldberg, 1989] .

### 2.3.2 Tournament Selection

*Tournament selection* is a much more recent development of GA that seeks to address the problem of premature convergence due to super-individuals [David, 1989]. In a Tournament Selection, the operating variable is the size of Tour. After fitness values are computed, Tour number of individuals are randomly selected and put into a *"tournament."* The individual with highest fitness in the tournament is selected to be the parent. As the Tour gets bigger, the more biased is the selection process towards better solutions, in other words, *higher elitism*, but it is still relatively easy for lesser individuals to get selected than *Roulette Wheel selection.* This is because even if there exist a super-dominant individual, that individual will not be selected for reproduction, if it is first not selected to participate in a tournament. Of course, it would win any tournament it joins, but there would still be many other tournaments in which it could not be selected. Therefore, it is important to note that tournament participants are selected randomly without bias. At the same time, this selection process discourages the very weak individuals, since they are very unlikely to win any "tournament" at all because the chance of the tournament filling with only weaker individuals is almost nonexistent. In Roulette Wheel selection, the weaker individuals are still assigned portions of the probability pie. In summary, Tournament Selection protects diversity by lessening the super-individual's chances of reproducing, while simultaneously lessening the chance of very weak individuals. It has the combined effect of elitism, because it kills weak individuals, and diversification, since super-individuals are not allowed to dominate.

This method matches the real evolutionary process more closely, because survival depends on competition rather than all the collective population. Also, nature kills off weaker individuals, rather than make stronger individuals rule over everyone else. In this sense, massive cloning is bad for a population, because lesser diversity may easily kill off a population of clones not resistant to a disease. Diversity is a way of a living system to survive virulent diseases which may easily kill dominant individuals. It is sad that in today's society, many individuals aspire to quite similar goals in life. It makes us less resistant to social diseases that may arise in the future.

It is shown from experiments discussed in next section just how much more effective tournament selection is over standard selection methods with or without fitness scaling.

## 2.4 Crossover and Recombination

Because the model in this paper uses non-contiguous integer sets as alphabets, simple crossovers do not generate enough new solutions for possible improvement of the population. For example, the two 8-bit binary individuals with integer values of 58 and 179 produced resulting integer values of 51 and 186. However, had these been integer values, such intermediary values cannot be created. It is also impractical to map the current chromosomes to binary for crossover and convert back to integer because we have non-contiguous alphabet space. Also, the chromosome is already 244 long, and if we use 4 bit per integer that would be 976 long! Therefore, we have to use *Recombination* techniques to bring about optimal diversity in the population and controls premature convergence due to selective pressure exerted by the super dominant solutions. In recombination, at least in the simple version, for each segment of chromosome, the value is randomly chosen from that corresponding segment of the parents. This is a very fast process and produces a more diverse population than crossover. Below is an illustration of a crossover operator and an integer recombination operator.

$$\begin{array}{cccccccccccccccccc} 0 & 0 & 1 & 1 & 1 & 0 & 1 & 0 & \longrightarrow & 0 & 0 & 1 & 1 & 1 & 0 & 1 & 1 \\ 1 & 0 & 1 & 1 & 0 & 0 & 1 & 1 & \longrightarrow & 1 & 0 & 1 & 1 & 0 & 0 & 1 & 0 \end{array}$$

$$\begin{array}{ccccccccc} 15 & 9 & 12 & 14 & \longrightarrow & 15 & 4 & 16 & 14 \\ 13 & 4 & 16 & 2 & \longrightarrow & 14 & 9 & 12 & 2 \end{array}$$

This can be likened to a multi-point crossover over binary space. However, in our situation, we have the additional problem of our chromosome being non-contiguous integers.

Blending or hybridizing algorithms are also possible wherein we get an integer truncated randomly weighted average of the two values given by parents. This would be like:

$$Child\ Value \ = \ (1-p) \times (parent\ a) \ + \ p \times (parent\ b)$$

where p is a random number between 0 and 1. However, since our alphabet set is non-contiguous, we have to first map the alphabet to an array and use the array index for blending. But this would be too computationally expensive.

## 2.5 Mutation

*Mutation* can also be done in various ways, but since we have a non-contiguous integer alphabet set, we are forced to use a customized *mutation operator* [Ross, 1994]. The mutation operator should be random, and yet it must not assign any time slot not allowed for that particular class. This is solved by first reading in all the allowable times for each class and mapping the random number to the length of that array and assigning the content of selected random integer number given as follows:

$$Mutated\ Value[i,j] \ = \ Alphabet[i, Int(Rnd() \times Length\ of\ Alphabet[i]) \ + 1]$$

where Alphabet[i,j] is a matrix element for ith individual and jth class and Rnd() generates random floating value between 0 and 1.

This is a very slow process, but fortunately, mutation is done rarely. Even in nature, mutation is quite rare. It is easy to think that higher mutation rate should generate better results, because it can generate more variety of solutions. Studies show that ideal mutation rate is about 1/N, where N is the size of chromosome. In our case, it is about 0.004 only. This fact can be verified in the experiments. With mutation of 0.04 or greater, the Genetic Algorithm will simply not converge.

### 2.6 Reinsertion

After Evaluation, Selection, Recombination and Mutation comes *Reinsertion*, where we insert the newly created individuals into the population. Many kinds of reinsertions are possible. The simplest way is to erase all the old population and replace it with new population. Another method, known as *Elitism*, is to retain $x$ number of top individuals or remove $x$ number of bad individuals. Actually, removing $x$ bad individuals is the same as retaining $N - x$ better individuals. Therefore, they are both variations of Elitism. Reinsertion can also be done with more children produced than necessary, so we only use the better of the produced children for reinsertion.

The concept of Elitism is to ensure that we will not have evolutionary digression. This sounds like a good idea, and in fact, small systems, which require relatively small number of iterations, behave well with Elitism. However, as the system complexity grows and the numbers of iterations also grow, Elitism gives undue bias towards the better individuals such that it hinders the newer solutions from being explored. This could lead to dangerous convergence or near convergence to sub-optimal points. Experiments also showed that, in general, GA worked better without Elitism.

### 2.7 On Population Size

Interestingly, there is no mention yet in the literatures of variable sized population, since that is how nature realistically works. In real life, when population has high average fitness rating, the number of individuals tend to increase due to prosperity, and when the population has weak average fitness rating, the number of individuals tend to decrease, since the weaker of the already weak population cannot withstand environmental difficulties. This is especially true when we consider many organisms at the same time. This makes sense for GA, because when population is weak, for example at the start, it is beneficial to have stronger selective pressure and having less population makes it easy for stronger individuals to propagate themselves, because they are less hindered by randomness to dominate the population. When the population gets stronger, we must avoid super-individuals from easily dominating to avoid premature convergence. We also need more diverse variety of individuals to get better solutions, because if current solution is much closer to optimal solution then improving it is very difficult. A larger population can also possibly help this.

## 3 Some Results and Analyses

This section discusses some of the results of our GAs implemented in the Ateneo setting. We discuss here experiments with several core GA parameters that influence the performance of GA. When a parameter is not being experimented on, they take some default settings given as follows:

| | |
|---|---|
| Population Size | 40 |
| Crossover/Recombination Rate | 0.70 |
| Mutation Rate | 0.004 |
| Selection Method | Tournament |
| Crossover Method | Recombination |
| Reinsertion Method | No Elitism |
| Constraint Set | Full |

For example, population size 40 is used in all experiments, except when population size itself is being tested. It is then changed to 20, 40, 100 and 200 for that experiment only.

### 3.1 Experimental Designs

The experiments aim to measure quantitatively the parameters that can yield better results. Better results are obtained if a GA converges to a solution with higher fitness value. Among models with similar converged fitness value, the model, which converges faster, and give more consistent output through many runs would be the better one.

Often times, the users of heuristics do not know what it the optimal outcome to be expected. In our timetabling situation, however, we already know that a schedule of zero conflicts and room usage of less than available rooms is desirable. Meaning, if the GA converges to a solution of less than desirable, the GA has failed in finding an acceptable timetable. The succeeding sections describe combinations of experiments being done here. Each GA model runs until there is no change in fitness value for 50 consecutive generations. Each model runs 7 times to give the Best, Worse and Average runs. Average run is given in terms of the median, instead of the mean, because comparing mean time across runs with different resultant fitness value and comparing resultant fitness value across runs with different speed are irrelevant. Mean fitness and run time with their standard deviations from the 7 runs are also given.

### 3.2 Selection Method

This is perhaps the most important parameter for our GA. Using standard Roulette Wheel Selection with or without Fitness Scaling enhancement yields very bad results. Median fitness of Roulette Wheel selection refers to a result of (24,17,5,0), meaning 17 faculty time slot conflicts, 5 occurrences of more than 3 hour lectures, 0 instances of student conflicts and 25 rooms utilized. Compared to this Tournament Selection yields (16,0,0,0), meaning 0 faculty time slot conflicts, 0 occurrence of more than 3 hour lectures, 0 instances of student conflicts and only 16 rooms utilized. Convergence time is also significantly improved from more than 100 seconds to less than 10 seconds.

| Selection | Best Run | Median Run | Worst Run | Mean | Std. Dev. |
|---|---|---|---|---|---|
| Roulette Wheel | 0.00156 | 0.00142 | 0.00124 | 0.00142 | 0.00011 |
| Fitness Scaling | 0.00150 | 0.00131 | 0.00123 | 0.00133 | 0.00009 |
| Tournament | 0.01235 | 0.01235 | 0.01235 | 0.01235 | 0 |

Table 3.2.1 : Fitness Values

| Selection | Best Run | Median Run | Worst Run | Mean | Std. Dev. |
|---|---|---|---|---|---|
| Roulette Wheel | 100.24 | 111.06 | 100.17 | 104.07 | 4.00 |
| Fitness Scaling | 100.18 | 100.30 | 99.08 | 99.72 | 0.58 |
| Tournament | 6.65 | 7.80 | 9.89 | 8.24 | 1.21 |

Table 3.2.2 : Convergence time in seconds

Scaled Roulette Wheel generally results in lower fitness value throughout the run. However, the fluctuations in fitness value are somewhat dampened. Tournament Selection shows significantly improved results by almost 10 times. As shown here, while, fluctuations still exists, fitness values generally increase while population matures, which is the desired result.

## 3.3 Crossover vs Recombination

| 244 classes | Best Run | Median Run | Worst Run | Mean | Std. Dev. |
|---|---|---|---|---|---|
| Crossover | 0.01235 | 0.01235 | 0.01024 | 0.01207 | 0.00073 |
| Recombination | 0.01235 | 0.01235 | 0.01235 | 0.01235 | 0 |

Table 3.3.1 : Fitness Values

| 244 classes | Best Run | Median Run | Worst Run | Mean | Std. Dev. |
|---|---|---|---|---|---|
| Crossover | 5.22 | 5.44 | 7.52 | 6.25 | 0.94 |
| Recombination | 6.65 | 7.80 | 9.89 | 8.24 | 1.21 |

Table 3.3.2 : Convergence time in seconds

Because it is easier to do crossover than recombination, crossover-based GA is on the average faster. However, recombination yields slightly more consistent results, as can be seen by the zero standard deviation in Recombination GA. Because the results are so close to each other, it is worthwhile to examine the results on a bigger problem. Results of 732 class model is given below:

| 732 classes | Best Run | Median Run | Worst Run | Mean | Std. Dev. |
|---|---|---|---|---|---|
| Crossover | 0.00442 | 0.00433 | 0.00406 | 0.00433 | 0.00013 |
| Recombination | 0.00442 | 0.00433 | 0.00406 | 0.00428 | 0.00016 |

Table 3.3.3 : Fitness Values

| 732 classes | Best Run | Median Run | Worst Run | Mean | Std. Dev. |
|---|---|---|---|---|---|
| Crossover | 39.66 | 34.21 | 27.02 | 36.18 | 10.65 |
| Recombination | 44.60 | 38.28 | 31.85 | 42.98 | 10.73 |

Table 3.3.4 : Convergence time in seconds

In smaller model, Recombination works better and in larger model, Crossover model works better in the first several iterations, but as the model matures, they converge to similar value. Recombination, however, shows a little smoother or consistent increase in fitness value. Crossover-based GA reaches moderately high fitness value rapidly, but takes much longer to improve to even higher fitness values. Recombination based GA has a more steady increase in its increase of fitness value. Since the actual problem at hand is only 244 classes and Recombination models show slightly smoother performance, while results are identical for both Crossover and Recombination models in their best and median runs, this paper uses Recombination technique by default.

## 3.4 Elitism vs No Elitism

| 244 classes | Best Run | Median Run | Worst Run | Mean | Std. Dev. |
|---|---|---|---|---|---|
| Crossover | 0.01235 | 0.01235 | 0.00990 | 0.01153 | 0.00107 |
| Recombination | 0.01235 | 0.01235 | 0.01235 | 0.01235 | 0 |

Table 3.4.1 : Fitness Values

| 244 classes | Best Run | Median Run | Worst Run | Mean | Std. Dev. |
|---|---|---|---|---|---|
| Crossover | 8.34 | 9.23 | 4.95 | 7.47 | 2.13 |
| Recombination | 6.65 | 7.80 | 9.89 | 8.24 | 1.21 |

Table 3.4.2 : Convergence time in seconds

Because Elitism lengthens the influence of good individuals, it can converge to a solution faster. This can be seen by the shorter mean time. However, since a good individual may not necessarily lead to the optimal solution, it can also make GA converge to the less than optimal solution. In other words, if we are lucky that the earlier strong solutions lead to optimal solution, then Elitism makes us get there faster. However, if it is in the foot of the wrong mountain to climb, it can have serious problem in venturing into another solution. Thus, we can observe from above that Elitism can generate fast results at times, but it can also make it very hard for GA to find the true global solution, as a kind of trap. Elitism model gets *"stuck"* at a suboptimal solution from around generations 70 to 120. In fact, Elitism eventually breaks out to find the true optimal solution, but takes quite some time for that to happen. In some runs with Elitism, model converges permanently to suboptimal solutions.

## 3.5 Constraint Sets

| | Best Run | Median Run | Worst Run | Mean | Std. Dev. |
|---|---|---|---|---|---|
| Basic | 1.00000 | 1.00000 | 1.00000 | 1.00000 | 0 |
| + 3 hour constr. | 1.00000 | 1.00000 | 1.00000 | 1.00000 | 0 |
| + student conf. | 1.00000 | 1.00000 | 1.00000 | 1.00000 | 0 |
| + min room | 0.01235 | 0.01235 | 0.01235 | 0.01235 | 0 |
| Basic + stud conf. | 1.00000 | 1.00000 | 1.00000 | 1.00000 | 0 |

Table 3.5.1 : Fitness Values

| | Best Run | Median Run | Worst Run | Mean | Std. Dev. |
|---|---|---|---|---|---|
| Basic | 2.69 | 2.91 | 3.35 | 2.94 | 0.20 |
| + 3 hour constr. | 3.29 | 3.90 | 4.34 | 3.83 | 0.36 |
| + student conf. | 2.97 | 3.41 | 3.96 | 3.36 | 0.34 |
| + min room | 6.65 | 7.80 | 9.89 | 8.24 | 1.21 |
| Basic + stud conf. | 2.64 | 2.69 | 3.18 | 2.81 | 0.22 |

Table 3.5.2 : Convergence time in seconds

This tests the behavior of the GA model as the more constraints are added to the system. Since GAs are robustly resistant to irregular solution space, additional constraints and complexity merely slow the evaluation process. In general, this model is shown to be quite resistant to increasing complexity. Aside from GAs being flexible, this result is also because of the way this particular GA is formulated and structured. Alphabet set construction takes much of the optimization out, but actually lessen the solution space.

Dealing with time and space (room) issues separately reduces the complexity of constraints greatly. Also, since the chromosome is sorted by faculty and each segment refers to a single class, evaluation functions can easily dissect and score the solutions. The structure also makes future additions of constraints easy to do. Interesting to note after adding the student conflict constraint is that the heuristic actually converges faster. From the very rapid time the solutions are found, we can surmise that the problem is not very difficult to solve when room constraints are taken out. Seemingly, the additional constraint of student conflict helps GA decide which solution to settle into. Without slowing the GA, student conflicts are a rather rare happening anyway. The constraint seems to act as a tie-breaker type of constraint.

To check if the order of introducing constraints matter, result of a model with basic and student conflict constraints is also shown. Introducing the student conflict constraint also shows increasing convergence rate compared to basic model. Therefore, the sequence of introducing the constraints seems to have little importance compared to the kind of constraint being introduced. In general, it may be possible that adding a little of easy-to-satisfy constraints may help convergence to a solution in this type of GA, where there are many solutions with equal fitness value towards the mature state of the run. Of course, introducing too many easy-to-solve constraints will compound itself to increase the complexity, and will slow the model down. This will have to be specifically studied in the future by another paper for it to be proven empirically.

## 3.6 Size of Population

| | Best Run | Median Run | Worst Run | Mean | Std. Dev. |
|---|---|---|---|---|---|
| 20 | 0.01235 | 0.01235 | 0.01042 | 0.01186 | 0.00072 |
| 40 | 0.01235 | 0.01235 | 0.01099 | 0.01205 | 0.00054 |
| 100 | 0.01316 | 0.01235 | 0.01235 | 0.01258 | 0.00040 |
| 200 | 0.01408 | 0.01235 | 0.01235 | 0.01283 | 0.00067 |

Table 3.6.1 : Fitness Values

| | Best Run | Median Run | Worst Run | Mean | Std. Dev. |
|---|---|---|---|---|---|
| 20 | 3.35 | 5.99 | 4.18 | 4.21 | 0.89 |
| 40 | 5.99 | 10.00 | 9.72 | 7.85 | 2.19 |
| 100 | 34.27 | 16.42 | 21.64 | 23.05 | 8.62 |
| 200 | 70.14 | 34.93 | 41.46 | 49.22 | 13.92 |

Table 3.6.2 : Convergence time in seconds

This analysis monitors the behavior of GA as the population size alters. Small size GA iterates very rapidly and converges to a solution very quickly. Often times, the solutions are good enough as evidenced by the fact that all the 4 models have the same median run fitness value. However, smaller sized GAs behave less consistently as seen by large standard deviations and the fact that it's best and worst run fitness are significantly lower than those of larger population models. Largely attributable to the lack of diversity a small population can pose, GA takes longer to converge in terms of time. Quite understandably, large populations are carrying many different individuals, who do not necessarily benefit the population in general. This is so true in our actual society too. As society grows larger, many individuals only want to just *"live off the society"*, being costly parasites without contributing any benefit to the over-all population. In nature, this is related to the carrying capacity of the environment for the survival of the whole population. Bigger population size strains the environmental capacity.

## 3.7 Size of Problem

One of the most crucial sensitivity analyses would be to test how the model fares when the problem size changes. Here, we consider problems of double and triple the original size. Original data set is arbitrarily doubled and tripled to produce 488 class problem and 732 class problem.

Because sufficiently larger problems generally need bigger population size to work properly, population size of 120 is used for this experiment. This must be a little big for 244 class problem and a little small for 732 problem, but should be a good enough average figure to test the differences of 244, 488 and 732 class size models.

As a very important analysis, more data is presented to understand the behavior much better. First table monitors the fitness value and time taken to converge to that solution. Next table monitors the iteration counts of the Best, Median, and Worst runs. Last table monitors the fitness in a matrix of (# or rooms used, # teacher conflict, # 3 hour violations, # student conflicts). We multiply this matrix by (5, 30, 15, 30) to compute for weighted penalty. Fitness is then simply 1/(1+Weighted Penalty).

| Fitness Measure | Best Run | Median Run | Worst Run | Mean | Std. Dev. |
|---|---|---|---|---|---|
| 244 | 0.01316 | 0.01235 | 0.01235 | 0.01246 | 0.00031 |
| 488 | 0.00662 | 0.00641 | 0.00585 | 0.00622 | 0.00031 |
| 732 | 0.00452 | 0.00433 | 0.00407 | 0.00432 | 0.00016 |

Table 3.7.1 : Fitness Values

| | Best Run | Median Run | Worst Run | Mean | Std. Dev. |
|---|---|---|---|---|---|
| 244 | 9.88 | 8.90 | 11.31 | 9.04 | 1.82 |
| 488 | 29.22 | 22.46 | 20.43 | 21.81 | 4.08 |
| 732 | 56.03 | 32.96 | 39.27 | 44.44 | 11.09 |

Table 3.7.2 : Convergence time in seconds

| | Best Run | Median Run | Worst Run | Mean | Std. Dev. |
|---|---|---|---|---|---|
| 244 | 185 | 147 | 71 | 11.71 | 43.33 |
| 488 | 214 | 196 | 180 | 182.14 | 30.44 |
| 732 | 321 | 179 | 252 | 278.14 | 68.50 |

Table 3.7.3 : Iteration Counts

| | Best Run | Median Run | Worst Run |
|---|---|---|---|
| 244 | (15,0,0,0) | (16,0,0,0) | (16,0,0,0,) |
| 488 | (30,0,0,0) | (31,0,0,0) | (31,0,1,0,) |
| 732 | (44,0,0,0) | (46,0,0,0) | (46,0,1,0,) |

Table 3.7.4 : Solution

Most significantly, this model's computational time does not increase very steeply as the problem gets bigger. As to be understood, the bigger models take longer to converge because of two reasons: One is that the number of iterations needed to converge increases; and the other is that each iteration takes longer due to the enlarged chromosome.

Looking at iteration counters, it is actually almost linear in increase. Increase in time is expected, but not to be worried too much about, because as long as the number of iterations it takes to solve the problem do not grow geometrically, the GA will definitely converge and can be solved in reasonable finite time. A problem is when the iterations grow geometrically, because whether the larger versions can be solved at all is uncertain. Or, even if it can be solved at all, to solve in reasonable finite time is unsure. Hence, the basic problem of computational complexity associated with combinatorial explosion with problem size becomes an affliction.

However, bigger models sometimes do not yield a conflict free schedule. One violation of the 3 hours constraint is common in experimental results. Although this does not have the highest penalty, this is actually the hardest constraint to satisfy, since the GA must not only check concurrent classes but adjacent classes as well. However, when the GA do find a conflict free schedule, the room utilization of bigger problems is actually better! For example, in the median run, 16 rooms are used for 244 class problem, and only 31 rooms are used for 488 class problem which is double the size. Furthermore, only 46 rooms are used in the 732 class problem. This means that the GA does not only solve the bigger problem, but actually takes advantage of the larger data set to find ways to allocate rooms to even smaller wastage.

This practical behavior of school timetabling problem is not surprising. After all, the system complexity does not increase significantly as the problem grows. It is just a matter of having more rooms, more teachers and more classes. So long as the ratios of these three elements remain relatively constant, there is no significant difficulty expected in taking down a larger model. This is because in timetabling, the primary cause of conflicts is due to the teachers seeking certain arrangements or permutations to deliver courses in their own desired time, so that other personal constraints can be balanced as well. In the 244 class model, a faculty teaches no more than 6 classes per week, and this is the same in the 732 class model. The problem would seriously become more difficult if a faculty would start teaching 10 subjects or more. But that would be teaching 30 units a semester! God forbid such unfortunate events from us. There exists a certain maximal teaching capacity both to ensure that quality education can be delivered and that the system become not overly complicated.

The behavior of GA in our given problem is because of the peculiar fact that it is a school timetabling. It cannot be safely generalized that GA will always be stable with the increase of problem size, especially when the interconnected type of constraints will grow in complexity with the growth of overall number of variables. No theoretical result is yet available to give definite conclusions. After all, this is heuristics, which are always customized to suit different kinds of problems that behave differently.

## References

[1] S. Abdennadher & M. Marte (2000): , *University Course Timetabling Using Constraint Handling Rules*, Journal of Artificial Intelligence, Special Issue on Constraint Handling Rules, 14(4):311-326

[2] D. Abransom & J. Abela (1992): *A Parallel Genetic Algorithm for Solving the School Timetabling Problem*, 15th Autstrailian Computer Science Conference, Hobart

[3] S. Adams (1996): *The Dilbert Principle*, Harper Collins Books

[4] G.E. Bullelock (1996): *Programming Parallel Algorithms*, Carnegie Mellon University

[5] B.P. Buckles (1992): *Genetic Algorithms*, IEEE Computer Society Press, CA.

[6] C. Breshears (1995): *A Beginner's Guide to PVM - Parallel Virtual Machine*, University of Tennessee

[7] E.K. Burke, D.B. Elliman & R. Weare (1994): *A University Timetabling System based on Graph Colouring and Constraint Manipulation*, University of Nottingham

[8] G. Chartrand & O.R. Oellermann (1993): *Applied and Algorithmic Graph Theory*, McGraw-Hill, Inc., Singapore

[9] L. Davis (1996): *Handbook of Genetic Algorithms*, International Thomson Computer Press

[10] W. Erben & J. Keppler (1995): *A Genetic Algorithm Solving a Weekly Course-Timetabling Problem*, Department of Computer Science, Fachhochschule Konstanz

[11] M. Fischer & J. Dongarra (1996): *Experiences with Windows 95/NT as a Cluster Computing Platform for Parallel Computing*, University of California, San Diego, Oak Ridge National Laboratory

[12] M. Fischer & J. Dongarra (1996): *Another Architecture: PVM on Windows 95/NT*, University of California, San Diego, Oak Ridge National Laboratory

[13] D.E. Goldberg (1989): *Genetic Algorithms in Search, Optimization and Machine Learning*, Addison-Wesley Publishing Co., Inc.

[14] J. Gross & J. Yellen (1998): *Graph Theory and Its Applications*, CRC Press, Florida

[15] J.H. Heizer & B. Render (1996): *Production and Operations Management : Strategic and Tactical Decisions*, Prentice Hall International, New Jersey

[16] A.Y. Hermosilla : *Optimization Heuristics*, Working Paper in Progress, Department of Mathematics, U.P. Diliman

[17] F. Hoffman (1997): *Genetic Algorithm, Berkeley University*, http://www.ang-physik.uni-kiel.de/ hoefi/ga.html

[18] J.R. Koza (1996): *Genetic Programming Tutorial*, CVPR-96 Conference, San Francisco

[19] D. Levine (1994): *A Parallel Genetic Algorithm for the Set Partitioning Problem*, Argonne National Laboratory

[20] Ludwig Maximilians Universitat Munchen (2000): *Constraint Based University Timetabling*, http://www.pms.informatik.uni-muenchen.de/ ifiplan/, LMU

[21] G.L. Nemhauser & L.A. Wolsey (1999): *Integer and Combinatorial Optimization*, John Wiley & Sons, Inc.

[22] M. Obitko (1998): *Introduction to Genetic Algorithms with Java Applets*, http://cs.felk.cvut.cz/ xobitko/ga/

[23] M. Pinedo (1995): *Scheduling: Theory, Algorithms and Systems*, Prentice Hall, New Jersey

[24] J.G. Pira (1999): *School Timetabling by Genetic Algorithm and Simulated Annealing*, MS Thesis, University of the Philippines-Diliman

[25] H. Pohlheim (1998): *Genetic and Evolutionary Algorithms: Principles, Methods and Algorithms*, Germany

[26] D. Quagliarella, J. Periaux, C. Poloni & G. Winter (1998): *Genetic Algorithms and Evolution Strategies in Engineering and Computer Science*, John Wiley & Sons Ltd.

[27] Queensland University of Technology (2000): *Overview of Timetabling at QUT*, http://www.qut.edu.au/admin/timetable/timetabling/overview.html, QUT

[28] P. Ross, D. Corne & H-L Fang (1994): *Successful Lecture Timetabling with Evolutionary Algorithms*, University of Edinburgh

[29] B.T. Tesfaldet & A.Y. Hermosilla (1999): *A Lamarckian Genetic Algorithm Applied to the Travelling Salesman Problem*, Advances in Complex Systems, Hermes Science Publications, Vol. # 4, pp. 431-457.

## APPENDIX: Sample Result

| Cat No. | Sec. | Instructor | Dept | Time | Room | Time | Days | Room |
|---|---|---|---|---|---|---|---|---|
| MA 11 | D | ABERIN, Ma. Alva | 2 | 15 | 8 | 2:30 PM | MWF | SEC 2 |
| MA 12 | H | ABERIN, Ma. Alva | 2 | 7 | 11 | 4:30 PM | TTh | SEC 5 |
| MA 19 | E | ABERIN, Ma. Alva | 2 | 1 | 9 | 7:30 AM | TTh | SEC 3 |
| MA 19 | E | ABERIN, Ma. Alva | 2 | 10 | 9 | 9:30 AM | MWF | SEC 3 |
| MA 12 | I | ADVIENTO, Jasper | 2 | 11 | 12 | 10:30 AM | MWF | SEC 6 |
| MA 19 | B | ADVIENTO, Jasper | 2 | 1 | 8 | 7:30 AM | TTh | SEC 2 |
| MA 19 | B | ADVIENTO, Jasper | 2 | 9 | 19 | 8:30 AM | MWF | K 203 |
| MA 19 | D | ADVIENTO, Jasper | 2 | 10 | 12 | 9:30 AM | MWF | SEC 6 |
| MA 19 | D | ADVIENTO, Jasper | 2 | 8 | 12 | 7:30 AM | MWF | SEC 6 |
| CS 21A | B | AGLORO, Jr. | 1 | 2 | 3 | 9:00 AM | TTh | F 208 |
| CS 21B | B | AGLORO, Jr. | 1 | 10 | 2 | 9:30 AM | MWF | F 207 |
| CS 21B | E | AGLORO, Jr. | 1 | 14 | 3 | 1:30 PM | MWF | F 208 |
| CS 21B | G | AGLORO, Jr. | 1 | 13 | 2 | 12:30 PM | MWF | F 207 |
| CS 21B | H | AGLORO, Jr. | 1 | 16 | 3 | 3:30 PM | MWF | F 208 |
| MA 12 | A | ANG, Olivia | 2 | 4 | 11 | 12:00 PM | TTh | SEC 5 |
| MA 20 | C | ANG, Olivia | 2 | 10 | 19 | 9:30 AM | MWF | K 203 |
| MA 20 | C | ANG, Olivia | 2 | 12 | 12 | 11:30 AM | MWF | SEC 6 |
| MA 20 | J | ANG, Olivia | 2 | 5 | 11 | 1:30 PM | TTh | SEC 5 |
| MA 20 | J | ANG, Olivia | 2 | 2 | 12 | 9:00 AM | TTh | SEC 6 |
| CS 122 | D | ANGLO, Elizabeth | 1 | 16 | 9 | 3:30 PM | MWF | SEC 3 |
| CS 122 | E | ANGLO, Elizabeth | 1 | 13 | 4 | 12:30 PM | MWF | F 209 |
| EG 121 | V | AQUINO, Crispin | 3 | 9 | 16 | 8:30 AM | MWF | SS 286 |
| EG 121 | W | AQUINO, Crispin | 3 | 8 | 15 | 7:30 AM | MWF | SS 284 |
| EG 185 | A | AQUINO, Crispin | 3 | 2 | 13 | 9:00 AM | TTh | SS 280 |
| MGT 63 | D | AQUINO, Crispin | 3 | 11 | 13 | 10:30 AM | MWF | SS 280 |
| MGT 63 | E | AQUINO, Crispin | 3 | 10 | 15 | 9:30 AM | MWF | SS 284 |
| MA 12 | D | BARCELON, Rey | 2 | 1 | 12 | 7:30 AM | TTh | SEC 6 |
| MGT 51 | A | BARRETO, Ma. Jennifer | 3 | 4 | 14 | 12:00 PM | TTh | SS 282 |
| MGT 53 | D | BARRETO, Ma. Jennifer | 3 | 6 | 20 | 3:00 PM | TTh | K 204 |

# A DECEPTIVE GAME WITH MODIFIED INFORMATION

K.T. LEE

*School of Engineering and Science, Monash University Malaysia*
*Bandar Sunway 46150 Petaling Jaya Selangor, Malaysia*
*E-mail: lee.king.tak@engsci.monash.edu.my*

We consider a two-person game where the first player tries to maximize a given objective based on certain information provided. The second player may modify this information in order to frustrate the first player. This deceptive act from the second player is aimed at minimizing what the first player can obtain.

## 1 Introduction

Consider the following two-person zero-sum game $\Gamma$ in which the two players are called $P_1$ and $P_2$. Both players are informed that four numbers have been drawn, each uniformly and independently from the interval $[0, 1]$. These numbers are then dropped into four numbered boxes, one in each box. The boxes are closed, and the contents (that is, the madnitude of the number inside) of each box is written on a label pasted to the top of each box. Without allowing to open any box to see what it contains, $P_1$ selects a box, and his objective is to maximize the contents of the box that he has selected. This does not present any difficulties since the labels indicate the actual contents of the boxes. Hence the information of the labels is valuable to him. The difficulties arise when $P_2$ is introduced into the picture. Now $P_2$'s objective is to make the contents of the box selected by $P_1$ to be as small as possible. Before $P_1$ selects a box, $P_2$ can modify the information of the labels in the following way. He can do nothing, or to interchange the labels of any two boxes. $P_1$ knows this fact, but not what have been actually done to the labels. Can $P_1$ still use the information of the labels to assist him to select a box profitably? The game is understood to be played many times, and a solution is seeked in terms of the expected values.

Suppose $P_1$ adopts the strategy $\varphi$ of selecting each box with probability of $\frac{1}{4}$. It is easy to see that he can ensure to obtain an expected payoff of $\frac{1}{2}$, irrespect of what $P_2$ will do. For this bottomline case, $P_1$ does not need or use the information provided by the labels. The implication here is the information has been so distorted by $P_2$ that it has become completely useless to $P_1$. Let $v$ denote the value of $\Gamma$. The above discussion shows $v \geq \frac{1}{2}$.

## 2 A Solution

We will show $v = \frac{1}{2}$. It suffices to construct a strategy for $P_2$ that will restrict $P_1$ from obtaining an expected payoff of more than $\frac{1}{2}$. Towards this end, rename the contents of the boxes $a$, $b$, $c$, and $d$ with $a > b > c > d$. The set where two or more of these variables are equal has zero probability measure in $[0,1]^4$, and hence may be ignored. Let $p_{ab}$ denote the probability $P_2$ interchanges the labels with contents $a$ and $b$, etc.. Consider the following strategy $\sigma$ of $P_2$:

If $c \leq (a+b+c+d)/4$, choose

$$\begin{aligned}
p_{ab} &= \frac{(a-b+c-d)\{(a-b)(a-c)+(b-d)(c-d)\}}{2(a-c)\{(a-b)^2+(a-d)^2+(b-d)^2\}}, \\
p_{ac} &= \frac{(a+b+d-3c)}{4(a-c)}, \\
p_{ad} &= \frac{(a-b+c-d)\{(a-c)(a-d)+(b-c)(b-d)\}}{2(a-c)\{(a-b)^2+(a-d)^2+(b-d)^2\}}, \\
p_{bd} &= \frac{(a-b)(a-c)(a-d)+(a-b)(c-d)^2+(a-c)(b-d)^2+(a-d)(b-c)^2}{2(a-c)\{(a-b)^2+(a-d)^2+(b-d)^2\}}
\end{aligned} \tag{1}$$

Otherwise, choose

$$\begin{aligned}
p_{ad} &= \frac{(a-b+c-d)\{(a-c)(b-c)+(a-d)(b-d)\}}{2(b-d)\{(a-c)^2+(a-d)^2+(c-d)^2\}}, \\
p_{bd} &= \frac{(3b-a-c-d)}{4(b-d)}, \\
p_{cd} &= \frac{(a-b+c-d)\{(a-b)(a-c)+(b-d)(c-d)\}}{2(b-d)\{(a-c)^2+(a-d)^2+(c-d)^2\}}, \\
p_{ac} &= \frac{(a-d)(b-d)(c-d)+(c-d)(a-b)^2+(b-d)(a-c)^2+(a-d)(b-c)^2}{2(b-d)\{(a-c)^2+(a-d)^2+(c-d)^2\}}
\end{aligned} \tag{2}$$

It is easy to verify the variables $p_{..}$ are nonnegative and sum to one. These probabilities may be obtained by applying the method in Lee[3] and Teo to $\Gamma$, and exploiting any symmetries involved. Next, we verify the strategy $\sigma$ does restrict $P_1$ from obtaining an expected payoff of more than $\frac{1}{2}$. We cosider only the case $c \leq (a+b+c+d)/4$.

Suppose $P_1$ sees the tuple of labels $(a, b, c, d)$. This tuple can only come from the tuples of contents $(b, a, c, d)$, $(c, b, a, d)$, $(d, b, c, a)$ and $(a, d, c, b)$, and these tuples arise equally likely. We can then compute the conditional expectation of $P_1$ if he selects any one of the four boxes. As an illustration, suppose he selects the box with label $a$. Thus his conditional expectation is $bp_{ab} + cp_{ac} + dp_{ad} + ap_{bd} = (a+b+c+d)/4$. In this way, for each tuple of labels, $P_1$ can compute a tuple of conditional expectatations. For the tuple of labels $(a, b, c, d)$, it turns out that the tuple of conditonal expectations is $(m, m, m, m)$ where $m = (a + b + c + d)/4$. This holds also for the case $c > (a + b + c + d)/4$. It follows that $P_1$'s conditional expectatation is always equal to $m$ no matter which box (or any mixture of the boxes) he selects. Hence if $P_2$ uses the strategy $\sigma$, $P_1$ will receive an expected payoff of $\frac{1}{2}$, computed from the integral of his conditional expectation $m$ over $[0, 1]^4$.

In summary, the value of $\Gamma$ is $\frac{1}{2}$. An optimal strategy of $P_1$ is $\varphi$, and an optimal strategy of $P_2$ is $\sigma$.

## 3 Some Variants

Allowing $P_2$ to interchange any two labels leads to the situation where the information of the labels can become completely useless to $P_1$. We now restrict slightly what $P_2$ can do to the labels. Note in $\Gamma$, $P_2$ can do one of the following seven things to the labels: leave the labels unchanged, interchange the labels of $a$ and $b$, interchange the labels of $a$ and $c$, interchange the labels of $a$ and $d$, interchange the labels of $b$ and $c$, interchange the labels of $b$ and $d$, or interchange the labels of $c$ and $d$.

Consider the game which is the same as $\Gamma$ except $P_2$ is not allowed to interchange the labels of $b$ and $c$. The value of this game is still $\frac{1}{2}$ because $P_2$ can still use the strategy $\sigma$ to restrict $P_1$ from obtaining an expected payoff of more than $\frac{1}{2}$.

Consider now the game $\Gamma_1$ which is the same as $\Gamma$ except $P_2$ is not allowed to interchange the labels of $c$ and $d$. We will show the value of $\Gamma_1 = \frac{61}{120}$ which is greater than $\frac{1}{2}$. Hence in this game, $P_2$ *cannot* make the information of the labels completely useless, and $P_1$ *can* construct a strategy, based on the labels, that will ensure obtaining an expected payoff of more than $\frac{1}{2}$.

For the box with label $c$, its contents is either $a$, $b$ or $c$. This box's contents cannot be $d$ since that means $P_2$ has interchanged the labels of the boxes with contents of $c$ and $d$. Hence, if $P_1$ selects the box with label $c$, his conditional expectation is $c$ or more. On the other hand, if he selects the boxes equiprobably, his conditonal expectation is $(a+b+c+d)/4 = m$. Since $P_1$ is maximizing, he will try to attain the larger of these two conditional expectations. This suggests the strategy $\varphi_1$ of $P_1$:

If $c \leq (a+b+c+d)/4$, select each of the four boxes with probability of $\frac{1}{4}$. Otherwise, select the box with label $c$.

The following observation will show the value of $\Gamma_1 > \frac{1}{2}$. Returning back to $\Gamma$ temporarily, we observe $P_1$ can only obtain a conditional expectation of $m$ over the whole set of $[0,1]^4$. The strategy $\varphi_1$ enables $P_1$ in $\Gamma_1$ to obtain a conditional expectation of more than $m$ over certain subset (that has a positive probability measure) in $[0,1]^4$ as well as a conditional expectation of $m$ over its complement in $[0,1]^4$. It then follows $P_1$ can ensure an expected payoff of more than $\frac{1}{2}$. We next show $P_1$ cannot improve upon $\varphi_1$; that is, $\varphi_1$ is an optimal strategy of $P_1$. Let $\sigma_1$ denote the following strategy of $P_2$:

If $c \leq (a+b+c+d)/4$, choose $p_{..}$ according to (1).
Otherwise, choose

$$p_{ab} = \frac{2\{(b-c)^2 + (a-c)(c-d)\}}{(a-b)^2 + (a-d)^2 + (b-d)^2},$$

$$p_{ad} = \frac{2\{(c-d)^2 + (a-c)(c-b)\}}{(a-b)^2 + (a-d)^2 + (b-d)^2},$$

$$p_{bd} = \frac{2\{(a-c)^2 + (b-c)(c-d)\}}{(a-b)^2 + (a-d)^2 + (b-d)^2}.$$

If $P_2$ uses $\sigma_1$, we can compute $P_1$'s tuple of conditonal expectations, given he sees the tuple of labels $(a,b,c,d)$. The results are summarized below:

If $c \leq (a+b+c+d)/4$, $P_1$'s tuple of conditonal expectations $= (m,m,m,m)$. Otherwise, $P_1$'s tuple of conditonal expectations $= (a+b+d-2c, c, c, c)$.

Consider the case $c > (a+b+c+d)/4$. Then $a+b+d-2c < c$. With the tuple of his conditional expectations $= (a+b+d-2c, c, c, c)$, $P_1$ can obtain at most $c$ for this case. It follows that by using $\sigma_1$, $P_2$ can prevent $P_1$ from obtaining a conditional expectation of more than $m$ if $c \leq (a+b+c+d)/4$, and from obtaining a conditional expectation of more than $c$ otherwise. This shows that for the game $\Gamma_1$, $\varphi_1$ is an optimal strategy of $P_1$, and $\sigma_1$ is an optimal strategy of $P_2$. By direct calculation, the value of $\Gamma_1 = \frac{61}{120}$.

Consider now the game $\Gamma_2$ which is the same as $\Gamma$ except $P_2$ is not allowed to interchange the labels of $a$ and $b$. It will be shown below the value of $\Gamma_2$ is just greater than $\frac{1}{2}$.

For the box with label $b$, its contents cannot be $a$ since that means $P_2$ has interchanged the labels of the boxes with contents of $a$ and $b$. Thus the '$a$' contents is in one of the three boxes with labels $a$, $c$ or $d$. Next examine the possible contents (not necessarily in the proper order of which contents in which box) in these three boxes. The only possibilities are $\{a, b, c\}$, $\{a, b, d\}$ and $\{a, c, d\}$. Hence, if $P_1$ selects each of the three boxes with labels $a$, $c$ or $d$ with probability $\frac{1}{3}$, his conditional expectation is $(a + c + d)/3$ or more. On the other hand, if he selects the boxes equiprobably, his conditonal expectation is $(a + b + c + d)/4 = m$. Since $P_1$ is maximizing, he will try to attain the larger of these two conditional expectations. Note the constraint $(a+b+c+d)/4 \geq (a+c+d)/3$ is equivalent to $b \geq (a+b+c+d)/4$. This suggests the strategy $\varphi_2$ of $P_1$:

If $b \geq (a + b + c + d)/4$, select each of the four boxes with probability $\frac{1}{4}$. Otherwise, select each of the three boxes with labels $a$, $c$ or $d$ with probability $\frac{1}{3}$.

We next show $P_1$ cannot improve upon $\varphi_2$; that is, $\varphi_2$ is an optimal strategy of $P_1$. Towards this end, let $\sigma_2$ denote the following strategy of $P_2$:

If $b \geq (a + b + c + d)/4$, choose $p_{..}$ according to (2).
Otherwise, choose $p_{ac} = p_{ad} = p_{cd} = \frac{1}{3}$.

If $P_2$ uses $\sigma_2$, we can compute $P_1$'s tuple of conditonal expectations, given he sees the tuple of labels $(a, b, c, d)$. The results are summarized below:

If $b \geq (a+b+c+d)/4$, $P_1$'s tuple of conditonal expectations $= (m, m, m, m)$.
Otherwise, $P_1$'s tuple of conditonal expectations $= (\frac{a+c+d}{3}, b, \frac{a+c+d}{3}, \frac{a+c+d}{3})$.

The process proceeds in a similar manner as described in $\Gamma_1$. In summary, the value of $\Gamma_2$ is $\frac{11521}{23040}$. An optimal strategy of $P_1$ is $\varphi_2$, and an optimal strategy of $P_2$ is $\sigma_2$.

## 4 Conclusions

The game $\Gamma$ and its variants dicussed here are generally known as deception games. The study of deception games was initiated by Mark Thompson in his Harvard undergraduate thesis in 1970. The proof that very general deception games always have a solution can be found in Baston[1] and Bostock. Some examples of deception games can be found in the remaining references.

## Acknowledgments

This research is partially supported by University of Malaya, Malaysia.

## References

1. V.J. Baston and F.A. Bostock, Deception games, *International Journal of Game Theory* **17** 129-134 (1988).
2. V.J. Baston and F.A. Bostock, A simple cover-up game, *American Mathematical Monthly* **95** 850-854(1988).
3. K.T. Lee and K.L. Teo, A game with distorted information, *Naval Research Logistics* **40** 993-1001 (1993).
4. K.T. Lee, On a deception game with three boxes, *International Journal of Game Theory* **22** 89-95(1993).
5. J. Spencer, A deception game, *American Mathematical Monthly* **80** 416-417 (1973).

# DUADIC CODES

SAN LING
*Department of Mathematics, National University of Singapore, Singapore 117543, Republic of Singapore*
*Email:* lings@math.nus.edu.sg

Binary duadic codes were introduced in the 1980s by Leon, Masley and Pless in [7]. They can be regarded as generalizations of binary quadratic residue codes. The base field restriction was later removed by Smid [14,15]. In particular, the $Q$-codes of Pless [12] are duadic codes over $\mathbf{F}_4$. In another direction, Camion [3], Ward [16], van Lint and MacWilliams [10] have defined generalized quadratic residue codes, which are ideals in algebras of finite abelian groups. The quadratic residue codes form a subclass where the groups used are cyclic. Unifying the various constructions using group algebras, a generalization of the construction of duadic codes was discussed in [4], where a broader splitting of a finite abelian group was used to construct a more general family of codes. The broader definition of splitting also turned out to be more appropriate when one tries to work over finite rings instead of fields. Duadic codes over the finite ring $\mathbf{Z}_4$ were defined and studied by Langevin and Solé in [6], and subsequently over other finite rings [8,9]. We highlight some of the recent developments in the study of duadic codes.

## 1 Introduction

Binary duadic codes were introduced by Leon, Masley and Pless [7] as a generalization of quadratic residue codes. These codes share a number of properties similar to those of quadratic residue codes. A few interesting new examples of binary codes were also found. The approach adopted by the authors made use of idempotent generators. Later, Smid [14,15] generalized the notion of duadic codes to other base fields. Using his definition, the $Q$-codes of Pless in [12] are duadic codes over $\mathbf{F}_4$.

In a different direction, various authors such as Camion [3], Ward [16] and van Lint and MacWilliams [10] generalized quadratic residue codes to abelian codes which are not necessarily cyclic codes. This is achieved by regarding quadratic residue codes as ideals in cyclic group algebras, and the generalizations are defined as ideals in abelian group algebras.

In recent years, some further generalizations have been studied and interesting results obtained. Split group codes were introduced by Ding, Kohel and Ling in [4]. By broadening the definition of partitions, or splittings, used in duadic codes, it was shown that the main theorems for duadic codes still remain valid in this larger setting. New good codes were also found as examples.

The interest in the past decade in codes over finite rings has also infected the study of duadic codes. Building on the generalized notion of splittings in [4], Langevin and Solé introduced and studied duadic codes over $\mathbf{Z}_4$ in [6]. Their work was further extended to the rings $\mathbf{Z}_{2k}$, where $k$ is a positive integer, and $\mathbf{F}_2[u]/(u^2)$, cf. [8,9]. Duadic codes over such rings turn out to be rather interesting objects. A number of extremal Type II codes are found among their augmented and extended codes.

In this talk, we attempt to survey some of the main developments and the key results in the study of duadic codes, focusing more on recent developments.

## 2 Basic Definitions and Notations

Let $q$ be a prime power and let $F = \mathbf{F}_q$. A linear code $C$ of length $n$ over $F$ is a vector subspace of $F^n$. It is customary to denote by $k$ the dimension of $C$. For every vector $\mathbf{c} \in C$, the Hamming weight of $\mathbf{c}$, denoted by $w(\mathbf{c})$, is the number of nonzero coordinates of $\mathbf{c}$. For every pair of vectors $\mathbf{c}, \mathbf{c}' \in C$, the Hamming distance $d(\mathbf{c}, \mathbf{c}')$ is the number of coordinates at which $\mathbf{c}$ and $\mathbf{c}'$ differ. The minimum Hamming distance $d$ of $C$ is defined as

$$d = \min\{d(\mathbf{c}, \mathbf{c}') \mid \mathbf{c}, \mathbf{c}' \in C \text{ and } \mathbf{c} \neq \mathbf{c}'\}.$$

It is well known that, for a linear code $C$,

$$d = \min\{w(\mathbf{c}) \mid \mathbf{c} \in C \backslash \{\mathbf{0}\}\}.$$

Let $\langle\ ,\ \rangle$ denote the usual inner product on $F^n$. The dual $C^\perp$ of $C$ is defined with respect to this inner product:

$$C^\perp = \{\mathbf{x} \in F^n \mid \langle \mathbf{x}, \mathbf{c} \rangle = 0\ \forall \mathbf{c} \in C\}.$$

We say that $C$ is self-orthogonal if $C \subseteq C^\perp$. If $C = C^\perp$, we say $C$ is self-dual. There is much interest in self-dual codes and there is a wealth of literature on this subject, cf. [13].

Two linear codes over $F$ are said to be equivalent if one can be obtained from the other through a combination of a permutation of the coordinates and multiplication of some coordinates by some scalars.

For a linear code $C$ and a nonzero scalar $\epsilon \in F\backslash\{0\}$, the extended code $\overline{C}_\epsilon$ is defined as

$$\overline{C}_\epsilon = \left\{(c_0, c_1, \ldots, c_{n-1}, c_{\infty,\epsilon}) \mid (c_0, c_1, \ldots, c_{n-1}) \in C \text{ and } c_{\infty,\epsilon} = \epsilon \sum_{i=0}^{n-1} c_i \right\}.$$

A cyclic code is defined to be a linear code $C$ that is invariant under the cyclic shift, i.e.,

$$(c_0, c_1, \ldots, c_{n-1}) \in C \Rightarrow (c_{n-1}, c_0, \ldots, c_{n-2}) \in C.$$

It is well known and easy to see that every cyclic code $C$ of length $n$ over $F$ may be identified as an ideal in $F[X]/(X^n - 1) \simeq F[\mathbf{Z}_n]$, where $\mathbf{Z}_n$ is the cyclic group of integers modulo $n$:

$$(c_0, c_1, \ldots, c_{n-1}) \leftrightarrow \sum_{i=0}^{n-1} c_i X^i \in F[X]/(X^n - 1).$$

The addition of two codewords in $C$ corresponds to the addition in the group algebra $F[\mathbf{Z}_n]$, while the cyclic shift corresponds to multiplication by $X$ in the group algebra. Every ideal in $F[\mathbf{Z}_n]$ is principal. Its generator may be chosen to be an *idempotent*.

## 3 Classical Duadic Codes and the Fourier Transform

We first give a brief discussion on the first (binary) duadic codes as introduced by Leon, Masley and Pless in [7]. Then we use the Discrete Fourier Transform to obtain a spectral interpretation of such codes.

### *3.1 Classical Duadic Codes*

Throughout this subsection, $F = \mathbf{F}_2$, so we deal only with binary codes. Let $n$ be an odd positive integer and let $\mathbf{Z}_n$ be the cyclic group of integers modulo $n$. A 2-cyclotomic coset in $\mathbf{Z}_n$ is a subset of the form $\{s, 2s, 2^2s, \ldots\}$. Since $n$ is odd, multiplication by 2 is a group action on $\mathbf{Z}_n$, so $\mathbf{Z}_n$ splits into a union of orbits under this group action. A 2-cyclotomic coset is simply an orbit under this map.

**Definition** *A* **splitting** *of $\mathbf{Z}_n$ is a triple $(\{0\}, S_0, S_1)$ such that*

*(i) $S_0, S_1$ are unions of orbits (2-cyclotomic cosets);*

*(ii) $\mathbf{Z}_n = \{0\} \cup S_0 \cup S_1$ is a partition of $\mathbf{Z}_n$; and*

*(iii) there exists $\alpha \in \mathbf{Z}_n^*$ such that $\alpha(S_0) = S_1$, $\alpha(S_1) = S_0$ (hence $\alpha(0) = 0$).*

**Remark**: Such a splitting exists if and only if all prime divisors of $n$ are congruent to $\pm 1$ mod 8.

Given an odd $n$ and a splitting of $\mathbf{Z}_n$, a **duadic code** associated to this splitting is a binary cyclic code of length $n$, generated as an ideal in $F[X]/(X^n - 1)$ by one of the following idempotents:

$$e_0 = ((n+1)/2) + \sum_{i \in S_0} X^i, \qquad 1 + e_0,$$

$$e_1 = ((n+1)/2) + \sum_{i \in S_1} X^i, \qquad 1 + e_1.$$

We denote the duadic codes as follows:

$$C_0 = \langle e_0 \rangle, \quad C_1 = \langle e_1 \rangle, \quad C_0' = \langle 1 + e_1 \rangle, \quad C_1' = \langle 1 + e_0 \rangle.$$

Since $F \backslash \{0\} = \{1\}$, each of these four codes has a unique extended code defined with $\epsilon = 1$. We denote the extended codes by $\overline{C_0}, \overline{C_1}, \overline{C_0'}, \overline{C_1'}$.

**Example** Let $n$ be a prime congruent to $\pm 1 \bmod 8$, let $S_0$ be the set of nonzero quadratic residues modulo $n$ and let $S_1$ be the set of quadratic non-residues. Let $\alpha$ be any quadratic nonresidue. Then $(\{0\}, S_0, S_1)$ is a splitting given by $\alpha$. The duadic codes constructed with this splitting are exactly the quadratic residue codes of length $n$. Therefore, duadic codes may be regarded as generalizations of quadratic residue codes.

The following theorem summarizes the properties of duadic codes (cf. [7]):

**Theorem 1** *Let $n$ be a positive integer all of whose prime factors are congruent to $\pm 1 \bmod 8$ and let $(\{0\}, S_0, S_1)$ be a splitting of $\mathbf{Z}_n$ given by the automorphism $\alpha$. Let $C_0, C_1, C_0', C_1'$ denote the duadic codes associated with this splitting. Then the following statements are true.*

1. *$C_0, C_1, C_0', C_1'$ are of length $n$ and, for $i = 0, 1$, $\dim(C_i) = (n+1)/2$ and $\dim(C_i') = (n-1)/2$.*

2. *$C_0, C_1$ are equivalent; $C_0', C_1'$ are equivalent.*

3. *$C_i' \subset C_i$ and $C_i = \langle \mathbf{h} \rangle \oplus C_i'$, where $\mathbf{h}$ is the all-one vector $(1, \ldots, 1)$. Moreover, $C_i'$ is the even weight subcode of $C_i$.*

4. *Let $d_0$ be the minimum weight in $C_i \backslash C_i'$. Then $d_0^2 \geq n$.*

5. *If $\alpha = -1$, then $C_i'^{\perp} = C_i$ and all the words in $C_i'$ have weights congruent to $0 \bmod 4$.*

6. *If $\alpha = -1$, then $d_0^2 - d_0 + 1 \geq n$.*

7. *If* $\alpha = -1$ *and* $C_i \backslash C_i'$ *contains a vector of minimum weight* $d_0$ *with* $d_0^2 - d_0 + 1 = n$, *then*

   *(i) the vectors of weight* $d_0$ *in* $C_i$ *hold a projective plane of order* $d_0 - 1$;

   *(ii) the minimum weight of* $C_i$ *is* $d_0$.

8. $\overline{C_0}$ *equivalent to* $\overline{C_1}$.

9. *If* $\alpha = -1$, *then* $\overline{C_0}, \overline{C_1}$ *are self-dual.*

10. *If* $\alpha = -1$ *and* $n + 1 \equiv 0 \bmod 8$, *then the weights of all the words in* $\overline{C_0}, \overline{C_1}$ *are congruent to* $0 \bmod 4$.

Compared to quadratic residue codes, there are some advantages in considering duadic codes. For example, quadratic residue codes only exist in prime lengths, but duadic codes can have composite lengths. At some lengths, while extended quadratic residue codes exist, they may not be self-dual. However, for some such lengths (e.g., $n = 74$), certain extended duadic codes can be shown to be self-dual. In certain cases, e.g., for length 113, some of the more flexibly-defined duadic codes also give a better minimum distance compared to the quadratic residue codes of the same length and dimension.

### *3.2 Spectral Interpretation*

Discrete Fourier Transform, also known as the Mattson-Solomon Transform, leads to a spectral interpretation of cyclic codes. Such an interpretation is convenient at times for the analysis of such codes. For example, we will see shortly that the spectral interpretation yields an easy way to describe the dual of a code. For the sake of the later part of this talk, our discussion here is done in a much more general setting.

Let $R$ be either a finite field or the ring $\mathbf{Z}_{p^m}$, where $p$ is a prime. Let $q$ be the order of the residue field of $R$. (When $R = \mathbf{F}_q$, it is also its residue field. When $R = \mathbf{Z}_{p^m}$, the residue field is $\mathbf{F}_p$, so $q = p$.) For each $d$, there is a unique Galois extension of $R$ of degree $d$, denoted by $R_d$. When $R$ is a finite field, $R_d$ is just the degree $d$ field extension of $R$. When $R = \mathbf{Z}_{p^m}$, $R_d$ is the Galois ring of degree $d$ over $R$ (cf. [11]).

Let $n$ be a positive integer relatively prime to the order of $R$ and let $G$ be an abelian group of order $n$. Then the map $x \mapsto qx$ defines a group action on $G$. Let $O_0, O_1, \ldots, O_s$ denote the orbits (or, equivalently, $q$-cyclotomic cosets) of this action. Let $d_j$ denote the size of the orbit $O_j$. Let $M$ be the order of $q$ modulo $N$, where $N$ is the exponent of $G$. Then $R_M$ is the smallest Galois extension of $R$ containing an $N^{\text{th}}$ root of unity. Let $\zeta$ denote such a primitive $N^{\text{th}}$ root of unity in $R_M$.

Since $G$ is a finite abelian group, we may write

$$G = \prod_{i=1}^{t} \mathbf{Z}_{n_i},$$

for some $n_1, \ldots, n_t$. Then every element $g \in G$ may be written as a $t$-uple $(g_1, \ldots, g_t)$. For an element $\mathbf{c} = \sum_{g=0}^{n-1} c_g X^g \in R[G]$, its Fourier Transform is defined to be $\hat{\mathbf{c}} = \sum_{h=0}^{n-1} \hat{c}_h X^h$, where

$$\hat{c}_h = \sum_{g=0}^{n-1} c_g \langle g, h \rangle.$$

Here, $\langle g, h \rangle = \zeta^{\sum_{i=1}^{t} g_i h_i (N/n_i)}$.

A few well known properties of the Fourier Transform include:

1. $\widehat{cc'}_h = \hat{c}_h \hat{c'}_h$;

2. if $\phi$ is the Frobenius map, then $\phi(\hat{c}_h) = \hat{c}_{qh}$;

3. $\sum_{g \in G} c_g c'_g = \frac{1}{n} \sum_{h \in G} \hat{c}_h \hat{c'}_{-h}$

Using the above properties, one may show that the Fourier Transform gives rise to the following ring isomorphism

$$R[G] \simeq R \times R_{d_1} \times \cdots \times R_{d_s}.$$

Consequently, every ideal $I$ in $R[G]$ may be decomposed as

$$I \simeq I_0 \times I_1 \times \cdots \times I_s,$$

where $I_j$ is an ideal in $R_{d_j}$.

The ideals $I$ in $R[G]$ are called **abelian codes**. When $G$ is cyclic, then $I$ is a cyclic code. In this sense, abelian codes are considered to be generalizations of cyclic codes.

The map $x \mapsto -x$ induces an action on the set of orbits $O_0, O_1, \ldots, O_s$. Let $\sigma$ be the map on $\{0, 1, \ldots, s\}$ induced by $x \mapsto -x$. (Note that $\sigma(0) = 0$). The dual of $I$ now has an easy description:

$$I^{\perp} \simeq I^o_{\sigma(0)} \times I^o_{\sigma(1)} \times \cdots I^o_{\sigma(s)},$$

where, for any ideal $J$ in a ring $S$, $J^o = \mathrm{Ann}(J)$ denotes the annihilator of $J$ in $S$.

The spectral interpretation of abelian codes now allows us to obtain an alternative, but equivalent, way to view a classical duadic code. Recall that a duadic code is one that is generated by one of the idempotents: $e_0$, $e_1$, $1+e_0$ and $1+e_1$. Let

$$X_0 = \{a \in \mathbf{Z}_n \mid e_0(\zeta^a) = 0\}$$

and let $X_1 = \mathbf{Z}_n \backslash (\{0\} \cup X_0)$. Then $(\{0\}, X_0, X_1)$ is a splitting of $\mathbf{Z}_n$ given by the same $\alpha$ that interchanges $S_0$ and $S_1$.

The four duadic codes $C_0, C_1, C_0'$ and $C_1'$ now have the following spectral description:

1. $C_0 = I_0 \times \cdots \times I_s$: $I_0 = (1)$, $I_j = (0)$ if $O_j \subseteq X_0$ and $I_j = (1)$ if $O_j \subseteq X_1$;
2. $C_0' = I_0 \times \cdots \times I_s$: $I_0 = (0)$, $I_j = (0)$ if $O_j \subseteq X_0$ and $I_j = (1)$ if $O_j \subseteq X_1$;
3. $C_1 = I_0 \times \cdots \times I_s$: $I_0 = (1)$, $I_j = (1)$ if $O_j \subseteq X_0$ and $I_j = (0)$ if $O_j \subseteq X_1$;
4. $C_1' = I_0 \times \cdots \times I_s$: $I_0 = (0)$, $I_j = (1)$ if $O_j \subseteq X_0$ and $I_j = (0)$ if $O_j \subseteq X_1$.

One advantage of the spectral interpretation is that a number of the properties as stated in Theorem 1 now afford simpler proofs.

As our discussion on the Fourier Transform was done in the general setting where $R$ is any finite field or $\mathbf{Z}_{p^m}$, the generalization of the classical binary duadic codes to other base fields is obvious. Likewise, as the discussion on the Fourier Transform used an arbitrary finite abelian group $G$ instead of just a cyclic group, the definition of a duadic code can also be extended to abelian codes in a straight forward manner. In fact, all these generalizations may be considered as special cases of the split group codes introduced in [4].

## 4 Split Group Codes

Now let $F = \mathbf{F}_q$ and let $G$ be a finite abelian group of order $n$. Suppose that $n$ and $q$ are relatively prime.

As in the previous section, we consider the group action on $G$ induced by the map $x \mapsto qx$. Let $O_0, O_1, \ldots, O_s$ denote the orbits and let $d_j$ denote the size of $O_j$. We first generalize the notion of a splitting of $G$ as follows:

**Definition** *A* **splitting** *of $G$ is a triple $(Z, X_0, X_1)$ such that*

*(i) $Z, X_0, X_1$ are unions of orbits;*

*(ii) $G = Z \cup X_0 \cup X_1$ is a partition of $G$; and*

*(iii) there exists* $\alpha \in \mathrm{Aut}(G)$ *such that* $\alpha(X_0) = X_1$, $\alpha(X_1) = X_0$ *(hence* $\alpha(Z) = Z$*).*

**Definition** *Given a splitting* $(Z, X_0, X_1)$ *of* $G$*, a* **split group code** *over* $F$ *associated to this splitting is one of the following ideals in* $F[G]$*:*

1. $C_0 = I_0 \times \cdots \times I_s$: $I_j = (1)$ *if* $O_j \subseteq Z$, $I_j = (0)$ *if* $O_j \subseteq X_0$ *and* $I_j = (1)$ *if* $O_j \subseteq X_1$;

2. $C_0' = I_0 \times \cdots \times I_s$: $I_0 = (0)$ *if* $O_j \subseteq Z$, $I_j = (0)$ *if* $O_j \subseteq X_0$ *and* $I_j = (1)$ *if* $O_j \subseteq X_1$;

3. $C_1 = I_0 \times \cdots \times I_s$: $I_0 = (1)$ *if* $O_j \subseteq Z$, $I_j = (1)$ *if* $O_j \subseteq X_0$ *and* $I_j = (0)$ *if* $O_j \subseteq X_1$;

4. $C_1' = I_0 \times \cdots \times I_s$: $I_0 = (0)$ *if* $O_j \subseteq Z$, $I_j = (1)$ *if* $O_j \subseteq X_0$ *and* $I_j = (0)$ *if* $O_j \subseteq X_1$.

**Remarks**

1. When $F = \mathbf{F}_2$, $G = \mathbf{Z}_n$ and $Z = \{0\}$, the codes defined above are exactly the classical binary duadic codes.

2. When $F = \mathbf{F}_q$, $G = \mathbf{Z}_n$ and $Z = \{0\}$, the above definition gives the generalizations due to Smid.

3. For arbitrary $F$ and $G$, with $Z = \{0\}$, we obtain the generalization of duadic codes to abelian codes, including the generalized quadratic residue codes studied in [3,10,16].

Almost all the properties of the classical duadic codes carry over to split group codes with very minor modifications.

Let $G'$ denote the subgroup of $G$ generated by $Z$. Let $h$ denote the order of the quotient $G/G'$.

**Theorem 2 (cf. [4])** *Let* $F = \mathbf{F}_q$ *and let* $G$ *be a finite abelian group whose order* $n$ *is relatively prime to* $q$*. Let* $(Z, X_0, X_1)$ *be a splitting of* $G$ *given by the automorphism* $\alpha$*. Let* $C_0, C_1, C_0', C_1'$ *denote the split group codes associated with this splitting. Then the following statements are true.*

1. $C_0, C_1, C_0', C_1'$ *are of length* $n$ *and, for* $i = 0, 1$, $\dim(C_i) = (n + |Z|)/2$ *and* $\dim(C_i') = (n - |Z|)/2$.

2. $C_0, C_1$ *are equivalent;* $C_0', C_1'$ *are equivalent.*

3. *$C_i' \subset C_i$ and $C_i = \langle \mathbf{h} \rangle \oplus C_i'$, where $\mathbf{h}$ is the all-one vector $(1, \ldots, 1)$. Moreover, $C_i'$ is the subcode of $C_i$ defined by*

$$C_i' = \{(c_0, c_1, \ldots, c_{n-1}) \in C_i \mid \sum c_i = 0\};$$

4. *If $\alpha = -1$, then $C_i'^{\perp} = C_i$. Moreover, when $q = 2$, all the words in $C_i'$ have weights congruent to $0 \bmod 4$.*

*For the next three assertions, assume that there is an integer $r$ such that the maps $x \mapsto \alpha x$ and $x \mapsto q^r x$ agree on $Z$.*

5. *Let $d_0$ be the minimum weight in $C_i \backslash C_i'$. Then $d_0^2 \geq h$.*
6. *If $\alpha = -1$, then $d_0^2 - d_0 + 1 \geq h$.*
7. *If $\alpha = -1$ and $C_i \backslash C_i'$ contains a vector of minimum weight $d_0$ with $d_0^2 - d_0 + 1 = n$, then*
   *(i) the vectors of weight $d_0$ in $C_i$ hold a projective plane of order $d_0 - 1$;*
   *(ii) the minimum weight of $C_i$ is $d_0$.*

**Remark** The notion of an extended code also exists in this context, but in order to obtain results parallel to those for classical extended duadic codes, one needs to add $|Z|$ coordinates instead of the usual one.

The flexibility in the choice of $Z$ in the splitting of $G$ is useful in producing codes of good parameters. We give one such example below:

**Example** Let $F = \mathbf{F}_2$ and let $G = \mathbf{Z}_{3m}$, where $m$ is prime. Let $Z = m\mathbf{Z}_{3m}$ and let

$$X_0 = \left\{a \in G^* \, : \, \left(\frac{a}{m}\right) = 1\right\} \cup \left\{a \in 3G^* \, : \, \left(\frac{a}{m}\right) = -1\right\},$$

where $G^*$ denotes the multiplicative group of integers modulo $3m$ and $\left(\frac{\cdot}{m}\right)$ denotes the Legendre symbol. The automorphism $\alpha$ may be chosen to be any element of $G^*$ that is a nonsquare mod $m$. Such a splitting gives some binary codes that are either the best known or, in the case of $n = 141$ and $k = 69$, better than all others known previously, cf. [2].

| $m$ | $C_0$ | $C_0'$ |
|---|---|---|
| 7 | $[21, 12, 5]$ | $[21, 9, 8]$ |
| 31 | $[93, 48, 14]$ | |
| 47 | | $[141, 69, 24]$ |

It would be interesting to explore among the noncyclic abelian codes to see if any good codes may be found.

## 5 Over Local Rings

### 5.1 Duadic Codes

We now leave the case where the underlying alphabet is a finite field. Duadic codes over finite rings were first considered by Langevin and Solé in [6], where they worked over $\mathbf{Z}_4$. They were later generalized to other rings in [8,9]. We begin with the case where the underlying alphabet is $\mathbf{Z}_{p^m}$.

Let $R = \mathbf{Z}_{p^m}$ and let the other notations be as in the earlier sections. For any positive integer $d$, the Galois ring $R_d$ has exactly $m + 1$ ideals: $(1), (p), \ldots, (p^{m-1}), (0)$.

Again, we let $G$ be a finite abelian group whose order is relatively prime to $p$. Let $O_0, O_1, \ldots, O_s$ be the orbits under the group action $x \mapsto px$. Suppose that the automorphism $\alpha$ induces a permutation $\tau$ on $\{0, 1, \ldots, s\}$

The notion of the splitting used in this context is exactly the same as that used for split group codes. Duadic codes will only be defined when $m$ is even. We also introduce the notion of generalized duadic codes.

**Definition** *Let $R = \mathbf{Z}_{p^m}$ with $m$ even. Let $G$ be a finite abelian group whose order $n$ is relatively prime to $p$. Let $(Z, X_0, X_1)$ be a splitting of $G$ given by $\alpha \in \mathrm{Aut}(G)$. A* **duadic code** *over $R$ associated to this splitting is defined to be an ideal $I = I_0 \times I_1 \times \cdots \times I_s \subseteq R[G]$ such that:*

- *if $O_j \subseteq X_0 \cup X_1$, then $I_j = (p^i)$ if and only if $I_{\tau(j)} = (p^{m-i})$,*
- *if $O_j \subseteq Z$, then $I_j = (p^{m/2})$.*

Without the restriction for $m$ to be even, we define a **generalized duadic code** in the same way as for the duadic code, except that $I_0$ can be any ideal.

**Remark** If there is a splitting with $Z = \{0\}$, then generalized duadic codes exist for odd $m$.

The above definition of a duadic code over $\mathbf{Z}_{p^m}$ leads to a characterization result on the self-dual abelian codes and the self-dual augmented and extended abelian codes over such an alphabet. This provides a tool for the study of self-dual codes over these rings.

**Theorem 3** *Let $R = \mathbf{Z}_{p^m}$, let $G$ be a finite abelian group whose order $n$ is relatively prime to $p$ and let $C$ be an abelian code in $R[G]$.*
*(1) Assuming $m$ is even, $C$ is self-dual if and only if $C$ is a duadic code attached to a splitting $(Z, X_0, X_1)$ of $G$ given by $\alpha = -1$.*
*(2) A generalized duadic code over $R$ attached to a splitting of $G$ with $\alpha = -1$ and $I_0 = (p^i)$ $(i \geq m/2)$ is self-orthogonal.*

Let $R^*$ be the multiplicative group of units in $R$.

**Definition** *Given* $R = \mathbf{Z}_{p^m}$*,* $G$ *a finite abelian group whose order* $n$ *is relatively prime to* $p$ *and* $C$ *an abelian code in* $R[G]$*, for* $\epsilon \in R^*$*, the* **augmented and extended** *abelian code, denoted by* $\overline{C}_\epsilon$*, is the code obtained by first augmenting* $C$ *by the all-one vector followed by extending* $(c_1, \ldots, c_n)$ *by an extra coordinate* $\epsilon \sum c_i$*.*

**Theorem 4** *Let* $R = \mathbf{Z}_{p^m}$*, let* $G$ *be a finite abelian group whose order* $n$ *is relatively prime to* $p$ *and let* $C$ *be an abelian code in* $R[G]$*. Then* $\overline{C}_\epsilon$ *is self-dual if and only if* $C$ *is a generalized duadic code over* $R$ *attached to a splitting with* $\alpha = -1$ *and* $\epsilon^2 n + 1 = 0 \in R$*.*

### *5.2 Codes over* $\mathbf{Z}_{2^m}$

Now let $p = 2$.

**Definition** *Let* $C$ *be a code over* $\mathbf{Z}_{2^m}$*. The* **Euclidean weight** *of* $\mathbf{c} = (c_1, \ldots, c_n) \in C$ *is defined to be* $\sum_{i=1}^{n} \min\{c_i^2, (2^{m+1} - c_i)^2\}$*.*

**Definition** *A code* $C$ *over* $\mathbf{Z}_{2^m}$ *is said to be of* **Type II** *if* $C$ *is self-dual and the Euclidean weights of all the words in* $C$ *are divisible by* $2^{m+1}$*.*

**Remark** When $m = 1$, the notion of the Euclidean weight coincides with that of the Hamming weight and the above definition of a Type II binary code is consistent with the usual definition.

We have the following characterization result on Type II codes over $\mathbf{Z}_{2^m}$.

**Theorem 5** *Let* $R = \mathbf{Z}_{2^m}$*, let* $G$ *be a finite abelian group of odd order. Let* $C$ *be an abelian code in* $R[G]$ *and let* $\epsilon \in R^*$ *be such that* $\overline{C}_\epsilon$ *is self-dual. Then* $\overline{C}_\epsilon$ *is of Type II if and only if* $\epsilon^2 n + 1 \equiv 0 \bmod 2^{m+1}$ .

## 6 Abelian Codes over $\mathbf{Z}_{2k}$

Codes over $\mathbf{Z}_{2k}$ were introduced and studied in [1] and subsequently in [5]. We recall here the notions of the Euclidean weight and a Type II code over $\mathbf{Z}_{2k}$.

**Definition** *Let* $C$ *be a code over* $\mathbf{Z}_{2k}$*. The* **Euclidean weight** *of* $\mathbf{c} = (c_1, \ldots, c_n) \in C$ *is defined to be* $\sum_{i=1}^{n} \min\{c_i^2, (2k - c_i)^2\}$*.*

**Definition** *A code* $C$ *over* $\mathbf{Z}_{2k}$ *is said to be of* **Type II** *if* $C$ *is self-dual and the Euclidean weights of all the words in* $C$ *are divisible by* $4k$*.*

Let $2k = p_1^{m_1} \cdots p_t^{m_t}$, where $p_1 = 2$ and $p_i$ $(2 \le i \le t)$ are distinct odd primes.

By the Chinese Remainder Theorem, $\mathbf{Z}_{2k} = \mathbf{Z}_{p_1^{m_1}} \times \cdots \mathbf{Z}_{p_t^{m_t}}$, so a code $C$ over $\mathbf{Z}_{2k}$ decomposes as a product of codes $C_i$ for $1 \leq i \leq t$, where $C_i$ is a code over $\mathbf{Z}_{p_i^{m_i}}$. In [5], this product is called the Chinese product.

It is known that [5,8]:

1. $C$ is self-dual if and only if each $C_i$ is self-dual;
2. $C$ is of Type II if and only if $C_1$ is of Type II and the other $C_i$ are self-dual; and
3. for $\epsilon \in \mathbf{Z}_{2k}^*$, $\overline{C}_\epsilon$ is the Chinese product of $\overline{(C_i)}_\epsilon$ $(1 \leq i \leq t)$.

We define duadic codes and generalized duadic codes over $\mathbf{Z}_{2k}$ as follows:

**Definition** *Let $G$ be a finite abelian group whose order $n$ is relatively prime to $2k$. An ideal $C$ of $\mathbf{Z}_{2k}[G]$ is a* **(generalized) duadic code** *over $\mathbf{Z}_{2k}$* **with** *$\alpha \in \mathrm{Aut}(G)$ if and only if each $C_i$ is a (generalized) duadic code over $\mathbf{Z}_{p_i^{m_i}}$ attached to a splitting $(X_i, A_i, B_i)$ given by $\alpha \in \mathrm{Aut}(G)$.*

Using the Chinese Remainder Theorem and Theorems 3, 4 and 5 above, characterization results may be obtained for self-dual abelian codes, self-dual augmented and extended abelian codes and Type II augmented and extended abelian codes over $\mathbf{Z}_{2k}$.

**Examples**

A number of *extremal* Type II codes are found among the augmented and extended codes $\overline{C}_\epsilon$, cf. [8]. We list some of them in the table below:

| $R$ | length | Euclidean $d$ |
|---|---|---|
| $\mathbf{Z}_{2^m}$ $(m \geq 2)$ | 8 | $2^{m+1}$ |
| $\mathbf{Z}_{2^{2m}}$ $(m \geq 1)$ | 16 | $2^{2m+1}$ |
| $\mathbf{Z}_4$ | 24 | 16 |
| $\mathbf{Z}_4$ | 32 | 16 |
| $\mathbf{Z}_4$ | 40 | 16 |
| $\mathbf{Z}_4$ | 48 | 24 |
| $\mathbf{Z}_6$ | 24 | 24 |
| $\mathbf{Z}_6$ | 48 | 36 |
| $\mathbf{Z}_8$ | 48 | 48 |

## Acknowledgments

This was an invited lecture at the Third Asian Mathematical Conference held at the University of the Philippines, Diliman. The author wishes to thank the organizers for the kind invitation and the warm hospitality. The author also thanks the Singapore Mathematical Society, the Lee Kong Chian Centre for Mathematical Research and the MOE-ARF research grant R-146-000-018-112 for financial support.

## References

1. E. Bannai, S.T. Dougherty, M. Harada & M. Oura, *Type II codes, even unimodular lattices, and invariant rings*, IEEE Trans. Inform. Theory **45** (1999), 1194–1205
2. A. Brouwer, *Bounds on the minimum distance of linear code*, http://www.win.tue.nl/~aeb/voorlincod.html
3. P. Camion, *Global quadratic abelian codes*, in *Information Theory*, G. Longo ed., Vienna: Springer-Verlag, 1975, vol. 219, pp. 293–310. CISM Courses and Lectures.
4. C. Ding, D.R. Kohel & S. Ling, *Split group codes*, IEEE Trans. Inform. Theory **46** (2000), 485–495
5. S.T. Dougherty, M. Harada & P. Solé, *Self-dual codes over rings and the Chinese remainder theorem*, Hokkaido Math J **28** (1999), 253–283
6. P. Langevin & P. Solé, *Duadic* $\mathbf{Z}_4$*-codes*, Finite Fields Their Appl. **6** (2000), 309–326
7. J.S. Leon, J.M. Masley & V. Pless, *Duadic codes*, IEEE Trans. Inform. Theory **30** (1984), 709–714
8. S. Ling & P. Solé, *Duadic codes over* $\mathbf{Z}_{2k}$, to appear in *IEEE Trans. Inform. Theory*
9. S. Ling & P. Solé, *Duadic codes over* $\mathbf{F}_2 + u\mathbf{F}_2$, preprint 2000.
10. J.H. van Lint & F.J. MacWilliams, *Generalized quadratic residue codes*, IEEE Trans. Inform. Theory **24** (1978), 730–737
11. B.R. McDonald, *Finite rings with identity*, Marcel Dekker, New York, 1974 .
12. V. Pless, *Q-codes*, J. Comb. Theory Ser. A **43** (1986), 258–276
13. E.M. Rains & N.J.A. Sloane, *Self-dual codes*, in *Handbook of Coding Theory*, V. Pless, W.C. Huffman eds, North Holland, 1998.
14. M. Smid, *On duadic codes*, Dept. Math., Tech. Univ., Eindhoven, The Netherlands, Tech. Rep. 86-WSK-04, May 1986.
15. M. Smid, *Duadic codes*, IEEE Trans. Inform. Theory **33** (1987), 432-433
16. H.N. Ward, *Quadratic residue codes and symplectic groups*, J. Algebra **29** (1974), 150–171

# THE THEORY OF BINARY NONLINEARIZATION AND ITS APPLICATIONS TO SOLITON EQUATIONS

WEN-XIU MA
*Department of Mathematics, City University of Hong Kong, Hong Kong, Kowloon, P. R. China*
*E-mail: mawx@cityu.edu.hk*

**Binary symmetry constraints are applied to the nonlinearization of spectral problems and adjoint spectral problems into so-called binary constrained flows, which provide candidates for finite-dimensional Liouville integrable Hamiltonian systems. The resulting constraints on the potentials of spectral problems give rise to a kind of involutive solutions to zero curvature equations, and thus the integrability by quadratures can be shown for zero curvature equations once the corresponding binary constrained flows are proved to be integrable. The whole process to carry out binary symmetry constraints is called binary nonlinearization. The principal task of binary nonlinearization is to expose the Liouville integrability for the resulting binary constrained flows, which can usually be achieved as a consequence of the existence of hereditary recursion operators. The theory of binary nonlinearization is applied to the multi-wave interaction equations associated with a $4 \times 4$ matrix spectral problem as an illustrative example. The Bäcklund transformations resulted from symmetry constraints are given for the multi-wave interaction equations, and thus a kind of involutive solutions is presented and the integrability by quadratures is shown for the multi-wave interaction equations.**

## 1 Introduction

The nonlinearization technique arose in the theory of soliton equations ten years ago [1-3]. One gradually realizes that it provides a powerful approach for analyzing soliton equations, in both continuous and discrete cases, especially for showing the integrability by quadratures for soliton equations. The manipulation of nonlinearization leads to finite-dimensional Liouville integrable systems (for example, see [3-8]) and also establish a bridge between infinite-dimensional soliton equations and finite-dimensional Liouville integrable systems [9-11]. What is more, it paves a method of separation of variables for soliton equations [12-15], and thus the determination of finite-gap solutions to soliton equations is transformed into solving the corresponding Jacobi inversions problems [14,15].

Mathematically speaking, much excitement in the theory of nonlinearization comes from a kind of specific symmetry constraints, engendered from the variational derivative of the spectral parameter of spectral problems [10,16], which has a close relation with eigenfunctions of hereditary recursion operators [17,18]. Symmetry constraints are themselves a common conceptional umbrella under which one can manipulate both mono-nonlinearization [1] and binary nonlinearization [10], and the study of symmetry constraints is a crucial part of the kernel of the mathematical theory of nonlinearization.

Although it is not clear how to determine pairs of canonical variables to obtain Hamiltonian structures of constrained flows in a general case of mono-nonlinearization, there exists a natural way for determining symplectic structures to exhibit Hamiltonian forms for binary constrained flows while doing binary nonlinearization [19]. Therefore, one can expect to establish a rigorous theory for binary nonlinearization. In this paper, we would only like to discuss the procedure of manipulating binary nonlinearization. More considerations on the topics related to binary nonlinearization will be made elsewhere.

The paper is structured as follows. In section 2, a procedure for doing binary nonlinearization is depicted for general matrix spectral problems. Then in section 3, an illustrative example is carried out to make an application of the procedure to the multi-wave interaction equations associated with a $4 \times 4$ matrix spectral problem. Finally in section 4, a summary is given along with some concluding remarks.

## 2 Binary Nonlinearization

In this section, we would like to describe the procedure of manipulating binary nonlinearization for soliton equations (see, for example, [10,19]). Let us start from a general matrix spectral problem

$$\phi_x = U\phi = U(u,\lambda)\phi,\ U = (U_{ij})_{r\times r},\ \phi = (\phi_1,\cdots,\phi_r)^T \tag{2.1}$$

where $\lambda$ is a spectral parameter and $u = (u_1,\cdots,u_q)^T$ is a $q$-dimensional vector potential. Associated with the spectral problem (2.1), suppose that for each integer $m \geq 0$, we have the evolution law for the eigenfunction $\phi$:

$$\phi_{t_m} = V^{(m)}\phi = V^{(m)}(u,u_x,\cdots;\lambda)\phi,\ V^{(m)} = (V_{ij}^{(m)})_{r\times r}. \tag{2.2}$$

If the Gateaux derivative $U'$ of $U$ with the potential $u$ is injective, then the isospectral ($\lambda_{t_m} = 0$) zero curvature equation

$$U_{t_m} - V_x^{(m)} + [U,V^{(m)}] = 0 \tag{2.3}$$

will determine [?] an evolution equation

$$u_{t_m} = X_m(u) = JG_m = J\frac{\delta \tilde{H}_m}{\delta u},\ u = u(x,t_m), \tag{2.4}$$

where $X_m$ depends on the potential $u$ and its spatial derivatives of $u$ up to some finite order, $J(u)$ is a Hamiltonian operator, and $\tilde{H}_m(u)$ is a Hamiltonian functional. Evidently, the compatability condition of the adjoint spectral problem and the adjoint associated spectral problem

$$\psi_x = -U^T(u,\lambda)\psi,\ \psi_{t_m} = -V^{(m)T}(u,\lambda)\psi,\ \psi = (\psi_1,\cdots,\psi_r)^T \tag{2.5}$$

is still the same as the zero curvature equation (2.3), guaranteed by the evolution equation $u_{t_m} = X_m(u)$ defined by (2.4).

It has been pointed out [10,16] that

$$J\frac{\delta\lambda}{\delta u} = E^{-1}J\psi^T\frac{\partial U(u,\lambda)}{\partial u}\phi,\ E = -\int_\Omega \psi^T\frac{\partial U(u,\lambda)}{\partial \lambda}\phi\,dx,$$

is a symmetry of the evolution equation (2.4), where $E$ is called the normalized constant, and $\Omega = (0,T)$ if $u$ is assumed to be periodic with period $T$ and $\Omega = (-\infty,\infty)$ if $u$ is assumed to belong to the Schwartz space. Let us now introduce $N$ distinct eigenvalues $\lambda_1,\cdots,\lambda_N$, and so we have

$$\phi_x^{(s)} = U(u,\lambda_s)\phi^{(s)},\ \psi_x^{(s)} = -U^T(u,\lambda_s)\psi^{(s)},\ 1\le s\le N; \tag{2.6}$$

$$\phi_{t_m}^{(s)} = V^{(m)}(u,\lambda_s)\phi^{(s)},\ \psi_{t_m}^{(s)} = -V^{(m)T}(u,\lambda_s)\psi^{(s)},\ 1\le s\le N; \tag{2.7}$$

where the corresponding eigenfunctions and adjoint eigenfunctions are denoted by $\phi^{(s)}$ and $\psi^{(s)}$, $1\le s\le N$. Suppose that the covariant $G_{m_0}$ is Lie point, and then the so-called binary Bargmann symmetry constraint reads as

$$X_{m_0} = \sum_{s=1}^N E_s\mu_s J\frac{\delta\lambda_s}{\delta u},\ \text{i.e.,}\ JG_{m_0} = J\sum_{s=1}^N \mu_s\psi^{(s)T}\frac{\partial U(u,\lambda_s)}{\partial u}\phi^{(s)}, \tag{2.8}$$

where $\mu_s$, $1\le s\le N$, are arbitrary non-zero constants, and $E_s$, $1\le s\le N$, are $N$ normalized constants. The right-hand side of the binary symmetry constraint (2.8) is a linear combination of $N$ symmetries $E_sJ\delta\lambda_s/\delta u$, $1\le s\le N$, and the requirement of $\mu_s\ne 0$, $1\le s\le N$, is natural, since the symmetry $J\delta\lambda_s/\delta u$ is just not involved in our symmetry constraint (2.8) if some constant $\mu_s = 0$. It is also worthy to mention that $\phi^{(s)}$ and $\psi^{(s)}$ can not be expressed in terms of $x$, $u$ and spatial derivatives of $u$ to some finite order, and thus such symmetries (or the corresponding covariants $\delta\lambda_s/\delta u$, $1\le s\le N$) are not Lie point, contact or Lie-Bäcklund symmetries.

Let us now assume that we can solve the symmetry constraint (2.8) for $u$:

$$u = \tilde{u}(\phi^{(1)},\cdots,\phi^{(N)};\psi^{(1)},\cdots,\psi^{(N)}). \tag{2.9}$$

Replacing $u$ with $\widetilde{u}$ in the system (2.6) and the system (2.7), we obtain the so-called spatial binary constrained flow:

$$\phi_x^{(s)} = U(\widetilde{u}, \lambda_s)\phi^{(s)},\ \psi_x^{(s)} = -U^T(\widetilde{u}, \lambda_s)\psi^{(s)},\ 1 \leq s \leq N, \tag{2.10}$$

and the so-called temporal binary constrained flow

$$\phi_{t_m}^{(s)} = V^{(m)}(\widetilde{u}, \lambda_s)\phi^{(s)},\ \psi_{t_m}^{(s)} = -V^{(m)T}(\widetilde{u}, \lambda_s)\psi^{(s)},\ 1 \leq s \leq N. \tag{2.11}$$

Note that the spatial and temporal binary constrained flows (2.10) and (2.11) are nonlinear with respect to eigenfunctions and adjoint eigenfunctions, but the original spectral problems (2.6) and (2.7) are linear. Moreover, the spatial binary constrained flow (2.10) is a system of ordinary differential equations, and the temporal binary constrained flow (2.11) is usually a system of partial differential equations but it can be transformed into a system of ordinary differential equations under the control of (2.10).

The principal task in the theory of binary nonlinearization is to show that the spatial binary constrained flow (2.10) and the temporal binary constrained flow (2.11) under the control of (2.10) are Liouville integrable, and thus the integrability by quadratures can be shown for evolution equations possessing zero curvature representations. The main tools to do this are Lax representations generated from the stationary zero curvature equations and $r$-matrix formulations of Lax operators [21]. These will form the important part of the theory of binary nonlinearization and will be shed right on in our example.

Now if $\phi^{(s)}$ and $\psi^{(s)}$, $1 \leq s \leq N$, solve two binary constrained flows (2.10) and (2.11) simultaneously, then $u = \widetilde{u}$ will present a solution to the evolution equation (2.4), since the evolution equation (2.4) with the potential $u = \widetilde{u}$ is the compatability of two binary constrained flows (2.10) and (2.11). It also follows that the evolution equation $u_{t_m} = X_m(u)$ is decomposed into two finite-dimensional Liouville integrable systems, and the constraint $u = \widetilde{u}$ gives rise to a Bäcklund transformation between the evolution equation and the resulting finite-dimensional Liouville integrable systems [22]. Furthermore, two binary constrained flows (2.10) and (2.11) can be solved by separation of varaibles and all what we need to do upon obtaining separated variables is to solve the Jacobi inversion problems [15]. This whole process to carry out symmetry constraints is called binary nonlinearization [17,19].

## 3 Applications to Soliton Equations

By a soliton hierarchy we means a hierarchy of evolution equations with a recursion relation as follows

$$u_{t_m} = X_m(u) = \Phi^m X_0 = JG_m = J\frac{\delta \widetilde{H}_m}{\delta u},\ m \geq 0, \tag{3.1}$$

where $u$ is the vector potential $u = u(x, t_m)$, $X_m$, $m \geq 0$, depend on the potentional $u$ and its spatial derivatives of $u$ up to some finite order, $\Phi$ is a herediatry common recursion operator, $J(u)$ is a Hamiltonian operator, and $\tilde{H}_m(u)$, $m \geq 0$, are the Hamiltonian functionals. Moreover, we have

$$[X_m, X_n] := \frac{\partial}{\partial \varepsilon}\Big(X_m(u + \varepsilon X_n - X_n(u + \varepsilon X_m)\Big)\Big|_{\varepsilon=0} = 0, \ m, n \geq 0, \quad (3.2)$$

$$\{\tilde{H}_m, \tilde{H}_n\}_J := \int (\frac{\delta \tilde{H}_m}{\delta u})^T J \frac{\delta \tilde{H}_n}{\delta u}\, dx = 0, \ m, n \geq 0, \quad (3.3)$$

where $[\cdot,\cdot]$ and $\{\cdot,\cdot\}_J$ are called the commutator of vector fields and the Poisson bracket assocaited with the Hamiltonian operator $J$, resprectively. It follows that each equation $u_{t_m} = X_m$ in the soliton hierarhy (3.1) has infinitely many symmetries $\{X_n\}_{n=0}^{\infty}$ and infinitely many conserved functionals $\{\tilde{H}_n\}_{n=0}^{\infty}$. This property can be obtained from a bi-Hamiltonian formaultion, and its natrural bi-Hamiltonian formaulation for the hierarchy (3.1) is formed by $J$ and $M = J\Psi$ if they are a Hamiltonian pair, where $\Psi$ denotes the adjoint operator of $\Phi$, i.e., $\Psi = \Phi^{\dagger}$. The above property can aslo be obtained from a weaker condition that $M$ is a symmetric operator ($M^{\dagger} = M$), i.e, $\Phi J = J\Psi$ [23].

A soliton hierarchy can usually be constructed from a hierarchy of zero curvature equations as the compatability conditions of a given matrix spectral problem (2.1) and its associated spectral porblems (2.2) [24,25]. The hereditary recursion operator can be generated from the matrix spectral problem (2.1), and the Hamilotnian formulations can be determined by a trace identity [23,26]. All this information shows that we can have everything for manipulating binary nonlinearization for our soliton hierarchy (3.1). Applications of binary nonlinearization have been made to the KdV, the MKdV, the AKNS, the Kaup-Newell, the Dirac, the WKI soliton hierarchies and so on [10,16-19] [27-32]. In what follows, an illustrative example of the multi-wave interaction equations associate with a $4 \times 4$ matrix spectral porblem will be carried out for making applications of the procedure of binary nonlinearization.

### *3.1 Multi-wave Interaction Equations*

Let us start from a 4×4 matrix spectral problem:

$$\phi_x = U(u,\lambda)\,\phi,\ U = \begin{pmatrix} \alpha_1\lambda & u_{12} & u_{13} & u_{14} \\ u_{21} & \alpha_2\lambda & u_{23} & u_{24} \\ u_{31} & u_{32} & \alpha_3\lambda & u_{34} \\ u_{41} & u_{42} & u_{43} & \alpha_4\lambda \end{pmatrix} = \lambda U_0 + U_1,\ \phi = \begin{pmatrix} \phi_1 \\ \phi_2 \\ \phi_3 \\ \phi_4 \end{pmatrix}, \quad (3.4)$$

where $U_0 = \mathrm{diag}(\alpha_1, \alpha_2, \alpha_3, \alpha_4)$, and $\alpha_1, \alpha_2, \alpha_3, \alpha_4$ are distinct constants, and the potential $u$ is defined by

$$u = \rho(U_1) = (u_{21}, u_{12}, u_{31}, u_{13}, u_{41}, u_{32}, u_{23}, u_{14}, u_{42}, u_{24}, u_{43}, u_{34})^T. \quad (3.5)$$

The associated spectral problem is chosen as

$$\phi_t = V(u,\lambda)\,\phi,\ V = \begin{pmatrix} \beta_1\lambda & v_{12} & v_{13} & v_{14} \\ v_{21} & \beta_2\lambda & v_{23} & v_{24} \\ v_{31} & v_{32} & \beta_3\lambda & v_{34} \\ v_{41} & v_{42} & v_{43} & \beta_4\lambda \end{pmatrix} = \lambda V_0 + V_1, \quad (3.6)$$

where $V_0 = \mathrm{diag}(\beta_1, \beta_2, \beta_3, \beta_4)$, and $\beta_1, \beta_2, \beta_3, \beta_4$ are distinct constants. Then the isospectral ($\lambda_t = 0$) compatibility condition $U_t - V_x + [U, V] = 0$ of the spectral problem (3.4) and the associated spectral problem (3.6) gives rise to

$$U_{1t} - V_{1x} + [U_1, V_1] = 0,\ [U_0, V_1] = [V_0, U_1]. \quad (3.7)$$

These two matrix equations lead to the following multi-wave interaction equations

$$u_{ij,t} = \frac{\beta_i - \beta_j}{\alpha_i - \alpha_j} u_{ij,x} + \sum_{\substack{k=1 \\ k\neq i,j}}^{4} \left(\frac{\beta_k - \beta_i}{\alpha_k - \alpha_i} - \frac{\beta_k - \beta_j}{\alpha_k - \alpha_j}\right) u_{ik} u_{kj},\ 1 \le i \ne j \le 4, \quad (3.8)$$

which contain a couple of physically important nonlinear models as special reductions [25]. Note that the compatability condition of the adjoint spectral problem and adjoint associated spectral problem:

$$\psi_x = -U^T(u,\lambda)\psi,\ \psi_t = -V^T(u,\lambda)\psi,\ \psi = (\psi_1, \psi_2, \psi_3, \psi_4)^T \quad (3.9)$$

still gives rise to the above multi-wave interaction equations. It is easy to find that the multi-wave interaction equations (3.8) have the Hamiltonian structure

$$u_{ij,t} = J \frac{\delta \tilde{H}}{\delta u_{ij}},\ 1 \le i \ne j \le 4, \quad (3.10)$$

where the Hamiltonian operator $J$ reads as

$$J = \mathrm{diag}\Big((\alpha_1 - \alpha_2)\sigma_0, (\alpha_1 - \alpha_3)\sigma_0, (\alpha_1 - \alpha_4)\sigma_0,$$
$$(\alpha_2 - \alpha_3)\sigma_0, (\alpha_2 - \alpha_4)\sigma_0, (\alpha_3 - \alpha_4)\sigma_0\Big),\ \sigma_0 = \begin{pmatrix} 0 & 1 \\ -1 & 0 \end{pmatrix}. \quad (3.11)$$

The Hamiltonian function $\tilde{H}$ in (3.10) can directly be computed but it is omitted here due to its complicated form. We remark that the soliton hierarchy containing higher-order multi-wave interaction equations can also be generated from the spectral problem (3.6), but in this report we focus on the multi-wave interaction equations (3.8).

### *3.2 Binary Constrained Flows*

Let us now construct binary constrained flows of the multi-wave interaction equations (3.8). Based on the general procedure of binary nonlinearization, the symmetry of (3.8) generated from the conserved functional $\lambda(u)$ reads as

$$EJ\frac{\delta\lambda}{\delta u} = E\rho([U_0, \rho^{-1}(\frac{\delta\lambda}{\delta u})]) = \rho([U_0, \rho^{-1}(\psi^T\frac{\partial U(u,\lambda)}{\partial u}\phi)]), \tag{3.12}$$

where $E$ is the normalized constant and $\rho$ is the mapping defined by (3.5). Upon introducing $N$ distinct eigenvalues $\lambda_1, \cdots, \lambda_N$, we obtain $N$ replicas of Lax systems:

$$\phi_x^{(s)} = U(u,\lambda_s)\phi^{(s)},\ \psi_x^{(s)} = -U^T(u,\lambda_s)\psi^{(s)},\ 1 \le s \le N; \tag{3.13}$$

$$\phi_t^{(s)} = V(u,\lambda_s)\phi^{(s)},\ \psi_t^{(s)} = -V^T(u,\lambda_s)\psi^{(s)},\ 1 \le s \le N; \tag{3.14}$$

where the eigenfunctions $\phi^{(s)}$ and the adjoint eigenfunctions $\psi^{(s)}$ are assumed to be denoted by

$$\phi^{(s)} = (\phi_{1s}, \phi_{2s}, \phi_{3s}, \phi_{4s})^T,\ \psi^{(s)} = (\psi_{1s}, \psi_{2s}, \psi_{3s}, \psi_{4s})^T,\ 1 \le s \le N. \tag{3.15}$$

These $N$ eigenvalues lead to a more general symmetry of (3.8)

$$Z_0 := \rho([U_0, \rho^{-1}(\sum_{s=1}^N \mu_s \psi^{(s)T}\frac{\partial U(u,\lambda_s)}{\partial u}\phi^{(s)})]) = \rho([U_0, \sum_{s=1}^N \mu_s \phi^{(s)}\psi^{(s)T}]),$$

where the $\mu_s$'s are arbitrary non-zero constants. On the other hand, we can find by inspection that the multi-wave interaction equations (3.8) have a Lie point symmetry

$$Y_0 := \rho([\Gamma, U_1]),\ \Gamma = \mathrm{diag}(\gamma_1, \gamma_2, \gamma_3, \gamma_4),\ 1 \le i \ne j \le 4,$$

where $\gamma_1, \gamma_2, \gamma_3, \gamma_4$ are distinct constants. In fact, we can directly prove that both $Y_0$ and $Z_0$ are two symmetries of the multi-wave interaction equations (3.8). Define

$$\delta U_1 = [\Gamma, U_1] \text{ or } [U_0, \sum_{s=1}^N \mu_s \phi^{(s)}\psi^{(s)T}],$$

then $[U_0, \delta V_1] = [V_0, \delta U_1]$ uniquely determines

$$\delta V_1 = [\Gamma, V_1] \text{ or } [V_0, \sum_{s=1}^{N} \mu_s \phi^{(s)} \psi^{(s)T}].$$

Now it follows from (3.7) that the symmetry problem only requires to verify that

$$(\delta U_1, \delta V_1) = ([\Gamma, U_1], [\Gamma, V_1]) \text{ or } ([U_0, \sum_{s=1}^{N} \mu_s \phi^{(s)} \psi^{(s)T}], [V_0, \sum_{s=1}^{N} \mu_s \phi^{(s)} \psi^{(s)T}])$$

satisfies the linearized system of the multi-wave interaction equations (3.8):

$$(\delta U_1)_t - (\delta V_1)_x + [\delta U_1, V_1] + [U_1, \delta V_1] = 0, \tag{3.16}$$

which can easily be proved.

Therefore, a binary symmetry constraint of (3.8) can be taken as

$$Y_0 = Z_0, \text{ i.e., } [\Gamma, U_1] = [U_0, \sum_{s=1}^{N} \mu_s \phi^{(s)} \psi^{(s)T}]. \tag{3.17}$$

Solving this equation for $u$, we obtain the required constraints on the potentials

$$u_{ij} = \tilde{u}_{ij} := \frac{\alpha_i - \alpha_j}{\gamma_i - \gamma_j} \langle \Phi_i, B\Psi_j \rangle, \; 1 \le i \ne j \le 4, \tag{3.18}$$

where $\langle \cdot, \cdot \rangle$ denotes the standard inner-product of $\mathbb{R}^N$, $B$ is defined by

$$B = \text{diag}(\mu_1, \mu_2, \cdots, \mu_N), \tag{3.19}$$

and $\Phi_i$, $\Psi_i$ are defined by

$$\Phi_i = (\phi_{i1}, \cdots, \phi_{iN})^T, \; \Psi_i = (\psi_{i1}, \cdots, \psi_{iN})^T, \; 1 \le i \le 4. \tag{3.20}$$

The substitution of $u$ with

$$\tilde{u} = (\tilde{u}_{21}, \tilde{u}_{12}, \tilde{u}_{31}, \tilde{u}_{13}, \tilde{u}_{41}, \tilde{u}_{32}, \tilde{u}_{23}, \tilde{u}_{14}, \tilde{u}_{42}, \tilde{u}_{24}, \tilde{u}_{43}, \tilde{u}_{34})$$

into the binary Lax systems (3.13) and (3.14) yields the so-called binary constrained flows:

$$\phi_x^{(s)} = U(\tilde{u}, \lambda_s)\phi^{(s)}, \; \psi_x^{(s)} = -U^T(\tilde{u}, \lambda_s)\psi^{(s)}, \; 1 \le s \le N; \tag{3.21}$$

$$\phi_t^{(s)} = V(\tilde{u}, \lambda_s)\phi^{(s)}, \; \psi_t^{(s)} = -V^T(\tilde{u}, \lambda_s)\psi^{(s)}, \; 1 \le s \le N; \tag{3.22}$$

which are two systems of ordinary differential equations, since $V$ doesn't involve any spatial derivative of $u$.

*3.3 Liouville Integrability*

In order to analyze the Liouville integrability of the above two binary constrained flows, let us first introduce a symplectic structure over $\mathbb{R}^{8N}$:

$$\omega^2 = \sum_{i=1}^{4} B d\Phi_i \wedge d\Psi_i = \sum_{i=1}^{4}\sum_{s=1}^{N} \mu_s d\phi_{is} \wedge d\psi_{is}, \tag{3.23}$$

where $B$ is defined by (3.19) and $\Phi_i$ and $\Psi_i$ are defined by (3.20). Then the associated Poisson bracket reads as

$$\begin{aligned} \{f,g\} &= \omega^2(Idg, Idf) = \sum_{i=1}^{4}(\langle \frac{\partial f}{\partial \Psi_i}, B^{-1}\frac{\partial g}{\partial \Phi_i}\rangle - \langle \frac{\partial f}{\partial \Phi_i}, B^{-1}\frac{\partial g}{\partial \Psi_i}\rangle) \\ &= \sum_{i=1}^{4}\sum_{s=1}^{N} \mu_s^{-1}\Big(\frac{\partial f}{\partial \psi_{is}}\frac{\partial g}{\partial \phi_{is}} - \frac{\partial f}{\partial \phi_{is}}\frac{\partial g}{\partial \psi_{is}}\Big), \ f,g \in C^\infty(\mathbb{R}^{8N}), \end{aligned} \tag{3.24}$$

where the vector field $Idf$ is uniquely determined by

$$\omega^2(X, Idf) = df(X), \ X \in T(\mathbb{R}^{8N}).$$

A Hamiltonian system with a Hamiltonian $H$ defined over this symplectic manifold $(\mathbb{R}^{8N}, \omega^2)$ is given by

$$\Phi_{it} = \{\Phi_i, H\} = -B^{-1}\frac{\partial H}{\partial \Psi_i}, \ \Psi_{it} = \{\Psi_i, H\} = B^{-1}\frac{\partial H}{\partial \Phi_i}, \ 1 \le i \le 4, \tag{3.25}$$

where $t$ is assumed to be the evolution variable.

In order to prove that two binary constrained flows are Liouville integrable, we need generate their Hamiltonian structures and integrals of motion. A direct computation can verify the following theorem.

**Theorem 3.1** *The spatial binary constrained flow (3.21) and the temporal binary constrained flow (3.22) are Hamiltonian systems with the evolution variables $x$, $t$ and the Hamiltonian functions*

$$H^x = -\sum_{k=1}^{4} \alpha_k \langle A\Phi_k, B\Psi_k\rangle - \sum_{1\le k<l\le 4} \frac{\alpha_k - \alpha_l}{\beta_k - \beta_l}\langle \Phi_k, B\Psi_l\rangle\langle \Phi_l, B\Psi_k\rangle, \tag{3.26}$$

$$H^t = -\sum_{k=1}^{4} \beta_k \langle A\Phi_k, B\Psi_k\rangle - \sum_{1\le k<l\le 4} \frac{\beta_k - \beta_l}{\gamma_k - \gamma_l}\langle \Phi_k, B\Psi_l\rangle\langle \Phi_l, B\Psi_k\rangle, \tag{3.27}$$

*respectively, where the matrix $A$ is given by*

$$A = diag(\lambda_1, \lambda_2, \cdots, \lambda_N). \tag{3.28}$$

It is well known that Lax representations can engender integrals of motion. For the binary constrained flows (3.21) and (3.22), let us introduce a Lax operator $L(\lambda)$ as follows

$$L(\lambda) = (L_{ij}(\lambda))_{4\times 4} = \Gamma + D(\lambda) = \mathrm{diag}(\gamma_1, \gamma_2, \gamma_3, \gamma_4) + D(\lambda), \quad (3.29)$$

$$D(\lambda) = (D_{ij}(\lambda))_{4\times 4}, \ D_{ij}(\lambda) = \sum_{l=1}^{N} \frac{\mu_l \phi_{il} \psi_{jl}}{\lambda - \lambda_l}, \ 1 \le i, j \le 4. \quad (3.30)$$

Now if we define

$$\widetilde{U}(\lambda) = U(\tilde{u}, \lambda), \ \widetilde{V}(\lambda) = V(\tilde{u}, \lambda), \quad (3.31)$$

then we obtain the following result.

**Theorem 3.2** *The binary constrained flows (3.21) and (3.22) have the following Lax representations:*

$$(L(\lambda))_x = [\widetilde{U}(\lambda), L(\lambda)], \ (L(\lambda))_t = [\widetilde{V}(\lambda), L(\lambda)], \quad (3.32)$$

*respectively.*

An $r$-matrix formulation can also be directly shown for the Lax operator $L(\lambda)$ defined by (3.29) and (3.30).

**Theorem 3.3** *The Lax operator $L(\lambda)$ defined by (3.29) and (3.30) has the following $r$-matrix formulation*

$$\{L(\lambda) \overset{\otimes}{,} L(\mu)\} = \left[\frac{1}{\mu - \lambda}\mathcal{P}, L_1(\lambda) + L_2(\mu)\right], \ \mathcal{P} = \sum_{i,j=1}^{4} E_{ij} \otimes E_{ji}, \quad (3.33)$$

*where $L_1(\lambda) = L(\lambda) \otimes I_4, \ L_2(\mu) = I_4 \otimes L(\mu)$, and*

$$(E_{ij})_{kl} = \delta_{ik}\delta_{jl}, \ \{L(\lambda) \overset{\otimes}{,} L(\mu)\}_{ij,kl} = \{L_{ik}(\lambda), L_{jl}(\mu)\}, \ 1 \le i, j, k, l \le 4.$$

Now first from the Lax representations in (3.32), we obtain

$$(\nu I_4 - L(\lambda))_x = [\widetilde{U}(\lambda), \nu I_4 - L(\lambda)], \ (\nu I_4 - L(\lambda))_t = [\widetilde{V}(\lambda), \nu I_4 - L(\lambda)],$$

where $\nu$ is a parameter, and thus [26]

$$(\det(\nu I_4 - L(\lambda)))_x = 0, \ (\det(\nu I_4 - L(\lambda)))_t = 0.$$

Second from the $r$-matrix formulation (3.33), we can obtain

$$\{\mathrm{tr} L^k(\lambda), \mathrm{tr} L^l(\mu)\} = 0, \ k, l \ge 1.$$

Expand the determinant of the matrix $\nu I_4 - L(\lambda)$ as

$$\det(\nu I_4 - L(\lambda)) = \nu^4 - \mathcal{F}_1(\lambda)\nu^3 + \mathcal{F}_2(\lambda)\nu^2 - \mathcal{F}_3(\lambda)\nu + \mathcal{F}_4(\lambda), \tag{3.34}$$

where by Newton's identities on elementary symmetric polynomials, we have

$$\begin{aligned}
&\mathcal{F}_1(\lambda) = \mathrm{tr}L(\lambda),\ \mathcal{F}_2(\lambda) = \tfrac{1}{2}((\mathrm{tr}L(\lambda))^2 - \mathrm{tr}L^2(\lambda)),\\
&\mathcal{F}_3(\lambda) = \tfrac{1}{6}(\mathrm{tr}L(\lambda))^3 + \tfrac{1}{3}\mathrm{tr}L^3(\lambda) - \tfrac{1}{2}(\mathrm{tr}L(\lambda))\mathrm{tr}L^2(\lambda),\\
&\mathcal{F}_4(\lambda) = \det L(\lambda) = \tfrac{1}{24}(\mathrm{tr}L(\lambda))^4 - \tfrac{1}{4}\mathrm{tr}L^4(\lambda) + \tfrac{1}{8}(\mathrm{tr}L^2(\lambda))^2\\
&\qquad -\tfrac{1}{4}(\mathrm{tr}L(\lambda))^2\mathrm{tr}L^2(\lambda) + \tfrac{1}{3}(\mathrm{tr}L(\lambda))\mathrm{tr}L^3(\lambda).
\end{aligned}$$

Therefore, it follows that

$$(\mathcal{F}_i(\lambda))_x = 0,\ (\mathcal{F}_i(\lambda))_t = 0,\ \{\mathcal{F}_i(\lambda), \mathcal{F}_j(\mu)\} = 0,\ 1 \le i,j \le 4. \tag{3.35}$$

Make a further expansion of $\mathcal{F}_i(\lambda)$ as follows

$$\mathcal{F}_i(\lambda) = \sum_{k\ge 0} F_{ik}\lambda^{-k},\ 1 \le i \le 4, \tag{3.36}$$

where obviously the $F_{i0}$'s are all constants. Then it follows from (3.35) that

$$(F_{ik})_x = 0,\ (F_{ik})_t = 0,\ \{F_{ik}, F_{jl}\} = 0,\ 1 \le i,j \le 4,\ k,l \ge 0, \tag{3.37}$$

Therefore, two binary constrained flows (3.21) and (3.22) have the common integrals of motion: $F_{il}$, $1 \le i \le 4$, $l \ge 1$, which are in involution in pairs under the Poisson bracket (3.24). Now it is direct to show the following theorem on the Liouville integrability of two binary constrained flows (3.21) and (3.22).

**Theorem 3.4** *Two binary constrained flows (3.21) and (3.22) have the common integrals of motion: $F_{il}$, $1 \le i \le 4$, $l \ge 1$, which are in involution in pairs under the Poisson bracket (3.24), and of which the functions $F_{is}$, $1 \le i \le 4$, $1 \le s \le N$, are functionally independent over a dense open subset of $\mathbb{R}^{8N}$. Thus, two binary constrained flows (3.21) and (3.22) are Liouville integrable.*

### *3.4 Involutive Solutions*

Note that under the constraints on the potentials (3.18), the compatability condition of (3.13) and (3.14) is still only the multi-wave interaction equations (3.8). Thus, solutions $(\Phi_i(x,t), \Psi_i(x,t))$ to two binary constrained flows (3.21) and (3.22) will present solutions to the multi-wave interaction equations (3.8), namely,

$$u_{ij}(x,t) = \frac{\alpha_i - \alpha_j}{\gamma_i - \gamma_j}\langle \Phi_i(x,t), B\Psi_j(x,t)\rangle,\ 1 \le i \ne j \le 4, \tag{3.38}$$

which also shows the integrability by quadratures for the multi-wave interaction equations (3.8) because $\Phi_i$ and $\Psi_i$, $1 \le i \le 4$, can be determined by quadratures.

It is also easy to observe that under the Poisson bracket (3.24), two Hamiltonians $H^x$ and $H^t$ defined by (3.27) and (3.27) commute, namely

$$\{H^x, H^t\} = \sum_{i=1}^{4}\Big(\langle\frac{\partial H^x}{\partial \Psi_i}, B^{-1}\frac{\partial H^t}{\partial \Phi_i}\rangle - \langle\frac{\partial H^x}{\partial \Phi_i}, B^{-1}\frac{\partial H^t}{\partial \Psi_i}\rangle\Big) = 0. \tag{3.39}$$

It follows that the above solutions determined by (3.38) give rise to a kind of involutive solutions to the multi-wave interaction equations (3.8). Let us denote two Hamiltonian flows of (3.21) and (3.22) by $g_x^{H^x}$ and $g_t^{H^t}$ respectively. Then the resulting involutive solutions can be written as

$$\begin{aligned} u_{ij}(x,t) &= \frac{\alpha_i - \alpha_j}{\gamma_i - \gamma_j}\langle g_x^{H^x} g_t^{H^t}\Phi_{i0}, Bg_x^{H^x} g_t^{H^t}\Psi_{j0}\rangle \\ &= \frac{\alpha_i - \alpha_j}{\gamma_i - \gamma_j}\langle g_t^{H^t} g_x^{H^x}\Phi_{i0}, Bg_t^{H^t} g_x^{H^x}\Psi_{j0}\rangle,\ 1 \le i \ne j \le 4, \end{aligned}$$

where the initial values $\Phi_{i0}$ and $\Psi_{i0}$ of $\Phi_i$ and $\Psi_i$, $1 \le i \le 4$, can be taken to be any arbitrary constant vectors of $\mathbb{R}^N$.

In a word, solutions of two binary constrained flows (3.21) and (3.22) can lead to involutive solutions of the multi-wave interaction equations (3.8). More importantly, such involutive solutions show us an explicit Bäcklund transformation from the multi-wave interaction equations (3.8) to two finite-dimensional Liouville integrable Hamiltonian systems determined by (3.26) and (3.27) and thus the integrability by quadratures for the multi-wave interaction equations (3.8).

## 4 Summary and Conclusion

The multi-wave interaction equations (3.8) have been taken as an illustrative example of binary nonlinearization. Binary symmetry constraints (3.17), containing arbitrary distinct constants $\gamma_1, \gamma_2, \gamma_3, \gamma_4$, and arbitrary non-zero constants $\mu_1, \mu_2, \mu_3, \mu_4$, were proposed for the multi-wave interaction equations (3.8). Two finite-dimensional Liouville integrable Hamiltonian systems determined by (3.26) and (3.27), resulted from two binary constrained flows, determine a kind of involutive solutions to the multi-wave interaction equations (3.8). If we take $B = I_N$, and

$$\Gamma = V_0, \text{ i.e., } \mathrm{diag}(\gamma_1, \gamma_2, \gamma_3, \gamma_4) = \mathrm{diag}(\beta_1, \beta_2, \beta_3, \beta_4),$$

then the result established here will cover all results presented in Ref. [30]. This also implies that our result leads to a larger class of finite-dimensional integrable Hamiltonian systems.

The theory of binary nonlinearization shows that binary symmetry constraints in the continuous case decompose soliton equations (PDEs) into binary constrained flows (ODEs). The $r$-matrix formulation can be used to expose the Liouville integrability of binary constrained flows. The resulting constraints on the potentials also present a kind of involutive solutions to soliton equations and thus eventually show the integrability by quadratures for soliton equations. The whole theory can also be applied to many other types of soliton equations.

We point out that in the case on $2 \times 2$ traceless matrix spectral problems, binary nonlinearization will be reduced to mono-nonlinearization, if we take the reduction

$$(\psi_{1j}, \psi_{2j})^T = (-\phi_{2j}, \phi_{1j})^T, \ 1 \le j \le N.$$

That is to say, in such a case, all results achieved through mono-nonlinearization can be obtained from the results achieved through binary nonlinearization. Therefore, in this sense, binary nonlinearization is much broader and more systematic.

We have seen that symmetry constraints yield nonlinear constraints on potentials of soliton equations, and put linear spectral problems (linear with respect to eigenfunctions) into nonlinear binary constrained flows (nonlinear again with respect to eigenfunctions), which seemly makes it more complicated to solve soliton equations. However, since spectral problems are overdetermined, to guarantee the existence of eigenfunctions of spectral problems, one needs compatability conditions which are often nonlinear partial differential equations. The symmetry property of the constraints brings us the Liouville integrability for nonlinear binary constrained flows. Therefore, the symmetry property makes up for the complexity of nonlinearization while manipulating binary nonlinearization.

Of particular interest in the study of binary symmetry constraints are to create new classical integrable Hamiltonian systems which supplement the known classes of classical integrable systems and to expose the integrability by quadratures for soliton equations by using binary constrained flows. We mention that high-order symmetry constraints with involved Lie-Bäcklund symmetries having non-degenerate Hamiltonians can also be carried out for soliton equations without much difficulty, but the case of degenerate Hamiltonians needs some particular consideration [32]. The rigorous considerations on binary symmetry constraints and two-binary symmetry constraints, especially on variable symplectic structures of binary constrained flows, are being expected to be made in a future publication.

## Acknowledgments

This work was supported by a grant from the Research Grants Council of Hong Kong Special Administrative Region, China (Project no. 9040466), and a grant from the City University of Hong Kong (Project no. 7001041).

## References

1. C.W. Cao, *Sci. China Ser.* A **33**, 528 (1990).
2. Y.B. Zeng and Y.S. Li, *Acta Math. Sinica* (N.S.) **6**, 257 (1990).
3. W.X. Ma, in *Nonlinear Physics*, eds. C.H. Gu, Y.S. Li and G.Z. Tu (Springer-Verlag, Berlin, 1990).
4. C.W. Cao and X.G. Geng, in *Nonlinear Physics*, eds. C.H. Gu, Y.S. Li and G.Z. Tu (Springer-Verlag, Berlin, 1990).
5. Y.B. Zeng and Y.S. Li, *J. Math. Phys.* **31**, 2835 (1990).
6. O. Ragnisco and S. Rauch-Wojciechowski, *Inverse Problems* **8**, 245 (1992).
7. W.X. Ma, *Acta Math. Appl. Sinica* (English Ser.) **9**, 92 (1993).
8. O. Ragnisco and S. Rauch-Wojciechowski, *J. Math. Phys.* **35**, 834 (1994).
9. M. Antonowicz and S. Rauch-Wojciechowski, *J. Math. Phys.* **33**, 2115 (1992).
10. W.X. Ma and W. Strampp, *Phys. Lett.* A **185**, 277 (1994).
11. C.W. Cao, Y.T. Wu and X. G. Geng, *J. Math. Phys.* **40**, 3948 (1999).
12. E.K. Sklyanin, *Progr. Theoret. Phys.* Suppl. No.118, 35 (1995).
13. M. Blaszak, *Multi-Hamiltonian Theory of Dynamical Systems* (Springer-Verlag, Berlin, 1998).
14. Y.B. Zeng and W.X. Ma, *J. Math. Phys.* **40**, 6526 (1999).
15. W.X. Ma and Y.B. Zeng, Binary Constrained Flows and Separation of Variables for Soliton Equations, accepted for publication in *Proceedings of the Conference on Integrable Systems* (Kruskal 2000), Adelaide, Australia, 2000
16. W.X. Ma, *J. Phys. Soc. Jpn.* **64**, 1085 (1995).
17. W.X. Ma, *Physica* A **219**, 467 (1995).
18. W.X. Ma, B. Fuchssteiner and W. Oevel, *Physica* A **233**, 331 (1996).

19. W.X. Ma and B. Fuchssteiner, in *Nonlinear Physics: Theory and Experiment*, eds. E. Alfinito, M. Boiti, L. Martina and F. Pempinelli (World Scientific, Singapore, 1996).
20. W.X. Ma, *J. Phys. A: Math. Gen.* **26**, 2573 (1993).
21. W.X. Ma, Binary Bargmann Symmetry Constraints of Soliton Equations, accepted for publication in *Proceedings of the 3rd World Congress of Nonlinear Analysts*, Catania, Italy, 2000
22. W.X. Ma and X.G. Geng, Bäcklund Transformations of Soliton Systems from Symmetry Constraints, accepted for publication in *Proceedings of the Workshop on Bäcklund & Darboux Transformations*, Halifax, Canada, 1999
23. G.Z. Tu, *J. Math. Phys.* **30**, 330 (1989).
24. S. Novikov, S.V. Manakov, L.P. Pitaevskii and V.E. Zakharov, *Theory of Solitons: The Inverse Scattering Method* (Consultants Bureau, New York, 1984).
25. M.J. Ablowitz and P.A. Clarkson, *Solitons, Nonlinear Evolution Equations and Inverse Scattering* (Cambridge University Press, Cambridge, 1991).
26. G.Z. Tu, *J. Phys. A: Math. Gen.* **22**, 2375 (1989).
27. X.X. Xu, *Chinese Phys. Lett.* **12**, 513 (1995).
28. W.X. Ma, Q. Ding, W.G. Zhang and B.G. Lu, *Il Nuovo Cimento* B **111**, 1135 (1996).
29. W.X. Ma, *Chinese Ann. Math.* Ser. B **18**, 79 (1997).
30. Q.Y. Shi and S.M. Zhu, *J. Math. Phys.* **41**, 2157 (2000).
31. W.X. Ma and R.G. Zhou, Nonlinearization of Spectral Problems for the Perturbation KdV Systems, accepted for publication in *Physica* A.
32. Y.S. Li and W.X. Ma, *Chaos, Solitons & Fractals* **11**, 697 (2000).

# Trigonometric identities and geometrical inequalities for links and knots *

ALEXANDER D. MEDNYKH

*Sobolev Institute of Mathematics, Novosibirsk, 630090, Russia*

*e-mail : mednykh@math.nsc.ru*

**Abstract**

In the present paper links and knots are investigated as a singular set of geometric cone–manifolds with the three-sphere as underlying space. Trigonometric identities between lengths of singular components and cone angles of these cone–manifolds (Sine, Cosine, and Tangent rules) are obtained. Geometrical inequalities between volumes and singular geodesic lengths of the cone–manifolds are also given. They can be considered as a sort of isoperimetric inequalities well-known for convex polyhedra.

Keywords: *hyperbolic orbifold, hyperbolic cone-manifold, complex length, Tangent Rule, Sine Rule, Cosine Rule, isoperimetric inequalities*

**Table of contents**



## 0. Introduction

Knot theory was born around the year 1867 in Scotland from the imagination of three physicists: J. C. Maxwell, P. G. Tait, and W. Thomson (Lord Kelvin). For more details see [Kn], [HKW]. Maxwell's interest for knots came from his theory of electromagnetism. For instance, he gave in [Ma] an important interpretation of Gauss integral formula for the linking coefficient of two knots in the 3-space: it is equal to the work required to move a magnet pole along one knot while the other knot is run by an electric current. Another curious fact is that Seifert surface whose boundary is a given knot being introduced by Tait via pure physical arguments. Due to efforts of J. B. Listing, K. Reidemeister, and M. Dehn knot theory was gradually embodied in the more general theory of 3-dimensional manifolds. The notion of the fundamental group was introduced and the group theory became one of the most powerful tools in the knot theory. In 1975 R. Riley [R] had found examples of hyperbolic structures on some knot and link complements in the three-sphere. Later, in the spring of 1977 W. P. Thurston had announced an existence

*Supported by the RFBR (grant 99-01-00630, 00-15-96165) and by DFG

theorem for Riemannian metrics of constant negative curvature on 3-manifolds. In particular, it turned out that knot complement of a simple knot (except for toral and satellite) admits a hyperbolic structure. This fact allowed to consider knot theory as a part of geometry and Kleinian group theory. Starting from Alexander's works polynomial invariants became a convenient instrument for knot investigation. A lot of different kinds of such polynomials were investigated in the last twenty years. Among these there are Jones-, Kaufmann-, HOMFLY-, A–polynomials and others ([Kauf], [CCGLS], [HLM2]). This relates the knot theory with algebra and algebraic geometry.

In the present paper we investigate knots and links as singular subsets of the 3-sphere endowed by Riemannian metric of constant curvature (negative, positive, or zero). More precisely, our aim is to investigate the structure of geometrical cone–manifolds whose underlying space is the three-sphere and the singular set is a given knot or link.

Section 1 contains a list of trigonometric identities (Sine, Cosine, and Tangent rules) relating the lengths of singular geodesics of geometrical cone-manifolds with their cone–angles. Cone-manifolds are supposed to be hyperbolic, spherical, or Euclidean. Similar results are known for the right–angled hexagons in the hyperbolic 3-space which can be considered as triangles with complex lengths and angles [Fench]. Related results can be also obtained for a class of knotted graphs. For example, they take place for the rational knots with bridges through their tunnels.

Section 2 is devoted to explicit calculation of volume of some cone-manifolds in hyperbolic and spherical geometries. In particular, simple volume formulas will be obtained for the figure-eight cone–manifold. Partially, these results are well-known and were given earlier in [HLM3], [MV], and [Kj].

Section 3 gives inequalities between volumes and singular geodesic lengths of the cone–manifolds under investigation. They can be consider as a sort of isoperimetric inequalities well-known for convex polyhedra [BZ].

## 1. Trigonometric identities for knots and links

### 1.1 Cone–manifolds, complex distances and lengths

We start with the definition of cone–manifold modeled in hyperbolic, spherical or Euclidean structure.

**Definition 1.1.1.** A 3–dimensional *hyperbolic cone–manifold* is a Riemannian 3–dimensional manifold of constant negative sectional curvature with cone-type singularity along simple closed geodesics. To each component of singular set we associate a real number $n \geq 1$ such that the cone-angle around the component is $\alpha = 2\pi/n$. The concept of the hyperbolic cone-manifold generalizes the hyperbolic manifold which appears in the partial case when all cone-angles are $2\pi$. The hyperbolic cone-manifold is also a generalization of the hyperbolic 3–orbifold which arises when all associated numbers $n$ are integers. Euclidean and spherical cone–manifolds are defined similarly.

In the present paper hyperbolic, spherical or Euclidean cone-manifolds $C$ are considered whose underlying space is the three-dimensional sphere and the singular set $\Sigma = \Sigma^1 \cup \Sigma^2 \cup \ldots \cup \Sigma^k$ is a link consisting of components $\Sigma^j = \Sigma_{\alpha_j}$, $j = 1, 2, \ldots, k$ with cone-angles $\alpha_1, \ldots, \alpha_k$ respectively.

Recall a few well-known facts from the hyperbolic geometry.

Let $\mathbb{H}^3 = \{(z, \xi) \in \mathbb{C} \times \mathbb{R} : \xi > 0\}$ be the upper half model of the 3 -dimensional hyperbolic space endowed by the Riemannian metric $ds^2 = \dfrac{dzd\overline{z} + d\xi^2}{\xi^2}$. We identify the group of orientation preserving isometries of $\mathbb{H}^3$ with the group $PSL(2, \mathbb{C})$ consisting of linear fractional transformations

$$A : z \in \mathbb{C} \to \frac{az + b}{cz + d}.$$

By the canonical procedure the linear transformation $A$ can be uniquely extended to the isometry of $\mathbb{H}^3$. We prefer to deal with the matrix $\widetilde{A} = \begin{pmatrix} a & b \\ c & d \end{pmatrix} \in SL(2, \mathbb{C})$ rather than the element $A \in PSL(2, \mathbb{C})$. The matrix $\widetilde{A}$ is uniquely determined by the element $A$ up to a sign. If there will be no confusions we shall use the same letter $A$ for both $A$ and $\widetilde{A}$.

Let $C$ be a hyperbolic cone–manifold with the singular set $\Sigma$. Then $C$ defines a nonsingular but incomplete hyperbolic manifold $N = C - \Sigma$. Denote by $\Phi$ the fundamental group of the manifold $N$.

The hyperbolic structure of $N$ defines, up to conjugation in $PSL(2, \mathbb{C})$, a holonomy homomorphism

$$\hat{h} : \Phi \to PSL(2, \mathbb{C}).$$

It is shown in [Zhou] that the monodromy homomorphism of a compact orientable cone-orbifold can be lifted to $SL(2, \mathbb{C})$. Denote by $h : \Phi \to SL(2, \mathbb{C})$ this lifting homomorphism. Chose an orientation on the link $\Sigma = \Sigma^1 \cup \Sigma^2 \cup \ldots \cup \Sigma^k$ and fix a meridian-longitude pair $\{m_j, l_j\}$ for each component $\Sigma_j = \Sigma_{\alpha_j}$. Then the matrices $M_j = h(m_j)$ and $L_j = h(l_j)$ satisfy the following properties:

$$\operatorname{tr}(M_j) = 2\cos(\alpha_j/2), \quad M_j L_j = L_j M_j, \; j = 1, 2, \ldots, k.$$

**Definition 1.1.2**. A *complex length* $\gamma_j$ of the singular component $\Sigma^j$ of the cone-manifold $C$ is defined as displacement of the isometry $L_j$ of $\mathbb{H}^3$ , where $L_j = h(l_j)$ is represented by the longitude $l_j$ of $\Sigma^j$.

Immediately from the definition we get [Fench, p.46]

$$2\cosh\gamma_j = \operatorname{tr}(L_j^2) \tag{1.1.1}$$

We note [BZie, p.38] that the meridian-longitude pair $\{m_j, l_j\}$ of the oriented link is uniquely determined up to a common conjugating element of the group $\Phi$. Hence, the complex length $\gamma_j = l_j + i\,\varphi_j$ is uniquely determined up to a sign and (mod $2\pi i$) by the above definition.

We need two conventions to choose correctly real and imaginary parts of $\gamma_j$. The first convention is the following. Since $\Sigma^j$ does not shrink to a point, $l_j \neq 0$. Hence, we choose $\gamma_j$ in such a way that $l_j = \Re\gamma_j > 0$. The second convention is concerned with the imaginary part $\varphi_j = \Im\gamma_j$. We want to choose $\varphi_j$ such that the following identity holds

$$\cosh\frac{\gamma_j}{2} = -\frac{1}{2}\,\mathrm{tr}(L_j) \tag{1.1.2}$$

By virtue of identity $\mathrm{tr}(L_j)^2 - 2 = \mathrm{tr}(L_j^2)$ equality (1.1.1) is a consequence of (1.1.2). The converse, in general, is true only up to a sign. Under the second convention (1.1.1) and (1.1.2) are equivalent. The two above conventions lead to convenient analytic formulas for calculation of $\gamma_j$ and $l_j$. More precisely, there are simple relations between these numbers and eigenvalues of matrix $L_j$. Recall that $\det L_j = 1$. Since matrix $L_j$ is loxodromic it has two eigenvalues $f_j$ and $1/f_j$. We choose $f_j$ so that $|f_j| > 1$. The case $|f_j| = 1$ is impossible because in this case the matrix $L_j$ is elliptic and $l_j = 0$. Hence

$$f_j = -e^{\frac{\gamma_j}{2}}, \quad |f_j| = e^{\frac{l_j}{2}}. \tag{1.1.3}$$

**1.2. Whitehead link cone–manifold**

Denote by $W(\alpha,\beta)$ the cone-manifold whose underlying space is the 3-sphere and whose singular set consists of two components of the Whitehead link with cone angles $\alpha = 2\pi/m$ and $\beta = 2\pi/n$ (see Fig.1). It follows from Thurston's theorem that $W(\alpha,\beta)$ admits a hyperbolic structure for all sufficiently small $\alpha$ and $\beta$. The region of hyperbolicity of $W(\alpha,\beta)$ was investigated in [HLM2] and [KM]. In particular, this cone–manifold is hyperbolic for $m, n > 2.507..$ The following theorems have been obtained in [M].

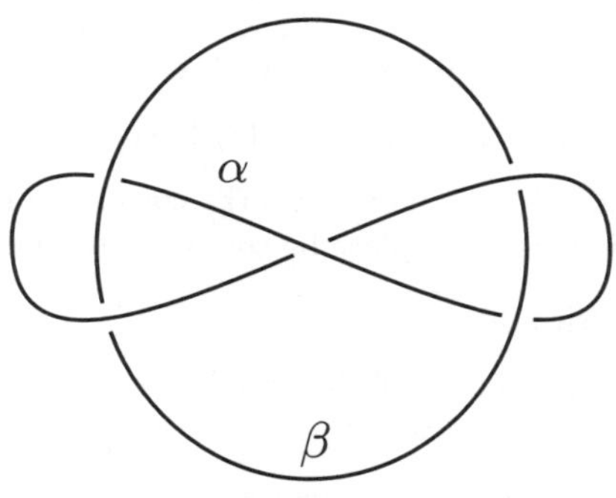


Figure 1: The Whitehead link cone–manifold $W(\alpha,\beta)$.

**Theorem 1.2.1 (The Tangent Rule).** *Suppose that cone–manifold $W(\alpha,\beta)$ is hyperbolic. Denote by $\gamma_\alpha$ and $\gamma_\beta$ complex lengths of the singular geodesics of*

*$W(\alpha,\beta)$ with cone angles $\alpha$ and $\beta$ respectively. Then*

$$\frac{\tanh\frac{\gamma_\alpha}{4}}{\tanh\frac{\gamma_\beta}{4}} = \frac{\tan\frac{\alpha}{2}}{\tan\frac{\beta}{2}}.$$

**Theorem 1.2.2 (The Sine Rule).** *Let $\gamma_\alpha = l_\alpha + i\,\varphi_\alpha$ (resp.$\gamma_\beta$) be a complex length of the singular geodesic of a hyperbolic cone-manifold $W(\alpha,\beta)$ with cone angle $\alpha$ (resp. $\beta$). Then*

$$\frac{\sin\frac{\varphi_\alpha}{2}}{\sinh\frac{l_\alpha}{2}} = \frac{\sin\frac{\varphi_\beta}{2}}{\sinh\frac{l_\beta}{2}}.$$

Euclidean analogues of Theorems 1.2.1 and 1.2.2 were obtained by R.N. Shmatkov (1999). The similar results can be stated also for spherical cone–manifold $W(\alpha,\beta)$.

**1.3. Cone–manifold $6_2^2(\alpha,\beta)$**

According to [Rolf] denote by $6_2^2$ two-bridge link with the rational slope 10/3. Consider a cone–manifold $6_2^2(\alpha,\beta)$ whose underlying space is the three-sphere and singular set is formed by two components of the link $6_2^2(\alpha,\beta)$ with cone angles $\alpha$ and $\beta$ (Fig.2). A canonical fundamental set for two-bridge cone–manifolds admitting hyperbolic, Euclidean, or spherical structure has been constructed in [MR1]. Applying this construction to the cone–manifold under consideration we get the following proposition.

**Proposition 1.3.1** *The cone manifold $6_2^2(\alpha,\beta)$ admits a hyperbolic, Euclidean, or spherical structure in regions $R_h, R_e$ and $R_s$ respectively, where*

$$(i)\quad R_h = \{(\alpha,\beta)\in\mathbb{R}^2 : 0\le\alpha,\beta<4\pi/3, \cos\alpha/2+\cos\beta/2>1/2\}$$

$$(ii)\quad R_e = \{(\alpha,\beta)\in\mathbb{R}^2 : 0\le\alpha,\beta<4\pi/3, \cos\alpha/2+\cos\beta/2=1/2\}$$

$$(iii)\quad R_h = \{(\alpha,\beta)\in\mathbb{R}^2 : 2\pi/3<\alpha,\beta<4\pi/3, |\cos\alpha/2\pm\cos\beta/2|<1/2\}$$

We recall that in the case $\alpha=0$ or $\beta=0$ the corresponding component of the cone–manifold $6_2^2(\alpha,\beta)$ should be replace by complete hyperbolic cusp.

| The cone–manifold $6_2^2(\alpha,\beta)$. Fig. 2 |
|---|

The next proposition gives an explicit formula for a real length of the singular component of the cone–manifold $6_2^2(\alpha,\beta)$.

**Proposition 1.3.2.** *Suppose that cone–manifold $6_2^2(\alpha,\beta)$, $(\alpha,\beta)\in R_h$ is hyperbolic. Then the length $l_\alpha$ of the singular geodesic of $6_2^2(\alpha,\beta)$ with cone angle $\alpha$ is defined by the formula*

$$\cosh\frac{l_\alpha}{2} = 2\frac{N+M\sqrt{1+K^2}}{(1+M^2)\sqrt{1+N^2}},$$

*where* $M = \cot\frac{\alpha}{2}, N = \cot\frac{\beta}{2}$, *and* $K^2 = (1+M^2)(1+N^2)/4$.

Proof is based on the following considerations.

Let $\Sigma = \Sigma^1 \cup \Sigma^2$ be the singular set of cone–manifold $C = 6_2^2(\alpha,\beta)$. Consider a nonsingular noncomplete hyperbolic manifold $C - \Sigma$ and denote by $\Phi = \pi_1(C - \Sigma)$ its fundamental group. Then $\Phi$ has the following presentation

$$\Phi =< s_\alpha, s_\beta : s_\alpha l_\alpha = l_\alpha s_\alpha,\, l_\alpha = s_\beta s_\alpha s_\beta {s_\alpha}^{-1} {s_\beta}^{-1} {s_\alpha}^{-1} s_\beta s_\alpha s_\beta >,$$

where $s_\alpha$ and $s_\beta$ are meridians of the components $\Sigma_\alpha$ and $\Sigma_\beta$ respectively, and $l_\alpha$ is a longitude of $\Sigma_\alpha$. Let $h : \Phi \to SL(2,\mathbb{C})$ be a holonomy homomorphism, $S = h(s_\alpha)$, and $T = h(l_\beta$.) After a suitable conjugation in the group $SL(2,\mathbb{C})$ homomorphism $h$ can be chosen in such a way that

$$S = \begin{pmatrix} \cos\mu & i\,e^{\frac{\rho}{2}}\sin\mu \\ i\,e^{-\frac{\rho}{2}}\sin\mu & \cos\mu \end{pmatrix}, \quad T = \begin{pmatrix} \cos\nu & i\,e^{-\frac{\rho}{2}}\sin\nu \\ i\,e^{\frac{\rho}{2}}\sin\nu & \cos\nu \end{pmatrix}, \qquad (1.3.1)$$

where $\mu = \alpha/2$, $\nu = \beta/2$, $\rho$ is a complex distance between axis of $S$ and $T$ in the hyperbolic space $\mathbb{H}^3$.

Set $L = h(l_\alpha)$. Then the matrix equation $SL = LS$ and a restriction that $L$ is loxodromic for all $\alpha > 0$ gives the following quadratic equation for $u = \cosh\rho$

$$u^2 - (MN + \sqrt{1+K^2})u + K^2 + MN\sqrt{1+K^2} = 0, \qquad (1.3.2)$$

where $M$, $N$ and $K$ are the same as in the statement of the proposition.

Routine calculation of the trace of the element $L = TSTS^{-1}T^{-1}S^{-1}TST$ modulo equation (1.3.2) gives

$$\mathrm{tr}(L) = 4\frac{N + M\sqrt{1+K^2}}{(1+M^2)\sqrt{1+N^2}}. \qquad (1.3.3)$$

Inside of region of hyperbolicity $R_h$ we have $\mathrm{tr}(L) > 2$. From(1.1.1) the complex length $l_\alpha = l_\alpha + i\varphi_\alpha$ is given by $\cosh\frac{\gamma_\alpha}{2} = -\frac{1}{2}\mathrm{tr}(L)$. Hence $\cosh\frac{\gamma_\alpha}{2} < -1$, $\frac{\gamma_\alpha}{2} = \frac{l_\alpha}{2} + i\pi$ and

$$\cosh\frac{l_\alpha}{2} = \frac{1}{2}\mathrm{tr}(L) = 2\frac{N + M\sqrt{1+K^2}}{(1+M^2)\sqrt{1+N^2}}.$$

**Theorem 1.3.3 (The Sine Rule).** *Let $l_\alpha$ and $l_\beta$ be lengths of the singular geodesics of a hyperbolic cone–manifold $6_2^2(\alpha,\beta)$ with cone angles $\alpha$ and $\beta$ respectively. Then*

$$\frac{\sin\frac{\alpha}{2}}{\sinh\frac{l_\alpha}{2}} = \frac{\sin\frac{\beta}{2}}{\sinh\frac{l_\beta}{2}}.$$

Proof. Immediately from Proposition 1.3.2 we get

$$\sinh\frac{l_\alpha}{2} = \frac{\sqrt{\Delta}}{(1+M^2)\sqrt{1+N^2}} = \sin\mu\frac{\sqrt{\Delta}}{\sqrt{(1+M^2)(1+N^2)}},$$

where $\Delta = 4M^2 + 4N^2 + 4MM\sqrt{4 + (1+M^2)(1+N^2)} - (1+M^2)(1+N^2)$ is a symmetric function of $M$ and $N$. Hence

$$\frac{\sinh \frac{l_\alpha}{2}}{\sin \mu} = \frac{\sqrt{\Delta}}{\sqrt{(1+M^2)(1+N^2)}} = \frac{\sinh \frac{l_\beta}{2}}{\sin \nu}$$

and the result follows.

**Theorem 1.3.4 (The Cosine Rule).** *Let $l_\alpha$ and $l_\beta$ be lengths of the singular geodesics of a hyperbolic cone–manifold $6_2^2(\alpha, \beta)$ with cone angles $\alpha$ and $\beta$ respectively. Then*

$$\frac{\cos \frac{\alpha}{2} \cosh \frac{l_\beta}{2} - \cos \frac{\beta}{2} \cosh \frac{l_\alpha}{2}}{\cos \alpha - \cos \beta} = -1.$$

Proof. By Proposition 1.3.2 we get

$$\cosh \frac{l_\alpha}{2} = 2\frac{N + M\sqrt{1+K^2}}{(1+M^2)\sqrt{1+N^2}}$$

and

$$\cosh \frac{l_\beta}{2} = 2\frac{M + N\sqrt{1+K^2}}{(1+N^2)\sqrt{1+M^2}}.$$

Hence

$$\frac{M}{\sqrt{1+M^2}} \cosh \frac{l_\beta}{2} - \frac{N}{\sqrt{1+N^2}} \cosh \frac{l_\alpha}{2} = 2\frac{M^2 - N^2}{(1+M^2)(1+N^2)} = \cos \beta - \cos \alpha.$$

Since $\frac{M}{\sqrt{1+M^2}} = \cos \frac{\alpha}{2}$ and $\frac{N}{\sqrt{1+N^2}} = \cos \frac{\beta}{2}$ the theorem is proved.

### 1.4 The Borromean cone–manifold

In this subsection we investigate geometric properties of a cone–manifold $B(\alpha, \beta, \gamma)$ with singular set the Borromean rings (Fig. 3). The cone angles of three components of the singular set are $\alpha, \beta, \gamma$. As above, the corresponding lengths of the singular set components will be denoted by $l_\alpha, l_\beta$, and $l_\gamma$.

The Borromean cone–manifold $B(\alpha, \beta, \gamma)$. Fig. 3

It is well-known fact that $B(\alpha, \beta, \gamma)$ can be obtained by gluing together eight copies of the Lambert cube $Q(\alpha/2, \beta/2, \gamma/2)$ with essential dihedral angles $\alpha/2, \beta/2, \gamma/2$. See [T] and [HLM1] for details. In particular, it is shown in [T] that $Q(\alpha/2, \beta/2, \gamma/2)$ (and hence $B(\alpha, \beta, \gamma)$) is hyperbolic if $0 \leq \alpha, \beta, \gamma < \pi$ and Euclidean if $\alpha/2 = \beta/2 = \gamma/2 = \pi$.

Moreover, if $L_\alpha, L_\beta, L_\gamma$ denote the edge lengths of $Q(\alpha/2, \beta/2, \gamma/2)$ with dihedral angles $\alpha/2, \beta/2, \gamma/2$ we get

$$L_\alpha = \frac{l_\alpha}{4}, \quad L_\beta = \frac{l_\beta}{4}, \quad L_\gamma = \frac{l_\gamma}{4}. \qquad (1.4.1)$$

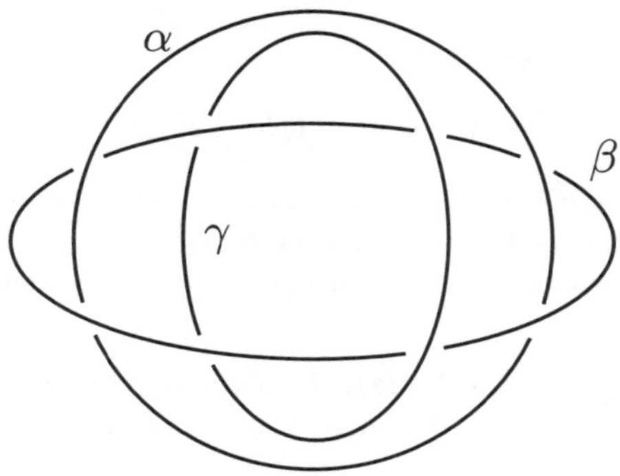


Figure 2: The Borromean cone–manifold $B(\alpha, \beta, \gamma)$.

The Lambert cube $Q(\alpha/2, \beta/2, \gamma/2)$. Fig. 4

As in the case of cone–manifolds $W(\alpha, \beta)$ and $6_2^2(\alpha, \beta)$ there are simple trigonometric identities relating the lengths $l_\alpha, l_\beta, l_\gamma$ of $B(\alpha, \beta, \gamma)$ with its cone angles $\alpha, \beta, \gamma$. We start with the following

**Theorem 1.4.1 (The Tangent Rule).** *Let $B(\alpha, \beta, \gamma)$, be a hyperbolic Borromean cone–manifold with cone angles $0 < \alpha, \beta, \gamma < \pi$ and the singular geodesic lengths $l_\alpha, l_\beta, l_\gamma$. Then*

$$\frac{\tan \frac{\alpha}{2}}{\tanh \frac{l_\alpha}{4}} = \frac{\tan \frac{\beta}{2}}{\tanh \frac{l_\beta}{4}} = \frac{\tan \frac{\gamma}{2}}{\tanh \frac{l_\gamma}{4}} = T,$$

*where $T$ is a positive number defined by* $T^2 = K + \sqrt{K^2 + L^2M^2N^2}$, $L = \tan \frac{\alpha}{2}$, $M = \tan \frac{\beta}{2}$, $N = \tan \frac{\gamma}{2}$, *and* $K = (L^2 + M^2 + N^2 + 1)/2$.

Proof. We prefer to deal with the Lambert cube $Q(\alpha/2, \beta/2, \gamma/2)$ rather then cone–manifold $B(\alpha, \beta, \gamma)$. It follows from the result of [K] that the edge lengths $L_\alpha, L_\beta$ and $L_\gamma$ are related with its angles by

$$\frac{\tan \frac{\alpha}{2}}{\tanh L_\alpha} = \frac{\tan \frac{\beta}{2}}{\tanh L_\beta} = \frac{\tan \frac{\gamma}{2}}{\tanh L_\gamma} = T, \tag{1.4.2}$$

where $T = \tan \theta$ for some angle $\theta$ such that $\alpha, \beta, \gamma \leq 2\theta \leq \pi$. The simple proof of this formula by means of Gram matrix techniques can be find also in [V]. The following equation for $T$ was obtained in ([K], p.564, eq. (II)) and ([HLM1], eq. (A.2)) in slightly different terms

$$T^2 = \frac{T^2 - L^2}{1 + L^2} \frac{T^2 - M^2}{1 + M^2} \frac{T^2 - N^2}{1 + N^2}, \tag{1.4.3}$$

The last equation is equivalent to

$$(T^2 + 1)(T^4 - (L^2 + M^2 + N^2 + 1)T^2 - L^2M^2N^2) = 0.$$

Since $T$ is a positive number we get

$$T^4 - (L^2 + M^2 + N^2 + 1)T^2 - L^2 M^2 N^2 = 0. \qquad (1.4.4)$$

Hence $T^2 = K + \sqrt{K^2 + L^2 M^2 N^2}$, and $K = (L^2 + M^2 + N^2 + 1)/2$. Taking into account (1.4.1) and (1.4.2) we finish the proof.

The next three theorems can be considered as a consequences of the Tangent Rule.

**Theorem 1.4.2 (The Sine Rule).** *Let $B(\alpha, \beta, \gamma)$, be a hyperbolic Borromean cone–manifold with cone angles $0 < \alpha, \beta, \gamma < \pi$ and the singular geodesic lengths $l_\alpha, l_\beta, l_\gamma$. Then*

$$\frac{\sin\frac{\alpha}{2}}{\sinh\frac{l_\alpha}{4}} \frac{\sin\frac{\beta}{2}}{\sinh\frac{l_\beta}{4}} \frac{\sin\frac{\gamma}{2}}{\sinh\frac{l_\gamma}{4}} = T,$$

*where $T$ is a positive number defined by* $T^2 = K + \sqrt{K^2 + L^2 M^2 N^2}$, $L = \tan\frac{\alpha}{2}$, $M = \tan\frac{\beta}{2}$, $N = \tan\frac{\gamma}{2}$, *and* $K = (L^2 + M^2 + N^2 + 1)/2$.

Proof. We rewrite the statement of the Tangent Rule in the form

$$\sinh^2 L_\alpha = \frac{L^2}{T^2 - L^2}, \ \sinh^2 L_\beta = \frac{M^2}{T^2 - M^2}, \ \sinh^2 L_\gamma = \frac{N^2}{T^2 - N^2}, \qquad (1.4.5)$$

We get also

$$\sin^2\frac{\alpha}{2} = \frac{L^2}{1 + L^2}, \quad \sin^2\frac{\beta}{2} = \frac{M^2}{1 + M^2}, \quad \sin^2\frac{\gamma}{2} = \frac{N^2}{1 + N^2}. \qquad (1.4.6)$$

By virtue of (1.4.3) we have from (1.4.5) and (1.4.6)

$$\frac{\sin^2\frac{\alpha}{2}}{\sinh^2 L_\alpha} \frac{\sin^2\frac{\beta}{2}}{\sinh^2 L_\beta} \frac{\sin^2\frac{\gamma}{2}}{\sinh^2 L_\gamma} = \frac{T^2 - L^2}{1 + L^2} \frac{T^2 - M^2}{1 + M^2} \frac{T^2 - N^2}{1 + N^2} = T^2.$$

By taking the square root we obtain the statement of the theorem.

By similar arguments the following theorems can be proved.

**Theorem 1.4.3 (The Cosine Rule).**

$$\frac{\cos\frac{\alpha}{2}}{\cosh\frac{l_\alpha}{4}} \frac{\cos\frac{\beta}{2}}{\cosh\frac{l_\beta}{4}} \frac{\cos\frac{\gamma}{2}}{\cosh\frac{l_\gamma}{4}} = \frac{1}{T^2},$$

**Theorem 1.4.4 (The Sine-Cosine Rule).**

$$\frac{\sin\frac{\alpha}{2}}{\sinh\frac{l_\alpha}{4}} \frac{\sin\frac{\beta}{2}}{\sinh\frac{l_\beta}{4}} \frac{\cos\frac{\gamma}{2}}{\cosh\frac{l_\gamma}{4}} = 1,$$

## 2. Explicit volume calculation

### 2.1. The Schläfli formula

In this section we will obtain explicit formulas for volume of some special cone–manifolds in the hyperbolic and spherical geometries. In the case of complete hyperbolic structure on the simplest knot and link complements such formulas in terms of Lobachevsky function are well-known and widely represented in [T]. In general situation, a hyperbolic cone–manifold can be obtained by completion of non-complete hyperbolic structure on a suitable knot or link complement. If the cone–manifold is compact explicit formulas are know just in a few cases [Hds] , [HLM3], [MV], [Kj]. In all these cases the starting point for the volume calculation is the Schläfli formula (see, for example [Hds] )

**Theorem 2.1.1. (The Schläfli volume formula)** *Suppose that $C_t$ is a smooth 1–parameter family of (curvature $K$) cone–manifold structures on a $n$-manifold, with singular locus $\Sigma$ of a fixed topological type. Then the derivative of volume of $C_t$ satisfies*

$$(n-1)K dV(C_t) = \sum_{\sigma} V_{n-2}(\sigma) d\theta(\sigma)$$

*where the sum is over all components $\sigma$ of the singular locus $\Sigma$, and $\theta(\sigma)$ is the cone angle along $\sigma$.*

In the present paper we will deal mostly with three-dimensional cone–manifold structures of negative constant curvature $K = -1$. The Schläfli formula in this case reduces to

$$dV = -\frac{1}{2}\sum_i l_{\theta_i} d\theta_i, \tag{2.1.1}$$

where the sum is taken over all components of the singular set $\Sigma$ with lengths $l_{\theta_i}$ and cone angles $\theta_i$.

Our aim is to obtain the volume formulas for cone–manifolds $W(\alpha,\beta), 6_2^2(\alpha,\beta)$ and $B(\alpha,\beta,\gamma)$ described in the above section. Since the figure eight cone–manifold $4_1(\alpha)$ is the two-fold covering of $6_2^2(\alpha,\pi)$ its volume is twice the volume of $6_2^2(\alpha,\pi)$. This leads to a simple volume formula for the figure eight cone–manifold obtained earlier in more complicated form in [HLM3], [MV] and [Kj].

### 2.2. Volume of the Whitehead link cone–manifold

First of all we consider the case of the hyperbolic Whitehead link with one complete cusp.

**Theoren 2.2.2.** Let $W(0,\alpha)$ be a hyperbolic Whitehead link cone–manifold with a complete hyperbolic structure on one cusp and cone angle $\alpha$, $0 \le \alpha < \pi$ on the another. Then the volume of $W(0,\alpha)$ is given by the formula

$$Vol\, W(0,\alpha) = \frac{1}{2}\int_{\alpha}^{\pi} arcosh\,(8 - 8\cos t + \cos 2t) dt.$$

Proof. By [KM] cone–manifold $W(0,\alpha)$ is hyperbolic for all $0 \le \alpha < \pi$. Denote by $V_\alpha$ the hyperbolic volume of $W(0,\alpha)$. By Schläfli formula we have $dV_\alpha = -\frac{1}{2} l_\alpha d\alpha$.

By calculation produced in [M] we obtain

$$\cosh l_\alpha = \frac{M^4 + 10M^2 + 17}{(M^2+1)^2}, \tag{2.2.1}$$

where $M = \cot \frac{\alpha}{2}$. Simplifying (2.2.1) we get $\cosh l_\alpha = 8 - 8\cos\alpha + \cos 2\alpha$ and

$$l_\alpha = arcosh\,(8 - 8\cos\alpha + \cos 2\alpha). \tag{2.2.2}$$

By integrating the Schläfli formula we have

$$V_\alpha = -\frac{1}{2}\int_\theta^\alpha arcosh\,(8 - 8\cos t + \cos 2t)dt + V_\theta, \tag{2.2.3}$$

for an arbitrary $\theta$, $0 \le \theta < \pi$. We note that the geometrical limit $W(0,\pi)$ of the cone–manifolds $W(0,\theta)$ as $\theta \to \pi - 0$ is not hyperbolic, since its two-fold covering branched over the $\pi$-component is the torus link 4/1. Also, $W(0,\pi)$ contains no two-dimensional suborbifolds of the type $S^2(\pi,\pi,\pi)$. Hence, by Theorem 7.1.2 of [Kj] we have $\lim_{\theta\to\pi-0} V_\theta = 0$. Going over to the limit we immediately get from (2.2.3) the statement of the theorem.

In the case of closed cone–manifold $W(\alpha,\beta)$ the volume function becomes more complicated and can be expressed in terms of roots of a cubic equation. See [M] and [KM] for details.

**2.3. Volumes of the $6_2^2(\alpha,\beta)$, the figure eight, and the Borromean rings cone–manifolds**

This subsection will be organized in the following way. First of all, by making use of length formula for a singular geodesic of the cone–manifold $6_2^2(\alpha,\beta)$ from section 1 and the Schläfli variation formula we obtain a simple expression for $Vol\, 6_2^2(\alpha,\beta)$. Then taking into account that the figure eight cone–manifold $4_1(\alpha)$ and the Borromean rings cone–manifold $B(\alpha,\alpha,\alpha)$ are, respectively two-fold and three-fold coverings of $6_2^2(\alpha,\beta)$ for $\beta = \pi$ and $\beta = \frac{2\pi}{3}$, we find the volume formulas for both of them. These formulas turn out to be more simple then the corresponding formulas obtained earlier in [HLM3], [MV], and [K].

**Theorem 2.3.1.** *Suppose that cone–manifold $6_2^2(\alpha,\beta)$, $(\alpha,\beta) \in R_e$ is hyperbolic. Then its volume is defined by the formula*

$$Vol\, 6_2^2(\alpha,\beta) = \int_\alpha^{\alpha^*} E(\frac{\alpha}{2},\frac{\beta}{2})d\alpha,$$

*where* $E(\mu,\nu) = arcosh\,(2\sin^2\mu\cos\nu + \cos\mu\sqrt{4\sin^2\mu\sin^2\nu + 1})$ *and* $\alpha^*$, $0 \le \alpha^* < \frac{2\pi}{3}$ *is uniquely determined by the equation* $\cos\frac{\alpha^*}{2} + \cos\frac{\beta}{2} = \frac{1}{2}$.

Proof. Recall (Proposition 1.3.1) that $6_2^2(\alpha,\beta)$ is hyperbolic for $(\alpha,\beta) \in R_h$, where the region $R_h$ is bounded by the coordinate axes and by the curve $R_e = \{(\alpha,\beta) \in \mathbb{R}^2 : 0 \le \alpha,\beta < 4\pi/3, \cos\frac{\alpha}{2} + \cos\frac{\beta}{2} = \frac{1}{2}\}$. Moreover, for all points

$(\alpha, \beta) \in R_e$ cone–manifold $6_2^2(\alpha, \beta)$ admits Euclidean structure. If $(\alpha, \beta) \in R_h$ then the lengths $l_\alpha$ and $l_\beta$ of are defined by Proposition 1.3.2. Hence

$$\cosh \frac{l_\alpha}{2} = 2\sin^2\mu\cos\nu + \cos\mu\sqrt{4\sin^2\mu\sin^2\nu + 1}, \qquad (2.3.1)$$

where $\mu = \frac{\alpha}{2}. \nu = \frac{\beta}{2}$ and the similar formula takes place for $\cosh \frac{l_\beta}{2}$. By the Schläfli formula for $V = V(\alpha, \beta) = Vol\, 6_2^2(\alpha, \beta)$ we get

$$dV = -\frac{l_\alpha}{2}d\alpha - \frac{l_\beta}{2}d\beta. \qquad (2.3.2)$$

Choose a path of integration $\gamma$ to be a segment with terminal points $(\alpha, \beta)$ and $(\alpha^*, \beta)$, where $(\alpha^*, \beta) \in R_e$ and note that $6_2^2(\alpha^*, \beta)$ is the Euclidean cone–manifold. Along the path $\gamma$ we have $\beta \equiv const$ and (2.3.2) reduces to $dV = -\frac{l_\alpha}{2}d\alpha$. By Theorem 7.1.2 in [Kj] we obtain $V(\alpha, \beta) \to 0$ as $\alpha \to \alpha^*$. Hence by (2.3.1)

$$V(\alpha, \beta) = \int_{\alpha^*}^{\alpha} -\frac{l_\alpha}{2}d\alpha = \int_{\alpha}^{\alpha^*} E(\frac{\alpha}{2}, \frac{\beta}{2})d\alpha,$$

where $E(\mu, \nu)$ is the same as in the statement of the theorem.

**Corollary 2.3.2.** The figure eight cone–manifold $4_1(\alpha)$ is hyperbolic for $0 \le \alpha < \frac{2\pi}{3}$. The hyperbolic volume of $4_1(\alpha)$ is given by the formula

$$Vol\, 4_1(\alpha) = \int_{\alpha}^{\frac{2\pi}{3}} arcosh\,(1 + \cos t - \cos 2t)dt.$$

Proof. We have $2E(\frac{\alpha}{2}, \frac{\pi}{2}) = arcosh\,(1+\cos\alpha-\cos 2\alpha)$ and $V(\alpha, \pi) = \int_{\alpha}^{\frac{2\pi}{3}} E(\frac{\alpha}{2}, \frac{\pi}{2})d\alpha$. Since the figure eight cone–manifold $4_1(\alpha)$ is two-fold covering of $V(\alpha, \pi)$, by Proposition 1.3.1 it is hyperbolic for $0 \le \alpha < \frac{2\pi}{3}$. We get

$$Vol\, 4_1(\alpha) = 2V(\alpha, \pi) = \int_{\alpha}^{\frac{2\pi}{3}} arcosh\,(1 + \cos t - \cos 2t)dt.$$

We remark that equivalent but more complicated formulas for $Vol\, 4_1(\alpha)$ were obtained in [HLM3], [MV], and [Kj].

**Corollary 2.3.3.** The hyperbolic volume of the Borromean rings cone–manifold $B(\alpha, \alpha, \alpha)$, $0 \le \alpha < \pi$ is given by the formula

$$Vol\, B(\alpha, \alpha, \alpha) = 12\int_0^{\cos\frac{\alpha}{2}} arcosh\,\frac{u + \sqrt{4 - 3u^2}}{2}\frac{du}{\sqrt{1 - u^2}}.$$

Proof is based on the fact that $B(\alpha, \alpha, \alpha)$ is the three-fold covering of the cone–manifold $6_2^2(\alpha, \frac{2\pi}{3})$ and on equality $E(\frac{\alpha}{2}, \frac{2\pi}{3}) = 2\, arcosh\, \frac{u+\sqrt{4-3u^2}}{2}$, where $u = \cos\frac{\alpha}{2}$.

It was noted in the subsection 1.4 that the volume of the hyperbolic cone–manifold $B(\alpha,\beta,\gamma)$, $0 \le \alpha,\beta,\gamma < \pi$ is eight times the volume of the Lambert cube $L(\frac{\alpha}{2},\frac{\beta}{2},\frac{\gamma}{2})$. Hence, according to [K], $Vol\, B(\alpha,\beta,\gamma)$ can be obtain as a linear combination of eight Lobachevsky functions. The analogues of Theorem 2.3.1 and its corollaries can be obtained also in the spherical geometry. We restrict ourselves by the statement of a spherical analog of the Corollary 2.3.2 ([MR2]).

**Theorem 2.3.4.** The figure eight cone–manifold $4_1(\alpha)$ is spherical for $\frac{2\pi}{3} < \alpha < \frac{4\pi}{3}$. The spherical volume of $4_1(\alpha)$ is given by the formula

$$Vol\, 4_1(\alpha) = \int_{\frac{2\pi}{3}}^{\alpha} arccos\,(1+\cos t - \cos 2t)dt, \; \frac{2\pi}{3} < \alpha \le \pi$$

and

$$Vol\, 4_1(\alpha) = 2\pi(\alpha - 0.9\pi) - \int_{\pi}^{\alpha} arccos\,(1+\cos t - \cos 2t)dt, \; \pi < \alpha < \frac{4\pi}{3}.$$

Earlier [HLM3] the existence of the spherical structure on $4_1(\alpha)$ was established only for $\frac{2\pi}{3} < \alpha \le \pi$.

**3. Geometrical inequalities**

As above denote by $V(\alpha)$ and $l(\alpha)$ the volume and the singular geodesic length of the figure eight cone–manifold $4_1(\alpha)$. It follows from Theorem 2.3.4 and Corollary 2.3.2 (see also [HLM3]) that $V(\alpha) \to 0$ as $\alpha \to \frac{2\pi}{3} \pm 0$. Hence "the Euclidean" volume $V(\frac{2\pi}{3}) = 0$. Certainly, this contradicts the geometric intuition. To avoid this phenomenon we introduce the notion of *specific volume* $v(\alpha)$. By definition $v(\alpha) = \frac{V(\alpha)}{l^3(\alpha)}$. In particular, in the Euclidean case for $\alpha = \alpha_0 = \frac{2\pi}{3}$ we get from [MR1]

$$v_0 = v(\alpha_0) = \frac{\sqrt{3}}{108}.$$

Explicit volume formulas obtained in Section 2 ensure that the specific volume function $v(\alpha)$ is continuous for all $0 < \alpha < \frac{4\pi}{3}$.

**Theorem 3.1.** *Let $4_1(\alpha)$ be the figure eight cone–manifold with cone angle $\alpha$. Denote by $V(\alpha)$ the volume and by $l(\alpha)$ the length of the singular geodesic of $4_1(\alpha)$. Then*

$$(i) \quad V(\alpha) > v_0\, l^3(\alpha), \; if\; 0 < \alpha < \alpha_0 = \frac{2\pi}{3} \; (hyperbolic\ case)$$

$$(ii) \quad V(\alpha) = v_0\, l^3(\alpha), \quad if\; \alpha = \alpha_0 \qquad (Euclidean\ case)$$

$$(iii) \quad V(\alpha) < v_0\, l^3(\alpha), \; if\; \alpha_0 < \alpha < 2\alpha_0 \quad (spherical\ case),$$

*where $v_0 = \frac{\sqrt{3}}{108}$ is the specific volume of $4_1(\alpha_0)$.*

Proof. Case (ii) immediately follows from the definition of $v_0 = v(\alpha_0)$.

To prove (i) we will show that

$$(V(\alpha) - v_0\, l^3(\alpha))' < 0,\ 0 < \alpha < \alpha_0.$$

Since $l(\alpha) > 0$ and by the Schläfli formula $V'(\alpha) = -\frac{1}{2}l(\alpha)$, the last inequality is equivalent to

$$1 + 6v_0\, l(\alpha)l'(\alpha) > 0,\ 0 < \alpha < \alpha_0.$$

By Corollary 2.3.2 we have $l(\alpha) = 2\,arcosh\,(1+\cos\alpha - \cos 2\alpha)$, and the inequality is verified by straightforward calculation.

In case (iii) we need to prove

$$(V(\alpha) - v_0\, l(\alpha))' > 0,\ \alpha_0 < \alpha < 2\alpha_0.$$

By the Schläfli formula and Proposition 2.3.4 we have

$$V'(\alpha) = \frac{1}{2}l(\alpha) = arccosh\,(1 + \cos\alpha - \cos 2\alpha).$$

Again, the inequality is a routine consequence of these formulas.

The next theorem can be proved by similar arguments.

**Theorem 3.2.** *The Borromean rings cone–manifold $B(\alpha,\alpha,\alpha)$ is hyperbolic for $0 < \alpha < \pi$, Euclidean for $\alpha = \pi$, and spherical for $\pi < \alpha < 2\pi$. Denote by $V(\alpha)$ the volume and by $L(\alpha)$ the half length of a singular geodesic of $B(\alpha,\alpha,\alpha)$. Then*

$$(i) \quad V(\alpha) > L^3(\alpha),\ 0 < \alpha < \pi$$

$$(ii) \quad V(\alpha) = L^3(\alpha), \qquad \alpha = \pi$$

$$(iii)\ V(\alpha) < L^3(\alpha),\ \pi < \alpha < 2\pi.$$

Different arguments are needed to obtain the following result

**Theorem 3.3.** *Let $B(\alpha,\beta,\gamma)$, $0 < \alpha,\beta,\gamma < \pi$ be a hyperbolic Borromean rings cone–manifold. Denote by $V(\alpha,\beta,\gamma)$ the volume and by $L(\alpha), L(\beta), L(\gamma)$ half lengths of singular geodesics of $B(\alpha,\beta,\gamma)$ with cone angles $\alpha,\beta,\gamma$ respectively. Then*

$$V(0,0,0) > V(\alpha,\beta,\gamma) > L(\alpha)L(\beta)L(\gamma).$$

Proof. In 1928 Grötsch [G] has proved the following theorem: *Let $\mathcal{S}$ be the conformal image of a square and let $A, B, C, D$ be the images of the sides of the square traversed in a clockwise direction. Suppose the distance between the "opposite" sides $A$ and $C$ of $\mathcal{S}$ is $a$, and between $B$ and $D$ is $b$. Then the area of $\mathcal{S}$ can be no smaller then $ab$.*

The $n$–dimensional analog of the Grötsch theorem for an arbitrary Riemannian metric was obtained in ([BZ], Theorem 8.2.1). In particular, it follows from [BZ] that hyperbolic volume of the Lambert cube $L(\frac{\alpha}{2}, \frac{\beta}{2}, \frac{\gamma}{2})$ with essential edges $L_\alpha, L_\beta, L_\gamma$ satisfies the inequality

$$Vol\, L(\frac{\alpha}{2}, \frac{\beta}{2}, \frac{\gamma}{2}) > L_\alpha L_\beta L_\gamma. \qquad (3.1)$$

In the Euclidean case $\alpha = \beta = \gamma = \pi$ we get the equality $Vol\, L(\frac{\pi}{2}, \frac{\pi}{2}, \frac{\pi}{2}) = L_\pi^3$.

Since by the Schläfli theorem $\frac{\partial}{\partial \alpha} V(\alpha, \beta, \gamma) = -L(\alpha) < 0$, $0 < \alpha < \pi$, the upper bound $V(\alpha, \beta, \gamma) < V(0, 0, 0)$ is established. We recall from section 2.4 that

$$V(\alpha, \beta, \gamma) = 8\, Vol\, L(\frac{\alpha}{2}, \frac{\beta}{2}, \frac{\gamma}{2}),\ L_\alpha = 2L(\alpha),\ L_\beta = 2L(\beta),\ L_\gamma = 2L(\gamma).$$

Hence, the lower bound follows from inequality (3.1).

No doubt the spherical analog of the theorem takes place too. It will be obtain anywhere more.

**Remark 3.4.** We note that the length $L(\alpha)$ in Theorem 3.2 is bounded above by $\log 3$. The equality $L(\alpha) = \log 3$ holds for $\cos \alpha = -\frac{1}{3}$. Contrary, the length $L(\alpha)$ in Theorem 3.3 is unbounded. More precisely, $L(\alpha) \to +\infty$, $L(\beta) \to 0$, $L(\gamma) \to 0$ as $\alpha \to \frac{\pi}{2}$, $\beta \to 0$, $\gamma \to 0$. The proof of these properties immediately follows from the Tangent Rule (Theorem 1.4.1).

# References


[BZ] Yu. D. Burago, V. A. Zalgaller, *Geometric inequalities,* Grundlehren der Mathematischen Wissenschaften, 285, Berlin ets., Springer-Verlag, 1988.

[BZie] G. Burde, H. Zieschang, *Knots,* De Gruyter Studies in Mathematics, Berlin - New-York, 1985.

[CCGLS] D. Cooper, M. Culler, H. Gillet, D. D. Long and P. B. Shalen, *Plane curves associated to character varieties of 3-manifolds,* Invent. Math., vol. 118 (1994), 47-84.

[G] H. Grötsch, *Extremalprobleme der Konformen Abbildung I,* Berichte Akad. Leipzig, 80, 1928, 367-376.

[Fench] W. Fenchel, *Elementary Geometry in Hyperbolic Space,* Walter de Gruyter, Berlin–New York 1989.

[HKW] P. de la Harpe, M. Kervaire and C. Weber, *On the Jones Polynomial,* L'Enseignement Mathématique, Vol.32 (1986), 271–335.

[HLM1] H. M. Hilden, M. T. Lozano, J. M. Montesinos-Amilibia, *On the Borromean orbifolds: geometry and arithmetic,* in: Topology'90, eds. B. A. Apanasov, W. Newmann, A. Reid, L. Siebenmann, de Gruyter, Berlin, 1992, 133–167.

[HLM2] H. M. Hilden, M. T. Lozano, J. M. Montesinos-Amilibia, *On the arithmetic 2-bridge knots and link orbifolds and a new knot invariant,* Journal of Knots and its Ramification, vol.4., no.1, 1995, 81-114.

[HLM3] H. M. Hilden, M. T. Lozano, J. M. Montesinos-Amilibia, *On a remarkable polyhedron geometrizing the figure eight cone manifolds,* J. Math. Sci. Univ. Tokyo, vol. 2, 1995, 501–561.

[Hds] C. D. Hodgson, *Schläfli revisited: Variation of volume in constant curvature spaces,* Preprint.

[Kauf] L. Kauffman, *New Invariants in the Theory of Knots,* Amer. Math. Monthly, vol. 95, no. 3, 195-242.

[K] R. Kellerhals, *On the volume of hyperbolic polyhedra,* Math. Ann., vol. 28 (1989), 245–270.

[KM] A.-Ch. Kim, A. Mednykh, *On the hyperbolic structure on the Whitehead link cone–manifold,* Preprint, 1999.

[Kj] S. Kojima, *Deformations of hyperbolic 3-cone-manifolds,* J. Differential Geometry, Vol. 49 (1998) 469-516.

[Kn] C. G. Knott, *Life and Scientific Work of P.G.Tait,* Cambridge University Press (1911).

[Ma] J. C. Maxwell *A treatise on eletricity and magmetism,* Oxford (1883).

[M] A. Mednykh, *On the the Remarkable Properties of the Hyperbolic Whitehead Link Cone–Manifolds,* Universität Bielefeld, Sonderforschungsbereich 343, "Discrete Structuren in der Mathematik", Preprint, 99–039.

[MR1] A. Mednykh, A. Rasskazov, *On the structure of the canonical fundamental set for the 2-bridge link orbifolds,* Universität Bielefeld, Sonderforschungsbereich 343, "Discrete Structuren in der Mathematik", Preprint, 98–062.

[MR2] A. Mednykh, A. Rasskazov, *On the degenaration of geometric structures on the figure eight cone–manifold,* Preprint, 1999

[MV] A. Mednykh, A. Vesnin, *Hyperbolic volumes of Fibonacci manifolds,* Siberian Math. J.36, 1995, 235–245.

[R] R. Riley, *An elliptical path from parabolic representations to hyperbolic structure,* in: Topology of Low-Dimension manifolds, LNM, 722, Springer-Verlag, 1979, 99-133.

[Rolf] D. Rolfsen, *Knots and links,* Publish of Perish Inc., Berkeley Ca., 1976.

[T] W. P. Thurston, *The geometry and topology of three-manifolds,* Princeton University Press (1979).

[Zhou] Qing Zhou, *The Moduli Space of Hyperbolic Cone Structures,* J. Diferential Geometry, Vol. 51 (1999), 517-550.

[V] E. B. Vinberg, *Geometry II,* Encyclopeadia of Mathematical Sciences, Vol.29, Springer Verlag, 1992.

# Generalized functions on the sphere and heat functions *

Mitsuo Morimoto

*Department of Mathematics, International Christian University,*

*Osawa 3-10-2, Mitaka-shi, Tokyo, 181-8585, Japan*

*E-mail: mitsuo@icu.ac.jp*

**Abstract**

First we shall review our results on the characterization of generalized functions on the sphere as initial values of heat functions ([5] and [6]). Second the space of initial values of heat functions will be determined (Theorem 6).

# 1 Hyperfunctions on the sphere

Let $\mathbb{S} = \{x = (x_1, x_2, \cdots, x_{n+1}) \in \mathbb{R}^{n+1}; x^2 = x_1^2 + x_2^2 + \cdots + x_{n+1}^2 = 1\}$ be the unit sphere of $n$ dimension. We denote by $\mathcal{B}(\mathbb{S})$ the space of hyperfunctions (or analytic functionals) on $\mathbb{S}$; i.e., $\mathcal{B}(\mathbb{S}) = \mathcal{A}'(\mathbb{S})$, where $\mathcal{A}(\mathbb{S})$ is the space of real analytic functions on $\mathbb{S}$ (see [4]).

Let $\mathcal{H}_k(\mathbb{S})$ be the space of $k$-th spherical harmonics on $\mathbb{S}$ and $N(k,n) = \dim \mathcal{H}_k(\mathbb{S})$. A real analytic function $f \in \mathcal{A}(\mathbb{S})$ can be expanded into the spherical harmonics:

$$f(x) = \sum_{k=0}^{\infty} f_k(x), \quad x \in \mathbb{S}. \tag{1}$$

The spherical harmonic component $f_k$ of $f$ is given by the integral over $\mathbb{S}$:

$$f_k(x) = N(k,n) \int_{\mathbb{S}} P_{k,n}(x \cdot y) f(y) dy,$$

*Ther Third Asian Mathematical Conference (AMC2000), October 23–27, 2000, Manila

where $P_{k,n}(t)$ is the Legendre polynomial of degree $k$.

**Theorem 1.** *For $f \in \mathcal{A}(\mathbb{S})$ the spherical harmonic expansion* (1) *converges in the topology of $\mathcal{A}(\mathbb{S})$. $f \in \mathcal{A}(\mathbb{S})$ if and only if* $\limsup_{k\to\infty} \sqrt[k]{\|u_k\|_{\mathbb{S}}} < 1$, *where $\|u_k\|_{\mathbb{S}}$ denotes the supremum norm on $\mathbb{S}$.*

Let $u \in \mathcal{B}(\mathbb{S})$. Then for $f \in \mathcal{A}(\mathbb{S})$ we have

$$\begin{aligned}\langle u, f\rangle &= \sum_{k=0}^{\infty} \langle u, f_k\rangle \\ &= \sum_{k=0}^{\infty} \langle u_x, N(k,n) \int_{\mathbb{S}} P_{k,n}(x\cdot y) f(y)dy\rangle \\ &= \sum_{k=0}^{\infty} N(k,n) \int_{\mathbb{S}} \langle u_x, P_{k,n}(x\cdot y)\rangle f(y)dy,\end{aligned}$$

which motivates us to define the $k$-spherical harmonic component $u_k$ of $u$ by

$$u_k(y) = N(k,n)\langle u_x, P_{k,n}(x\cdot y)\rangle. \tag{2}$$

Then we have $u_k \in \mathcal{H}_k(\mathbb{S})$ and $\langle u, f\rangle = \sum_{k=0}^{\infty} \int_{\mathbb{S}} u_k(y) f(y)dy$, which we denote by

$$u = \sum_{k=0}^{\infty} u_k. \tag{3}$$

**Theorem 2.** *For $u \in \mathcal{B}(\mathbb{S})$ the spherical harmonic expansion* (3) *converges in the topology of $\mathcal{B}(\mathbb{S}) = \mathcal{A}'(\mathbb{S})$. $u \in \mathcal{B}(\mathbb{S})$ if and only if*

$$\limsup_{k\to\infty} \sqrt[k]{\|u_k\|_{\mathbb{S}}} \le 1. \tag{4}$$

For the proof of Theorems 1 and 2 see [4].

A spherical harmonic function $u_k$ of degree $k$ is an eigenfunction of the Laplace-Beltrami operator $\Delta_{\mathbb{S}}$ on $\mathbb{S}$:

$$\Delta_{\mathbb{S}} u_k = -k(k+n-1)u_k \text{ for } u_k \in \mathcal{H}_k(\mathbb{S}). \tag{5}$$

Given a hyperfunction (3), consider the function

$$U(x,t) = \sum_{k=0}^{\infty} e^{-k(k+n-1)t} u_k(x), \quad x \in \mathbb{S},\ t > 0. \tag{6}$$

Thanks to (5) $U(x,t)$ satisfies the heat equation

$$\left(\frac{\partial}{\partial t} - \Delta_{\mathbb{S}}\right) U(x,t) = 0, \tag{7}$$

for $x \in \mathbb{S}$, $t > 0$, with the initial condition $U(x,0) = u(x)$. A solution of the heat equation (7) will be called a heat function.

If we define the heat kernel for $\mathbb{S}$ by

$$E(x,y,t) = \sum_{k=0}^{\infty} N(k,n) e^{-k(k+n-1)t} P_{k,n}(x \cdot y),$$

we can write $U(x,t) = \int_{\mathbb{S}} E(x,y,t) u(y) dy$.

We can prove the following theorem by means of (4) and (6):

**Theorem 3 ([6]).** *Let $u(x)$ be given by (3) and $U(x,t)$ by (6). The followings are equivalent:*
(a) $u(x) \in \mathcal{B}(\mathbb{S})$.
(b) *For all $\epsilon > 0$ there exists $C \geq 0$ such that*

$$\|U(x,t)\|_{\mathbb{S}} \leq C e^{\epsilon/t}, \quad 0 < t \leq 1.$$

This fact for the Euclidean space $\mathbb{R}^n$ was proved by Matsuzawa [3].

# 2 Ultradistributions on the sphere

For $u = \sum_{k=0}^{\infty} u_k(x) \in \mathcal{B}(\mathbb{S})$ we define the fractional power of $(-\Delta_{\mathbb{S}})$ as follows:

$$(-\Delta_{\mathbb{S}})^{(m/2)} u = \sum_{k=0}^{\infty} (k(k+n-1))^{(m/2)} u_k,$$

where $m = 0, 1, 2, \cdots$. For $s > 0$ and $h > 0$ we put

$$\|u\|_{s,h} = \sup_m \frac{1}{(m!)^s h^m} \|(-\Delta_{\mathbb{S}})^{(m/2)} u\|_{L^2}$$

The space $\mathcal{E}_{s,h}(\mathbb{S}) = \{u \in \mathcal{B}(\mathbb{S}); \|u\|_{s,h} < \infty\}$ is a Banach space with the norm $\|u\|_{s,h}$. The space

$$\mathcal{E}_{\{s\}}(\mathbb{S}) = \text{ind} \lim_{h\to\infty} \mathcal{E}_{s,h}(\mathbb{S}) = \bigcup_{h>0} \mathcal{E}_{s,h}(\mathbb{S})$$

is a DFS space. If $s > 1$, then $\mathcal{E}_{\{s\}}(\mathbb{S})$ is a space of non quasi-analytic functions of Gevrey class. If $s = 1$, then we have $\mathcal{E}_{\{1\}}(\mathbb{S}) = \mathcal{A}(\mathbb{S})$. If $0 < s < 1$, then $\mathcal{E}_{\{s\}}(\mathbb{S}) \subset \mathrm{Exp}(\tilde{\mathbb{S}}; (0))$, where $\mathrm{Exp}(\tilde{\mathbb{S}}; (0))$ is the space of entire functions on the complex sphere $\tilde{\mathbb{S}}$ of exponential type 0. Note that the complex sphere $\tilde{\mathbb{S}}$ is defined by $\tilde{\mathbb{S}} = \{z \in \mathbb{C}^{n+1}; z^2 = z_1^2 + z_2^2 + \cdots + z_{n+1}^2 = 1\}$.

**Theorem 4 ([2]).** *The followings are equivalent:*
(a) $f = \sum_{k=0}^{\infty} f_k \in \mathcal{E}_{\{s\}}(\mathbb{S})$.
(b) *For some* $M \geq 0$ *and* $C \geq 0$ *we have* $\|f_k\|_{\mathbb{S}} \leq C \exp(-Mk^{1/s})$ *for* $k = 0, 1, 2, \cdots$.
(c) *For some* $M \geq 0$ *we have* $\lim_{k\to\infty} \|f_k\|_{\mathbb{S}} \exp(Mk^{1/s}) = 0$.

Denote by $\mathcal{E}'_{\{s\}}(\mathbb{S})$ the dual space of $\mathcal{E}_{\{s\}}(\mathbb{S})$. An element of $\mathcal{E}'_{\{s\}}(\mathbb{S})$ is called a ultradistribution if $s > 1$, a hyperfunction if $s = 1$ and an entire functional if $0 < s < 1$.

Let $f \in \mathcal{E}_{\{s\}}(\mathbb{S})$ with $f = \sum_{k=0}^{\infty} f_k$. The $k$-th spherical component $u_k$ of $u \in \mathcal{E}'_{\{s\}}(\mathbb{S})$ is defined by (2). Then we have $\langle u, f\rangle = \sum_{k=0}^{\infty} \int_{\mathbb{S}} u_k(y) f(y) dy$, which we shall denote by $u = \sum_{k=0}^{\infty} u_k$ and call the spherical harmonic expansion of $u$.

**Theorem 5 ([2]).** *The followings are equivalent:*
(a) $u = \sum_{k=0}^{\infty} u_k \in \mathcal{E}'_{\{s\}}(\mathbb{S})$.
(b) *For any* $M > 0$ *there exists* $C_M \geq 0$ *such that* $\|u_k\|_{\mathbb{S}} \leq C_M \exp(Mk^{1/s})$ *for* $k = 0, 1, 2, \cdots$.
(c) *For any* $M \geq 0$ *we have* $\lim_{k\to\infty} \|u_k\|_{\mathbb{S}} \exp(-Mk^{1/s}) = 0$.

Let $B = \{x = (x_1, x_2, \cdots, x_{n+1}); x^2 < 1\}$ be the unit open ball in $\mathbb{R}^{n+1}$. The sphere $\mathbb{S}$ is the boundary of $B$. A function $F \in \mathcal{A}(B)$ is called harmonic if $\Delta F = 0$, $\Delta$ being the Laplacian. For a harmonic function $F$ on $B$ we can define the hyperfunction boundary value $F|_{\mathbb{S}} \in \mathcal{B}(\mathbb{S})$. Conversely, for any hyperfunction $u \in \mathcal{B}(\mathbb{S})$ there is a unique harmonic function $F$ with $F|_{\mathbb{S}} = u$. We can summerize this fact saying that there is a one-to-one correspondence between harmonic functions on $B$ and hyperfunctions on $\mathbb{S}$.

The following theorem answer the similar question about heat functions and hyperfunctions.

**Theorem 6.** *Let* (3) *be a formal sum of spherical harmonics* $u_k \in \mathcal{H}_k(\mathbb{S})$ *and* $U(x,t)$ *be given by* (6).

*If* (3) *is a spherical harmonic expansion of* $u \in \mathcal{E}'_{\{1/2\}}(\mathbb{S})$, *then* (6) *converges and defines a heat function* $U(x,t)$ *on* $\mathbb{S} \times \{t > 0\}$.

*Conversely, suppose a heat function $U(x,t)$ on $\mathbb{S}\times\{t>0\}$ is given. Then it can be expanded in* (6) *with some $u_k \in \mathcal{H}_k(\mathbb{S})$, which defines $u \in \mathcal{E}'_{\{1/2\}}(\mathbb{S})$ by* (3).

In sum, we can say that $u \in \mathcal{E}'_{\{1/2\}}(\mathbb{S})$ if and only if $U(x,t)$ is a heat function on $\mathbb{S}\times\{t>0\}$.

*Proof.* Suppose $u \in \mathcal{E}'_{\{1/2\}}(\mathbb{S})$. Then by Theorem 5, for any $M>0$ there is $C_M>0$ such that $\|u_k\|_{\mathbb{S}} \le C_M \exp(Mk^2)$. If $t>M$, then we have

$$\sum_{k=0}^{\infty} e^{-k(k+n-1)t}\|u_k\|_{\mathbb{S}} \le C_M \sum_{k=0}^{\infty} \exp(-k(k+n-1)t+Mk^2) < \infty.$$

Therefore, (6) converges absolutely for $x \in \mathbb{S}$ and $t>M$ and defines a heat function there. $M>0$ being arbitrary, (6) defines a heat function $U(x,t)$ on $\mathbb{S}\times\{t>0\}$ with the unitial condtion $u(x)$.

Conversely, let $U(x,t)$ be defined for $x\in\mathbb{S}$ and $t>0$ and satisfies the heat equation (7). For $t>0$ fixed $U(x,t)$ can be expanded in spherical harmonics:

$$U(x,t)=\sum_{k=0}^{\infty} u_k(x,t), \tag{8}$$

where $u_k(x,t)\in\mathcal{H}_k(\mathbb{S})$ is given by

$$u_k(x,t)=N(k,n)\int_{\mathbb{S}} P_{k,n}(x\cdot y)U(y,t)dy.$$

Because $U(x,t)$ satisfies the heat equation (7), we have, for fixed $x\in\mathbb{S}$,

$$\begin{aligned}\frac{\partial}{\partial t}u_k(x,t) &= N(k,n)\int_{\mathbb{S}} P_{k,n}(x\cdot y)\frac{\partial}{\partial t}U(y,t)dy\\ &= N(k,n)\int_{\mathbb{S}} P_{k,n}(x\cdot y)\Delta_{\mathbb{S},y}U(y,t)dy\\ &= N(k,n)\int_{\mathbb{S}} \Delta_{\mathbb{S},y}P_{k,n}(x\cdot y)U(y,t)dy\\ &= -k(k+n-1)N(k,n)\int_{\mathbb{S}} P_{k,n}(x\cdot y)U(y,t)dy\\ &= -k(k+n-1)u_k(x,t).\end{aligned}$$

Solving this ordinary differential equation we can find a function $u_k(x)$ such that $u_k(x,t) = e^{-k(k+n-1)t}u_k(x)$. Fixing $t > 0$ we can see $u_k(x) \in \mathcal{H}_k(\mathbb{S})$. Then (8) becomes (6). Because for $t > 0$ fixed the function $x \mapsto U(x,t)$ is real analytic on $\mathbb{S}$, we have $\lim_{k\to\infty} \|u_k\|_{\mathbb{S}} \exp(-k(k+n-1)t) = 0$. By Theorem 5, we can conclude $u(x) = \sum_{k=0}^{\infty} u_k(x) \in \mathcal{E}'_{\{1/2\}}(\mathbb{S})$. □

Theorem 3 on hyperfunctions can be generalized as follows:

**Theorem 7 ([6]).** *Suppose $s > 1/2$. Let (3) be a formal sum of spherical harmonics and $U(x,t)$ be given by (6). Then the followings are equivalent:*
(a) $u \in \mathcal{E}'_{\{s\}}(\mathbb{S})$.
(b) *For any $\epsilon > 0$ there exists $C \geq 0$ such that*

$$\|U(x,t)\|_{\mathbb{S}} \leq C \exp((\epsilon/t)^{1/(2s-1)}), \quad 0 < t \leq 1.$$

For the Euclidean case S-Y. Chung and D. Kim ([1]) conjectured that Theorem would be valid without the condition "$s > 1/2$". But 1/2 is the best value for $s$.

# References

[1] S-Y. Chung and D. Kim, Representation of quasianalytic ultradistributions, Arkiv for mathematik, 31(1993), 51 – 60.

[2] K. Fujita and M. Morimoto, Gevrey classes on compact real analytic Riemannian manifolds, Tokyo J. Math., 18(1995), 341 – 355.

[3] T. Matsuzawa, A calculus approach to hyperfunctions II, Trans. Amer. Math. Soc., 313(1989), 619-654.

[4] M. Morimoto, Analytic Functionals on the Sphere, Translations of Mathematical Monographs, Volume 178, AMS, 1998.

[5] M. Morimoto and M. Suwa, A characterication of analytic functionals on the sphere I, Finite or Infinite Diemnsional Complex Analysis, Lecture Notes in Pure & Applied Mathematics, Marcel Dekkar, 214(2000), pp 331 – 336.

[6] M. Morimoto and M. Suwa, A characterization of analytic functionals on the sphere II, Proceedings of the Second ISAAC Congress, Vol. 2, Kluwer Academic Publ., 2000, pp 799 – 807.

# ESSENTIAL SPECTRUM AND NULLITIES PRESERVED UNDER AN ADMISSIBLE PERTURBATION

MARIE REDINA L. MUMPAR-VICTORIA
Department of Mathematics
University of the Philippines
Diliman, Quezon City, Philippines
eden@math01.cs.upd.edu.ph

**Abstract**

This paper will show the derivation of an admissible perturbation on Special Expressions. This perturbation which will turn out to be relatively compact, preserves the essential spectrum and the nullities of the Special Expression referred to.

## 1 Special Expressions

Consider the differential expression of the form

$$\mathbf{M_o}f \;:=\; \sum_{\sigma=0}^{r} a_\sigma\, t^{\alpha_\sigma}\, f^{(\rho_\sigma)} \tag{1}$$

with $r \in \mathbb{N}$, $\rho_0, \cdots, \rho_r \in \mathbb{N}_0$, and $\alpha_\sigma \in \mathbb{R}$ $(\sigma = 0, \cdots, r)$ such that

$$\alpha_0 = 0, \;\; \alpha_1 \leq \rho_1 \tag{2}$$

and

$$1 \geq \frac{\alpha_\sigma - \alpha_{\sigma-1}}{\rho_\sigma - \rho_{\sigma-1}} \geq \frac{\alpha_{\sigma+1} - \alpha_\sigma}{\rho_{\sigma+1} - \rho_\sigma} \tag{3}$$

for $1, \cdots, r-1$ if $r > 1$

Denote by $\sigma_1 < \sigma_2 < \cdots < \sigma_{s-1}$ those indices $\sigma$ $(\sigma = 1, \cdots, r-1)$ for which the strict inequality holds in (3).

Then together with $\sigma_0 := 0$ and $\sigma_s := r$, we have

$$\frac{\alpha_{\sigma_j} - \alpha_{\sigma_{j-1}}}{\rho_{\sigma_j} - \rho_{\sigma_{j-1}}} = \frac{\alpha_\sigma - \alpha_{\sigma-1}}{\rho_\sigma - \rho_{\sigma-1}} \tag{4}$$

for $\sigma_{j-1} < \sigma < \sigma_j$ $(j = 1, \cdots, s)$ if $s \geq 1$ and

$$\frac{\alpha_{\sigma_j} - \alpha_{\sigma_{j-1}}}{\rho_{\sigma_j} - \rho_{\sigma_{j-1}}} > \frac{\alpha_{\sigma_{j+1}} - \alpha_{\sigma_j}}{\rho_{\sigma_{j+1}} - \rho_{\sigma_j}} \tag{5}$$

for $j = 1, \cdots, s-1$ if $s > 2$

We assume the constants $\alpha_\sigma \in \mathbb{C} \setminus \{0\}$ to satisfy the following conditions:

$$a_\sigma \in \mathbb{R} \setminus \{0\} \text{ for } \sigma = \sigma_1, \cdots, r \text{ and}$$

for each $k = \rho_{\sigma_1}, \cdots, n$ we have

$$c_k := \sum_{\substack{\rho_k + \rho_\lambda = 2k \\ \sigma_i \leq k,\ \lambda \leq \sigma_{i+1}}} (-1)^{\rho_k + k} \; a_k \, a_\lambda \geq 0 \tag{6}$$

$$(\rho_{\sigma_i} \leq k \leq \rho_{\sigma_{i+1}}, i = 1, \cdots, s-1)$$

The $a_\sigma$ $(\sigma = 0, \cdots, \sigma_1 - 1)$ may be arbitrary complex constants.

A condition sufficient for (6) to hold is the following simpler condition:

$$\begin{aligned} \operatorname{sgn}((-1)^{\rho_\sigma/2} a_\sigma) &= \text{constant for all } \sigma \geq \sigma_1 \text{ such that } \rho_\sigma \text{ is even} \\ \operatorname{sgn}((-1)^{(\rho_\sigma+1)/2} a_\sigma) &= \text{constant for all } \sigma \geq \sigma_1 \text{ such that } \rho_\sigma \text{ is odd} \end{aligned} \tag{7}$$

**Definition 1.1** Differential equations of the form (1), satisfying (2), (3), and (6) is called a *special expression of order* $n$ defined on $[1, \infty)$

The expression given by

$$\mathbf{M_{o,o}} f = \sum_{\sigma=0}^{\sigma_1} a_\sigma t^{\alpha_\sigma} f^{(\rho_\sigma)}$$

is called the *essential part* of the special expression $\mathbf{M_o}$.

If we plot the points $(\rho_\sigma, \alpha_\sigma) \in \mathbb{R}^2$ $(\sigma = 0, \cdots, r)$ in the cartesian plane $\mathbb{R}^2$, and connect $(\rho_\sigma, \alpha_\sigma)$ and $(\rho_{\sigma+1}, \alpha_{\sigma+1})$ $(\sigma = 0, \cdots, r-1)$ by a line segment, then we will obtain a polygonal path with kink (corner) points $(\rho_{\sigma_i}, \alpha_{\sigma_i})$ $(i = 0, \cdots, s-1)$. We will call the $\sigma_i$'s the *kink indices*, and the polygonal path constructed, the *polygonal path generated by* $\mathbf{M_o}$.

The polygonal path generated by $\mathbf{M_{o,o}}$ lies on or below the bisectrix while the polygonal path generated by $\mathbf{M_o} - \mathbf{M_{o,o}}$ lies below the bisectrix.

**Definition 1.2** Define the function $\gamma : [0, n] \to \mathbb{R}$ by

$$\gamma(l) := \frac{2}{\rho_{\sigma_{i+1}} - \rho_{\sigma_i}} \{(l - \rho_{\sigma_i})\alpha_{\sigma_{i+1}} + (\rho_{\sigma_{i+1}} - l)\alpha_{\sigma_i}\}$$

where $\rho_{\sigma_i} \leq l \leq \rho_{\sigma_{i+1}}$.

Then the graph of $\gamma$ is precisely the *polygonal path generated by* $\mathbf{M_o}$.

# 2 Perturbations of Special Expressions

The perturbation given in this paper was derived from a lemma in Goldberg[1] specialized in the $\mathcal{L}_2-$ Hilbert space setting.

**Lemma 2.1** *Given the interval $I = [1, \infty)$ and $\epsilon > 0$, there exists a $\tilde{K}$, depending only on $\epsilon$, such that for all $r_l$ locally in $\mathcal{L}_2(I)$ and for all $f$ in the domain of the maximal operator $\mathbf{D}^{(l+1)}$ ($\mathbf{D} = \frac{d}{dt}, l \in \mathbb{N}_0$)*

$$\| r_l f^{(l)} \|_{2,I}^2 \;\leq\; (\epsilon \| f^{(l+1)} \|_{2,I}^2 + \tilde{K} \| f^{(l)} \|_{2,I}^2) \sup_{[x,x+1]\subset I} \int_x^{x+1} |r_l(t)|^2 \, dt \tag{8}$$

Let us now define an admissible perturbation of special expressions.

**Definition 2.2** Let $\mathbf{M}$ be a differential expression of the form

$$\mathbf{M}f \;=\; \sum_{l=0}^{n-1} r_l(t)\; f^{(l)} \tag{9}$$

We say that $\mathbf{M}$ is an *admissible perturbation* of the differential expression $\mathbf{M}_o$ if there exists a $B$ such that the coefficients $r_l$ satisfy the following:

$$\sup_{[x,x+1]\subset I} \int_x^{x+1} \left| \frac{r_l(t)}{s_l(t)} \right|^2 dt \;<\; B \tag{10}$$

where $r_l \in \mathbf{C}^l(I)$ for $l = 0, \cdots, n-1$ and $0 < s_l$ is an auxilliary function $s_l \in C^\infty(I)$ satisfying

$$s_l(t) = o\left(t^{\frac{1}{2}\gamma(l+1)}\right) \;\text{ and }\; s_l(t) = o\left(t^{\frac{1}{2}\gamma(l)}\right) \tag{11}$$

For the invariance of nullities, we can only admit a somewhat less general class of perturbations consisting of expressions (9) satisfying:

$$\sup_{[x,x+1]\subset I} \int_x^{x+1} \left| \frac{r_l^{(l-b)}(t)}{s_b(t)} \right|^2 dt \;<\; \tilde{B} \tag{12}$$

for $b = 0, \cdots, n-l, \;\; l = b, \cdots, n-l.$

**Remark 2.3** In the succeeding discussions, we will see that this admissible perturbation will turn out to be a relatively compact perturbation.

The following proposition of Schultze[4] will be used to prove the main result.

**Proposition 2.4** *If* $\mathbf{M_o}$ *is a special expression, then there are constants* $b_l > 0$ $(l = 0, \cdots, n)$, $K \geq 0$, *and* $\eta \in I$ *such that for all* $f \in C_0^\infty(\eta, \infty)$, *we have*

$$\| \mathbf{M_o} f \|_2^2 \geq \sum_{l=0}^{n} \int_I b_l \, t^{\gamma(l)} \, |f^{(l)}|^2 + (b_o - K) \, \| f \|_2^2 \tag{13}$$

With the following lemma we can show that (9)is a relatively bounded perturbation of the special expression $\mathbf{M_o}$ as in form (1).

**Lemma 2.5** *Let* $\mathbf{M_o}$ *be a special expression and* $\mathbf{M}$ *a corresponding expression, i.e., of the form* (9) *satisfying* (10) *and* (11). *Then there exist* $\eta \in I$, $0 < \alpha < 1$, $\beta \geq 0$ *such that for all* $f \in C_o^\infty(\eta, \infty)$, *we have*

$$\| \mathbf{M} f \|_2 \leq \alpha \, \| \mathbf{M_o} f \|_2 \; + \; \beta \, \| f \|_2 \tag{14}$$

**Remark 2.6** If $\mathbf{M}$ and $\mathbf{M_o}$ satisfy (14), then $\mathcal{D}\,(\mathbf{T_o}(\mathbf{M_o} + \mathbf{M})) = \mathcal{D}\,(\mathbf{T_o}(\mathbf{M_o}))$

The theorem below that is due to Kauffman[2] is the basic perturbation theorem that we will apply.

**Theorem 2.7** *Let* $\mathbf{M_o}$ *be given as in* (1) *and* $\mathcal{R}(\mathbf{T_o}(\mathbf{M}))$ *closed. Let* $\mathbf{M}$ *be another expression of the form* (9) *with order* $\mathbf{M}$ < *order* $\mathbf{M_o}$ *satisfying the following condition:*

*There is a* $g \in C([1, \infty))$, $g > 0$, *and* $\lim_{t \to \infty} g(t) = \infty$ *such that* $g\mathbf{M}f \in \mathcal{L}_2([1, \infty])$ *for all* $f \in \mathcal{D}(\mathbf{T_o}(\mathbf{M_o}))$. *Then the operator, defined as the restriction of* $\mathbf{M}$ *on* $\mathcal{D}(\mathbf{T_o}(\mathbf{M_o}))$, *is relatively compact with respect to* $\mathbf{T_o}(\mathbf{M_o})$, *and we have*

$$\mathcal{D}\,(\mathbf{T_o}(\mathbf{M_o} + \mathbf{M})) \; = \; \mathcal{D}\,(\mathbf{T_o}(\mathbf{M_o})),$$

$$\text{nul}\,(\mathbf{M}^+ + \mathbf{N}^+) \; = \; \text{nul}\,\mathbf{M}^+$$

$$\mathcal{R}(\mathbf{T_o}(\mathbf{M_o} + \mathbf{M})) \;\; \text{is closed.}$$

# 3 Main Results

The crucial lemma that will show the main result is the following:

**Lemma 3.1** *There exist* $0 < g \in C([1, \infty))$ *with* $\lim_{t \to \infty} g(t) = \infty$, *and constants* $\eta \in I$, $0 < \alpha < 1$, $\beta \geq 0$ *such that for* $f \in C_0^\infty(\eta, \infty)$ *and* $l = 0, \cdots, n-1$ *we have*

$$\left\| r_l \, g \, f^{(l)} \right\|_2 \; \leq \alpha \, \| \mathbf{M_o} f \|_2 \; + \; \beta \, \| f \|_2,$$

Proof:

From (11), we have

$$s_l(t) = o\left(t^{\frac{1}{2}\gamma(l+1)}\right) \text{ and } s_l(t) = o\left(t^{\frac{1}{2}\gamma(l)}\right)$$

for $l = 0, \cdots, n-1$.

Set

$$g(t) := \left[\frac{1}{t} + \sum_{l=0}^{n-1} |s_l(t)| \left(t^{-\frac{1}{2}\gamma(l)} + t^{-\frac{1}{2}\gamma(l+1)}\right)\right]^{-\frac{1}{2}}.$$

Then,

$$g(t)|s_l(t)| = o\left(t^{\frac{1}{2}\gamma(l)}\right) \text{ and } g(t)|s_l(t)| = o\left(t^{\frac{1}{2}\gamma(l+1)}\right)$$

for $l = 0, \cdots, n-1$.

Let $\tilde{\gamma}(t) = \min\{\gamma(l+1), \gamma(l)\}$. From Proposition 2.4, we have for $f \in C_0^\infty(\eta, \infty)$,

$$\begin{aligned}
\left\| r_l\, g\, f^{(l)} \right\|_2^2 &= \left\| \tfrac{r_l}{s_l}\, g\, (s_l\, f^{(l)}) \right\|_2^2 \\
&= \int_I \left|\frac{r_l}{s_l}\right|^2 |g(t)\, s_l(t)|^2\, |f^{(l)}|^2 \\
&\le \mu \int_I \left|\frac{r_l}{s_l}\right|^2 t^{\tilde{\gamma}(l)}\, |f^{(l)}|^2 \\
&\qquad \text{for any } \mu > 0 \text{ making } \eta \text{ large enough,} \\
&= \mu \left\| \tfrac{r_l}{s_l}\, t^{\frac{1}{2}\tilde{\gamma}(l)}\, f^{(l)} \right\|_2^2.
\end{aligned}$$

By using the argument as in the proof of Proposition 2.4, we obtain

$$\begin{aligned}
\left\| r_l\, g\, f^{(l)} \right\|_2^2 &\le \mu B \left(\epsilon \left\| \left(t^{\frac{1}{2}\tilde{\gamma}(l)}\, f^{(l)}\right)' \right\|_2^2 + \tilde{K} \left\| \left(t^{\frac{1}{2}\tilde{\gamma}(l)}\, f^{(l)}\right) \right\|_2^2\right) \\
&\qquad \text{with } \epsilon = 1 \text{ we get,} \\
&\le \mu B \left(2\left(\int_I |t^{\tilde{\gamma}(l)}|\, |f^{(l+1)}|^2 + \int_I |(t^{\frac{1}{2}\tilde{\gamma}(l)})'|^2\, |f^{(l)}|^2\right) + \right. \\
&\qquad \left. \tilde{K} \int_I |t^{\tilde{\gamma}(l)}|\, |f^{(l)}|^2\right) \\
&\le \mu B \left(2 \int t^{\gamma(l+1)}\, |f^{(l+1)}|^2 + (\delta + \tilde{K}) \int t^{\gamma(l)} |f^{(l)}|^2\right). \\
&\qquad \text{for arbitrary } \delta > 0 \\
&\le \mu B \left((\delta + \tilde{K})\, \| f \|_2^2 + (2 + \delta + \tilde{K}) \textstyle\int_I t^{\gamma(l)}\, |f^{(l)}|^2\right) \\
&\qquad \text{with } \gamma(0) = 0.
\end{aligned}$$

Choose $\mu > 0$ so small that $\mu < \frac{\alpha^2\, b_l}{B(2+\delta+\tilde{K})}$.

$$\begin{aligned}
\left\| r_l\, g\, f^{(l)} \right\|_2^2 &\le \alpha^2 \int b_l\, t^{\gamma(l)}\, |f^{(l)}|^2 \\
&\le [\, \alpha \, \| \mathbf{M_o} f \|_2 + \beta \, \| f \|_2 \,]^2.
\end{aligned}$$

Equivalently, we have

$$\left\| \, r_l \, g \, f^{(l)} \, \right\|_2 \;\le \alpha \parallel \mathbf{M_o} f \parallel_2 \; + \; \beta \parallel f \parallel_2 .$$

Theorem 3.2 give the invariance of the essential spectrum and the nullities under the relatively compact perturbation we have obtained.

**Theorem 3.2** *Let* $\mathbf{M_o}$ *be a special expression and* $\mathbf{M}$ *a corresponding perturbation, i.e., an expression of the form* (9) *satisfying* (10) *and* (11). *Then*

$$\sigma_e(\mathbf{M_o} + \mathbf{M}) \;=\; \sigma_e(\mathbf{M_o})$$

*In addition, if* $\mathbf{M}$ *satisfies* (3.9), *then for every* $\lambda \in \mathbb{C} \setminus \sigma_e(\mathbf{M_o})$,

$$\text{nul}\,(\mathbf{M_o} + \mathbf{M} - \lambda) = \text{nul}\,(\mathbf{M_o} - \lambda).$$

Outline of the Proof:

From the proof of Lemma 3.1, we have

$$s_l(t) = o\left(t^{\frac{1}{2}\gamma(l+1)}\right) \;\text{ and }\; s_l(t) = o\left(t^{\frac{1}{2}\gamma(l)}\right)$$

for $l = 0, \cdots, n-1$.

$$g(t) := \left[\frac{1}{t} + \sum_{l=0}^{n-1} |s_l(t)| \left(t^{-\frac{1}{2}\gamma(l)} + t^{-\frac{1}{2}\gamma(l+1)}\right)\right]^{-\frac{1}{2}}.$$

and,

$$g(t)|s_l(t)| = o\left(t^{\frac{1}{2}\gamma(l)}\right) \;\text{ and }\; g(t)|s_l(t)| = o\left(t^{\frac{1}{2}\gamma(l+1)}\right)$$

for $l = 0, \cdots, n-1$

then

$$\left\| \, r_l \, g \, f^{(l)} \, \right\|_2 \;\le \alpha \parallel \mathbf{M_o} f \parallel_2 \; + \; \beta \parallel f \parallel_2, \;\text{ for } f \in C_0^\infty(\eta, \infty)$$

With Theorem 2.7 and with $\mathbf{D} = \frac{d}{dt}$, it follows that

$$\begin{aligned}
\mathcal{D}\left(\mathbf{T_o}(\mathbf{M_o})\right) &= \mathcal{D}\left(\mathbf{T_o}(\mathbf{M_o} - \lambda)\right) \\
&= \mathcal{D}\left(\mathbf{T_o}(\mathbf{M_o} - \lambda) + \sum_{l=0}^{n-1} r_l(t) D^{(l)}\right) \\
&= \mathcal{D}\left(\mathbf{T_o}(\mathbf{M_o} + \mathbf{M} - \lambda)\right)
\end{aligned}$$

For h $\in \mathcal{D}\left(\mathbf{T_o}(\mathbf{M_o})\right)$, there exists a sequence $\{f_n\} \in C_0^\infty(\eta, \infty)$ such that $f_n \to h$ and $\mathbf{M_o} f_n \to \mathbf{M_o} h$. (This follows from the closedness of $\mathbf{T_o}$).

Then since $\{f_n\}$ and $\{\mathbf{M_o} f_n\}$ are Cauchy sequences and $f_n - f_m \in C_0^\infty(\eta, \infty)$ for all n and m, we have

$$\left\| \, g \, r_l \, (f_n^{(l)} - f_m^{(l)}) \, \right\|_2 \;\le\; \alpha \parallel \mathbf{M_o}(f_n - f_m) \parallel_2 + \beta \parallel f_n - f_m \parallel_2 \to 0$$

as n,m $\to \infty$

Hence $\left\{g\, r_l\, f_n^{(l)}\right\}$ converges in $\mathcal{L}_2(I)$ and therefore is also pointwise convergent almost everywhere. This implies that as n $\to \infty$, $\left\{g\, r_l\, f_n^{(l)}(t)\right\} \to g\, r_l\, h^{(l)}(t)$ almost everywhere, and so $g\, r_l\, h \in \mathcal{L}_2(I)$.

Theorem 2.7 gives for closed $\mathcal{R}(\mathbf{T_o}(\mathbf{M_o}-\lambda))$ that $\mathcal{R}(\mathbf{T_o}(\mathbf{M_o}+\mathbf{M}-\lambda))$ is closed and nul $((\mathbf{M_o}-\lambda)^+)$ $=$ nul $((\mathbf{M_o}+\mathbf{M}-\lambda)^+)$.

And if $\mathcal{R}(\mathbf{T_o}(\mathbf{M_o}+\mathbf{M}-\lambda))$ is closed, it follows in the same way that $\mathcal{R}(\mathbf{T_o}(\mathbf{M_o}-\lambda))$ is closed. Hence we have proven the first assertion, that is,

$$\sigma_e(\mathbf{M_o}+\mathbf{M}) \;=\; \sigma_e(\mathbf{M_o}).$$

However for the invariance of the nullities, what we have shown is that

$$\text{nul}\,(\mathbf{M_o}-\lambda)^+ \;=\; \text{nul}\,(\mathbf{M_o}+\mathbf{M}-\lambda)^+,$$

then to complete the proof of the theorem we must show that $\mathbf{M}^+$ is a relatively compact perturbation of a certain special expression.

Let us consider the constants $K_{\sigma,\tau}$ $(K_{\sigma,0}=1)$, we have

$$\begin{aligned}
\mathbf{M_o}^+ f &= \sum_{\sigma=0}^{r}(-1)^{\rho_\sigma}\,\bar{a}_\sigma\,(t^{\alpha_\sigma} f)^{(\rho_\sigma)} \\
&= \sum_{\sigma=0}^{r}(-1)^{\rho_\sigma}\,\bar{a}_\sigma \sum_{\tau=0}^{\rho_\sigma} K_{\sigma,\tau}\, t^{\alpha_\sigma-\tau}\, f^{(\rho_\sigma-\tau)} \\
&= \sum_{\sigma=0}^{r}(-1)^{\rho_\sigma}\,\bar{a}_\sigma\, t^{\alpha_\sigma}\, f^{(\rho_\sigma)} + \sum_{\sigma=0}^{\sigma_1}(-1)\bar{a}_\sigma \sum_{\tau=1}^{\rho_\sigma} K_{\sigma,\tau}\, t^{\alpha_\sigma-\tau}\, f^{(\rho_\sigma-\tau)} + \\
&\qquad \sum_{\sigma=\sigma_1+1}^{r}(-1)^{\rho_\sigma}\,\bar{a}_\sigma \sum_{\tau=1}^{\rho_\sigma} K_{\sigma,\tau}\, t^{\alpha_\sigma-\tau}\, f^{(\rho_\sigma-\tau)}
\end{aligned}$$

Defining

$$\tilde{\mathbf{M}}_{\mathbf{o}}^{+} f = \begin{cases} \displaystyle\sum_{\sigma=0}^{r}(-1)^{\rho_\sigma}\bar{a}_\sigma t^{\alpha_\sigma} f^{(\rho_\sigma)}, & \text{if } \alpha_1 < \rho_1 \\ \displaystyle\sum_{\sigma=0}^{r}(-1)^{\rho_\sigma} t^{\alpha_\sigma} f^{(\rho_\sigma)} + & \\ \displaystyle\sum_{\sigma=0}^{\sigma_1}(-1)^{\rho_\sigma}\bar{a}_\sigma \sum_{\sigma=1}^{\rho_\sigma} K_{\sigma,\tau} t^{\alpha_\sigma-\tau} f^{(\rho_\sigma-\tau)}, & \text{if } \alpha_1 = \rho_1, \end{cases}$$

we see that $\tilde{\mathbf{M}}_{\mathbf{o}}^{+}$ is a special expression with the same polygonal path (same $\gamma(l)$ as $\mathbf{M_o}$). Since (12) holds for the $s_l$, then the coefficients of $\mathbf{M_o}^+ - \tilde{\mathbf{M}}_{\mathbf{o}}^{+}$ satisfy (10), (11).

If $\mathcal{R}(\mathbf{T_o}(\tilde{\mathbf{M}}_{\mathbf{o}}^{+}-\lambda))$ is closed, then

$$\text{nul}\,((\tilde{\mathbf{M}}_{\mathbf{o}}{}^{+}-\bar{\lambda})^+) \;=\; \text{nul}\,(\mathbf{M_o}-\lambda) \;=\; \text{nul}\,(\mathbf{M_o}+\mathbf{M}-\lambda).$$

But $\mathcal{R}(\mathbf{T_o}(\tilde{\mathbf{M}}_\mathbf{o}{}^+ - \bar{\lambda}))$ is closed if and only if $\mathcal{R}(\mathbf{T_o}(\mathbf{M_o} - \lambda)^+)$ is closed. And this holds if and only if $\mathcal{R}(\mathbf{T_o}(\mathbf{M_o} - \lambda))$ is closed. Thus proving the second assertion.

## References

[1] S. GOLDBERG, "Unbounded Linear Operators," Mc-Graw Hill, New York, 1966.

[2] R. M. KAUFFMAN, On the limit-$n$ classification of ordinary differential operators with positive coefficients, *Proc. London Math. Soc. (3)* **35** (1977), 496-526.

[3] R. Mumpar-Victoria, Perturbations of a class of ordinary differential expressions preserving the essential spectrum and the nullities, PhD Thesis, Dept. of Math, Univ. of the Phil., (1999).

[4] B. SCHULTZE, Spectral properties of not necessarily self-adjoint linear operators, *Advances in Mathematics* **83**, No. 1 (1990), 75-95.

# AN EXTRAPOLATION TECHNIQUE FOR A SINGULARLY PERTURBED PROBLEM ON A SHISHKIN MESH

M.C. NATIVIDAD
*University of the Philippines, Diliman*
*Quezon City, Philippines 1101*
*E-mail: joanne@math01.cs.upd.edu.ph*

M. STYNES
*National University of Ireland*
*Cork, Ireland*
*E-mail: stynes@ucc.ie*

We consider a singularly perturbed reaction-diffusion problem on a piecewise uniform mesh, called the Shishkin mesh, and show how Richardson Extrapolation improves the accuracy of the computed solution. In particular, we shall show how error improves from $O(N^{-2}\ln^2 N)$ to $O(N^{-4}\ln^4 N)$. Numerical tests are provided to support the theoretical results.

## 1 Introduction

Consider the linear problem $(P_\varepsilon)$

$$(P_\varepsilon) \qquad \begin{cases} L_\varepsilon u_\varepsilon(x) := -\varepsilon^2 u_\varepsilon''(x) + b(x)u_\varepsilon(x) = f(x), & x \in (0,1) \\ \text{where } u_\varepsilon(0) = u_0, \quad u_\varepsilon(1) = u_1, \end{cases}$$

where $b(x) > 2\beta^2 > 0$, $\quad b(x), f(x) \in C^8([0,1])$, $\quad 0 < \varepsilon \ll 1$ and $u_0$ and $u_1$ are given.

When $\varepsilon$ is small, boundary layers may occur and standard numerical methods may exhibit inaccurate results. Numerical methods that behave uniformly for all values of $\varepsilon$ have been used in the literature (e.g., see Stynes and Roos [1], Stynes and Tobiska [2], and Sun and Stynes [3]).

Richardson extrapolation is an error-reduction procedure where two computed solutions approximating a particular quantity are combined to provide a third and better approximation (for more details see Johnson and Riess [4] and Ralston [5]). This technique has been used in improving the accuracy of the solution of a regular perturbation problem on a uniform mesh (see Keller [6,7]). Furthermore, it has also been used in solving singularly perturbed problems using a special non-uniform mesh (see Vulanović *et al.* [8]). In this paper we shall show how Richardson extrapolation improves the accuracy of the numerical solution of $(P_\varepsilon)$ on a piecewise uniform mesh, called the Shishkin mesh.

This mesh is simple and it resolves the boundary layer (Miller *et al.* [9] gives a more detailed explanation of the mesh). In particular, we will show how the error is improved from $O(N^{-2}\ln^2 N)$ to $O(N^{-4}\ln^4 N)$.

The paper is presented as follows. Section 2 contains bounds on the error of the smooth and layer components of the solution. Section 3 shows why extrapolation improves the accuracy of the computed solution. Lastly, Section 4 gives a numerical example that confirms the theoretical results. Proofs of results stated in the paper will appear in Natividad and Stynes [10].

Throughout our analysis, $C$ (sometimes subscripted) is a generic constant that is independent of $\varepsilon$ and of the mesh. Note that $C$ may take different values in different places.

## 2 Bounds on the error in the Smooth and Singular components of the solution

To obtain $\varepsilon$-uniform convergence results, the solution $u_\varepsilon$ can be decomposed into $u_\varepsilon = v_\varepsilon + w_\varepsilon$ where $v_\varepsilon$ and $w_\varepsilon$ are the smooth and layer components respectively. Furthermore, the layer part $w_\varepsilon$ is decomposed into $w_\varepsilon = w_l + w_r$ where $w_l$ and $w_r$ correspond to layer part in the left and right boundary layers, respectively. For $0 \le k \le 8$, the bounds of the derivatives of $v_\varepsilon, w_l$ and $w_r$ for $x \in (0,1)$ are given by:

$$
\begin{aligned}
|v_\varepsilon^{(k)}(x)| &\le C(1+\varepsilon^{-(k-4)}e(x,\beta)) \\
|w_l^{(k)}(x)| &\le C\varepsilon^{-k}e^{-\beta x/\varepsilon} \\
\text{and}\quad |w_r^{(k)}(x)| &\le C\varepsilon^{-k}e^{-\beta(1-x)/\varepsilon}
\end{aligned}
$$

for some constant $C$ where $e(x,\beta) = e^{-\beta x/\varepsilon} + e^{-b(1-x)/\varepsilon}$.

The method used to solve $(P_\varepsilon)$ uses a piecewise uniform mesh called the *Shishkin mesh.* To construct this mesh, we first specify its transition points $\tau$ and $1-\tau$ by

$$
\tau = \min\{\frac{1}{4}, \frac{4\varepsilon}{\beta}\ln N\},
$$

where we assume that $N$ is divisible by 4. We assume further that $\tau = (4\varepsilon \ln N)/\beta$. Otherwise, $N$ is very large compared to $1/\varepsilon$ and the analysis can be carried out using a standard consistency/stability analysis.

The subintervals $[0,\tau]$ and $[1-\tau,1]$ are divided into $N/4$ equal subdivisions with mesh lengths $h_i = x_i - x_{i-1} = h = 4\tau/N$. On the other hand, the subinterval $[\tau, 1-\tau]$ is divided into $N/2$ equal subdivisions with mesh lengths

$H = 2(1-2\tau)/N$. Now we form the discretization of the problem $(P_\varepsilon)$:

$$(P_\varepsilon^N) \qquad \begin{cases} \text{Find the mesh function } U_\varepsilon^N \text{ such that} \\ L_\varepsilon^N U_\varepsilon^N(x_i) = -\varepsilon^2 \delta^2 U_\varepsilon^N(x_i) + b(x_i) U_\varepsilon^N(x_i) = f(x_i) \\ \text{where } U_\varepsilon^N(0) = u_0, \quad U_\varepsilon^N(1) = u_1 \end{cases}$$

and

$$\delta^2 U_\varepsilon^N(x_i) = \frac{2}{h_{i+1} + h_i} \left\{ \frac{U_\varepsilon^N(x_{i+1}) - U_\varepsilon^N(x_i)}{h_{i+1}} - \frac{U_\varepsilon^N(x_i) - U_\varepsilon^N(x_{i-1})}{h_i} \right\}.$$

The solution $U_\varepsilon^N$ of $(P_\varepsilon^N)$ is decomposed into smooth and layer components in a manner analogous to the decomposition of the solution $u_\varepsilon$ of $(P_\varepsilon)$. Thus $U_\varepsilon^N = V_\varepsilon^N + W_\varepsilon^N$ where $V_\varepsilon^N$ is the solution of the nonhomogeneous problem

$$L_\varepsilon^N V_\varepsilon^N = f, \quad V_\varepsilon^N(0) = v_\varepsilon(0), \quad V_\varepsilon^N(1) = v_\varepsilon(1)$$

and therefore $W_\varepsilon^N$ must satisfy

$$L_\varepsilon^N W_\varepsilon^N = 0, \quad W_\varepsilon^N = w_\varepsilon(0), \quad W_\varepsilon^N(1) = w_\varepsilon(1).$$

Furthermore, $W_\varepsilon^N = W_l^N + W_r^N$ where $W_l^N$ and $W_r^N$ are defined by

$$L_\varepsilon^N W_l^N = 0, \quad W_l^N(0) = w_\varepsilon(0), \quad W_l^N(1) = 0$$
$$\text{and} \quad L_\varepsilon^N W_r^N = 0, \quad W_r^N(0) = 0, \quad W_r^N(1) = w_\varepsilon(1).$$

The error can be expressed as $U_\varepsilon^N - u_\varepsilon = (V_\varepsilon^N - v_\varepsilon) - (W_\varepsilon^N - w_\varepsilon)$ and the errors in the smooth and layer components can be estimated separately. The following theorem bounds the error between the solution $u_\varepsilon(x)$ and the computed solution $U_\varepsilon^N(x)$ in the whole interval $[0,1]$.

**Theorem 2.1** *Recall that $u_\varepsilon$ is the solution to $(P_\varepsilon)$ and $U_\varepsilon^N$ is the computed solution to $(P_\varepsilon^N)$. Then for all $x_i \in [0,1]$ we have*

$$|U_\varepsilon^N(x_i) - u_\varepsilon(x_i)| \le CN^{-2} \ln^2 N$$

*for some constant $C$.*

## 3 Extrapolation of $U_\varepsilon^N(x_i)$

To extrapolate the computed solution $U_\varepsilon^N(x_i)$ , we solve $(P_\varepsilon^N)$ on two different Shishkin meshes, one with $N$ subintervals and the other with $2N$ subintervals. We denote the solutions of these discrete problems by $U_\varepsilon^N$ and $\tilde{U}_\varepsilon^{2N}$, respectively. Here $U_\varepsilon^N = V_\varepsilon^N + W_\varepsilon^N$ and $\tilde{U}_\varepsilon^{2N} = \tilde{V}_\varepsilon^{2N} + \tilde{W}_\varepsilon^{2N}$. As before, the first computed solution $U_\varepsilon^N$ is solved with the transition points $\tau$ and $1-\tau$. Next, $\tilde{U}_\varepsilon^{2N}$ is solved with the *same* transition points $\tau$ and $1-\tau$.

Note that in the standard Shishkin mesh, the transition point $\tau$ is dependent on the number of subintervals $N$. But in solving $\tilde{U}_\varepsilon^{2N}(x_i)$ where the number of subintervals is doubled and thus a change in the value of $\tau$ is expected, we modify the mesh by retaining the positions of the transition points $\tau$ and $1-\tau$. This is done so that the two meshes be nested, that is, $\{x_i^N\} \subset \{\tilde{x}_i^{2N}\}$, where $\{x_i^N\}$ are the mesh points of the Shishkin mesh defined previously while $\{\tilde{x}_i^{2N}\}$ are the mesh points of our modified Shishkin mesh. Thus when solving $\tilde{U}_\varepsilon^{2N}(x_i)$, we compute approximations at the same mesh points as in $U_\varepsilon^N(x_i)$.

The following theorem shows that after extrapolation the accuracy of the computed solution is improved to $O(N^{-4}\ln^4 N)$.

**Theorem 3.1** *For all $x_i \in (0,1)$, the error between $u_\varepsilon$ and the extrapolated value of $U_\varepsilon^N$ is bounded as follows:*

$$\left| u_\varepsilon(x_i) - \frac{4\tilde{U}_\varepsilon^{2N}(x_i) - U_\varepsilon^N(x_i)}{3} \right| \le CN^{-4}\ln^4 N$$

*for some constant $C$.*

## 4 Numerical Results

Consider the linear reaction-diffusion problem:

$$\begin{aligned} -\varepsilon^2 u''(x) + (1+x)^2 u(x) &= f(x), \quad x \in (0,1) \\ u(0) = u(1) &= 0 \end{aligned}$$

where $f(x)$ is chosen in such a way that the real solution $u(x)$ is:

$$\begin{aligned} u(x) &= e^{-x(1+x)/\varepsilon} + e^{-(1-x^2)/\varepsilon} + \cos^2 \pi x \\ &+ \quad (e^{-1/\varepsilon} - e^{-2/\varepsilon})x - 2 - e^{-1/\varepsilon}. \end{aligned}$$

Table 1 shows the maximum error between the real and computed solutions before and after extrapolation and the corresponding rates of convergence with $\varepsilon = 10^{-6}$. As can be seen in the table, extrapolation decreases the error and increases the rates of convergence. Furthermore, Figure 1 shows that before extrapolation, the error between the real and computed solution is $O(N^{-2}\ln^2 N)$ while the error after extrapolation is now $O(N^{-4}\ln^4 N)$.

From the theory and from the results of the example given above, we can conclude that the form of extrapolation discussed here improved the accuracy of the computed solution from $O(N^{-2}\ln^2 N)$ to $O(N^{-4}\ln^4 N)$ of singularly perturbed reaction-diffusion problems using Shishkin meshes.

Table 1. Maximum errors before and after extrapolation with the corresponding rates of convergence; $\varepsilon = 10^{-6}$

| N/rate | error before extrapolation | error after extrapolation |
|---|---|---|
| 64 | 3.1807e-002 | 1.5022e-003 |
| rate | 1.4739e+000 | 2.9151e+000 |
| 128 | 1.4371e-002 | 3.1214e-004 |
| rate | 1.9170e+000 | 3.8760e+000 |
| 256 | 4.9157e-003 | 3.5672e-005 |
| rate | 1.9485e+000 | 3.8617e+000 |
| 512 | 1.6021e-003 | 3.8671e-006 |
| rate | 1.9936e+000 | 3.9599e+000 |
| 1024 | 4.9636e-004 | 3.7716e-007 |
| rate | 1.9912e+000 | 3.9851e+000 |

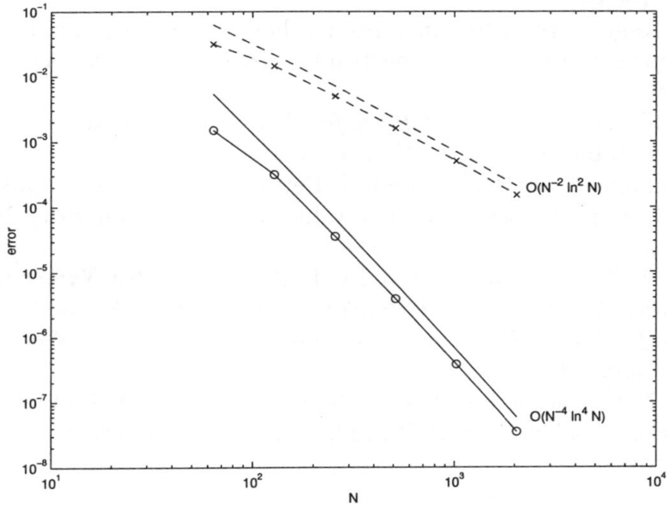


Figure 1. Comparison of errors before and after extrapolation

## Acknowledgements

The work of M.C. Natividad was supported by the Center of Excellence Grant awarded to the Department of Mathematics, University of the Philippines, Diliman by the Commission on Higher Education of the Philippine Government.

**References**


1. M. Stynes and H.G. Roos, The midpoint upwind scheme, *Applied Numer. Math.* **23**, 361 – 374 (1997).
2. M. Stynes and L. Tobiska, A finite difference analysis of a streamline diffusion method on a Shishkin mesh, *Numer. Algorithms* **18**, 337 – 360 (1998).
3. G. Sun and M. Stynes, A finite difference analysis of a streamline diffusion method on a Shishkin mesh, *Numer. Algorithms* **18**, 337 – 360 (1998).
4. L. Johnson and R.D. Riess, *Numerical Analysis* (2nd Edition) (Addison – Wesley, Reading, Mass, 1982).
5. A. Ralston, *A First Course in Numerical Analysis* (McGraw – Hill, New York, 1965).
6. H.B. Keller, Accurate difference methods for linear ordinary differential systems subject to linear constraints, *SIAM J. Numer. Anal.* **6**, 8 – 30 (1969).
7. H.B. Keller, *Numerical Methods for Two-Point Boundary Value Problem* (Dover Publications, Inc., New York, 1992).
8. R. Vulanović, D. Herceg, and N. Petrović, On the extrapolation for a singularly perturbed boundary value problem, *Computing* **36**, 69 – 79 (1986).
9. J.J.H Miller, E. O'Riordan and G.I. Shishkin, *Fitted Numerical Methods for Singular Perturbation Problems, Error Estimates in the Maximum Norm for Linear Problems in One and Two Dimension* (World Scientific, Singapore, 1996).
10. M.C. Natividad and M. Stynes, Richardson extrapolation for a reaction-diffusion problem using a Shishkin mesh (in preparation).

# Decomposition-Metaheuristic Method Applied to a Capacitated Facility Location Problem

MARRICK C. NERI[1] and AUGUSTO Y. HERMOSILLA[2]
*Department of Mathematics, University of the Philippines*
*Diliman, Quezon City, Philippines*
[1] *marrick@math01.cs.upd.edu.ph* and [2] *auggie@math01.cs.upd.edu.ph* *

**Abstract**

A *capacitated facility location problem* (CFLP) with simulated random data is formulated as a mixed integer linear program. Benders decomposition is used to generate an optimal solution for this problem. Instead of the usual branch and bound method, Simulated Annealing and some of its variants are used to generate binary integer solutions for the relaxed master problem of the decomposition. Objective function value results are comparable with, if not, sometimes better than GAMS's using the branch and bound method.

## 1 Introduction

*Mixed integer problems* (MIPs) involve integer and continuous variables in a linear objective function and linear constraints. The simplex method, though widely and popularly used in solving linear problems involving continuous variables, cannot however solve to global optimality an MIP because of the inclusion of integer variables. Some MIP cases are large scale and their solutions require computations that exceed the computational limits of existing commercial linear programming codes. An example of a computationally complex MIP is the *capacitated facility location problem* (CFLP).

In the CFLP, a decision maker is faced with the problem of which plants and warehouses to operate, how many of each commodity to produce in a plant, and what routes to take in shipping the commodities from the plants to the warehouses to the demand centers-at minimum cost. If the company produces several kinds of commodities, the number of constraints may become very large [Geoffrion, 1974].

Algorithms, such as decomposition techniques, have been developed to counter the computational complexity of the problem with large-scale dimensionality. These algorithms were shown to be effective in generating optimal solutions to large-scale MIPs. A large-scale problem is decomposed into a number of smaller and easier problems, each of which consequently uses smaller subsets of constraints or variables in the computations. These algorithms that decompose a problem into smaller problems are grouped as *decomposition techniques*, and the most familiar ones are the

*This study was granted financial support by the *Office of the Vice Chancellor for Research and Development-University of the Philippines*, Grant No. 00009.1 NSET

- **Dantzig-Wolfe decomposition (DWD)**, and
- **Benders decomposition (BD)** .

Dantzig-Wolfe decomposition is primarily a delayed column generation algorithm applied on a problem with an excessive number of columns. The problem is initially solved using a smaller number of columns representing the variables in its coefficient matrix. An improving column is generated only if its reduced cost is negative and the solution approaches optimality. On the other hand, Benders decomposition uses delayed row generation and the cutting plane method on a problem with a large number of constraints. A constraint is added only when it is violated by the current solution [Bertsimas, 1997]. Solving the full master problem using BD is the same as solving the dual of the problem using DWD [Eiselt, 1987]. In effect, BD is the *"dual"* of DWD [Bertsimas, 1997].

A problem is *NP-hard*, meaning that it has a non-deterministic polynomial time complexity, if the optimal solution can only be obtained in a non-polynomial, usually exponential, function of execution time. As the problem size increases in terms of the number of variables and constraints, the time needed by an optimal algorithm to solve the problem could explode exponentially. As a result, only small sized problems can be solved by classic methods within reasonable time. It is commonly believed that no efficient algorithm could ever be possibly found to solve this class of inherently hard problems [NIST]. Problems that are usually NP-hard are pure integer combinatorial problems, such as the *travelling salesman problem, knapsack problem, boolean satisfiability problems*, and *bin packing*, to mention a few [Osman, *et.al.*, 1996].

Researchers are left with two alternatives to solve NP-hard problems: one is to look for *heuristics* that merely find *near-optimal solutions*, but significantly much more quickly, another is to develop dynamic optimization algorithms that work well on real-life, rather than worst-case yet probably unlikely, instances [Johnson, *et.al.*, 1995].

*Heuristic methods*, or simply heuristics (meaning *approximation algorithms*) are processes that quickly identify good-quality solutions to a given problem. They are usually iterative and empirically developed through human intuition, experience, and judgment. Although they are not *exact algorithms*, they nonetheless could provide global optimal solutions in a finite time. For more complex problems, they oftentimes provide near-optimal or acceptable approximate solutions within reasonable time. Heuristics usually work more efficiently than an exact algorithm in providing acceptable approximate solutions to real life large combinatorial integer problems. Although an exact algorithm is certain and assured to work theoretically every time, in practice, it may perform poorly due to the computational complexity associated with large-scale, real-life problems. The purpose of heuristics is therefore simply to identify acceptable approximate solutions to problems where time for manageable and meaningful decisionmaking is more important than solution quality in terms of perfect global optimality [Barr, *et.al.*, 1995].

The most recent advancement in heuristic search algorithms to solve pure integer combinatorial problems is *metaheuristics.* Metaheuristics use concepts in the physical and mathematical sciences, neural nets, biological evolution, among others, to guide search heuristics in finding optimal solutions or near-optimal solutions on limited time to problems that are usually NP-hard. One metaheuristic is the *Simulated Annealing* (SA) algorithm that is patterned after the annealing process of metals and crystals [Osman, *et.al.*, 1996].

SA has been used extensively in solving the *Travelling Salesman Problem* (TSP). The TSP is an *NP-hard* binary integer linear problem with open or closed variables applied on the routes of tour a salesman travels as he visits each of n cities once [Telfar, 1994]. SA has recently been making crossroads in solving other integer problems, such as the cutting stock problem, previously solved by using branch and bound methods only [Lutfiyya, *et.al.*, 1991].

# 2 More on the CFLP and BD

Consider a company that produces a commodity at plants with known production capacities. The commodity is stored at warehouses before shipment to customer zones. The demand for the commodity at each customer zone is known. Also, fixed charges are associated with the possession and operation of the plants and warehouses. There are lower and upper bounds on a warehouse's throughput capacity, and holding costs are charged at the commodities. The possible locations for plants and warehouses are given, but the sites to be used should be selected so as to result in the minimum production, distribution, holding and operating costs. All costs are taken to be linear [Geoffrion, 1974].

A single-period *capacitated facility location problem* (CFLP) is an MIP that involves both continuous and integral variables. The decision-maker is faced with the problem of

- which plants and warehouses to operate,
- how many of each commodity to produce in a plant, and
- what routes to take in shipping the commodities from the plants to the warehouses to the demand centers

at **minimum cost**. If the company produces several kinds of commodities, the number of constraints included in the formulation of the problem can become exponentially very large [Bertsimas, 1997] .

## 2.1 A Mathematical Formulation

For this paper, we formulate the CFLP as an MIP as follows:

Minimize

$$(1)\quad \sum_{jkl}(\, c_j \,+\, t_{jkl})\, x_{jkl} \,+\, \sum_j z_j^p Fp_j \,+\, \sum_k \left( \, z_k^w Fw_k \,+\, \sum_l \, h_k D_l y_{kl} \right)$$

Subject to

$$\begin{array}{lll}
(2) & \sum_{kl}\; x_{jkl} \;\le\; z_j^p S_j & \forall\, j \in J \\
(3) & \sum_j\; x_{jkl} \;=\; y_{kl} D_l & \forall\, k \in K\;,\; l \in L \\
(4) & \sum_k\; y_{kl} \;=\; 1 & \forall\, l \in L \\
(5) & \underline{V}_k z_k^w \;\le\; \sum_{il}\; y_{kl} D_{il} \;\le\; \overline{V}_k z_k^w & \forall\, k \in K \\
(6) & \textit{Other linear configuration constraints on} & \\
 & y\;,\; z^p \text{ and } z^w & \\
(7) & x \;\ge\; 0\;;\; z \in \{\, 0, 1\,\} &
\end{array}$$

which uses the following notation [Eiselt, 1987]:

| | |
|---|---|
| $J$ | index set of plants, |
| $K$ | index set of warehouses, |
| $L$ | index set of customer centers, |
| $D_l$ | demand for the commodity at customer zone $l$, |
| $S_j$ | supply at plant $j$, |
| $\underline{V}_k$ | minimum allowed throughput for a warehouse at site $k$, |
| $\overline{V}_k$ | maximum allowed throughput for a warehouse at site $k$, |
| $Fp_j$ | fixed charges for the possession and operation of plant $j$, |
| $Fw_k$ | fixed charges for the possession and operation of warehouse $l$, |
| $c_j$ | cost of producing the commodity at plant $j$, |
| $h_k$ | holding cost for the commodity at warehouse $k$, |
| $t_{jkl}$ | transportation cost for the commodity from plant $j$ through warehouse $k$ to customer zone $l$, |
| $x_{jkl}$ | a variable denoting the quantity of the commodity shipped from plant $j$ through warehouse $k$ to customer zone $l$, |
| $y_{kl}$ | a $0-1$ variable that will be 1 if warehouse $k$ serves customer zone $l$, |
| $z_j^p$ | $0-1$ variables for plant $j$, |
| $z_k^w$ | $0-1$ variables for warehouse $k$. |

All summations are understood to run over all possible meaningful combinations for $i$, $j$ , $k$ , and $l$ , since some permutations of the indices may either be physically impossible or uneconomical. The $x$ variables are continuous and the binary variables $y$ and $z$ must either be zero or one. Other linear configuration constraints could be, for example, a) a limit on the number of customer zones served by a particular plant, or b) a necessity to operate a particular plant due to political or social implications. However, this research is limited to handling only constraints (2) to (5). A CFLP is infeasible only when the total demand exceeds the total supply in all the open plants. It is assumed that the available supply is at least equal to the demand. The CFLP as an MIP, is traditionally solved using the *simplex* and integer programming methods, such as BB.

With our MIP formulation, Benders decomposes the CFLP into two problems:

**The Subproblem (Continuous)**

$$\text{Maximize} \quad \sum_j (-S_j)\, z_j^p v_j \;+\; \sum_{kl} y_{kl} D_l u_{kl}$$

$$\text{Subject to} \quad \begin{array}{l} -v_j \;+\; u_{kl} \;\leq\; c_j \;+\; t_{jkl}\;, \quad \forall\, j\;,\; k\;,\; l \\ v_j \;\geq\; 0\;,\; u_{kl} \in \mathbb{R} \end{array}$$

**The Relaxed Master Problem (Integer)**

$$\text{Minimize} \quad \sum_j z_j^p Fp_j \;+\; \sum_k \left( z_k^w Fw_k \;+\; \sum_l h_k D_l y_{kl} \right) \;+\; Y_0$$

Subject to

$$Y_0 + \sum_j S_j z_j^p v_j^m - \sum_{kl} y_{kl} D_l u_{kl}^m \geq 0 , \quad m = 1 \; to \; r$$
$$CFLP \; constraints \; (4) , (5) , and \; (6) \quad and$$
$$y , z \in \{ 0,1 \}$$

## 2.2 Computational Complexity of the CFLP

The problem of determining the subset of sites for locating the plants and warehouses and that of setting up the necessary transportation routes involves $0 - 1$ constraints, and hence, is a **NP-hard** combinatorial optimization problem. Once the optimal solution of this combinatorial problem is obtained, the problem of determining the optimal amounts of commodities to be shipped along the open transportation routes is a simple transportation problem [Osman, *et.al.*, 1996].

The model formulated above has $jkl$ continuous variables and $j + k + kl$ binary variables. There are $j$ supply constraints, $kl$ demand constraints, $l$ one-warehouse-per-customer zone constraints, and $2k$ constraints on the warehouses' configurations, since constraint (5) can be split into two inequality constraints. The total number of constraints is at least $j + kl + l + 2k$ , not yet including other possible configuration constraints. If, for instance, the company has 10 plants, 20 warehouses, and 100 customer zones, the number of variables is 20000 continuous and 2030 binary. There are 2150 constraints. The number of configurations for the open plants is $2^{10}$ ; for the warehouses, $2^{20}$; and for the shipping routes $2^{2000}$. The number of possible solutions, some of which may yet be infeasible, for the binary integer variables alone is $2^{10}2^{20}2^{2000}$ , a very large combinatorial number to cause computational complexity. If the problem involves only binary integer solutions, and as the problem size increases, the problem becomes NP-hard of at least order $2^N$, where $N$ is the number of binary variables. Deterministic algorithms of polynomial time complexity, such as *branch and bound* (BB), may not solve this problem to global optimality.

At this point, worthwhile to discuss briefly are the difficulties of NP-complete problems. Given a large-scale problem, it is not enough that we are able to come up with a mathematical formulation of the problem. We must also make sure that the problem can be solved within reasonable computational time. There are certain problems that can be solved in polynomial times. This means that the computational time required as the size of the problem increase will grow in a polynomial fashion. Such problems are indeed wonderful to us, since it means that we can formulate problems of enormous size without having to worry about whether the problem can be solved within our lifetime. Linear programs in continuous variables are one of the most beloved polynomial problems and this is the reason why we are not scared of linear problems that deal with thousands of variables. In fact, even a student package LP optimizer can often take several hundreds of variables and constraints.

However, there are other types of difficult-to-solve problems which do not possess such polynomial time structure but rather exponential time structures. These problems are computed in fairly comparable time frame when the problem is small, but its computation time structure makes it impossible to be solved, as the problem size becomes larger. There are also problems called *NP-complete*, or *completely nondeterministic polynomial*, or *the hardest (in the complete sense)*, which are much harder than the regular exponential time problems.

A fine illustration of this principle is the brute force enumeration of all possible paths in a graph in order to find the shortest path. When the graph is small with, say 7 vertices, the number of paths is only 25, or only 32 paths all in all. Even the relatively slow PC's can crunch those possibilities and come up with the best path in seconds. However, as the problem size gets bigger, say 32 vertices, to enumerate all the possible paths, we have to go through all the possibilities of up to 230 paths. If we can enumerate and compare 1000 paths in a second, we will still need more than 12 days to come up with optimal path. Given 50 cities, the problem will easily go beyond our lifetimes. In short, we will be long dead and gone before we can get the solution to the exponentially timed problems.

To illustrate the point further, consider a problem of size N. Suppose the following functions represent the computational time in microseconds, and we will examine by the following table the behavior of each type of functions, as the size N enlarges:

| Time | | Problem | Size | N | | |
|---|---|---|---|---|---|---|
| Function | 10 | 20 | 30 | 40 | 50 | 60 |
| $N$ | .00001 second | .00002 second | .00003 second | .00004 second | .00005 second | .00006 second |
| $N^2$ | .00001 second | .00004 second | .00009 second | .00016 second | .00025 second | .00036 second |
| $N^5$ | .1 second | 3.2 second | 24.3 second | 1.7 minutes | 5.2 minutes | 13.0 minutes |
| $2^N$ | .001 second | 1.0 second | 17.9 minutes | 12.7 days | 35.7 years | 366 centuries |

As shown by the table, the polynomial $N$ , $N^2$ , and $N^5$ time functions require computation times of seconds in small problems to minutes in large problems. However, the non-polynomial $2^N$ problem's time requirement will grow from seconds to days and even centuries, as the problem size increases. In other words, the computational difficulty explodes as problem gets bigger and bigger.

# 3 Statement of this Paper's Problem

In BD, an MIP is decomposed into two smaller problems: a *relaxed master problem* that involves *only integer* variables, and a *subproblem* that involves only the *continuous variables* [Murty, 1976]

The usual approach in BD is to use the BB method to solve the integer master problem. However, this method may no longer perform very well when the problem is NP-hard or afflicted with *computational complexity.* A primary feature of computational complexity is *time complexity*, which is the execution time of the algorithm to solve completely the problem. No exact algorithm is to known to exist that can solve to optimality the NP-hard problems within reasonable time limits.

A Benders decomposed CFLP may not be solved to optimality using BB on the integer master problem. This paper, therefore, seeks to provide other alternative methods that can solve the integer master problem while addressing specific computational complexity issues.

## 3.1 Objectives

The basic objectives of this paper are as follows:

1. *Compile comprehensively available literature on BD and SA.*
2. *Formulate a CFLP with simulated random data and solve it using a modified BD, with a proposed SA algorithm to solve the relaxed master problem of the decomposed problem.*
3. *Compare the results of a BD using a pure or classic SA algorithm against a decomposition using variants of SA and against GAMS's BB method.*
4. *Develop a program using Matlab 5.3 to solve a mixed-integer and linear CFLP, which can be useful for future researchers with interest along this line of study.*

In other words, BD is to be applied on a CFLP with randomly generated data. However, instead of using BB to solve the master problem, the metaheuristic SA is used. Variants of SA, e.g. *SA with Restart* and *SA with Systematic Exploration*, are also tested and the results are compared with the results of pure SA. Also, the results of the modified BD are compared with the results of GAMS. Lastly, the proposed algorithm given in a later section is implemented in MATLAB.

## 3.2 Methodology

In this paper, a CFLP is solved using BD. However, instead of the usual BB method incorporated in the decomposition method, this study considers it as a contribution to investigate the usefulness of SA in finding an optimal solution or near-optimal solution to the relaxed master problem. Part of this study is to compare the results of solving a CFLP using variants of SA incorporated into the decomposition method. The various results studied in comparable settings are used to compare the relative performance of the proposed methodology using variants of SA.

The methodology of this study is described as follows:

1. *Research through available background literature.*
2. *Formulate the CFLP and construct an iterative scheme for the BD using SA to solve the integer master problem.*
3. *Develop appropriate computer programs, and in particular, a program that yield simulated random data.*
4. *Design and execute the experiments using the randomly simulated data.*

5. *Analyze and compare the experimental results of using BD with classic SA and BD with a variant SA, and also with GAMS.*

6. *Report the computational results, together with some analyses, and make conclusions and relevant recommendations.*

## 3.3 Limitations

The scope of this study is limited in the following manner:

1. The authors do not have the opportunity to avail of commercial optimization solvers where the BD could be programmed. The only software available to the authors is MATLAB 5.3. Using GAMS205, checks have been done also on the consistency of the solutions obtained from the modified BD.

2. There is a dearth of available and usable real data, especially in the Philippine setting. Attempts to obtain real-life data via the Internet have been futile. The data used to test the program in this study are only randomly generated using the random function *rand*() built inside MATLAB 5.3. All programs used in this research are coded and implemented in MATLAB 5.3.

3. Comparisons between the results of a decomposition method using SA and a decomposition method using BB are not performed due to some difficulties in coding a BB algorithm in MATLAB 5.3. Comparisons of the results of the modified BD embedded with a pure or variant SA with those of GAMS are limited to the optimal objective values, iterations, and running time.

4. This study does not include a comprehensive comparison of the program developed with that of existing commercial softwares in terms of computing resource utilization efficiency issues. As a benchmarking exploration, the aim of this paper is simply to exploit the use of SA in BD.

## 3.4 Significance of the Study

This study is significant in the following areas:

1. Unlike the classical CFLP model which simply determines what warehouses to operate, the CFLP version proposed in this paper also includes the selection of plants to operate. The proposed CFLP model allows plants to be opened or closed.

2. Large-scale MIPs are often hard to solve using exact algorithms. This paper adequately highlights the advantage of the use of decomposition methods in solving MIPs, in particular, CFLPs.

3. The current trend in combinatorial optimization is the use of metaheuristics. This study incorporates SA into a BD algorithm, and thereby offers an improvement to the existing decomposition algorithm by combining it with a currently popular search method of SA in combinatorial optimization.

4. This study demonstrates the versatility of MATLAB 5.3 in solving MIPs.

5. This is a pioneering attempt to unify the capabilities of a metaheuristic and a widely used decomposition algorithm in providing alternative near-optimal solutions to problems in the NP-hard context.

The realization of this paper's objectives could well contribute to the development of BD and SA as combined and hybridized powerful tools in heuristic combinatorial optimization. Likewise, this paper would hopefully ignite interest in other researchers to explore its potentials in offering further enhancements and continuous improvements, such as hybridizations of heuristics and distributive parallelizations, of methods and theories along this line of study, as a way to conquer the gargantuan problem of computational complexity.

## 3.5 Summary of Algorithms Used

### 3.5.1 Benders Decomposition (BD)

Benders decomposes a MIP into two problems:

- A *relaxed master problem*: over the integer variables, pure integer constraints, and complicating constraints with the dual costs of the subproblem as coefficient vectors of the constraints.

  Usual solution method: *Branch and Bound (BB).*

- A *subproblem*: over the continuous variables and complicating constraints using fixed integer variables obtained from the master problem.

  Usual solution method: *Simplex.*

### 3.5.2 Simulated Annealing (SA)

Pure combinatorial integer problems, such as the master problem of BD, are usually solved by BB. However, most combinatorial integer problems are NP-hard: the optimal solution cannot be obtained in polynomial time using deterministic approaches such as the classic BB. *Metaheuristics* have been alternatively developed to solve to optimality or near-optimality combinatorial integer problems. One widely used metaheuristic is the *Simulated Annealing* (SA) algorithm [Lutfiyya, *et.al.*, 1991], [Telfar, 1994], [Johnson, *et.al.*, 1995].

SA is a metaheuristic that

- is patterned after the annealing process of crystals and metals,

- uses local search heuristics to determine optimal solution at a given temperature, and

- allows *randomized uphill moves* if $\theta < e^{(-\Delta/T)}$ , where $\theta$ is a random number, $\Delta$ is the change in the cost, and T is the current temperature. *Entrapment* in a local optimum is thus avoided.

### 3.6 Proposed Decomposition-Metaheuristic Algorithm

To fulfill the objectives and aims of this paper, the authors propose the following modification of the earlier algorithms, and is given as follows:

Step 1. Input CFLP case data.

Step 2. Initialize starting facility and routing binary solutions (which may be infeasible), and set these to be the best integer solutions. The set of extreme point solutions $ep$ is currently empty. For the objective function value, initialize the upper bound $UB$ and the lower bound $LB$.

Step 3. Solve the integer relaxed master problem using pure or variant SA. The best or near-optimal objective cost of the master problem becomes the new lower bound $LB$. Let $F$ be the facility and holding costs associated with the best integer solutions. If $UB \leq LB$ , proceed to Step 5. Else, go to Step 4.

Step 4. Solve the continuous subproblem using the simplex method and with the integer facility and routing solutions from the relaxed master problem as vectors of fixed parameters. Let $Z$ be the optimal objective cost and let $ep_m$ be the extreme point solution of the subproblem. If $Z + F < UB$ , then $UB = Z + F$ . If $UB \leq LB$ , proceed to Step 5. Else, let $ep = ep \cup \{ ep - m \}$ and repeat Step 3.

Step 5. **Stop.** Optimality (or near-optimality) is reached. Solve the corresponding simple transportation problem using the two-phase simplex method, with the best facility and routing binary configurations as fixed parameters.

**Note:** The decision maker could terminate the running of the algorithm at any time. Results, however, may not be the best.

The following are the modified BD algorithms used in the research.

- BD embedded with pure SA.
- BD embedded with SA with Restart (R).
- BD embedded with SA with Systematic Exploration (SE).
- BD embedded with SA with SE and R.

## 4 Sample Results and Analyses

In this paper, all programs are coded in MATLAB. Three randomly generated data sets for the CFLP are used in the experiments. These are DATA 346 (3 plants, 4 warehouses, 6 customer zones), DATA 777(7 plants, 7 warehouses, 7 customer zones), and DATA 6610 (6 plants, 6 warehouses, 10 customer zones). The programs are run in two microcomputers, with Pentium III 500Mhz processors and 256 MB memory.

| DATA 346 | Objective Value | Iterations | Elapsed Time in Seconds |
|---|---|---|---|
| BD w/ SA | 30,858.68 | 392 | 55316 |
| BD w/ SA w/ R. | 28,432.12 | 398 | 57187 |
| BD w/ SA w/ SE | 33,492.92 | 16 | 811 |
| BD w/ SA w/ SE & R | 28,419.97 | 411 | 25946 |
| GAMS | 28,925.43 | 4,911 | 32 |
| DATA 777 | | | |
| BD w/ SA | 1,494,497.43 | 215 | 22,330 |
| BD w/ SA w/ R. | 1,446,982.06 | 412 | 51,013 |
| BD w/ SA w/ SE | 1,485,434.15 | 277 | 15,007 |
| BD w/ SA w/ SE & R | 1,510,406.49 | 1251 | 136,420 |
| GAMS | 1,609,140.31 | 4,807 | 63 |
| DATA 6610 | | | |
| BD w/ SA | 3,197,696.50 | 125 | Interrupted |
| BD w/ SA w/ R. | 3,189,052.76 | 125 | Interrupted |
| BD w/ SA w/ SE | 3,195,697.27 | 6 | 5969 |
| BD w/ SA w/ SE & R | 3,139,930.15 | 96 | Interrupted |
| GAMS | 3,096,961.51 | 4,824 | 70 |

*NOTE: GAMS OPTIONS: ITERLIM = 5000, OPTCR = 0.0*
*"BD" Benders Decomposition* ; "R" Restart ; "SE" Systematic Exploration

BD with pure or variant SA yielded acceptable approximate solutions for all the data sets comparable with, if not sometimes better than, the results of GAMS. BD with SA with SE and R yielded the best result for the CFLP with test data DATA346, with the objective value of 28,419,97. The fastest algorithm to converge is the one with SA with SE. However, it also yielded the least best solution.

For the CFLP with the primary reference test data DATA777, BD with SA with R yielded the best solution with an objective function value of 1,446,982.0634, after 412 iterations. The execution time is 51,013 seconds. The next best solution output is by BD with SE, although this result was obtained more quickly, requiring only 277 iterations in 15,007 seconds.

The algorithm that performed poorly with regards to the objective function value is BD with SA with SE and R. Slightly better is BD with SA.

The same problem with simulated data DATA777 is solved using GAMS. GAMS generated solutions using MIP methods, particularly the BB method. Four runs of GAMS yielded the same result.

The CFLP with test data DATA6610 is more difficult to solve. Only BD with SA with SE converged; the rest were terminated (or interrupted) after they exceeded the time limit of 36 hours. The length of time needed to solve this CFLP could be largely due to the numerical complexity of the solution space for this data.

The algorithm with variant SA with SE and R, yielded the best objective value of 3,139,930.15. This result was obtained more quickly, using only 96 iterations against the

125 iterations of the two other interrupted algorithms. BD with SA with R gives the next best objective result of 3,189,052.76.

The quick convergence of the algorithm with SA with SE is unusual and unexpected. The reason behind this result could be the intrinsic nature of SA in generating random solutions. The algorithm may have randomly *"hit"* the converging solution quickly.

The CFLP with test data DATA6610 was also solved using GAMS, and it is in this larger-scale CFLP where GAMS produced the best result. However, it is yet premature to conclude that GAMS performs better than the modified algorithm in solving bigger CFLPs due to time restrictions on the algorithms; it may even prove otherwise.

It should be noted that, in solving the CFLP, GAMS has an iteration limit of ITERLIM = 5000, and OPTCR = 0.0, meaning the program will strive to get the best, if not the optimal, result within a neighborhood approximately 0.0% distance from the predicted global optimum. In all test cases of the CFLP, the authors even increased the iteration limit of GAMS to 50,000. All results remained the same except for DATA346 where GAMS reached the same best solution of Benders with SA with Systematic Exploration with Restart.

GAMS usually output its solution in just a matter of seconds. The *best"* solution of BD with SA is frequently obtained after hours of computer running time. Although GAMS output its result much faster than the program developed in this research, the quality of the results of the proposed modified algorithms are at least comparable with, if not sometimes better than, the results of GAMS.

The *"time-weakness"* in the decomposition-metaheuristic program may be largely due to the SA parameters, such as the cooling factor of 90% , and the degree of uncertainty in obtaining the best results randomly, or the complexity of the solution space. Also, the software itself may be a cause of this slowness - the quickness of MATLAB 5.3 in running this kind of algorithms may need further investigation.

Nevertheless, the time factor - if not beyond tolerable magnitude - should not sway the decision-maker into deciding against a result which is by far better than a result quickly obtained, especially if this result could save thousands of pesos, if not millions, for the company. The longer time spent on the waiting to obtain the near-optimal solution may be offset by the larger savings of the company. Further improvements could be made on the programs to enhance their performance, specifically in terms of running time.

Best results are obtained from an algorithm that contains a variant of SA with a restart, even though this algorithm reaches the best result more slowly. In terms of optimal objective value, the results of Benders decomposition algorithm with an embedded pure or variant SA are comparable, if not sometimes better, than the results of GAMS. As the problem size increases (from DATA346 to DATA6610), the proposed algorithm continue to output *"good enough* solutions. However, due to the time complexity of the solution space, the programs are usually terminated after approximately 36 hours, thereby yielding not exactly the best solution. GAMS performance for DATA6610 is better than the proposed algorithms.

# 5 Summary, Conclusion, and Recommendation

As mathematical programming techniques evolve and computer capabilities improve, problems once considered large-scale and computationally intractable are now becoming amenable to practical mathematical programming solutions. Robust and efficient search algorithms now contest traditional methods of solving MIPs. Problems whose structures involve a large quantity of variables and constraints are now being broken down to inter-related smaller subproblems and solved by decomposition techniques.

In particular, the CFLP, an MIP, had once been solved using BB methods. But in keeping up with advancements in combinatorial search methods, this research endeavors to solve a CFLP using BD with an embedded SA algorithm. The optimal results here are shown to be better than the result obtained from a commercial program using BB.

This study shows the applicability of the proposed decomposition method (BD) with an embedded metaheuristic (SA) working hand in hand in solving a CFLP. This also shows the decomposition-metaheuristic algorithm's potential in solving other MIPs which have been solved largely and solely under the auspices of traditional programming methods, thus opening new avenues for further research.

The optimal solutions for the CFLP test data obtained using BD with SA are shown to be comparable with, if not sometimes better than, those in GAMS. The program developed in this paper used the capabilities of a MATLAB version without a compiler. Predictably, better results could have been obtained at a faster rate had the SA subprogram been compiled.

The BD presented in this research is the classic type that terminates once the lower bound and the upper bound on the objective value converge, i.e., $UB \leq LB$. Geoffrion and Graves altered this criterion a bit by including an error $e$. Their variant of BD terminates on the condition $UB \leq LB + e$. Studies on this variant of BD - with SA or any metaheuristic to solve the integer master problem as given in this paper - are highly recommended.

Since this research has focused on a field yet in its infant stage, more improvements could still be done to the algorithm and program codes that this author has developed to accommodate the many possible modifications and adaptations, such as:

1. Further research could be done on BD with SA by experimenting on different values for the parameters. For example, the initial temperature could be 90% or 80% instead of 100%. The cooling schedule could be slower or faster. More trials or moves could be made at each temperature level.

2. For a SA algorithm with either Restart or Descent, the restart or descent action could be triggered after a variable amount of moves.

3. Metaheuristics are very much adaptable to NP-hard combinatorial integer problems. Other metaheuristics, such as genetic algorithm or tabu search, could be experimented upon to solve the integer master problem of BD.

4. Using the decomposition-metaheuristic algorithm developed in this research on multi-period and multicommodity CFLP problems is an additional area of future research.

5. Another open problem is to compare the results of a decomposition method using SA and a decomposition method using BB.

6. Expanding the time-limit environment may allow better performance of the proposed metaheuristic.

Much more could be done to explore the other possibilities in this line of research, with the end in view of shedding more light in the exploration and ultimate solution of MIPs and NP-hard combinatorial optimization problems.

## References


[1] Barr, R. S., Golden, B. L., Kelly, J. P., Resende, M. G.C., Stewart, W. R. Jr. (1995): "Designing and Reporting on Computational Experiments with Heuristic Methods," *Journal of Heuristics*, Kluwer Academic Publishers

[2] Bertsimas, D., Tsitsiklis, J.N.(1997): "Introduction to Linear Optimization," *Athena Scientific*

[3] Brown, G. G., Graves, G. W., Honczarenko, M. D. (1987): "Design and Operation of a Multicommodity Production/Distribution System Using Primal Goal Decomposition," *Management Science*, No. 11, November, 1987

[4] Cerny, V. (1985): "A Thermodynamical Approach to the Travelling Salesman Problem: an Efficient Simulated Annealing Algorithm," *Journal of Optimization Theory and Applications*, 45

[5] Charon, I., Hudry, O. (1995): "Mixing Different Components of Metaheuristics," *Proceedings of the Metaheuristics International Conference (MIC) 95*," Kluwer Academic Publishers, Norwell, MA Etats-Unis, pp. 350-355

[6] Eiselt, N. A., Pederzoli, G., Sandblom, C. -L. (1987): "Continuous Optimization Models," Walter de Gruyter & Co., Berlin 30

[7] Fang, S. -C., Puthenpura, S. (1993): *Linear Optimization and Extensions (Theory and Algorithms)*, Prentice-Hall International, Inc.

[8] Geoffrion, A. M., Graves, G. W. (1974): "Multicommodity Distribution System Design by Benders Decomposition," *Management Science*, Vol. 20, No. 5, January, 1974.

[9] Glover, F., Laguna, M. (1997): "General Purpose Heuristics for Integer Programming - Part 1," *Journal of Heuristics*, Vol. 2, No. 4, pp. 343-358

[10] Greenberg, H.J., "Mathematical Programming Glossary", *World Wide Web*, http://www.cudenver.edu/ hgreenbe/glossary/glossary.html, 1996-2000.

[11] Hermosilla, A.Y. : *Optimization Heuristics*, Working Paper in Progress, Department of Mathematics, U.P. Diliman

[12] Holmberg, K. (1994): "Cross Decomposition Applied to Integer Programming Problems: Duality Gaps and Convexification in Parts," *Operations Research*, Vol. 42

[13] Infanger, G. (1992): "Monte Carlo (Importance) Sampling within a Benders Decomposition Algorithm for Stochastic Linear Programs," *Annals of Operations Research*, Vol. 39

[14] Jackson, L. A., *Facility Location Using Cross Decomposition*, MS Thesis, B.A.,Cameron University, June, 1990.

[15] Johnson, D. S., McGeoch, L. A. (1995): *The Travelling Salesman Problem: A Case Study in Local Search in Combinatorial Optimization*, edited by Aarts, E., et al, John Wiley & Sons Inc., pp. 215-311

[16] Kirkpatrick, S., Gelatt, C. D., and Vecchi, P. M. (1983): "Optimization by Simulated Annealing," *Science*, 220

[17] Lawler, E.L. (1976): *Combinatorial Optimization: Networks and Matroids*, Holt, Rinehart and Winston, New York

[18] Lutfiyya, H., McMillin, B., Poshyanonda, P., Dagli, C. (1991): " Composite Stock Cutting Through Simulated Annealing," Report Numbers CSC 91-09 and ISC 91-04, *Journal of Mathematical and Computer Modeling*, Pergamon Press, New York, Vol. 16, No. 1

[19] Martin, O. C., Otto, S. W. (1993): *Combining Simulated Annealing with Local Search Heuristics*, CS/E 94-016, Oregon Graduate Institute

[20] Mashford, J. S. (1990): *A Method for the Solution of Fixed Charge Problems by Benders' Decomposition*, Akademie-Verlag, Berlin, 1990.

[21] Metropolis, W., Roenbluth, A., Rosenbluth, M., Teller, A., and Teller, E. (1953): "Equation of the State Calculations by Fast Computing Machines," *Journal of Chemical Physics*, 21

[22] Murty, K. G. (1976): *Linear and Combinatorial Programming*, John Wiley & Sons, Inc.

[23] Nash, S. G., Sofer, A. (1996): *Linear and Nonlinear Programming*, McGraw-Hill, New York

[24] National Institute of Standards and Technology (NIST), *Dictionary of Algorithms, Data Structures, and Problems*, http://hissa.nist.gov/dads/

[25] Osman, I. H., Kelly, J. P. (1996): *Meta-Heuristics: Theory and Applications*, Kluwer Academic Publishers, Boston, MA

[26] Taha, H. A. (1992): *Operations Research, an Introduction*, 5th ed.,Macmillan Publishing Company

[27] Telfar, G. (1994): *Generally Applicable Heuristics for Global Optimisation: An Investigation of Algorithm Performance for the Euclidean Traveling Salesman Problem*, MS Reading Paper, Institute of Statistics and Operations Research, Victoria University of Welington

[28] Van Roy, T. J. (1986): "A Cross Decomposition Algorithm for Capacitated Facility Location," *Operations Research*, Vol. 34, pp.145-163

[29] Vilkko, M. *New Approach to Short-Term Planning of Hydro-Thermal Power Production*, Doctor Thesis Presented by Weiss, R.,
http://autinfo.hut.fi/kussit/as-116-140/sems00/Weiss/weiss/htm

# Homogeneous Extensions Of Shifts With Hölder Continuous Skewing Functions

Mohd. Salmi Md. Noorani
School of Mathematical Sciences
Faculty of Science and Technology
Universiti Kebangsaan Malaysia
43600 Bangi, Selangor
Malaysia
e-mail: msn@pkrisc.cc.ukm.my

**Abstract**

Let $\sigma: X \to X$ be an aperiodic subshift of finite type and $H$ a closed subgroup of a compact metric group $G$. Also let $\phi: X \to G$ be Hölder continuous. In this paper we consider a homogeneous extension of $\sigma$ with skewing function $\phi$ given by $\sigma_\phi: X \times G/H \to X \times G/H$, $\sigma_\phi(x, gH) = (\sigma(x), \phi(x)gH)$, $\forall(x, gH) \in X \times G/H$. We equip $\sigma$ with an equilibrium state measure and $\sigma_\phi$ with the (restricted) Haar extension measure. With respect to this extension system, we give the conditions for $\sigma_\phi$ to be weak-mixing or ergodic in terms of a certain functional equation. In particular these conditions strenghthen the previously known result in the case of an arbitrary measurable $\phi$. We also show that generically $\sigma_\phi$ is ergodic and weak-mixing. When $\sigma_\phi$ is not ergodic then we give a description of the ergodic components of $\sigma_\phi$. All these results are inspired by a key and recent result of Parry, where he showed that the measurable eigenfunctions of a certain transfer operator can essentially be replaced by Hölder continuous functions.

## 1 Introduction

Let $\sigma: X \to X$ be an ergodic Markov shift and $G$ a compact metric group with a closed subgroup $H$. Also let $\phi: X \to G$ be measurable. Define the homogeneous extension of $\sigma$ (with skewing function $\phi$) as $\sigma_\phi: X \times G/H \to X \times G/H$, $\sigma_\phi(x, gH) = (\sigma(x), \phi(x)gH)$, $\forall(x, gH) \in X \times G/H$. Then $\sigma_\phi$ is measure-preserving with respect to the (restricted) Haar extension measure $\mu \times \lambda'$ where $\mu$ is the associated Markov measure and $\lambda'$ is the Borel probability measure defined on $G/H$ by $\lambda'(A) = \lambda(p^{-1}(A))$ ($p$ is the natural map from $G$ to $G/H$ and $\lambda$ is the Haar measure). For this extension system, we showed that (see Noorani (1997)), a necessary and sufficient condition for $\sigma_\phi$ to be ergodic is that for any non-trivial irreducible $d$-dimensional unitary representation $R$ of $G$ satisfying $bR(h) = b$, $\forall h \in H$, for some $0 \neq b \in \mathbb{C}^d$, the only measurable solution $w: X \to \mathbb{C}^d$ satisfying the functional equation

$$w(\sigma(x)) = R(\phi(x))w(x) \qquad a.e \;\; x$$

is $w = 0$ *a.e.*

There also exists a similar criterion for $\sigma_\phi$ to be weak-mixing, namely: $\sigma_\phi$ is weak-mixing iff the only measurable solution $w: X \to \mathbb{C}$ satisfying the equation

$$w(\sigma(x)) = \alpha\chi(\phi(x))w(x) \qquad a.e \;\; x$$

where $1 \neq \alpha \in \mathbb{C}$ and $\chi$ a non-trivial one-dimensional representation of $G$ such that $\chi(h) = 1$, for all $h \in H$ is $w = 0$ *a.e* (see Noorani (1997), for details).

In this note we shall again be interested in the aforementioned criteria but in the special case where $\sigma$ is an aperiodic subshift of finite type equipped with an equilibrium state measure and the associated skewing function $\phi$ is assumed to be Hölder continuous. Our main results gives a more precise condition for the homogeneous extension to be ergodic and also weak-mixing. The main tool in proving these results comes from a recent work of Parry (1997) where he considered problems of the same nature for group extension of shifts. (For related results, see Parry & Pollicott (1997) and Walters (1996)). In fact, the approach is to relate the above functional equations to eigenfunctions of a certain transfer operator. Then we can invoke the key result of Parry which says that measurable eigenfunctions of the transfer operator can essentially be replaced by Hölder continuous functions. Along the way we also show that generically $\sigma_\phi$ is ergodic and weak-mixing. When $\sigma_\phi$ is not ergodic, we give a description of the ergodic components of $\sigma_\phi$. As in the main results, these latter results are also modelled along the line of Parry.

## 2 S.s.f.t, Hölder Function Spaces and a Transfer Operator

Let $A$ be a $k \times k$ aperiodic 0-1 matrix and define

$$X^+ = \left\{ x = (x_n)_0^\infty \in \prod_0^\infty \{1, 2, \ldots, k\}: \quad A(x_n, x_{n+1}) = 1 \ \forall \ n \geq 0 \right\}.$$

Define the shift map on $X^+$ as $\sigma: X^+ \to X^+$, $\sigma(x) = y$, $y_n = x_{n+1}$, $n \geq 0$. This means all sequences are shifted one place to the left, with the first term being deleted. The set $\{1, 2, \ldots, k\}$ is equipped with the discrete topology so that $X^+$ is compact with the Tychonov product topology and $\sigma$ is a local homeomorphism. Then, as is well-known, this dynamical system is called a *(one-sided) subshift of finite type* with *transition matrix* $A$.

Now let $0 < \theta < 1$. We shall associate $X^+$ with the metric $d_\theta(x, y) = \theta^n$ where $n = \max\{m: x_i = y_i, 0 \leq i < m\}$. Let $f: X^+ \to U(d)$ be continuous where $U(d)$ is the group of unitary $d \times d$ matrices equipped with the Euclidean matrix norm $\|\cdot\|$. For such a function $f$, define for $n \geq 0$,

$$\text{var}_n f = \sup\{\|f(x) - f(y)\|: d_\theta(x, y) \leq \theta^n\}$$

and set $|f|_\theta = \sup_{n \geq 0}\{\text{var}_n f / \theta^n\}$. Also set $|f|_\infty = \sup\{\|f(x)\|: x \in X^+\}$. Then, define

$$F_\theta^+(U(d)) = \{f: X^+ \to U(d): f \text{ is continuous } \& \ |f|_\theta < \infty\}.$$

Then $F_\theta^+(U(d))$ together with the norm $\|f\|_\theta = |f|_\theta + |f|_\infty$ is complete. It is easy to see that $F_{\theta_1}^+(U(d)) \subset F_{\theta_2}^+(U(d))$ when $\theta_1 \leq \theta_2$.

Now recall that a function $k$ from metric spaces $(Y_1, d_1)$ to $(Y_2, d_2)$ is said to be $\alpha$-Hölder continuous, $0 < \alpha < 1$, if $d_2(k(x), k(y)) \leq M d_1(x, y)^\alpha$ for all $x, y \in Y_1$, for some $M > 0$. Thus if $f: X^+ \to U(d)$ is $\alpha$-Hölder continuous with respect to the metric $d_\theta$ then $f \in F_{\theta^\alpha}^+(U(d))$. And the converse is also true. This follows from the easily proven fact: $\|f(x) - f(y)\| \leq C d_\theta(x, y)$ if and only if $\text{var}_n f \leq C\theta^n, n = 0, 1, \ldots$.

In particular if we set

$$F^+ = \bigcup_{0 < \theta < 1} F_\theta^+(U(d))$$

then we see that $F^+$ is the space of all Hölder continuous functions from $X^+$ to the unitary group $U(d)$. One can check that locally constant functions defined on $X^+$ which in turn includes functions depending on a finite number of coordinates are in $F^+$. In fact , they are in

$$\bigcap_{0<\theta<1} F_\theta^+(U(d)).$$

All the above set up is fairly standard and can be found in for e.g., Parry and Pollicott (1990). We shall also be interested in the space of all real-valued Hölder continuous functions on $X^+$,

$$F_\theta^+(\mathbb{R}) = \{f: X^+ \to \mathbb{R} : f \text{ is continuous } \& \ |f|_\theta < \infty\}$$

where $|\cdot|_\theta$ is defined similarly as before. It is well-known (Bowen (1975)) that for a given $g \in F_\theta^+(\mathbb{R})$ there exists a unique $\sigma$-invariant probability measure $m$ – called the *equilibrium state* of $g$ – so that

$$h_m(\sigma) + \int g \, dm \geq h_\nu(\sigma) + \int g \, d\nu$$

for all other $\sigma$-invariant probability measure $\nu$. (Here, $h_\nu(\sigma)$ denotes the metric entropy of $\sigma$ w.r.t $\nu$ (see, for instance, Walters (1982)). From here onwards, we shall assume that $m$ is the equilibrium state for some $g \in F_\theta^+(\mathbb{R})$ which is *normalised*, i.e.

$$\sum_{\sigma y = x} e^{g(y)} = 1, \qquad \forall \, x \in X^+.$$

Let us denote by $F_\theta^+(\mathbb{C}^d)$ the Banach space of all $\mathbb{C}^d$-valued Holder continuous functions defined on $X^+$ equiped with the (analogouesly defined) norm $\|\cdot\|_\theta$. For a normalised $g \in F_\theta^+(\mathbb{R})$ and $f \in F_\theta^+(U(d))$ we define the transfer operator $L: F_\theta^+(\mathbb{C}^d) \to F_\theta^+(\mathbb{C}^d)$, $L = L_{g,f}$, by

$$(Lw)(x) = \sum_{\sigma y = x} e^{g(y)} f(y) w(y), \quad x \in X^+$$

for all $w \in F_\theta^+(\mathbb{C}^d)$. Then it is easy to see that $L$ is a continuous linear operator on $F_\theta^+(\mathbb{C}^d)$. For completeness, we shall give the proof of the key result due to Parry which is to do with the spectral properties of the operator $L$. We shall need the following two lemmas.

**Lemma 1** *Consider the operator* $L = L_{g,f}$ *with respect to some* $f \in F_\theta^+(U(d))$ *and normalised* $g \in F_\theta^+(\mathbb{R})$*. Then there exists a constant* $C > 0$ *such that*

$$\|L^n w\|_\theta \le C|w|_\infty + \theta^n |w|_\theta$$

*for each* $n \ge 0$ *and* $w \in F_\theta^+(\mathbb{C}^d)$*. Moreover* $C$ *only depends on* $f$*,* $g$ *and* $\theta$*.*

The proof of the above lemma is essentially a simple modification of the analogous result due to Parry & Pollicott (1990) for real-valued Hölder continuous functions. The proof of the next lemma can also be found in Parry & Pollicott (1990).

**Lemma 2** *Let* $1 > \theta' > \theta$*. Then the function space* $F_\theta^+(U(d))$ *can be written as*

$$F_\theta^+(U(d)) = V \oplus W$$

*where* $V$ *and* $W$ *are* $L$*-invariant subspaces of* $F_\theta^+(U(d))$ *such that the spectral radius of* $L$ *when restricted to* $W$ *is at most* $\theta'$ *and the subspace* $V$ *is finite-dimensional.*

The following result is due to Parry and is the key to everything else that follows.

**Theorem 1 (Parry (1997))** *Let* $\sigma$ *be equiped with an equilibrium state measure* $m$ *corresponding to a normalised* $g \in F_\theta^+(\mathbb{R})$ *and let* $f \in F_\theta^+(U(d))$*. Also let* $w\colon X^+ \to \mathbb{C}^d$ *be integrable with respect to* $m$*. If the operator* $L = L_{g,f}$ *satisfy*

$$Lw(x) = w(x) \quad a.e.\ [m],$$

*then there exists* $w' \in F_\theta^+(\mathbb{C}^d)$ *such that* $w = w'$ *a.e* $[m]$ *and*

$$Lw'(x) = w'(x) \quad everywhere\ x.$$

**Proof** Suppose $w\colon X^+ \to \mathbb{C}^d$ is integrable and $Lw(x) = w(x)$ $a.e.$ $[m]$. We shall approximate $w$ in the $L^1$-sense by locally constant functions $w_k$ such that

$$\int |w - w_k| dm \le \frac{1}{k}.$$

Note that $w_k \in F_\theta^+(\mathbb{C}^d)$.

For each $k = 1, 2, \ldots$, one can apply Lemma 1 to conclude that there exists a subsequence $\{m_n\}$ such that $L^{m_n}w_k$ converges uniformly to some $w_k^* \in F_\theta^+(\mathbb{C}^d)$. Now write $v_k = w - w_k$ and $v_k^* = w - w_k^*$. Then since $g$ is normalised and $f$ is unitary valued, we have

$$\begin{aligned}
\int |v_k^*|\, dm &= \int |\lim_{m_n\to\infty} L^{m_n}v_k|\, dm \\
&\leq \int \lim_{m_n\to\infty} |L^{m_n}v_k|\, dm \\
&= \int |v_k|\, dm \\
&\leq \frac{1}{k}
\end{aligned}$$

Now choose any $\theta'$ such that $\theta < \theta' < 1$. Then by Lemma 2, we can write $F_\theta^+(\mathbb{C}^d) = V_{\theta'} \oplus V$ where each subspace $V_{\theta'}$ and $V$ is $L$-invariant. Moreover, the spectral radius of $L|V_{\theta'}$ is at most $\theta' < 1$ and $V$ is finite-dimensional. Write $w_k = w'_k + w''_k$ where $w'_k \in V_{\theta'}$ and $w''_k \in V$. By the spectral radius formula, $L^m w'_k \to 0$ so that $L^{m_n}w''_k \to w_k^*$ uniformly. Since $V$ is finite-dimensional, it is closed. Thus $w_k^* \in V$. In short, we have $w = w_k^* + v_k^*$ where $w_k^*$ is in a finite-dimensional subspace and $\int |v_k^*|\, dm \leq 1/k$. In particular, this means that $w_k^*$ is $L^1$-Cauchy in $V$. Hence an $L^1$-limit point, say $w'$, exists and $w = w'$ *a.e* $[m]$. Also since all norms on $V$ are equivalent, this convergence also holds with respect to $\|\cdot\|_\theta$ and hence $w' \in F_\theta^+(\mathbb{C}^d)$.
When $w = w'$ *a.e* it is clear that $Lw' = w'$ *a.e.* Next we show that $Lw' = w'$ $\forall x \in X^+$. Consider the set $A = \{x \in X^+ : Lw' - w' \neq 0\}$. By continuity, $A$ is open. Since equilibrium state measures are positive on open sets (see Walters (1982)), we must have $A = \emptyset$. Hence $Lw' = w'$ $\forall x$ as required. □

The following result, which appears as a remark in Parry (1997), relates eigenfunctions of the transfer operator and the functional equation mentioned in the introduction.

**Proposition 1** *Let $f \in F_\theta^+(U(d))$. If $w$ is measurable and $w \circ \sigma = f \cdot w$ a.e then $Lw = w$ a.e $[m]$. On the other hand if $w$ is continuous and $Lw = w$ for all $x$ then $w \circ \sigma = f \cdot w$ everywhere.*

**Proof** Let $w$ be measurable and $w \circ \sigma = f \cdot w$ *a.e.* Then, since $g$ is normalised, it is straight-forward to check that $Lw = w$ *a.e*

Now suppose $w: X^+ \to \mathbb{C}^d$ is continuous and

$$L_f w(x) = \sum_{\sigma(y)=x} e^{g(y)} f(y) w(y) = w(x).$$

Then since $f$ is unitary valued, we have $\sum_{\sigma(y)=x} e^{g(y)} |w(y)| \geq |w(x)|$. Observe that since $g$ is normalised, the right-hand side of this inequality is a convex combination of the $|w(y)|$ where $\sigma(y) = x$. By continuity, we can choose some $x'$ so that $|w(x)|$ is maximum at this point and this in turn implies that the previous inequality must be an equality whenever $x' = \sigma(y)$, i.e $|w(y)| = |w(x')|$. By considering the equation $L^n w = w$, we have $|w(y)| = |w(x')|$ if $x' = \sigma^n(y)$. The set $\{y: \sigma^n y = x', \text{some } n\}$ is dense in $X^+$. Thus $|w|$ is constant. Now

$$\sum_{\sigma(y)=x} e^{g(y)} f(y) w(y) = w(x)$$

and the left-hand side is a convex combination of the $f(y)w(y)$ which has the same norm as $w(x)$. Thus it follows that $w(x) = f(y)w(y)$ whenever $\sigma(y) = x$. In particular, $w \circ \sigma(x) = f(x) \cdot w(x)$ for all $x$ and this completes the proof. □

We also have

**Proposition 2** *Let $m$ be the equilibrium state of a normalised $F_\theta^+(\mathbb{R})$ function. Also let $f \in F_\theta^+(U(d))$ and $w: X^+ \to \mathbb{C}^d$ be measurable. Then $w \circ \sigma(x) = f(x)w(x)$ a.e $[m]$ implies $w = w'$ a.e $[m]$ and $w' \circ \sigma(x) = f(x)w'(x)$ $\forall x$ for some $w' \in F_\theta^+(\mathbb{C}^d)$.*

**Proof** By proposition 1, we have $L_f w = w$ *a.e.* So that by Parry's theorem, $w = w'$ *a.e* for some $w' \in F_\theta^+(\mathbb{C}^d)$. In particular $L_f w' = w'$ $\forall x$. The result follows by invoking once again proposition 1. □

# 3 Applications to Homogeneous Extensions

We now apply Parry's result (Theorem 1) to homogeneous extensions of shifts. As in the introduction, let $\sigma: X^+ \to X^+$ be a subshift of finite type such that the associated transition matrix is aperiodic. And let $m$ be an equilibrium state corresponding to a $F_\theta^+(\mathbb{R})$ function. For example, the Markov measure arising from a stochastic matrix $P$ which is compatible with $A$ is an equilibrium state for the locally constant function $g(x) = \log P(x_0, x_1)$ (see Parry & Pollicott (1990)). With respect to $m$, $\sigma$ possess a very strong ergodic property, namely it is *exact* (see Bowen (1975)).

As in the introduction, let $G$ be a compact metric group (with metric $d_G$) and let $\phi: X \to G$ be Hölder continuous, i.e $d_G(\phi(x), \phi(y)) \leq M d_\theta(x, y)^\alpha$, for some $M, 0 < \alpha < 1$. Observe that when $R$ is an irreducible unitary representation of $G$ (of dimension $d$, say), then $R \circ \phi \in F_\theta^+(U(d))$. Also let $H$ be a closed subgroup of $G$. The homogeneous extension $\sigma_f: X^+ \times G/H \to X^+ \times G/H$ given by

$$\sigma_\phi(x, gH) = (\sigma x, \phi(x) gH)$$

is then measure-preserving with respect to the the (restricted) Haar extension measure $m \times \lambda'$. One should observe that, there is a compact group extension $\tilde{\sigma}_\phi: X \times G \to X \times G$ defined by $\tilde{\sigma}_\phi(x, g) = (\sigma x, \phi(x) g)$ associated with the homogeneous extension $\sigma_\phi$. Note also that both systems share the same skewing-function $\phi$ (see, Noorani (1997), for details). Hence, by replacing $f$ with $R \circ \phi$ in Proposition 2 , our criterion for the ergodicity of the homogeneous extension (see introduction) simplifies to:

**Theorem 2** *Let $R$ be a non-trivial irreducible $d$-dimensional unitary representation of $G$ such that $bR(h) = b$, $\forall h \in H$, for some non-zero $b \in \mathbb{C}^d$. Then $\sigma_\phi$ is ergodic (with respect to $m \times \lambda'$) if and only if the only solutions $w \in F_\theta^+(\mathbb{C}^d)$ satisfying the functional equation*

$$R \circ \phi(x) \cdot w(x) = w \circ \sigma(x) \qquad (\textit{everywhere})$$

*is $w \equiv 0$ everywhere.*

Similarly we also have

**Theorem 3** *Let the associated group extension of $\sigma_\phi$ be ergodic. Also let $1 \neq \alpha \in \mathbb{C}$ and $\chi$ a non-trivial one-dimensional irreducible character of $G$ such that $\chi(h) = 1$ $\forall h \in H$. Then $\sigma_\phi$ is weak-mixing (with respect to $m \times \lambda'$) if and only if the only solutions $w \in F_\theta^+(\mathbb{C})$ satisfying the functional equation*

$$\alpha \chi \circ \phi(x) \cdot w(x) = w \circ \sigma(x) \qquad (\textit{everywhere } x)$$

*is $w \equiv 0$ everywhere.*

The above results can also be extended to cover the case where $X$ is an aperiodic two-sided subshift of finite type. Recall that this means, for a given $k \times k$ aperiodic 0-1 matrix $A$,

$$X = \left\{ x = (x_n)_{-\infty}^{\infty} \in \prod_{-\infty}^{\infty} \{1, \ldots, k\} \colon A(x_n, x_{n+1}) = 1 \ \forall \ n \in \mathbb{Z} \right\}.$$

The shift map $\sigma \colon X \to X$ is then defined as $\sigma(x) = y$, $y_n = x_{n+1}$, $n \in \mathbb{Z}$. The (two-sided) Hölder function spaces $F_\theta(U(d)), F_\theta(\mathbb{C}^d)$ and $F_\theta(\mathbb{R})$ are defined analogously as in the previous section (see Parry & Pollicott (1990)). We shall need the following result which relates functions on the two-sided shifts to the one-sided version.

**Proposition 3 ( Sinai (1972), Parry & Pollicott (1990))** *If $f \in F_\theta(U(d))$ then there exists $h \in F_{\theta^{1/2}}(U(d))$ and $f' \in F^+_{\theta^{1/2}}(U(d))$ such that*

$$f' = (h \circ \sigma)^{-1} \cdot f \cdot h.$$

Strictly speaking, the function $f'$ in the above proposition is an element of $F_{\theta^{1/2}}(U(d))$. But since it depends only on future coordinates, i.e $f(x) = f(y)$ whenever $x_i = y_i, i = 0, 1, 2, \ldots$, $f'$ is then viewed as an element of $F^+_{\theta^{1/2}}(U(d))$. We shall also denote by $m$ the equilibrium state measure arising from a normalised $F_\theta(\mathbb{R})$ function.

**Proposition 4** *Let $f \in F_\theta(U(d))$ and $w \colon X \to \mathbb{C}^d$ be measurable. If $w \circ \sigma = f \cdot w \quad$ a.e $[m]$ then there exists some $w' \in F^+_{\theta^{1/2}}(\mathbb{C}^d)$ such that $w = w' \quad$ a.e $[m]$. Moreover $w' \circ \sigma = f \cdot w \quad$ (everywhere)*

**Proof** By Proposition 3, we have $f = (h \circ \sigma)^{-1} \cdot f' \cdot h$ for some $h \in F_{\theta^{1/2}}(U(d))$ and $f' \in F^+_{\theta^{1/2}}(U(d))$. Hence

$$(hw) \circ \sigma = f' \cdot (hw) \quad a.e.$$

Now it can be shown that (see, e.g Noorani (1997)) the function $hw$ depends only future coordinates, so that both $hw$ and $f'$ are defined on $X^+$. We are now in the same situation as before. In particular by applying Proposition 1, we have $L_{f'}(hw) = hw \quad a.e$ and Parry's theorem implies there exists some $w'' \in F^+_{\theta^{1/2}}(\mathbb{C}^d)$ such that $hw = w'' \quad a.e\ [m]$ and $L_{f'} w'' = w$ everywhere. We take $w' = h^{-1} w''$ to complete the proof. □

Observe that, with respect to $m$, $\sigma$ is Bernoulli . Then the analogue of theorems 2 and 3 for the two-sided case is as follows:

**Theorem 4** *Let $\sigma$ be an aperiodic two-sided s.s.f.t equipped with an equilibrium state measure corresponding to a normalised $F_\theta(\mathbb{R})$ function. Also let $\sigma_\phi$ be the corresponding homogeneous extension.*

*a) Let $R$ be a non-trivial irreducible d-dimensional unitary representation of $G$ such that $bR(h) = b$, $\forall h \in H$, for some non-zero $b \in \mathbb{C}^d$. Then $\sigma_\phi$ is ergodic (with respect to $m \times \lambda'$) if and only if the only solutions $w \in F_{\theta^{1/2}}(\mathbb{C}^d)$ satisfying the functional equation*

$$R \circ \phi(x) \cdot w(x) = w \circ \sigma(x) \qquad (everywhere)$$

*is $w \equiv 0$ everywhere.*

*b) Let the associated group extension of $\sigma_\phi$ be ergodic. Also let $1 \neq \alpha \in \mathbb{C}$ and $\chi$ a non-trivial one-dimensional irreducible character of $G$ such that $\chi(h) = 1$ $\forall h \in H$. Then $\sigma_\phi$ is weak-mixing (with respect to $m \times \lambda'$) if and only if the only solutions $w \in F_{\theta^{1/2}}(\mathbb{C})$ satisfying the functional equation*

$$\alpha\chi \circ \phi(x) \cdot w(x) = w \circ \sigma(x) \qquad (everywhere\ x)$$

*is $w \equiv 0$ everywhere.*

From here onwards, for ease of presentation, we shall concentrate on the two-sided case. In analogy with Parry's Corollary 4.5 and 4.6 (Parry (1997)), we have

**Corollary 1** *The homogeneous extension $\sigma_\phi$ is ergodic if and only if $\sigma_\phi$ has a dense orbit.*

**Proof** Suppose $\sigma_\phi$ is ergodic with respect to $m \times \lambda'$. Then since $m \times \lambda'$ is positive on non-empty open sets we deduce that $\sigma_\phi$ has a dense orbit (see Walters (1982), Theorem 5.16).

Conversely suppose $\sigma_\phi$ has a dense orbit. Let $R$ be a non-trivial irreducible $d$-dimensional representation of $G$ satisfying $bR(h) = b$, $\forall h \in H$ for some non-zero $b \in \mathbb{C}^d$ and

$$R \circ \phi(x) \cdot w(x) = w \circ \sigma(x) \qquad (everywhere)$$

where $w \in F_{\theta^{1/2}}(\mathbb{C}^d)$.

Define $K: X \times G/H \to \mathbb{C}^d$ by $K(x, gH) = bR(g^{-1})w(x)$. Then it is straightforward to check that $K$ is $\sigma_\phi$-invariant. In particular $K$ is constant on the dense orbit of $\sigma_\phi$ and by continuity $K$ is constant on the whole of $X \times G/H$. Since $R$ is non-trivial and irreducible, $w = 0$. Ergodicity of $\sigma_\phi$ then follows by applying theorem 4. □

**Corollary 2** *Let the associated group extension of $\sigma_\phi$ be ergodic. If for any two distinct periodic points $x, y$ (of $\sigma$) of period $p$ we have $\chi(\phi^p(x)) \neq \chi(\phi^p(y))$ for all non-trivial one-dimensional characters $\chi$ with $\chi(h) = 1\ \forall h \in H$ then $\sigma_\phi$ is weak-mixing.*
*(Here, and throughout, $\phi^n(x)$ means $\phi(\sigma^{n-1}x) \cdots \phi(\sigma x)\phi(x)$).*

**Proof** Let $1 \neq \alpha \in \mathbb{C}$ and $\chi$ be a non trivial one-dimensional (irreducible) character of $G$ satisfying $\chi(h) = 1\ \forall h \in H$. For a contradiction, suppose there exists some $w \in F_{\theta^{1/2}}(\mathbb{C})$ such that $w \neq 0$ and

$$\alpha\chi \circ \phi \cdot w = w \circ \sigma \qquad (\textit{everywhere}).$$

But this functional equation says for any two distinct periodic points $x, y$ of period $p$ we have $\chi(\phi^p(x)) = \alpha^{-p} = \chi(\phi^p(y))$. Hence $w \equiv 0$ and Theorem 4 then gives us $\sigma_\phi$ is weak-mixing. □

### 3.1 Genericity

We now move on to the question of genericity. We say that a compact metric group $G$ has the *general position property* if $G$ is infinite and there exists a positive integer $k$ such that the set

$$\{(g_1, \ldots, g_k) \in G \times \ldots \times G : < g_1, \ldots, g_k >= G\}$$

is dense in $\underbrace{G \times \ldots \times G}_{k \text{ times}}$.
(Here $< g_1, \ldots, g_k >= G$ means the smallest closed subgroup containing $\{g_1, \ldots, g_k\}$ is $G$.) We shall always assume that $G$ is identifiable with some subgroup of $U(d)$ (some $d$). The prime example is when $G$ is a compact Lie group.

We shall need the following result of Parry (see Parry (1997), Prop. 6.4).

**Lemma 3** *Let $G$ be a compact metric group and $R$ be an irreducible d-dimensional unitary representation of $G$. Then the set*

$$A_R = \{\phi \in F_\theta(G) \colon R(\phi(x))w(x) = w(\sigma(x)) \text{ a.e for some integrable } w \colon X \to \mathbb{C}^d\}$$

*is closed in $F_\theta(G)$.*

The proofs of the following two results follows closely the proof provided by Parry in the setting of group extensions.

**Theorem 5** *Let $G$ be a compact metric group and suppose that $G$ has the general position property. Then*

$$\{\phi \in F_\theta(G) \colon \sigma_\phi \text{ is ergodic}\}$$

*is a dense $G_\delta$ set in $F_\theta(G)$.*

**Proof** Let $A = \{\phi \in F_\theta(G) \colon \sigma_\phi \text{ is ergodic}\}$ and let

$$B = \bigcup_R \{\phi \in F_\theta(G) \colon R(\phi)w = w \circ \sigma \text{ a.e, } w \text{ integrable and } w \neq 0\}$$

where the union is taken over all non-trivial and irreducible represention $R$ of $G$ such that $bR(h) = b$, $\forall h \in H$, for some non-zero $b \in \mathbb{C}^d$. Then the complement of $B$,

$$\begin{aligned} B^c &= \bigcap_R \{\phi \in F_\theta(G) \colon R(\phi)w = w \circ \sigma \text{ a.e.} \Rightarrow w = 0\} \\ &= \bigcap_R \{\phi \in F_\theta(G) \colon \sigma_\phi \text{ ergodic w.r.t } R\}. \end{aligned}$$

Since $A = B^c$ and the set $\{\phi \in F_\theta(G) \colon \sigma_\phi \text{ ergodic w.r.t } R\}$ is open by Lemma 3, we deduce that $A$ is a countable intersection of open sets, i.e $A$ is $G_\delta$. We next show that $A$ is dense in $F_\theta(G)$. In particular, we show that given $\phi \in A$ we can construct some $\phi' \in A$, arbitrary close to $\phi$ such that $\sigma'_\phi$ is ergodic. By the general position property, we may assume that there exists some $\ell$ such that

$$\{(g_1, \ldots, g_\ell) \in G \times \ldots \times G \colon < g_1, \ldots, g_\ell >= G\}$$

is dense in $\underbrace{G \times \ldots \times G}_{\ell \text{ times}}$. Now, let $p_0$ be a fixed point of $\sigma$ and let $p_1, \ldots, p_\ell$ be $\ell$ distinct points such that

$$\lim_{n\to\infty} \sigma^n p_i = p_0 = \lim_{n\to\infty} \sigma^{-n} p_i$$

that is the forward and backward iterates of each $p_i$ converges to $p_0$, for each $i = 1, \dots, \ell$. For $i = 1, \dots, \ell$, consider the sequences

$$\phi(\sigma p_i), \quad \phi(\sigma^2 p_i)\phi(\sigma p_i), \dots, \phi(\sigma^n p_i) \cdots \phi(\sigma p_i), \dots$$
$$\phi(\sigma^{-1} p_i), \quad \phi(\sigma^{-1} p_i)\phi(\sigma^{-2} p_i), \dots, \phi(\sigma^{-1} p_i) \cdots \phi(\sigma^{-n} p_i)$$

Each of these sequences are in $G$ and by the compactness of $G$, there exists convergence subsequences. Thus w.l.o.g write

$$\lim_{n\to\infty} \phi(\sigma^n p_i) \cdots \phi(\sigma p_i) = g_i^+ \ \& \ \lim_{n\to\infty} \phi(\sigma^{-1} p_i) \cdots \phi(\sigma^{-n} p_i) = g_i^-$$

Again by the general position property, we may choose some $g_i'$ which is arbitrary close to the identity of $G$ so that if $g_i = g_i^+ \phi(p_i) g_i' g_i^-$, $i = 1, \dots, \ell$ then $g_1, \dots, g_\ell$ generates $G$. We are now ready to define $\phi'$: Let $C_1, \dots, C_\ell$ be (small) cylindrical neighbourhoods of $p_1, \dots, p_\ell$ such that $C_i$ are pairwise disjoint and $\sigma^n p_i \notin C_1 \cup \dots \cup C_\ell$ for $n \neq 0$. Then take $\phi'(x)$ as follows:

$$\phi'(x) = \begin{cases} \phi(x) & \text{if } x \notin \bigcup_{i=1}^{\ell} C_i \\ \phi(x) g_i' & \text{if } x \in C_i, \ i = 1, \dots, \ell \end{cases}$$

Then $\phi' \in F_\theta(G)$ and $\phi'$ is $F_\theta(G)$-close to $\phi$. Finally we show that $\sigma'_\phi$ is ergodic. Thus let $R$ be a non-trivial irreducible representation of $G$, $w \in F_{\theta^{1/2}}(\mathbb{C}^d)$ and $R(\phi')w = w \circ \sigma$. Hence $w$ satifies

$$R(\phi(\sigma^n p_i)) \cdots R(\phi(p_i)) R(g_i') \cdots R(\phi(\sigma^{-n} p_i)) w(\sigma^{-n} p_i)$$

so that, taking limits as $n \to \infty$ (as before), we have

$$w(p_0) = R(g_i^+ \phi(p_i)) g_i' g_i^- w(p_0) = R(g_i) w(p_0)$$

for each $i = 1, \dots, \ell$. Since $\{g_1, \dots, g_\ell\}$ generates $G$, we have $R(g)w(p_0) = w(p_0)$ $\forall g \in G$. Since $R$ is irreducible and non-trivial, necessarily $w(p_0) = 0$. Now, consider $< w, w >$. It is easy to see that $< w, w >$ is $\sigma$-invariant. Hence by the ergodicity of $\sigma$ we have $< w, w >$ is constant. In particular, $< w(x), w(x) > = < w(p_0), w(p_0) > = 0$, $\forall x$. Thus $w = 0$ and this completes the proof. □

**Theorem 6** *Let $G$ be a compact metric group and suppose that $G$ has the general position property. Also, given a homogeneous extension $\sigma_\phi$, let $\tilde{\sigma}_\phi$ be the associated group extension. Then the set*

$$K = \{\phi \in F_\theta(G) : \tilde{\sigma}_\phi \text{ is ergodic } \& \ \sigma_\phi \text{ is weak-mixing}\}$$

*contains a dense $G_\delta$ set of $F_\theta(G)$.*

**Proof** Let $A = \{\phi \in F_\theta(G): \sigma_\phi \text{ is ergodic}\}$. Then $K \subset A$ and the previous result tells us that $A$ is a dense $G_\delta$ set. Suppose $\phi \in A$ and $\phi$ is not weak-mixing. Then there exists a 1-dimensional non-trivial (irreducible) character $\chi$ satisfying $\chi(h) = 1$, $\forall h \in H$ and $\alpha \in \mathbb{C}, |\alpha| = 1$ such that $w(\sigma x) = \alpha\chi(\phi(x))w(x)$ $\forall x \in X$. Now, choose 2 periodic points $x$ and $y$ having the same least period, say $p \geq 1$ such that their orbits are disjoint. Then $\alpha^p\chi(\phi^p(x)) = \alpha^p\chi(\phi^p(y))$. Let $L_\chi = \{\phi \in F_\theta(G): \chi(\phi^p(x)) = \chi(\phi^p(y))\}$ so that $L_\chi \neq \emptyset$. Moreover $L_\chi$ is closed and nowhere dense. Hence the Baire category theorem implies $\cap_\chi L_\chi^c$ is a dense $G_\delta$ set in $F_\theta(G)$. In particular, $A \cap \cap_\chi L_\chi^c$ is also dense $G_\delta$. This set, by corollary 2, is contained in $K$ and this completes the proof. □

### 3.2 Ergodic Components

Let $G$ be a compact metric group. Then $G$ is said to have the decreasing chain property if every decreasing sequence of closed subgroups of $G$ has a non-trivial minimal element. For example, any compact Lie group will satisfy this property.

The following result is due to Parry (1997).

**Proposition 5** *Let $G$ be a compact metric group having the decreasing chain property. Also, for each $\phi \in F_\theta(G)$ let $\tilde{\sigma}_\phi$ be the corresponding group extension with skewing-function $\phi$. Suppose $\tilde{\sigma}_\phi$ is not ergodic. Then there exists a non-trivial closed subgroup $\overline{H}$ of $G$ and $\phi' \in F_\theta(\overline{H})$ such that $\tilde{\sigma}_\phi$ is topologically conjugate to $\tilde{\sigma}_{\phi'}$ and the restriction of $\tilde{\sigma}_{\phi'}$ to $X \times \overline{H}$ is ergodic. Moreover, the decomposition of $X \times G$ into sets of the form $X \times \overline{H}g$, $g \in G$ are the ergodic components of $\tilde{\sigma}_{\phi'}$.*

**Corollary 3** *Let $G$ be a compact metric group having the decreasing chain property. Suppose the homogeneous extension $\sigma_\phi$ is not ergodic. Then there exists a non-trivial closed subgroup $\overline{H}$ of $G$ and $\phi' \in F_\theta(\overline{H})$ such that $\sigma_\phi$ is topologically conjugate to $\sigma_{\phi'}$ and the restriction of $\sigma_{\phi'}$ to $\pi(X \times \overline{H})$ is ergodic. Moreover, the ergodic components of $\sigma_{\phi'}: X \times G/H \to X \times G/H$ are sets of the form $\pi(X \times \overline{H}g)$, $g \in G$.*
*(Here, $\pi$, is the natural projection map from $X \times G$ to $X \times G/H$.)*

**Proof** Suppose the homogeneous extension $\sigma_\phi$ is not ergodic. Then the associated group extension $\tilde{\sigma}_\phi$ is also not ergodic. In particular, the previous result implies the existence of the subgroup $\overline{H}$ and the function $\phi' \in F_\theta(\overline{H})$ such that $\tilde{\sigma}_\phi$ is topologically conjugate to $\tilde{\sigma}_{\phi'}$. It is then an easy matter to show that the corresponding homogeneous extensions $\sigma_\phi$ and $\sigma_{\phi'}$ are also topologically conjugate. The distinct elements of $\pi(X \times \overline{H}g)$, $g \in G$ are then the ergodic components of $\sigma_{\phi'}: X \times G/H \to X \times G/H$ and the proof is thus completed. □

## References


Bowen, R., *Equilibrium states and the ergodic theory of anosov diffeomorphisms*, Lecture Notes in Mathematics **470**, Springer, Berlin, 1975.

Katok, A. and Hasselblatt, B., *Introduction to the Modern Theory of Dynamical Systems.* Encylopedia of Mathematics and its Applications **54**, Cambridge University Press, Cambridge, 1995.

Noorani, M.S.M., Ergodicity and weak-mixing of homogeneous extensions of measure-preserving transformations with applications to Markov shifts. *Mh. Math.*, 1997, **123**, 149-170.

Parry, W., Skew-products of shifts with a compact Lie group. *Journal of the London Mathematical Society*, 1997, **56**(2), 395-404.

Parry, W. and Pollicott, M., *Zeta functions and the periodic orbit structure of hyperbolic dynamics.* Astérisque **187-188**, Société Mathématique de France, 1990.

Parry, W. and Pollicott, M., The Livšiv cocycle equation for compact Lie group extensions of hyperbolic systems. *Journal of the London Mathematical Society*, 1997, **56**(2), 405-416.

Sinai, Ya. G., Gibbs measures in ergodic theory. *Russian Math. Surveys*, 1972, **166**, 21-69.

Walters, P., *An Introduction to Ergodic Theory*, Springer, Berlin, 1982.

Walters, P., Topological Wiener–Wintner ergodic theorems and a random $L^2$ ergodic theorem. *Ergod. Th. & Dynam. Sys.*, 1996, **16**, 179–206.

# The Abelian Index of finite groups

Kenneth K. Nwabueze,
Norhayati Hamzah,
Saiful A. Husain

**Abstract**

Let $G$ be a finite $p$-group of order $p^k$ and let $\mathcal{I}(G)$ denote the numerical measure (which we call the abelian index) of deviation of $G$ from being an abelian group. Using the Burnside ring theoretic method, we demonstrate that $I(G) = 1$ if and only if $G$ is abelian whereas in all other cases $I(G) = p$. This result seems to suggest that $\mathcal{I}(G)$ is a blunt measure of deviation of a finite $p$-group from being abelian.

AMS Subject Classification: 20A11
Key words; abelian index, burnside rings, groups,

# 1 Introduction

In [2] and [3], a numerical measure of deviation for a finite $p$ -group $G$ from being cyclic and elementary abelian respectively was provided. The method used in those papers were Burnside ring theoretic and this clears the corridor for similar computations. Here we shall be interested in finding such numerical deviations induced from abelian subgroups of a finite $p$-group $G$. The main result in this paper is the following:

**Main result**: Let $G$ denote a finite $p$-group of order $p^k$ and let $\mathcal{I}(G)$ denote its Abelian index (see section 2 for the definition of Abelian index). Then one has:
(a) $\mathcal{I}(G) = 1$ if and only if $G$ is abelian.
(b) In all other cases $\mathcal{I}(G) = p$.

This result seems to suggest that $\mathcal{I}(G)$ is a blunt measure of the deviation for a finite $p$-group $G$ from being abelian, since this depends only on two values 1 and $p$. Numerical measures of this sort have found various applications in some induction theorems for functors on the category of finite groups (see, for instance, [5]). We have arranged this paper as follows: In section 2, we collect some useful results concerning the Burnside rings and give the Burnside ring theoretic formulation of the Abelian Index. This section, also provides some elementary useful algebraic techniques used in section 3. Finally, in section 3 we give the proof of the main result.

**Notation 1.1** *Unless otherwise stated we shall adopt the following notation:*

| | |
|---|---|
| $\{1_G\}$ : | *the trivial subgroup of a group G* |
| $p$ : | *a prime number* |
| $\|G\|$ : | *the order of G* |
| $Z$ : | *the ring of integers* |
| $< T >$: | *subgroup of G generated by T* |

# 2 Preliminaries

Let $G$ be a finite group and consider $\mathcal{S}$, sets of subgroups of a finite group $G$ which is closed with respect to conjugation and such that $G \in \mathcal{S}$. We define a $(G\ ,\ \mathcal{S})$-set as a finite left $G$-set $S$ with $G_s \in \mathcal{S}$ for all $s$ in $S$. This implies that for any subgroup $U \in \mathcal{S}$ the set $G/U = \{gU \mid g \in G\}$ of left cosets of $U$ in $G$ is a $(G\ ,\ \mathcal{S})$-set; the $G$-action on $G/U$ is defined by left multiplication $G \times G/U \longrightarrow G/U\ :\ (h\ ,\ gU) \longrightarrow hgU$. Furthermore for any two $(G\ ,\ \mathcal{S})$-sets $S_1$ and $S_2$ one has $S_1 \times S_2$ and $S_1 \cup S_2$ are also $(G\ ,\ \mathcal{S})$-sets. Observe that the isomorphism classes of $(G\ ,\ \mathcal{S})$-sets form a commutative semiring $B^+(G\ ,\ \mathcal{S})$ because one has the obvious natural isomorphisms

$$\begin{aligned} S_1 \cup S_2 &\cong S_2 \cup S_1 \\ (S_1 \cup S_2) \cup S_3 &\cong S_1 \cup (S_2 \cup S_3) \\ S_1 \times S_2 &\cong S_2 \times S_1 \\ (S_1 \times S_2) \times S_3 &\cong S_1 \times (S_2 \times S_3) \\ S_1 \times (S_2 \cup S_3) &\cong (S_1 \times S_2) \cup (S_1 \times S_3). \end{aligned}$$

Moreover $1 \in B^+(G\ ,\ \mathcal{S})$ exists, namely $G/G$. Therefore any map from the set $B^+(G\ ,\ \mathcal{S})$ into a ring $R$ which commutes with sums and products and sends $1 \in B^+(G\ ,\ \mathcal{S})$ onto $1_R$ is a homomorphism from the semiring $B^+(G\ ,\ \mathcal{S})$ into $R$ and this factors uniquely through the universal ring associated to $B^+(G\ ,\ \mathcal{S})$, *the Burnside rings* $B(G\ ,\ \mathcal{S})$ *of* $G$ *with respect to* $\mathcal{S}$. Observe that when $\mathcal{S}$ is assumed to be the set $\mathcal{S}(G)$, the set of all subgroups of $G$, then $B(G\ ,\ \mathcal{S})$ coincides with the usual Burnside ring $B(G)$ of $G$, constructed from the semiring $B^+(G)$ of isomorphism classes of all finite $G$-sets. We state the following useful results (see [2]).

**Theorem 2.1** *(i)* $B(G\ ,\ \mathcal{S})$ *is generated freely as an additive group by the isomorphism classes of transitive* $(G\ ,\ \mathcal{S})$*-sets, i.e., G-sets of the form* $G/U$ *with* $U \in \mathcal{S}$*: so its rank equals the number* $k = k_{\mathcal{S}}$ *of G-conjugacy classes of subgroups in* $\mathcal{S}$.
*(ii) For any subgroup* $V \leq G$ *of G, the mapping*

$$\chi_V : S \longmapsto \#\{s \in S \mid V \leq G_s\}$$

*which associates with any* $(G\ ,\ \mathcal{S})$*-set* $S$ *the number of elements in* $S^V := \{s \in S \mid V \leq G_s\}$*, the set of* $V$*-invariant elements in* $S$ *– induces a homomorphism also denoted by* $\chi_V$ *or more precisely* $\chi_V^{\mathcal{S}}$ *– from* $B(G\ ,\ \mathcal{S})$ *into* $Z$*.*
*(iii) Any homomorphism from* $B(G\ ,\ \mathcal{S})$ *into* $Z$ *takes the form given in (ii).*

As a further remark, note that we have (see[1]) that for $V,\ W \leq G$ one has $\chi_V^{\mathcal{S}} = \chi_W^{\mathcal{S}}$ if and only if $\overline{V} := \bigcap_{V \leq U \in \mathcal{S}} U$ is conjugate to $\overline{W} := \bigcap_{W \leq U \in \mathcal{S}} U$. So, in view of the fact that $\overline{V} \in \mathcal{S}$ for all $V \in \mathcal{S}(G)$ and $V \in \mathcal{S}$ if and only if $V = \overline{V}$, one has $k = k_{\mathcal{S}}$ different homomorphisms from $B(G\ ,\ \mathcal{S})$ into $Z$ which – after choosing a system $\mathcal{S}' = \{U_1,\ U_2,\ \ldots\ldots, U_k\}$ of representatives of conjugacy classes of subgroups of $\mathcal{S}$ with $|U_1| \geq |U_2| \geq \ldots\ldots \geq |U_k|$ – may be denoted by $\chi_1 = \chi_{U_1},\ \chi_2 = \chi_{U_2},\ \ldots, \chi_k = \chi_{U_k}$.

**Theorem 2.2** *The product map* $\chi := \prod_{i=1}^{k} \chi_i :\ B(G\ ,\ \mathcal{S}) \longrightarrow \prod_{i=1}^{k} Z$ *of* $k$ *different homomorphisms from* $B(G\ ,\ \mathcal{S})$ *into* $Z$ *is injective and maps* $B(G\ ,\ \mathcal{S})$ *onto a subring of finite index* $\prod_{i=1}^{k}(N_G(U_i) : U_i)$ *of* $\prod_{i=1}^{k} Z$ *– this way identifying* $\prod_{i=1}^{k} Z$ *with the integral closure* $\tilde{B}(G\ ,\ \mathcal{S})$ *of* $B(G\ ,\ \mathcal{S})$ *in its total quotient ring* $\tilde{B}(G\ ,\ \mathcal{S}) \cong Q \otimes_Z B(G\ ,\ \mathcal{S}) \cong \prod_{i=1}^{k} Q$*.*

**Proof:** see [2]

We note from above that for $B(G\ ,\ \mathcal{S}) = B(G)$ and for every subgroup $U$ of $G$ there exists a canonical homomorphism $\chi_U : B(G) \longrightarrow Z$, which maps every finite $G$-set $S$ onto the cardinality $\chi_U(S) := \#S^U$ of its subset $S^U = \{s \in S \mid us = s \text{ for all } u \in U\}$ of $U$ invariant elements – in particular $\chi_1(S) = \#S$ if $1 = \{1_G\}$ denotes the trivial subgroup of $G$. One also has for $U,\ V \leq G$ that $\chi_U = \chi_V$ if and only if $U \overset{G}{\sim} V$ (where $U \overset{G}{\sim} V$ denotes that $U$ and $V$ are conjugate in $G$) and $\chi_U(x) = \chi_U(x')$ for all $U \leq G$ if and only if $x = x'$ for all $x,\ x' \in B(G)$. So identifying each $x \in B(G)$ with the associated map $U \rightarrow \chi_U(x)$ – also denoted by $x$ – from the set $\mathcal{S}(G)$ of all subgroups of $G$ into $Z$, we can consider $B(G)$ in a canonical way as a subring of $\tilde{B}(G) := \prod'_{U \leq G} Z$, consisting of all maps from $\mathcal{S}(G)$ into $Z$ which are constant on each conjugacy class of subgroups.

Next consider the isomorphism classes of the transitive $G$-sets of the form $G/U := \{gU \mid g \in G\}$. These isomorphism classes form a $Z$ basis of $B(G)$ and for $U, V \leq G$ we have $G/U \cong G/V$ if and only if $U \overset{G}{\sim} V$. This then implies that every $x \in B(G)$ can be expressed uniquely in the form $x = \sum'_{U \leq G} \mu_U(x) \cdot G/U$ where " $\sum'$" indicates that the sum extends over just one subgroup out of every $G$-conjugacy class of subgroups. That is, every $x \in B(G)$ can be expressed as a linear combination of the isomorphism classes of transitive $G$-sets of type $G/U$ with uniquely determined integral coefficients $\mu_U(x) \in Z$ subject to the relation $\mu_U(x) = \mu_V(x)$ if $U \overset{G}{\sim} V$. Now recall that for $U, V \leq G$ one has $\chi_V(G/U) \neq 0$ if and only if $V \preceq U$ in which case one has

$$\chi_V(G/U) \quad = \quad \#\{gU \in G/U \mid VgU = gU\}$$

$$
\begin{aligned}
&= \frac{1}{|U|} \cdot \#\{g \in G \mid VgU = gU\} \\
&= \frac{1}{|U|} \cdot \#\{g \in G \mid V \leq gUg^{-1}\} \\
&= (N_G(U) : U) \cdot \#\{U' \leq G \mid V \leq U' \overset{G}{\sim} U\},
\end{aligned}
$$

where as usual

$$
\begin{aligned}
N_G(U) &= \{g \in G \mid g^{-1}Ug = U\} \\
&= \{g \in G \mid UgU = gU\} \\
&= \{g \in G \mid gU \in (G/U)^U\}
\end{aligned}
$$

is the normalizer of $U$ in $G$. We note that given any $x \in B(G)$, a subgroup $U \leq G$ is a maximal subgroup (relative to "$\preceq$") with $\mu_U(x) \neq 0$ if and only if it is a maximal subgroup with $\chi_U(x) \neq 0$ because if $U \leq G$ is maximal with $\mu_V(x) \neq 0$ then

$$
\chi_U(x) = \sum_{U \preceq V \subseteq G}{}' \mu_V(x)\chi_U(G/V) = \mu_U(x)\chi_U(G/U).
$$

By assumption $\mu_U(x) \neq 0$ and we know that $\chi_U(G/U) \neq 0$. Therefore $\chi_U(x) \neq 0$. In addition, for any $U' \in \mathcal{S}(G)$ with $U \preceq U'$ but $U \not\sim U'$ we have $\chi_{U'}(x) = 0$ because $\mu_V(x).\chi_{U'}(G/V) = 0$ for all $V \leq G$ in view of the fact that $\chi_{U'}(G/V) \neq 0$ implies $U' \preceq V$ and therefore $\mu_V(x) = 0$. The converse is proved by reversing the argument.

Note that in the foregone case one has $\chi_U(x) = \mu_U(x).\chi_U(G/U) = \mu_U(x) \cdot (N_G(U) : U)$. Because as observed earlier, every $x \in B(G)$ can be expressed uniquely in the form $x = \sum_{U \leq G}' \mu_U(x) \cdot (G/U)$, it follows that in the case where $G$ is a $p$-group one has

$$
\chi_1(x) = \sum_{U \leq G} \mu_U(x) \cdot (G : U) \equiv \mu_G(x) = \chi_G(x)(mod p).
$$

Hence, if $V$ is a $p$-subgroup of an arbitrary finite group $G$ and if $U$ is a subgroup of $G$ with an index $(G : U)$ which is prime to $p$ then $\chi_V(G/U) \equiv \chi_1(G/U) = (G : U) \not\equiv 0(mod\, p)$, and therefore $V \preceq U$. We now state a useful result (see [2]).

**Lemma 2.3** *Let $x \in \tilde{B}(G)$. Then $x \in B(G)$ if and only if for every $U \trianglelefteq V \leq G$ with $(V : U)$ a power of a prime one has $\sum_{vU \in V/U} x(< v\, ,\, U >) \equiv 0(mod\ (V : U))$.*

Next we formulate the definition of the Abelian index in terms of the Burnside ring. This arises in the following context.

**Definition 2.4** *Let $\mathcal{S}$ be a family of all abelian subgroups of $G$ and let $(U)$ denote the conjugacy class of the subgroup $U$ of $G$. Define $\mathcal{C}_{\mathcal{S}}(U) \in \tilde{B}(G)$ by,*

$$
\mathcal{C}_{\mathcal{S}}(U) := \begin{cases} 1, & \text{if } U \in \mathcal{S} \\ 0, & \text{otherwise.} \end{cases}
$$

*We call the integer $\mathcal{I}(G) := min(n \in N \mid n \cdot \mathcal{C}_{\mathcal{S}}(U) \in B(G))$ the Abelian index of $G$.*

Observe that $|G| \cdot \mathcal{C}_{\mathcal{S}}(U) \in B(G)$ and this implies that $|G| \geq \mathcal{I}(G)$ – more precisely, it implies that $\mathcal{I}(G)$ divides $|G|$.

# 3 Proof of main result

Before we prove the main result we state the following lemmata

**Lemma 3.1** *Let $\mathcal{S}$ be a family of all abelian subgroups of a finite group $G$ and let $\mathcal{I}(G)$ denote the abelian index with respect to the family $\mathcal{S}$. Suppose that $\mathcal{I}(G) = 1$, then for all $U$, $V \leq G$ with $U \trianglelefteq V$ and $(V : U)$ a power of a prime, one has that $U \in \mathcal{S}$ if and only if $V \in \mathcal{S}$.*

**Proof:** Let $(V : U) = p^\alpha$ for some $\alpha \geq 0$. Then $x \in B(G)$ implies $x_{(U)} \equiv x_{(V)} (mod\ p)$. Therefore $U \in \mathcal{S}$ implies $x_{(V)} \equiv 1 (mod\ p)$, so $x_{(V)} \neq 0$. We have $x_{(V)} = 1$, and so $V \in \mathcal{S}$. Conversely, $U \notin \mathcal{S}$ implies $x_{(U)} \equiv 0 (mod\ p)$, hence $x_{(V)} \neq 1$, which implies $x_{(V)} = 0$, that is $V \notin \mathcal{S}$.

**Lemma 3.2** *If $G$ is a finite $p$-group then $\mathcal{I}(G) = 1$ if and only if $\mathcal{S} = \mathcal{S}(G)$.*

**Proof:** Assume $\mathcal{I}(G) = 1$. Since $G$ is a $p$ group one has for $x \in B(G)$ that

$$\sum_{gU \in G/U} x_{(<g,U>)} \equiv (mod\ (G : U))$$
$$\equiv (mod\ p^{\alpha - i}), \quad \text{where } |U| = p^i.$$

But

$$\sum_{gU \in G/U} x_{(<g,U>)} = \#\{gU \in G/U \mid <g,U> \in \mathcal{S}\} \equiv 0 (mod\ (G : U)).$$

This implies

$$|G| \mid \#\{g \in G \mid <g,U> \in \mathcal{S}\}.$$

Because

$$0 \leq \#\{g \in G \mid <g,U> \in \mathcal{S}\} \leq |G|,$$

we must either have that

$$\#\{g \in G \mid <g,U> \in \mathcal{S}\} = 0, \text{ that is } <g,U> \notin \mathcal{S} \text{ for all } g \in G$$

or

$$\#\{g \in G \mid <g,U> \in \mathcal{S}\} = |G|, \text{ that is } <g,U> \in \mathcal{S} \text{ for all } g \in G.$$

In the second case we have $U \in \mathcal{S}$, and in the first case $U \notin \mathcal{S}$. Now by assumption, there is some $U \in \mathcal{S}$. For that $U$ we have necessarily $<g,U> \in \mathcal{S}$ for all $g \in U$. It then follows that $G$ is in $\mathcal{S}$. Now assume that there is also some $U \leq G$ with $U \notin \mathcal{S}$. Then one has that

$< g, U > \notin \mathcal{S}$ for all $g \in G$. It follows also that $G \notin \mathcal{S}$, contradicting the conclusion that $G \in \mathcal{S}$. Hence $\mathcal{S} = sub(G)$.

We now prove the main result.

**Theorem 3.3** *Suppose that $G$ is a finite p-group. Then one has that $\mathcal{I}(G) = 1$ if and only if $G$ is abelian. If $G$ is nonabelian, then $\mathcal{I}(G) = p$.*

**Proof:** The first part follows from [1] and lemma 3.2 – in particular if $G$ is abelian then $\mathcal{I}(G) = 1$. For the second part, suppose $\mathcal{S}$ is the family of all abelian subgroups of a nonabelian group $G$ and let $U \in \mathcal{S}$. Let $\mathcal{I}(G)$ and $\mathcal{C}_\mathcal{S}(U)$ be as in Definition (2.4). Then $\mathcal{I}(G) \cdot \mathcal{C}_\mathcal{S}(U) \in B(G)$ and $\sum_{gU \in G/U} c_\mathcal{S}(< g, U >) \equiv 0(mod\ |G : U|)$. To compute $\mathcal{I}(G)$, we fix $|U| = p$. So

$$\begin{aligned}
&\sum_{gU \in G/U} c_\mathcal{S}(< g, U >) \\
&= \#\{gU \in G/U \mid < g, U > \text{ is abelian }\} \\
&= \#\{gU \in G/U \mid < g, U > \text{ is abelian of order } p\} \\
&+ \sum_{k \geq 2}^{n} \#\{gU \in G/U \mid < g, U > \text{ is abelian of order } p^k\} \\
&= p^k + (p-1) \cdot \#\{A \leq G \mid A \text{ is abelian of order } p^2,\ A \supseteq U\} \\
&+ \sum_{k \geq 3}^{n} (p^{k-1} - p^{k-2}) \cdot \#\{A \leq G \mid A \text{ is abelian of order } p^k,\ A \supseteq U\}
\end{aligned}$$

Obviously $p$ divides

$$\mathcal{K}(G) := \#\{A \leq G \mid A \text{ is abelian of order } p^2,\ A \supseteq U\}$$

if $G$ is nonabelian (see [4]). This implies that $\mathcal{I}(G) = p$, is the desired minimal positive integer such that $\mathcal{I}(G) \cdot (1 + \mathcal{K}(G)) \equiv 0(mod\ (G : U))$, that is, $\mathcal{I}(G) = p$.

## REFERENCES

[1] **Dress, A** (1969): A characterization of solvable groups, *Math. Z. 110, 213 – 217.*

[2] **Nwabueze, K.K.** (1998 ): Certain applications of the Burnside Ring and its ghost ring, *Bull. Belg. Math. Soc. 5, 101–114.*

[3] **Nwabueze, K.K.**(1997): The Artin exponent of finite groups, *Acta Math. et inf., uni ost. 5, 71–79.*

[4] **Nwabueze, K.K.** (1996): A Burnside ring theoretic proof of T.Y. Lam's results concerning the Artin exponent of finite groups, *C.R. Maths. Rep Acad. Sci. Canada, vol. XVIII, No. 1, 17 – 21.*

[5] **Swan, R.**(1960): Induced Representations and Projective Modules, *Ann. Math. 71, 552–578.*

[6] **Zassenhaus, H.**(1958): Theory of Groups, *Chelsea, New York.*

*Dept. of Mathematics,*
*Faculty of Science,*
*Universiti Brunei Darussalam,*
*Bandar Seri Begawan, BRUNEI DARUSSALAM.*

# DIRICHLET FORMS AND SYMMETRIC MARKOVIAN SEMIGROUPS ON VON NEUMANN ALGEBRAS

YONG MOON PARK

*Department of Mathematic, Yonsei University, Seoul 120-749, Korea*
*E-mail: ympark@yonsei.ac.kr*

We give a brief survey on the theory of Dirichlet forms on standard forms of von Neumann algebras, and then discuss the construction of Dirichlet forms which generate symmetric Markovian semigroups. More precisely, let $\mathcal{M}$ be a von Neumann algebra acting on a Hilbert space $\mathcal{H}$ with a cyclic and separating vector $\xi_0$. we give an explicit expression of Dirichlet forms on the standard form $(\mathcal{M}, \mathcal{H}, \mathcal{P}, J)$ associated with the pair $(\mathcal{M}, \xi_0)$. As an application, we describe the construction of Dirichlet forms and symmetric Markovian semigroups on CCR algebras with respect to quasi-free states.

## 1 Introduction

In this lecture we would like to give a brief survey on the theory of noncommutative Dirichlet forms and associated symmetric Markovian semigroups on standard forms of von Neumann algebras, and then describe a construction method of Dirichlet forms. In order to show an application of the general construction method, we describe the construction of symmetric Markovian semigroups associated with quasi-free states on CCR algebras. The semigroups we constructed are ergodic.

The theory of commutative Dirichlet forms and associated Markovian semigroups, originated with work of Beurling and Deny, [7] has been developed especially by Fukushima [16] and Silverstein. [31] The theory has been intensively investigated in connection with their important applications to the theory of Markov processes and to various branches of mathematical physics. See Ref. 20 and references therein.

The efforts in constructing and studying of noncommutative (quantum) Markovian semigroups are as old as the equilibrium description of quantum systems. [12] The study of noncommutative Dirichlet forms was pioneered by Albeverio and Høegh-Krohn, [4] Sauvageot [28,29,30] and extensively developed by Davies and Lindsay, [14] and Guido, Isola and Scarlatti. [17] All these authors considered Markovianity of forms and semigroups only with respect to a tracial state $\phi_0$. Recently, Goldstein and Lindsay, [18,19] and Cipriani [9,10] have extended the abstract theory to a faithful normal state $\phi_0$ in the context of standard forms.

The need to construct Markovian semigroups on von Neumann algebras,which are symmetric with respect to a non-tracial state, is clear for various applications to open systems, [12] quantum statistical mechanics, [8] and quantum probability theory. [1,3,26] Although on an abstract level we have quite well-developed theory as mentioned above, the progress in concrete application is very slow. We would like to mention a few recent works in this direction. The completely positive Hamiltonian semigroup for quantum spin chains in the ground state representation has been considered in Refs. 15, 23 and 24. In Refs. 21 and 22, Majewski and Zegarlinski used the generalized conditional expectation to construct generators of spin-flip type dynamics for quantum spin systems. We should mention that in Refs. 21 and 22 the author have also constructed Markovian semigroups of a diffusion type generators for quantum spin systems under the condition of hyper-asymptotic Abelianess which is not proven yet. In Ref. 25, we gave a general construction method of Dirichlet forms on standard forms of von Neumann algebras and applied the method to construct translation invariant Markovian semigroups for quantum spin systems. In Ref. 11, quantum Ornstein-Uhlenbeck semigroups were constructed by means of noncommutative Dirichlet forms. Extending the method proposed in Ref. 25, symmetric Markovian semigroups associated with quasi-free states on CCR algebras are constructed. [6]

In section 2, we give a brief survey on the abstract theory of Dirichlet forms on standard forms of von Neumann algebras. In section 3, we describe a general construction method of Dirichlet forms on von Neumann algebras proposed in Ref. 25. We will generalize the result of Ref. 25 slightly and give its proof. In section 4, we discuss how we can apply the general construction method to construct symmetric Markovian semigroups on CCR algebras with respect to quasi-free states.

## 2 Review on the Theory of Dirichlet Forms on von Neumann Algebras

The theory of Dirichlet forms and symmetric Markovian semigroups on von Neumann algebras with respect to (nontracial) faithful normal states has been developed by Goldstein and Lindsay, [18,19] and Cipriani. [9,10] In this section, we give a brief review on the theory of noncommutative Dirichlet forms developed by Cipriani. [10]

In order to understand the basis structure of the noncommutative theory, let us first consider the commutative situation. The basic definitions of the commutative theory rely on the structure of the ordered Hilbert space $L^2(X, m)$, given by the closed,convex cone of non-negative functions $L^2_+(X, m)$. These objects form a so called standard form of the Abelian von Neumann algebra $L^\infty(X,m)$. Next, consider a sesquilinear form $\mathcal{E}(\cdot,\cdot) : D(\mathcal{E}) \times D(\mathcal{E}) \to \mathbb{C}$ defined on a dense subset $D(\mathcal{E}) \subset L^2(X,m)$, and associated quadratic form $\mathcal{E}[\cdot] : D(\mathcal{E}) \to \mathbb{C}$, $\mathcal{E}[u] := \mathcal{E}(u,u)$. Assume that the quadratic form $(\mathcal{E}, D(\mathcal{E}))$ is real, bounded below and closed. Let $\{T_t\}_{t\geq 0}$ be the semigroup associated to $(\mathcal{E}, D(\mathcal{E}))$. The first Beurling - Deny criterion ( Theorem 1.3.2 of Ref. 13) characterized that $\{T_t\}_{t\geq 0}$ is positive preserving if and only if

$$u \in D(\mathcal{E}) \cap L^2_{\mathbb{R}}(X,m) \Rightarrow u_+, u_- \in D(\mathcal{E}), \quad \mathcal{E}(u_+, u_-) \leq 0,$$

where $u_+ := \sup(u,0)$ and $u_- := \sup(-u,0)$. The second Beurling-Deny criterion(Theorem1.3.2 of Ref. 13) establishes that $\{T_t\}_{t\geq 0}$ is sub-Markovian($0 \leq u \leq 1 \Rightarrow 0 \leq T_t u \leq 1$) if and only if

$$u \in D(\mathcal{E}) \cap L^2_{\mathbb{R}}(X,m) \Rightarrow u \wedge 1 \in D(\mathcal{E}), \quad \mathcal{E}[u \wedge 1] \leq \mathcal{E}[u],$$

where $u \wedge 1 = \inf(u,1)$.

To generalized the above characterization to the noncommutative setting,we introduce the concept of a positive cone in a Hilbert space. Let $\mathcal{M}$ be a $\sigma$-finite von Neumann algebra acting on a complex Hilbert space $\mathcal{H}$. A self-dual cone $\mathcal{P}$ in $\mathcal{H}$ is a subset satisfying the property

$$\{\xi \in \mathcal{H} : (\xi,\eta) \geq 0, \forall \eta \in \mathcal{P}\} = \mathcal{P}.$$

$\mathcal{P}$ is then a closed convex cone and $\mathcal{H}$ is the complexification of the real subspace $\mathcal{H}^J := \{\xi \in \mathcal{H} : (\xi,\eta) \in \mathbb{R} \ \ \forall \eta \in \mathcal{P}\}$, which elements are called *J-real* : $\mathcal{H} = \mathcal{H}^J + i\mathcal{H}^J$. Such a $\mathcal{P}$ gives rise to a structure of ordered Hilbert space $\mathcal{H}^J$ (denoted by $\leq$) and to an anti-unitary involution $J$ on $\mathcal{H}$, which preserves $\mathcal{H}^J$ : $J(\xi + i\eta) := \xi - i\eta$ for all $\xi, \eta \in \mathcal{H}^J$. Any $J$-real element $\xi \in \mathcal{H}^J$ can be decomposed uniquely as a difference, $\xi = \xi_+ - \xi_-$, of two orthogonal, positive elements, called the positive and the negative part of $\xi : \xi_+, \xi_- \in \mathcal{P}, \quad (\xi_+, \xi_-) = 0$.

A standard form $(\mathcal{M}, \mathcal{H}, \mathcal{P}, J)$ of the von Neumann algebra $\mathcal{M}$ acting faithfully on the Hilbert space $\mathcal{H}$ consists of self-dual, closed, convex cone $\mathcal{P}$ in $\mathcal{H}$ and the anti-unitary involution $J$ satisfying the properties:

(a) $J\mathcal{M}J = \mathcal{M}'$

(b) $JxJ = x^*, \quad \forall x \in \mathcal{M} \cap \mathcal{M}'$

(c) $J\xi = \xi, \quad \forall \xi \in \mathcal{P}$

(d) $xJxJ(\mathcal{P}) \subset \mathcal{P}, \quad \forall x \in \mathcal{M}$, where $\mathcal{M}'$ is the commutant of $\mathcal{M}$. An concrete example of a standard form will be given in the next section.

A bounded operator $A$ on $\mathcal{H}$ is called *J-real* if $AJ = JA$ and *positive preserving* if $A\mathcal{P} \subset \mathcal{P}$. The semigroup $\{T_t\}_{t\geq 0}$ is said to be *J-real* if $T_t$ is $J$-real for any $t \geq 0$ and it is called *positive preserving* if $T_t$ is positive preserving for any $t \geq 0$.

Let us fix a cyclic and separating vector $\xi_0$ in $\mathcal{P}$. A bounded operator $A : \mathcal{H} \to \mathcal{H}$ is called *sub-Markovian* (with respect to $\xi_0$) if $0 \leq \xi \leq \xi_0$ implies $0 \leq A\xi \leq \xi_0$. $A$ is called *Markovian* if it is sub-Markovian and also $A\xi_0 = \xi_0$. A semigroup $\{T_t\}_{t\geq 0}$ is said to be *sub-Markovian* (with respect to $\xi_0$) if $T_t$ is sub-Markovian for every $t \geq 0$. The semigroup $\{T_t\}_{t\geq 0}$ is called *Markovian* if $T_t$ is Markovian for every $t \geq 0$.

Next, we consider a sesquilinear form on some linear manifold of $\mathcal{H}$ : $\mathcal{E}(\cdot,\cdot) : D(\mathcal{E}) \times D(\mathcal{E}) \to \mathbb{C}$. We also consider the associated quadratic form: $\mathcal{E}[\cdot] : D(\mathcal{E}) \to \mathbb{C}$, $\mathcal{E}[\xi] := \mathcal{E}(\xi,\xi)$. A real quadratic form $\mathcal{E}[\cdot]$ is said to be *semi bounded* if $\inf\{\mathcal{E}[\xi] : \xi \in D(\mathcal{E}),\ ||\xi|| = 1\} = -b > -\infty$. A quadratic form $(\mathcal{E}, D(\mathcal{E}))$ is said to be *J-real* if $JD(\mathcal{E}) \subset D(\mathcal{E})$ and $\mathcal{E}[J\xi] = \overline{\mathcal{E}[\xi]}$ for any $\xi \in D(\mathcal{E})$. For a given semi bounded quadratic form $\mathcal{E}$, one consider the inner product given by $(\xi,\eta)_\lambda := \mathcal{E}(\xi,\eta) + \lambda(\xi,\eta)$, for $\lambda > b$. The form $\mathcal{E}$ is *closed* if $D(\mathcal{E})$ is a Hilbert space for some of the above norms. The form $\mathcal{E}$ is called *closable* if it admits a closed extension.

Associated to a semi bounded closed form $\mathcal{E}$, there is a self-adjoint operator $(H, D(H))$ and a strongly continuous, symmetric semigroup $\{T_t\}_{t\geq 0}$. Each of the above objects determines uniquely the others according to well known relations (see Section 3.1 of Ref. 8 and Section 1.3 of Ref. 16).

From now on we will consider only $J$-real, real-valued, semi bounded, densely defined quadratic forms. It is easy to check that these forms satisfy the relation : $\mathcal{E}[\xi + i\eta] = \mathcal{E}[\xi] + \mathcal{E}[\eta]$ for all $\xi + i\eta \in D(\mathcal{E})^J + iD(\mathcal{E})^J = D(\mathcal{E})$.

Let us denoted by $\mathrm{Proj}\,(\xi, \mathcal{Q})$ the projection of the vector $\xi \in \mathcal{H}^J$ onto the closed, convex cone $\mathcal{Q} \subset \mathcal{P}$. For $\xi, \eta \in \mathcal{H}^J$, define

$$\xi \vee \eta := \mathrm{Proj}\,(\xi, \eta + \mathcal{P})$$
$$\xi \wedge \eta := \mathrm{Proj}\,(\xi, \eta - \mathcal{P})$$

**Definition 2.1** A $J$-real, real-valued, densely defined quadratic form $(\mathcal{E}, D(\mathcal{E}))$ is called Markovian with respect to $\xi_0 \in \mathcal{P}$ if

$$\xi \in D(\mathcal{E})^J \text{ implies } \xi \wedge \xi_0 \in D(\mathcal{E}) \text{ and } \mathcal{E}[\xi \wedge \xi_0] \leq \mathcal{E}[\xi].$$

A closed Markovian form is called a Dirichlet form.

Next, we state main results without the proofs. For the proofs we refer to Ref. 10.

**Proposition 2.1** *[Prop.4.5(b) of Ref. 10]. Let $(\mathcal{E}, D(\mathcal{E}))$ be a $J$-real, real-valued, densely defined quadratic form. The followings are equivalent:*

*(a) $\mathcal{E}[\cdot] \geq 0$ and $\xi \in D(\mathcal{E})^J$ implies $\xi_\pm \in D(\mathcal{E})$, $\quad \mathcal{E}(\xi_+, \xi_-) \leq 0$.*

*(b) $\xi \in D(\mathcal{E})^J$ implies $\xi_+ \in D(\mathcal{E})$ and $\mathcal{E}[\xi_+] \leq \mathcal{E}[\xi]$.*

The following is the first Beurling-Deny Type criterion:

**Proposition 2.2** *[Theorem 4.7 (b) of Ref. 10]. Let $\{T_t\}_{t\geq 0}$ be a $J$-real, strongly continuous, symmetric semigroup on $\mathcal{H}$ and let $(\mathcal{E}, D(\mathcal{E}))$ be the associated $J$-real, real-valued quadratic form. The followings are equivalent:*

*(a) $\xi \in D(\mathcal{E})^J$ implies $\xi_+ \in D(\mathcal{E})$ and $\mathcal{E}[\xi_+] \leq \mathcal{E}[\xi]$.*

*(b) $\{T_t\}$ is positive preserving and contractive.*

**Proposition 2.3** *[Proposition 4.10 (ii) of Ref. 10] Let $(\mathcal{E}, D(\mathcal{E}))$ be a $J$-real, real-valued, densely defined closed form. If $\xi \in D(\mathcal{E})^J$ implies $\xi_+ \in D(\mathcal{E})$ and $\mathcal{E}[\xi_+] \leq \mathcal{E}[\xi]$ and $\xi_0 \in D(\mathcal{E})$, then $\mathcal{E}$ is Dirichlet if and only if $\mathcal{E}(\xi_0, \xi) \geq 0$ for all $\xi \in D(\mathcal{E}) \cap \mathcal{P}$.*

The following is the second Beurling-Deny Type criterion which is the characterization of sub-Markovian semigroups in terms of Dirichlet forms with respect to the fixed vector $\xi_0 \in \mathcal{P}$ :

**Theorem 2.1** *[Theorem 4.11 of Ref. 10]. Let $\{T_t\}$ be a $J$-real, strongly continuous, symmetric semigroup on $\mathcal{H}$ and let $(\mathcal{E}, D(\mathcal{E}))$ be the associated densely defined $J$-real, real-valued quadratic form. The followings are equivalent :*

*(a) $\{T_t\}_{t\geq 0}$ is sub-Markovian.*

*(b) $\mathcal{E}$ is a Dirichlet form.*

## 3 Construction of Dirichlet Forms

In this section we describe a general construction method of Dirichlet forms on the natural standard forms of von Neumann algebras associated with the Tomita-Takesaki theory. [8,32] The method has been developed in Ref. 25. However we generalizes the result of Ref. 25 slightly and produce the proof. See also Ref. 6.

Let $\mathcal{M}$ be a $\sigma$-finite von Neumann algebra acting on a Hilbert space $\mathcal{H}$ and let $\xi_0 \in \mathcal{H}$ be a cyclic and separating vector for $\mathcal{M}$. We use $\Delta$ and $J$ to denote respectively, the modular operator and the modular conjugation associated with the pair $(\mathcal{M}, \xi_0)$. [5,8] The associated modular automorphism group is denoted by $\sigma_t : \sigma_t(A) = \Delta^{it} A \Delta^{-it}, A \in \mathcal{M}$. Finally, $j : \mathcal{M} \to \mathcal{M}'$ is the anti linear $*$-isomorphism defined by $j(A) = JAJ$, $A \in \mathcal{M}$. The natural positive cone $\mathcal{P}$ associated with $(\mathcal{M}, \xi_0)$ is the closure of the set

$$\{Aj(A)\xi_0; A \in \mathcal{M}\}.$$

By a general result the cone $\mathcal{P}$ can be obtained by the closure of the set

$$\{\Delta^{1/4} AA^* \xi_0; A \in \mathcal{M}\}.$$

The natural cone $\mathcal{P}$ is self-dual. [5,8] For the detailed properties of $\mathcal{P}$, we refer to Section 2.5 of Ref. 8. The form $(\mathcal{M}, \mathcal{H}, \mathcal{P}, J)$ is the standard form associated with the pair $(\mathcal{M}, \xi_0)$.

In order to express Dirichlet forms, let us introduce the notion of admissible functions:

**Definition 3.1** An analytic function $f : D \to \mathbb{C}$ on a domain $D$ containing the strip $Im\ z \in [-1/4, 1/4]$ is said to be admissible if the following properties hold;

(a) $f(t) \geq 0$ for $\forall t \in \mathbb{R}$,

(b) $f(t + i/4) + f(t - i/4) \geq 0$ for $\forall t \in \mathbb{R}$,

(c) there exist $M > 0$ and $p > 1$ such that the bound

$$|f(t + is)| \leq M(1 + |t|)^{-p}$$

holds uniformly in $s \in [-1/4, 1/4]$.

Let us give an example of an admissible function. Using the residue integration method, it is easy to check that

$$\int_{-\infty}^{\infty} (\cosh k)^{-1} e^{ikt}\, dk = 2\pi (e^{\frac{\pi}{2}t} + e^{-\frac{\pi}{2}t})^{-1}.$$

See also the expression in p. 94 of Ref. 8. Using the above, one may check that the function given by

$$f(t) = \frac{2}{\sqrt{2\pi}} \int (e^{k/4} + e^{-k/4})^{-1} e^{-\frac{1}{2}k^2} e^{-ikt}\, dk \tag{3.1}$$

is an admissible function.

We are ready to give a construction of Dirichlet forms on the standard form $(\mathcal{M}, \mathcal{H}, \mathcal{P}, J)$ associated with the pair $(\mathcal{M}, \xi_0)$. Denote by $\mathcal{M}_0$ the dense subset of $\mathcal{M}$ consisting of every $\sigma_t$-analytic element with a domain containing the strip $I_{1/2} := \{z; |Im\ z| \leq \frac{1}{2}\}$. [8] By Prop. 2.5.21 of Ref. 8, any $A \in \mathcal{M}_0$ is strongly analytic. In the following, the inner product $\langle\cdot,\cdot\rangle$ on $\mathcal{H}$ is conjugate linear in the first and linear in the second variable. For given admissible function $f$ and $x \in \mathcal{M}_0$, define a sesquilinear form $\mathcal{E} : \mathcal{H} \times \mathcal{H} \to \mathbb{C}$ by

$$\begin{aligned}\mathcal{E}(\eta,\xi) = &\int \langle \big(\sigma_{t-i/4}(x) - j\,(\sigma_{t-i/4}(x^*))\big)\eta, \\ &\qquad \big(\sigma_{t-i/4}(x) - j\,(\sigma_{t-i/4}(x^*))\big)\xi\rangle f(t)\,dt \\ &+ \int \langle \big(\sigma_{t-i/4}(x^*) - j\,(\sigma_{t-i/4}(x))\big)\eta, \\ &\qquad \big(\sigma_{t-i/4}(x^*) - j\,(\sigma_{t-i/4}(x))\big)\xi\rangle f(t)\,dt \\ \equiv\ & \mathcal{E}^{(1)}(\eta,\xi) + \mathcal{E}^{(2)}(\eta,\xi)\end{aligned} \tag{3.2}$$

Then the associated quadratic form is given by

$$\begin{aligned}\mathcal{E}[\xi] = &\int ||\,[\sigma_{t-i/4}(x) - j(\sigma_{t-i/4}(x^*))]\xi||^2 f(t)\,dt \\ &+ \int ||\,[\sigma_{t-i/4}(x^*) - j(\sigma_{t-i/4}(x))]\xi||^2 f(t)\,dt \\ \equiv\ & \mathcal{E}^{(1)}[\xi] + \mathcal{E}^{(2)}[\xi]\end{aligned} \tag{3.3}$$

In Ref. 25, we have considered the case $x = x^* \in \mathcal{M}_0$. The following is the result corresponding to Theorem 3.1 of Ref. 25.

**Theorem 3.1** *For given admissible function $f$ and $x \in \mathcal{M}_0$, let $(\mathcal{E},\mathcal{H})$ be defined as in (3.2). Let $H$ be the self-adjoint operator associated with $(\mathcal{E},\mathcal{H})$. Assume that there exists a constant $M > 0$ such that the bound*

$$\sup_{s\in[-1/4,1/4]} ||\,\sigma_{t+is}(x)\,|| \le M$$

*holds uniformly in $t \in \mathbb{R}$. Then the following properties hold:*

*(a)* $H\xi_0 = 0$,

*(b)* $\mathcal{E}$ *is* $J$*-real,*

*(c)* $\mathcal{E}(\xi_+,\xi_-) \le 0 \quad \forall \xi \in \mathcal{H}^J$.

*Furthermore the form $(\mathcal{E},\mathcal{H})$ is a Dirichlet form.*

We will produce the proof of Theorem 3.1 at the end of this section. The following is a consequence of Theorem 3.1:

**Theorem 3.2** *Let $\{T_t\}_{t\geq 0}$ be the semigroup generated by the form $(\mathcal{E}, \mathcal{H})$ in Theorem 3.1. Then $\{T_t\}_{t\geq 0}$ is a J-real, strongly continuous, symmetric Markovian semigroup.*

**Proof.** It follows from Theorem 3.1(a) that $T_t(\xi_0) = \xi_0$ for any $t \geq 0$. Thus the theorem follows from Theorem 2.1. □

We now produce the proof of Theorem 3.1.

**Proof of Theorem 3.1.** (a) Notice that $JA\xi_0 = \Delta^{1/2}A^*\xi_0$ for any $A \in \mathcal{M}$. Thus we have that

$$\begin{aligned}\left(\sigma_{t-i/4}(x) - j\left(\sigma_{t-i/4}(x^*)\right)\right)\xi_0 &= \Delta^{1/4}(\sigma_t(x))\xi_0 - J\Delta^{1/4}\sigma_t(x)^*\Delta^{-1/4}\xi_0 \\ &= \Delta^{1/4}(\sigma_t(x))\xi_0 - \Delta^{1/4}\sigma_t(x)\xi_0 \\ &= 0.\end{aligned}$$

Replacing $x$ by $x^*$ in the above, we get

$$\left(\sigma_{t-i/4}(x^*) - j\left(\sigma_{t-i/4}(x)\right)\right)\xi_0 = 0.$$

Thus (a) follows from (3.2) and the above fact.

(b) A direct estimate shows that

$$\begin{aligned}&\|\left(\sigma_{t-i/4}(x) - j\left(\sigma_{t-i/4}(x^*)\right)\right)J\xi\|^2 \\ &= \|J\left(\left[\sigma_{t-i/4}(x^*) - j\left(\sigma_{t-i/4}(x)\right)\right]\right)\xi\|^2 \\ &= \|\left(\sigma_{t-i/4}(x^*) - j\left(\sigma_{t-i/4}(x)\right)\right)\xi\|^2,\end{aligned}$$

which implies $\mathcal{E}^{(1)}[J\xi] = \mathcal{E}^{(2)}[\xi]$. The method used in the above also implies that $\mathcal{E}^{(2)}[J\xi] = \mathcal{E}^{(1)}[\xi]$.

(c) By the expression of $\mathcal{E}(\eta, \xi)$ in (3.2), $\mathcal{E}(\xi_+, \xi_-)$ can be written as

$$\begin{aligned}\mathcal{E}(\xi_+, \xi_-) &= \mathcal{E}^{(1)}(\xi_+, \xi_-) + \mathcal{E}^{(2)}(\xi_+, \xi_-) \\ &= (\mathrm{I}^{(1)} + \mathrm{II}^{(1)}) + (\mathrm{I}^{(2)} + \mathrm{II}^{(2)}) \qquad (3.4)\end{aligned}$$

where

$$\begin{aligned}\mathrm{I}^{(1)} &= \int \Big(\langle\sigma_{t-i/4}(x)\xi_+, \sigma_{t-i/4}(x)\xi_-\rangle \\ &\quad + \langle\sigma_{t-i/4}(x^*)\xi_-, \sigma_{t-i/4}(x^*)\xi_+\rangle\Big) f(t)\, dt, \qquad (3.5) \\ \mathrm{II}^{(1)} &= -\int \Big(\langle\sigma_{t-i/4}(x)\xi_+, j\left(\sigma_{t-i/4}(x^*)\right)\xi_-\rangle \\ &\quad + \langle j\left(\sigma_{t-i/4}(x^*)\right)\xi_+, \sigma_{t-i/4}(x)\xi_-\rangle\Big) f(t)\, dt.\end{aligned}$$

and $\mathrm{I}^{(2)}$ and $\mathrm{II}^{(2)}$ are obtained from $\mathrm{I}^{(1)}$ and $\mathrm{II}^{(1)}$ respectively replacing $x$ by $x^*$ in the above.

As a consequence of Theorem 4(7) of Ref. 5, $\mathcal{M}\xi_+ \perp \mathcal{M}\xi_-$, which implies $\mathrm{I}^{(1)} = 0$ and $\mathrm{I}^{(2)} = 0$. See also the proof of Theorem 5.3 (ii) of Ref. 10. Next, we first consider $\mathrm{II}^{(1)}$. It can be checked that $\sigma_{t-i/s}(x)^* = \sigma_{t+i/s}(x^*)$, for any $x \in \mathcal{M}_0$ and $s \in \mathbb{R}$, and so

$$\begin{aligned} &\langle \sigma_{t-i/4}(x)\xi_+ \, , \, j\,(\sigma_{t-i/4}(x^*))\xi_- \rangle \\ &\qquad = \langle\, \xi_+ \, , \, \sigma_{t+i/4}(x^*) j((\sigma_{t+i/4}(x))^*)\xi_- \rangle \end{aligned} \tag{3.6}$$

and

$$\begin{aligned} &\langle j\,(\sigma_{t-i/4}(x^*))\xi_+ \, , \, (\sigma_{t-i/4}(x))\xi_- \rangle \\ &\qquad = \langle \xi_+ \, , \, \sigma_{t-i/4}(x) j((\sigma_{t-i/4}(x^*))^*)\xi_- \rangle. \end{aligned} \tag{3.7}$$

It follows from (3.5)-(3.7) that

$$\begin{aligned} \mathrm{II}^{(1)} = &- \int \langle\, \xi_+ \, , \, \sigma_{t+i/4}(x^*) j((\sigma_{t+i/4}(x))^*)\xi_- \rangle \, f(t)\, dt \\ &- \int \langle \xi_+ \, , \, \sigma_{t-i/4}(x) j((\sigma_{t-i/4}(x^*))^*)\xi_- \,\rangle f(t)\, dt. \end{aligned}$$

Notice that the map $A \mapsto j(A^*)$ from $\mathcal{M}$ to $\mathcal{M}'$ is linear. It can be shown that for any $x \in \mathcal{M}_0$ and $\xi \in \mathcal{H}$, the map

$$z \mapsto j((\sigma_z(x))^*)\xi$$

is analytic on a domain containing the strip $I_{1/2}$. In fact, the analyticity follows from the facts that $\langle \eta \, , \, j(\sigma_z(x)^*)\xi \rangle = \langle \sigma_z(x)^* \, J\xi \, , \, J\eta \,\rangle = \langle J\xi, \sigma_z(x) J\eta \,\rangle$ for any $\eta, \xi \in \mathcal{H}$, and that weak analyticity implies strong analyticity ( see Theorem VI.4 of Ref. 27).

Using the Cauchy integral theorem, the assumption in the theorem, the property (c) in the Definition 3.1 and $\sigma_t(x)^* = \sigma_t(x^*)$, we obtain that

$$\begin{aligned} \mathrm{II}^{(1)} = &- \int \langle \xi_+, \sigma_t(x)^* j((\sigma_t(x))^*)\xi_- \rangle f(t - i/4)\, dt \\ &- \int \langle \xi_+, \sigma_t(x) j(\sigma_t(x))\xi_- \rangle f(t + i/4)\, dt. \end{aligned}$$

Replacing $x$ and $x^*$ in the above, we obtain the expression of $\mathrm{II}^{(2)}$. Thus we get

$$\begin{aligned} \mathrm{II} = - \int &\langle \xi_+, [\sigma_t(x) j(\sigma_t(x)) + \sigma_t(x)^* j((\sigma_t(x)^*)]\xi_- \rangle \\ &\cdot (f(t - i/4) + f(t + i/4))\, dt. \end{aligned}$$

Since $\sigma_t(x)j(\sigma_t(x))\xi_- \in \mathcal{P}$, $\langle \xi_+, [\sigma_t(x)j(\sigma_t(x)) + \sigma_t(x)^* j((\sigma_t(x)^*)]\xi_- \rangle \geq 0$ for $t \in \mathbb{R}$. By the property (b) in the Definition 3.1, we conclude that II $\leq 0$. This proved the part (c) of the theorem.

Clearly $\mathcal{E}[\cdot] \geq 0$. Note that $\mathcal{E}(\xi_0, \xi) = 0$, $\forall \xi \in D(\mathcal{E})$. By Theorem 3.1 (a)-(c), Proposition 2.2, and Proposition 2.3, $(\mathcal{E}, \mathcal{H})$ is a Dirichlet form. □

## 4 Dirichlet Forms and Symmetric Markovian Semigroups on CCR Algebras with respect to Quasi-free States

As an application of the general construction method explained in Section 3, we describe the construction of Dirichlet forms and associated symmetric Markovian semigroups on the algebras of the canonical commutation relations (CCRs) with respect to quasi-free states. For the details, we refer to Ref. 6.

Let $\mathfrak{h}_0$ be a complex separable pre-Hilbert space. Denote by $\mathfrak{h}$ the completion of $\mathfrak{h}_0$. The inner product $(\cdot, \cdot)$ on $\mathfrak{h}$ is conjugate linear in the first and linear in the second variable. Let $\mathcal{A}(\mathfrak{h}_0)$ be the $C^*$-algebra generated by the Weyl operators $W(f)$, $f \in \mathfrak{h}_0$, satisfying

$$\begin{aligned} W(-f) &= W(f)^*, \\ W(f)W(g) &= e^{-\frac{i}{2}\mathrm{Im}(f,g)} W(f+g), \quad \forall f, g \in \mathfrak{h}_0 \end{aligned} \tag{4.1}$$

For the abstract properties of $\mathcal{A}(\mathfrak{h}_0)$, see Theorem 5.2.2 of Ref. 8.

Next, we describe quasi-free states on $\mathcal{A}(\mathfrak{h}_0)$. Let $A$ be a bounded and non-negative operator on $\mathfrak{h}$. Recall that $\varphi \in \mathfrak{h}$ is an analytic vector for an operator $B$ on $\mathfrak{h}$ if $\varphi \in D(B^n)$, $n \in \mathbb{N}$, and if

$$\sum_{n=0}^{\infty} \frac{\|B^n \varphi\|}{n!} t^n < \infty$$

for some $t > 0$. In the following, we assume that $A$ satisfies the following properties.

**Assumption 4.1** (a) There exists $\alpha \in (0, 1)$ such that

$$0 < A \leq \alpha \mathbf{1} < \mathbf{1}.$$

(b) The inverse $A^{-1}$ of $A$ exists as a (unbounded) self-adjoint and positive operator on $\mathfrak{h}$.

(c) For any $z \in \mathbb{C}$, $A^z$ leaves $\mathfrak{h}_0$ invariant, i.e., $A^z \mathfrak{h}_0 \subset \mathfrak{h}_0$. Moreover, $z \mapsto A^{iz}\varphi$ is entire analytic for any $\varphi \in \mathfrak{h}_0$.

(d) Any $\varphi \in \mathfrak{h}_0$ is an analytic vector for $A^{-1/2}$.

We remark that a dense submanifold $\mathfrak{h}_0$ of $\mathfrak{h}$ satisfying the assumption exists by the spectral theorem.

**Example 4.1 (Ideal Base gases)** Let $\mathfrak{h}$ be $L^2(\mathbb{R}^d, dx)$ and $\Delta$ the Laplacian operator on $L^2(\mathbb{R}^d, dx)$. Let $A$ be given by

$$A = \exp\{-\beta(-\frac{1}{2}\Delta + \mu\mathbf{1})\},$$

where $\beta > 0$ and $\mu > 0$. Then $\alpha = \exp(-\beta\mu)$. For $f \in L^2(\mathbb{R}^d, dx)$, denote by $\hat{f}$ the Fourier transform of $f$. Choose $\mathfrak{h}_0 = \{f \in L^2(\mathbb{R}^d, dx) | \hat{f} \in C_c(\mathbb{R}^d)\}$. Clearly Assumption 4.1 satisfied.

For given $A \in \mathcal{L}(\mathfrak{h})$ satisfying Assumption 4.1, the gauge invariant quasi-free state $\omega$ on $\mathcal{A}(\mathfrak{h}_0)$ is defined by

$$\omega(W(f)) = \exp\{-(f, \frac{1}{4}(\mathbf{1} + A)(\mathbf{1} - A)^{-1}f)\}, \quad f \in \mathfrak{h}_0 \tag{4.2}$$

Let$(\mathcal{H}_\omega, \pi_\omega, \Omega_\omega)$ be the $GNS$ representation [8] of $(\mathcal{A}(\mathfrak{h}_0), \omega)$, and let $\mathcal{M} = \pi_\omega(\mathcal{A}(\mathfrak{h}_0))''$. Notice that the representation $(\mathcal{H}_\omega, \pi_\omega, \Omega_\omega)$ of the CCR algebra $\mathcal{A}(\mathfrak{h}_0)$ is regular [8] in the sense that the unitary group $t \in \mathbb{R} \longmapsto \pi_\omega(W(tf))$ are strongly continuous for all $f \in \mathfrak{h}_0$. We use the notation $\Phi_\omega(f)$ to denote the infinitesimal generator of the unitary group $\pi_\omega(W(tf))$;

$$\pi_\omega(W(tf)) = \exp(it\Phi_\omega(f)). \tag{4.3}$$

Notice also that, since $\omega$ is an entire analytic state [8] by (4.2) and Assumption 4.1, the cyclic vector $\Omega_\omega$ is an entire vector for all $\Phi_\omega(f)$, $f \in \mathfrak{h}_0$. The annihilation and creation operators defined for each $f \in \mathfrak{h}_0$ by

$$D(a_\omega(f)) = D(\Phi_\omega(f)) \cap D(\Phi_\omega(if)) = D(a^*_\omega(f))$$

and

$$\begin{aligned} a_\omega(f) &= 2^{-1/2}(\Phi_\omega(f) + i\Phi_\omega(if)), \\ a^*_\omega(f) &= 2^{-1/2}(\Phi_\omega(f) - i\Phi_\omega(if)) \end{aligned} \tag{4.4}$$

are densely defined, closed and $a_\omega(f)^* = a^*_\omega(f)$, and satisfy the canonical commutation relations( CCRs): for $\forall f, g \in \mathfrak{h}_0$

$$[a_\omega(f), a_\omega(g)] = 0,$$

$$[a_\omega(f), a^*_\omega(g)] = (f, g)\mathbf{1}. \tag{4.5}$$

See section 5.2.3 of Ref. 8 for the details.

In the rest of this section, we suppress $\omega$ and $\pi_\omega$ from the notations. Thus $\mathcal{H} = \mathcal{H}_\omega$, $W(f) = \pi_\omega(W(f))$, $\Phi(f) = \Phi_\omega(f), a\,(f) = a_\omega(f)$ and $a^*(f) =$

$a^*_\omega(f)$ for any $f \in \mathfrak{h}_0$. We also use the notation $\xi_0 = \Omega_\omega$. Let $\alpha_t : \mathcal{M} \to \mathcal{M}$ be the group of automorphism on $\mathcal{M}$ defined by

$$\alpha_t(W(f)) = W(A^{it} f), \quad f \in \mathfrak{h}_0.$$

It can be check that the state $\omega$ defined by (4.2) is a $\alpha$-KMS state. [8] Thus $\alpha_t$, $t \in \mathbb{R}$, is the modular automorphism by Theorem 5.3.10 of Ref. 8. Let $\Delta$ and $J$ be the modular operator and modular conjugation respectively. Then $(\mathcal{M}, \mathcal{H}, \mathcal{P}, J)$ is the natural standard form associated to $(\mathcal{M}, \xi_0)$.

Next, we introduce several dense manifolds of $\mathcal{H}$. For any $B \in \mathcal{M}$, define

$$B_n = \left(\frac{n}{\pi}\right)^{\frac{1}{2}} \int \sigma_t(B) e^{-nt^2} dt, \quad n \in \mathbb{N}. \tag{4.6}$$

Then $B_n$ is an entire analytic element for $\sigma_t$, $\|B_n\| \leq \|B\|$ for all $n \in \mathbb{N}$ and $B_n \to B$ strongly. See the proof of proposition 2.5.22 of Ref. 8. Put

$$\begin{aligned} \mathcal{M}_0 &:= \text{ the algebra generated by } B_n, \quad B \in \mathcal{M}, n \in \mathbb{N} \\ \mathcal{W}_0 &:= \text{ the algebra generated by } W_n(f), \quad f \in \mathfrak{h}_0, n \in \mathbb{N}, \\ \mathcal{W} &:= \text{ the algebra generated by } W(f), \quad f \in \mathfrak{h}_0. \end{aligned} \tag{4.7}$$

Denote by $\mathcal{H}_{fin}$ the subset of finite linear combinations of the vectors of the following type:

$$\psi_n = \left[\prod_{j=1}^{n} \Phi(f_j)\right] \xi_0, \quad f_j \in \mathfrak{h}_0, \, j = 1, \cdots, n.$$

Clearly $\mathcal{M}_0\xi_0$, $\mathcal{W}_0\xi_0$, $\mathcal{W}\xi_0$ and $\mathcal{H}_{fin}$ are dense in $\mathcal{H}$.

We denote by $a^\#(f)$ either $a(f)$ or $a^*(f)$, for any $f \in \mathfrak{h}_0$. Notice that $a^\#(f)$ and $j(a^\#(f))$, $f \in \mathfrak{h}_0$, are affiliated with $\mathcal{M}$ and $\mathcal{M}'$ respectively. For any $f \in \mathfrak{h}_0$, we write

$$\begin{aligned} \sigma_z(a(f)) &:= a(A^{i\bar{z}} f) \\ \sigma_z(a^*(f)) &:= a^*(A^{iz} f). \end{aligned} \tag{4.8}$$

We are ready to describe Dirichlet forms. An admissible function $f$ is said to be *normalized* if $\int f(t)dt = 1$. For given normalized admissible function $f$ and a complete orthonormal system(CONS) $\{g_n\}_{n=1}^\infty \subset \mathfrak{h}_0$ of $\mathfrak{h}$, define a sesquilinear form $\mathcal{E} : D(\mathcal{E}) \times D(\mathcal{E}) \to \mathbb{C}$ as follows:

$$D(\mathcal{E}) = \mathcal{W}_0\xi_0 (\text{ or } \mathcal{M}_0\xi, \, \mathcal{W}\xi_0)$$

and

$$\mathcal{E}(\eta, \xi) = \sum_{n=1}^{\infty} \mathcal{E}^{(n)}(\eta, \xi), \quad \eta, \xi \in D(\mathcal{E}), \tag{4.9}$$

where for each $n \in \mathbb{N}$

$$\mathcal{E}^{(n)}(\eta, \xi) = \int \langle \big(\sigma_{t-i/4}(a(g_n)) - j\,(\sigma_{t-i/4}(a^*(g_n)))\big)\eta, \\ \big(\sigma_{t-i/4}(a(g_n)) - j\,(\sigma_{t-i/4}(a^*(g_n)))\big)\xi\rangle f(t)\,dt \\ + \int \langle \big(\sigma_{t-i/4}(a^*(g_n)) - j\,(\sigma_{t-i/4}(a(g_n)))\big)\eta, \\ \big(\sigma_{t-i/4}(a^*(g_n)) - j\,(\sigma_{t-i/4}(a(g_n)))\big)\xi\rangle f(t)\,dt. \tag{4.10}$$

We also defined the associated quadratic forms by

$$\mathcal{E}^{(n)}[\xi] = \mathcal{E}^{(n)}(\xi, \xi), \quad \xi \in D(\xi),\ n \in \mathbb{N}$$

$$\mathcal{E}[\xi] = \sum_{n=1}^{\infty} \mathcal{E}^{(n)}[\xi], \quad \xi \in D(\mathcal{E}).$$

We remark that the expression $\mathcal{E}^{(n)}(\eta, \xi)$ in (4.10) can be obtained from $\mathcal{E}(\eta, \xi)$ in (3.2) by replacing $x$ by $a(g_n)$.

We state main results. It turns out that the form defined in (4.9) and (4.10) is independent of the normalized admissible function $f$ and the CONS $\{g_n\}$ we have chosen:

**Proposition 4.1** *Let $D(\mathcal{E})$ be either $\mathcal{W}_0\xi_0$ or else $\mathcal{W}\xi_0$ and let $(\mathcal{E}, D(\mathcal{E}))$ be defined as in (4.9) and (4.10). The form $(\mathcal{E}, D(\mathcal{E}))$ is closable. Moreover,$(\mathcal{E}, D(\mathcal{E}))$ is independent of the normalized admissible function $f$ and the CONS $\{g_n\} \subset \mathfrak{h}_0$ we have chosen.*

**Proposition 4.2** *For each $n \in \mathbb{N}$, $(\mathcal{E}^{(n)}, \mathcal{M}_0\xi_0)$ is closable. The closure $(\overline{\mathcal{E}}^{(n)}, D(\overline{\mathcal{E}}^{(n)}))$ is a Dirichlet form.*

**Theorem 4.1** *Let us consider the following form :*

$$D(\overline{\mathcal{E}}) = \{\xi \in \bigcap_{n=1}^{\infty} D(\overline{\mathcal{E}}^{(n)}) : \sum_{n=1}^{\infty} \overline{\mathcal{E}}^{(n)}[\xi] < \infty\},$$

$$\overline{\mathcal{E}}[\xi] = \sum_{n=1}^{\infty} \overline{\mathcal{E}}^{(n)}[\xi],$$

*where for each $n \in \mathbb{N}$, $(\overline{\mathcal{E}}^{(n)}, D(\overline{\mathcal{E}}^{(n)}))$ is the closed Dirichlet form obtained in Proposition 4.2. Then $(\overline{\mathcal{E}}, D(\overline{\mathcal{E}}))$ is a densely defined Dirichlet form.*

**Theorem 4.2** *Let $\{T_t\}$ be the symmetric Markovian semigroup associated to the Dirichlet form $(\overline{\mathcal{E}}, D(\overline{\mathcal{E}}))$, and let $H$ be the Dirichlet operator, i.e., $T_t = e^{-tH}, t \geq 0$. Then the following results hold:*

(a) *H is essentially self-adjoint on* $\mathcal{H}_{.fin}$.

(b) *H is independent of the normalized admissible function f and the CONS* $\{g_n\} \subset \mathfrak{h}_0$ *we have chosen.*

(c) *The zero is a simple eigenvalue of H with eigenvector* $\xi_0$. *Moreover* $(0, \alpha^{-1/2} - \alpha^{1/2}) \cap \sigma(H) = \phi$. *Thus* $\{T_t\}$ *is ergodic.*

The complete proofs of the above results are lengthy and required several limiting processes to take care of the unboundedness of $a(f), f \in \mathfrak{h}_0$. We refer to Ref. 6 for the proofs. Since the proof of proposition 4.1 is not hard, we give the proof.

**Proof of proposition 4.1.** We first consider the case of $D(\mathcal{E}) = W\xi_0$. Notice that

$$\begin{aligned}
&\{\sigma_{t-i/4}(a(g_n)) - j(\sigma_{t-i/4}(a^*(g_n)))\}W(h_1)\xi_0 \\
&= [\sigma_{t-i/4}(a(g_n)), W(h_1)]\xi_0 \\
&= [a(A^{it-1/4}g_n), W(h_1)]\xi_0 \\
&= i2^{-1/2}(A^{it-1/4}g_n, h_1)W(h_1)\xi_0.
\end{aligned}$$

Here we have used (4.8),(4.3),(4.4) and (4.5) in that order. The method used in the above yields

$$\begin{aligned}
&\{\sigma_{t-i/4}(a^*(g_n)) - j(\sigma_{t-i/4}(a(g_n)))\}W(h_1)\xi_0 \\
&= -i2^{-1/2}(h_1, A^{it+1/4}g_n)W(h_1)\xi_0.
\end{aligned}$$

Thus it follows from the definition of $\mathcal{E}(\eta, \xi)$ in (4.9) and (4.10) that

$$\begin{aligned}
&\mathcal{E}(W(h_2)\xi_0, W(h_1)\xi_0) \\
&= 2^{-1}\sum_{n=1}^{\infty}\int (h_2, A^{it-1/4}g_n)(A^{it-1/4}g_n, h_1)\langle W(h_2)\xi_0, W(h_1)\xi_0\rangle f(t)\,dt \\
&+2^{-1}\sum_{n=1}^{\infty}\int (A^{it+1/4}g_n, h_2)(h_1, A^{it+1/4}g_n)\langle W(h_2)\xi_0, W(h_1)\xi_0\rangle f(t)\,dt.
\end{aligned}$$

Using the Lebesgue dominated convergence theorem and the Parseval relation, we get

$$\begin{aligned}
&\mathcal{E}(W(h_2)\xi_0, W(h_1)\xi_0) \\
&= 2^{-1}\{(h_2, A^{-1/2}h_1) + (A^{1/2}h_1, h_2)\}\langle W(h_2)\xi_0, W(h_1)\xi_0\rangle.
\end{aligned} \tag{4.11}$$

Thus $(\mathcal{E}, \mathcal{W}\xi_0)$ is independent of the normalized admissible function $f$ and the CONS $\{g_n\} \subset \mathfrak{h}_0$.

Define an operator $H : \mathcal{W}\xi_0 \to \mathcal{H}$ by

$$H(W(h_1)\xi_0) \\ = i2^{-1/2}\{(a(A^{-1/2}h_1) - j(a^*(h_1)))^* - (a^*(A^{1/2}h_1) - j(a(h_1)))^*\}W(h_1)\xi_0.$$

Notice that

$$\begin{aligned}(a(A^{-1/2}h_1) - j(a^*(h_1)))W(h_2)\xi_0 &= [a(A^{-1/2}h_1), W(h_2)]\xi_0 \\ &= i2^{-1/2}(A^{-1/2}h_1, h_2)W(h_2)\xi_0,\end{aligned}$$

and

$$(a^*(A^{1/2}h_1) - j(a(h_1)))W(h_2)\xi_0 = -i2^{-1/2}(h_2, A^{1/2}h_1)W(h_2)\xi_0.$$

Using the above relations and (4.11), we have

$$\mathcal{E}(W(h_2)\xi_0, W(h_1)\xi_0) = \langle W(h_2)\xi_0, HW(h_1)\xi_0\rangle.$$

It follows that $H$ is a positive symmetric operator on $\mathcal{W}\xi_0$. Thus $(\mathcal{E}, \mathcal{W}\xi_0)$ is closable (Theorem X.23 of Ref. 27). The proof for $D(\mathcal{E}) = \mathcal{W}_0\xi_0$ follows from the argument similar to that used for $D(\mathcal{E}) = \mathcal{W}\xi_0$. This completes the proof. □

## References


1. L. Accardi, Topics in quantum probability, *Phys. Rep.* **77**, 169-192 (1981).
2. L. Accardi and Cecchini, C.: Conditional expectations in von Neumann algebras and a theorem of Takesaki, *J. Funct. Anal.* **45**, 245-273 (1982).
3. L. Accardi, A. Frigerio, and J.T. Lewis, Quantum stochastic processes, *Publ. RIMS* **18**, 97-133 (1982).
4. S. Albeverio and R. Høegh-Krohn, Dirichlet forms and Markovian semigroups on $C^*$-algebras, *Commun. Math. Phys.* **56**, 173-187 (1977).
5. H. Araki, Some properties of modular conjugation operator of von Neumann algebras and noncommutative Radon-Nikodym theorem with chain rule, *Pacific J. Math.* **50** (2), 309-354 (1974).
6. C. Bahn, C. K. Ko and Y. M. Park, Dirichlet forms and symmertic Markovian semigroups on CCR algebras with respect to quasi-free states : preprint preparation.
7. A. Beurling and J. Deny, Dirichlet spaces, *Proc. Nat. Acad. Sci.* **45**, (1959).
8. O. Bratteli and D.W. Robinson, *Operator algebras and quantum statistical mechanics*, Springer-Verlag, New York-Heidelberg-Berlin, vol I (1979), vol. II (1981).

9. F. Cipriani, *Dirichlet forms and Markovian semigroups on standard forms of von Neumann algebras*, (Thesis Trieste, 1992).
10. F. Cipriani, Dirichlet forms and Markovian semigroups on standard forms of von Neumann algebras, *J. Funct. Anal.* **147**, 259-300 (1997).
11. F. Cipriani, F. Fagnola and J. M. Lindsay, Spectral Analysis and Feller Properties for Quantum Ornstein-Uhlenbeck Semigroups, *Comm. Math. Phys.* **210**, 85-105 (2000).
12. E.B. Davies, *Quantum theory of open systems*, Academic Press, London-New York-San Francisco (1976).
13. E.B. Davies, *Heat-kernel and spectral theory*, Vol. 92, Cambridge Univ. Press, Academic Press (1989).
14. E.B. Davies and J.M. Lindsay, Non commutative symmetric Markov semigroups, *Math. Zeit.* **210**, 379-411 (1992).
15. M. Fanes and B. Nachtergaele, R.F. Werner, Finitely correlated states on quantum spin chains, *Commun. Math. Phys.* **144**, 443-490 (1992).
16. M. Fukushima, *Dirichlet Forms and Markov Processes*, North-Holland, Kodansha (1980).
17. D. Guido, T. Isola and S. Scarlatti, Non-commutative Dirichlet forms on semifinite von Neumann algebras, *J. Funct. Anal.* **135**, 50-75 (1996).
18. S. Goldstein and J.M. Lindsay, Beuring-Deny conditions for KMS-symmetric dynamical semigroups, *C. R. Acad. Sci. Paris. Ser. I* **317**, 1053-1057 (1993).
19. S. Goldstein and J.M. Lindsay, KMS-symmetric Markov semigroups, *Math. Zeit.* **219**, 590-608 (1995).
20. Z.M. Ma and M. Röckner, *An introduction to the thorey of (non-symmetric) Dirichlet forms*, Springer-verlag (1992).
21. A.W. Majewski and B. Zegarlinski, Quantum stochastic dynamics I: Spin systems on a lattice, *MPEJ* **1**, Paper 2 (1995).
22. A.W. Majewski and B. Zegarlinski, Quantum stochastic dynamics II, *Rev. Math. Phys.* **8** (5), 689-713 (1996).
23. T. Matsui, Markov semigroups on UHF algebras, *Rev. Mod. Phys.* **5**, 587-600 (1993).
24. B. Nachtergaele, The spectral gap for some spin chains with discrete symmetry breaking, *Commun. Math. Phys.* **175**, 565-606 (1996).
25. Y. M. Park, Construction of Dirichlet forms on standard forms of von Neumann algebras, *Infinite Dimensional Analysis, Quantum Probability and Related Topics*, Vol. 3, No. 1, 1-14 (2000).
26. K.R. Parthasarathy, *An Introduction to Quantum Stochastic Calculus*, Birkhäuser, Basel (1992).
27. M. Reed and B. Simon, *Method of modern mathematical physics II:*

*Fourier analysis, Self-adjointness*, Academic Press, New York (1975).

28. J.L. Sauvageot, Semi-groupe de la chaleur transverse sur la $C^*$ algebre d'un feuilletage riemannien, *C. R. Acad. Sci. Paris. Ser. I* **310**, 531-536 (1990).
29. J.L. Sauvageot, Tangent bimodule and locality for dissipative operators on $C^*$-algebras, in *Quantum probability and applications IV* (L. Accardi and W. von Waldenfels, Eds.), Lecture Notes in Math. vol. 1396, Springer-Verlag, Berlin-Heidelberg-New York, (1987).
30. J.L. Sauvageot, Quantum Dirichlet forms, differential calculus and semigroups, in *Quantum probability and applications V* (L. Accardi and W. von Waldenfels, Eds.), Lecture Notes in Math. vol. 1442, Springer-Verlag, Berlin-Heidelberg-New York, (1988).
31. M. L. Silverstein, *Symmetric Markov Process* (Springer-verlag, 1974).
32. M. Takesaki, *Tomita's Theory of Modular Hilbert Algebras and its applications*, Lecture Notes in Mathematics, vol. 128, Springer-verlag, Berlin-Heidelberg-New York (1970).

# Symbols for Semilinear Wave Equation with Conormal Cauchy Datas Singular at a Point

**A. Piriou**
Laboratoire J.A. Dieudonné
Université de Nice Sophia - Antipolis, UMR N$^o$ 6621 du CNRS
Parc Valrose, 06108 Nice, Cedex 2, France

**Abstract**

The singularities of the solution $u$ of a semilinear wave equation with conormal Cauchy datas singular at a point $S$ were studied by N. Ritter, J.M. Bony (see also J.M. Chemin for the fully nonlinear case). They proved that the wave front set of $u$ is included in the union of the conormal bundle to the light cone issued from $S$ and the conormal bundle to the vertex $S$. Outside the intersection of these two conormal bundles, $u$ is conormal in Hörmander sense, and our goal is to study the corresponding principal symbols, which give a good description of the leading part of the singularities of $u$. We state general conjectures, which are proven in some particular cases, and we present remarks which should be useful for the general case.

# 1 Introduction

We consider the semilinear Cauchy Problem

$$\left\{ \begin{array}{rcll} \dfrac{\partial^2 u}{\partial x_1^2} - \displaystyle\sum_{j=2}^{n} \dfrac{\partial^2 u}{\partial x_j^2} & = & f(x, u, \nabla u) & \text{in } \mathbb{R}^n \\ \gamma_k u & = & g_k \qquad (k = 0, 1) & \text{in } \mathbb{R}^{n-1} \end{array} \right. \tag{1}$$

where $x = (x_1, x')$, $x_1 \in \mathbb{R}$ is the time variable, $x' = (x_2, \ldots, x_n) \in \mathbb{R}^{n-1}$ is the space variable, $\nabla u = \left( \dfrac{\partial u}{\partial x_1}, \dfrac{\partial u}{\partial x_2}, \ldots, \dfrac{\partial u}{\partial x_n} \right)$, $f = f(x, u, v)$ is a real-valued $C^\infty$-function in $\mathbb{R}^n \times \mathbb{R} \times \mathbb{R}^n$, and $\gamma_0 u = u(0, \cdot)$, $\gamma_1 = \gamma_0 \dfrac{\partial}{\partial x_1}$. (For simplicity, we consider only real-valued solutions of (1), but the case of complex-valued solutions can be handled with slight modifications).

Existence and uniqueness properties for (1) are well-known: If $g_k$ belongs to the Sobolev space $H^{s-k}_{loc}(\mathbb{R}^{n-1})$, $s > n/2+1$, then (1) admits a unique solution in $H^s_{loc}(\mathbb{R}^n)$. A similar result holds if the Sobolev spaces $H^{s-k}_{loc}(\mathbb{R}^n)$, $H^s_{loc}(\mathbb{R}^n)$ are respectively replaced by the Besov spaces ${}^\infty H^{s-k}_{loc}(\mathbb{R}^{n-1})$, ${}^\infty H^s_{loc}(\mathbb{R}^n)$.

We are concerned with the propagation of singularities: If we know the singularities of the Cauchy datas $g_k$, can we determine directly the singularities of the solution $u$? In the linear case (that is when $f$ is linear with respect to $u, v$), we know that the wave front set of $u$ (denoted $WF(u)$) is included in the union of the bicharacteristic strips issued from the points $(0, x', \xi)$ such that $\xi^2 - |\xi'|^2 = 0$, $(x', \xi')$ on $WF(g_0)$ or $WF(g_1)$. In the semilinear case, interactions of microlocal singularities arise, and the situation is very intricate (see M. Beals, [1]). But results can be given if we restrict to the class of conormal singularities, which are recalled in the next section.

# 2 Conormal and Lagrangian singularities

Let $X$ be a smooth manifold of dimension $n$, $T^*X$ its conormal bundle, and $\Lambda$ a smooth closed conical Lagrangian submanifold of $T^*X \setminus O$. We denote by $\Psi^d(X)$ the space of proper, classical $\Psi$DO(pseudodifferential operators) of order $d$ in $X$, and we set

$$\mathcal{M}(\Lambda) = \{L \in \Psi^1(X) : \sigma(L) = 0 \text{ on } \Lambda\} \tag{2}$$

where $\sigma$ denotes the principal symbol.

**Definition 1 (See Hörmander, [7])** *The space $I^m(X, \Lambda)$ of Lagrangian distributions (with respect to $\Lambda$) of order $m$ is defined as the set of all $u \in {}^\infty H^s_{loc}(X)$ such that $L_1 \dots L_N u \in {}^\infty H^s_{loc}(X)$ for all $N$ and all $L_j \in \mathcal{M}(\Lambda)$. Here $s = -m - n/4$.*

It is clear that $\mathcal{M}(\Lambda)$ is a $\Psi^0(X)$-module, and that:

$$u \in I^m(X, \Lambda) \Longrightarrow WF(u) \subset \Lambda.$$

**Definition 2 (See Hörmander, [7])** *Let $M$ be a smooth closed submanifold of $X$. The space $I^m(X, M) = H^s(X, M)$ of conormal distributions with respect to $M$ and of order $m$ (or index $s = -m - n/4$) is the space $I^m(X, N(M) \setminus O)$, where $N(M)$ is the conormal bundle of $M$.*

When $u \in I^m(X, M)$, then $WF(u) \subset N(M) \setminus O$, and $\operatorname{sing\,supp}(u) \subset M$.

Elements of $I^m(X, \Lambda)$ can also be described by means of Fourier type integrals and this leads to the definition of principal symbols. For instance, in the conormal case, a distribution $u \in \mathcal{D}'(X)$ belongs to $H^s(X, M)$ if and only if:

1. $u$ is $C^\infty$ in $X \setminus M$.

2. close to each point of $M$, taking local coordinates $(x_1, \ldots, x_n)$ such that $M = \{x_1 = \ldots = x_n = 0\}$, $p = \text{codim } M$

then

$$u(x) = \int_{\mathbb{R}^p} e^{i(x_1\xi_1+\cdots+x_p\xi_p)} \sigma(x; \xi_1, \ldots, \xi_p) d\xi_1 \ldots d\xi_p \tag{3}$$

where $\sigma \in S^\mu(\mathbb{R}^n \times \mathbb{R}^n)$ with $\mu = -s - p/2$. (see [7] for the definition of the space $S^\mu$ of symbols of degree $\mu$.) The equivalence class of $\sigma|_{x\in M}$ in $S^\mu \setminus S^{\mu-1}$ defines the principal symbol of $u$ as a class of symbol sections of a convenient vector bundle over $N(M) \setminus \{0\}$ (see [7]).

# 3 Conormal Cauchy problem

We consider the Cauchy problem (1), assuming up to now that the Cauchy datas are conormal with respect to the submanifold $\{0\}$ of $\mathbb{R}^{n-1}$:

$$g_k \in H^{s-k}(\mathbb{R}^{n-1}, \{0\}), \qquad k = 0, 1.$$

According to (3), we have

$$g_k(x') = \int_{\mathbb{R}^{n-1}} e^{x'\cdot\xi'} \sigma_k(\xi') d\xi', \qquad \sigma_k \in S^{\mu+k}, \quad \mu = -s - \frac{n-1}{2}. \tag{4}$$

Firstly, we recall the linear case, that is the case where

$$f(x, u, v) = \sum_{j=1}^{n} a_j(x) \frac{\partial u}{\partial x_j} + b(x)u + c(x)$$

with $a_j, b, c \in C^\infty(\mathbb{R}^n)$. By standard WKB method, the solution $u$ is given, modulo $C^\infty(\mathbb{R}^n)$, by

$$u(x) \equiv \sum_{\pm} \int_{\mathbb{R}^{n-1}} e^{i(x'\cdot\xi' \pm x_1|\xi'|)} a_\pm(x, \xi') d\xi' \tag{5}$$

with $a_\pm \in S^\mu(\mathbb{R}^n \times \mathbb{R}^{n-1})$.

The phase functions $x' \cdot \xi' \pm x_1|\xi'|$ parametrize the Lagrangian manifold

$$\Lambda_1^\pm = \Big\{(x, \xi) \in \mathbb{R}^n \times (\mathbb{R}^n \setminus \{0\}) \Big| \; x_1 \in \mathbb{R}, \;\; \xi' \in \mathbb{R}^{n-1} \setminus \{0\}, \;\; \xi_1 = \pm|\xi'|, \;\; x' = \mp x_1 \frac{\xi'}{|\xi'|} \Big\}$$

thus $u \in I^m(\mathbb{R}^n, \Lambda_1)$ where $m = -s - n/4$ and $\Lambda_1 = \Lambda_1^+ \cup \Lambda_1^-$ is the conormal bundle to the (non-smooth) light cone $\Gamma = \{x \in \mathbb{R}^n \,||x_1| = |x'|\}$. The principal symbol of $u$

is defined by the principal part $a^0_\pm$ of $a_\pm$in $S^\mu$ modulo $S^{\mu-1}$, which is determined by the transport equation

$$\left[2\left(\xi_1\frac{\partial}{\partial x_1}-\sum_{j=2}^{n}\xi_j\frac{\partial}{\partial x_j}\right)-\sum_{j=1}^{n}a_j(x)\xi_j\right]a_0^\pm\equiv 0 \qquad \text{on } \Lambda_1^\pm \tag{6}$$

and the initial condition

$$a_0^\pm(x,\xi)\Big|_{\begin{array}{lcl} x_1 & = & 0 \\ \xi_1 & = & \pm|\xi'|\end{array}} \equiv \frac{1}{2}\left(\sigma_0(\xi')\pm\frac{1}{i|\xi'|}\sigma_1(\xi')\right) \tag{7}$$

where $\sigma_k(\xi')$ was introduced at (4).

For studying the nonlinear case, the algebra properties of the Lagrangian and conormal spaces are crucial. Returning to Definition 1, and recalling that the Besov space ${}^\infty H^s_{loc}(X)$ is an algebra when $s > \dim X/2$, it is clear, by Leibniz rule, that $I^m(X,\Lambda)$ is also an algebra (and more generally stable under the action of $C^\infty$ functions) if $s > \dim X/2$ and if $\mathcal{M}(\Lambda)$ is generated by vector fields. This last condition is satisfied in the conormal case $\Lambda = N(M)\setminus O$, $M$ smooth, thus $H^s(X,M)$ is an algebra when $s > \dim X/2$ (in fact, from (3), it can be proven that $s > (\text{codim } M)/2$ is sufficient). But even when $s = -m - n/4$ is large, the Lagrangian space $I^m(X,\Lambda_1)$ of linear solutions $u$ given by (5) is not an algebra, due to the singularity of the light cone $\Gamma$ at the point 0: by nonlinear operations, new microlocal singularities $(0,\xi)$ with $\xi \in \mathbb{R}^n\setminus\{0\}$ are expected. More precisely, we have the following result:

**Proposition 1** *Let $\Lambda_0 = \{(0,\xi)|\xi\in\mathbb{R}^n\setminus\{0\}\}$ be the conormal bundle to $\{0\}\subset\mathbb{R}^n$. If $u_1$, $u_2$ are of the form (5) with $\mu < -(n-1)$ (which corresponds to $s > \frac{n-1}{2}$), then, microlocally in $\Lambda_0\setminus\Lambda_1 = \{(0,\xi)|\xi\in\mathbb{R}^n\setminus\{0\}, |\xi_1|\neq|\xi'|\}$, $u_1u_2$ is in $H^s(\mathbb{R}^n,\{0\})$ with $s' = 2s-\frac{n}{2}+1$, and this index is optimal.*

Now set $\Lambda = \Lambda_1\cup\Lambda_0$, define $\mathcal{M}(\Lambda)$ as in (1) and $I^m(X,\Lambda)$ as in Definition 1. A remarkable fact, proven by N. Ritter [14], is that $\mathcal{M}(\Lambda)$ is now generated by vector fields, and thus $I^m(X,\Lambda)$ is an algebra when $s > n/2\,(s = -m-n/4)$. Thanks to commutation arguments and linear energy estimates, this allowed to prove the following result:

**Theorem 1 (N. Ritter [14], J. M. Bony [2])** *If $s > n/2+2$ and $g_k \in H^{s-k}(\mathbb{R}^{n-1},\{0\})$, $(k=0,1)$, then the solution $u$ of the semilinear Cauchy problem (1) belongs to $I^m(\mathbb{R}^n,\Lambda_1\cup\Lambda_0)$, where $m = -s-n/4$.*

# 4 Principal symbols and optimal conormal regularity

As a consequence of Theorem 1, we have $WF(u) \subset \Lambda_1 \cup \Lambda_0$, singsupp $u \subset \Gamma$, the light cone issued from 0, the restriction of $u$ to $\{x \neq 0\}$ belongs to the conormal space $H^s(\mathbb{R}^n \setminus \{0\}, \Gamma \setminus \{0\})$, and microlocally in $\Lambda_0 \setminus \Lambda_1$, $u$ is in the conormal space $H^s(\mathbb{R}^n, \{0\})$.

In this work, our concern is the following:

> Is the index $s$ optimal for the two conormal regularities stated above, and more precisely, what can be said about the two corresponding principal symbols?

The linear case shows that $s$ is optimal for $\Gamma \setminus \{0\}$, and the following conjecture is relative to the principal symbol:

**Conjecture 1** *The principal symbol of $u \in H^s(\mathbb{R}^n \setminus \{0\}, \Gamma \setminus \{0\})$ is obtained by solving the transport equation (6) formally associated to the linearization of (1) (that is with $a_j = \frac{\partial f}{\partial v_j}(x, u, \nabla u)\Big|_\Gamma$) with the initial condition (7).*

For the microlocal conormal regularity in $\Lambda_0 \setminus \Lambda_1$, it is clear that $s$ is not optimal. Indeed, thanks to the algebra property of

$$\{w \in {}^\infty H^s_{loc}(\mathbb{R}^n) | w \text{ is microlocally in } {}^\infty H^r_{loc}(\mathbb{R}^n) \text{ at a given point } (\bar{x}, \bar{\xi})\}$$

when $\frac{n}{2} < s \leq r < 2s - \frac{n}{2}$ (see M. Beals [1]) and to the ellipticity of the wave operator in $\Lambda_0 \setminus \Lambda_1$, it is not hard to prove that, microlocally in $\Lambda_0 \setminus \Lambda_1$, $u \in H^{2s-n/2+2}(\mathbb{R}^n, \{0\})$. The following conjecture is motivated by our Proposition 1:

**Conjecture 2** *Microlocally in $\Lambda_0 \setminus \Lambda_1$, $u$ is in $H^{s'}(\mathbb{R}^n, \{0\})$ with $s' = 2s - n/2 + 3$, and this index is optimal.*

We conclude by giving some partial results. Conjectures 1 and 2 are true when $n = 2$: see A. Piriou, J.P. Varenne [13], H. Bougrini, A. Piriou, J.P. Varenne [3] and G. Laschon [8] for the more general case of Cauchy datas conormal with respect to a smooth hypersurface in any dimension. Conjecture 1 is true in the weakly semilinear case:

**Theorem 2** *We consider the semilinear Cauchy problem (1) with $f(x, u, \nabla u) = \phi(x, u)$ depending only on $x$ and $u$, and $g_k \in H^{s-k}(\mathbb{R}^{n-1}, \{0\})$, $k = 0, 1$, $s > n/2 + 1$. Then in $\{x \neq 0\}$, the principal symbol of $u$ as an element of $H^s(\mathbb{R}^n \setminus \{0\}, \Gamma \setminus \{0\})$ is the same as in the linear case $f = 0$, and thus it is determined by the transport equation (6) with $a_j = 0$ and the initial condition (7).*

Sketch of the proof. Denoting by $P$ the wave operator, we have $u = u_0 + u_1$ where $u_0$ is the solution of the linear Cauchy problem $Pu_0 = 0$, $\gamma_k u_0 = g_k$ $(k = 0, 1)$, and $u_1$ satisfies $Pu_1 = F$, $\gamma_k u_1 = 0$ $(k = 0, 1)$ with $F = f(x, u)$. Since $u \in I^m(\mathbb{R}^n, \Lambda_1 \cup \Lambda_0)$ we have also $F \in I^m(\mathbb{R}^n, \Lambda_1 \cup \Lambda_0)$. By adapting the Bony's proof of Theorem 1, we can deduce that $u_1 \in I^{m-1}(\mathbb{R}^n, \Lambda_1 \cup \Lambda_0)$, thus $u_1 \in H^{s+1}(\mathbb{R}^n \setminus \{0\}, \Gamma \setminus \{0\})$ in $\mathbb{R}^n \setminus \{0\}$, and gives no contribution to the principal symbol of $u$ considered in Theorem 2.

For studying Conjectures 1 and 2 in more general cases, Lagrangian spaces $I^{m,m'}(\mathbb{R}^n, \Lambda_1, \Lambda_0)$ allowing different regularities on $\Lambda_1$ and $\Lambda_0$ are needed. Such spaces were introduced by R.B. Melrose and G. Uhlmann [11], V. Guillemin and G. Uhlmann [6], G. Mendoza [12], A. Greenleaf and G. Uhlmann [5] in a linear context. In particular, G. Mendoza [12] considers distributions which are, microlocally close to $\Lambda_1^{\pm}$, defined by oscillatory integrals

$$u(x) = \int e^{i(x\cdot\xi + s(\xi_1 \mp |\xi'|) + s\sigma)} a(x, s; \xi, \sigma) ds\, d\xi\, d\sigma$$

where $a(x, s; \xi, \sigma)$ is a product-type symbol with $\xi, \sigma$ as frequency variables. This form seems well adapted to our problems, and a nonlinear study of Mendoza spaces should be performed to go further in the analysis of Conjectures 1 and 2.

# References


[1] **M. Beals**, *Propagation and Interaction of Singularities in Nonlinear Hyperbolic Problems*, Birkhäuser (1989)

[2] **J.M. Bony**, *Singularités des solutions des problèmes de Cauchy hyperboliques non linéaires*, Advances in Microlocal Analysis, H.G. Garner (ed), D. Reidel, p. 15-39 (1986)

[3] **H. Bougrini, A. Piriou, J.P. Varenne**, *Propagation et interaction des symboles principaux pour les ondes conormales semi-linéares*, Comm. in PDE, 23(1), p. 333-370 (1998)

[4] **J.Y. Chemin**, *Régularité de la solution d'un problème de Cauchy fortement non linéaire à données singulières en un point*, Ann. Inst. Fourier, 39(1), p. 101 - 123 (1989)

[5] **A. Greenleaf, G. Uhlmann**, *Estimates for Singular Radon Transforms and Pseudo Differential Operators with Singular Symbols*, J. of Funct. Anal., 89, p .202 - 232 (1990)

[6] **V. Guillemin, G. Uhlmann**, *Oscillatory Integrals with Singular Symbols*, Duke Math. J, 48, p. 251 - 267, (1981)

[7] **L. Hörmander**, *The Analysis of Linear Partial Differential Operators III, IV*, Springer Verlag (1985)

[8] **G. Laschon**, *Thèse de Doctorat*, Nice (1998)

[9] **R.B. Melrose, N. Ritter**, *Interaction of Nonlinear Progressing Waves*, Annals of Math. 121, p. 187 - 213 (1985)

[10] **R.B. Melrose, N. Ritter**, *Interaction of Progressing Waves for Semilinear Wave Equation II*, Arkiv for Math. 25, p. 91 - 114 (1987)

[11] **R.B. Melrose, G. Uhlmann**, *Lagrangian Intersection and the Cauchy Problem*, Comm. in Pure and Appl. Math, Vol 32, p. 483 - 519 (1979)

[12] **G. Mendoza**, *Symbol Calculus associated with Intersecting Lagrangians*, Comm. in PDE, 7(9), p. 1035 - 1116 (1982)

[13] **A. Piriou, J.P. Varenne**, *Problème de Cauchy semi-linéaire à donnée initiale conormale: régularité à l'arête et symbole principal*, J. Math. Pures Appl., 75, p. 9 - 49 (1996)

[14] **N. Ritter**, *Progressing Wave Solutions to Nonlinear Hyperbolic Cauchy Problems*, PhD Thesis, MIT (1986)

# TRIANGLE GRAPHS WITH MAXIMUM DEGREE AT MOST 3

**Romulo G. Pizaña**
Department of Mathematics
College of Science, University of the Philippines, Diliman
e-mail: pizaña@math.upd.edu.ph

**Rolando E. Ramos**
Institute of Mathematical Sciences and Physics
College of Arts and Sciences, University of the Philippines, Los Baños, Laguna

**ABSTRACT**

The triangle graph $T(G)$ of a graph $G$ with triangles is the graph whose vertices are the triangles of $G$ in which two vertices are adjacent if and only if the corresponding triangles have a common edge in $G$. A graph $H$ is a triangle graph if there exists a graph $G$ with triangles such that $T(G) \cong H$. In this paper, we describe triangle graphs with maximum degree at most three.

# 1 INTRODUCTION

Let $G$ be a graph with triangles and $S(G)$ denote the set of all triangles in $G$. The *triangle graph of* $G$, denoted by $T(G)$, is the graph with vertex set $S(G)$ in which two vertices are joined if and only if they have a common edge in $G$.

A graph $G$ is called a *triangle graph* if there exists a graph $H$ with triangles such that $T(H)$ is isomorphic to $G$.

If for a graph $G$ there exists a graph $H$ with triangles such that $T(H) \cong G$, then $H$ is a *triangle configuration* of $G$. We say that an edge of a triangle is *free* if there is no triangle edge-adjacent to it.

A *chord* of the cycle $C$ is a line connecting two non-adjacent vertices of $C$. Two chords of $C$ are said to be *parallel* if the chords do not cross each other and they have no common end vertex. An *elementary cycle* $C$ is a cycle where no two non-adjacent vertices are connected by a path.

A cycle $C$ is said to be a *cycle with parallel chords* (CPC) if any two chords $k_i$ , $k_j$, $i \neq j$, of $C$, are parallel. We say that a cycle $C$ is *CPC of Type I* if $C$ can be decomposed into consecutively adjacent chordless cycles $C^1$, $C^2$, ..., $C^n$ such that $C^i \cap C^{i+1} = k_i$ for $i = 1, 2, \ldots, n-1$ and $C^i \cap C^j = \Phi$ for $j \neq i+1, j > i$ and for $j \neq i-1, j > i$. We say that a cycle $C$ is *CPC of Type II* if $C$ can be decomposed into chordless cycles $C^1$, $C^2$, ..., $C^n$ such that there exists a chordless cycle $C^m$ for some m, $1 \leq m \leq n$ such that $C^m \cap C^n = k_s$ for $s = 1, 2, \ldots, q \geq 3$ $(q < n)$, but $C^i \cap C^j = \Phi$ for $j \neq i$, $m$. We will call the cycles $C^s$ as *boundary cycles.* We say that a cycle $C$ is *CPC of Type III* if $C$ can be decomposed into chordless cycles $C^1$, $C^2$, ..., $C^n$ such that $C^i \cap C^j \neq \Phi$ implies that $C^i \cap C^j = k_r$, $r = 1, 2, 3, \ldots t$ and $C$ can be further decomposed into $Q^i$ which are CPC's of Type I or Type II.

A *cut-vertex* of a graph is a vertex whose removal increases the number of components and a *bridge* is such a line. A *nonseparable* graph is connected, nontrivial and has no cut-vertex. A *block* of a graph is a maximal nonseparable subgraph. A block of order $p \geq 3$ is called a *cyclic block* . A graph is acyclic if it has no cycle.

For $n \geq 4$, the *wheel*, $W_n$, is defined to be the graph $K_1 + C_{n-1}$.

A theta graph $T$ is isomorphic to

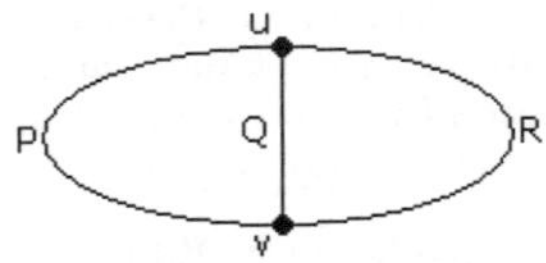


where $P$, $Q$, $R$, are internally disjoint $u-v$ paths.

# 2 MAIN RESULTS

The following results aim to describe triangle graphs $G$, with maximum degree at most three. Paths, cycles and trees have been shown to be triangle graphs. Hence, focus will be given to cyclic blocks $G$. The following propositions will be used to prove a theorem describing triangle graphs $G$ with maximum degree at most 3.

**Proposition 1**. Any cycle with parallel chord(s), except $K_4 - e$, is a triangle graph.

**Sketch of Proof**. Any cycle with parallel chords of Type I can be decomposed into elementary cycles consecutively edge-adjacent to each other. Each elementary cycle has a particular corresponding triangle configuration in the form of a wheel for $p \geq 4$ or $K(1,1,3)$ for $p = 3$. Hence, CPC's of type I are triangle graphs. For CPC's of Type II, there exists a central elementary cycle to which the boundary elementary cycles are correspondingly attached. The same will be the case for Type III CPC's as they are just combinations of Type I and Type II CPC's.

**Remark:** We will call graphs which are cycles with parallel chords CPC's of type I, II, III as $G$-I graphs.

**Lemma 1**. Theta graphs $T$ such that $|P|$, $|Q|$, $|R| \geq 7$ are triangle graphs.

**Proof.** To show that $T$ is a triangle graph, we show that we can find a triangle configuration for $T$. WLOG, take the elementary cycle $C$ composed of the paths P and Q. We know that there exists a wheel $W_n$, $n \geq 12$ which is the triangle configuration for $C$. Then we can easily construct a sequence of consecutive edge-adjacent triangles starting with $T_w$, the triangle corresponding to vertex $u$ in $T$ and ending with $T_v$, the triangle corresponding to vertex $v$ in $T$. This is the triangle configuration for the path $R$.

**Proposition 2.** Let $G$ be a block with the following properties:

i. $K_4 - e$ free

ii. deg $v = 2$ or $3$ for all $v \in V(G)$

iii. For any induced theta subgraph $T \subseteq G$, $|P|$, $|Q|$, $|R| \geq 7$.

Then $G$ is a triangle graph.

**Sketch of Proof.** The proof will be by induction on the number $n$ of elementary cycles in $G$. For $n = 1$, $G$ is an elementary cycle and hence $G$ is a triangle graph. The induction hypothesis is as follows: if $G$ is a block satisfying i, ii, iii with $n \leq k$ elementary cycles, then $G$ is a triangle graph. Suppose that $G$ has $n = k + 1$ elementary cycles. Then $G - P_o$ for some path $P_o$ has $n = k$ elementary cycles and by the induction hypothesis, $G - P_o = G^*$ is a triangle graph, hence, having a triangle configuration. Construct some path $P_r$ where $|P| \geq 7$ connecting any two vertices $v$ and $u$ in $G^*$ such that properties i, ii, iii are still satisfied. Without loss of generality, let vertices $u$, $v$ be on the boundary of $G^*$. By Lemma 1, $G^*$ with the newly-constructed path is a triangle graph.

**Remark:** We will call graphs described in Proposition 2 above, as $G$-II graphs.

**Proposition 3**. A graph $G$ with $\Delta(G) \leq 3$ is a triangle graph if and only if every cyclic block of $G$ is a triangle graph.

**Proof.** The necessity is clear. For the sufficiency, suppose that every cyclic block of $G$, where $\Delta(G) \leq 3$, is a triangle graph. To show that $G$ is a triangle graph. Without loss of generality, let $G$ be connected. $G$ can be decomposed into blocks which are either acyclic or cyclic. Clearly, every acyclic block is a triangle graph. It is given that every cyclic block is a triangle graph. Hence, for every block $B_i \subseteq G$, there exists a graph $H_i$ with triangles such that $T(H_i) \cong B_i$. The adjacency of the $H_i$'s can be easily determined following the adjacency of the $B_i$'s in $G$ because $\Delta(G) \leq 3$. Therefore, $T(H) \cong G$ where $H \cong \cup H_i$. That is, $G$ is a triangle graph.

The following theorem then describes triangle graphs $G$ where $\Delta(G) \leq 3$.

**Theorem.** Suppose a graph $G$ satisfies the following properties:

1. $\Delta(G) \leq 3$
2. the cyclic blocks of $G$ are either $G$-I or $G$-II graphs

Then $G$ is a triangle graph.

Note that the above theorem follows immediately from Propositions 1, 2 nd 3.

**REFERENCES:**

Behzad, M., Chartrand, G., and Lesniak, L., *"Graphs and Diagraphs"*, $1^{st}$ edition, Belmont, California: Wadsworkth, Inc., 1979.

Beineke, L. and Wilson, R., eds., *"Selected Topics in Graph Theory"*, London: Academic Press, 1978.

Berge, Claude, *"Graphs"* , $2^{nd}$ revised editions, Amsterdam: North-Holland, 1985

Harary, Frank, ed., *"Proof Techniques in Graph Theory"*. New York and London: Academic Press, 1969.

Ramos, R.E., Dorado, C.A., *"On Triangle Graphs"* Proceedings, Annual Symposium of the Division of Mathematical Sciences, UP Alumni Center, Diliman, Quezon City, 21-22 October 1988, 165-172.

Pizaña, R.G., *"On Triangle Graphs"*, Doctoral Dissertation, University of the Philippines-Diliman, Quezon City, 1992.

# Algebraic Geometry and Physics

**Shi-shyr Roan**

*Institute of Mathematics , Academia Sinica , Taipei , Taiwan*
*E-mail: maroan@ccvax.sinica.edu.tw*

**Abstract**

This article is an interdisciplinary review and an on-going progress report over the last few years made by myself and collaborators in certain fundamental subjects on two major theoretic branches in mathematics and theoretical physics: algebraic geometry and quantum physics. I shall take a practical approach, concentrating more on explicit examples rather than formal developments. Topics covered are divided in three sections: (I) Algebraic geometry on two-dimensional exactly solvable statistical lattice models and its related Hamiltonians: I will report results on the algebraic geometry of rapidity curves appeared in the chiral Potts model, and the algebraic Bethe Ansatz equation in connection with quantum inverse scattering method for the related one-dimensional Hamiltonion chain, e.g., XXZ, Hofstadter type Hamiltonian. (II) Infinite symmetry algebras arising from quantum spin chain and conformal field theory: I will explain certain progress made on Onsager algebra, the relation with the superintegrable chiral Potts quantum chain and problems on its spectrum. In conformal field theory, mathematical aspects of characters of N=2 superconformal algebra are discussed, especially on the modular invariant property connected to the theory. (III). Algebraic geometry problems on orbifolds stemming from string theory: I will report recent progress on crepant resolutions of quotient singularity of dimension greater than or equal to three. The direction of present-day research of engaging finite group representations in the geometry of orbifolds is briefly reviewed, and the mathematical aspect of various formulas on the topology of string vacuum will be discussed.

# 1 Introduction

In this report, I would like to summarize our recent work and on-going progress we have made in the interdisciplinary area of algebraic geometry and quantum physics. A brief historic account relevant to the subjects will also be presented. We will discuss some fundamental problems in solvable theories in these fields. Here the term "solvable" is loosely used in the following sense: in physics, it means some physical quantities of the theory, e.g., the partition function, can be computed exactly in a closed form; in mathematics, it involves certain mathematical formulae or equations expressed in an explicit form whose solution can be effectively calculated. The physical models we are concerned with here are those among integrable statistical systems or string theory, where techniques or problems in algebraic geometry naturally arise, in particular, those related to 2-dimensional exactly solvable statistical lattice models and orbifold models of superstring compactification. It is known that a 2-dimensional statistical lattice model associates a 1-dimensional quantum Hamiltonian chain, whose eigenvalues determine the partition function of the statistical system. Incorporating the symmetries of a quantum chain could serve as a useful means of gaining solutions of the original model. Accordingly, the representation theory of symmetrical algebras was brought in within the context of solvable models. Recently many such theories have been explored along this scheme in literature with spectacular progress. But, just to keep things simple, we will in this article restrict our attention only to a few such kind of models we have been involved with at the present day. Let us address the main topics in this report with a comparison list of theories involved in the areas of physics and mathematics:

| **Physics** | **Mathematics** |
|---|---|
| (Integrable) Chiral Potts Model, | Algebraic Curve; |
| Superintegrable Chiral Potts Model, | Onsager Algebra, $sl_2-$loop Algebra; |
| Hofstadter Type Hamiltonian, | Bethe Equation on Riemann Surface; |
| $\mathrm{N} = 2$ Conformal Field Theory, | Elliptic Theta Form; |
| Orbifold Model in String Theory, | Crepant Resolution, $G-$Hilbert Scheme . |

Here I will summarize the research that we undertook on these issue over the past five years. The presentation is divided in three sections with the references based upon our work in the area:

- Exactly solvable integrable model: [33, 48].
- Symmetry algebra of quantum spin chain, superconformal field theory: [13, 22, 31, 49].
- Orbifold, finite group representation and string theory: [11, 46, 47]

**Convention .** In this article, $\mathbf{R}$, $\mathbf{C}$ will denote the field of real, complex numbers respectively; $\mathbf{Z}$ the ring of integers. For positive integers $n$ and $N$, we will denote $\overset{n}{\otimes}\ \mathbf{C}^N$ the tensor product of $n$-copies of the complex $N$-dimensional vector space $\mathbf{C}^N$, and $\mathbf{Z}_N$ the quotient ring $\mathbf{Z}/N\mathbf{Z}$.

## 2 Exactly Solvable Integrable Model

In this section, we discuss a class of 2-dimensional lattice models, called solvable integrable models, in which certain interesting physical quantities, in particular the partition function, can be explicitly computed. The models are built by assigning spins $\sigma_i$ on sites of a (finite or infinite) planar lattice with adjacent spins interacting along the edges by a rule of Boltzmann weights. The spins $\sigma_i$ take values of $\pm 1$; or $N$ different values, $0, \ldots, N-1 \in \mathbf{Z}_N$. For a such system, the Yang-Baxter (or star-triangle) relation of Boltzmann weights plays a central role in the theory of an exactly solvable model in statistical mechanics and field theory. For statistical lattice models, it is well known that a model is solvable by techniques of commuting transfer matrices if the Boltzmann weights satisfy the Yang-Baxter equation. There are a number of notable Yang-Baxter solvable lattice models which have been extensively studied in both literature of physics and mathematics (see, e.g., [3]). Here I will consider only a few of such kind of models in which there exists a certain intricate relation with problems in algebraic geometry.

In the 1980s, a two-dimensional lattice integrable model with phenomenological origin in physics, known as the chiral Potts $N$-state model, was found in statistical mechanics where the "rapidity" (or "spectral variable") lies on a algebraic curve of genus higher than one,

( A brief history account can be found in Sect. 4.1 of [34]). The model takes $N(\geq 3)$ values for the spin per site, which is a generalization of the renowned Ising model first solved by L. Onsager in 1944 [39]. The Boltzmann weights, $W_{p,q}, \overline{W}_{p,q}$, have the form,

$$\frac{W_{p,q}(n)}{W_{p,q}(0)} = \prod_{j=1}^{n} \frac{d_p b_q - a_p c_q \omega^j}{b_p d_q - c_p a_q \omega^j}, \qquad \frac{\overline{W}_{p,q}(n)}{\overline{W}_{p,q}(0)} = \prod_{j=1}^{n} \frac{\omega a_p d_q - d_p a_q \omega^j}{c_p b_q - b_p c_q \omega^j}$$

where $\omega = e^{\frac{2\pi i}{N}}$, $n \in \mathbf{Z}_N$, and $p, q$ are two rapidities represented by the ratio of four-vectors, $a : b : c : d$, satisfying the following set of equations,

$$\begin{aligned} k a^N + k' c^N = d^N, \quad k b^N + k' d^N = c^N, \\ a^N + k' b^N = k d^N, \quad k' a^N + b^N = k c^N. \end{aligned} \tag{1}$$

Here $k, k'$ are the real parameter with the relation $k^2 + k'^2 = 1$. The periodicity of $n$ modulo $N$ of the Boltzmann weights follows from the above constraints of rapidities, which form a curve of genus $N^3 - 2N^2 + 1$ in the projective 3-space $\mathbf{P}^3$. For a lattice with the horizontal size $L$ and periodic boundary condition, the combined weights of intersection between two consecutive rows give rise to the transfer matrix,

$$T_{\{\sigma, \sigma'\}}(p, q) = \prod_{l=1}^{L} \overline{W}_{p,q}(\sigma_l - \sigma'_l) W_{p,q}(\sigma_l - \sigma'_{l+1}) \ ,$$

where $\sigma = (\sigma_1, \ldots, \sigma_L)$ , $\sigma' = (\sigma'_1, \ldots, \sigma'_L)$ with $\sigma_l, \sigma'_l \in \mathbf{Z}_N$. The transfer matrix $T(p, q)$ is an operator acting on the $L$-tensor vector space of $\mathbf{C}^N$, $\overset{L}{\otimes}$ $\mathbf{C}^N$. Here the coordinates of a vector in $\mathbf{C}^N$ will be indexed by elements in $\mathbf{Z}_N$, equivalently to say, it is represented by a sequence of coordinates, $\{v_k\}_{k \in \mathbf{Z}}$, with the $N$-periodic condition, $v_k = v_{k+N}$. The complete spectrum of the operator $T(p, q)$ is one of the main mathematical problems of physical interest. By the star-triangle relation satisfied by Boltzmann weights, it ensures the commuting relation of transfer matrices for a fixed rapidity $p$:

$$[T(p, q), \ \ T(p, q')] = 0 \ .$$

Furthermore, $T(p, q)$ also satisfies a certain functional equation involving an automorphism of the Riemann surface defined by $R(a_q, b_q, c_q, d_q)$

$= (b_q, \omega a_q, d_q, c_q)$. Note that for $N = 2$, it is the Ising case where (1) defines an elliptic curve, known to carry a canonical additive group structure. Solutions of the functional equation can be obtained by using elliptic theta function representations of $a, b, c, d$ in the rapidity. However, the "difference property" of rapidities no longer holds for the spectral curve (1) as $N \geq 3$. This creates the peculiar character which makes the chiral Potts model distinct with all previously known models, hence inspired the algebro-geometry approach on studying integrable models with a such kind of nature. Note that the curve (1) possesses a large finite number of symmetries, and the normalized Boltzmann weights are meromorphic functions of that Riemann surface. In various problems of computating the physical quantities, e.g., free energy, order parameter etc., the spectral curve can be reduced to some lower genus one by modulating certain symmetry subgroup. For the spectrum of the transfer matrix, the curves are the ones obtained from (1) quotiented by a certain order $N^2$ symmetry subgroup, with the variables, $t = \frac{ab}{cd}, \lambda = (\frac{d}{c})^N$, satisfying the relation,

$$\mathcal{C}_{N,k'} : \quad t^N = \frac{(1 - k'\lambda)(1 - k'\lambda^{-1})}{k^2} \ . \tag{2}$$

This is one-parameter curves with the characterization of the genus $N-1$ hyperelliptic curves carrying the dihedral symmetrical group $D_N$ structure. With the functional equation derived in [8, 9], the computation was made in the infinite lattice, (i.e., $L \to \infty$), to obtain a closed form for the "ground state" energy in [5] and the excitation energy in [35, 20], by an ingenious method devised by R. J. Baxter bypassing the algebraic geometry of the hyperelliptic curves. These have been the only few of progresses made on the physical computations of spectrum to this day. As the solutions involved are rational functions on Riemann surfaces, which in principal could be constructed from its zeros and poles, techniques in algebraic geometry would be expected of great help to such problems. Through the classical work of S. Kowalevski, R. J. Baxter first found the expression of hyperelliptic function parametrization of Boltzmann weights years ago [6]; subsequently the Baxter's parametrization was explored in [43] by an algebraic geometry method where the prime form expression of the Boltzmann weights was identified. In an recent work [48], differential equations for the periods and the expression of the modulus of the hyperellpitic family (2) in terms of the Schwarz's triangle function were obtained. However all these findings in algebraic geometry, though interesting in a certain sense, haven't yet provided any significant insight in solving the original eigenvalue problem. Similar difficulties occur on the role of algebraic geometry when applying to solutions of other physical quantities, among which a notable one is the strong conjecture made in [2] on the order parameter with a Ising-like form:

$$\langle Z_0^j \rangle = (1 - k'^2)^{\frac{j(N-j)}{2N^2}} \ , \quad 0 \leq j \leq N \ .$$

After many unsuccessful attempts of employing algebraic geometry into problems of the chiral Potts model, B.M. McCoy made the following epigram in [34] which goes as :

*The nineteen century saw many brilliant creations of the human mind. Among them are algebraic geometry and Marxism. In the late twentieth century Marxism has been shown to be incapable of solving any practical problem but we still do not know about algebraic geometry.*

Nevertheless, the study of chiral Potts model does raise some of the most intractable problems in algebraic geometry at the present day; the quest of rigorous mathematical theorems with physical implication should be an interesting, though difficult, program in algebraic geometry. On the other hand, in the physical content there are some notable 1-dimensional Hamiltonian chains which are speculated to have certain intimate connection with the chiral Potts model; though the exact formulation is not apparently clear right now. Among these Hamiltonions, the XXZ spin chain, Hofstadter type Hamiltonian, discrete sine-Gordon model are included. It is well known that the XXZ chain is the one corresponding to the 2-dimensional six-vertex statistical model [32, 55], a renowned solvable system whose structure was the primary source of the mathematical development of quantum group theory (for a pedagogical account, see e.g. [18, 30]). By the work of Bazhanov and Stroganov [9], it proposed that the chiral Potts model can be regarded as a certain new algebraic structure related to the six-vertex model. By this suggestion, the quantum inverse method approach developed by the Leningrad school in the early eighties [51] could be brought into the study of those related spin chains, which would possibly provide some valuable insights to the chiral Potts model. Indeed it is the case with certain discrete integrable models -with discretized space and time variables-involved. The Hofstadter type Hamiltonian is a such example with progress we recently made.

The Hofstadter Hamiltonian has the following form for parameters $\mu, \nu \in \mathbf{R}$,

$$H_{Hof} = \mu(U + U^{-1}) + \nu(V + V^{-1}),$$

where $U, V$ are unitary operators satisfying the Weyl commutation relation: $UV = \omega VU$, $\omega := e^{2\pi\sqrt{-1}\Phi}$ with the phase $\Phi \in \mathbf{R}$. The Hamiltonion can be interpreted as a 2-dimensional Bloch system in the presence of constant electric and magnetic fields, which has been a model with a long and rich history in physics. The phase $\Phi$ represents the magnetic flux per plaquette. This problem has attracted an unceasing interest in both the physical and mathematical communities since the 1976 numerical work of Hofstadter [27] exhibited a beautiful fractal structure of butterfly spectral band versus the magnetic flux. For a rational flux $\Phi$, a surprising connection was recently found between problems of diagonalizing the Hofstadter type Hamiltonians and solving the 1-dimensional spin chain of a finite site by the quantum transfer matrix method. The development was motivated by the work of Wiegmann and Zabrodin [54] on the appearance of quantum $U_q(sl_2)$ symmetry in magnetic translation problems. Subsequently in [19], Faddeev and Kashaev considered the following Hamiltonian, including the Hamiltonian $H_{Hof}$ as a certain special limit,

$$H_{FK} = \mu(U + U^{-1}) + \nu(V + V^{-1}) + \rho(W + W^{-1}), \qquad (3)$$

where $U, V, W$ satisfy the pair-wise Weyl commutation relations: $UV = \omega VU$, $VW = \omega WV$, $WU = \omega UW$. For a rational flux $\Phi = P/N$ with $\omega$ a primitive $N$-th root of unity, by the well known result on representations of the Weyl algebra, one can rewrite the operators $U, V, W$ in $H_{FK}$ as $\alpha U, \beta V, \gamma W$ with $\alpha, \beta, \gamma \in \mathbf{C}$, plus the $N$-power identity of $U, V, W$: $U^N = V^N = W^N = 1$. The Hamiltonian $H_{FK}$ can be realized in the transfer matrix of a Yang-Baxter model for the six-vertex $R$-matrix; its Yang-Baxter solution $\mathcal{L}$-operator, appeared also in the study of chiral Potts $N$-state model [8], has the following form:

$$\mathcal{L}_h(x) = \begin{pmatrix} aZX & xbX \\ xcZ & d \end{pmatrix} , \quad x \in \mathbf{C} ,$$

where $X, Z$ are operators of $\mathbf{C}^N$ defined by

$$X, Z : \mathbf{C}^N \longrightarrow \mathbf{C}^N, \quad X|m\rangle := |m+1\rangle, \quad Z|m\rangle := \omega^m |m\rangle \tag{4}$$

where $m \in \mathbf{Z}_N$ and the parameter $h$ represents the ratio of four complex numbers, $a : b : c : d$, i.e., an element in $\mathbf{P}^3$. It follows that the $\mathcal{L}$-operator of a size $L$ ( with operators of $\overset{L}{\otimes} \mathbf{C}^N$ as its entries),

$$\mathcal{L}_{\vec{h}}(x) = \mathcal{L}_{h_0}(x) \otimes \mathcal{L}_{h_1}(x) \otimes \ldots \otimes \mathcal{L}_{h_{L-1}}(x) ,$$

again satisfies the Yang-Baxter relation. The transfer matrices,

$$T_{\vec{h}}(x) := \mathrm{tr}(\mathcal{L}_{\vec{h}}(x), x \in \mathbf{C},$$

form a family of commuting operators of $\overset{L}{\otimes} \mathbf{C}^N$, hence with the common eigenvectors. The exact spectra of $T_{\vec{h}}(x)$ will be the main problem we are concerned with. A powerful algebraic Bethe-ansatz-technique in the investigation of this type of problems was introduced in [19] as follows. By applying a gauge transformation of the $\mathcal{L}$-operator at each site $j$ by introducing a set of new variables $\xi_j$s,

$$\widetilde{\mathcal{L}}_{h_j}(x, \xi_j, \xi_{j+1}) = \begin{pmatrix} 1 & \xi_j - 1 \\ 1 & \xi_j \end{pmatrix} \mathcal{L}_{h_j}(x) \begin{pmatrix} 1 & \xi_{j+1} - 1 \\ 1 & \xi_{j+1} \end{pmatrix}^{-1} ,$$

for $0 \leq j \leq L-1$, and $\xi_L := \xi_0$. it remains the same for the transfer matrix, i.e., $T_{\vec{h}}(x) = \mathrm{tr}(\bigotimes_{j=0}^{L-1} \widetilde{\mathcal{L}}_{h_j}(x, \xi_j, \xi_{j+1}))$. However, there exists a family of vectors $|p\rangle$ in $\overset{L}{\otimes} \mathbf{C}^N$ with the parameter $p$ in a "spectral curve" $\mathcal{C}_{\vec{h}}$, which is defined in $(x, \xi_0, \ldots, \xi_{L-1})$-space by the relations,

$$\mathcal{C}_{\vec{h}} : \quad \xi_j^N = (-1)^N \frac{\xi_{j+1}^N a_j^N - x^N b_j^N}{\xi_{j+1}^N x^N c_j^N - d_j^N} \quad , \quad j = 0, \ldots, L-1 ,$$

such that $T_{\vec{h}}(x)$ acts on those selected vectors $|p\rangle$ in a "geometrical" manner,

$$T_{\vec{h}}(x)|p\rangle = |\tau_- p\rangle \Delta_-(p) + |\tau_+ p\rangle \Delta_+(p) .$$

Here $\tau_\pm$ are automorphisms, and $\Delta_\pm$ the algebraic functions of $\mathcal{C}_{\vec{h}}$ defined by

$$\begin{array}{l} \tau_\pm : (x, \xi_0, \ldots, \xi_{L-1}) \mapsto (q^{\pm 1}x, q^{-1}\xi_0, \ldots, q^{-1}\xi_{L-1}), \qquad q := \omega^{\frac{1}{2}} \\ \Delta_-(x, \xi_0, \ldots, \xi_{L-1}) = \prod_{j=0}^{L-1}(d_j - x\xi_{j+1}c_j), \\ \Delta_+(x, \xi_0, \ldots, \xi_{L-1}) = \prod_{j=0}^{L-1} \frac{\xi_j(a_jd_j - x^2b_jc_j)}{\xi_{j+1}a_j - xb_j} . \end{array}$$

For a common eigenvector $\langle\varphi|$ of $T_{\vec{h}}(x)$ with the eigenvalue $\Lambda(x) \in \mathbf{C}[x]$, the function $Q(p) = \langle\varphi|p\rangle$ of $\mathcal{C}_{\vec{h}}$ satisfies the following relation, called the Bethe equation (of the model),

$$\Lambda(x)Q(p) = Q(\tau_-(p))\Delta_-(p) + Q(\tau_+(p))\Delta_+(p) \ , \quad \text{for } p \in \mathcal{C}_{\vec{h}} \ . \tag{5}$$

The family of vectors $|p\rangle$ is called the "Baxter vector" or "Baxter vacuum state" in literature, which was explained in [33] from the aspect of algebraic geometry. As the Bethe solutions $Q(p)$ of (5) are rational functions of the algebraic curve $\mathcal{C}_{\vec{h}}$, the algebraic geometry of $\mathcal{C}_{\vec{h}}$ would play a vital role in the solvability of the Bethe equation in this scheme. The complexity of the problem heavily depends on the size $L$ and the function theory of the corresponding curve $\mathcal{C}_{\vec{h}}$. For the diagonalization of Hofstadter type Hamiltonion (3), it is related to the case $L = 3$, the simplest one with respect to the size. However even in this situation, the general solution of (5) remains an unsolved problem due to the high genus of the curve $\mathcal{C}_{\vec{h}}$ though a primitive qualitative investigation was conducted in [33].

When the curve $\mathcal{C}_{\vec{h}}$ degenerates into a finite collection of rational curves for certain special values of $\vec{h}$, algebraic geometry of the spectral curve becomes a trivial one . With a detailed analysis of the form of Baxter vector and a certain "diagonalization" procedure of eliminating the effects of variables $\xi_j$, the equation (5) is reduced to a finite number of polynomial Bethe equations, which possess a form of "Strum-Liouville like" difference equation on the complex plane. Even with a simple expression of these rational Bethe equations, the determination of its solutions $\Lambda(x), Q(x)$ for an arbitrary given size $L$ still posts a difficult mathematical problem, from both the quantitative and qualitative aspects. For $L = 3$, we made a thorough mathematical study of the rational Bethe equations in [33], and derived its complete solutions.

The results have revealed some certain mathematical characters common with other standard intergable models as the XXX-chain by using the algebraic Bethe-ansatz-technique; one can discuss the "degeneracy" of eigenstates of the transfer matrix with respect to the Bethe solutions. However, inspired by the butterfly fractal structure of the Hofstadter Hamiltonian, we adopt the infinity procedure by letting $N \to \infty$ while keeping the size $L$ fixed as the "thermodynamic" limit, a consideration corresponding to the irrational flux in the Block problem. We have made the detailed discussion of these structures with possibly physical interest. It is worth noting that the results obtained for rational Bethe solutions are corresponding to those in the spectrum problem of the Hamilition (3) with a special choice of representations of Weyl operators $U, V, W$. For a higher size $L$, some mathematical subtleties involved in solutions of the rational Bethe equation are difficulty to overcome at this moment. However, we have now conducted a similar analysis on the case $L = 4$, due to its connection with discrete quantum pendulum, discrete sine-Gordon model. The progress we now obtain has indicated a certain new feature unseen in the case $L = 3$. A deeper study would enhance our understanding of the efficiency of this algebraic method on diagonalizing integrable Hamiltonians. But ultimately, the quest of possible interplay of these analysis with the chiral Potts model would be our primary concern for the programme undertaken by now.

# 3 Symmetry Algebra of Quantum Spin Chain, Superconformal Field Theory

The principle of symmetries has been one of the main ideas of modern physics. If a model possesses certain symmetries, its analysis becomes much simpler by employing the representation theory of the symmetrical algebra. In this section, we present two such models which we have made certain progress in recent years.

In the family of the chiral Potts $N$-state model, a peculiar integrable structure occurs to enhance on solving the model when the vertical rapidity $p$ takes a special value. This special case is called "super-integrable", which means the vertical rapidity satisfies the relations: $a_p = b_p, c_p = d_p$. The corresponding element in $\mathcal{C}_{N,k'}$ of (2) has the coordinate $\lambda_p = 1$, i.e., $p$ lies above a branched point of the hyperelliptic curve $\mathcal{C}_{N,k'}$. In this way, the superintegrable case can be regarded as a general chiral Potts model with a special symmetry condition on the rapidity. It turns out the integrabilty of the model has a stronger sense than the usual ones. In this case, the corresponding $N$-state quantum chain has the following Hamiltonian form,

$$\begin{array}{c} H(k') = H_0 + k' H_1 \ , \\ H_0 := -2\sum_{l=1}^{L}\sum_{n=1}^{N-1} \frac{X_l^n}{1-\omega^{-n}}, \quad H_1 := -2\sum_{l=1}^{L}\sum_{n=1}^{N-1} \frac{Z_l^n Z_{l+1}^{N-n}}{1-\omega^{-n}}, \end{array} \quad (6)$$

where $X_l, Z_l$ are the operators on $\overset{L}{\otimes} \mathbf{C}^N$ with the Weyl operators $X, Z$ of (4) at the $l$-th site: $X_l = I \otimes \ldots \otimes \overset{l\text{th}}{X} \otimes \ldots \otimes I$, $Z_l = I \otimes \ldots \otimes \overset{l\text{th}}{Z} \otimes \ldots \otimes I, (Z_{L+1} = Z_1)$. Note that the operator $H(k')$ is Hermitian for real $k'$, hence with the real eigenvalues. When $N = 2$, $H(k')$ is the Ising quantum chain ; for $N = 3$, one obtains the $\mathbf{Z}_3$-symmetrical self-dual chiral clock model with the chiral angles $\varphi = \phi = \frac{\pi}{2}$ in [28]. For a general $N$, $H(k')$ was first constructed in a paper of G. von Gehlen and R. Rittenberg [21], in which the following Dolan-Grady condition [15] was shown to hold for the operators $A_0, A_1$ with $A_0 = -2N^{-1}H_0, A_1 = -2N^{-1}H_1$,

$$[A_1, [A_1, [A_1, A_0]]] = 16[A_1, A_0] \ , \quad [A_0, [A_0, [A_0, A_1]]] = 16[A_0, A_1] \ .$$

The above relation is equivalent to the Onsager algebra [14], which is composed of operators $A_m, G_m$ ($m \in \mathbf{Z}$) through the recurrent relation via $G_1 := \frac{1}{4}[A_1, A_0]$,

$$A_{m-1} - A_{m+1} = \frac{1}{2}[A_m, G_1] \ , \quad G_m = \tfrac{1}{4}[A_m, A_0] \ .$$

This Lie algebra satsifies following relations

$$[A_m, A_l] = 4G_{m-l} \ , \quad [A_m, G_l] = 2(A_{m-l} - A_{m+l}) \ , \quad [G_m, G_l] = 0 \ ,$$

which appeared in the seminal paper [39] of L. Onsager in 1944 on the solution of two-dimensional Ising model. Ever since then, an intimate relationship has been known to exist between Onsager algebra and $sl_2$, the exact connection was clarified in [42] through a mathematical formulation by regarding the Onsager algebra as the fixed subalgebra of

$sl_2$-loop algebra, $sl_2[t, t^{-1}]$, by the canonical involution. In a recent joint work with E. Date [13], we made a systematic study of the algebraic structure of the Onsager algebra; obtained the complete classification of "closed" Lie-ideals[1] and the structure of quotient algebras. The key ingredient is to employ the $sl_2$-loop variable $t$ by working on the Laurent polynomials on the relevant problems which we concern. The ideal theory of the Onsager algebra arisen from the finite dimension representation of the algebra has revealed a simple and elegant structure, which provides a clear link between representations of Onsager algebra and $sl_2$-loop algebra. With the $t$-evaluation, the finite dimensional unitary representations of Onsager algebra are derived in a canonical manner. Meanwhile, the essence differentiating Onsager algebra from $sl_2[t, t^{-1}]$ has indicated the structure of certain solvable algebras and a combinatorial nature of the content, though the full explanation is still lacking by now. The combinatoric implication of Onsager algebra would be expected by recent progress made on subjects related to a deformed Dolan-Grady relation (see, e.g., [52]). However, the mathematical analysis on Onsager algebra only ensures a certain special Ising-like form for the eigenavlues of the Hamiltonian (6), the precise expression could not yet be derived even for the "ground-state" energy, known in physical literature [1, 4]. The expression involves the Chebychev type polynomials for $N$=2 [3]. For higher $N$ , by a recent study on specific cases, we are able to achieve the equations satisfied by the Baxter-Albertini-McCoy-Perk polynomials, and understand the nature and the distribution of their zeros [22]. Nevertheless, the complete knowledge of the spectrum of (6) still remains a theoretical mathematical challenge in the context of study of Onsager algebra.

Another type of infinite dimensional algebras we have been working on is the superconformal algebra, which describes the underling symmetries of superstring theory. It is known in physics that certain conformal field theories can be identified with the continuum scaling limit of some critical integrable lattice models, a fact which is served as a link between 2-dimensional statistical mechanics and string theory. The supersymmetric extensions of spacetime symmetries can enhance profound structures in physical content; in fact, supersymmetry is more or less mandatory in string theory in which quantities of physical in-

[1]Here a Lie-ideal $I$ of a Lie algebra $L$ is called closed if $L/I$ has the trivial center.

terest could be calculated exactly. Though the subject has been studied in physical literature extensively since long (see [12] and references therein ), the systematic mathematical study of conformal superalgebras has been taken up by mathematicians only until recent years [10]. Mathematically, the superconformal fields are described by a simple Lie superalgebra, called a superconformal algebra. It is spanned by modes of a finite family of local fields, containing the Virasoro and some other even, odd ones, such that coefficients of the operator product expansions are linear combinations of fields in the family and their derivatives. Among the many conformal field theories, the N=2 conformal theory possesses a peculiar nature which defines a class of its own. The reason is partly due to the existence of its connection with physical spacetime described by Kähler geometry, which is the central core in the analytic study of algebraic geometry. The N=2 conformal algebra, denoted by SCA, consists of the Virasoro field $L(z)$, two super-current $G^{\pm}(z)$ and a $U(1)$-current $J(z)$, whose coefficients $L_m, J_n, G_p^{\pm}, (m, n \in \mathbf{Z}, p \in \frac{1}{2}+\mathbf{Z})$, form a super-Lie algebra with a central element $c$ :

$$
\begin{aligned}
&[L_m, L_n] = (m-n)L_{m+n} + \tfrac{c(m^3-m)}{12}\delta_{m,-n}, \qquad &&[J_m, J_n] = \tfrac{cm}{3}\delta_{m,-n} \ , \\
&[L_m, J_n] = -nJ_{m+n} \ , &&[L_m, G_p^{\pm}] = (\tfrac{m}{2}-p)G_{m+p}^{\pm} \ , \\
&[J_m, G_p^{\pm}] = \pm G_{m+p}^{\pm} \ , &&\{G_p^+, G_q^+\} = \{G_p^-, G_q^-\} = 0 \ , \\
&\{G_p^+, G_q^-\} = 2L_{p+q} + (p-q)J_{p+q} + \tfrac{c(p^2-\frac{1}{4})}{3} \ \delta_{p,-q} \ .
\end{aligned}
$$

The algebra has a standard Cartan decomposition with its Cartan subalgebra spanned by $c, L_0, J_0$. An irreducible highest weight module (HWM) is characterized by the central charge $c$, and the quantum numbers $h, Q$ for $L_0, J_0$ respectively, by which one defines the character of a HWM as the following Laurent series,

$$
\mathrm{NS}(z,\tau) = \mathrm{Tr}(y^Q q^{h-\frac{c}{24}}) \ , \quad y := e^{2\pi\sqrt{-1}z} \ , \ q := e^{2\pi\sqrt{-1}\tau} \ .
$$

The discrete series of the unitary irreducible HWMs has the central charge $c < 3$, parametrized by

$$
c = \frac{3k}{k+2} \ , \quad h = \frac{l^2+2l-m^2}{k+2} \ , \quad Q = \frac{m}{k+2} \ ,
$$

where $1 \le k \in \mathbf{Z}$, $l, m \in \mathbf{Z}$ with $l - m \in 2\mathbf{Z}, |m| \le l \le k$. In rational conformal field theory, the representations in the discrete series form the building blocks of the Gepner's models of string compactification [23], which gives rise to manifolds with SU($n$) holonomy. The topological invariants, e.g., Euler numbers, Hodge numbers and elliptic genus, of such manifolds can be expressed by characters of HWMs involved, (for this aspect, see e.g. [17, 44] and references therein). Amid the many features special for the N=2 conformal algebra, the modular invariant property related to the theory has drawn the constant attention from geometers and mathematical physicists because of its involvement with the theory of classical elliptic modular form. As the algebra SCA has one parameter family of equivalent generators, which form the twisting currents, $L^a(z), J^a(z), G^{\pm a}(z)$, one can also consider the "twisted" character,

$$\mathrm{Ch}^{(a,b)}(z,\tau) = \mathrm{Tr}(q^{L_0^a - \frac{c}{8}}(e^{2\pi\sqrt{-1}b}y)^{J_0^a}) \ , \quad \frac{-1}{2} \leq a, b \leq \frac{1}{2} \ .$$

The NS$(z,\tau)$ is corresponding to $(a,b) = (0,0)$, called the Neveu-Schwarz sector. The Ramon sector is the one for $(a,b) = (\frac{-1}{2}, \frac{-1}{2})$. In a recent work [49], we have obtained an expression of the HWM characters in the discrete series in terms of Jacobi-theta form,

$$\mathrm{Ch}_{l,m}^{(a,b)}(z,\tau) = \frac{e^{\pi\sqrt{-1}(\frac{1}{2}-\frac{l+1}{k+2})}\Theta^{(\frac{1}{2}+\frac{l+1}{k+2},\frac{1}{2})}(0,(k+2)\tau)\Theta^{(a,b)}(z,\tau)}{\Theta^{(\frac{1}{2}+\frac{l-m+1+2a}{2(k+2)},b)}(z,(k+2)\tau)\Theta^{(\frac{1}{2}-\frac{l+m+1-2a}{2(k+2)},b)}(z,(k+2)\tau)}$$

where $\Theta^{(s,t)}(z,\tau)$ is the following expression given by the elliptic theta function $\vartheta^{s,t)}(z,\tau)$ and the Dedekind eta function $\eta(\tau)$:

$$\Theta^{(s,t)}(z,\tau) = \frac{\vartheta^{(s,t)}(z,\tau)}{\eta(\tau)^3} \ , \qquad s,t \in \mathbf{R} \ .$$

Note that for the Ramon sector, the characters are involved with the Jacobi-theta form (of weight $-1$ and index $\frac{1}{2}$). With the help of modular form theory, we clarify the symmetries among HWMs in the discrete series for a fixed central charge, then obtain the Heisenberg and modular invariant property of these characters in an explicit manner. As an ongoing program, with the knowledge obtained in N=2 conformal field theory, we shall apply it to the geometrical study of $c_1 = 0$ manifolds as an alternative way to understand the topological invariants of such manifolds, especially on the elliptic genus, of which the mathematical structure has now become clearer after the work of Hirzebruch school [25]. On the other hand, the fundamental vital role of modular group has been a well-known fact in different areas of mathematics; the importance of modular data has also arisen in many physical theories other that conformal field theory, e.g. quantum Hall effect. For the reason that is apparent from the modular invariant property, another problem would be the connection between N=2 conformal algebra and affine algebras from the perspective of representation theory. It is known that different theories can realize the same modular data, even within the content of conformal field theories only. Inside SCA there is a finite dimensional super-Lie algebra $sl_2(2|1)$, which is geometrically realized as super-vector fields of the one-dimensional projective superspace $\mathbf{P}^{1,2}$ with 2-component Grassmann variables. Meanwhile characters of HWM in the discrete series of N=2 conformal theory have shown a remarkable similarity to those of admissible highest weight representations of the affine algebra $\widehat{sl_2}(2|1)$ in [31]. The highlight of our present analysis is that first, the characters of two sides are found to match in a perfect way, and second, both algebras share analogous features on the structure. A deeper connection is now under our investigation, and partial results are promising.

# 4 Orbifold, Finite Group Representation and String Theory

There has been spectacular progress in the development of string theory since its inception thirty years ago. Development in this area has never been impeded by the lack of experimental confirmation. Indeed, numerous bold and imaginative strides have been taken and the sheer elegance and logical consistency by the arguments have served as a primary motivation for string theorists to push their information ahead. Meanwhile, in the process of developing the theory, other than employing powerful mathematical machinery, string theorists have proposed many elegant mathematical formulae and conjectures, which had never been anticipated by mathematicians before. However these predictions were explained and justified only at the level for the need of the physical theory, or through the physical intuition. Yet even when definite predictions exist, an explanation is an explanation in the physical sense, still not a mathematical proof. With this attitude, a rigorous mathematical justification would be desired for further understanding the structure so that results obtained would be expected to make feedback to the original theory in physics. Amid the many such examples occurring in the past few decades, a notable one would be the Vafa's formula on string vacuum of the orbifoldized Laudau-Ginzburg models in N=2 superconformal theory [53]. Subsequently, by using known results and techniques in topology and algebraic geometry, the formula which Vafa proposed was justified mathematically in [41] as the expected form of the Euler number, Hodge number of a crepant resolution of Calabi-Yau (possibly singular) hypersurface in a weighted projective 4-space. Relying on this formula, striking numerical evidences were produced in the early 90s, subsequently stemmed the quest of Calabi-Yau mirror symmetry [56]. In the remaining section, I shall focus on the continuum of such a interdisciplinary work with progress recently made on the orbifold theory.

By an orbifold, we mean an algebraic (or analytic) variety with at most (finite group) quotient singularities. This geometrical object was introduced in the 1950s by I. Satake, which he called a V-manifold, on his course of study of the Siegel's modular forms [50]. It is a simple and natural construction, however provides a powerful tool in the study of various mathematical areas, including algebraic geometry, topology, infinite dimensional algebra representation, etc. As the procedure of an orbifold is viewed as an identification of certain "known" data by a discrete twist group, one would expect the orbifold construction will eventually play an equally important role in physics. Indeed it has been shown to be the case in various physical theories, e.g. conformal field theory, quantum Hall effect and string theory. From the geometrical aspect, the essence of an orbifold has always been related to the study of singularity arising from the construction. In the mid-80s, string theorists studied a special type of physical model built upon an orbifold $M/G$ where $M$ is a manifold quotiented by a finite group $G$. It required the following description of the string vacuum in terms of the topology of target space [16], which was called the orbifold Euler characteristic in [26],

$$\chi(M,G) = \frac{1}{|G|} \sum_{g,h\in G,\ gh=hg} \chi(M^{g,h})$$

where $M^{g,h}$ is the simultaneous fixed-point set of $g, h$. For the local situation where $G$ acts linearly on $\mathbf{C}^n$, one has the following equivalent formulation in terms of representations of the group $G$,

$$\chi(\mathbf{C}^n, G) = \#\{\text{irreducible representations of } G\}\ ,\ G \subset \mathrm{GL}_n(\mathbf{C})\ . \quad (7)$$

For the special interest in string theory, $M$ is a complex manifold with a $G$-invariant holomorphic volume form, equivalently to say, locally the group $G$ in (7) is a subgroup of $\mathrm{SL}_n(\mathbf{C})$. So the $n$-dimensional orbifold $X(= M/G)$ has at most Gorenstein singularities. For the consistency of physical theory, it became immediately desirable to demand a crepant resolution $\widehat{X}$ of $X$ with the compatible Euler number: $\chi(\widehat{X}) = \chi(M,G)$. In the local version, it states that there exists a resolution $\widehat{\mathbf{C}^n/G}$ of $\mathbf{C}^n/G$ for a finite group $G$ in $\mathrm{SL}_n(\mathbf{C})$, with the canonical bundle of $\widehat{\mathbf{C}^n/G}$ trivial and $\chi(\widehat{\mathbf{C}^n/G}) = \chi(\mathbf{C}^n, G)$ . For $n = 2$, these groups were classified into the renowned *A-D-E* series by F. Klein in 1872 (or so) in his work of invariant theory of regular solids in $\mathbf{R}^3$ [29]. The minimal resolution of a surface singularity provides the correct solution $\widehat{X}$ [26]. However it has long been understood that the nature of "minimality" of singularities of algebraic (or analytical) varieties differs significantly as the dimension of the variety is larger than two. A notable example would be the minimal model program carried out by S. Mori and others on the 3-dimensional birational geometry, which, as the prime achievement in algebraic geometry of the 1980s, has provided an effective tool for the study of algebraic 3-folds, (see [37] and references therein). However, the crepant resolution of Gorenstein 3-dimensional orbifolds was still missing in the theory. For the existence of crepant resolution of 3-dimensional oridifolds stemming from string theory, the complete mathematical solution of the conjecture was obtained in the mid-90s (see [46] and references therein); the proof was based upon techniques in toric geometry and singularity theory, but heavily depending on the classical work of Miller-Blichfeldt-Dickson classification of finite groups in $\mathrm{SL}_3(\mathbf{C})$ [36] and invariant theory of two simple groups, $I_{60}$ (icosahedral group), $H_{168}$ (Klein group) [29]. The result implies that "terminal" singularities of 3-dimensional Gorenstein orbifolds are in fact "smooth" in the minimal model theory of 3-folds. Nevertheless, due to the computational nature of methods in the proof, the qualitative understanding on the these crepant resolutions was still lacking on certain aspects from a mathematical viewpoint. Recently the $G$-Hilbert scheme $\mathrm{Hilb}^G(\mathbf{C}^n)$, developed by I. Nakamura *et al.* [38], has provided a conceptual approach and understanding of crepant resolutions of Gorenstein 3-orbifolds. Along this line, a plausible method has been proposed on problems of resolutions of higher dimensional orbifolds by engaging the representation theory of finite groups. Even for the abelian group $G$ in the dimension $n = 3$, the conclusion on the trivial canonical bundle of $\mathrm{Hilb}^G(\mathbf{C}^3)$ would lead to a mathematical subtlety that could not be ignored by both physicists and mathematicians working on the mirror symmetry problem of Calabi-Yau 3-folds in string theory. A standard well known example in Calabi-Yau mirrors is the Fermat quintic in $\mathbf{P}^4$ with the special marginal deformation family:

$$X : \sum_{j=1}^{5} z_i^5 + \lambda z_1 z_2 z_3 z_4 z_5 = 0 \ .$$

With the maximal diagonal group $SD$ of $z_i$s preserving the family $X$, the mirror $X^*$ is constructed by "the" crepant resolution of $X/SD$, $X^* = \widehat{X/SD}$ [24], by which the roles of $H^{1,1}, H^{2,1}$ are interchangeable in the "quantum" sense. On the one-dimensional space $H^{1,1}(X) \sim H^{2,1}(X^*)$, the choice of crepant resolution $\widehat{X/SD}$ makes no difference on the conclusion. While working on $H^{2,1}(X)$ and $H^{1,1}(X^*)$, many topological invariants like Euler characteristic, elliptic genus, have been known indifferent to the choice of crepant resolutions, hence the same for $X^*$. However, the topological triple intersection of cohomologies differs for two crepant resolutions (see, e.g., [45]), hence the choice of crepant resolution as the mirror $X^* = \widehat{X/SD}$ leads to the different effect on the topological cubic form of $H^{1,1}(X^*)$, which as the "classical" level, the quantum triple product of the physical theory will be built upon (see, e.g., articles in [56]). The question of the "good" model for $X^*$ has rarely been raised in the past, partly due to the lack of mathematical knowledge on the issue. However, with the $G$-Hilbert scheme now as the mirror $X^*$, it seems to have left some fundamental open problems on its formalism of mirror Calabi-Yau spaces and the question of the arbitrariness of the choice of crepant resolutions remains a mathematical question to be answered. For $n \geq 4$, it is known that a quotient orbifold $\mathbf{C}^n/G$ does not possess any crepant resolution in general, even for a cyclic group $G$ in the case $n = 4$, (for a selection of examples, see [47]). For simplicity, in the discussion of $G$-Hilbert scheme, the quotient singularities we shall only concern in higher dimensions are of hypersurface type, among which certain simple group quotient singularities are included. For the abelian case, the group of the $A$-type hypersurface quotient singularity of dimension $n$ is defined by

$$A_r(n) := \{g \in \mathrm{SL}_n(\mathbf{C}) \mid g : \text{ diagonal }, \ g^{r+1} = 1\}, \quad r \geq 1 ,$$

and the orbifold can be realized as the following hypersurface in $\mathbf{C}^{n+1}$,

$$x^{r+1} = \prod_{j=1}^{n} y_j , \qquad (x, y_1, \cdots, y_n) \in \mathbf{C}^{n+1} .$$

One can use the toric techniques to study the orbifold geometry for $A_r(n)$. For $n = 4$, we obtained the structure of $\mathrm{Hilb}^{A_r(4)}(\mathbf{C}^4)$ in [11], which is a smooth toric variety with the non-trivial canonical bundle,

$$\omega = \mathcal{O}(\sum_{k=1}^{m} E_k) , \quad m = \frac{r(r+1)(r+2)}{6} ,$$

where $E_k$s are disjoint smooth exceptional divisors in $\mathrm{Hilb}^{A_r(4)}(\mathbf{C}^4)$ , each of them is isomorphic to $\mathbf{P}^1 \times \mathbf{P}^1 \times \mathbf{P}^1$ whose normal bundle restricting to each factor $\mathbf{P}^1$ is the $(-1)$-hyperplane bundle. Hence one can blow down a family of $\mathbf{P}^1$s of $E_k$ along one of the three projections of the triple product of $\mathbf{P}^1$. By this means, one obtains different crepant resolutions of $\mathbf{C}^4/A_r(4)$. The relation between those different crepant resolutions is interpreted as the "flop" of 4-folds. Note that in Mori's minimal model program of birational geometry of 3-folds, one of the important concept introduced is the flip operation. Amid flip operations in the 3-fold theory, an invertible one, called flop, was known since long among geometers; it had played a vital role in the study of degeneration of K3 surfaces (see e.g. [40]). In our searching crepant resolutions of $A$-type hypersurface singularity through the $G$-Hilbert scheme, the flop operation of dimension 4 appeared again naturally in the process. The construction would be expected by the common nature of varieties with the trivial canonical bundle. The explicit description of the "flop" of folds is defined as the connection between three different "small"[2] resolutions of the isolated singularity of an four-dimensional affine variety $S$ with the following realization in $\mathbf{C}^8$,

$$S = \{(x_i, y_i)_{1 \leq i \leq 4} \in \mathbf{C}^8 \mid x_i y_i = x_j y_j, \ x_i x_j = y_{i'} y_{j'} \ , \ 1 \leq i \neq j \leq 4 \}$$

where the indices $i', j'$ denotes the complimentary pair of $i, j$. All these three small resolutions are 4-folds with the trivial canonical bundle. In fact, all the three small ones are dominated by a smooth variety $\widetilde{S}$ over the singular variety $S$ such that the exceptional locus of $\widetilde{S}$ over $S$ is a divisor isomorphic to the triple product of $\mathbf{P}^1$. Those small crepant resolutions are the smooth varieties obtained by blowing-down the family $\mathbf{P}^1$s along one of three projections. The toric description of these relations can be found in [33]. On the other hand, elements of the $G$-Hilbert scheme parametrize certain $G$-regular representations, which are easy to obtain through the data of toric variety for the case $G = A_r(4)$. Nevertheless, the analysis on these $G$-representations displays a common pattern for all $r$. For a simple group $G$, the understanding of $\mathrm{Hilb}^G(\mathbf{C}^n)$ in terms of $G$-representations could be an interesting problem from both the geometrical and representation theoretical point of view. Several examples are now under investigation along this scheme of interpretation; the programme is now under progress.

# 5 Conclusion Remarks

In this article, a comprehensive analysis and the direction of our present-day research are briefly reviewed on certain topics related to algebro-geometry approach to solvable theories in physics. Though the rudimentary knowledge of a complete mathematical theory has yet unknown, partly for that reason most of the topics are explained through examples, we hope the presentation would help to give a feeling of how the current research is oriented. Even in such a formulation, many problems

[2]Here the "smallness" for a resolution means one with the exceptional locus of codimension $\geq 2$.

which emerged with an empirical ground have still been a difficult mathematical task for the solution, and remain to be completely understood on their fundamental nature. However, the growing important role of algebraic geometry in certain physical problems would be in evidence through our discussion on recent achievement and newly progress in 2-dimensional statistical physics and string theory. Hopely, this consideration will lead to impute new insights to certain unsolved problems of common interest in algebraic geometry and solvable quantum physics. Meanwhile, the development of computer science in the last decade has made possible the use of numerical data to extract correct information in a wide variety of problems, both in mathematics and physics. The power of its calculation has served us well and continues to act as the underlying driving force of our search of a theoretical conceptual framework at a whole new level of complexity. These experimental investigations led us to foresee the vital role of quantitative research as a significant part in advancing the mathematical knowledge for decades to come. In comparison to the development of physics, it is of interest to note a short remark describing the career of 1913 Nobel Laureate in physics, H. K. Onnes, related to his version on experimental physics around the turn of centuries from 19th to 20th:

*In 1881 he published a paper, "Algemeene theorie de vloeistoffen" (General theory of liquids), which dealt with the kinetic theory of liquid state, approaching Van der Waals' law of corresponding states from a mechanistic point of view. This work can be considered as the beginning of his life-long investigations into the properties of matter at low temperatures. In his inaugural address, " De beteekenis van het quantitatief onderzoek in de natnurkunde" (The importance of quantitative research in physics), he arrived at his well-known motto "Door meten tot weten" (knowledge through measurement), an appreciation of the value of measurements which concerned him through his scientific career.*

(This passage is quoted from "Biography of Heike Kamerlingh Onnes, Nobel Lectures, Physics 1901-1921", Elsevier Publishing Company, Amsterdam.) H. K. Onnes' valuing on quantitative research could serve one primary factor behind his laboratory success of discovering superconductivity; it would also provide a permanent symbolic reminder as a scientific discipline in physics as well. In the science of mathematics, "knowledge through computation" could be the one philosophically more in tune with H. K. Onnes' view on the quantitative research. Along this line at the dawn at the new millennium, the great challenge of how to embrace the spectacular computing capability provided by the successful computer technology in advancing our mathematical knowledge beneficial to physical problems will confront algebraic geometers in the years to come. In concert with this principle in the context of solvable models, the quest of mathematical insights and rigorous theorems to the physical theory will still deserve a great deal of merits from the mathematical point of view.

## Acknowledgements

I would like to thank L. Chiang, E. Date, G. von Gehlen, V.G. Kac, S.S. Lin and M. Wakimoto for collaboration of the research presented in this article. I also acknowledge the support by research programs of the National Science Council of Taiwan over the years when most of this work was carried out.

## References

[1] G. Albertini, B. M. McCoy, and J. H. H. Perk, Commensurate-incommensurate transition in the ground state of superintegrable chiral Potts model, Phys. Lett. A 135 (1989) 159-166; Level crossing transitions and the massless phases of the superintegrable chiral Potts chain, Phys. Lett. A 139 (1989) 204-212; Eigenvalue spectrum of the superintegrable chiral Potts model, Adv. Stud. Pure Math., vol. 19, Kinokuniya Academic (1989) 1-55.

[2] G. Albertini, B. M. McCoy, J. H. H. Perk and S. Tang, Excitation spectrum and order parameter for the integrable $N$-state chiral Potts model, Nucl. Phys. B 314 (1989) 741-763.

[3] R. J. Baxter, *Exactly solved models in statistical mechanics*, Academic Press, 1982; Solving models in statistical mechanics, Adv. Stud. Pure Math., vol. 19, Kinokuniya Academic (1989) 95-116.

[4] R. J. Baxter, The superintegrable chiral Potts model, Phys. Letts. A 133 (1988) 185-189.

[5] R. J. Baxter, Chiral Potts model: eigenvalues of transfer matrix, Phys. Lett. A 146 (1990) 110 -114.

[6] R. J. Baxter, Hyperelliptic function parametrization for chiral Potts model, Proc. ICM Kyoto, Japan 1990.

[7] R. J. Baxter, J. H. H. Perk and H. Au-yang, New solutions of star-triangle relations for the chiral Potts model, Phys. Letts. A 128 (1988) 138-142 .

[8] R. J. Baxter, V. V. Bazhanov and J. H. H. Perk, Functional relations for transfer matrices of the chiral Potts model, Intern. J. Mod. Phys. B 4 (1990) 803-870.

[9] V.V. Bazhanov and Yu.G. Stroganov, Chiral Potts model as a descendant of the six-vertex model, J. Stat. Phys. 59 (1990) 799-817.

[10] S.J. Cheng and V. G. Kac, Conformal modules, Asian J. Math. 1 (1997) 181-193.

B.L. Feigin, A.M. Semikhatov, V.A. Sirota, and I. Yu. Tipunin, Resolutions and characters of irreducible representations of the $N = 2$ superconformal algebra, hep-th/9805179.

[11] L. Chiang and S. S. Roan, Orbifolds and finite group representations, math.AG/0007072 ; On hypersurface quotient singularity of dimension 4, math.AG/0011151.

[12] V. K. Dobrev, Characters of the unitarizable highest weight modules over $N = 2$ superconformal algebras, Phys. Lett. B 186 (1987) 43-51;

E. B. Kiritsis, Character formulae and the structure of the presentations of the $N = 1, N = 2$ superconformal algebras, Int. J. Mod. Phys. A 3 (1988) 1871-1906;

Y. Matsuo, Character formula of $C < 1$ unitary representation of $N = 2$ superconformal algebra, Prog. Theor. Phys. 77 (1987) 793-797;

F. Ravanini and S. K. Yang, Modular invariance in $N = 2$ superconformal field theories, Phys. Lett. B 195 (1987), 202-208.

[13] E. Date and S. S. Roan, The structure of quotients of the Onsager algebra by closed ideals, J. Phys. A: Math. Gen. 33 (2000) 3275-3296, math.QA/9911018 ; The algebraic structure of the Onsager algebra, Czech. J. Phys. 50 (2000) 37-44, cond-mat/0002418.

[14] B. Davies, Onsager's algebra and superintegrability, J. Phys. A: Math. Gen. 23 (1990) 2245-2261; Onsager's algebra and the Dolan-Grady condition in the non-self case, J. Math. Phys. 32 (1991) 2945-2950.

[15] L. Dolan and M. Grady, Conserved charges from self-duality, Phys. Rev. D 25 (1982) 1587-1604.

[16] L. Dixon, J. Harvey, C. Vafa and E. Witten, Strings on orbifolds, Nucl. Phys. B 261 (1985) 678-686, Strings on orbifolds II, Nucl. Phys. B 274 (1986) 285-314.

[17] T. Eguchi, H. Ooguri, A. Taormina, and S. K. Yang, Superconformal algebras and string compactification on manifolds with SU(n) holonomy, Nucl. Phys. B 315 (1989) 193-221.

[18] L. D. Faddeev, *Integrable Models, Quantum Groups and Conformal Field Theory*, Lectures in Technical University, Berlin 1992, preprint no. SFB 288 N1.

[19] L. D. Faddeev and R. M. Kashaev, Generalized Bethe Ansatz equations for Hofstadter problem, Comm. Math. Phys. 155 (1995) 181-191, hep-th/9312133.

[20] G. von Gehlen, Integrable $Z_n$-chiral Potts model: The missing rapidity-momentum relation, in *Statistical Models, Yang-Baxter Equation and Related Topics* (eds. M. L. Ge and F. Y. Wu, Nankai Inst. Math. Tianjin ) 1996, 102-117, hep-th/9601001.

[21] G. von Gehlen and R. Rittenberg, $Z_n$-symmetric quantum chains with infinite set of conserved charges and $Z_n$ zero modes, Nucl. Phys. B 257 (1985) 351-370.

[22] G. von Gehlen and S. S. Roan, Superintegrable chiral Potts model and generalized Chebychev polynomials (in preparation).

[23] D. Gepner, Space-time supersymmtry in compactified string theory and superconformal models, Nucl. Phys. B 296 (1988) 757-778.

[24] B. R. Greene and M. R. Plesser, Duality in Calabi-Yau moduli space, Nucl. Phys. B 338 (1990) 15-37.

S. S. Roan, The mirror of Calabi-Yau orbifold, Intern. J. Math. 2 (1991) 439-455.

[25] F. Hirzebruch, T. Berger and R. Jung, *Manifolds and modular forms*, Aspects of Mathematics, Friedr. Vieweg & Sohn Verlagsgesellschaft mbH, Braunschweig/Wiesbaden, 1992;

F. Hirzebruch, Complex cobordism and the elliptic genus, (notes by A. Langer), Contemporary Mathematics Vol. 241, 1999 Amer. Math. Soc. 9-20.

[26] F. Hirzebruch and T. Höfer, On the Euler number of an orbifold, Math. Ann. 286 (1990) 255-260.

[27] D. R. Hofstadter, Energy levels and wave functions of Bloch electrons in rational and irrational magnetic fields, Phys. Rev. B 14 (1976) 2239-2249.

[28] S. Howes, L.P. Kadanoff and M. den Nijs, Quantum model for commensurate-incommensurate transitions, Nucl. Phys. B 215 (1983) 169-208.

[29] F. Klein, *Gesammelte Mathematische Abhandlungen Bd. I-III*, Springer-Verlag, Berlin 1922 ( reprint 1973 ).

[30] M. Jimbo, Introduction to the Yang-Baxter equation, Intern. J. Mod. Phys. A 4 (1989) 3759-3777; Solvable lattice models and quantum groups, Intern. Congress Math. Kyoto (1990) 1343-1351 .

[31] V. G. Kac, S. S. Roan and M. Wakimoto, Characters of affine $sl_2(2|1)$ and N=2 superconformal algebra, (in preparation).

[32] E. H. Lieb, Exact solution of the problem of the entropy of two-dimensional ice, Phys. Rev. Lett. 18 (1967) 692-694; Exact solution of F model of an antiferroelectric, Phys. Rev. lett. 18 (1967) 1046-1048; Exact solution of the two-dimensional Slater KDP model of a ferroelectric, Phys. Rev. Lett. 19 (1967) 108-110; The residual entropy of square ice, Phys. Rev. 162 (1967) 162-172.

[33] S.S. Lin and S. S. Roan, Algebraic geometry approach to the Bethe equation for Hofstadter type models, cond-mat/9912473; Bethe ansatz of discrete quantum pendulum and discrete sine-Gordon, ( in preparation).

[34] B. M. McCoy, Integrable models in statistical mechanics: The hidden field with unsolved problems, (The 1999 Heineman Prize address), Intern. J. Mod. Phys. A, 14 (1999) 3921-3933, math-ph/9904003.

[35] B. M. McCoy and S. S. Roan, Excitation spectrum and phase structure of chiral Potts model, Phys. Lett. A 150 (1990) 347-354;

B. M. McCoy, The chiral Potts model: From physics to mathematics and back, in *Special functions*, ICM 90, Satellite Conf. Proc. (eds. M Kashiwara and T. Miwa) Springer (1991) 245.

[36] G. A. Miller, H. F. Blichfeldt and L. E. Dickson, *Theory and applications of finite groups*, John Wiley and Sons. Inc., 1916.

[37] S. Mori, Birational classification of algebraic threefolds, Intern. Congress Math. Kyoto (1990) 235-248.

[38] I. Nakamura, Hilbert schemes of abelian group orbits, preprint 1998.

Y. Gomi, I. Nakamura and K. Shinoda, HIlbert scheme of G-orbits in dimension three, preprint, (to appear in Asian J. Math.).

Y. Ito and H. Nakajima, McKay correspondence and Hilbert schemes in dimension three, math.AG/9803120.

T. Bridgeland, A. King and M. Reid, Mukai implies McKay, math.AG/9908027.

[39] L. Onsager, A two-dimensional model with an order-disorder transition (crystal statistics I) Phys. Rev. 65 (1944) 117-149 ; The Ising model in two dimensions, in *Critical Phenomena in Alloys, Magnets and Superconductors* (eds. R. E. Mills, E. Ascher and R. I. Jeffe) McGraw-Hill, New York, 1971 .

[40] S. S. Roan, *Degeneration of* K3 *and abelian surfaces*, Ph.D. thesis of Brandeis Univ. USA, 1974;

V. Kulikov, Degeneration of K3 and Enriques surfaces, Math. USSR Izvestija 11 (1977) 957-989;

U. Persson and H. Pinkham, Degeneration of surfaces with trivial canonical divisor, Annals of Math. 113 (1981) 45-66.

[41] S. S. Roan , On the Calabi-Yau orbifolds in weighted projective spaces, Intern. J. Math. 1 (1990) 221-232.

[42] S. S. Roan, Onsager's algebra, loop algebra and chiral Potts model, Max-Planck-Institut für Mathematik, Bonn, 1991 Preprint MPI 91-70.

[43] S. S. Roan, A characterization of "rapidity" curve in chiral Potts model, Comm. Math. Phys. 145 (1992) 605-634.

[44] S. S. Roan, Modular invariance of manifolds with $SU(n)$ holonomy, Intern. J. Math. 3 (1992) 359-395.

[45] S. S. Roan, Topological coupling of Calabi-Yau orbifold, J. Group Theory in Physics 1 (1993) 83-103.

[46] S. S. Roan, Minimal resolution of Gorenstein orbifolds in dimension three, Topology 35 (1996) 489-508.

[47] S. S. Roan, Crepant resolution and fibred CY manifolds, Preprint MIAS 97-1, Inst. of Math. Acad. Sinica, Taiwan, 1997.

[48] S. S. Roan, Schwarz's function and chiral Potts model, Helv. Phys. Acta 71 (1998) 242-261; Mirror symmetry of elliptic curves and Ising model, J. Geom. Phys. 20 (1996) 273-296.

[49] S. S. Roan, Heisenberg and modular invariance of N=2 conformal field theory, Intern. J. Mod. Phys. A 15 (2000) 3065-3094, hep-th/9902198.

[50] I. Satake, On a generalization of the notion of manifold, Proc. Nat. Acad. Sci. U.S.A. 42 (1956) 359-363; The Gauss-Bonnet theorem for V-manifolds, J. Math. Soc. Japan 9 (1957) 464-492.

[51] L. A. Takhtadzhan and L. D. Faddeev , The quantum method of the inverse problem and the Heisenberg XYZ model, Russ. Math. Surveys 34 ( 1979 ) 11-68.

[52] P. Terwlliger, An introduction to Leonard pairs and Leonard systems, Sūrikaisekikenkyūsho Kōkyūroku 1109, Algebraic combinatorics, Kyoto 1999, 67-79.

[53] C. Vafa, String vacua and orbifoldized LG models, Mod. Phys. Lett. A 4 (1989) 1169-1185.

[54] P. B. Wiegmann and A. V. Zabrodin, Bethe-ansatz approach to Hofstadter problem, quantum group and magnetic translations, Nucl. Phys. B 422 (1994) 495-514.

[55] C. Y. Yang and C. P. Yang, One-dimensional chain of anisotropic spin-spin interactions. I. Proof of Bethe's hypothesis for ground state in a finite system, Phys. Rev. 150 (1966) 321-327; II. Properties of the ground-state energy per lattice site for an infinite system, Phys. Rev. 150 (1966) 327-339;

B. Sutherland, C. N. Yang and C. P. Yang, Exact solution of a model of two-dimensional ferroelectric in an arbitrary external electric field, Phy. Rev. Lett. 19 (1967) 588-591.

[56] S. T. Yau (ed.) *Essays on Mirror Manifolds,* International Press, 1992 ;

R. Penner and S. T. Yau, (eds. ), *Perspectives in Mathematical Physics,* International Press, 1994 ;

B. Greene and S. T. Yau (eds.), *Mirror Symmetry II* American Mathematical Society, International Press, 1997.

# WAVE BLOCKING IN PARALLEL COUPLED NERVE FIBERS

M. RODRIGO, H. IKEDA, AND M. MIMURA

ABSTRACT. In this paper, we shall consider a system of reaction-diffusion equations arising from a model of nerve impulse propagation along two parallel coupled nerve fibers. We will study the phenomenon of wave blocking by showing the existence and stability of two classes of stationary (weak) solutions of our model system.

In theoretical studies of electrical impulse propagation along an unmyelinated nerve fiber, uniform geometric properties are usually assumed. If a uniform nerve fiber is stimulated at one end and the potential across the membrane exceeds a certain threshold, then ions move through the membrane stimulating adjacent parts and so a wave travelling down the nerve fiber is initiated.

It is possible that if there is a sudden increase of the membrane area ahead of the wave, then the membrane current density falls. As a result, there is a temporary decrease of action potential and if this potential falls below the threshold, the wave may be stopped.

This phenomenon was studied analytically by Pauwelussen [7]. He considered a nerve fiber with only one branching point at $x = 0$, where $x$ denotes the distance along an infinitely long nerve fiber. Furthermore, he assumed that there is one branch of radius 1 for $x < 0$ and $k$ branches of radius $r$ for $x > 0$ and the membrane potentials on the $k$ branches are identical for any positive time $t$. Under these assumptions, he gave the following equations:

$$
(1) \quad \begin{cases} \left.\begin{aligned} u_t &= e_\epsilon(x)u_{xx} + u(1-u)(u-a) - v \\ v_t &= \delta(u - \gamma v) \end{aligned}\right\} , \quad x \in \mathbb{R}, \quad x \neq 0, \quad t > 0, \\ (u,v)(x,0) = (u_0,v_0)(x), \quad x \in \mathbb{R}, \quad x \neq 0, \end{cases}
$$

where $u(x,t)$ represents the transmembrane potential and $v(x,t)$ the recovery mechanism of the cell membrane. The constants appearing in (1) satisfy $0 < a < 1/2$, $0 < \delta \ll 1$, and $0 < \gamma$. The function $e_\epsilon(x)$ is defined by

$$
e_\epsilon(x) \equiv \begin{cases} 1, & x < 0, \\ \epsilon = \dfrac{1}{r^3 k^2} \in (0,1], & x > 0. \end{cases}
$$

When $\epsilon = 1$, system (1) reduces to the well-known FitzHugh–Nagumo equations. It is known that this system admits travelling pulse solutions, i.e., solutions which are of the form $(u,v)(x,t) = (u_P, v_P)(z)$ $(z = x - c_P t,\ c_P > 0)$ such that $(u_P, v_P)(\pm\infty) = (0,0)$. Actually, there are two such pulse solutions but we shall consider only the stable, fast pulse solution.

Suppose that the initial condition of (1) is given by $(u_0, v_0)(x) = (u_P, v_P)(x + x_0)$ for some $x_0 \gg 1$. If $\epsilon_1^* < \epsilon < 1$, for some $\epsilon_1^* > 0$, then the solution of (1) continues to propagate as a travelling pulse solution with speed $c_P$ until it approaches the origin. At $x = 0$, the speed of propagation changes but the pulse wave proceeds beyond this point. However, if $0 < \epsilon \le \epsilon_1^*$, then the pulse wave is blocked at the origin and fails to propagate.

In the absence of a recovery mechanism ($v \equiv 0$), system (1) reduces to the scalar equation

$$(2) \qquad \begin{cases} u_t = e_\epsilon(x)u_{xx} + u(1-u)(u-a), & x \in \mathbb{R}, \quad x \neq 0, \quad t > 0, \\ u(x,0) = u_0(x), & x \in \mathbb{R}, \quad x \neq 0. \end{cases}$$

When $\epsilon = 1$, it is known that this equation admits a unique (up to horizontal translations) monotonic decreasing travelling front solution $u(x,t) = u_F(z)$ ($z = x - c_F t$, $c_F > 0$) satisfying $u_F(-\infty) = 1$, $u_F(+\infty) = 0$. This solution can be interpreted as the upstroke (or front part) of the FitzHugh–Nagumo pulse solution $(u_P, v_P)$.

On the other hand, when $\epsilon \neq 1$, it was found in [6] that there exists a critical value $\epsilon_2^* > 0$ such that

(i) if $0 < \epsilon < \epsilon_2^*$, then there exist exactly two stationary monotonic decreasing solutions $q_\pm(x)$ of (2) satisfying $q_\pm(-\infty) = 1$, $q_\pm(+\infty) = 0$;
(ii) if $0 < \epsilon < \epsilon_2^*$, then the solution of (2) tends to $q_-(x)$ as $t \to +\infty$;
(iii) if $\epsilon_2^* < \epsilon < 1$, then the solution of (2) tends to the unique travelling front solution corresponding to the case when $e_\epsilon(x) \equiv \epsilon$.

The previous results were for the case of a single non–uniform nerve fiber. However, in some instances, the density of nerve fibers increases sharply, e.g., in the myocardium. Thus, it is natural to consider the phenomenon of wave blocking in an array of coupled nerve fibers.

We propose the following system of equations which can be interpreted as a model for nerve impulse propagation along two non–uniform parallel nerve fibers coupled electrotonically:

(3)

$$\begin{cases} \left. \begin{aligned} u_{1t} &= e_1(x)u_{1xx} + f(u_1) - v_1 + d(u_2 - u_1) \\ u_{2t} &= e_2(x)u_{2xx} + f(u_2) - v_2 + d(u_1 - u_2) \\ v_{1t} &= \delta(u_1 - \gamma v_1) \\ v_{2t} &= \delta(u_2 - \gamma v_2) \end{aligned} \right\} , \quad x \in \mathbb{R}, \quad x \neq 0, \quad t > 0, \\ (u_1, u_2, v_1, v_2)(x, 0) = (u_{10}, u_{20}, v_{10}, v_{20})(x), \quad x \in \mathbb{R}, \quad x \neq 0, \end{cases}$$

where $u_i, v_i$ ($i = 1, 2$) represent transmembrane potentials and recovery mechanisms, respectively. The constants $\delta$ and $\gamma$ are the same as in (1), while $f \in C^1(\mathbb{R})$ is assumed to satisfy the following:

$$f(0) = f(a) = f(1) = 0, \quad 0 < a < 1,$$
$$f(s) < 0 \quad \text{for} \quad 0 < s < a, \quad f(s) > 0 \quad \text{for} \quad a < s < 1,$$
$$f'(0) < 0, \quad f'(1) < 0, \quad \int_0^1 f(s)\,ds > 0.$$

A typical example is $f(s) = s(1-s)(s-a)$ ($0 < a < 1/2$).

Physically speaking, it is reasonable to assume that $d$ is a sufficiently small positive number but in this paper, we assume that

$$0 < |d| \ll 1.$$

The functions $e_1(x)$ and $e_2(x)$ are defined by

$$e_1(x) \equiv \begin{cases} 1, & x < 0, \\ \epsilon_1 \in (0,1], & x > 0, \end{cases} \quad \text{and} \quad e_2(x) \equiv \begin{cases} 1, & x < 0, \\ \epsilon_2 \in (0,1], & x > 0. \end{cases}$$

Suppose for the moment that $\epsilon_1 = \epsilon_2 = 1$. Then, system (3) reduces to the model proposed by Keener [5] for nerve impulse propagation along two uniform parallel nerve fibers coupled electrotonically. It is known that in this case, there exist at least two types of travelling pulse solutions of (3). The first one (so–called symmetric pulse solution) is of the form $(u_1, u_2, v_1, v_2)(x,t) = (u_P, u_P, v_P, v_P)(z)$, where $(u_P, v_P)(z)$ is a travelling pulse solution of the FitzHugh–Nagumo equations. The second one (so–called antisymmetric pulse solution) is of the form $(u_1, u_2, v_1, v_2)(x,t) = (U_1, U_2, V_1, V_2)(z; d)$ such that $(U_1, V_1)(z; 0) = (u_P, v_P)(z)$ and $(U_2, V_2)(z; 0) = (0,0)$.

The symmetric pulse solution is asymptotically stable if $d > 0$ and unstable if $d < 0$. On the other hand, the antisymmetric pulse solution is always asymptotically stable for $0 < |d| \ll 1$ ([1], [2], [8]).

Let the initial condition of (3) be $(u_{10}, u_{20}, v_{10}, v_{20})(x) = (u_P, u_P, v_P, v_P)(x + x_0)$ for some $x_0 \gg 1$. When $\epsilon_1 \neq 1$ or $\epsilon_2 \neq 1$, we observed the following numerically: if $\epsilon_3^* < \epsilon_1, \epsilon_2 \leq 1$, for some critical $\epsilon_3^*$, then the speed of propagation changes at $x = 0$ but the symmetric pulse wave proceeds beyond this point. The same situation happens if $0 < \epsilon_1 \leq \epsilon_3^* < \epsilon_2 \leq 1$ or $0 < \epsilon_2 \leq \epsilon_3^* < \epsilon_1 \leq 1$. However, if $0 < \epsilon_1, \epsilon_2 \leq \epsilon_3^*$, then the symmetric pulse wave is blocked at the origin and fails to propagate.

On the other hand, for the antisymmetric case, we assign the initial condition $(u_{10}, u_{20}, v_{10}, v_{20})(x) = (U_1, U_2, V_1, V_2)(x + x_0)$ for some $x_0 \gg 1$. Numerically, we obtained the following results: for some $\epsilon_4^* > 0$, if $\epsilon_4^* < \epsilon_1 \leq 1$ and $\epsilon_2 > 0$, then the speed of propagation changes at $x = 0$ but the antisymmetric pulse wave proceeds beyond this point, provided $|d|$ is sufficiently small. If $0 < \epsilon_1 \leq \epsilon_4^*$ and $\epsilon_2 > 0$, then the antisymmetric pulse wave is blocked at the origin and fails to propagate.

In the absence of recovery ($v_1 \equiv v_2 \equiv 0$), system (3) reduces to the following system:

(4)

$$\begin{cases} \left.\begin{aligned} u_{1t} &= e_1(x) u_{1xx} + f(u_1) + d(u_2 - u_1) \\ u_{2t} &= e_2(x) u_{2xx} + f(u_2) + d(u_1 - u_2) \end{aligned}\right\}, \quad x \in \mathbb{R}, \quad x \neq 0, \quad t > 0, \\ (u_1, u_2)(x, 0) = (u_{10}, u_{20})(x), \quad x \in \mathbb{R}, \quad x \neq 0, \end{cases}$$

When $\epsilon_1 = \epsilon_2 = 1$, it is known that this system allows at least two types of travelling front solutions. The first one (so–called symmetric front solution) is of the form $(u_1, u_2)(x,t) = (u_F, u_F)(z)$, where $u_F(z)$ is the unique monotonic travelling front solution of $u_t = u_{xx} + f(u)$ and satisfies $u_F(-\infty) = 1$, $u_F(+\infty) = 0$.

The second solution (so–called antisymmetric front solution) can be represented in the form $(u_1, u_2)(x,t) = (U_{1F}, U_{2F})(z; d)$ such that $U_{1F}(z; 0) = u_F(z)$ and $U_{2F}(z; 0) = 0$. This solution satisfies the boundary conditions $(U_{1F}, U_{2F})(-\infty) = (P_1, P_2)$, $(U_{1F}, U_{2F})(+\infty) = (0,0)$. Here, $(P_1, P_2)$ is a constant steady state of (4) and is expressible as

$$P_1 = 1 + d\,a_1 + d^2 g(d), \quad P_2 = d\,b_1 + d^2 h(d),$$

for some functions $g(d), h(d)$, some constants $a_1, b_1$, and $|d|$ sufficiently small.

The symmetric front solution is asymptotically stable if $d > 0$ and unstable if $d < 0$. On the other hand, the antisymmetric front solution is always asymptotically stable for $0 < |d| \ll 1$ ([1], [2], [4]).

Let the initial condition of (4) be $(u_{10}, u_{20})(x) = (u_F, u_F)(x + x_0)$ for some $x_0 \gg 1$. When $\epsilon_1 \neq 1$ or $\epsilon_2 \neq 1$, we observed the following numerically: if $\epsilon_5^* < \epsilon_1, \epsilon_2 \leq 1$, for some $\epsilon_5^* > 0$, then the speed of propagation changes at $x = 0$ but the symmetric front wave proceeds beyond this point. The same situation happens if $0 < \epsilon_1 \leq \epsilon_5^* < \epsilon_2 \leq 1$ or $0 < \epsilon_2 \leq \epsilon_5^* < \epsilon_1 \leq 1$. However, if $0 < \epsilon_1, \epsilon_2 \leq \epsilon_5^*$, then the symmetric front wave is blocked at the origin.

On the other hand, for the antisymmetric case, we assign the initial condition $(u_{10}, u_{20})(x) = (U_{1F}, U_{2F})(x + x_0)$ for some $x_0 \gg 1$. Numerically, we obtained the following results: if $\epsilon_5^* < \epsilon_1 \leq 1$ for any $\epsilon_2 > 0$, then the speed of propagation changes at $x = 0$ but the antisymmetric front wave proceeds beyond this point, provided $|d|$ is sufficiently small. If $0 < \epsilon_1 \leq \epsilon_5^*$ for any $\epsilon_2 > 0$, then the antisymmetric front wave is blocked at the origin.

In this paper, as a first step, we shall study the phenomenon of wave blocking in the system (4). We shall show the existence of two classes of stationary solutions of (4), corresponding to the blocking of the symmetric and antisymmetric front wave solutions of (4) when $\epsilon_1 = \epsilon_2 = 1$.

By a (weak) stationary solution of (4) we mean an ordered pair of functions $(U, V)(x)$ which have continuous second derivatives everywhere except at $x = 0$ (where it is continuously differentiable), and satisfies

$$\begin{array}{l} e_1(x)\, U'' + f(U) + d(V - U) = 0 \\ e_2(x)\, V'' + f(V) + d(U - V) = 0 \end{array}, \quad x \in \mathbb{R}, \quad x \neq 0. \tag{5}$$

Define $\epsilon^* \equiv \int_0^a f(\tau)\, d\tau / \int_1^a f(\tau)\, d\tau$. Our results are as follows:

**Theorem 1.** *For every $\epsilon_1, \epsilon_2 \in (0, \epsilon^*)$ and $|d|$ sufficiently small, there exist at least four solutions $(U^i, V^i)(x; d)$ $(i = 1, 2, 3, 4)$ of (5) satisfying the boundary conditions*

$$(U^i, V^i)(-\infty) = (1, 1), \quad (U^i, V^i)(+\infty) = (0, 0).$$

*For $|d| \ll 1$, the solution $U^1(x; d)$ is stable while $U^i(x; d)$ $(i = 2, 3, 4)$ are unstable.*

**Theorem 2.** *For every $\epsilon_1 \in (0, \epsilon^*)$, $\epsilon_2 > 0$, and $|d|$ sufficiently small, there exist at least two solutions $(U^i, V^i)(x; d)$ $(i = 1, 2)$ of (5) satisfying the boundary conditions*

$$(U^i, V^i)(-\infty) = (P_1, P_2), \quad (U^i, V^i)(+\infty) = (0, 0).$$

*For $|d| \ll 1$, the solution $U^1(x; d)$ is stable while $U^2(x; d)$ is unstable.*

The method of proof for the existence is to expand the solution as powers of $d$ and show the existence of the remainder terms using an implicit function theorem due to Fife [3]. For the stability, we construct an Evans' function whose zeroes correspond to the isolated eigenvalues of the linear operator obtained by linearizing about the stationary solution.

## References


[1] A. Bose and C. K. R. T. Jones, *Stability of the in-phase travelling wave solution in a pair of coupled nerve fibers*, Indiana Univ. Math. J., 44 (1995) pp. 189-220.

[2] A. Chmaj, *Travelling Waves in multicomponent reaction–diffusion systems: existence, uniquenesss and stability*, Ph.D. Thesis, The University of Utah, 1995.

[3] P. C. Fife, ***Boundary and interior transition layer phenomena for pairs of second-order differential equations***, J. Math. Anal. Appl., 54 (1976) pp. 497-521.
[4] H. Ikeda and M. Rodrigo, ***Travelling front waves in parallel coupled nerve fibers***, Math. J. Toyama Univ., 19 (1996) pp. 211-252.
[5] J. Keener, ***Frequency decoupling of parallel excitable fibers***, SIAM J. Appl. Math., 49 (1989), pp. 211-230.
[6] J. P. Pauwelussen, ***Nerve impulse propagation in a branching nerve system: a simple model***, Physica 4D, (1981) pp. 67-88.
[7] J. P. Pauwelussen, ***One way traffic of pulses in a neuron***, J. Math. Biol., 15 (1982) pp. 151-171.
[8] M. Rodrigo, ***Travelling pulses in parallel coupled nerve fibers***, M. Sc. Thesis, Toyama University, 1997.

DEPARTMENT OF MATHEMATICAL AND LIFE SCIENCES, INSTITUTE FOR NONLINEAR SCIENCES AND APPLIED MATHEMATICS, GRADUATE SCHOOL OF SCIENCE, HIROSHIMA UNIVERSITY, HIGASHI-HIROSHIMA, 739-8526, JAPAN
*E-mail address*: mrodrigo@math.sci.hiroshima-u.ac.jp

DEPARTMENT OF MATHEMATICS, FACULTY OF SCIENCE, TOYAMA UNIVERSITY, GOFUKU, TOYAMA 930-8555, JAPAN
*E-mail address*: ikeda@sci.toyama-u.ac.jp

DEPARTMENT OF MATHEMATICAL AND LIFE SCIENCES, INSTITUTE FOR NONLINEAR SCIENCES AND APPLIED MATHEMATICS, GRADUATE SCHOOL OF SCIENCE, HIROSHIMA UNIVERSITY, HIGASHI-HIROSHIMA, 739-8526, JAPAN
*E-mail address*: mimura@math.sci.hiroshima-u.ac.jp

# NULLITIES OF DIFFERENTIAL EXPRESSIONS WITH LOGARITHMICALLY DECAYING COEFFICIENTS IN WEIGHTED SPACE

M.P. ROQUE
*Department of Mathematics*
*University of the Philippines*
*Diliman, Quezon City*
*email: roque@math01.cs.upd.edu.ph*

Consider the differential expression of the form $\mu_0 = \sum_{k=0}^{r} b_k (\log t)^{\beta_k} t^{\rho_k} D_t^{\rho_k}$ where $b_k \in \mathbb{C}$, $\beta_k \in \mathbb{R}$, and $\rho_k \in \mathbb{N}_0$ for every k, satisfying certain conditions. We shall classify the nullities of the maximal operator generated by differential expressions of the form $\mu_0$ together with their relatively compact perturbations in the weighted space $\mathcal{L}_2^w[1,\infty)$. These expressions have known nullities in the usual $\mathcal{L}_2$-setting. It shall be shown that if the weight function $w$ satisfies certain conditions, the nullities of $\mu_0$ are determined by the coefficients $b_k$.

## 1 Introduction

We define the weighted $\mathcal{L}_2$ space on an interval $I$ with weight $w$ as

$$\mathcal{L}_2^w(I) = \{\ f : I \to \mathbb{C} \mid f \text{ measurable and } \int_I |f(t)|^2 w(t)dt < \infty\}$$

where $w$ is a positive measurable function on $I$. For $f,\ g \in \mathcal{L}_2^w(I)$, we can define the inner product

$$< f,\ g >_w = \int_I f(t)\overline{g(t)}w(t)dt.$$

Together with this inner product, the weighted space $\mathcal{L}_2^w(I)$ is a Hilbert space and is equivalent to $\mathcal{L}_2(I)$ under the linear isometry $W : \mathcal{L}_2^w(I) \to \mathcal{L}_2(I)$ defined by $Wf = w^{\frac{1}{2}} f$.

The weighted spaces come from eigenvalue problems of the form $\tau \cdot y = \lambda w(x) \cdot y$ where $\tau$ is a differential expression and $w$ is a positive measurable function on a given interval $I$. These problems arise as unweighted ones in boundary-initial value problems of partial differential expressions using the method of separation of variables. The factor $w(x)$ can be considered as part of the differential expression $\tau$ in the ordinary Hilbert space. The same problem may be considered in the Hilbert space with weight $w$, where the factor cancels in the Dirichlet form. These two methods yield different spectral theories and the nullities of the differential expression may depend on the weight function. This was shown in [1] for a general class of expressions of the form $\sum_{k=0}^{r} b_k t^{\alpha_k} D_t^{\rho_k}$ and in [2] for expressions of the form $\sum_{k=0}^{n} b_k e^{t\beta_k} D_t^{\rho_k}$.

The nullities of the maximal operator generated by differential expressions of the form $\mu_0 = \sum_{k=0}^{r} b_k (\log t)^{\beta_k} t^{\rho_k} D_t^{\rho_k}$ where $b_k$, $\beta_k$ and $\rho_k$ satisfy certain conditions have been classified in $\mathcal{L}_2[1,\infty)$ ([3]). The spectral properties of these expressions in $\mathcal{L}_2[1,\infty)$ are completely determined by the coefficients and this was also established in [4] for the essential spectrum of $\mu_0$ in $\mathcal{L}_2^w[1,\infty)$.

## 2 Preliminaries

We now define the maximal operator generated by a differential expression in the weighted space $\mathcal{L}_2^w(I)$.

**Definition 1.** Let $\tau$ be a differential expression given by $\tau = \sum_{k=0}^{n} p_k D^k$, where $p_k$ is a complex-valued function on an interval $I$ for every $k$,

1. the **maximal operator** $T_1^w(\tau)$ generated by $\tau$ in $\mathcal{L}_2^w(I)$ is defined as follows:

$$\text{dom } (T_1^w(\tau)) = \{ f \mid f \in AC(I) \cap \mathcal{L}_2^w(I) \quad \text{and} \quad \tau f \in \mathcal{L}_2^w(I)\}$$
$$T_1^w(\tau) f = \tau f = \sum_{k=0}^{n} p_k D^k f.$$

2. the **essential resolvent of** $\tau$, denoted by $\rho_e^w(\tau)$, and **nullity** of $\tau$, denoted by $\mathbf{nul}^w(\tau)$ are defined as:

$$\rho_e^w(\tau) = \{ \lambda \in \mathbb{C} \mid \text{range of} \quad (\lambda I - T_1^w(\tau)) \quad \text{is closed}\}$$
$$\text{nul}^w(\tau - \lambda) = \text{dim ker } (T_1^w(\tau - \lambda)).$$

The **essential spectrum** $\sigma_e^w$ of a differential expression is the complement of its essential resolvent.

The problem of classifying the nullities of a differential expression $\tau$ relative to a weighted space $\mathcal{L}_2^w(I)$ can be translated into that of obtaining the nullities of the expression $w^{\frac{1}{2}} \tau w^{-\frac{1}{2}}$ in $\mathcal{L}_2(I)$ ([1]). The following lemma presents this method.

**Lemma 2.1** *Consider the weighted space $\mathcal{L}_2^w(I)$ and let $\tau$ be a differential expression defined on $I$. Then,*

$$nul^w(\tau) = nul(w^{\frac{1}{2}} \tau w^{-\frac{1}{2}})$$

We now consider differential expressions of the form :

$$\mu_0 = \sum_{k=0}^{r} b_k (\log t)^{\beta_k} t^{\rho_k} D_t^{\rho_k} \tag{1}$$

where $D_t = \frac{d}{dt}$, $\rho_k \in \mathbf{N}_0$ for every $k$, such that $0 = \rho_0 < \rho_1 < \cdots < \rho_r = n$ and $\beta_k \in \mathbb{R}$, satisfying

$$\beta_0 = 0 \quad \text{and} \quad 0 \geq \frac{\beta_k - \beta_{k-1}}{\rho_k - \rho_{k-1}} \geq \frac{\beta_{k+1} - \beta_k}{\rho_{k+1} - \rho_k} \tag{2}$$

for $k = 1, \ldots, r-1$ if $r > 1$. If we denote by $\sigma_1 < \sigma_2 < \cdots < \sigma_{s-1}$ those indices $k$ ($k = 1, 2, \ldots, r-1$), for which strong inequality holds in (2), the coefficients $b_k \in \mathbb{C}$ satisfy the following conditions for $k = \sigma_1, \ldots, r$, $b_k \in \mathbb{R} \setminus \{0\}$ and

$$c_\sigma = \sum_{\substack{\rho_\delta + \rho_\lambda = 2\sigma \\ \sigma_i \leq \delta,\, \lambda \leq \sigma_{i+1}}} (-1)^{\rho_\delta + \sigma} b_\delta b_\lambda \geq 0 \tag{3}$$

where $\rho_{\sigma_i} \leq \sigma \leq \rho_{\sigma_{i+1}}$, $i = 1, \ldots, s-1$. For $k = 0, \ldots, \sigma_1 - 1$, $b_k$ may be an arbitrary complex constant. We call $\sigma_1, \ldots, \sigma_s$ the **kink indices**.

The points $(\beta_k, \rho_k)$, $k = 0, 1, \ldots, r$ on the Cartesian plane determine a polygonal path starting from the origin with corner points at the kink indices and with the property $0 \geq m_{\sigma_1} \geq \cdots \geq m_{\sigma_s}$, where $m_{\sigma_i}$ is the slope of the $i^{th}$ polygonal path which is the line connecting $(\beta_{\sigma_{i-1}}, \rho_{\sigma_{i-1}})$ and $(\beta_{\sigma_i}, \rho_{\sigma_i})$.

**Definition 2.** The polygonal path generated by $\mu_0$ is given by the graph of the function $\gamma$: $[0, n] \to \mathbb{R}$ defined by:

$$\tilde{\gamma}(k) = \frac{1}{\rho_{\sigma_{i+1}} - \rho_{\sigma_i}} \left\{ (k - \rho_{\sigma_i}) \beta_{\sigma_{i+1}} + (\rho_{\sigma_{i+1}} - k) \beta_{\sigma_i} \right\}$$

for $k \in [\rho_{\sigma_i}, \rho_{\sigma_{i+1}}]$.

**Remark 1.** For every $k$, $\gamma(\rho_k) = \beta_k$. If $k > j$, then $\tilde{\gamma}(k) \leq \tilde{\gamma}(j)$. Strict inequality holds if $\beta_k < \beta_{k-1}$ for every $k$.

We now define the relatively compact perturbations of the above expressions.

**Definition 3.** A relatively compact perturbation of the differential expression $\mu_0$ is a differential expression $\mu_1$ of the form :

$$\mu_1 = \sum_{j=0}^{n} r_j(t) t^j D_t^j \tag{4}$$

where $r_j \in C^j([1,\infty))$, and $r_j(t) = o[(\log t)^{\tilde{\gamma}(j)}]$. as $t \to \infty$ for every $j = 0,\ldots,n$. We shall consider a smaller class of perturbations for the invariance of the nullities which consists of expressions $\mu_1$ of the form (4) satisfying :

$$r_j^{(m)}(s) = o[t^{-m}(\log t)^{\tilde{\gamma}(j-m)}] \tag{5}$$

for $j = 0,\ldots,n$, and $m = 0,\ldots,j$.

In [3], the nullities of differential expressions of the form (1) satisfying the above conditions together with their relatively compact perturbations in $\mathcal{L}_2[1,\infty)$ were classified.

**Theorem.** Let $\mu_0$ be a differential expression of the form (1) satisfying (2) and (3) and let $\mu_1$ be a relatively compact perturbation of $\mu_0$ of the form (4) satisfying (5). Then if $\beta_k < \beta_{k-1}$ for every $k$, for every $\lambda \in \rho_e(\mu_0)$,

$$nul(\mu_0+\mu_1-\lambda) \quad = \quad nul(\mu_0-\lambda) \quad =$$

$$\#\left\{z \mid \sum_{k=0}^{\sigma_1} b_k z^{\rho_k} = \lambda,\ \mathrm{Re}\, z<0\right\} + \sum_{i=1}^{s-1} \#\left\{z \mid \sum_{k=\sigma_i}^{\sigma_{i+1}} b_k z^{\rho_k} = 0,\ \mathrm{Re}\, z<0\right\}.$$

If $\beta_k = 0$ for $k = 0,1,\ldots,\sigma_1$, for every $\lambda \in \rho_e(\mu_0)$,

$$nul(\mu_0+\mu_1-\lambda) \quad = \quad nul(\mu_0-\lambda) \quad =$$

$$\#\left\{z \mid \sum_{k=0}^{\sigma_1} b_k \prod_{j=0}^{\rho_k-1}(z-(j+\frac{1}{2})) = \lambda,\ \mathrm{Re}\, z<0\right\} + \sum_{i=1}^{s-1} \#\left\{z \mid \sum_{k=\sigma_i}^{\sigma_{i+1}} b_k z^{\rho_k} = 0,\ \mathrm{Re}\, z<0\right\}.$$

## 3 Main Results

Lemma 2.1 relates the nullities of a differential expression $\tau$ relative to a weighted space with weight function $w$ to the spectral properties of $w^{\frac{1}{2}}\tau w^{-\frac{1}{2}}$ in $\mathcal{L}_2$ space. In the following results, it will be assumed that $w \in C^n[1,\infty)$. Also, we say that $t^j w^{(j)}(t) = \mathcal{O}[w(t)]$ if and only if $\frac{t^j w^{(j)}(t)}{w(t)}$ is bounded. A characterization of $w^{\frac{1}{2}}\mu_1 w^{-\frac{1}{2}}$ in $\mathcal{L}_2[1,\infty)$ for $\mu_1 \in \mathcal{L}_2^w[1,\infty)$ is given by the following lemma.

**Lemma 3.1** *Let $\mu_0$ be a differential expression of the form (1) satisfying (2) and let $\mu_1$ be a relatively compact perturbation of $\mu_0$ of the form (4) in $\mathcal{L}_2^w[1,\infty)$ satisfying (5). If $t^j w^{(j)}(t) = \mathcal{O}[w(t)]$ then $w^{\frac{1}{2}}\mu_1 w^{-\frac{1}{2}}$ is a relatively compact perturbation of $\mu_0$ in $\mathcal{L}_2[1,\infty)$ satisfying (5).*

**Proof:** We have

$$(w^{\frac{1}{2}}\mu_1 w^{-\frac{1}{2}})f = \left[\sum_{j=0}^{n}\sum_{i=j}^{n}\binom{i}{j} r_i(t) w^{\frac{1}{2}}(w^{-\frac{1}{2}})^{(i-j)} D_t^j\right] f.$$

For $j = 0, 1, \ldots, n$, let $R_j(t) = \sum_{i=j}^{n} \binom{i}{j} r_i(t) w^{\frac{1}{2}} (w^{-\frac{1}{2}})^{(i-j)}$. We need to show that $R_j^{(m)}(t) = o[t^{-m}(\log t)^{\gamma(j-m)}]$ as $t \to \infty$ for every $j$ and $m = 0, 1, \ldots, j$.

For $m = 1, \ldots, j$,

$$R_j^{(m)}(t) = \sum_{i=j}^{n} \sum_{l=0}^{m} \sum_{q=0}^{l} \binom{i}{j} \binom{m}{l} \binom{l}{q} r_i^{(l-q)}(t)(w^{\frac{1}{2}})^{(q)}(w^{-\frac{1}{2}})^{(i-j+m-l)}.$$

Since $(l-q) \leq m$, for every $i = j, \ldots, n$ we have $i - (l-q) \geq (j-m)$ and by Remark 1 it follows that $\gamma(i-(l-q)) \leq \gamma(j-m)$. Hence, the assertion follows since

$$\lim_{t\to\infty} \frac{r_i^{(l-q)}(t)(w^{\frac{1}{2}})^{(q)}(w^{-\frac{1}{2}})^{(i-j+m-l)}}{(\log t)^{\gamma(j-m)}} = 0.$$

It has been shown in [4] that if $\mu_0$ is a differential expression of the form (1) satisfying (2), then $(w^{\frac{1}{2}}\mu_0 w^{-\frac{1}{2}}) = \mu_0 + \mu$ where

$$\mu = \sum_{k=1}^{r} \sum_{j=0}^{\rho_k - 1} b_k \binom{\rho_k}{j} (\log t)^{\beta_k} t^{\rho_k} w^{\frac{1}{2}} (w^{-\frac{1}{2}})^{(\rho_k - j)} D_t^j. \tag{6}$$

This expression $\mu$ may or may not be a relatively compact perturbation of $\mu_0$, depending on $w$. We consider two cases for the expression $\mu_0$: 1) $\beta_k < \beta_{k-1}$ for every $k$; and 2) $\beta_k = 0$ for $k = 0, 1, \ldots, \sigma_1$.

**Lemma 3.2** *Let $\mu_0$ be an expression of the form (1) satisfying (2) with $\beta_k < \beta_{k-1}$ for every $k$ and let $\mu$ be given by (6). If $t^j w^{(j)}(t) = \mathcal{O}[w(t)]$ for $j = 1, \ldots, n$, then $\mu$ is a relatively compact perturbation of $\mu_0$ satisfying (5).*

**Proof:** Let $\mu$ be the expression given by (6). Then $\mu$ can also be expressed as

$$\mu = \sum_{i=0}^{r-1} \sum_{j=\rho_i}^{\rho_{i+1}-1} \sum_{k=i+1}^{r} b_k \binom{\rho_k}{j} (\log t)^{\beta_k} t^{\rho_k} w^{\frac{1}{2}} (w^{-\frac{1}{2}})^{(\rho_k - j)} D_t^j \tag{7}$$

or $\mu = \sum_{j=0}^{n-1} h_j(t) t^j D_t^j$ where

$$h_j(t) = \sum_{k=i+1}^{r} b_k \binom{\rho_k}{j} (\log t)^{\beta_k} t^{\rho_k - j} w^{\frac{1}{2}} (w^{-\frac{1}{2}})^{(\rho_k - j)}$$

for $j \in [\rho_i, \rho_{i+1})$. For $m = 0, 1, \ldots, j$, we have

$$h_j^{(m)}(t) = \sum_{k=i+1}^{r} \sum_{l=0}^{m} \sum_{p=0}^{l} \sum_{q=0}^{m-l} C_k [(\log t)^{\beta_k}]^{(q)} [t^{\rho_k - j}]^{(l-p)} (w^{\frac{1}{2}})^{(m-l-q)} (w^{-\frac{1}{2}})^{(\rho_k - j + p)}$$

where $C_k = b_k \binom{\rho_k}{j}\binom{m}{l}\binom{l}{p}\binom{m-l}{q}$. Since $t^j w^{(j)}(t) = \mathcal{O}[w(t)] \quad \forall j \le n$, and $v+\gamma(j-m)-\gamma(\rho_k) > 0$ for $j \in [\rho_i, \rho_{i+1})$, $k = i+1, \ldots, r$ and $m = 0, 1, \ldots, j$. Hence $h_j^{(m)}(t) = o(t^{-m}(\log t)^{\gamma(j-m)})$ for $m = 0, 1, \ldots, j$ which implies that $\mu$ is a relatively compact perturbation of $\mu_0$ satisfying (5).

**Lemma 3.3** *Let $\mu_0$ be a differential expression of the form (1) satisfying (2) with $\beta_k = 0$ for $k = 0, \ldots, \sigma_1$ and let $\mu$ be given by (6). If $t^j w^{(j)}(t) = o[w(t)]$ as $t \to \infty$ for $j = 1, \ldots, \rho_{\sigma_1}$ and $t^j w^{(j)}(t) = \mathcal{O}[w(t)]$ for $j \le n$ then $\mu$ is a relatively compact perturbation of $\mu_0$ satisfying (5).*

**Proof:** Consider the following terms in $\mu$ given by (7) in Lemma 3.2:

$$\tau_1 = \sum_{i=0}^{\sigma_1-1}\sum_{j=\rho_i}^{\rho_{i+1}-1}\sum_{k=i+1}^{\sigma_1} b_k \binom{\rho_k}{j} t^{\rho_k-j} w^{\frac{1}{2}}(w^{-\frac{1}{2}})^{(\rho_k-j)} t^j D_t^j \tag{8}$$

Clearly, $w$ satisfies the conditions stated in Lemma 3.2 and it follows directly from the proof of this lemma that the terms in $\mu$ which are not in $\tau_1$ are relatively compact perturbations of $\mu_0$ satisfying (6). Let $r_j(t) = \sum_{k=i+1}^{\sigma_1} b_k \binom{\rho_k}{j} t^{\rho_k-j} w^{\frac{1}{2}}(w^{-\frac{1}{2}})^{(\rho_k-j)}$, then for $m = 0, 1, \ldots, j$,

$$r_j^{(m)}(t) = \sum_{k=i+1}^{\sigma_1}\sum_{l=0}^{m}\sum_{p=0}^{l} b_k \binom{\rho_k}{j}\binom{m}{l}\binom{l}{p} [t^{\rho_k-j}]^{(p)} (w^{\frac{1}{2}})^{(l-p)} (w^{-\frac{1}{2}})^{(\rho_k-j+m-l)}.$$

Since we are taking $j \in [0, \rho_{\sigma_1})$, we have $\gamma(j-m) = 0$.Furthermore, since $t^s w^{(s)}(t) = o[w(t)]$ for $s = 0, \ldots, \sigma_1$, $r_j^{(m)}(t) = o[t^{-m}(\log t)^{\gamma(j-m)}]$ and $\tau_1$ is a relatively compact perturbation of $\mu_0$ satisfying (6).

The following theorem follows directly from Lemmas 3.1, 3.2, 3.3, and 3.4 and the results presented in [3].

**Theorem 3.1** *Let $\mu_0$ be a differential expression of the form (1) satisfying (2) and (3) and let $\mu_1$ be a relatively compact perturbation of $\mu_0$ satisfying (5):*

1. *if $\beta_k < \beta_{k-1}$ for every $k$ and $t^j w^{(j)}(t) = \mathcal{O}[w(t)]$ for $j = 1, \ldots, n$; or*

2. *if $\beta_k = 0$ for $k = 0, \ldots, \sigma_1$ and $t^j w^{(j)}(t) = o[w(t)]$ as $t \to \infty$ for $j = 1, \ldots, \rho_{\sigma_1}$ and $t^j w^{(j)}(t) = \mathcal{O}[w(t)]$for $j \le n$,*

*then for every $\lambda \in \rho_e^w(\mu_0)$*

$$nul^w(\mu_0 + \mu_1 - \lambda) = nul(\mu_0 - \lambda) =$$

$$\#\left\{z \mid \sum_{k=0}^{\sigma_1} b_k z^{\rho_k} = \lambda,\ Re\, z < 0\right\} + \sum_{i=1}^{s-1} \#\left\{z \mid \sum_{k=\sigma_i}^{\sigma_{i+1}} b_k z^{\rho_k} = 0,\ Re\, z < 0\right\}.$$

**Acknowledgement**

Part of this research was completed while the author was a visiting researcher at the Department of Applied Physics, Waseda University, Tokyo, Japan under the JSPS-DOST Long-Term Invitation Program. The author wishes to thank her host scientist, Prof. Mitsuharu Otani, for his comments and suggestions.

**References**

1. J.C. Agapito, *On relatively compact perturbations of special expressions in* $L_2^w(I)$,Ph.D. Thesis, Department of Mathematics, University of the Philippines, 1993.
2. M.P. Roque, Spectral properties of differential expressions with exponentially decaying coefficients in weighted space, **Matimyas Matematika**, vol. 21, no. 2, pp. 26–36, (1998).
3. M.P. Roque, Spectral properties of differential expressions with exponentially and logarithmically growing coefficients, **Functional Analysis and Global Analysis: proc. of the conference held in Manila, Philippines**, Springer-Verlag, pp. 239–249 (1997).
4. M.P. Roque, On the essential spectrum of differential expressions with logarithmically decaying coefficients in weighted space, **Matimyas Matematika**, 21, pp. 158–166 (2000).
5. B. Schultze, Spectral properties of not necessarily self-adjoint linear operators, **Advances in Mathematics,**, 83, No. 1, pp. 75–95 (1990).
6. S. Goldberg, *Unbounded Linear Operators*(McGraw-Hill)

# PROBLEMS CONCERNING THE DEFICIENCY INDICES OF SINGULAR SELF-ADJOINT ORDINARY DIFFERENTIAL OPERATORS

BERND SCHULTZE

*University of Essen, Fachbereich 6, 45117 Essen, Germany*
*E-mail: bernd.schultze@uni-essen.de*

Classical and new problems and conjectures concerning the deficiency indices of singular self-adjoint ordinary differential operators on $[a, \infty)$ are presented as well as the results leading to them.

The deficiency index problem started with H.Weyl's [15] remarkable theory on singular Sturm-Liouville problems, i.e. he considered 2nd order linear formally self-adjoint differential expressions in $L^2(I)$, where $I = [a, \infty), a \in \mathbb{R}$, of the form:

$$My = -(p_1 y')' + p_0 y \tag{1}$$

with $p_1 \in C^1(I, \mathbb{R}), p_0 \in C(I, \mathbb{R})$ and $p_1 > 0$ on $I$.
For the construction of self-adjoint operators generated by $M$, H.Weyl imposed boundary conditions and found out that one has to distinguish two cases for $M$. He called them $limit - point$ resp. $limit - circle$ case. In the limit-point case, one boundary condition has to be imposed in $a$, in the limit-circle case a second one in infinity.
This alternative seemed to have been so impressive that it was falsely adopted in publications that tried to generalize Weyl's theory to $2n$-th order formally self-adjoint expressions of the form:

$$My = \sum_{j=0}^{n} (-1)^j (p_j y^{(j)})^{(j)} \tag{2}$$

with $p_j \in C^j(I, \mathbb{R})$ for $j = 0, ..., n$ and $p_n > 0$ on $I$.
This has been corrected in papers of K. Kodaira [8] and I. M. Glazman [4].
Defining $D(M) := \{f \in L^2(I) : f \in AC^{2n}, Mf \in L^2(I)\}$, $M$ generates the

so called *maximal* operator:

$$T_1(M) : D(M) \longrightarrow L^2(I), \qquad T_1(M)f := Mf.$$

It turns out that for $\lambda \in \mathbb{C} \setminus \mathbb{R}$

$$d(M) := \dim \operatorname{Ker}(T_1(M) - \lambda) = constant \tag{3}$$

and satisfies:

$$n \leq d(M) \leq 2n. \tag{4}$$

The integer $d(M)$ is called the *deficiency index* of $M$ and Glazman showed - giving explicit examples - that actually all the integers according to (4) occur. The cases $d(M) = n$ resp. $d(M) = 2n$ are still called limit-point resp. limit-circle case. In the latter it even holds (see[2],chapter XIII.6):

$$\dim \operatorname{Ker}(T_1(M) - \lambda) = 2n \quad \text{for all} \quad \lambda \in \mathbb{C} \tag{5}$$

So, for the $2n$-th order expressions one has not to consider 2 cases but actually $n+1$. The case in which an explicit operator falls seems to depend in a very sophisticated way on the behaviour of the coefficient functions $p_j$ $(j = 0, ..., n)$ at infinity.
This is also the case for the *essential spectrum* of $M$

$$\sigma_e(M) := \{\lambda \in \mathbb{C} : \text{range}(T_1(M) - \lambda) \text{is not closed}\}. \tag{6}$$

Among the numerous deficiency index criteria, one stands out by the simplicity of its assumption. It has been given already by H. Weyl:

**Theorem 1** *If* $n = 1$ *and* $p_0 \geq 0$, *then* $d(M) = 1$.

Proof:[7]

If we show that $My = 0$ has a solution $y \notin L^2(I)$, then (5) gives the assertion.
Consider the solution $y$ of the initial value problem:

$$My = 0, \;\; y(a) = 0, \;\; y'(a) = 1$$

and assume that $y'$ has a zero. If $x_1 := \min\{x \in I : y'(x) = 0\}$ then

$$0 < p_1(a) \leq \int_a^{x_1} My dx = 0.$$

Therefore $y' > 0$ on $I$ implying that $y \notin L^2(I)$ . ◇

This result, of course, waited to be generalized for larger $n$ and N. Everitt made the following

**Conjecture 1** *If* $p_j \geq 0$ *for all* $j = 0, ..., n$ *, then* $d(M) = n$.

This was disproved by R.M.Kauffman [6] and in [13] it was shown:

**Theorem 2** *For every integer* $k$ *with* $0 \leq k < \frac{n}{2}$ *there exists an expression* $M$ *as in (2) with nonnegative coefficients* $p_j$, $j = 0, ...n$ *, such that*

$$d(M) = n + 2k.$$

Proof:(sketch)[13]
Consider

$$My = \sum_{j=0}^{n} (-1)^j (a_j x^{\beta \cdot j} e^x y^{(j)})^{(j)} \tag{7}$$

where $0 < \beta < 1$ is arbitrary and $a_j$, $j = 0, ...n$, are positive constants ($a_n = 1$) that are chosen in such a way that the polynomial $p(z) = \sum_{j=0}^{n} a_j z^j$ has exactly $2k$ zeros with positive real part and $n - 2k$ zeros with negative real part, counting the zeros according to their multiplicities.
If one carries out the differentiation in (7) and factors out $e^x$ , one obtains

$$My = e^x (M_0 y + M_1 y) \tag{8}$$

with

$$M_0 y = \sum_{j=0}^{n-1} (-1)^j a_j x^{\beta \cdot j} y^{(j)} + (-1)^n \sum_{j=n}^{2n} \binom{n}{j-n} x^{\beta \cdot j} y^{(j)}. \tag{9}$$

$M_0$ is of a special non self-adjoint form, for which the essential spectral data are known, $M_1$ determines a relatively compact perturbation of it and has therefore no impact in them (see[12]). One has:

$$\sigma_e(M_0) = \{\sum_{j=0}^{n} a_j z^j : z \in i\mathbb{R}\}$$

and for $\lambda \notin \sigma_e(M_0)$ :

$$\mathrm{nul}(M_0 - \lambda) := \dim(\mathrm{Ker}(T_1(M_0) - \lambda) =$$

$$\sharp\{z \in \mathbb{C} : \sum_{j=0}^{n} a_j z^j = \lambda, \mathrm{Re}\ z < 0\} + \underbrace{\sharp\{z \in \mathbb{C} : \sum_{j=n}^{2n} \binom{n}{j-n} z^j = 0, \mathrm{Re}\ z < 0\}}_{=n}.$$

Choosing the $a_j$, $j = 0, ..., n$, in the indicated way one obtains $0 \notin \sigma_e(M_0)$ and nul( $M_0) = n + 2k$ . This gives $\sigma_e(M) = \emptyset$ and the assertion. ◇

In this negative result two new conjectures are hidden that are unanswered until now:

**Conjecture 2** *If* $p_j \geq 0$ *for all* $j = 0, ..., n$ , *then* $d(M) < 2n$.

**Conjecture 3** *If* $p_j \geq 0$ *for all* $j = 0, ..., n$ , *then* $d(M)$ *makes jumps of 2.*

This last conjecture stems from the following observation:
If one drops in (7) the condition that the $a_j$, $j = 0, ...n$ , are positive, then one can construct in this way examples for the whole range of examples for (4).

There is the following result for the limit-circle case (see[2],chapter XIII.6):

$$d(M) = 2n \quad \Longrightarrow \quad \sigma_e(M) = \emptyset \tag{10}$$

and until recently there have even no examples been known of expressions $M$ having essential spectrum and being not in the limit-point case.
The first such example was given in [11] and looks quite simple:

$$My = y^{(4)} + (xy')'. \tag{11}$$

It has: $d(M) = 3$ and $(-\infty, 0] \subset \sigma_e(M)$.
Despite the fact that many people tried to classify all fourth order expressions with polynomial coefficients, the deficiency index of this expression was first given only in 1980 in [5] and its complete essential spectrum is still unknown though there are some hints that it is the interval $(-\infty, 0]$ .

For $n = 3$ we have also such examples with nonempty essential spectrum:

$$My = -y^{(6)} - (xy'')'' \quad \text{it has} \quad d(M) = 4 \tag{12}$$

and

$$My = -y^{(6)} - 4(xy'')'' - (x^2y')' \quad \text{it has} \quad d(M) = 5. \tag{13}$$

These examples provoke:

**Conjecture 4** *For all $n, k \in \mathbb{N}$ with $n \leq k < 2n$ there exist expressions $M$ of order $2n$ with $\sigma_e(M) \neq \emptyset$ and $d(M) = k$.*

Now, $M$ as given in (2) is not the most general formally self-adjoint expression with smooth coefficients. Such an expression of order $m \in \mathbb{N}$ can be brought into the form:

$$My = \sum_{j=0}^{[\frac{m}{2}]} (-1)^j (p_j y^{(j)})^{(j)} + \frac{i}{2} \sum_{j=0}^{[\frac{m-1}{2}]} (-1)^j \{(q_j y^{(j+1)})^{(j)} + (q_j y^{(j)})^{(j+1)}\} \tag{14}$$

with $p_j \in C^j(I, \mathbb{R})$, $j = 0, ..., [\frac{m}{2}]$, $q_j \in C^{j+1}(I, \mathbb{R})$, $j = 0, ..., [\frac{m-1}{2}]$ and the coefficient of the highest derivative is positive on $I$.
For the nullities it turns out, that they are no longer constant on all of $\mathbb{C} \setminus \mathbb{R}$ but in general only in the upper and lower halfplane. So, there are now 2 deficiency indices:

$$(d_+(M), d_-(M)) := (\dim \operatorname{Ker}(T_1(M) - i), \dim \operatorname{Ker}(T_1(M) + i)) \tag{15}$$

and one has (cp.[7],[1],[14]):

**Theorem 3** *The following hold:*

*a)* $[\frac{m}{2}] \leq d_\pm(M) \leq m$ *with*

$d_\pm(M) = \frac{m-1}{2} \Rightarrow d_\mp \geq \frac{m+1}{2}$ *and* $d_+(M) = m \Leftrightarrow d_-(M) = m.$

*b)* $d_\pm(M) = m \Rightarrow \sigma_e(M) = \emptyset$.

*c) If* $p_j^{(k)}(x) = O(x^{2j-k})$ *for* $k = 0, ..., j,\ j = 0, ..., [\frac{m}{2}]$, *and*

$q_j^{(k)}(x) = O(x^{2j+1-k})$ *for* $k = 0, ..., j+1,\ j = 0, ..., [\frac{m-1}{2}]$, *as* $x \longrightarrow \infty$, *then* $\sigma_e(M) \neq \emptyset$.

Soon after Glazman's construction, part a) of this theorem suggested:

**Conjecture 5** *All pairs of integers according to the restrictions in part a) of theorem 3 actually occur as deficiency indices of some expression $M$ as in (14).*

Since this conjecture is open for a long time, easier questions of this type were asked, e.g.:
Are there even order expressions $M$ of type (14) for which $d_+(M) \neq d_-(M)$?
This was answered positively by McLeod [10], and Kogan and Rofe-Beketov [9], but a much simpler expression than theirs is:

$$My = y^{(4)} + \frac{i}{2}\{(xy')'' + (xy'')'\} \quad \text{it has} \quad (d_+(M), d_-(M) = (3,2). \tag{16}$$

Gilbert [3] showed that $|d_+(M) - d_-(M)|$ can become arbitrarily large. His examples of lowest degree, for which the deficiency indices differ by 2, has degree 7 and the simplest is:

$$\begin{aligned} My = &-\tfrac{i}{2}\{(x^3y^{(4)})^{(3)} + (x^3y^{(3)})^{(4)}\} - 11(x^3y^{(3)})^{(3)} \\ &+\tfrac{9i}{2}\{((5x^3+3x)y^{(2)})^{(3)} + ((5x^3+3x)y^{(3)})^{(2)}\} \\ &+9((9x^3+22x)y^{(2)})^{(2)} \\ &-27i\{((x^3+10x)y')'' + ((x^3+10x)y'')'\} \\ &-486(xy')' + 81i\{(xy)' + xy'\}. \end{aligned} \tag{17}$$

It has $(d_+(M), d_-(M)) = (3,5)$. One other remarkable fact for this expression is its susceptibility to perturbations. You change only one of the coefficients above and it looses the claimed property.
The lowest order at which the deficiency indices could differ by 2 according to theorem 3 a) is 5, and therefore we have another unanswered

**Conjecture 6** *There exists $M$ as in (14) of order 5 with* $(d_+(M), d_-(M)) = (2,4)$.

## References

1. M. Castillo, On the existence of the essential spectrum for a class of differential expressions, *Ph. D. Thesis, Department of Mathematics, Univ. of the Philippines, Diliman*, 1999.

2. N. Dunford and J. T. Schwartz, Linear operators II, *Wiley, New York*, 1963.
3. R. C. Gilbert, A class of symmetric ordinary differential operators whose deficiency numbers differ by an integer, *Proc. Roy. Soc. Edinburgh,* 82A, 117-134, 1978.
4. M. Glazman, On the deficiency indices of differential operators, *Doklad. Akad. Nauk SSSR*,64,151-154, 1949 (Russ).
5. C. G. M. Grudniewicz, Asymptotic methods for fourth-order differntial equations, *Proc. Roy. Soc. Edinburgh,* 87A, 53-63, 1980.
6. R. M. Kauffman, On the limit-n classification of ordinary differential operators with positive coefficients, *Proc. Roy. Soc. Edinburgh,* 90A, 209-236, 1977.
7. R. M. Kauffman, T. T. Read and A. Zettl, The deficiency index problem for powers of ordinary differential expressions, *Lecture Notes in Math. No. 621, Springer-Verlag, Berlin*, 1977.
8. K. Kodaira, On ordinary differential equations of any even order and the corresponding eigenfunction expansions, *Amer. J. of Math.*72, 502-544, 1950.
9. V. L. Kogan and F.S. Rofe-Beketov, On the question of the deficiency indices of differential operators with complex coefficients, *Proc. Roy. Soc. Edinburgh,* 72A, 281-298, 1973/74.
10. J.B. McLeod, The number of square-integrable solutions of ordinary differential equations, *Quarterly Journal Math.*, 17, 285-290, 1966.
11. B. Schultze, A fourth order limit-3 expression with nonempty essential spectrum, *Proc. Roy. Soc. Edinburgh,* 106A, 233-235, 1987.
12. B. Schultze, Spectral properties of not necessarily self-adjoint linear differential operators, *Advances in Mathematics* 83, No. 1, 75-95, 1990.
13. B. Schultze, On singular differential operators with positive coefficients, *Proc. Roy. Soc. Edinburgh,* 120A, 361-365, 1992.
14. B. Schultze, Regions of spectral stability and instability for singular ordinary differential operators, *Proc. of the conference on "Functional Analysis and Global Analysis", Manila 1996, Springer Verlag*, 250-257, 1997.
15. H. Weyl, On ordinary differential equations with singularities and the associated expansions of arbitrary functions, *Math. Annal.* 68, 222-269, 1910(Ger).

# ATTAINABILITY OF A TWO CHARACTERISTIC MANPOWER STRUCTURE INCLUDING DEMOTION

A. SRINIVASAN
READER IN MATHEMATICS, BISHOP HEBER COLLEGE,
TRICHY - 620 0017 TAMIL NADU INDIA
E-mail: mathsrinivas@yahoo.com
P. MARIAPPAN [1]
LECTURER[SS] IN MATHEMATICS,
BISHOP HEBER COLLEGE,
TRICHY - 620 017 TAMIL NADU INDIA
E-mail: mathmari@hotmail.com

In this paper the stochastic behaviour of a two-characteristic manpower model including demotion is discussed and using multinomial distribution, the probabilities of grade structure being attainable in one-step, under control of recruitment are calculated. Numerical results are also given to illustrate the model. In addition, the geometric description of some structures is also analysed.

## 1 Introduction

Bartholomew [2], Davies [5] and Abdallaovi [1] have analysed in detail, the problem of stochastic control of a graded manpower system and the probability of attaining a structure in one step for one dimensional manpower flow. Normally, in any organization, promotion or wastage for any grade depends on the length of service, time spent in that grade or physical location. Hence, the problem of maintaining a structure has been considered for a two-characteristic flow by Chandra [3]. Further the attainability of a two characteristic manpower structure also has been analysed by Chandra [4] by considering promotion aspect only using Trinomial distrubution. In this paper we have extended the results of Chandra [4] by conisdering the aspect of demotion along with promotion.

## 2 Notations And Assumptions

$i$ : represents the grade [First Characteristic (FC)],
$1 \leqslant i \leqslant N$.
$j$ : represents the length of service [Second Characteristic (SC)],
$l(i) \leqslant j \leqslant u(i)$

[1]Research Supported By UGC / FIP / IX Plan 31.05.1999.

$q_{i,j-1}$ : probability that a person is moving from state$(i, j-1)$ to state$(i, j)$ in one time period.
$p_{i-1,j-1}$ : probability that a person is moving from state $(i-1, j-1)$ to state$(i, j)$in one time period.
$f_{i+1,j-1}$ : probability that a person is moving from$(i+1, j-1)$ to state $(i, j)$ in one time period.
$r_{ij}$ : probability of recruitment to state$(i, j)$ in one time period.
$s_{ij}(t)$ : expected stock in state $(i, j)$ at a time t.
$R_i(t)$ : the recruitment vector for grade i at time t.
$m_{ij}$ : the goal structure of state$(i, j)$
$s_{ij}$ : the given structure of state $(i, j)$
$x_{1ij}$ : number of persons who move from state$(i, j-1)$ to$(i, j)$
$x_{2ij}$ : number of persons who move from state$(i-1, j-1) to (i, j)$
$x_{3ij}$ : number of persons who move from state $(i+1, j-1)$ to$(i, j)$
$x_{4ij}$ : number of recruits to state$(i, j)$
$P_j(m_{ij})$ : probability of attaining the structure $m_{ij}$
$P(m_i)$ : probability of attaining the structure $m_i$
$m_i = [\, m_{i,l(i)}, \ldots, m_{ij}, \ldots, m_{i,u(i)}]$
$R_i(t) = [R_{i,l(i)}(t), \ldots, R_{ij}(t), \ldots, R_{i,u(i)}(t)]$
$s_i = [\, s_{i,l(i)}, \ldots, s_{ij}, \ldots, s_{i,u(i)}]$

We make the following assumptions :

1. Time is considered in discrete units.
2. Movement in one time period from states with FC i can only to states with FC $(i-1)$ or $(i)$ or $(i+1)$ or out of the system.
3. Service rendered in other systems before joining the present one is also counted.
4. Since only a single period $(t, t+1)$ is considered the suffix t can be omitted.
5. Grade sizes are constant.

In the next section, the use of multinomial distribution to obtain the probability of a structure being attained in one step in a two-characteristic flow along with promotion including demotion is analysed.

## 3 Multinomial Method

The probability of attaining a structure in one step is of fundamental importance. For a one dimensional manpower flow Abdallaovi [1] used trinomial distribution to evaluate such probability. Chandra [4] evaluated the same for a two characteristic manpower flow by considering only promotion using trinomial distribution in a different way. We use the multinomial distribution approach to evaluate the same for a two-characteristic manpower flow by considering demotion and promotion .

We discuss the problem of attainability for grade $i$.

Without loss of generality, we take

$$s_{i-1} = [s_{(i-1),1}, \quad ......., \quad s_{(i-1),u(i-1)}]$$

$$s_i = [s_{i,1}, \quad ......., \quad s_{i,u(i)}]$$

$$m_i = [m_{i,1}, \quad ..........., \quad m_{i,u(i)}]$$

Since the grade sizes are taken to be constant, at least one $m_{ij}$, say $m_{ik}$, is greater than $s_{ij}$.Obviously $m_{ik}$ is attainable by promotion, demotion and recruitment, with probability one. Consider any other $m_{ij}$ ($m_{ij}$ $> s_{ij}$), then $m_{ij}$ consists of four components namely $x_{1ij}$, $x_{2ij}$, $x_{3ij}$ and $x_{4ij}$.

Then

$$x_{1ij} + x_{2ij} + x_{3ij} + x_{4ij} = m_{ij} \text{ ; } \forall \text{ j} \qquad (1)$$

If any person in state (i, j) is considered, he / she has been either continuing in the same grade or promoted to grade i or demoted to grade i or recruited to grade i.

Let,

$a = s_{i,j-1}{}^{*} \; q_{i,j-1} / \; m_{ij}$
$b = s_{i-1,j-1}{}^{*} p_{i-1,j-1} / \; m_{ij}$
$c = s_{i+1,j-1}{}^{*} \; f_{I+1,j-1} / \; m_{ij}$ and
$d =$ 1 - a - b - c (2)

Here $x_{1ij}$, $x_{2ij}$, $x_{3ij}$ and $x_{4ij}$ are random variables occurring with the probabilities $a$, $b$, $c$ and$d$ respectively.

Then the joint probability mass function of these random variables is given by the multinomial distribution

$$P\left[x_{1ij}, x_{2ij}, x_{3ij}, x_{4ij}\right] \quad = \quad \frac{m_{ij}!}{x_{1ij}! \quad x_{2ij}! \quad x_{3ij}! x_{4ij}!} a^{x_{1ij}} b^{x_{2ij}} c^{x_{3ij}} d^{x_{4ij}} \qquad (3)$$

As $x_{ij}$ takes the value from 0 to $s_{i,j-1} * q_{i,j-1}$, $x_{2ij}$ takes the value from 0 to $s_{i-1,j-1} * p_{i-1,j-1}$, $x_{3ij}$ takes the value from 0 to $s_{i+1,j-1} * f_{i+1,j-1}$ and $x_{4ij}$ takes the value from 0 to $m_{ij}$ subject to the condition given by (1),

$$P_j[m_{ij}] = \sum_{x_{1ij}=0}^{s_{i,j-1}*q_{i,j-1}} \quad \sum_{x_{2ij}=0}^{s_{i-1,j-1}*p_{i-1,j-1}} \quad \sum_{x_{3ij}=0}^{s_{i+1,j+1}*f_{i+1,j-1}} \quad \sum_{x_{4ij}=0}^{m_{ij}} P\left[x_{1ij}, x_{2ij}, x_{3ij}, x_{4ij}\right] \; (4)$$

subject to be condition that the above sum is less than or equal to 1.
The structure $m_i$ is attained only when all $m_{ij's}$ (j = 1, ... u(i)) are attained.

$$\therefore \quad P(m_i) = \prod_{j=1}^{u(i)} P_j\left(m_{ij}\right) \qquad (5)$$

When $s_{i,j-1}$, $s_{i-1,j-1}$ and $s_{i+1,j-1}$ are less than $m_{ij}$ for all j, it is clear that the structure $m_i$ is attainable with probability 1. For a better understanding of the method an illustration is given in the next section.

## 4 Illustration :

Let $m_i$ = (40, 25, 25); ie., $m_{i1}$=40, $m_{i2}$ = 25 and $m_{i3}$ = 25.

Taking $P(m_{i1})$ = 1, we have computed $P(m_{i2})$, $P(m_{i3})$ and $P(m_i)$ by considering different set of values for $s_{i,j-1}$, $s_{i-1,j-1}$, $s_{i+1,j-1}$, $q_{i,j-1}$, $p_{i-1,j-1}$ and $f_{i+1,j-1}$; (j = 1,2,3). The results are tabulated using C programming and are given in the following tables.

**Table - 1**

| $s_{i1}$ | $s_{i-1,1}$ | $s_{i+1,1}$ | $q_{i,1}$ | $p_{i-1,1}$ | $f_{i+1,1}$ | $P(m_{i2})$ | $s_{i,2}$ | $s_{i-1,2}$ | $s_{i+1,2}$ | $q_{i,2}$ | $p_{i-1,2}$ | $f_{i+1,2}$ | $P(m_{i3})$ | $P(m_i)$ |
|---|---|---|---|---|---|---|---|---|---|---|---|---|---|---|
| 30 | 24 | 20 | 0.005 | 0.01 | 0.005 | 0.610 | 25 | 22 | 22 | 0.005 | 0.01 | 0.05 | 0.489 | 0.298 |
| 30 | 24 | 22 | 0.005 | 0.01 | 0.005 | 0.603 | 28 | 22 | 22 | 0.005 | 0.01 | 0.05 | 0.482 | 0.291 |
| 30 | 24 | 22 | 0.010 | 0.01 | 0.050 | 0.399 | 28 | 22 | 22 | 0.010 | 0.01 | 0.05 | 0.416 | 0.166 |
| 25 | 20 | 22 | 0.700 | 0.20 | 0.050 | 0.087 | 26 | 19 | 19 | 0.700 | 0.20 | 0.05 | 0.019 | 0.002 |
| 25 | 23 | 22 | 0.700 | 0.20 | 0.050 | 0.042 | 26 | 23 | 19 | 0.700 | 0.20 | 0.05 | 0.014 | 0.001 |
| 25 | 23 | 22 | 0.710 | 0.21 | 0.060 | 0.011 | 26 | 23 | 19 | 0.710 | 0.21 | 0.06 | 0.012 | 0 |

From the table it is clear that $P(m_{i2})$ decreases when $s_{i,1}$,$s_{i-1,1}$, and $s_{i+1,1}$ increases and also when the transition probabilities increases. In a very similar manner $P(m_{i3})$ decreases when $s_{i,2}$, $s_{i-1,2}$ and $s_{i+1,2}$ increases and also when the transition probabilities increases. In other words, the probability of attaining a structure is more when the goal structure is nearer to the given structure or when the probability of a person being recruited to state $(i, j)$ is more. So, given a goal structure for grade $i$, it is possible to find the probability of attaining it from the given structure.

**Table - 2**

| $s_{i1}$ | $s_{i-1,1}$ | $s_{i+1,1}$ | $q_{i,1}$ | $p_{i-1,1}$ | $f_{i+1,1}$ | $P_2(m_{i2})$ | $s_{i2}$ | $s_{i-1,2}$ | $s_{i+1,2}$ | $q_{i,2}$ | $p_{i-1,2}$ | $f_{i+1,2}$ | $P_3(m_{i3})$ | $P(m_i)$ |
|---|---|---|---|---|---|---|---|---|---|---|---|---|---|---|
| 30 | 24 | 20 | 0.4 | 0.21 | 0.05 | 0.172 | 40 | 20 | 21 | 0.4 | 0.21 | 0.05 | 0.129 | 0.022 |
| 35 | 24 | 20 | 0.4 | 0.22 | 0.05 | 0.132 | 40 | 20 | 21 | 0.4 | 0.22 | 0.05 | 0.109 | 0.014 |
| 35 | 24 | 20 | 0.4 | 0.23 | 0.05 | 0.110 | 40 | 20 | 22 | 0.4 | 0.23 | 0.05 | 0.087 | 0.010 |
| 35 | 24 | 20 | 0.4 | 0.24 | 0.05 | 0.091 | 40 | 20 | 23 | 0.4 | 0.24 | 0.05 | 0.067 | 0.006 |
| 40 | 25 | 20 | 0.4 | 0.25 | 0.05 | 0.090 | 41 | 24 | 24 | 0.4 | 0.25 | 0.05 | 0.058 | 0.005 |
| 40 | 25 | 20 | 0.4 | 0.26 | 0.05 | 0.067 | 42 | 24 | 24 | 0.4 | 0.26 | 0.05 | 0.019 | 0.001 |
| 40 | 25 | 20 | 0.4 | 0.27 | 0.05 | 0.047 | 43 | 24 | 24 | 0.4 | 0.27 | 0.05 | 0.009 | 0 |
| 42 | 24 | 23 | 0.4 | 0.28 | 0.05 | 0.004 | 42 | 24 | 24 | 0.4 | 0.28 | 0.05 | 0.004 | 0 |

From the above table we observe that keeping the retention and demotion probabilities fixed, the probability of attaining the goal structure for grade $i$ decreases when the probability of promotion from grade $(i-1)$ to grade $i$ increases.

**Table - 3**

| $s_{i1}$ | $s_{i-1,1}$ | $s_{i+1,1}$ | $q_{i,1}$ | $p_{i-1,1}$ | $f_{i+1,1}$ | $P_2(m_{i2})$ | $s_{i,2}$ | $s_{i-1,2}$ | $s_{i+1,2}$ | $q_{i,2}$ | $p_{i-1,2}$ | $f_{i+1,2}$ | $P_3(m_{i3})$ | $P(m_i)$ |
|---|---|---|---|---|---|---|---|---|---|---|---|---|---|---|
| 30 | 24 | 20 | 0.4 | 0.2 | 0.05 | 0.115 | 28 | 24 | 20 | 0.4 | 0.2 | 0.05 | 0.111 | 0.013 |
| 30 | 24 | 20 | 0.41 | 0.2 | 0.05 | 0.098 | 28 | 24 | 20 | 0.41 | 0.2 | 0.05 | 0.095 | 0.009 |
| 30 | 24 | 20 | 0.42 | 0.2 | 0.05 | 0.082 | 28 | 24 | 20 | 0.42 | 0.2 | 0.05 | 0.081 | 0.007 |
| 30 | 24 | 20 | 0.43 | 0.2 | 0.05 | 0.067 | 40 | 24 | 20 | 0.43 | 0.2 | 0.05 | 0.053 | 0.004 |
| 35 | 24 | 20 | 0.44 | 0.2 | 0.05 | 0.066 | 41 | 24 | 20 | 0.44 | 0.2 | 0.05 | 0.045 | 0.003 |
| 35 | 24 | 20 | 0.45 | 0.2 | 0.05 | 0.049 | 42 | 24 | 20 | 0.45 | 0.2 | 0.05 | 0.003 | 0 |
| 36 | 24 | 20 | 0.46 | 0.2 | 0.05 | 0.047 | 42 | 23 | 21 | 0.46 | 0.2 | 0.05 | 0.002 | 0 |
| 37 | 24 | 20 | 0.47 | 0.2 | 0.05 | 0.042 | 40 | 24 | 24 | 0.47 | 0.2 | 0.05 | 0.002 | 0 |

From the above table, we observe that keeping the promotion and demotion probabilities fixed, the probability of attaining the goal structure for grade $i$ decreases when the probability of retention in grade $i$ increases.

**Table - 4**

| $s_{i1}$ | $s_{i-1,1}$ | $s_{i+1,1}$ | $q_{i,1}$ | $p_{i-1,1}$ | $f_{i+1,1}$ | $P_2(m_{i2})$ | $s_{i2}$ | $s_{i-1,2}$ | $s_{i+1,2}$ | $q_{i,2}$ | $p_{i-1,2}$ | $f_{i+1,2}$ | $P_3(m_{i3})$ | $P(m_i)$ |
|---|---|---|---|---|---|---|---|---|---|---|---|---|---|---|
| 30 | 24 | 20 | 0.4 | 0.21 | 0.05 | 0.172 | 32 | 24 | 20 | 0.4 | 0.21 | 0.05 | 0.112 | 0.019 |
| 30 | 24 | 20 | 0.4 | 0.21 | 0.06 | 0.146 | 32 | 24 | 20 | 0.4 | 0.21 | 0.06 | 0.093 | 0.014 |
| 30 | 24 | 20 | 0.4 | 0.21 | 0.07 | 0.122 | 32 | 24 | 20 | 0.4 | 0.21 | 0.07 | 0.076 | 0.009 |
| 30 | 24 | 20 | 0.4 | 0.21 | 0.08 | 0.101 | 32 | 24 | 20 | 0.4 | 0.21 | 0.08 | 0.061 | 0.006 |
| 30 | 24 | 20 | 0.4 | 0.21 | 0.09 | 0.082 | 32 | 24 | 20 | 0.4 | 0.21 | 0.09 | 0.048 | 0.004 |
| 45 | 24 | 20 | 0.4 | 0.21 | 0.10 | 0.052 | 24 | 24 | 65 | 0.4 | 0.21 | 0.10 | 0.044 | 0.002 |
| 45 | 24 | 20 | 0.4 | 0.21 | 0.11 | 0.031 | 24 | 24 | 75 | 0.4 | 0.21 | 0.11 | 0.034 | 0.001 |
| 45 | 24 | 20 | 0.4 | 0.21 | 0.12 | 0.012 | 24 | 24 | 75 | 0.4 | 0.21 | 0.12 | 0.032 | 0 |

From the above table, we observe that keeping the promotion and retention probabilities fixed, the probability of attaining the goal structure for grade i decreases when the probability of demotion from grade (i + 1) to grade i increases.

In the next section, the geometric behaviour of the set of goal structures, attainable from $(s_{i-1}, s_i)$ is discussed.

## 5 Geometric Relationship

In this section, let grade be FC and physical location be SC,

Let $1 \leqslant j \leqslant b(i)$.

Let $s_{i-1}$ and $s_i$ be the given structures.

Let $s_i(s_{i-1}, s_i)$ be the set of structures for grade i, which are attainable with non-zero probability of one-step, from the structures $s_{i-1}$ and $s_i$.

$$b_i = s_i\ Q_i + s_{i-1}\ P_{i-1} + s_{i+1}\ F_{i+1} + R_i \qquad (6)$$

$m_i(t) = [\ m_{i,l}, \ldots, m_{ij}, \ldots, m_{i,b(i)}]$ is a member of the set $S_i(s_{i-1}, s_i)$ if and only if

i.e., $s_i\ Q_i + s_{i-1}\ P_{i-1} + s_{i+1}\ F_{i+1} \leqslant m_i$ (7)

The boundary of the region is determined by

$$Y_i = s_i\ Q_i + s_{i-1}\ P_{i-1} + s_{i+1}\ F_{i+1} \quad (8)$$

Thus $Y_i$ is the structure obtained for grade i, in one-step from $(s_{i-1},\ s_i)$, when leavers from each state are replaced in that state by equal number of recruits. Therefore, $Y_i$ is attainable with probability one.

Consider the points

$X_{ir} = Y_i + L\ [\ R_{i,l},\ \dots\ ,\ R_{ij},\ \dots\ ,\ R_{i,b(i)}]'E_r$

where r = 1,2,..,b(i); L = (1,1, ... ,1) $_{1xb(i)}$and $E_r$is the 1xb(i) vector with 1 in the $r^{th}$ place and zero elsewhere . Now it can be shown that

$X_{ir}$'s ( r = 1,2,..,b(i)) are the vertices of the convex hull $S_i(s_{i-1}, s_i)$. Obviously,

$$X_{ir} = Y_i + A_r \quad (9)$$

where $A_r$ is the 1xb(i) vector with { $\sum_{j=1}^{b(i)}$ $R_{ij}$ } in the r$^{th}$ place and zero elsewhere.

Using (6) and (7) we have,

$$b_i = Y_i \quad + \quad R_i$$

$$= \quad \frac{R_{i1}}{\sum_j R_{ij}} x_{i1} \quad + \quad \frac{R_{i2}}{\sum_j R_{ij}} x_{i2} \quad + \quad \dots\dots\dots \quad + \quad \frac{R_{i,b(i)}}{\sum_j R_{ij}} x_{i,b(i)}$$

$$= \sum_{r=1}^{b(i)} C_{ir} X_{ir}; \quad where C_{ir} = \frac{R_{ir}}{\sum_j R_{ij}} \quad and \quad \sum_{r=1}^{b(i)} C_{ir} = 1. \tag{10}$$

Thus $x_{i1}, x_{i2}, \ldots, x_{ib(i)}$ are the convex vertices of a convex hull, of which $b(i)$ is a member.

Consider any other point in that convex hull. Let

$$b_i' = s_i Q_i + s_{i-1} P_{i-1} + s_{i+1} F_{i+1} + R_i' \tag{11}$$

Since the grade sizes are assumed to be constant

$$b_i' = Y_i + R_i'$$

Thus, $x_{i1}, x_{i2},, \ldots, x_{i,b(i)}$. are the vertices of a Convex hull, of which $b_i$ is a member. Consider any other point in that convex hull

$$Let \quad b_i' = s_i Q_i + s_{i-1} P_{i-1} + s_{i+1} F_{i+1} + R_i' \tag{12}$$

Since the grade sizes are assumed to be constant

$$b_i' = Y_i + R_i'$$

$$b_i^I = \sum_{r=1}^{b(i)} \frac{C_{ir}'}{\sum_j R_{ij}} x_{ir} \quad ;$$

$$where \quad C_{ir}' = \frac{R_{ir}'}{\sum_j R_{ij}} \quad and \quad \sum_{r=1}^{b(i)} C_{ir}' = 1$$

Thus $b_i'$s is also expressed as a convex combination of $x_{i1}, x_{i2},..x_{i,b(i)},$ which are the vertices of the Convex hull $S_i(s_{i-1}, s_i)$.

Thus it is not only shown that the set of all attainable structures for grade i forms a convex hull for which $x_{i-1}, x_{i-2}, \ldots, x_{i-b(i)}$, are the vertices, but also that only that structure which can be expressed as a convex combination of those vertices, ie., attainable from $s_{i-1}$ and $s_i$ in one step. The abo e result explains the geometric behaviour of the attainable structures.

## 6 Conclusion

In this paper we found that :

1. The probability of attaining a structure in grade i is more when the goal structure is nearest to the given structure or when the probability of a person having recruited to state(i,j) is more.
2. The probability of attaining the goal structure for grade i decreases when the probability of promotion from grade (i-1) to grade i increases, keeping the retension probability and demotion probability fixed.
3. The probability attaining the goal structure for grade i decreases when the probability of retension from grade i increases, keeping the promotion probability and demotion probability fixed.
4. The probability attaining the goal structure for grade i decreases when the probability of demotion from grade i increases, keeping the retension probability and promotion probability fixed.
5. The geometric behaviour of the attainable structure is also explained.

As a matter of fact, when demotion factor is taken into account, our paper subsumes the results of Chandra[4].

**ACKNOWLEDGEMENT**

The authors wish to express their gratefulness and thanks to UGC, India for its financial assistance during the preparation of this paper.

**REFERENCES**

1. G. Abdallaoui, J. Appl. Prob., V24, 1006(1987).
2. D.J. Batholomew, Adv. Appl. Prob., V11,603(1979).
3. S. Chandra , IAPQR Trans., j. of the Indian Society for Productivity Quality and reliability, Vol.16, No.1, 23(1991).
4. S. Chandra, OPSEARCH, Vol.27, No.3(1990).
5. G.S. Davies, J. Appl. Prob., V20, 423(1983).

# EXPECTED TIME FOR RECRUITMENT IN MANPOWER PLANNING ASSOCIATED WITH CORRELATED RENEWAL SEQUENCIES - A SHOCK MODEL APPROACH

**A. SRINIVASAN**
**READER IN MATHEMATICS, BISHOP HEBER COLLEGE,**
**TRICHY - 620 0017 TAMIL NADU INDIA**
**E-mail: mathsrinivas@yahoo.com**
**P. MARIAPPAN** [1]
**LECTURER[SS] IN MATHEMATICS, BISHOP HEBER COLLEGE,**
**TRICHY - 620 017 TAMIL NADU INDIA**
**E-mail: mathmari@hotmail.com**

In any organization the exit or wastage of personnel is very common. The exit of personnel in marketing organization is very frequent, especially when policies regarding incentives, targets are revised. In this paper, a mathematical model is considered under the above setup, wherein the expected time to recruitment and its variance are obtained using shock model approach when the sequences associated with the exodus of personnel and the inter decision times taken by the organization form a correlated pair of renewal sequences. A numerical example has been given to illustrate the model, assuming specific forms of distributions involved.

## 1.Introduction

Consider a sequence of independent and identically distributed pairs of non-negative random variables $(X_n, Y_n)$; $n = 0, 1, 2, \cdots$ where the variates $(X_n, Y_n)$ are pair wise independent, but $X_n$ and $Y_n$ may be correlated for a given n. We call $(X_n, Y_n)$ as a correlated pair of renewal sequence. Such a paired process appears in the process of stochastic system subject to random shocks at random intervals. For a detailed study of shock models one may refer Esary, Prochan and Marshall [3]. Wastage of personnel due to retirement, death and resignations is a common phenomenon in administrative as well as production-oriented organizations. A number of models are discussed in Bartholomew and Forbes [2]. Optimal appointment policies using Markov chain model is discussed by Grinold [4]. In [7] Sathiyamoorthy and Elangovan have considered the problem of breakdown of an organization due to staff depletion which takes place whenever the organization announces revised policies regarding sales targets, revision of wages, incentives and perquisites etc., and obtained the expected time to recruitment and its variance using shock model approach when the magnitude $X_n$ of the staff depletion at the $n^{th}$ decision epoch are independent of $Y_n$, the time between two consecutive decision making epochs.

In this paper, we have studied the general manpower model governed by the correlated pairs $(X_n, Y_n)$

[1]Research Supported By UGC / FIP / IX Plan 31.05.1999.

of renewal sequences, where the organization reaches an uneconomical status which other wise be called the breakdown point if the magnitude $X_n$ exceeds a pre specified threshold level 'z'. More specifically we have considered two different models depending on whether $X_n$ is correlated to the length of the interval since the last decision or to the length of subsequent interval until the next decision is made. The former is called Model-I and the latter is called Model-II. For these two models we have obtained the expected duration to the breakdown of the organization or the expected time to recruitment due to staff depletion and its variance using shock model approach of Shanthikumar and Sumitha [6], without using any cost structure. Our paper extends the results of Sathiyamoorthy and Elangovan [7] for the correlated case when the threshold level is pre specified.

In Section-2, we have listed the notations. In Section-3, we have given all the assumptions made. In Section-4, the formula to evaluate mean and variance of time to recruitment for Model-I and Model-II are given. In Section-5, the formula to evaluate mean and variance of time to recruitment for Model-I and Model-II are derived based on the Bivariate Poisson distribution. In Section-6, a numerical illustration is discussed for Model-I and Model-II.

## 2.Assumptions Of The Model

1. The organization takes decisions at n random epochs, and at every decision making epoch a random number of persons quit the organization.
2. There is an associated loss of manpower to the organization if a person quits.
3. If the total number of persons who leave the organization exceeds a pre specified threshold level, the organization faces a break down and so recruitment is necessary.
4. The magnitude of the staff depletion at the $n^{th}$ decision making epoch are correlated to the time between the corresponding two consecutive decision making epochs.

## 3.Notations

| | | |
|---|---|---|
| $X_j$ | : | a discrete random variable denoting the number of persons who leave the organization at the jth decision making epoch |
| Y | : | time between the two consecutive decision making epochs |

| | | |
|---|---|---|
| N(t) | : | number of decisions made in [0, t] |
| M(t) | : | $\underset{0 \leqslant j \leqslant N(t)}{Max} \{X_j\}$ |
| z | : | pre specified threshold level which is positive |
| $T_z$ | : | time to recruitment when the threshold level is z; $T_z(\omega) = \inf \{t; M(\omega, t) \geqslant z\}$ for sample path $\omega$ |
| $F_{XY}$ (x, y) | : | Joint distribution of X and Y |
| $F_X(z)$ | : | Marginal distribution of X |
| $\overline{F}_x(z) : 1 - F_x(z)$ | | |
| $E(T_z)$ | : | mean time to recruitment |
| $V(T_z)$ | : | Variance of the time to recruitment |

**4.Results**

**Model - I**

The organization starts at time t = 0. The total number of persons $X_n$ leaving the organization at the $n^{th}$ decision making epoch is correlated to the time interval $Y_n$, since the (n -1)$^{th}$ decision making epoch and does not affect the future events. For this model, from Santhikumar and Sumita [7] we have the following result.

**Theorem - 4.1**

The mean time to recruitment and its variance are given by

$$E(T_z) = \frac{E(Y)}{\bar{F}_X(z)} \tag{1}$$

$$V(T_z) = \left\{\frac{E(Y^2)}{\bar{F}_X(z)}\right\} + \left\{\frac{E(Y)}{(\bar{F}_X(z))^2}\right\}\{2F_X(z) \quad E(Y/X<z) - E(Y)\} \tag{2}$$

**Model - II**

It differs significantly from Model-I in that the number of persons $X_n$ leaving of the $n^{th}$ decision making epoch affects the future events $Y_n$, the time interval until the $(n+1)^{th}$ decision making epoch. For this model, from Santhikumar and Sumita [7] we have the following results.

**Theorem - 4.2** The mean time to recruitment and its variance are given by

$$E(T_z) = E(Y/X<z)\left\{\frac{F_X(z)}{\bar{F}_X(z)}\right\} \tag{3}$$

$$V(T_z) = E(Y^2/X<z) + \left\{\frac{F_X(z)}{\bar{F}_X(z)}\right\} + [E(T_z)]^2 \tag{4}$$

**5.Special Case**

In this section we assume that the number X of people who leave the organization at any decision making epoch and the time Y between two consecutive decision making epochs are correlated and follow a Bivariate Poisson Distribution with parameter $a_1$, $a_2$ and $a_{12}$. Here $a_{12}$ refers the mean dependence rate between X and Y. In [5, 8, 9], the joint distribution of X and Y of a Bivariate Poisson distribution is given by

$$P_{XY}(x,y) = e^{-(a_1+a_2+a_{12})} \sum_{i=0}^{min(x, \quad y)} \frac{a_{12}^i \quad a_1^{x-i} \quad a_2^{y-j}}{i! \quad (x-i)! \quad (y-j)!}$$

where $a_{12} < \min(a_1, a_2)$, $a_1, a_2 > 0$ and $x, y = 0, 1, 2, \ldots$,..

Note that $E(X) = a_1 + a_{12}$, $E(Y) = a_2 + a_{12}$ and $Cov(X, Y) = a_{12}$

In [5,9], the marginal distributions are found to be

$$p_X(x) = e^{-(a_1+a_{12})} \frac{(a_1+a_{12})^x}{x \;!} \;; \quad x = 0, 1, 2, \ldots.. \tag{5}$$

$$p_Y(y) = e^{-(a_1+a_{12})} \frac{(a_1+a_{12})^y}{y \;!} \;; \quad y = 0, 1, 2, \ldots.. \tag{6}$$

Write

$$L_1 = a_1 + a_{12}; L_2 = \quad a_2 + a_{12} \quad ; \quad L_3 = L_2\,(L_2+1)\,; \quad K\,(z) = \sum_{x=0}^{z-1} \frac{L_1^x}{x!}$$

$$G_1\,(z) \quad = \quad \sum_{x=0}^{z-1} \frac{L_1^x}{x!} \left[a_2 + a_{12} \quad \frac{x}{L_1}\right] \frac{1}{K\,(z)}$$

$$G_2\,(z) \quad = \quad \sum_{x=0}^{z-1} \frac{[}{L_1^x} x \quad ! \left[a_2 + a_1 a_{12} \quad \frac{x}{L_1^2} + \left[a_2 + \frac{a_{12} x}{L_1}\right]^2\right] \frac{1}{K\,(z)}]$$

$$G_3\,(z) \quad = \quad e^{-L_1} \quad \left[K\,(z) + \frac{L_1^z}{z!}\right]$$

$$G_4\,(z) \quad = \quad 1 \quad - \quad G_3\,(z)$$

From Theorems 4.1 and 4.2 and using the above results, we find the following results for the special case.

**MODEL - I**

$$E\,(T_z) \quad = \quad \frac{L_2}{G_4\,(z)} \tag{7}$$

$$V\,(T_z) \quad = \quad \left[\frac{L_3}{G_4\,(z)}\right] + \left[\frac{L_2}{(G_4\,(z))^2}\right] \quad [2 * G_3\,(z) * G_1\,(z) \quad - \quad L_2] \tag{8}$$

**MODEL- II**

$$E\,(T_z) \quad = \quad G_1\,(z) \quad \left\{\frac{G_3\,(z)}{G_4\,(z)}\right\} \tag{9}$$

$$V\,(T_z) \quad = \quad G_2\,(z) \quad \left\{\frac{G_3\,(z)}{G_4\,(z)}\right\} \quad + \quad [E\,(T_z)]^2 \tag{10}$$

## 6. Numerical Illustration

### MODEL - I

$a_1$, $a_2$ and $a_{12}$ are kept fixed, variations are with respect to z.

**Table - I**

$\mathbf{a_1 = 0.5, a_2 = 0.6}$ **and** $\mathbf{a_{12} = 0.4}$

| Value of z | $\mathbf{E(T_z)}$ | $\mathbf{V(T_z)}$ |
|---|---|---|
| 1 | 4.40 | 7.38 |
| 2 | 15.91 | 118.28 |
| 3 | 74.30 | 3521.73 |
| 4 | 426.61 | 132025.00 |
| 5 | 2911.70 | 6393952.00 |

The parameters $a_1$, $a_2$ and $a_{12}$ are kept fixed and variations are with respect to z. As the value of z increases, the mean and variance of the time to recruitment increase. It is quite reasonable that as the value of z, the threshold value increases, the time to recruitment also increases and hence the result that the mean time to next recruitment also increases.

### MODEL - II

$a_1$, $a_2$ and $a_{12}$ are kept fixed, variations are with respect to z.

**Table - II**

$a_1 = 0.5$, $a_2 = 0.6$ and $a_{12} = 0.4$

| Value of z | $\mathbf{E(T_z)}$ | $\mathbf{V(T_z)}$ |
|---|---|---|
| 1 | 2.04 | 7.41 |
| 2 | 10.67 | 135.98 |
| 3 | 59.85 | 3717.33 |
| 4 | 367.04 | 135582.50 |
| 5 | 2552.83 | 6523004.00 |

The parameters $a_1$, $a_2$ and $a_{12}$ are kept fixed and variations are with respect to z. As the value of z increases, the mean and variance of the time to recruitment increase. It is quite reasonable that as the value of z, the threshold value increases, the time to recruitment also increases and hence the result that the mean time to next recruitment also increases.

From Tables I and II it is interesting to note that for each fixed z, the expected time to recruitment is greater for Model - I than for Model - II. This may be due to the different interpretation given for $Y_n$ in Models I and II.

**Models I and II**

$a_1$, $a_2$ and z are kept fixed, variations are with respect to $a_{12}$.

**Table - III**

$a_1 = 1$, $a_2 = 2$ and $z = 8$

| Value of $a_{12}$ | $E(T_z)$ for Model I | $E(T_z)$ for Model II |
|---|---|---|
| 0.45 | 114496.30 | 59039.16 |
| 0.50 | 90006.52 | 48039.61 |
| 0.55 | 71661.48 | 39548.04 |
| 0.60 | 57170.07 | 32587.29 |
| 0.65 | 46215.82 | 27179.02 |
| 0.70 | 37654.60 | 22821.78 |
| 0.75 | 30861.10 | 19255.82 |
| 0.80 | 25544.43 | 16391.00 |
| 0.85 | 21185.23 | 13965.33 |

We observe that the mean time to recruitment decreases as $a_{12}$ increases for both Model I and Model II.

**Acknowledgement**

The authors wish to express their gratefulness and thanks to UGC, India for its financial assistance during the preparation of this paper.

**References**

1. D.J. Bartholomew, Statistical Model for Social Process ($3^{rd}$ Ed.,), John Wiley and Sons, New York(1973).

2. D.J. Bartholomew, and Andrew F. Forbes Statistical Techniques for Manpower Planning, John Wiley and Sons, New York[1979].

3. J.D. Esary, A.W. Marshall and F. Prochan Ann. Probability, Vol.1, No.4, p 627-650 [1973].

4. R.C. Grinold, Operations Research, Vol.24, No.3, p. 387-399 [1976].

5. Normal L. Johnson and Samuel Kotz Discrete Distributions, John Wiley and Sons, New York [1969].

6. J.G. Santhikumar, and U. Sumita, J. Appl. Prob., 20, p.660-614[1983].

7. R. Sathiyamoorthy and R. Elangovan, Journal of Decision and Mathematika Sciences, Vol.3, No. 1-3[1998].

8. C.V.R.S. Vijayakumar and K. Srinivasa Rao, International Journal of Management and Systems, Vol. 16, No.2, p.157-168[ 2000].

9. Yuhan Shih Chow and Henry Teicher, Probability Theory, Narosa Publishing House, New Delhi [1979].

# An Orthogonally Additive Functional on a Space of Banach Lattice Valued Functions

Supama
Department of Mathematics, Gadjah Mada University
Yogyakarta 55281, Indonesia
Email:maspomo@yahoo.com

Soeparna Darmawijaya
Department of Mathematics, Gadjah Mada University
Yogyakarta 55281, Indonesia
Email:soeparna@astaga.com

**Abstract**

Let $m_L[a,b]$ be the set of all strongly measurable functions from $[a,b]$ into a Banach lattice $L$ and $\phi$ be an $N$-function defined on $L$ satisfying the $\Delta_2$-condition. For any real number $a > 0$, the space $W_\phi[a,\infty)$ will be defined as follows :

$$W_\phi[a,\infty) = \{f \in m_L[a,b]; \lim_{x\to\infty} \frac{1}{x}\int_a^x \phi(f(t))d\mu = 0\}$$

In this paper, we present a sufficient and necessary condition for wich the functional $F : W_\phi[a,\infty) \to \Re$ is ortoghonally additive and continuous.

# 1 Introduction

A representation for orthogonally additive functionals on sequence spaces have been given by Chew and Paredes in [1] and [4] respectively. In [1], Chew also gave a representation of those in function spaces. Supama and Soeparna [5,6] succesfully constructed a space such that the space in [1] and [4] are as the models. A representation for orthogonally additive functionals on those space are also given in [5]. By defining the (*)-condition, Supama and Soeparna [6] proved the representation theorem given in [5] by another approach.

The theory of Banach lattices and with linear operator defined on, or with values in Banach lattices were given in [3]. Then, Groenewegen [2] talked about Banach lattices valued function. The dual of any space and representable operators are also given in [2].

$\mathcal{N}$ and $\Re$ denote the set of all positive integers and real numbers system respectively. Let $L$ be a Banach lattice. A Banach lattice valued function $s : [a, b] \to L$ is called a *step function* if there exist measurable sets $A_1, A_2, ..., A_n \subset [a, b]$ and $\alpha_1, \alpha_2, ..., \alpha_n \in L$ such that :

$$s(x) = \sum_{i=1}^{n} \alpha_i \chi_{A_i}(x); x \in [a, b]$$

Further, a function $f : [a, b] \to L$ is said to be *strongly measurable* or *measurable* for the short, if there exists a sequence of step functions $s_n$ such that :

$$\lim_{n\to\infty} \|s_n(x) - f(x)\| = 0,$$

for almost all $x \in [a, b]$. A collection of all measurable functions from $[a, b]$ into $L$ will be denoted by $m_L[a, b]$. It is easily verified that $m_L[a, b]$ is a Riesz space.

**Example 1.1** *Let* $f : [0, 1] \to \Re$ *be a function defined by* $f(x) = x$. *For* $n \in \mathcal{N}$, *we define a function* $s_n : [0, 1] \to \Re$ *as follows :*

$$\begin{aligned} s_1(x) &= 1 \ , x \in [0, 1] \\ s_2(x) &= \begin{cases} \frac{1}{2} & , x \in [0, \frac{1}{2}] \\ 1 & , x \in [\frac{1}{2}, 1] \end{cases} \\ \ldots, &\ \text{etc.} \\ s_n(x) &= \begin{cases} \frac{1}{n} & , x \in [0, \frac{1}{n}] \\ \frac{2}{n} & , x \in [\frac{1}{n}, \frac{2}{n}] \\ & \ldots \\ \frac{n-1}{n} & , x \in [\frac{n-2}{n}, \frac{n-1}{n}] \\ 1 & , x \in [\frac{n-1}{n}, 1] \end{cases} \end{aligned}$$

*It is clearly that* $s_n$ *is a step function for every* $n \in \mathcal{N}$ *and :* $\lim_{n\to\infty} \|s_n(x) - f(x)\| = 0$.

*Hence,* $f : [0, 1] \to \Re$ *is measurable.*

Let $s : [a,b] \to L$ be a step function. It means there exists a collection of measurable sets $A_1, A_2, ..., A_n \subset [a,b]$ and $\alpha_1, \alpha_2, ..., \alpha_n \in L$ such that :

$$s(x) = \sum_{i=1}^{n} \alpha_i \chi_{A_i}(x); x \in [a,b]$$

For any measurableset $A \subset [a,b]$ we define the integral of $s$ on $A$ as follows :

$$\int_A s(x)d\mu = \sum_{i=1}^{n} \mu(A_i \cap A)\alpha_i$$

It is easily verified that this definition does not depend on the representation chosen for the step function $s$. Now we define an integrable function as follows.

**Definition 1.2** *Let $A \subset [a,b]$ be a measurable set. A measurable function $f : [a,b] \to L$ is said to be integrable on $A$ if there exists $\alpha \in L$ such that for every sequence of step functions $s_n, s_n : [a,b] \to L$, with : $\lim_{n\to\infty} \|s_n(x) - f(x)\| = 0$ implies :*

$$\lim_{n\to\infty} \int s_n(x)d\mu = \alpha$$

*Further, the integral of $f$ on $A$ is given by :*

$$\int_A f(x)d\mu = \alpha.$$

**Example 1.3** *Let $f : [0,1] \to \Re$ be a function defined by $f(x) = x$ and $s_n$ be a sequence of step functions as given in example* 1.1. *Since* $\lim_{n\to\infty} \|s_n(x) - f(x)\| = 0$ *and :*

$$\begin{aligned} \lim_{n\to\infty} \int_0^1 s_n(x)d\mu &= \lim_{n\to\infty} \frac{1}{n}(\frac{1}{n} + \frac{2}{n} + ... + \frac{n-1}{n} + 1) \\ &= \lim_{n\to\infty} (\frac{1}{2} + \frac{1}{2n^2}) = \frac{1}{2}, \end{aligned}$$

*then $f$ is integrable on $[0,1]$ and :*

$$\int_0^1 f(x)d\mu = \lim_{n\to\infty} \int_0^1 s_n(x)d\mu = \frac{1}{2}$$

Let $\Im$ denotes the set of all $f \in m_L[a,b]$ which is integrable on a measurable set $A \subset [a,b]$. It is easy to prove that $\Im$ is a linear space over $\Re$. Further, for any $f \in m_L[a,b]$ and $x \in [a,b]$, we define the truncated function $f_x$ as follows :

$$f_x(t) = \begin{cases} f(t) & ,t \le x \\ 0 & ,t > x \end{cases}$$

It is clearly that $f_x \in m_L[a,b]$ for every $f \in m_L[a,b]$ and $x \in [a,b]$. Arbitrary two functions $f, g \in m_L[a,b]$ is said to be *orthogonal (disjoint)*, denoted by $f \perp g$, if $|f(x)| \wedge |g(x)| = 0$ for almost all $x \in [a,b]$. Furthermore, an operator $F : m_L[a,b] \to m_L[a,b]$ is said to be *orthogonally additive* if for every $f, g \in m_L[a,b]$, $f \perp g$, we have :

$$F(f+g) = F(f) + F(g)$$

**Definition 1.4** *Let $X \subset m_L[a,b]$ be a linear space over $\Re$. A non-negative function $\|.\| : X \to [0,\infty)$ is called an F-norm iff :*

*i.* $\|f\| = 0 \Leftrightarrow f = 0$ *a.e on* $[a,b]$

*ii.* $\|-f\| = \|f\|$ *for every* $f \in X$

*iii.* $\|f+g\| \le \|f\| + \|g\|$ *for every* $f, g \in X$

*iv. If* $a_n \to a$ *in* $\Re$ *and* $f_n \to f$ *in* $X$ *then* $a_n f_n \to af$

*Further, $(X, \|.\|)$ is called an F-normed Space.*

Let $(X, \|.\|)$ be an $F$-normed space. If the function $d : X \times X \to \Re$ is defined by $d(f,g) = \|f-g\|$ for any $f, g \in X$, then $d$ is a metric on $X$. If $X$ is complete then it is called a *Frechet Space* or an *F-space* for the short.

**Definition 1.5** *An F-space X is called an FK-space if the canonical mapping $p_x : X \to L$,*

$$p_x(f) = f(x),$$

*is continuous for every $x \in [a,b]$.*

**Definition 1.6** *An F-space $X$ is called an AK-space if $f_x, \chi_A \in X$ for every $f \in X$, $x \in [a,b]$ and measurable set $A \subset [a,b]$and $\lim_{x \to b} \|f_x - f\| = 0$. The AK-space $X \subset m_L[a,\infty]$ is defined similarly. An F-space $X \subset m_L[a,\infty]$ is called an AK-space if $f_x, \chi_A \in X$ for every $f \in X$, $x \in [a,b]$ and measurable set $A \subset [a,\infty]$ with $\mu(A) < \infty$ and $\lim_{x\to\infty} \|f_x - f\| = 0$.*

# 2 Main Results

We begin by proving the following lemma which are required to prove the next theorems.

**Lemma 2.1** *Let $X$ be an FK-space satisfying $f_x \in X$ for every $f \in X$ and $x \in [a,b]$. If $f \in X$ and $x \in [a,b]$ then for every real number $\epsilon > 0$ there exists $\delta > 0$ such that for every $g \in X$ we have :*

$$\|f(x) - g(x)\| < \epsilon,$$

*whenever $\|f_x - g_x\| < \delta$.*

**Proof :** Let any $f \in X$ and $x \in [a,b]$ be given. Since $X$ is an $FK$-space then the canonical mapping $p_x : X \to L$, where :

$$p_x(g) = g(x) \quad , g \in X$$

is continuous at $f$. It means for any $\epsilon > 0$ there exists $\delta > 0$ such that for every $g \in X$ with $\|f - g\| < \delta$ implies $\|p_x(f) - p_x(g)\| < \epsilon$. Especially, if $\|f_x - g_x\| < \delta$ then we have :

$$\|f(x) - g(x)\| = \|p_x(f_x) - p_x(g_x)\| < \epsilon.$$

These completes the proof. □

Now, we assume that the Banach lattice $L$ is order complete.In the next theorem, we give sufficient conditions for which the operator $F : X \to Y$ is orthogonally additive and continuous.

**Theorem 2.2** *Let $(X, \|.\|)$ be an FK-space such that $f_x \in X$ and $\|f_x\| \le \|f\|$ for every $x \in [a,b]$ and $f \in X$. Suppose $Y \subset m_L[a,b]$ is an F-space respect with to the norm :*

$$\|h\| = \sup\{\|h(t)\|; t \in [a,b]\} \quad , h \in Y$$

*If the function $g(x,t) : [a,b] \times \Re \to Y$ satisfies the condition $g(x,0) = 0$ and $g(x,.)$ is continuous on $\Re$ for every $x \in [a,b]$ then the operator $F : X \to Y$,*

$$F(f)(x) = \int_a^x g(t, f(t))d\mu \quad , f \in X, x \in [a,b],$$

*is orthogonally additive and continuous on $X$.*

**Proof :** It is clear that $F$ is orthogonally additive. It follows from the condition $g(x,0) = 0$ for all $x \in [a,b]$.
Now, we are going to prove that $F$ is continuous on $[a,b]$. If we suppose the contrary then there exist $f \in X$ and $\epsilon > 0$ such that for every $\delta > 0$ there exists $k \in X$ such that $\|k - f\| < \delta$ and :

$$\sup_{x \in [a,b]} \| \int_a^x g(t, k(t))d\mu - \int_a^x g(t, f(t))d\mu\| \ge \epsilon.$$

For any $\delta_1 \le \frac{1}{2}$, there exists $k_1$ such that $\|k_1 - f\| < \delta_1$ and :

$$\sup_{x \in [a,b]} \| \int_a^x g(t, k_1(t))d\mu - \int_a^x g(t, f(t))d\mu\| \ge \epsilon.$$

So, we can choose $y_1 \in [a,b]$ such that :

$$\| \int_a^{y_1} g(t, k_1(t))d\mu - \int_a^{y_1} g(t, f(t))d\mu\| \ge \frac{\epsilon}{2}.$$

Further, the continuity of $g(x,.)$ implies that there exists $\delta > 0$ such that for every $w \in X$ with $\|w(t) - f(t)\| < \delta$ we have :

$$\|g(t, w(t)) - g(t, f(t))\| < \frac{\epsilon}{4|y_1 - a| + 1}$$

By Lemma 2.1 there exists $\delta_2 \le \frac{1}{2^2}$ such that for every $t \in [a, y_1]$ we have :

$$\|w(t) - f(t)\| < \delta,$$

whenever $\|w_{y_1} - f_{y_1}\| \leq \delta_2$. And hence,

$$\begin{aligned}\|\int_a^{y_1} g(t,w(t))d\mu - \int_a^{y_1} g(t,f(t))d\mu\| &\leq \int_a^{y_1} \|g(t,w(t)) - g(t,f(t))\|d\mu \\ &< \frac{\epsilon}{4|y_1 - a| + 1}\int_a^{y_1} d\mu < \frac{\epsilon}{4}\end{aligned}$$

On the other hand, there exists $k_2 \in X$ such that $\|k_2 - f\| < \delta_2$ and :

$$\sup_{x\in[a,b]} \|\int_a^x g(t,k_2(t))d\mu - \int_a^x g(t,f(t))d\mu\| \geq \epsilon.$$

By letting $w = k_2$ then we have :

$$\|\int_a^{y_1} g(t,k_2(t))d\mu - \int_a^{y_1} g(t,f(t))d\mu\| < \frac{\epsilon}{4}$$

Hence,

$$\|\int_{y_1}^{b} g(t,k_2(t))d\mu - \int_{y_1}^{b} g(t,f(t))d\mu\| \geq \frac{3\epsilon}{4}$$

Further, we can choose $y_2 \in [y_1, b]$ such that :

$$\|\int_{y_1}^{y_2} g(t,k_2(t))d\mu - \int_{y_1}^{y_2} g(t,f(t))d\mu\| \geq \frac{\epsilon}{2}$$

Since $g(x,.)$ is continuous on $\Re$ for every $x \in [a,b]$ then there exists $\delta' > 0$ such that for every $t \in [y_1, y_2]$ we have :

$$\|g(t,w(t)) - g(t,f(t))\| < \frac{\epsilon}{4|y_2 - y_1| + 1},$$

whenever $\|w(t) - f(t)\| < \delta'$. By Lemma 2.1 there exists $\delta_3 \leq \frac{1}{2^3}$ such that for every $t \in [y_1, y_2]$,

$$\|w(t) - f(t)\| < \delta',$$

whenever $\|w_{y_2} - f_{y_2}\| \leq \delta_3$. Thus, there exists $\delta_3 \leq \frac{1}{2^3}$ such that :

$$\begin{aligned}\|\int_{y_1}^{y_2} g(t,w(t))d\mu - \int_{y_1}^{y_2} g(t,f(t))d\mu\| &\leq \int_{y_1}^{y_2} \|g(t,w(t)) - g(t,f(t))\|d\mu \\ &< \frac{\epsilon}{4|y_2 - y_1| + 1}\int_{y_1}^{y_2} d\mu < \frac{\epsilon}{4}\end{aligned}$$

On the other hand, the discontinuity of $F$ implies that there exists $k_3 \in X$ such that $\|k_3 - f\| < \delta_3$ and :

$$\sup_{x\in[a,b]} \|\int_a^x g(t,k_3(t))d\mu - \int_a^x g(t,f(t))d\mu \geq \epsilon.$$

By choosing $w = k_3$ we have :

$$\| \int_a^{y_2} g(t, k_3(t))d\mu - \int_a^{y_2} g(t, f(t))d\mu \| < \frac{\epsilon}{4}.$$

Hence,

$$\| \int_{y_2}^{b} g(t, k_3(t))d\mu - \int_{y_2}^{b} g(t, f(t))d\mu \| \geq \frac{3\epsilon}{4}.$$

So, we can choose $y_3 \in [y_2, b]$ such that :

$$\| \int_{y_2}^{y_3} g(t, k_3(t))d\mu - \int_{y_2}^{y_3} g(t, f(t))d\mu \| \geq \frac{\epsilon}{2}.$$

Since $g(x, .)$ is continuous on $\Re$ for every $x \in [a, b]$ then there exists $\delta'' > 0$ such that :

$$\|g(t, w(t)) - g(t, f(t))\| < \frac{\epsilon}{4|y_3 - y_2| + 1},$$

whenever $\|w(t) - \|f(t) < \delta''$. Moreover, by Lemma 2.1 there exists $\delta_4 \leq \frac{1}{2^4}$ such that for every $t \in [y_2, y_3]$ we have :

$$\|w(t) - f(t)\| < \delta'',$$

whenever $\|w_{y_3} - f_{y_3}\| \leq \delta_4$. So, if $\|w_{y_3} - f_{y_3}\| \leq \delta_4$ we have :

$$\begin{aligned} \| \int_{y_2}^{y_3} g(t, w(t))d\mu - \int_{y_2}^{y_3} g(t, f(t))d\mu \| &\leq \int_{y_2}^{y_3} \|g(t, w(t)) - g(t, f(t))\|d\mu \\ &< \frac{\epsilon}{4|y_3 - y_2| + 1} \int_{y_1}^{y_2} d\mu < \frac{\epsilon}{4} \end{aligned}$$

etc. If we continue the process then we'll get a sequence $k_n \subset X$, a sequence $\delta_n \subset \Re$ with $\delta_n \leq \frac{1}{2^n}$, and an increasing sequence $y_n \subset [a, b]$ such that :

$$\begin{aligned} \|(k_{n,y_n} - k_{n,y_{n-1}}) - (f_{y_n} - f_{y_{n-1}})\| &\leq \|k_{n,y_n} - f_{y_n}\| + \|k_{n,y_{n-1}} - f_{y_{n-1}}\| \\ &\leq 2\|k_n - f\| < 2\delta_n \leq \frac{1}{2^{n-1}} \end{aligned}$$

where :

$$k_{n,y_n}(t) = \begin{cases} k_n(t) & , t \leq y_n \\ 0 & , t > y_n \end{cases}$$

for $n = 1, 2, 3, \ldots$, and :

$$\| \int_{y_{n-1}}^{y_n} g(t, k_n(t))d\mu - \int_{y_{n-1}}^{y_n} g(t, f(t))d\mu \| \geq \frac{\epsilon}{2},$$

for $n = 1, 2, 3, \ldots$, where $y_0 = a$. Suppose :

$$h_n = \sum_{i=1}^{n}(k_{i,y_i} - k_{i,y_{i-1}}) - f_{y_n},$$

where $k_{1,y_0} = 0$. Then for every $n \geq m$ we have :

$$\begin{aligned}
\|h_n - h_m\| &= \| \sum_{i=m+1}^{n} (k_{i,y_i} - k_{i,y_{i-1}}) - (f_{y_n} - f_{y_m})\| \\
&= \| \sum_{i=m+1}^{n} \{(k_{i,y_i} - k_{i,y_{i-1}}) - (f_{y_i} - f_{y_{i-1}})\}\| \\
&\leq \sum_{i=m+1}^{n} \{(k_{i,y_i} - f_{y_i}) - (k_{i,y_{i-1}} - f_{y_{i-1}})\} \\
&\leq 2 \sum_{i=m+1}^{n} \|k_i - f\| < 2 \sum_{i=m+1}^{n} \delta_i = 2 \sum_{i=m+1}^{n} \frac{1}{2^{i-1}} \\
&< 2 \sum_{i=m+1}^{\infty} \frac{1}{2^{i-1}} = \frac{1}{2^{m-1}} \to 0,
\end{aligned}$$

whenever $m \to \infty$. Thus, $h_n$ is a Cauchy sequence in a complete space $X$. Hence, there exists $h \in X$ such that $\lim_{n\to\infty} \|h_n - h\| = 0$. Since $X$ is an $FK$-space and by Lemma 2.1 we have :

$$\lim_{n\to\infty} \|h_n(t) - h(t)\| = 0,$$

for every $t \in [a, b]$. It follows that $h = k - f$, where :

$$k = k_{1,y_1} + \sum_{n=2}^{\infty} k_{n,y_{n-1},y_n},$$

$$k_{n,y_{n-1},y_n}(t) = \begin{cases} k_n(t) & , y_{n-1} < t \leq y_n \\ 0 & , \text{anyelse} \end{cases}$$

So, $k = (h + f) \in X$. It means that $\int_a^b g(t, k(t))d\mu = \sum_{n=1}^{\infty} \int_{y_{n-1}}^{y_n} g(t, k_n(t))d\mu$ exists. And hence,

$$\sum_{n=1}^{\infty} \int_{y_{n-1}}^{y_n} g(t, k_n(t)d\mu - \sum_{n=1}^{\infty} \int_{y_{n-1}}^{y_n} g(t, f(t)d\mu =$$
$$\sum_{n=1}^{\infty} \{\int_{y_{n-1}}^{y_n} g(t, k_n(t)d\mu - \int_{y_{n-1}}^{y_n} g(t, f(t)d\mu\}$$

exists, i.e. there exists $N \in \mathcal{N}$ such that :

$$\| \int_{y_{N-1}}^{y_N} g(t, k_N(t))d\mu - \int_{y_{N-1}}^{y_N} g(t, f(t))d\mu \| < \frac{\epsilon}{2}$$

This leads to contradiction. These completes the proof of the theorem.□

An $F$-normed space $X \subset m_L[a, b]$ is said to be *solid* if for every $f \in X$ and $g \in m_L[a, b]$ with $|g| \leq |f|$ a.e. on $[a, b]$ implies $g \in X$. Then we have :

**Theorem 2.3** *Let $X$ be a solid FK- and AK-space such that $\|f_x\| \leq \|f\|$ for every $x \in [a, b]$ and $f \in X$. Suppose $Y \subset m_L[a, b]$ is an F-space respect with to the norm :*

$$\|h\| = \sup\{\|h(t)\|; t \in [a, b]\} \quad , h \in Y$$

*The operator $F : X \to Y$ is orthogonally additive and continuous on $X$ if and only if there exists a function $g(x, t) : [a, b] \times \Re \to Y$ satisfying the condition $g(x, 0) = 0$ and $g(x, .)$ is continuous on $\Re$ for every $x \in [a, b]$ such that,*

$$F(f)(x) = \int_a^x g(t, f(t))d\mu(t)$$

*exists for every $f \in X$*

**Proof :** The necissity follows from Theorem 2.2.
For the sufficient, let $f \in X$ and $x, z \in [a, b]$ such that $z \leq x$. Denote $[a, z] = \cup_{i=1}^n E_i$, where $E_i$ is measurable, $\mu(E_i) \neq 0$ and $\mu(E_i \cap E_j) = 0$ for every $i \neq j$, $i, j = 1, 2, \ldots, n$. Then $f_z = \sum_{i=1}^n f\chi_{E_i}$. Since $F$ is orthogonally additive, then we have :

$$F(f_z) = F(\sum_{i=1}^n f\chi_{E_i}) = \sum_{i=1}^n F(f\chi_{E_i})$$

Further, let $\epsilon > 0$ be given. Since $f$ is measurable on $E_i$, for $i = 1, 2, \ldots, n$, then there exists a step function $s$ on $E_i$ such that :

$$\|f(t) - s(t)\| < \epsilon, \text{ for every } t \in E_i.$$

It means that for any $i = 1, 2, \ldots, n$ there exists $\delta' > 0$ and $\alpha_i \in E_i$ such that for every $t \in E_i$ we have :

$$\|f(t) - \alpha_i\| < \epsilon,$$

whenever $\mu(E_i) < \delta'$. Further, the function $g(x,t) : [a,b] \times \Re \rightarrow Y$ is defined such that :

$$\begin{aligned} &i. \quad g(t,0) = 0 \text{ for every } t \in [a,b] \\ &ii. \quad g(t_i,\alpha_i) = \frac{1}{\mu(E_i)} F(f\chi_{E_i}) \text{ for every } t_i \in E_i \end{aligned}$$

If we let $max\mu(E_i) \rightarrow 0$ then we have :

$$\begin{aligned} F(f_z) &= \sum_{i=1}^{n} F(f\chi_{E_i}) = \sum_{i=1}^{n} g(t_i,\alpha_i)\mu(E_i) \\ &= \lim_{max\mu(E_i)\rightarrow 0} \{\sum_{i=1}^{n} g(t_i, f(t_i))\mu(E_i)\} \\ &= \int_a^z g(t,f(t))d\mu \end{aligned}$$

Further, since $F$ is continuous and $X$ is an $AK$-space then :

$$F(f_x) = \lim_{z\rightarrow x} F(f_z) = \int_a^x g(t,f(t))d\mu$$

It can easily be proved that for every $x \in [a,b]$ the sets $\{F(f_x); f \in X, x \in [a,b]\}$ and $\{F(f)(x); f \in X, x \in [a,b]\}$ are isomorphic. It is followed by :

$$F(f)(x) = \int_a^x g(t,f(t))d\mu$$

Now we are going to prove that $g(t,.)$ is continuous on $\Re$ for every $t \in [a,b]$. Let $\{c_n\}$ be a sequence on $\Re$ such that $c_n \rightarrow c$ for some $c \in \Re$. Since $X$ is an $F$-space and $\chi_A \in X$ for every measurable set $A \subset [a,b]$ then for every $i = 1,2,\ldots,n$ :

$$\|c_m\chi_{E_i} - c\chi_{E_i}\| \rightarrow 0,$$

for large enough $m$. And the continuity of $F$ implies :

$$\|F(c_m\chi_{E_i}) - F(c\chi_{E_i})\| \rightarrow 0,$$

whenever $m \rightarrow \infty$. So, for every $t \in [a,b]$ we have :

$$\|g(t,c_m) - g(t,c)\| \rightarrow 0,$$

whenever $m \to \infty$. These complete the proof.□

Let $L$ be a Banach lattice. We write $L_+ = \{x \in L; x \geq 0\}$. A function $\phi : L \to \Re$ is called an $N-$ *function* if it satisfies the following conditions :

i. $\phi(x) = 0 \Longleftrightarrow x = 0$

ii. $\phi$ is increasing on $L_+$

iii. $\phi$ is continuous on $L$

iv. $\phi(-x) = \phi(x)$ for every $x \in L$

The $N$-function $\phi$ is said to satisfy the $\Delta_2 -$ *condition* if there exists a real number $M > 0$ such that : $\phi(2x) \leq M\phi(x)$ for all $x \in L_+$.

Let $\phi$ be an $N$-function satisfying the $\Delta_2$-condition. For any positive real number $a$, we define the modulared function space $W_\phi[a, \infty)$ as follows :

$$W_\phi[a, \infty) = \{f \in m_L[a, \infty); \quad \lim_{x \to \infty} \frac{1}{x} \int_a^x \phi(f(t))d\mu = 0\}$$

where the modulars $\rho_x$ and $\rho$ are defined as follows :

$$\begin{aligned} \rho_x(f) &= \frac{1}{x} \int_a^x \phi(f(t))d\mu \\ \rho(f) &= sup\{\rho_x(f); \quad x \in [a, \infty)\} \end{aligned}$$

It can easily be poved that $W_\phi[a, \infty)$ is a solid, $FK$- and $AK$-space respect with to the $F$-norm :

$$\|f\| = inf\{\epsilon > 0; \quad \rho(\frac{f}{\epsilon}) \leq \epsilon\}$$

and satisfies all conditions required in Theorem 2.3. We've also known that $\Re$ is a Banach lattice with the norm $|.|$ (absolute value). Hence, we have :

**Corollary 2.4** *The functional $F : W_\phi[a, \infty) \to \Re$ is orthogonally additive and continuous on $W_\phi[a, \infty)$ if and only if there exists a function $g(x, t) : [a, b] \times \Re \to \Re$ satisfying the condition $g(x, 0) = 0$ and $g(x, .)$ is continuous on $\Re$ for every $x \in [a, b]$, such that :*

$$F(f) = \int_a^\infty g(t, f(t))d\mu$$

*exists for every $f \in W_\phi[a, \infty)$.*

# References


[1] Chew Tuan Seng, *Orthogonally Additive Functionals*, 1985, Ph.D disertation, National University of Singapore.

[2] Groenewegen, G.L.M., *On Space of Banach Lattice Valued Functions and Measures*, 1982, Ph.D disertation, Nijmegen.

[3] Meyer, P. and Nieberg, *Banach Lattices*, 1991, Springer-Verlag.

[4] Paredes, L.I., *Orthogonally Additive Functionals and Superposition Operators on* $w_0(\Phi)$, 1993, Ph.D disertation, University of the Philippines.

[5] Supama and Soeparna D., *Orthogonally Additive Functionals on the Modulared Function Space* $W_0(\phi)$, 1998, Berkala Penelitian MIPA Th. VIII no.1, Gadjah Mada University.

[6] Supama and Soeparna D., *Orthogonally Additive Functionals on* $W_0^0(\phi)$, 1999, Journal of Indonesian Mathematical Society, Vol.5 no.2.

# BOUNDS ON DOUBLING RATES OF INVESTMENT CAPITAL DUE TO INCORRECT DISTRIBUTIONS

CHOON PENG TAN

*Institute of Mathematical Sciences, University of Malaya, 50603 Kuala Lumpur, Malaysia*
*E-mail: tcp@mnt.math.um.edu.my*

We consider an investment in a market of $m$ stocks using an optimal portfolio strategy based on estimating the probability distribution of the stock-price vector. Due to inaccuracies in estimating the stock-price distribution, it is not possible to achieve the optimal doubling rate of the investment capital. Barron and Cover (1988) obtained a bound on the difference in doubling rates in the form of the relative entropy of the corresponding distributions. We shall consider distributions belonging to the exponential class and determine the relative entropy, including the relative entropy of multivariate normal distributions.

## 1 Introduction

Consider a market of $m$ stocks where the $m$-dimensional stock-price vector $\mathbf{X}$ has components $X_i$, $i = 1, 2, \ldots, m$ and $X_i$ is the ratio of the price of the $i$th stock at the end of a trading day to the price at the beginning. Let $\mathbf{X}_1, \mathbf{X}_2, \ldots, \mathbf{X}_n$ be a sequence of stock-price vectors corresponding to $n$ trading days and a common investment portfolio $\mathbf{b} = (b_i)$ is used each trading day, where $b_i \geq 0$ for $i = 1, 2, \ldots, m$ and $\sum_{i=1}^{m} b_i = 1$. Assuming an initial investment capita of 1 unit, the capital after $n$ trading days $S_n$ is given by $S_n = \prod_{i=1}^{n} \mathbf{b}^t \mathbf{X}_i$. If $\mathbf{X}_1, \mathbf{X}_2, \ldots, \mathbf{X}_n$ are independent and identically distributed random vectors, then

$$\frac{1}{n} \log S_n \to W(\mathbf{b}) = E[\log \mathbf{b}^t \mathbf{X}]$$

with probability 1 as $n \to \infty$. Thus $S_n$ behaves asymptotically as $e^{nW(\mathbf{b})}$, where logarithms are to base $e$ and $W(\mathbf{b})$ is called the doubling rate of investment capital. The log-optimum portfolio vector $\mathbf{b}^*$ maximizes $W(\mathbf{b})$ over all possible portfolio vectors $\mathbf{b}$. Cover (1984) gives an algorithm that generates a sequence of vectors $\mathbf{b}^{(n)}$ that converges to $\mathbf{b}^*$ as $n \to \infty$. Barron and Cover (1988) gives a bound on the difference in the doubling rates $W(\mathbf{b}^*)$ caused by the error in estimating the true distribution $f(\mathbf{x})$ of $\mathbf{X}$ by $g(\mathbf{x})$. To be precise, let $\mathbf{X}_1, \mathbf{X}_2, \ldots, \mathbf{X}_n$ be a sequence of independent, identically distributed stock-price vectors from a distribution with probability density function $f(\mathbf{x})$. Let $\mathbf{b}_f^*$ be the log-optimum portfolio corresponding to $f(\mathbf{x})$ and $\mathbf{b}_g^*$ be the log-optimum portfolio corresponding to a probability density function $g(\mathbf{x})$. Then the difference in the doubling rates $W(\mathbf{b}^*)$ is bounded by $D(f \parallel g)$, i.e.

$$\Delta W = W(\mathbf{b}_f^*) - W(\mathbf{b}_g^*) \leq D(f \parallel g) \tag{1}$$

where

$$D(f \parallel g) = \int f(\mathbf{x}) \log \frac{f(\mathbf{x})}{g(\mathbf{x})}\, d\mathbf{x} \geq 0 \tag{2}$$

is the relative entropy of $f(\mathbf{x})$ and $g(\mathbf{x})$. We shall determine the relative entropy of two multivariate normal distributions and the relative entropy of two univariate distributions from the exponential class. We also assume that the distribution of the stock-price vector $\mathbf{X} \geq \mathbf{0}$ can be approximated by a multivariate normal distribution or a product of univariate distributions from the exponential class.

## 2 Main Results

First, we have the following result:

**Theorem 1.** *Let the stock-price vector $\mathbf{X} = (X_1, \ldots, X_m)$ have the joint distribution $N(\boldsymbol{\mu}_1, K_1)$ with p.d.f.*

$$f(\mathbf{x}) = \frac{1}{(\sqrt{2\pi})^m |K_1|^{1/2}}\, e^{-1/2(\mathbf{x}-\boldsymbol{\mu}_1)^t K_1^{-1}(\mathbf{x}-\boldsymbol{\mu}_1)}$$

*where the mean vector $\boldsymbol{\mu}_1 = (\mu_{11}, \ldots, \mu_{m1})^t$ and $K_1$ is the covariance matrix (with the notation $|A|$ for the determinant of $A$). If $f(\mathbf{x})$ is wrongly estimated as $g(\mathbf{x})$, where $g(\mathbf{x})$ is the p.d.f. of $N(\boldsymbol{\mu}_2, K_2)$, i.e.*

$$g(\mathbf{x}) = \frac{1}{(\sqrt{2\pi})^m |K_2|^{1/2}}\, e^{-1/2(\mathbf{x}-\boldsymbol{\mu}_2)^t K_2^{-1}(\mathbf{x}-\boldsymbol{\mu}_2)}$$

*with mean $\boldsymbol{\mu}_2 = (\mu_{12}, \ldots, \mu_{m2})^t$ and covariance matrix $K_2$, then the bound on the difference in the doubling rates of the capital is:*

$$D(f \parallel g) = \frac{1}{2} \log \left[ e^{-m} \frac{|K_2|}{|K_1|} \right] + \frac{1}{2} (\log e) tr(L^{(21)} K_2^{-1})$$

*where $tr(\cdot)$ denotes the trace of a matrix and $L^{(21)}$ is the interaction matrix between $\boldsymbol{\mu}_2$ and $\boldsymbol{\mu}_1$ with the $(i,j)$ element defined by:*

$$L_{ij}^{(21)} = \int (x_i - \mu_{i2})(x_j - \mu_{j2}) f(\mathbf{x}) d\mathbf{x}$$

*for* $i, j = 1, 2, \ldots, m$. *Furthermore,*

$$L^{(21)} = K_1 + (\boldsymbol{\mu}_2 - \boldsymbol{\mu}_1)(\boldsymbol{\mu}_2 - \boldsymbol{\mu}_1)^t.$$

**Proof.** First, assume that logarithms are to base $e$, we have

$$\begin{aligned} D(f \parallel g) &= \int f(\mathbf{x}) \ln f(\mathbf{x}) d\mathbf{x} - \int f(\mathbf{x}) \ln g(\mathbf{x}) d\mathbf{x} \\ &= -h(f) + \frac{1}{2} \ln[(2\pi)^m |K_2|] \\ &\quad + \frac{1}{2} \int (\mathbf{x} - \boldsymbol{\mu}_2)^t K_2^{-1} (\mathbf{x} - \boldsymbol{\mu}_2) f(\mathbf{x}) d\mathbf{x} \end{aligned} \tag{3}$$

where

$$h(f) = -\int f(\mathbf{x}) \ln f(\mathbf{x}) d\mathbf{x} = \frac{1}{2} \ln[(2\pi e)^m |K_1|] \tag{4}$$

is known (see Cover and Thomas, 1991). From (3) and (4),

$$D(f \parallel g) = \frac{1}{2} \ln \left[ e^{-m} \frac{|K_2|}{|K_1|} \right] + \frac{1}{2} \int (\mathbf{x} - \boldsymbol{\mu}_2)^t K_2^{-1} (\mathbf{x} - \boldsymbol{\mu}_2) f(\mathbf{x}) d\mathbf{x}. \tag{5}$$

Now,

$$\begin{aligned} &\int (\mathbf{x} - \boldsymbol{\mu}_2)^t K_2^{-1} (\mathbf{x} - \boldsymbol{\mu}_2) f(\mathbf{x}) d\mathbf{x} \\ &= \int \left[ \sum_{i,j} (x_i - \mu_{i2})(K_2^{-1})_{ij} (x_j - \mu_{j2}) \right] f(\mathbf{x}) d\mathbf{x} \\ &= \sum_{i,j} \left[ \int (x_j - \mu_{j2})(x_i - \mu_{i2}) f(\mathbf{x}) d\mathbf{x} \right] (K_2^{-1})_{ij} \\ &= \sum_{i,j} L_{ji}^{(21)} (K_2^{-1})_{ij} \\ &= \operatorname{tr}(L^{(21)} K_2^{-1}). \end{aligned} \tag{6}$$

From (5) and (6), we have

$$D(f \parallel g) = \frac{1}{2} \log \left[ e^{-m} \frac{|K_2|}{|K_1|} \right] + \frac{1}{2} (\log e) \operatorname{tr}(L^{(21)} K_2^{-1}) \tag{7}$$

for logarithms to any base. Now

$$\begin{aligned}L_{ij}^{(21)} &= \int (x_i - \mu_{i2})(x_j - \mu_{j2}) f(\mathbf{x}) d\mathbf{x} \\ &= \int [(x_i - \mu_{i1}) - (\mu_{i2} - \mu_{i1})][(x_j - \mu_{j1}) - (\mu_{j2} - \mu_{j1})] f(\mathbf{x}) d\mathbf{x} \\ &= E_{\boldsymbol{\mu}_1, K_1}[(X_i - \mu_{i1})(X_j - \mu_{j1})] + (\mu_{i2} - \mu_{i1})(\mu_{j2} - \mu_{j1}) \\ &= (K_1)_{ij} + (\mu_{i2} - \mu_{i1})(\mu_{j2} - \mu_{j1}).\end{aligned}$$

Hence,

$$L^{(21)} = K_1 + (\boldsymbol{\mu}_2 - \boldsymbol{\mu}_1)(\boldsymbol{\mu}_2 - \boldsymbol{\mu}_1)^t. \quad \square \tag{8}$$

Next, we consider the stock price $X$ having a probability density function $f(x;\boldsymbol{\theta})$ belonging to a regular exponential class. The domain of $f(x;\boldsymbol{\theta})$ does not depend on the parameter $\boldsymbol{\theta}$ and hence the order of differentiation with respect to $\theta$ and integration with respect to $x$ can be interchanged. In the first case, we assume that the parameter $\boldsymbol{\theta} = (\theta_1, \ldots, \theta_k)$ is wrongly estimated as $\boldsymbol{\theta}' = (\theta_1', \ldots, \theta_k')$.

**Theorem 2.** *Let the $m$ stock prices $X_1, X_2, \ldots, X_m$ be mutually independent with common p.d.f.*

$$f(x;\boldsymbol{\theta}) = a(\boldsymbol{\theta}) b(x) e^{\sum_{j=1}^{k} c_j(\boldsymbol{\theta}) d_j(x)}$$

*belonging to a regular exponential class, where $\boldsymbol{\theta} = (\theta_1, \ldots, \theta_k)$. If $f(x;\boldsymbol{\theta})$ is wrongly estimated as $f(x;\boldsymbol{\theta}')$ where $\boldsymbol{\theta}' = (\theta_1', \ldots, \theta_k')$, then the bound on the difference in the doubling rates of the capital is $mD(f(x;\boldsymbol{\theta}) \parallel f(x;\boldsymbol{\theta}'))$ where*

$$D(f(x;\boldsymbol{\theta}) \parallel f(x;\boldsymbol{\theta}')) = \log\left[\frac{a(\boldsymbol{\theta})}{a(\boldsymbol{\theta}')}\right] - \frac{(\log e)}{a(\boldsymbol{\theta})}[\nabla a(\boldsymbol{\theta})] G(\boldsymbol{\theta})^{-1}[\mathbf{c}(\boldsymbol{\theta}) - \mathbf{c}(\boldsymbol{\theta}')],$$

$\nabla a(\boldsymbol{\theta}) = \left(\frac{\partial a(\boldsymbol{\theta})}{\partial \theta_1}, \ldots, \frac{\partial a(\boldsymbol{\theta})}{\partial \theta_k}\right)^t$, $\mathbf{c}(\boldsymbol{\theta}) = (c_1(\boldsymbol{\theta}), \ldots, c_k(\boldsymbol{\theta}))$, *$G(\boldsymbol{\theta})$ is the nonsingular matrix corresponding to the Jacobian determinant* $\frac{\partial(c_1(\boldsymbol{\theta}), \ldots, c_k(\boldsymbol{\theta}))}{\partial(\theta_1, \ldots, \theta_k)}$, *namely*

$$G(\boldsymbol{\theta}) = \begin{bmatrix} \nabla c_1(\boldsymbol{\theta}) \\ \nabla c_2(\boldsymbol{\theta}) \\ \vdots \\ \nabla c_k(\boldsymbol{\theta}) \end{bmatrix}. \tag{9}$$

**Proof.** By a straightforward integration with respect to the components of $\mathbf{x}$, it can be shown that the bound on the difference in doubling rates is:

$$D(f(x;\boldsymbol{\theta}) \parallel f(x;\boldsymbol{\theta}')) = mD(f(x;\boldsymbol{\theta}) \parallel f(x;\boldsymbol{\theta}')). \tag{10}$$

Thus it suffices to determine $D(f(x;\boldsymbol{\theta}) \parallel f(x;\boldsymbol{\theta}'))$. Now, assuming logarithms are to base $e$,

$$\begin{aligned} D(f(x;\boldsymbol{\theta}) \parallel f(x;\boldsymbol{\theta}')) &= \int f(x;\boldsymbol{\theta}) \ln f(x;\boldsymbol{\theta}) dx - \int f(x;\boldsymbol{\theta}) \ln f(x;\boldsymbol{\theta}') dx \\ &= \ln a(\boldsymbol{\theta}) - \ln a(\boldsymbol{\theta}') + \sum_{i=1}^{k} [c_i(\boldsymbol{\theta}) - c_i(\boldsymbol{\theta}')] E_{\boldsymbol{\theta}}[d_i(X)]. \end{aligned} \tag{11}$$

To determine $E_{\boldsymbol{\theta}}[d_i(X)]$, we use the property

$$\int a(\boldsymbol{\theta}) b(x) e^{\sum_{j=1}^{k} c_j(\boldsymbol{\theta}) d_j(x)} dx = 1. \tag{12}$$

Differentiating (12) with respect to $\theta_i$, we have

$$\int \left[ \frac{\partial a(\boldsymbol{\theta})}{\partial \theta_i} + a(\boldsymbol{\theta}) \sum_{j=1}^{k} \frac{\partial c_j(\boldsymbol{\theta})}{\partial \theta_i} d_j(x) \right] b(x) e^{\sum_{j=1}^{k} c_j(\boldsymbol{\theta}) d_j(x)} dx = 0.$$

This is equivalent to

$$\frac{\partial a(\boldsymbol{\theta})}{\partial \theta_i} \frac{1}{a(\boldsymbol{\theta})} = - \sum_{j=1}^{k} \frac{\partial c_j(\boldsymbol{\theta})}{\partial \theta_i} E_{\boldsymbol{\theta}}[d_j(X)] \quad \text{for } i = 1, 2, \ldots, k.$$

This system of linear equations can be rewritten as:

$$\frac{1}{a(\boldsymbol{\theta})} [\nabla a(\boldsymbol{\theta})]^t = -G(\boldsymbol{\theta})^t E_{\boldsymbol{\theta}}[\mathbf{d}(X)]$$

where $\mathbf{d}(X) = (d_1(X), \ldots, d_k(X))$. Hence

$$E_{\boldsymbol{\theta}}[\mathbf{d}(X)] = -\frac{1}{a(\boldsymbol{\theta})} [G(\boldsymbol{\theta})^t]^{-1} [\nabla a(\boldsymbol{\theta})]^t. \tag{13}$$

From (11) and (13),

$$D(f(x;\boldsymbol{\theta}) \parallel f(x;\boldsymbol{\theta}')) = \log \left[ \frac{a(\boldsymbol{\theta})}{a(\boldsymbol{\theta}')} \right] - \frac{(\log e)}{a(\boldsymbol{\theta})} [\nabla a(\boldsymbol{\theta})] G(\boldsymbol{\theta})^{-1} [\mathbf{c}(\boldsymbol{\theta}) - \mathbf{c}(\boldsymbol{\theta}')].$$

□

Note that in the above, we assume the existence and continuity of the partial derivatives with respect to $\theta$ and that $G(\boldsymbol{\theta})$ is nonsingular. Next, we assume that the p.d.f. $f(x;\boldsymbol{\theta})$ of the stock price $X$ is wrongly estimated as another p.d.f. $g(x;\boldsymbol{\theta})$ of the regular exponential class, but from a different family.

**Theorem 3.** *Let the $m$ stock prices $X_1, X_2, \ldots, X_m$ be mutually independent with common p.d.f.*

$$f(x;\boldsymbol{\theta}) = a_1(\boldsymbol{\theta})b_1(x)e^{\sum_{j=1}^{k} c_j(\boldsymbol{\theta})d_j(x)}$$

*belonging to a regular exponential class where $\boldsymbol{\theta} = (\theta_1, \ldots, \theta_k)$. Suppose that $f(x;\boldsymbol{\theta})$ is wrongly estimated as $g(x;\boldsymbol{\theta})$ from a different family of the regular exponential class where*

$$g(x;\boldsymbol{\theta}) = a_2(\boldsymbol{\theta})b_2(x)e^{\sum_{j=1}^{k} r_j(\boldsymbol{\theta})s_j(x)}$$

*and the support set of $g(x;\boldsymbol{\theta})$ contains the support set of $f(x;\boldsymbol{\theta})$. Then the bound on the difference in the doubling rates of the capital is $mD(f(x;\boldsymbol{\theta}) \parallel g(x;\boldsymbol{\theta}))$ where*

$$\begin{aligned}
D(f(x;\boldsymbol{\theta}) \parallel g(x;\boldsymbol{\theta})) &= \log\left[\frac{a_1(\boldsymbol{\theta})}{a_2(\boldsymbol{\theta})}\right] + E_f\left[\log\frac{b_1(X)}{b_2(X)}\right] \\
&\quad -\frac{(\log e)}{a_1(\boldsymbol{\theta})}[\nabla a_1(\boldsymbol{\theta})][G(\boldsymbol{\theta})^{-1}]\mathbf{c}(\boldsymbol{\theta}) \\
&\quad -(\log e)\sum_{j=1}^{k} r_j(\boldsymbol{\theta})E_f[s_j(X)] \\
&= -\Bigg\{h(f) + \log[a_2(\boldsymbol{\theta})] + E_f[\log b_2(X)] \\
&\quad +(\log e)\sum_{j=1}^{k} r_j(\boldsymbol{\theta})E_f[s_j(X)]\Bigg\}
\end{aligned}$$

*where $G(\boldsymbol{\theta})$ and $c(\boldsymbol{\theta})$ are defined in Theorem 2, $h(f) = -\int f(x;\boldsymbol{\theta}) \times \log f(x;\boldsymbol{\theta})dx$ is the differential entropy of $f(x;\boldsymbol{\theta})$.*

**Proof.** Consider logarithms to base $e$ first.

$$\begin{aligned}D(f(x;\boldsymbol{\theta}) \parallel g(x;\boldsymbol{\theta})) &= \int f(x;\boldsymbol{\theta}) \ln f(x;\boldsymbol{\theta}) dx - \int f(x;\boldsymbol{\theta}) \ln g(x;\boldsymbol{\theta}) dx \\ &= [\ln a_1(\boldsymbol{\theta}) - \ln a_2(\boldsymbol{\theta})] + E_f[\ln b_1(X) - \ln b_2(X)] \\ &\quad + \sum_{j=1}^{k} c_j(\boldsymbol{\theta}) E_f[d_j(X)] - \sum_{j=1}^{k} r_j(\boldsymbol{\theta}) E_f[s_j(X)] \\ &= \ln\left[\frac{a_1(\boldsymbol{\theta})}{a_2(\boldsymbol{\theta})}\right] + E_f\left[\ln \frac{b_1(X)}{b_2(X)}\right] \\ &\quad + \mathbf{c}(\boldsymbol{\theta})^t E_f[\mathbf{d}(X)] - \sum_{j=1}^{k} r_j(\boldsymbol{\theta}) E_f[s_j(X)].\end{aligned}$$

From (13),

$$\begin{aligned}\mathbf{c}(\boldsymbol{\theta})^t E_f[\mathbf{d}(X)] &= -\frac{1}{a_1(\boldsymbol{\theta})} \mathbf{c}(\boldsymbol{\theta})^t [G(\boldsymbol{\theta})^{-1}]^t [\nabla a_1(\boldsymbol{\theta})]^t \\ &= -\frac{1}{a_1(\boldsymbol{\theta})} [\nabla a_1(\boldsymbol{\theta})][G(\boldsymbol{\theta})^{-1}] \mathbf{c}(\boldsymbol{\theta})\end{aligned}$$

and hence for logarithms to any base,

$$\begin{aligned}D(f(x;\boldsymbol{\theta}) \parallel g(x;\boldsymbol{\theta})) &= \log\left[\frac{a_1(\boldsymbol{\theta})}{a_2(\boldsymbol{\theta})}\right] + E_f\left[\log \frac{b_1(X)}{b_2(X)}\right] \\ &\quad - \frac{(\log e)}{a_1(\boldsymbol{\theta})} [\nabla a_1(\boldsymbol{\theta})][G(\boldsymbol{\theta})^{-1}] \mathbf{c}(\boldsymbol{\theta}) \\ &\quad - (\log e) \sum_{j=1}^{k} r_j(\boldsymbol{\theta}) E_f[s_j(X)].\end{aligned}$$

The second expression for $D(f(x;\boldsymbol{\theta}) \parallel g(x;\boldsymbol{\theta}))$ in terms of $h(f)$ is straightforward. □

We give a few examples of the expressions for relative entropy of gamma and normal distributions.

**Example 1.** Let $f(x)$ and $g(x)$ be p.d.f.'s having the distributions gamma$(\alpha_1, \beta_1)$ and gamma$(\alpha_2, \beta_2)$ respectively, i.e.

$$f(x) = \frac{x^{\alpha_1 - 1} e^{-x/\beta_1}}{\beta_1^{\alpha_1} \Gamma(\alpha_1)} \quad \text{for } 0 < x < \infty, \ \alpha_1 > 0, \ \beta_1 > 0$$

$$g(x) = \frac{x^{\alpha_2 - 1} e^{-x/\beta_2}}{\beta_2^{\alpha_2} \Gamma(\alpha_2)} \quad \text{for } 0 < x < \infty, \ \alpha_2 > 0, \ \beta_2 > 0.$$

Then

$$D(f \parallel g) = \log \left[ \left( \frac{\beta_2}{\beta_1} \right)^{\alpha_2} \frac{\Gamma(\alpha_2)}{\Gamma(\alpha_1)} \right] + (\alpha_1 - \alpha_2)\psi(\alpha_1)(\log e) + \alpha_1 \left( \frac{\beta_1}{\beta_2} - 1 \right) (\log e)$$

where $\psi(\alpha_1) = \frac{d}{d\alpha} \ln \Gamma(\alpha) \Big|_{\alpha = \alpha_1} = \frac{\Gamma'(\alpha_1)}{\Gamma(\alpha_1)}$.

**Example 2.** Let $f(x)$ and $g(x)$ be p.d.f.'s from the normal distributions $N(\mu_1, \sigma_1^2)$ and $N(\mu_2, \sigma_2^2)$ respectively. Then

$$D(f \parallel g) = \log \left( \frac{\sigma_2}{\sigma_1} \right) + \frac{(\log e)}{2\sigma_2^2} [(\sigma_1^2 - \sigma_2^2) + (\mu_1 - \mu_2)^2].$$

**Example 3.** Let $f(x)$ and $g(x)$ be p.d.f.'s from the distributions gamma$(\alpha, \beta)$ and $N(\mu, \sigma^2)$ respectively. Then

$$D(f \parallel g) = \frac{1}{2} \log(2\pi e \sigma^2) - \log[\beta \Gamma(\alpha)] + (\log e) \left[ \frac{\alpha\beta}{2\sigma^2} \{(\alpha + 1)\beta - 2\mu\} + \frac{\mu^2 - \sigma^2}{2\sigma^2} + (\alpha - 1)\psi(\alpha) - \alpha \right]$$

where $\psi(\alpha) = \frac{\Gamma'(\alpha)}{\Gamma(\alpha)}$.

Finally, if $f(\mathbf{x}) = f_1(x_1) f_2(x_2) \ldots f_m(x_m)$ is the joint p.d.f. of $m$ mutually independent stock prices $X_1, X_2, \ldots, X_m$ where $f_i(x_i)$, the p.d.f. of $X_i$ is wrongly estimated as $g_i(x_i)$ for $i = 1, 2, \ldots, m$ and $g(\mathbf{x}) = g_1(x_1) g_2(x_2) \ldots g_m(x_m)$, then the bound on the difference in the doubling rates of the capital is:

$$D(f(\mathbf{x}) \parallel g(\mathbf{x})) = \sum_{i=1}^{m} D(f_i(x_i) \parallel g_i(x_i)). \tag{14}$$

## References


1. A. Barron and T.M. Cover, "A bound on the financial value of information", IEEE Trans. Inform. Theory, **34**, 1097–1100 (1988).
2. T.M. Cover, "An algorithm for maximizing expected log investment return", IEEE Trans. Inform. Theory, **30**, 369–373 (1984).
3. T.M. Cover and J.A. Thomas, *Elements of Information Theory*, (John Wiley, New York, 1991).

# TAU ERGODIC COEFFICIENTS GENERATED BY THE $C$-NORM

CHOON PENG TAN AND PEK WAI LIEW
*Institute of Mathematical Sciences, University of Malaya, 50603 Kuala Lumpur, Malaysia*
*E-mail: tcp@mnt.math.um.edu.my*

We consider a new class of tau coefficients generated by the $c$-norm and study the associated maximization of a convex functional over a certain convex set $E$ of $n$-dimensional vectors in order to derive the explicit functional forms of these coefficients. The class of extreme points of $E$ obtained is a natural generalization of the class of extreme points associated with the tau coefficients generated by the trimmed $\ell_1$ norm, which is a special case of a $c$-norm. The main results derived may also be regarded as further extensions of the work of Chang and Li (1992) on characterizing the set of all extreme points of the unit ball of the $c$-norm in $\mathbb{R}^n$. Here we have derived a class of extreme points of the entire subset of the unit ball of the $c$-norm lying on a given hyperplane.

## 1 Introduction

The tau coefficient of a nonnegative, irreducible matrix $P$ generated by a given vector norm $\|\cdot\|$ is defined as:

$$\tau_{\|\cdot\|}(P) = \max\{\|x'P\| : x \in E\} \tag{1}$$

where

$$E = \{x : \|x\| \leq 1,\ x' \cdot w = 0,\ x \in \mathbb{R}^n\} \tag{2}$$

and $w = (w_i) > 0$ is the Perron vector of $P$ (i.e. $w$ is a positive eigenvector of $P$ corresponding to its spectral radius which is an eigenvalue). The well-known property of $\tau_{\|\cdot\|}(P)$ in bounding the subdominant eigenvalues of $P$ is studied in Rothblum and Tan (1985). For a Markov chain with transition matrix $P$, the maximum modulus of the subdominant eigenvalues of $P$ is also known as the geometric rate of convergence of the chain to stationarity. Thus $\tau_{\|\cdot\|}(P)$ provides an upper bound on the rate of convergence of a Markov chain. In order to use $\tau_{\|\cdot\|}(P)$, we must know its explicit functional form in terms of the entries of $P$. This is equivalent to solving the maximization problem (1) by maximizing the convex functional $\|x'P\|$ over the set of vectors $x$ in a compact and convex set $E$. It is well-known that the maximum of $\|x'P\|$ is achieved at an extreme point of $E$. Thus, characterizing the set of all extreme points of $E$ will lead to a solution of (1) provided that this set is not extensive.

This approach is not fruitful when $\|\cdot\|$ is an $\ell_2$ norm or a trimmed $\ell_p$ norm for $1 < p < \infty$ because the corresponding sets of extreme points of $E$ are very extensive (see Tan (1995)). Rothblum (1984) obtained the set of all extreme points of $E$ when $\|\cdot\|$ is an $\ell_1$ or $\ell_\infty$ norm. Further studies of the extreme points of $E$ were conducted by Gooi (1991). Tan (1992) and Tan (1995) for the case where $\|\cdot\|$ is a trimmed $\ell_1$ norm. Our objective here is to study the extreme points of $E$ when $\|\cdot\|$ is a $c$-norm which is a generalization of the trimmed $\ell_1$ norm. The main results obtained in the next section may also be regarded as extensions of the work of Chang and Li (1992) on characterizing the set of all extreme points of the unit ball of the $c$-norm in $\mathbb{R}^n$.

We begin with some definitions. Let $c = (c_1, c_2, \dots, c_n)$ be a nonnegative vector in $\mathbb{R}^n$, where the elements $c_i$ are ordered in nonincreasing order, i.e. $c_1 \geq c_2 \geq \cdots \geq c_n \geq 0$. For $x \in \mathbb{R}^n$, the $c$-norm of $x$ denoted by $\|x\|_c$, is defined as:

$$\|x\|_c = \max\left\{\sum_{k=1}^{n} c_k |x_{i_k}| : (x_{i_1}, x_{i_2}, \dots, x_{i_n}) \text{ is any permutation of the components of } x\right\}. \tag{3}$$

If the components of $x \in \mathbb{R}^n$ are ordered such that $|x_{(1)}| \geq |x_{(2)}| \geq \cdots \geq |x_{(n)}|$, then it is well-known that

$$\|x\|_c = \sum_{i=1}^{n} c_i |x_{(i)}|. \tag{4}$$

## 2 Main Results

First, we have the following result:

**Proposition 1.** *Suppose that $\|\cdot\|_c$ is a given $c$-norm on $\mathbb{R}^n$ where $c$ is a nonnegative vector with ordered elements $c_1 \geq c_2 \geq \cdots \geq c_n \geq 0$. Let $f = \lambda x + (1-\lambda) y \in \mathbb{R}^n$ for some $\lambda$ such that $0 < \lambda < 1$ where $\|x\|_c \leq 1$ and $\|y\|_c \leq 1$. If $\|f\|_c = 1$, then $\|x\|_c = \|y\|_c = 1$.*

**Proof.** Suppose that the elements of $f = (f_1, f_2, \dots, f_n)$ are ordered such that $|f_{(1)}| \geq |f_{(2)}| \geq \cdots \geq |f_{(n)}|$. Then

$$\begin{aligned}1 = \|f\|_c &= \sum_{i=1}^{n} c_i |f_{(i)}| \\ &= \sum_{i=1}^{n} c_i |\lambda x_{(i)} + (1-\lambda) y_{(i)}| \\ &\leq \lambda \sum_{i=1}^{n} c_i |x_{(i)}| + (1-\lambda) \sum_{i=1}^{n} c_i |y_{(i)}| \\ &\leq \lambda \|x\|_c + (1 - \lambda \|y\|_c \\ &\leq 1\end{aligned}$$

which implies that

$$\|x\|_c = \sum_{i=1}^{n} c_i |x_{(i)}| = 1 \tag{5}$$

and

$$\|y\|_c = \sum_{i=1}^{n} c_i |y_{(i)}| = 1. \qquad \square \tag{6}$$

**Proposition 2.** *Suppose that* $\|\cdot\|_c$ *is a given c-norm on* $\mathbb{R}^n$ *where* $c$ *is a nonnegative vector with ordered elements* $c_1 \geq c_2 \geq \cdots \geq c_n \geq 0$. *Let* $f \in \mathbb{R}^n$ *be a vector with ordered magnitudes* $|f_1| \geq |f_2| \geq \cdots \geq |f_n|$ *and* $f = \lambda x + (1-\lambda) y$ *for some* $\lambda$ *such that* $0 < \lambda < 1$, *where* $\|x\|_c \leq 1$ *and* $\|y\|_c \leq 1$.

(i) *If* $\|f\|_c = \sum_{i=1}^{n} c_i |f_i| = 1$, *then* $|f_i| = \lambda |x_i| + (1-\lambda)|y_i|$ *for all* $c_i > 0$, *i.e.* $f_i$, $x_i$ *and* $y_i$ *all have the same sign if* $f_i \neq 0$ *and* $f_i = x_i = y_i = 0$ *otherwise.*

(ii) *If* $\|f\|_c = \sum_{i=1}^{n} c_i |f_i| = 1$, *then* $|x_i| \geq |x_j|$ *and* $|y_i| \geq |y_j|$ *if* $c_i > c_j \geq 0$ *for* $i, j = 1, 2, \ldots, n$.

(iii) *If* $\|f\|_c = \sum_{i=1}^{n} c_i |f_i| = 1$ *and* $|f_i| = |f_j|$ *for 2 distinct indices* $i$ *and* $j$, *then* $|x_i| = |x_j|$ *and* $|y_i| = |y_j|$ *if* $c_i > c_j \geq 0$, *for* $i, j = 1, 2, \ldots, n$.

*Furthermore,* $\|f\|_c = 1$ *and* $|f_i| = |f_j|$ *for* $i \neq j$ *will imply that* $|f_j| = \lambda|x_j| + (1-\lambda)|y_j|$ *if* $c_i > c_j = 0$. *(In particular, if* $\|f\|_c = 1$ *and* $f_i = f_j = 0$ *for* $i \neq j$*, then* $x_i = x_j = 0$ *and* $y_i = y_j = 0$ *if* $c_i > c_j \geq 0$*, for* $i, j = 1, 2, \ldots, n$*.)*

**Proof.** (i) It is clear that

$$\begin{aligned} 1 = \|f\|_c &= \sum_{i=1}^{n} c_i|f_i| \\ &\leq \sum_{i=1}^{n} c_i(\lambda|x_i| + (1-\lambda)|y_i|) \\ &\leq \lambda\|x\|_c + (1-\lambda)\|y\|_c \leq 1 \end{aligned}$$

implies that $|f_i| = \lambda|x_i| + (1-\lambda)|y_i|$ for all $c_i > 0$.

(ii) From (5) and the definition (3),

$$\|x\|_c = \sum_{k=1}^{n} c_k|x_k| \geq c_i|x_j| + c_j|x_i| + \sum_{\substack{k=1 \\ k\neq i,j}}^{n} c_k|x_k|$$

which implies that $(|x_i| - |x_j|)(c_i - c_j) \geq 0$. Hence $|x_i| \geq |x_j|$ if $c_i > c_j \geq 0$. Similarly, using the same argument as before, $|y_i| \geq |y_j|$ if $c_i > c_j \geq 0$.

(iii) We consider 2 separate cases. In the first case, if $c_i > c_j > 0$, then from the result in (i),

$$|f_i| = \lambda|x_i| + (1-\lambda)|y_i| \quad \text{and } |f_j| = \lambda|x_j| + (1-\lambda)|y_j|.$$

Given that $|f_i| = |f_j|$, we have

$$\lambda(|x_i| - |x_j|) + (1-\lambda)(|y_i| - |y_j|) = 0. \tag{7}$$

From the result in (ii) where $|x_i| \geq |x_j|$ and $|y_i| \geq |y_j|$, we conclude that $|x_i| = |x_j|$ and $|y_i| = |y_j|$ from (7). In the second case, if $c_i > c_j = 0$, then $|f_i| = \lambda|x_i| + (1-\lambda)|y_i|$ from the result in (i) and $|f_j| \leq \lambda|x_j| + (1-\lambda)|y_j|$ is always true.

Given that $|f_i| = |f_j|$, we have

$$\lambda(|x_i| - |x_j|) + (1-\lambda)(|y_i| - |y_j|) \leq 0. \tag{8}$$

Again $|x_i| \geq |x_j|$ and $|y_i| \geq |y_j|$ will imply that $|x_i| = |x_j|$ and $|y_i| = |y_j|$ from (8). Furthermore $|f_j| = \lambda|x_j| + (1-\lambda)|y_j|$. $\square$

Before we proceed to introduce the extreme points of the convex set $E$ in (2) for the $c$-norm, we introduce the notation used to describe certain vectors in $\mathbb{R}^n$. Let $S$ be a set of indices where $S \subseteq \{1, 2 \ldots, n\}$. We denote an $n$-dimensional vector whose $i$th component is 1 if $1 \in S$ and 0 if $1 \notin S$ by $e^S$. The complement of a set $S$ will be denoted by $S^c$, i.e. $S^c = \{1, 2, \ldots, n\} \backslash S$. The number of distinct elements in a set $S$ is denoted by $|S|$. If $|S| = 1$ and $S = \{k\}$, we use the simplified notation $e_k$ for the vector $e^{(k)}$. In other words, $e_k$ is the coordinate vector with a 1 in the $k$th position and 0 elsewhere. If $S = \emptyset$, the empty set, we define $e^S = \mathbf{0}$.

First, we shall consider a $c$-norm on $\mathbb{R}^n$ where the nonnegative vector $c$ has at least 3 distinct elements. Let the elements of $c$ be ordered such that $c_1 \geq c_2 \geq \cdots \geq c_n \geq 0$. We assume that the largest element $c_1$ is repeated $m$ times and zero is repeated $q$ times. For $c$ to have at least 3 distinct elements, we assume that $n - m - q \geq 1$. Thus, the elements of $c$ are ordered as follows:

$$c_1 = c_2 = \cdots = c_m > c_{m+1} \geq \cdots \geq c_{n-q} > c_{n-q+1} = \cdots = c_n = 0. \tag{9}$$

**Theorem 1.** *Let $\|\cdot\|_c$ be a given c-norm on $\mathbb{R}^n$ where the nonnegative vector c has at least 3 distinct elements, with the largest element repeated m times, the element zero is repeated q times and the elements are ordered in nonincreasing order. Let $E = \{x : \|x\|_c \leq 1,\ x' \cdot w = 0,\ x \in \mathbb{R}^n\}$ and $w \in \mathbb{R}^n$. Then, for $m \geq 2$, $q \geq 1$ and $n - m - q \geq 3$, $F_2 \cup F_n \bigcup_{r=m+2}^{n-q-1} F_r \cup H_{m+1}$ is a set of extreme points of the convex set E where the sets $F_r$ and $H_{m+1}$ are defined as follows:*

$$\begin{aligned} F_r = \Bigg\{ \alpha^{-1}(e^S + \beta e_k - e^T) &: S,\ T \subseteq \{1, 2, \ldots, n\},\ S \cap T = \emptyset, \\ & |S| + |T| = r - 1,\ k \in (S \cup T)^c, \\ & \beta = w_k^{-1} \left( \sum_{i \in T} w_i - \sum_{i \in S} w_i \right), \\ & \alpha = \sum_{i=1}^{r-1} c_i + |\beta| c_r \quad \text{if } |\beta| \leq 1, \\ & \alpha = |\beta| c_1 + \sum_{i=2}^{r} c_i \quad \text{if } |\beta| > 1 \Bigg\} \end{aligned} \tag{10}$$

*for $r = 2, 3, \ldots, n$ and*

$$H_{m+1} = \{f : f \in F_{m+1},\ |\beta| \geq 1\}. \tag{11}$$

*The vectors* $\pm c_1^{-1}e_k$ *are extreme points of* $E$ *if* $w_k = 0$ *and* $k \in \{1, 2, \ldots, n\}$.

**Proof.** We consider 4 separate cases.

(i) Let $f = \alpha^{-1}(e^S + \beta e_k - e^T) \in F_r$ where $m+2 \leq r \leq n-q-1$. Then $f$ is a permutation of the vector $\alpha^{-1}(1, 1, \ldots, 1, \beta, -1, -1, \ldots, -1, 0, \ldots, 0)$ where 1 and $-1$ are repeated $|S|$ and $|T|$ times respectively. The vector $f \in F_r$ has $r$ nonzero elements if $\beta \neq 0$. Without loss of generality, we assume that $f$ has its elements ordered according to magnitude, i.e. $|f_i| = \alpha^{-1}$ for $i = 1, 2, \ldots, (r-1)$, $|f_r| = \alpha^{-1}|\beta|$ and $f_i = 0$ for $i = r+1, r+2, \ldots, n$ if $|\beta| \leq 1$; otherwise $|f_1| = \alpha^{-1}|\beta|$, $|f_i| = \alpha^{-1}$ for $i = 2, 3, \ldots, r$ and $f_i = 0$ for $i = r+1, r+2, \ldots, n$ if $|\beta| > 1$. We assume that $w \in \mathbb{R}^n$ is permutated to correspond with the permutation of the elements of $f$. Suppose $f = \lambda x + (1-\lambda)y$ for some $\lambda$ such that $0 < \lambda < 1$ where $x, y \in E$. Let $U = \{r+1, r+2, \ldots, n-q\}$ and $V = \{n-q+1, n-q+2, \ldots, n\}$ be 2 disjoint nonempty sets of indices, where $n - q \geq r + 1$ and $n - q + 1 \leq n$. For $i \in U$ and $j \in V$, we have $c_i \geq c_{n-q} > c_{n-q+1} = c_j$ by (9). Now $\|f\|_c = 1$ and $f_i = f_j = 0$ for $i \in U$, $j \in V$ will imply that $x_i = x_j = 0$ and $y_i = y_j = 0$ by Proposition 2(iii) and the fact that $c_i > c_j$. Hence

$$x_k = y_k = 0 \qquad \text{for } k = r+1, r+2, \ldots, n. \tag{12}$$

We consider 2 separate cases, namely $|\beta| \leq 1$ and $|\beta| > 1$.

Suppose $|\beta| \leq 1$. Let $W = \{1, 2, \ldots, m\}$ and $z = \{m+1, m+2, \ldots, r-1\}$. Then $W$ and $Z$ are 2 disjoint nonempty sets since $m \geq 2$ and $m+1 \leq r-1$. By (9), $c_i = c_m > c_{m+1} \geq c_j$ for $i \in W$, $j \in Z$. By Proposition 2(iii), $\|f\|_c = 1$ and $|f_i| = |f_j| = \alpha^{-1}$ for $i \in W$, $j \in Z$ imply that $|x_i| = |x_j|$ and $|y_i| = |y_j|$ since $c_i > c_j > 0$. Furthermore, $|f_k| = \lambda|x_k| + (1-\lambda)|y_k|$ for $k = 1, 2, \ldots, r-1$ by Proposition 2(i), i.e. $f_k$, $x_k$ and $y_k$ all have the same sign. Thus

$$|x_k| = \gamma \quad \text{and} \quad |y_k| = \delta \qquad \text{for } k = 1, 2, \ldots, r-1. \tag{13}$$

Now $f_i = \alpha^{-1}$ for $i \in S$ and $f_j = -\alpha^{-1}$ for $j \in T$ where $S \cup T = \{1, 2, \ldots, r-1\}$ and $S \cap T = \emptyset$ will imply that

$$x_i = \gamma,\ y_i = \delta \qquad \text{for } i \in S \tag{14}$$

$$\text{and} \qquad x_j = -\gamma,\ y_j = -\delta \quad \text{for } j \in T. \tag{15}$$

From (12), (14), (15) and the condition $x' \cdot w = 0$, we have

$$\gamma \sum_{i \in S} w_i - \gamma \sum_{j \in T} w_j + x_r w_r = 0 \text{ and hence } x_r = \gamma w_r^{-1} \left( \sum_{j \in T} w_j - \sum_{i \in S} w_i \right) = \gamma\beta.$$

Similarly, $y_r = \delta\beta$ from $y' \cdot w = 0$. Now $\|x\|_c = \gamma\left(\sum_{i=1}^{r-1} c_i + |\beta| c_r\right) = \gamma\alpha = 1$ from Proposition 1 implies that $\gamma = \alpha^{-1}$. Similarly, $\delta = \alpha^{-1}$ from the fact that $\|y\|_c = 1$. Hence $f_k = x_k = y_k$ for $k = 1, 2, \ldots, n$, i.e. $f = x = y$. We conclude that $f$ is an extreme point of $E$.

For the case $|\beta| > 1$, the method of proof is similar. Replace the sets $W$ and $Z$ defined previously by the nonempty sets $W = \{2, 3, \ldots, m\}$ and $Z = \{m+1, m+2, \ldots, r\}$. From $|f_i| = |f_j|$ for $i \in W$, $j \in Z$, we deduce that $|x_i| = |x_j| = \gamma$ and $|y_i| = |y_j| = \delta$. Using the orthogonality conditions $x' \cdot w = y' \cdot w = 0$, we can show that $x_1 = \gamma\beta$ and $y_1 = \delta\beta$. Finally, we use the fact that $\|x\|_c = \|y\|_c = 1$ to deduce that $\gamma = \delta = \alpha^{-1}$ and hence $f = x = y$ which shows that $f$ is an extreme point of $E$ for $|\beta| > 1$.

(ii) Before we show that $F_2$ is a set of extreme points of $E$, we note that

$$\begin{aligned} F_2 = \Bigg\{ \pm\alpha^{-1}\left(e_i - \frac{w_i}{w_k} e_k\right) &: i, k = 1, 2, \ldots, n,\ i \neq k, \\ \alpha = c_1 + \left|\frac{w_i}{w_k}\right| c_2 &\text{ if } |w_i| \leq |w_k|, \\ \alpha = \left|\frac{w_i}{w_k}\right| c_1 + c_2 &\text{ if } |w_i| > |w_k| \Bigg\} \end{aligned} \tag{16}$$

can be written as:

$$\begin{aligned} F_2 = \{\gamma^{-1}(w_k e_i - w_i e_k) &: i, k = 1, 2, \ldots, n,\ i \neq k, \\ &\gamma = \|w_k e_i - w_i e_k\|_c\}. \end{aligned} \tag{17}$$

Define the nonempty sets of indices: $U = \{3, 4, \ldots, n-q\}$ and $V = \{n-q+1, n-q+2, \ldots, n\}$ where $n - q \geq 3 + m \geq 5$ and $n \geq n - q + 1$. For $i \in U$, $j \in V$, we have $c_i \geq c_{n-q} > c_{n-q+1} = c_j$. Assume that the elements of $f \in F_2$ are ordered according to magnitude, i.e. $|f_1| \geq |f_2|$, $f_i = 0$ for $i = 3, 4, \ldots, n$. Without loss of generality, we can further assume that $|w_2| \geq |w_1|$ for $f = \gamma^{-1}(w_2 e_1 - w_1 e_2) \in F_2$ by permutating the elements of $f$ with the corresponding permutation of $w \in \mathbb{R}^n$, where $\gamma = \|w_2 e_1 - w_1 e_2\|_c$. Now suppose that $f = \lambda x + (1-\lambda) y$ for some $\lambda$ such that $0 < \lambda < 1$ where $x, y \in E$. By Proposition 2(iii), $\|f\|_c = 1$ and $f_i = f_j = 0$ for $i \in U$, $j \in V$ imply that $x_i = x_j = 0$ and $y_i = y_j = 0$ since $c_i > c_j$. From the fact that $c_k > 0$ for $k = 1, 2$, we deduce that $f_k$, $x_k$ and $y_k$ all have the same sign from Proposition 2(i). The conditions $x' \cdot w = y' \cdot w = 0$ imply that $x_1 = -x_2(w_2/w_1)$ and $y_1 = -y_2(w_2/w_1)$. Hence we can write $x_1 = \delta w_2$, $x_2 = -\delta w_1$ for some $\delta > 0$ and $y_1 = \eta w_2$, $y_2 = -\eta w_1$ for some $\eta > 0$. From $\|x\|_c = \|y\|_c = 1$, we obtain $\delta\|w_2 e_1 - w_1 e_2\|_c = \delta\gamma = 1$ and

$\eta\|w_2e_1 - w_1e_2\|_c = \eta\gamma = 1$. Hence $\delta = \eta = \gamma^{-1}$ and $f_k = x_k = y_k$ for $k = 1, 2$. In other words, $f = x = y$ and this shows that $f$ is an extreme point of $E$. Similarly, $-f$ is an extreme point of $E$.

(iii) Given $f \in F_n$, we assume without loss of generality that the elements of $f$ are ordered according to their magnitudes. We consider 2 separate cases, namely $|\beta| \leq 1$ and $|\beta > 1$. First consider $|\beta| \leq 1$. Here $|f_i| = \alpha^{-1}$ for $i = 1, 2, \ldots, n-1$ and $f_n = \alpha^{-1}\beta$. Define 2 disjoint, nonempty sets of indices as follows: $W = \{1, 2, \ldots, m\}$ and $Z = \{m+1, m+2, \ldots, n-1\}$, where $n - 1 \geq m + 3$ from $n - m \geq q + 3 \geq 4$. Now $\|f\|_c = 1$ and $|f_i| = |f_j|$ for $i \in W$, $j \in Z$ imply that $|x_i| = |x_j| = \gamma$ and $|y_i| = |y_j| = \delta$ by Proposition 2(iii) since $c_i > c_j$. Moreover, $f_k$, $x_k$ and $y_k$ all have the same sign for $k = 1, 2, \ldots, n-1$ by Proposition 2(i) and (iii). Hence if $f_i = \alpha^{-1}$ for $i \in S \subseteq \{1, 2, \ldots, n-1\}$, then $x_i = \gamma$ and $y_i = \delta$, and if $f_j = -\alpha^{-1}$ for $j \in T \subseteq \{1, 2, \ldots, n-1\}$, then $x_j = -\gamma$ and $y_j = -\delta$, where $S \cap T = \emptyset$ and $S \cup T = \{1, 2 \ldots, n-1\}$. From the orthogonality conditions $x' \cdot w = y' \cdot w = 0$ we obtain $x_n = \gamma w_n^{-1}\left(\sum_{j\in T} w_j - \sum_{i\in S} w_i\right) = \gamma\beta$ and $y_n = \delta\beta$. Again $\|f\|_c = 1$ will imply that $\|x\|_c = \gamma\alpha = 1$ and $\|y\|_c = \delta\alpha = 1$. Hence $\gamma = \delta = \alpha^{-1}$ and $f = x = y$ which shows that $f \in F_n$ is an extreme point of $E$, where $|\beta| \leq 1$. The proof for the second case where $f \in F_n$ and $|\beta| > 1$ is similar. In this latter case, we replace the sets $W$ and $Z$ defined previously by the nonempty sets $W = \{2, 3, \ldots, n-q\}$ and $Z = \{n-q+1, n-q+2, \ldots, n\}$ where $n - q \geq 3 + m \geq 5$ and $n \geq n - q + 1$.

(iv) We omit the proof that $f \in H_{m+1}$ is an extreme point of $E$ which is similar to the proof in (i) for $|\beta| \geq 1$. $\square$

**Corollary 1.** *Let $c$ be a positive vector with at least 2 distinct elements, with the largest element repeated $m$ times and the elements are ordered in nonincreasing order. Then for $m \geq 2$ and $n - m \geq 2$,*

$$F_2 \bigcup_{r=m+2}^{n} F_r \cup H_{m+1}$$

*is a set of extreme points of $E$.*

**Remarks.** We observe that the extreme points of $E$ depends on the structure of $c \geq \mathbf{0}$. There are more sets of extreme points of the type $F_r$ or subsets of $F_r$ if $c$ contains more distinct elements. In particular, we define the following subsets of $F_r$ which may be sets of extreme points of $E$ depending on the structure of $c$:

$$G_r = \{f : f \in F_r,\ |\beta| \leq 1\} \tag{18}$$

and
$$H_r = \{f : f \in F_r,\ |\beta| \geq 1\} \tag{19}$$

for $r = 2, 3, \ldots, n$.

**Corollary 2.** *Let $c$ be a nonnegative vector with at least 4 distinct elements, where the largest element is unique (i.e. $m = 1$), the second largest element is repeated $p$ times, the element zero is repeated $q$ times and the elements are ordered in nonincreasing order. Then, for $p \geq 2$, $q \geq 1$ and $n - p - q \geq 3$,*

$$F_2 \cup F_n \bigcup_{r=p+2}^{n-q-1} F_r \bigcup_{r=3}^{p+1} G_r$$

*is a set of extreme points of $E$. For the special case $p = 1$, $q \geq 1$ and $n-q \geq 4$,*

$$F_2 \cup F_n \bigcup_{r=3}^{n-q-1} F_r$$

*is a set of extreme points of $E$.*

**Corollary 3.** *Let $c$ be a nonnegative vector with at least 3 distinct elements where the largest element is unique (i.e. $m = 1$), the element zero is repeated $q$ times and the elements are ordered in nonincreasing order. Then for $q \geq 1$ and $n - q \geq 4$,*

$$F_2 \cup F_n \bigcup_{r=3}^{n-q-1} G_r$$

*is a set of extreme points of $E$.*

**Corollary 4.** *Let $c$ be a positive vector with at least 3 distinct positive elements, where the largest element is unique (i.e. $m = 1$), the second largest element is repeated $p$ times and the elements are ordered in nonincreasing order. Then for $p \geq 2$ and $n - p \geq 2$,*

$$F_2 \bigcup_{r=p+2}^{n} F_r \bigcup_{r=3}^{p+1} G_r$$

*is a set of extreme points of* $E$. *For the special case* $p = 1$ *and* $n \geq 3$, $\bigcup_{r=2}^{n} F_r$ *is a set of extreme points of* $E$.

**Remarks.** We note that the trimmed $\ell_1$ norm is a $c$-norm where $c$ has only 2 distinct elements, namely 1 and 0 where 1 is repeated $m$ times and 0 is repeated $n - m$ times. Hence, the $c$-norm with only 2 distinct elements is a direct generalization of the trimmed $\ell_1$ norm. For this case, we have the following theorem:

**Theorem 2.** (a) *Let* $c$ *be a nonnegative vector with only 2 distinct elements, namely* $c_1 > 0$ *and 0, where* $c_1$ *is repeated* $m$ *times, 0 is repeated* $n - m$ *times, and the elements are ordered in nonincreasing order. Then* $A_1 = F_2 \cup F_n$, $A_2 = F_n$, $A_3 = G_n$, $A_4 = H_3$, $A_5 = F_2 \cup H_n$ *are sets of extreme points of* $E = \{x : \|x\|_c \leq 1,\ x' \cdot w = 0,\ x \in \mathbb{R}^n\}$ *subject to the conditions:*

(i) $m \geq 3,\ n - m \geq 2$, (ii) $m = 2,\ n - m \geq 2$,
(iii) $m = 1,\ n - m \geq 2$, (iv) $m = 2,\ n - m = 1$,
(v) $m \geq 3, n - m = 1$, *respectively.*

(b) *Let* $c$ *be a positive vector with only 2 distinct elements, namely* $c_1 > 0$ *and* $c_n > 0$ *where* $c_1$ *is repeated* $m$ *times,* $c_n$ *is repeated* $n - m$ *times, and the elements are ordered in nonincreasing order. Then*

$$B_1 = F_2 \bigcup_{r=m+2}^{n} F_r \cup H_{m+1},\ B_2 = F_2 \bigcup_{r=3}^{n} G_r,\ B_3 = F_2 \cup H_n,\ B_4 = F_2$$

*are sets of extreme points of* $E = \{x : \|x\|_c \leq 1,\ x' \cdot w = 0,\ x \in \mathbb{R}^n\}$ *subject to the conditions:*

(i) $m \geq 2,\ n - m \geq 2$, (ii) $m = 1,\ n - m \geq 2$,
(iii) $m \geq 2,\ n - m = 1$, (iv) $m = 1,\ n - m = 1$, *respectively.*

We omit the proof of Theorem 2 which is similar to that of Theorem 1. Finally for the case where $c$ is a positive constant vector, we note that the tau coefficient generated by this $c$-norm is identical to the tau coefficient generated by the $\ell_1$ norm. We expect the structure of the extreme points of $E$ under the 2 norms to be similar, namely $F_2$ is the set of extreme points of $E$ in both cases (Rothblum, 1984).

## References

1. S.S. Chang and C.K. Li, "Certain isometries on $\mathbb{R}^n$", *Linear Algebra Appl.*, **Vol. 165**, 251–265 (1992).

2. K.M. Gooi, "Estimating the rate of convergence of Markov chains", *M.Sc. Thesis*, University of Malaya (1991).
3. U.G. Rothblum, "Explicit solutions to optimization problems on the intersections of the unit ball of the $\ell_1$ and $\ell_\infty$ norms with a hyperplane", Siam J. Alg. Disc. Meth., **Vol. 5, No. 4**, 619–632 (1984).
4. U.G. Rothblum and C.P. Tan, "Upper bounds on the maximum modulus of subdominant eigenvalues of nonegative matrices", Linear Algebra Appl., **Vol. 66**, 45–86 (1985).
5. C.P. Tan, "On the extreme points of the intersection of the unit ball of the trimmed $\ell_1$ norm with a hyperplane", in *Optimization Techniques and Applications*, ed. K.H. Phua et al., 124–131 (World Scientific, Singapore, 1992).
6. C.P. Tan, "On some optimization problems involving the enumeration of the extreme points of convex sets", in *Optimization Techniques and Applications: ICOTA' 95*, ed. G. Liu et al., 195–202 (World Scientific, Singapore, 1995).

# FINDING THE SPAN OF GRAPHS USING SMALL CIRCLE UNIT REPRESENTATIONS

JOSELITO A. UY
Mathematics Department
MSU-Iligan Institute of Technology
9200 Iligan City, Philippines
E-mail: *jauy@math1.msuiit.edu.ph*

A technique called *small circle unit representations* is a useful tool in determining the span of graphs containing a cycle. This paper illustrates this technique by finding the span of cycles, wheels and generalized wheels.

## 1 Preliminaries

Graphs considered here are simple graphs. If $G$ is a graph, denote its vertex set by $V(G)$ and its edge set by $E(G)$. The complete graph of order $n$ will be denoted by $K_n$. Reader of this paper may refer to [4] for other terms and concepts whose definitions are not given here.

$\mathbf{R}^n$ will be referred to as the Euclidean $n$-space with the usual distance formula. The Euclidean 0-space shall be understood to contain only the zero point.

**Definition 1.1** A graph $G$ is a *unit graph* in $\mathbf{R}^n$ if there is a one-to-one mapping $f: V(G) \to \mathbf{R}^n$ such that $|f(x) - f(y)| = 1$ whenever $[x,y] \in E(G)$. If $f$ is such a mapping, we shall refer to the set $\{f(x) \in \mathbf{R}^n : x \in V(G)\}$, together with the set $\{[f(x), f(y)] : [x,y] \in E(G)\}$, as a *unit representation* of $G$ in $\mathbf{R}^n$.

**Definition 1.2** Let $G$ be a unit graph in $\mathbf{R}^n$. The *span* of $G$ in $\mathbf{R}^n$, denoted by $span_n(G)$, is the real number $s$ such that the following conditions are satisfied:

1. No unit representation of $G$ in $\mathbf{R}^n$ is contained in any open ball of diameter $s$.
2. For every $\varepsilon > 0$ there exists a unit representation of $G$ in $\mathbf{R}^n$ which

is contained in some open ball of diameter $s+\varepsilon$.

The span of a graph is defined only in a Euclidean space where it has a unit representation. Thus the next concept is vital in the study of span of a graph.

**Definition 1.3** The *dimension* of a graph $G$, denoted by $dim(G)$, is the smallest integer $n$ such that $G$ is a unit graph in $\mathbf{R}^n$.

**Theorem 1.1** [2] $dim(K_n) = n-1$.

**Theorem 1.2** [7] $span_d(K_n) = \sqrt{\frac{2(n-1)}{n}}$ *for* $d \geq n-1$.

## 2 Span of Cycles

This section will define what we mean by small circle unit representations from the work of Buckley and Harary in [1]. It will also find the span of cycles.

**Lemma 2.1** [1] *For each integer* $n \geq 5$ $(n \neq 6)$ *there exists an integer* $k$, $n/4 < k < n/2$, *such that* $k$ *and* $n$ *are relatively prime.*

**Remark 2.1** In the proof of Lemma 2.1, the following values of $k$ are given:

$$k = \begin{cases} (n-1)/2 & \text{if } n \text{ is odd,} \\ (n-4)/2 & \text{if } n \text{ is even } \& \ n \not\equiv 0 \,(\text{mod} 4), \\ (n-2)/2 & \text{if } n \text{ is even } \& \ n \equiv 0 \,(\text{mod} 4). \end{cases}$$

These values of $k$ are the largest integers satisfying the conditions of Lemma 2.1.

**Definition 2.1** The *cycle* $C_n$, $n \geq 3$, is the graph with vertices 1, 2, …, $n$ and edges [1, 2], [2, 3], …, [$n$-1, $n$], [$n$, 1]. It is called an *even* or *odd cycle* depending on whether $n$ is even or odd.

**Lemma 2.2** [1] *The cycle* $C_n$ *has a unit representation in the plane with its vertices on a circle. When* $n = 6$, *the radius of the circle must be* 1; *for* $n \neq 6$, *there is such a circle with radius* $r < 1$.

The unit representations used in Lemma 2.2 are as follows. For $n = 3$ or 4, the cycle $C_n$ is embedded as a regular unit $n$-gon. Thus the radii are $\sqrt{3}/3$ and $\sqrt{2}/2$, respectively. For $n = 6$, the cycle $C_6$ is embedded as a regular hexagon. Since the sides have unit lengths, the radius is also 1. For $n \geq 5$, $n \neq 6$, let $k$ be the largest integer described in Remark 2.1. Let $C_n = [1, 2, \ldots, n, 1]$ and $C$ a circle on a plane with radius $r$. Take $n$ points $p_i$ from $C$ which are equally spaced on the circumference of $C$. The vertices of $C_n$ are associated with the points $p_i$ as follows: $p_i \leftrightarrow 1 + (i-1)k$, where the sum is taken modulo $n$. The central angle determined by $p_1$ and $p_{1+k}$ has measure strictly between $90^\circ$ and $180^\circ$. But the chord $[p_1, p_{1+k}]$ is the longest side of its central angle; hence the radius is less than 1. Thus

**Remark 2.2** For $n \geq 5$ $(n \neq 6)$, the cycle $C_n$ has a unit representation in the plane with its vertices on a circle with radius $r = [\csc(\pi k/n)]/2$, where $k$ and $n$ are relatively prime and $n/4 < k < n/2$.

**Definition 2.2** The unit representations of the cycle $C_n$ used in Lemma 2.2 shall be called *small circle unit representations.*

**Theorem 2.1** $span_2(C_{2t}) = \begin{cases} \sqrt{2} & \text{if } t = 2, \\ 1 & \text{if } t > 2. \end{cases}$

*Proof.* First, consider the case $t = 2$. A unit representation of $C_4$ in $\mathbf{R}^2$ is a unit square whose diagonal has length $\sqrt{2}$. Decreasing the length of a diagonal of a unit square would mean increasing the length of other diagonal. This means that the other unit representations of $C_4$ in $\mathbf{R}^2$ cannot be inscribed in an open disk of diameter $\varepsilon + \sqrt{2}$ for each $\varepsilon > 0$.

Second, consider the case $t > 2$. Let $\delta_1, \delta_2, \ldots, \delta_t, d, w$ be real numbers such that $0 < \delta_1 < \delta_2 < \ldots < \delta_t < 1$ and $0 < w < d < 1$. Consider two cases:

<u>Case 1</u>. $t$ is even. Let $p_1=(\delta_1,0)$, $p_2=(\delta_2,d)$, $p_3=(\delta_3,w)$, $p_4=(\delta_4,d)$, $p_5=(\delta_5,w)$, ... , $p_{t-1}=(\delta_{t-1},w)$, $p_t=(\delta_t,1)$, $p_{t+1}=(\delta_t,0)$, $p_{t+2}=(\delta_{t-1},d)$, $p_{t+3}=(\delta_{t-2},w)$, ... , $p_{2t-4}=(\delta_5,d)$, $p_{2t-3}=(\delta_4,w)$, $p_{2t-2}=(\delta_3,d)$, $p_{2t-1}=(\delta_2,w)$ and $p_{2t}=(\delta_1,1)$. Then $\delta_1,\delta_2,...,\delta_t,d,w$ can be chosen so that $A=[p_1,p_2,...,p_{2t},p_1]$ is a unit cycle in $\mathbf{R}^2$.

<u>Case 2</u>. $t$ is odd. Let $p_1=(\delta_1,0)$, $p_2=(\delta_2,d)$, $p_3=(\delta_3,w)$, $p_4=(\delta_4,d)$, $p_5=(\delta_5,w)$, ... , $p_{t-1}=(\delta_{t-1},d)$, $p_t=(\delta_t,0)$, $p_{t+1}=(\delta_t,1)$, $p_{t+2}=(\delta_{t-1},w)$, $p_{t+3}=(\delta_{t-2},d)$, ... , $p_{2t-4}=(\delta_5,d)$, $p_{2t-3}=(\delta_4,w)$, $p_{2t-2}=(\delta_3,d)$, $p_{2t-1}=(\delta_2,w)$ and $p_{2t}=(\delta_1,1)$. Then $\delta_1,\delta_2,...,\delta_k,d,w$ can be chosen so that $B=[p_1,p_2,...,p_{2t},p_1]$ is a unit cycle in $\mathbf{R}^2$.

Now, let $\delta_t\to\delta_1$. Then $d\to 1$, $w\to 0$ and $\delta_i\to 0$ for each $i$. Thus, a unit $C_{2t}$ can be inscribed in an open disk of diameter $1+\varepsilon$ for each $\varepsilon>0$.

**Theorem 2.2** *For odd cycle* $C_n$, $1<span_2(C_n)\le sec\frac{\pi}{2n}$.

*Proof.* By definition, $span_2(C_n)\ge 1$ for each $n$. Suppose that $span_2(C_n)=1$ for some $n$. Then there exists a unit representation of cycle $C_n$ whose edges are almost parallel to a diameter of an open disk of radius $1+\varepsilon$ where $\varepsilon$ is an arbitrary small positive number. This means that $C_n$ is bipartite, and so $n$ is even. This is a contradiction. Thus $span_2(C_n)>1$.

Since $n$ is odd, $k=(n-1)/2$ by Remark 2.1. Now, $C_n$ has a small circle unit representation in $\mathbf{R}^2$. The radius of this circle according to Remark 2.2 is $r=\frac{1}{2}\csc\frac{\pi k}{n}=\frac{1}{2}\csc\left(\frac{n-1}{n}\right)\frac{\pi}{2}=\frac{1}{2}\sec\frac{\pi}{2n}$.

Theorem 2.3 is proven in [3]. We include its proof here to provide the reader a way of constructing unit representations of even cycles whose span in $\mathbf{R}^3$ is equal to 1.

**Theorem 2.3** [3] *For even cycle* $C_n$, $span_3(C_n)=1$.

*Proof.* Take a circle $C$ of radius $\delta$ lying on a plane parallel to the $xy$-plane and $d$ units above the $xy$-plane. Let $C'$ be the projection of $C$ on the $xy$-plane. Also, let $t = n/2$. Take $t$ points $p_i$ on the circumference of $C$ such that the length of the arcs $p_1p_2, p_2p_3, \ldots, p_{t-1}p_t, p_tp_1$ are equal. Do the same thing with the circle $C'$ choosing $t$ points $q_i$. Assume further that the projection of the midpoint of the arc $p_ip_{i+1}$ on the $xy$-plane is the point $q_i$. Then the sequence $[p_1, q_1, p_2, q_2, \ldots, p_t, q_t, p_1]$ is a unit representation of the cycle $C_n$ in $\mathbf{R}^3$ by a proper choice of $d$ for each $\delta$, $0 < \delta < 1$. By making $\delta$ arbitrarily small, $d$ becomes arbitrarily close to 1. Clearly, the cycle can be inscribed in a sphere of diameter $1+\varepsilon$ for each $\varepsilon > 0$.

## 3 Span of Wheels

The wheel $W_n$ contains the cycle $C_n$. This section will illustrate how to find the span of $W_n$ by finding first the small circle unit representation of $C_n$.

**Definition 3.1** The *wheel* $W_n$ is the graph obtained from the cycle $C_n = [a_1, a_2, \ldots, a_n, a_1]$ by adding a new vertex $a_o$ and joining $a_o$ by an edge to each of the $n$ vertices of $C_n$.

**Lemma 3.1** [2] $dim(W_n) = \begin{cases} 2 & if \quad n = 6 \\ 3 & if \quad n \neq 6. \end{cases}$

**Theorem 3.1** $span_2(W_6) = 2.$

*Proof.* Embed the cycle $C_6$ as a regular hexagon with unit sides. It is clear that the hexagon has a unit radius. Place a vertex at the center of the hexagon and join the vertex by an edge to each of the vertices of the hexagon. The resulting graph is a unit is a unit representation of $W_6$ in $\mathbf{R}^2$. Observe that other unit representations of $W_6$ can be obtained from this by translation and/or rotation. Hence, $span_2(W_6) = 2$.

**Theorem 3.2** $span_3(W_n) = \begin{cases} \frac{\sqrt{6}}{2} & \text{if } n=3 \\ \sqrt{2} & \text{if } n=4. \end{cases}$

*Proof.* The wheel $W_3$ is the complete graph $K_4$. By Theorem 1.2, $span_3(W_3) = span_3(K_4) = \sqrt{6}/2$. In the case of $W_4$, take a unit square in the plane to represent the cycle $C_4$. For the fifth vertex, take a point $p$ that is $\sqrt{2}/2$ unit above the center of the square. Then we obtain a unit representation of $W_4$ in $\mathbf{R}^3$ which is contained in a ball of diameter $\sqrt{2}$. Other unit representations of $W_4$ in $\mathbf{R}^3$ can be obtained from this by increasing the distance of two non-consecutive vertices of $C_4$. This means that other unit representation of $W_4$ in $\mathbf{R}^3$ cannot be contained in a ball of diameter $\sqrt{2}$. Therefore, $span_3(W_4) = \sqrt{2}$.

**Theorem 3.3** *For $n \geq 5$ and $n \neq 6$,*

$$span_3(W_n) \leq \frac{2}{\sqrt{4 - csc^2 \alpha}},$$

*where*

$$\alpha = \begin{cases} (n-1)\pi/2n & \text{if } n \text{ is odd,} \\ (n-4)\pi/2n & \text{if } n \text{ is even and } n \not\equiv 0 \,(mod\, 4), \\ (n-2)\pi/2n & \text{if } n \text{ is even and } n \equiv 0 \,(mod\, 4). \end{cases}$$

*Proof.* Begin with a small circle unit representation of the cycle $C_n$ in $\mathbf{R}^3$ with the circle in the *xy*-plane centered at (0,0,0). By Remark 2.2, the radius of this circle is $r = \frac{1}{2}\csc\frac{\pi k}{n}$. By Remark 2.1, we have

$$\alpha = \frac{\pi k}{n} = \begin{cases} (n-1)\pi/2n & \text{if } n \text{ is odd,} \\ (n-4)\pi/2n & \text{if } n \text{ is even and } n \not\equiv 0 \,(mod\, 4), \\ (n-2)\pi/2n & \text{if } n \text{ is even and } n \equiv 0 \,(mod\, 4). \end{cases}$$

Observe that $3\pi/10 \le \alpha < \pi/2$. Since $r < 1$, there exists a positive real number $z$ such that the point $(0,0,z)$ is at a unit distance from each point on the circle. (This circle is a cross section of the unit sphere centered at $(0,0,z)$.) Thus we have obtained a unit representation of $W_n$ in $\mathbf{R}^3$ when $n \ne 6$.

Note that $z = \sqrt{1-r^2}$. Consider the triangle in the plane formed by the points $A = (r,0)$, $B = (-r,0)$ and $E = (0,z)$. Let us find the radius $d$ of the circle $C$ containing these points on its circumference. Clearly, the center of $C$ is of the form $F = (0,k)$. From the expressions $d = |A-F| = |E-F|$, we can obtain $r^2 + k^2 = (z-k)^2$. Solving for $k$ in the last equation yields $k = (z^2 - r^2)/2z = (1-2r^2)/2\sqrt{1-r^2}$. Thus

$$d = |A-F| = \sqrt{r^2+k^2} = \frac{z^2+r^2}{2z} = \frac{1}{2\sqrt{1-r^2}}.$$

Therefore, $span_3(W_n) \le 2d$.

## 4 Span of Generalized Wheels

The *generalized wheel* $W_{m,n}$ is the graph obtained from the cycle $C_n = [a_1, a_2, ..., a_n, a_1]$ and adding $m$ new vertices $b_1, b_2, ..., b_m$ and joining each $b_i$ by an edge to each of the vertices of $C_n$. Its span can be found by considering a small circle unit representation of $C_n$.

**Lemma 4.1** [1] *For* $n \ne 6$, $dim(W_{2,n}) = 3$.

**Definition 4.1** Let $A_1, A_2, ..., A_n$ be disjoint nonempty sets of vertices. For all $i \ne j$, let each vertex of $A_i$ be adjacent to each vertex of $A_j$. The resulting graph is called the *complete n-partite graph* and is denoted by $K(p_1, p_2, ..., p_n)$, where $p_i = |A_i|$ for all $i$.

**Lemma 4.2** [7] $span_3(K(2,2,2) = \sqrt{2}$.

**Theorem 4.1** $span_3(W_{2,n}) = \begin{cases} \dfrac{2\sqrt{6}}{3} & if\ n=3, \\ \sqrt{2} & if\ n=4. \end{cases}$

*Proof.* The points $p_1 = (1/2, -\sqrt{3}/6, 0)$, $p_2 = (-1/2, -\sqrt{3}/6, 0)$, $p_3 = (0, \sqrt{3}/3, 0)$, $p_4 = (0,0,\sqrt{6}/3)$, and $p_5 = (0,0,-\sqrt{6}/3)$ form a unit representation of $W_{2,3}$ in $\mathbf{R}^3$. Note that other unit representations of $W_{2,3}$ in $\mathbf{R}^3$ can be obtained from this by translation and/or rotation. Now, $[p_1, p_2, p_3, p_1]$ is a unit equilateral triangle in the *xy*-plane. Its centroid is at (0,0,0) while its span in the plane is $2\sqrt{3}/3$ (see Theorem 1.2). Observe that $p_4$ and $p_5$ are on the *z*-axis and $|p_4 - p_5| = 2\sqrt{6}/3$. Since $2\sqrt{3}/3 < 2\sqrt{6}/3$, $span_3(W_{2,3}) = 2\sqrt{6}/3$. On the other hand, $W_{2,4} = K(2,2,2)$; hence $span_3(W_{2,4}) = \sqrt{2}$ by Lemma 4.2.

**Theorem 4.2** *For* $n \geq 5$ *and* $n \neq 6$,

$$span_3(W_{2,n}) \leq \sqrt{4 - csc^2\alpha},$$

*where*

$$\alpha = \begin{cases} (n-1)\pi/2n & if\ n\ is\ odd, \\ (n-4)\pi/2n & if\ n\ is\ even\ and\ n \not\equiv 0\,(mod\ 4), \\ (n-2)\pi/2n & if\ n\ is\ even\ and\ n \equiv 0\,(mod\ 4). \end{cases}$$

*Proof.* Begin with a small circle unit representation of $C_n$ in $\mathbf{R}^3$ with the circle in the *xy*-plane centered at (0,0,0). By Remark 2.2 the radius of the circle is $r = \frac{1}{2}\csc\frac{\pi k}{n}$. By Remark 2.1, we have

$$\alpha = \frac{\pi k}{n} = \begin{cases} (n-1)\pi/2n & if\ n\ is\ odd, \\ (n-4)\pi/2n & if\ n\ is\ even\ and\ n \not\equiv 0\,(mod\ 4), \\ (n-2)\pi/2n & if\ n\ is\ even\ and\ n \equiv 0\,(mod\ 4). \end{cases}$$

Observe that $3\pi/10 \le \alpha < \pi/2$. Since $r < 1$, there exists a positive real number $z$ such that each of the points $(0,0,z)$ and $(0,0,-z)$ is at distance 1 from each point on the circle. Thus we can form a unit representation of $W_{2,n}$ in $\mathbf{R}^3$ when $n \ne 6$. Note that $z = \sqrt{1-r^2}$. Next we show that $z > r$ for each $n \ge 5$ $(n \ne 6)$. Suppose that $z \le r$ for some $n \ge 5$ $(n \ne 6)$. Then $z^2 \le r^2 \Leftrightarrow 1 - r^2 \le r^2 \Leftrightarrow 1 \le 2r^2 = \frac{1}{2}\csc^2\alpha \Leftrightarrow 2 \le \csc^2\alpha \Leftrightarrow \frac{1}{2} \ge \sin^2\alpha \Leftrightarrow \frac{\sqrt{2}}{2} \ge \sin\alpha$. Hence $\alpha \le \pi/4$, a contradiction since $\pi/4 < 3\pi/10$. Therefore, $span_3(W_{2,n}) \le 2z$.

**Theorem 4.3** $span_4(W_{m,3}) = \dfrac{\sqrt{6}}{2}$.

*Proof.* Place a unit equilateral triangle $C_3 = [p_1, p_2, p_3, p_1]$ on the $xy$-plane with centroid at (0,0,0,0). Each of the points on the circle

$$C = \{(0,0,a,b) \in R^4 : a^2 + b^2 = 2/3\}$$

is at a unit distance from each of the vertices of $C_3$. Let $\delta > 0$ and consider an arc of $C$ of length $\delta$. Take $m$ points $z_k$ from this arc, where $z_1 = (0,0,\sqrt{6}/3,0)$. The points $p_i$ and $z_k$ form a unit representation of $W_{m,3}$ in $\mathbf{R}^4$. Note that the points $p_i$ and $z_1$ form a unit $K_4$ in $\mathbf{R}^4$. Thus $span_4(W_{m,3}) \ge \sqrt{6}/2$. Letting $\delta \to 0$, we see that a unit $W_{m,3}$ in $\mathbf{R}^4$ can be inscribed by an open ball of diameter $\sqrt{6}/2 + \varepsilon$ for each $\varepsilon > 0$.

**Theorem 4.4** *For* $n \ge 4$ $(n \ne 6)$ *and* $m \ge 2$,

$$span_4(W_{m,n}) \le \frac{2}{\sqrt{4 - csc^2\alpha}},$$

*where*

$$\alpha = \begin{cases} (n-1)\pi/2n & \text{if } n \text{ is odd,} \\ (n-4)\pi/2n & \text{if } n \text{ is even and } n \not\equiv 0 \,(mod\, 4), \\ (n-2)\pi/2n & \text{if } n \text{ is even and } n \equiv 0 \,(mod\, 4). \end{cases}$$

*Proof.* Let $C_n = [p_1, p_2, \ldots, p_n, p_1]$ be a small circle unit representation of $C_n$ in $\mathbf{R}^3$ with the circle in the *xy*-plane centered at (0,0,0,0) and radius $r_n$. Since $r_n < 1$, there exists real number $z_n > 0$ such that each point on the circle $C = \{(0,0,a,b) \in R^4 : a^2 + b^2 = z_n^2\}$ is at a unit distance from each vertex of $C_n$. Let $\delta > 0$ and consider an arc of $C$ of length $\delta$. Take $m$ points $q_k$ from this arc, where $q_1 = (0,0,z_n,0)$. The projections of the points $p_i$ and $q_1$ into $\mathbf{R}^3$ form a unit $W_{1,n}$ in $\mathbf{R}^3$. Letting $\delta \to 0$, we see that $W_{m,n}$ can be inscribed in an open ball of diameter $\varepsilon + 2/\sqrt{4 - csc^2\alpha}$ for each $\varepsilon > 0$ (see Theorem 3.3).

**References**


1. Fred Buckley and Frank Harary, *On the Euclidean Dimension of a Wheel*, Graphs and Combinatorics 4(1988), 23-30.
2. Severino V. Gervacio, *Unit Embedding of Graphs in the Euclidean n-Space*, Plenary Paper, AMC '95, Suranaree University, Thailand.
3. Severino V. Gervacio & Blessilda P. Raposa, *On the Dimension and Span of Graphs*, NRCP Research Project, 1996-97.
4. Frank Harary, *Graph Theory*, Addison-Wesley, Massachusetts, 1972.
5. Rowena T. Isla, *On the Dimension of Graphs*, Ph.D. Thesis, MSU-Iligan Institute of Technology, March 2000.
6. Hiroshi Maehara, *Distance Graphs in Euclidean Space*, Ryukyu Math Journal **5**(1992), 33-51.
7. Joselito A. Uy, *On the Span of Graphs in the Euclidean n-Space*, Matimyas Matematika, Vol. 24, No. 1, January 1999, 25-32.

# On the Regularity of a Trust Region-CG Algorithm for Nonlinear Ill-posed Inverse Problems*

Yan-fei Wang Ya-xiang Yuan
State Key Laboratory of Scientific and Engineering Computing,
Institute of Computational Mathematics and Scientific/Engineering computing,
Academy of Mathematics and System Sciences,
Chinese Academy of Sciences, P.O.BOX 2719, Beijing, 100080, China
Email: wyf@lsec.cc.ac.cn yyx@lsec.cc.ac.cn

**Abstract.** In this paper we consider the regularity of the trust region-cg algorithm, when it is applied to nonlinear ill-posed iverse problems. The trust region algorithm can be viewed as a regularization method, but it differs from the traditional regularization method, because no penalty term is need. Thus, the determing of the so-called regularization parameter in a standard regularization method is avoided. Theoretical analysis of the trust region-cg method is presented, convergency and regularity of the trust region algorithm are proved, and numerical tests are also given.

**Key Words.** Nonlinear ill-posed problems, trust region-cg method, convergence, regularity

**AMS Subject Classifications:** 65J15, 65J20, 65K10

## 1 Introduction

In scientific and engineering computing, we are often encountered with nonlinear inverse problems. An inverse problem consists of a direct problem and some unknown function(s) or parameters. Inverse problems are usually ill-posed in the sense of J. Hadmard, i.e., at leat one term of the *existence, uniqueness, stability* of the solution is vilated. Particularly we are concerned with the stability, since in many applications the solution does not depend continuously on the unknown quantities and the problem is ill-posed. A typical ill-posed problems is to determine these unknowns given measured, or condaminated data.

We can outline the nonlinear ill-posed problems into an abstract operator equations

$$F(x) = y, \tag{1}$$

where $F : D(F) \subset X \longrightarrow Y$ is a nonlinear mapping, $X$ and $Y$ are both seperable Hilbert spaces. We assume that $F$ is continuous and compact for fixed $x \in D(F)$.

Problem (1) is typically ill-posed in the sense that a solution $x^+$ does not depend continuously on the obervation data $y$. Since in practice only approximate data with some error level

*Partially supported by Chinese NSF grant 19731010 and the Knowledge Innovation Program of CAS. The first author also partially supported by Hebei Province NSF grant 101001.

$\delta$, i.e.,

$$\|y_\delta - y\| \le \delta \tag{2}$$

are available, problem (1) has to be regularized (see e.g. [3, 7, 23]). Through out this paper we assume that a solution $x^+$ of (1) exists, i.e.

$$F(x^+) = y. \tag{3}$$

Regularization methods are such kind of methods which replace the ill-posed problem with a stabilized problem whose solution depends on a parameter, named as the regularization parameter. The regularized problem is well-posed in the sense of J. Hadamard. For a complete theoretical analysis of such kind of method, please see some well-written books [23, 3, 13, 7, 14, 17].

Certainly the most well-known and most widely used regularization method for nonlinear ill-posed problems is the method of Tikhonov regularization. In which one solves the unconstrained minimization problem

$$\min_{x\in X} J_\alpha[x, y] := \|F(x) - y_\delta\|^2 + \alpha\theta(x). \tag{4}$$

$\alpha > 0$ is the regularization parameter. $\theta(x)$ serves as the stabilizer, i.e., stablizes the minimization process and provides a priori information about the solution.

Replacing $F(x)$ by first order Taylor's expansion, i.e., (4) turns into

$$\min_{\xi\in X} J_\alpha[\xi, y] := \|y_\delta - F(x_k) - F'(x_k)\xi\|^2 + \alpha\theta(\xi). \tag{5}$$

If an approximate solution $\xi_k$ of (5) is computed, we can let $x_{k+1} = x_k + \xi_k$.

Assume that $x_k$ is some approximation of the solution $x^+$, then

$$F(x^+) - F(x_k) = F'(x_k)(x^+ - x_k) + r(x^+; x_k), \tag{6}$$

where $r(x^+; x_k)$ is the Taylor remainder. Denoting $\xi^+ = x^+ - x_k$ and solving for it leads to

$$F'(x_k)\xi^+ = y - F(x_k) - r(x^+; x_k). \tag{7}$$

The above relation can be rewritten as

$$F'(x_k)\xi^+ = y_\delta - F(x_k) + y - y_\delta - r(x^+; x_k). \tag{8}$$

Thus, the linearized problem

$$F'(x_k)\xi = y_\delta - F(x_k) \tag{9}$$

is an approximation to the original problem up to up to an error $err = y_\delta - y + r(x^+; x_k)$ with

$$\|err\| \le \delta + \|r(x^+; x_k)\|. \tag{10}$$

Clearly (5) is equivalent to applying Tikhonov regularization to problem (9).

**Assumption 1.1** *Assume that after k iterations, $\xi^+ = x^+ - x_k$ satisfies*

$$\|y_\delta - F(x_k) - F'(x_k)\xi^+\| \leq \omega\|y_\delta - F(x_k)\|, \;\; 0 < \omega < 1.$$

Apart from the above analysis, we need the following assumption:

**Assumption 1.2** *For a certain ball $\mathcal{B} \subset D(F)$ around the exact solution $x^+$ of (1), and some $1 > d > 0$ let*

$$\|F(x) - F(\hat{x}) - F'(\hat{x})(x - \hat{x})\| \leq d\|F(x) - F(\hat{x})\| \tag{11}$$

*for all $x, \; \hat{x} \in \mathcal{B}$.*

This assumption is helpful for analyzing the properties of the trust region algorithm which we presented in this paper.

Recently, optimization methods are becoming popular for solving nonlinear ill-posed inverse problems, for example, Gauss-Newton method ([1, 12]), Broyden's method ([11]), and Levenberg-Marquardt method ([9]), which have been well developed in nonlinear programming.

Trust region method has been used in parameter identification problem and image restoration problem (see [25, 26]) and seems promising. This paper will consider trust region method for nonlinear ill-posed inverse problems.

## 2 A Trust Region-CG Algorithm

Considering the unconstrained optimization problem

$$\min_{x \in X} J[x, y_\delta] := \|F(x) - y_\delta\|^2. \tag{12}$$

We denote by $g(x)$ the gradient of the functional $J$, $Hess(x)$ the approximate Hessian of $J$, i.e.,

$$g(x) = F'(x)^T(F(x) - y_\delta), \;\; Hess(x) = F'(x)^T F'(x).$$

At the $k$-th iteration, a trust region subproblem (TRS) for (12) is

$$\min_{x \in R^n} g_k^T \xi + \frac{1}{2}(Hess_k \xi, \xi) := \psi_k(\xi), \tag{13}$$

$$s.\ t.\ \|\xi\| \leq \Delta, \tag{14}$$

where $g_k = g(x_k), \;\; Hess_k = Hess(x_k)$ and $\Delta > 0$ is the trust region bound. (13)–(14) is solved exactly or inexactly to obtain a trial step $\xi_k$. The ratio

$$r_k = \frac{Ared_k}{Pred_k} \tag{15}$$

is used to decide whether the trial step $\xi_k$ is acceptable and to adjust the trust region bound.

$$Ared_k = J[x_k, y_\delta] - J[x_k + \xi_k, y_\delta] \tag{16}$$

is called the actual reduction in the objective model, and

$$Pred_k = \psi_k(0) - \psi_k(\xi_k) \tag{17}$$

is the predicted reduction. We outline the general trust region algorithm for unconstrained optimization as follows.

**Algorithm 2.1** *(Trust region algorithm for nonlinear ill-posed problem)*

*STEP 1* *Given the initial guess value* $x_1 \in R^n$, $\Delta_1 > 0$, $0 < \tau_3 < \tau_4 < 1 < \tau_1$, $0 \leq \tau_0 \leq \tau_2 < 1$, $\tau_2 > 0$, $k := 1$;

*STEP 2* *If some stopping rule is satisfied then STOP; Else, solve (13)-(14) giving* $\xi_k$;

*STEP 3* *Compute* $r_k$;

$$x_{k+1} = \begin{cases} x_k & if \quad r_k \leq \tau_0, \\ x_k + \xi_k & otherwise; \end{cases} \tag{18}$$

*Choose* $\Delta_{k+1}$ *that satisfies*

$$\Delta_{k+1} = \begin{cases} [\tau_3\|\xi_k\|, \tau_4\Delta_k] & if \quad r_k < \tau_2, \\ [\Delta_k, \tau_1\Delta_k] & otherwise; \end{cases} \tag{19}$$

*STEP 4* *Evaluate* $g_k$ *and* $Hess_k$; *k:=k+1; GOTO STEP 2.*

The constant $\tau_i$ $(i = 0, \cdots, 4)$ can be chosen by users. Typical values are $\tau_0 = 0$, $\tau_1 = 2$, $\tau_2 = \tau_3 = 0.25$, $\tau_4 = 0.5$. For other choices of those constants, please see [4], [5], [15], [19], etc.. The parameter $\tau_0$ is usually zero (see [4], [20]) or a small positive constant (see [2] and [21]). The advantage of using zero $\tau_0$ is that a trial step is accepted whenever the objective function is reduced. Hence it would not throw away a "good point", which is a desirable property especially when the function evaluations are very expensive.

In STEP 2, the stopping rule is based on some kind of so-called discrepancy principle, i.e., once the inequality

$$\|F(x_k) - y_\delta\| \leq \tilde{\omega}\delta, \text{ with } \tilde{\omega} > 1$$

is satisfied, no further iteration is needed.

The following lemma is well known (for example, see [16] and [6]):

**Lemma 2.2** *A vector* $\xi^* \in R^n$ *is a solution of (13)-(14) if and only if there exists* $\lambda^* \geq 0$ *such that*

$$(Hess_k + \lambda^* I)\xi^* = -g_k \tag{20}$$

*and that* $Hess_k + \lambda^* I$ *is positive semi-definite,* $\|\xi^*\| \leq \Delta$ *and*

$$\lambda^*(\Delta - \|\xi^*\|) = 0. \tag{21}$$

It is shown by Powell [20] that trust region algorithms for (12) is convergent if the trust region step satisfies

$$Pred(\xi) \geq c\|g\| \min\{\Delta, \|g\|/\|Hess\|\} \tag{22}$$

and some other conditions on $Hess$ are satisfied. It is easy to see that

$$\psi(0) - \min_{\|\xi\| \leq \Delta, \xi \in span\{g\}} \psi(s) \geq \frac{1}{2} \quad \|g\| \min\{\Delta, \|g\|/\|Hess\|\}. \tag{23}$$

Therefore it is quite common that in practice the trial step at each iteration of a trust region method is computed by solving the TRS (13)-(14) inexactly. One way to compute an inexact solution of (13)-(14) was the truncated conjugate gradient method proposed by Toint [24] and Steihaug [22] and analyzed by Yuan [30].

The conjugate gradient method for (13) generates a sequence as follows:

$$\xi_{l+1} = \xi_l + \alpha_l d_l, \tag{24}$$

$$d_{l+1} = -g^{\psi}_{l+1} + \beta_l d_l, \tag{25}$$

where $g_l^{\psi} = \nabla\psi_k(\xi_l) = Hess_k\, \xi_l + g_k$ with $g_k = g(x_k) = F'(x_k)^T(F(x_k) - y_\delta)$, $Hess_k = Hess(x_k) = F'(x_k)^T F'(x_k)$ and

$$\alpha_l = -g_l^{\psi^T} d_l / d_l^T Hess_k\, d_l, \quad \beta_l = \|g^{\psi}_{l+1}\|^2/\|g_l^{\psi}\|^2, \tag{26}$$

with the initial values $\xi_1 = 0, \;\; d_1 = -g_1^{\psi} = -g_k$.

Toint [24] and Steihaug [22] were the first to use the conjugate gradient method to solve the general trust region subproblem (13)-(14). Even without assuming the positive definiteness of $Hess$, we can continue the conjugate gradient method provided that $d_l^T Hess\, d_l$ is positive. If the iterate $\xi_l + \alpha_l d_l$ computed is in the trust region ball, it can be accepted, and the conjugate gradient iterates can be continued to the next iteration. Whenever $d_l^T Hess\, d_l$ is not positive or $\xi_l + \alpha_l d_l$ is outside the trust region, we can take the longest step along $d_l$ within the trust region and terminate the calculations.

**Algorithm 2.3** *(Truncated conjugate gradient method for TRS)*

*STEP 1* Given $\xi_1 = 0$, $0 < \tau < 1$, $\epsilon$ *(tolerance)* $> 0$ *and compute* $g_1^{\psi} = \nabla\psi(\xi_1)$, *set* $l := 1$, $d_1 = -g_1^{\psi} = -g_k$;

*STEP 2* *If* $\|A_k\xi_l - \tilde{u}_k\| \leq \tau\|\tilde{u}_k\|$, *stop, output* $\xi^* = \xi_l$;

*Compute* $d_l^T Hess\, d_l$: *if* $d_l^T Hess_k\, d_l \leq 0$ *then goto step 4;*

*Calculate* $\alpha_l$ *by (26).*

*STEP 3* *If* $\|\xi_l + \alpha_l d_l\| \geq \Delta_k$ *then goto step 4;*

*Set* $\xi_{l+1}$ *by (24) and* $g^{\psi}_{l+1} = g_l^{\psi} + \alpha_l Hess_k\, d_l$;

*Compute* $\beta_l$ *by (26) and set* $d_{l+1}$ *by (25);*

*l:=l+1, goto step 2.*

*STEP 4 Compute $\alpha_l^* \geq 0$ satisfying $\|\xi_l + \alpha_l^* d_l\| = \Delta$;*

*Set $\xi^* = \xi_l + \alpha_l^* d_l$, and stop.*

Note that $\alpha_l^*$ can be computed by choosing the positive root of the quadratic equation in $\alpha$:

$$\|d_l\|^2\alpha^2 + 2(\xi_l, d_l)\alpha + \|\xi_l\|^2 - \Delta_k^2 = 0. \tag{27}$$

Let $\xi^*$ be the inexact solution of (13)-(14) obtained by the above truncated CG method and $\hat{\xi}$ be the exact solution of (13)-(14). Recently Yuan [30] shows that

$$\frac{\psi(0) - \psi(\xi^*)}{\psi(0) - \psi(\hat{\xi})} \geq \frac{1}{2}, \tag{28}$$

which can be written as the following theorem:

**Theorem 2.4** *For any $\Delta > 0$, $g \in R^n$ and any positive definite matrix $Hess \in R^{n\times n}$, let $\hat{s}$ be the global solution of the trust region subproblem (13)-(14), and let $\xi^*$ be the solution obtained by the truncated CG method, then*

$$\psi(\xi^*) \leq \frac{1}{2}\psi(\hat{s}). \tag{29}$$

This theorem tells us that the reduction in the approximate model is at least half of the maximum reduction if we use the truncated conjugate gradient method for solving the subproblem (13)-(14).

Applying Algorithm 2.3 to compute the trial step $\xi_k$ in Step 2 of Algorithm 2.1, we obtain a trust region-cg algorithm for nonlinear ill-posed inverse problems. The algorithm consists of two stage iterations: the inner loop and the outer loop. The inner loop is the truncated conjugate gradient method, the outer loop is the trust region method.

To avoid too many inner loop iterations in one out loop iteration, we terminate the inner loop iteration if *itermax* cg steps have been taken, where *itermax* is a given positive number. We also terminate the inner look iteration if a progress in function reduction in the cg step is smaller than $\epsilon$. However, in our numerical tests, this termination rule was not activated.

# 3 Properties of Algorithm 2.3

In this section we give some properties of the truncated conjugate gradient method. The main result is the monotonicity of the iterates.

First, we present an equivalent form of the conjugate gradient method. Denote

$$A_k = F'(x_k),\ u_k = y - F(x_k) - r(x^+; x_k),\ \tilde{u}_k = y_\delta - F(x_k),$$

we have

$$Hess_k = A_k^* A_k,\ g_k = -A_k^* \tilde{u}_k,\ g_l^\psi = A_k^*(A_k \xi_l - \tilde{u}_k),$$

$$\begin{aligned} d_{l+1} &= -g_{l+1}^{\psi} + \beta_l d_l = -A_k^*(A_k\xi_{l+1} - \tilde{u}_k) + \beta_l d_l \\ &= -A_k^* A_k \xi_l - \alpha_l A_k^* A_k d_l + A_k^* \tilde{u}_k + \beta_l d_l. \end{aligned}$$

If we let $d_l = A_k^* z_l$ provided that such $z_l$ exists, then

$$d_{l+1} = A_k^*(\tilde{u}_k - A_k\xi_l - \alpha_l A_k d_l + \beta_l z_l).$$

Further if we denote $r_l = \tilde{u}_k - A_k\xi_l$, then clearly

$$r_{l+1} = \tilde{u}_k - A_k\xi_{l+1} = r_l - \alpha_l A_k d_l$$

and

$$d_{l+1} = A_k^*(r_l - \alpha_l A_k d_l + \beta_l z_l)$$

hold. We can generate the next search direction by $d_{l+1} = A_k^* z_{l+1}$ with $z_{l+1} = r_l - \alpha_l A_k d_l + \beta_l z_l = r_{l+1} + \beta_l z_l$. Hence, the conjugate gradient iterates can be genereated in the following way:

$$\xi_{l+1} = \xi_l + \alpha_l d_l \tag{30}$$

$$d_l = A_k^* z_l \tag{31}$$

$$z_{l+1} = r_{l+1} + \beta_l z_l \tag{32}$$

$$r_{l+1} = r_l - \alpha_l A_k d_l \tag{33}$$

$$\alpha_l = -\frac{g_l^{\psi^T} d_l}{d_l^T Hess\, d_l} = \frac{\|A_k^* r_l\|^2}{\|A_k d_l\|^2} \tag{34}$$

$$\beta_l = \frac{\|A_k^* r_{l+1}\|^2}{\|A_k^* r_l\|^2} \tag{35}$$

$$\sigma_{l+1} = 1 + \beta_l \sigma_l \tag{36}$$

The initial values are $\xi_1 = 0,\ d_1 = -g_k,\ z_1 = r_1,\ r_1 = \tilde{u}_k = y_\delta - F(x_k),\ \sigma_1 = 1$. Here another scalar $\sigma_l$ is added, which will be used for the analysis of the truncated conjugate gradient method.

One tool for the analysis of the truncated conjugate gradient method is the so-called residual polynomials (see [10, 3]). Let $\Pi_l$ be the set of all polynomials of degree $l$ or less, and set

$$\Pi_l^0 := \{p \in \Pi_l : p(0) = 1\}.$$

Then there is an 1-1 relation between elements $\xi \in \mathcal{K}_l(A_k^*\tilde{u}_k; A_k^* A_k)$ and $p \in \Pi_l^0$ via the representation

$$\tilde{u}_k - A_k\xi = p(A_k A_k^*)\tilde{u}_k \tag{37}$$

of the corresponding residual, where $\mathcal{K}_l(A_k^*\tilde{u}_k; A_k^* A_k)$ is the $l$-th Krylov subspace

$$\mathcal{K}_l(A_k^*\tilde{u}_k; A_k^* A_k) = \operatorname{span}\{A_k^*\tilde{u}_k, (A_k^* A_k)A_k^*\tilde{u}_k, \cdots, (A_k^* A_k)^{l-1} A_k^*\tilde{u}_k\}.$$

For simplicity, $p_l \in \Pi_l^0$ denotes the residual polynomial associated with $\xi_l$, the $l$-th CG iterate.

The bilinear form

$$< p, q >:= (p(A_k A_k^*)\tilde{u}_k, q(A_k A_k^*)\tilde{u}_k)$$

defines the inner product for $p,\ q \in \Pi_l$. If $q \in \Pi_{l-1}$ is an arbitrary polynomial of degree $l-1$ then the polynomial $p$ given by $p(\lambda) = p_l(\lambda) + t\lambda q(\lambda)$ belongs to $\Pi_l^0$ for every $t \in R$. Noticing that $p_l$ solves the minimization problem

$$< p, p > \longrightarrow \min \quad \text{for } p \in \Pi_l^0.$$

Hence,

$$< p_l, \lambda q > = \frac{1}{2}\frac{d}{dt} < p, p > |_{t=0} = 0, \text{ for all } q \in \Pi_{l-1}. \tag{38}$$

If we define $q = q_{l-1}$ by $p_l = 1 - \lambda q_{l-1}$, then clearly we have that,

$$< p_l, 1 > = < p_l, p_l >, \tag{39}$$

which will be used for later analysis.

From (30)-(36) and the above definitions, we have that

$$z_l = s_l(A_k A_k^*)\tilde{u}_k, \ s_l(\lambda) := \frac{p_l(\lambda) - p_{l+1}(\lambda)}{\alpha_l \lambda} \in \Pi_l. \tag{40}$$

It was pointed out by [8] that in general $s_l$ does not belong to $\Pi_l^0$. Instead, since the vectors $z_l$ are updated by $z_{l+1} = r_{l+1} + \beta_l z_l$ with $r_{l+1} = \tilde{u}_k - A_k \xi_{l+1}$, it follows from (40) that

$$s_{l+1}(\lambda) = p_{l+1}(\lambda) + \beta_l s_l(\lambda), \tag{41}$$

and hence, $s_l(0)$ and $\sigma_l$ of Algorithm 2.3 share the same recurrence relation (this in fact has been observed by [8]), i.e.,

$$s_l(0) = \sigma_l. \tag{42}$$

With the above analysis, we can now present the monotonicity of the iterates for perturbed right-hand side.

**Theorem 3.1** *Let $\gamma \geq 2$, $l^* \in N$. If Assumption 1.1 holds and if*

$$\|\tilde{u}_k - A_k \xi_l\|^2 + \|\tilde{u}_k - A_k \xi_{l+1}\|^2 > \omega\gamma \frac{\|z_l\| \|\tilde{u}_k\|}{\sigma_l}, \ 0 < \omega < 1, \ l = 1, 2, \cdots, l^*, \tag{43}$$

*then $\|\xi^+ - \xi_l\|$ is strictly monotonically decreasing for $l = 1, 2, \cdots, l^*$, and*

$$\|\xi^+\|^2 - \|\xi^+ - \xi_{l^*+1}\|^2 > (\gamma - 2)\omega \|\tilde{u}_k\| \sum_{l=1}^{l^*} \alpha_l \|z_l\|. \tag{44}$$

*Proof.* By induction, we obtain

$$\begin{aligned}
\|\xi^+ - \xi_{l+1}\|^2 &= \|\xi^+ - \xi_l - \alpha_l d_l\|^2 \\
&= \|\xi^+ - \xi_l\|^2 - 2(\xi^+ - \xi_l, \alpha_l A_k^* z_l) + (\alpha_l A_k^* z_l, \alpha_l A_k^* z_l) \\
&= \|\xi^+ - \xi_l\|^2 - \alpha_l(2A_k\xi^+ - 2A_k\xi_l - \alpha_l A_k d_l, z_l) \\
&= \|\xi^+ - \xi_l\|^2 - \alpha_l(\tilde{u}_k - A_k\xi_l, z_l) - \alpha_l(\tilde{u}_k - A_k\xi_{l+1}, z_l) \\
&\quad +2\alpha_l(\tilde{u}_k - A_k\xi^+, z_l).
\end{aligned}$$

From the definitions of $p_l$ and $z_l$, we have

$$\|\xi^+ - \xi_l\|^2 - \|\xi^+ - \xi_{l+1}\|^2 = \alpha_l < p_l, s_l > + \alpha_l < p_{l+1}, s_l > -2\alpha_l < \tilde{u}_k - A_k\xi^+, z_l > .$$

By (42), $s_l(\lambda) = \sigma_l + \lambda q(\lambda)$ for some polynomial $q \in \Pi_{l-1}$, and hence from (38) and (39) we find that

$$\begin{aligned}
&\|\xi^+ - \xi_l\|^2 - \|\xi^+ - \xi_{l+1}\|^2 = \alpha_l\sigma_l(< p_l, 1 > + < p_{l+1}, 1 >) - 2\alpha_l < \tilde{u}_k - A_k\xi^+, z_l > \\
&= \alpha_l\sigma_l(< p_l, p_l > + < p_{l+1}, p_{l+1} >) - 2\alpha_l < \tilde{u}_k - A_k\xi^+, z_l > .
\end{aligned}$$

Since $< p_l, p_l >= \|\tilde{u}_k - A_k\xi_l\|^2$, hence it follows from the above relation and (43) that

$$\|\xi^+ - \xi_l\|^2 - \|\xi^+ - \xi_{l+1}\|^2 > \omega\gamma\alpha_l\|z_l\|\|\tilde{u}_k\| - 2\omega\alpha_l\|\tilde{u}_k\|\|z_l\|, \text{ for all } l = 1, \cdots, l^*. \tag{45}$$

Thus, the monotonicity of $\|\xi^+ - \xi_l\|$ follows from the above inequality and the assumption $\gamma > 2$. Relation (44) follows by taking the sum of (45) for $l = 1, \cdots, l^*$ and observing the fact that $\xi_1 = 0$. Q.E.D

**Remark 3.2** *We have noted that in general $s_l$ will not belong to $\Pi_l^0$, however, $\hat{s}_l := s_l/\sigma_l \in \Pi_l^0$. Hence from the minimization property of the truncated CG we obtain*

$$\|\tilde{u}_k - A_k\xi_{l+1}\| \le \|\tilde{u}_k - A_k\xi_l\| =< p_l, p_l >^{\frac{1}{2}} \le < \hat{s}_l, \hat{s}_l >^{\frac{1}{2}} = \frac{1}{\sigma_l} < s_l, s_l >^{\frac{1}{2}} = \frac{1}{\sigma_l}\|z_l\|. \tag{46}$$

*This together with (43) yield that*

$$\begin{aligned}
2\|\tilde{u}_k - A_k\xi_{l^*}\|^2 &\ge \|\tilde{u}_k - A_k\xi_{l^*}\|^2 + \|\tilde{u}_k - A_k\xi_{l^*+1}\|^2 \\
&> \omega\gamma\frac{\|z_{l^*}\|\|\tilde{u}_k\|}{\sigma_{l^*}} > \omega\gamma\|\tilde{u}_k\|\|\tilde{u}_k - A_k\xi_{l^*}\|,
\end{aligned}$$

*which indicates that*

$$\|\tilde{u}_k - A_k\xi_{l^*}\| \ge \frac{\omega\gamma\|\tilde{u}_k\|}{2} > \omega\|\tilde{u}_k\|.$$

*Since $\|\tilde{u}_k - A_k\xi^+\| \le \omega\|\tilde{u}_k\|$ according to Assumption 1.1, this shows that $l^*$ can not exceed $l_k$, the smallest index of the inner iteration.*

# 4 Convergence of Trust Region-CG for Exact Data

Before presenting the proposition in the following paragraph, we first give an assumption:

**Assumption 4.1** *Assume that in each inner iteration, $\xi_l$ satisfies*

$$\|\tilde{u}_k - A_k \xi_l\| \geq \omega_1 \|\tilde{u}_k\|, \text{ with } \omega_1^2 = \gamma\omega \tag{47}$$

*until convergence, where* $0 < \omega < \gamma^{-1}$, $\gamma > 2$.

Assumption 4.1 is closely related with the termination rule of Algorithm 2.3. Where, $\tau$ in STEP 2 serves as the number $\omega_1$ here. Once the opposite inequality of Assumption 4.1 is satisfied, we stop the inner iteration.

**Proposition 4.2** *Suppose that Assumption 4.1 holds. The inequality (47) indicates that (43) is true if $l > 0$. Furthermore, there are only finitely many $l$ for which (47) holds.*

*Proof.* From the Remark 3.2 we know

$$\frac{1}{\sigma_l}\|z_l\| = < \hat{s}_l, \hat{s}_l >^{\frac{1}{2}} .$$

It is proved by [10] that $< \hat{s}_l, \hat{s}_l >$ is strictly monotonically decreasing with $l$, consequently

$$\frac{1}{\sigma_l}\|z_l\| < \frac{1}{\sigma_1}\|z_1\| = \|r_1\| = \|\tilde{u}_k\|.$$

The above inequality together and (47) indicates that (43) is true for $l > 0$. From Remark 3.2, we see that (47) holds for only finitely many indices $l$. Q.E.D

Now assume that $y_\delta = y$, we first prove the monotonicity of the trust region-cg algorithm, i.e., $x_k + \xi_{l_k}$ is a better approximation of $x^+$ than $x_k$. We also assumed that $F(x) = y$ has a solution $x^+ \in \mathcal{B} \subset D(F)$.

**Proposition 4.3** *Assume that Assumptions 1.2 and 4.1 holds, then the iteration error $\|x^+ - x_k\|$ is monotonically decreasing.*

*Proof.* According to Assumption 1.2, (11) holds for $x = x^+$, $\hat{x} = x_k$, i.e.

$$\|y - F(x_k) - F'(x_k)(x^+ - x_k)\| \leq d\|y - F(x_k)\|.$$

Note that $y_\delta = y$, the above expression indicates that Assumption 1.1 is fulfilled with $0 < d < 1$.

Due to Assumption 4.1,

$$\|\tilde{u}_k - A_k \xi_l\|^2 \geq \gamma\omega\|\tilde{u}_k\|^2$$

is satisfied. Hence from Proposition 4.2, the requirement of Theorem 3.1 is fulfilled.

From our notations, we have that $x_{k+1} = x_k + \xi_{l_k}$. Thus, from $\xi^+ = x^+ - x_k$, $\xi^+ - \xi_{l_k} = x^+ - x_{k+1}$ and Theorem 3.1, we see that $\|x^+ - x_{k+1}\| < \|x^+ - x_k\|$. Q.E.D

**Remark 4.4** *Propostion 4.3 also implies two inequalities:*

$$\|u_k\|\|w_k\| < \frac{1}{d(\gamma-2)}(\|x^+ - x_k\|^2 - \|x^+ - x_{k+1}\|^2) \tag{48}$$

*and*

$$\|u_k\|^2 < \frac{\|A_k\|^2}{d(\gamma-2)}(\|x^+ - x_k\|^2 - \|x^+ - x_{k+1}\|^2). \tag{49}$$

(48) is straightforward as $\|u_k - A_k\xi^+\| \le d\|u_k\|$ with $0 < d < 1$. (49) follows from relations

$$\|u_k\|\|w_k\| > \alpha_1\|u_k\|\|z_1\| = \alpha_1\|u_k\|^2$$

and

$$\alpha_1 = \frac{\|A_k^* r_1\|^2}{\|A_k d_1\|^2} = \frac{\|A_k^* u_k\|^2}{\|A_k A_k^* u_k\|^2} \ge \|A_k\|^{-2}.$$

**Theorem 4.5** *Given the exact data $y_\delta = y$ and suppose that Assumptions 1.2 and 4.1 hold. Then the iterates $\{x_k\}$ generated by Algorithms 2.1 and 2.3 converge to a solution of (1) as $k \to \infty$.*

*Proof.* First we prove that $\{x_k\}$ forms a Cauchy sequence. Let us denote the iteration errors by $e_k = x^+ - x_k$. Given $k$, $j \in N$ with $k > j$, let $\nu \in \{j, \cdots, k\}$ be chosen in such a way that

$$\|y - F(x_\nu)\| \le \|y - F(x_i)\|, \; i = j, \cdots, k.$$

Consider now

$$\|e_\nu - e_j\|^2 = \|e_j\|^2 - \|e_\nu\|^2 + 2(e_\nu, e_\nu - e_j). \tag{50}$$

Note that

$$e_j - e_\nu = x_\nu - x_j, \; x_\nu - x_j = \sum_{i=j}^{\nu-1} A_i^* w_i,$$

where $A_i = F'(x_i)$, $w_i = \sum_{l=1}^{l_i-1} \alpha_l z_l$. Hence we obtain

$$|(e_\nu, e_\nu - e_j)| = |\sum_{i=1}^{\nu-1}(A_i^* w_i, e_\nu)| \le \sum_{i=1}^{\nu-1} \|w_i\|\|A_i e_\nu\|.$$

We can estimate that

$$\begin{aligned} \|A_i e_\nu\| &= \|A_i e_i - A_i(e_\nu - e_i)\| \\ &\le \|y - F(x_i) - F'(x_i)e_i\| + \|F(x_\nu) - F(x_i) - F'(x_i)(e_\nu - e_i)\| \\ &\quad + \|y - F(x_\nu)\| \\ &\le d\|y - F(x_i)\| + d\|F(x_\nu) - F(x_i)\| + \|y - F(x_\nu)\| \\ &\le 2d\|y - F(x_i)\| + (1+d)\|y - F(x_\nu)\| \le (1+3d)\|y - F(x_i)\|. \end{aligned}$$

Relation (48) and the above expression give that

$$\begin{aligned} |(e_\nu, e_\nu - e_j)| &\le (1+3d)\sum_{i=j}^{\nu-1} \|w_i\|\|y - F(x_i)\| \\ &\le \frac{1+3d}{d(\gamma-2)}(\|x^+ - x_j\|^2 - \|x^+ - x_\nu\|^2), \end{aligned}$$

which, together with (50), yields

$$\|e_\nu - e_j\|^2 \le C(\|x^+ - x_j\|^2 - \|x^+ - x_\nu\|^2)$$

with $C = \frac{2(1+3d)}{d(\gamma-2)} + 1$ independent of $\nu$, $j$, $k$. Similarly one can obtain

$$\|e_j - e_\nu\|^2 \le C(\|x^+ - x_\nu\|^2 - \|x^+ - x_j\|^2),$$

hence

$$\begin{aligned} \|x_k - x_j\|^2 &= \|e_j - e_k\|^2 \le 2(\|e_k - e_\nu\|^2 + \|e_\nu - e_j\|^2) \\ &\le 2C(\|x^+ - x_j\|^2 - \|x^+ - x_k\|^2). \end{aligned} \tag{51}$$

Therefore, $\{x_k\}$ form a Cauchy sequence because the monotonicity of $\{\|x^+ - x_k\|\}$.

Denote the limit of $x_k$ by $x$. From (49) we know $\sum_{k=1}^{\infty}\|u_k\|^2$ converges, and therefore $F(x_k) \to y$ as $k \to \infty$. This indicates that $x$ is a solution of (1). Q.E.D

# 5 Regularity of the Algorithm for Inexact Data

Now we consider the case where inexact date $y_\delta$ instead of $y$. It is assumed that (2) is satisfied.

Our stopping rule is based on the discrepancy principle, i.e., we terminate the calculations at the smallest iteration index $k_D$ such that the discrepancy inequality

$$\|y_\delta - F(x_k^\delta)\| \le \tilde{\omega}\delta, \text{ with } \tilde{\omega} > 1 \tag{52}$$

holds.

We denote $x_k^\delta$ the corresponding iterates and consider the regularity of the trust region-cg algorithm.

**Theorem 5.1** *Assume that Assumptions 1.2 and 4.1 hold. Let $x$ be a solution of (1) with $F$ satisfies (11) for some $1 > d > 0$ in a ball $\mathcal{B} \subset D(F)$ around $x$. Let $\tilde{\omega}$ in (52) be chosen that $\tilde{\omega} > \frac{1+d}{1-d}$. Then $\|x - x_k^\delta\|$ is monotonically decreasing. Moreover, Algorithm 2.1 terminates after $k_D < \infty$ iterations.*

*Proof.* We prove that

$$\|x - x_{k+1}^\delta\| \le \|x - x_k^\delta\| \tag{53}$$

with $x$ a solution of (1).

Using Assumption 1.2, we estimate that

$$\begin{aligned}\|y_\delta - F(x_k^\delta - F'(x_k^\delta)(x - x_k^\delta)\| &\leq \delta + \|F(x) - F(x_k^\delta - F'(x_k^\delta)(x - x_k^\delta)\| \\ &\leq \delta + d\|y - F(x_k^\delta)\| \\ &\leq (1+d)\delta + d\|y_\delta - F(x_k^\delta)\|.\end{aligned}$$

According to the discrepancy principle, $\|y_\delta - F(x_k^\delta)\| > \tilde{\omega}\delta$ as $k < k_D$, hence

$$\delta < \frac{1}{\tilde{\omega}}\|y_\delta - F(x_k^\delta)\|$$

and

$$\|y_\delta - F(x_k^\delta - F'(x_k^\delta)(x - x_k^\delta)\| \leq \frac{1+d+\tilde{\omega}}{\tilde{\omega}}\|y_\delta - F(x_k^\delta)\|.$$

By assumption, $0 < (1 + d + \tilde{\omega})/\tilde{\omega} < 1$, hence Assumption 1.1 is fulfilled. Consequently Propostion 4.3 applies and the monotonicity assertion (53) follows as in the proof of Proposition 4.3.

Next we show that there are only finite number of iterations. In fact as the same as in the proof of (49), we have

$$\|y_\delta - F(x_k^\delta)\|^2 < \frac{L}{d(\gamma-2)}(\|x - x_k^\delta\|^2 - \|x - x_{k+1}^\delta\|^2) \tag{54}$$

with $L = \sup\{\|F'(x_k^\delta)\|^2\}$ for all $k < k_D$.

Assume that (53) holds for $x = x^+$. Now taking the sum of (54) for $k = 1, 2, \cdots, k_D - 1$ we obtain

$$(k_D - 1)\tilde{\omega}^2\delta^2 \leq \sum_{k=1}^{k_D-1} \|y_\delta - F(x_k^\delta)\|^2 \leq \frac{L}{d(\gamma-2)}\|x^+ - x_1\|^2 < \infty.$$

This indicates that $k_D$ is a finite number. Q.E.D

**Theorem 5.2** *Assume that $F(x_k^\delta) \to F(x_k)$ as $\delta \to 0$. If $k \leq k_D$ for all $\delta$ sufficiently small, then $x_k^\delta \to x_k$ for $k \leq k_D$ as $\delta \to 0$.*

*Proof.* Given sufficiently small number $\epsilon$, we want to prove $\|x_k^\delta - x_k\| \leq \epsilon$ as $\delta \to 0$ for $k \leq k_D$. We proceed by induction.

Assume that $x_k^\delta \to x_k$ as $\delta \to 0$, and that $k + 1 \leq k_D$. Note that

$$x_{k+1} = x_k + \xi_{l_k}, \;\; x_{k+1}^\delta = x_k^\delta + \xi_{l_k}^\delta,$$

$$\xi_{l_k} = F(x_k)^* \sum_{i=1}^{l_k-1} \alpha_i z_i, \;\; \xi_{l_k}^\delta = F(x_k^\delta)^* \sum_{i=1}^{l_k-1} \alpha_i^\delta z_i^\delta,$$

we can estimate that

$$\begin{aligned}
\|x_{k+1}^{\delta} - x_{k+1}\| &\leq \|x_k^{\delta} - x_k\| + \|\xi_{l_k}^{\delta} - \xi_{l_k}\| \\
&\leq \|x_k^{\delta} - x_k\| + \|F(x_k^{\delta})^* \sum_{i=1}^{l_k-1} (\alpha_i^{\delta} z_i^{\delta} - \alpha_i z_i)\| \\
&\quad + \|(F(x_k^{\delta})^* - F(x_k)^*) \sum_{i=1}^{l_k-1} \alpha_i z_i\| \\
&\leq \|x_k^{\delta} - x_k\| + \|F(x_k^{\delta})\|(l_k - 1) \max_i\{\|\alpha_i^{\delta} z_i^{\delta} - \alpha_i z_i\|\} \\
&\quad + \|F(x_k^{\delta}) - F(x_k)\|\|\sum_{i=1}^{l_k-1} \alpha_i z_i\| \qquad (55)
\end{aligned}$$

By the induction assumption $x_k^{\delta} \to x_k$, we have that

$$F(x_k^{\delta}) \to F(x_k).$$

Therefore, it follows that $\alpha_i^{\delta} \to \alpha_i$, $z_i^{\delta} \to z_i$. Consequently, it from (55) that $x_{k+1}^{\delta} \to x_{k+1}$. Q.E.D

**Theorem 5.3** *Assume that $F$ satisfies (11) in some ball $\mathcal{B} \subset D(F)$ and let $y_\delta$, $x_\delta$ as before. Then the iterates $x_k^{\delta}$ generated by Algorithms 2.1 and 2.3 converge to a solution of (1) as $k \to \infty$ and $\delta \to 0$.*

*Proof.* For simplicity, We use $k(\delta)$ instead of $k_D$ in the following analysis.

From Theorem 4.5 we know that iterates $x_k$ converge to a solution of (1). Combining this fact with Theorem 5.2, we find that the iterates $x_k^{\delta}$ converge to a solution of (1) for $k \leq k(\delta)$ as $\delta \to 0$.

Now assume that $k(\delta) \to \infty$ as $\delta \to 0$, and denote $x^+$ the limit of the iterates $x_k$, $x^+$ is a solution of (1). It suffieces to consider subsequences $\{k(\delta_n)\}_n$ which are monotinically increasing to infinity as $n \to \infty$ and $\delta_n \to 0$. Without loss of generality, let us consider $k(\delta_m) > k(\delta_n)$ for $m > n$. By the monotonicity of $x_k^{\delta}$, i.e., Theorem 5.1, we have

$$\|x_{k(\delta_m)}^{\delta_m} - x^+\| \leq \|x_{k(\delta_n)}^{\delta_m} - x^+\| \leq \|x_{k(\delta(n)}^{\delta(m)} - x_k(\delta_n)\| + \|x_{k(\delta_n)} - x^+\|.$$

Given a sufficiently small number $\epsilon > 0$ and for some sufficiently large number $n$, $\|x_{k(\delta_n)} - x^+\| \leq \epsilon/2$ by Theorem 4.5. On the other hand, for sufficiently large number $m$ and fixed $n$, $\|x_{k(\delta(n)}^{\delta(m)} - x_k(\delta_n)\| \leq \epsilon/2$ by Theorem 5.1. This proves that $\|x_{k(\delta_m)}^{\delta_m} - x^+\| \leq \epsilon$ for all $m$ sufficiently large, and thereafter $x_{k(\delta_m)}^{\delta_m} \to x^+$ as $m \to \infty$. Hence, we see that $x_k^{\delta} \to x^+$ as $k \to \infty$ and $\delta \to 0$. Q.E.D.

# 6 Numerical Test

In this section, we give an example to test our algorithm. The example is the inverse Gravimetry problem (see [23]). We write it as

$$F(x) : X \longrightarrow Y$$
$$F(x)(t) = \int_a^b k(t,s,x(s))ds = y(t), \ t \in [c,d]. \tag{56}$$

with $k(t,s,x(s)) = \ln \frac{(t-s)^2+H^2}{(t-s)^2+(x(s)-H)^2}$. Clearly the kernel $k$ is defined on the set $\Pi = \{[c,d] \times [a,b] \times R\}$ and $k(t,s,x(s)) \in C^1(\Pi)$. The first derivative $F'(x) : X \longrightarrow Y$ is defined by

$$[F'(x)u](t) = \int_a^b \frac{\partial k}{\partial x}(t,s,x(s))u(s)ds, \ t \in [c,d], \tag{57}$$

where the kernel $\frac{\partial k}{\partial x}(t,s,x(s))$ can be evaluated by

$$\frac{\partial k}{\partial x}(t,s,x(s)) = \frac{2(H-x(s))}{(t-s)^2+(x(s)-H)^2}.$$

$F'(x)$ is compact, since the kernel is square integrable.

Now, we will set up the problem of approximate determination of normal pseudosolution to the equation (56).

For simplicity, two equidistant grids on intervals $[a,b]$ and $[c,d]$ are applied:

$$\Sigma_n(s) = \{s_j : s_j = a + h_s(j-1), j = 1,2,\cdots,n\}, \ h_s = \frac{b-a}{n-1},$$
$$\Sigma_m(t) = \{t_i : t_i = c + h_t(i-1), i = 1,2,\cdots,n\}, \ h_t = \frac{d-c}{m-1}.$$

In this way, the spaces of all grid functions defined on $\Sigma_n(s)$ and $\Sigma_m(t)$, respectively, are treated as $X_n$ and $Y_m$.

The integral operator $F$ gives rise to an operator $F_{mn} : X_n \longrightarrow Y_m$ by

$$[F_{mn}(x)]_i = \int_a^b k(t_i,s,x(s))ds, \ 1 \le i \le m.$$

Similarly the derivative operator $F'(x)$ yields an $m \times n$ matrix:

$$[F'_{mn}(x)]_{ij} = \int_a^b \frac{\partial k}{\partial x}(t_i,s,x(s))\phi_j(s)ds, \ 1 \le i \le m, \ 1 \le j \le n.$$

Where $\phi_j(s)$ we used is the standard linear basic functions

$$\phi_j(s) = \begin{cases} \frac{s-s_{j-1}}{h}, & if \ s \in [s_{j-1}, s_j], \\ \frac{s_{j+1}-s}{h}, & if \ s \in [s_j, s_{j+1}], \\ 0, & else. \end{cases}$$

In which, $s_j = jh, \ h = \frac{1}{n}, \ j = 1,2,\cdots,n$. The integral can be computed numerically.

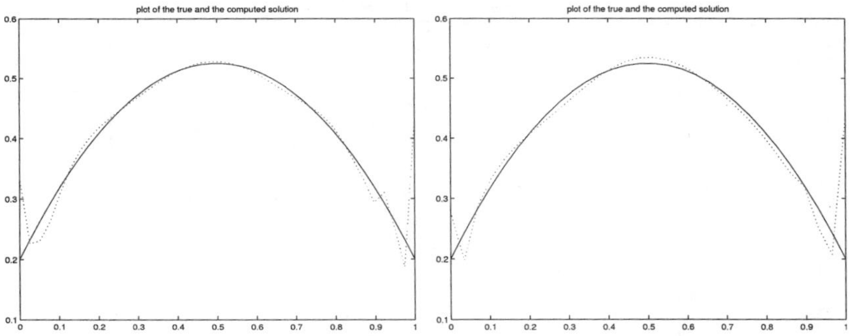


Figure 1: Solution of the inverse problem: nonlinear Fredholm equation

We take $[a, b] = [c, d] = [0, 1]$, $H = 0.1$ and different $m$, $n$ to give a discretization. Our true function is $x_{true}(s) = 1.3s(1-s) + 0.2$, and it is discretized by evaluating it at the points $s_i$ to give the components $x_i$ of $x$. The right-hand side $y$ is generated by integral (56).

The numerical results are shown in figures 1 and 2. In all of these figures, the true solution is denoted by solid line, the approximate solution is denoted by dotted line.

First we choose: $n = 30$, $m = 30$, $\tilde{\omega} = 4.8$, $\tau = 0.8$, $\Delta_0 = 0.1$ with perturbation error level $\delta = 0.01$. The results are shown in the left of Figure 1; It needs 18 inner iterations and 16 outer iterations to generate convergence.

Then we choose $n = 40$, $m = 40$, $\tilde{\omega} = 6.7$, $\tau = 0.8$, $\Delta_0 = 0.1$, with small perturbation $\delta = 0.005$, The results are shown in the right of Figure 1. It needs 20 inner loops and 18 outer loops to generate convergence.

Finally we choose $n = m = 50$, $\Delta_0 = 0.1$ with large perturbation $\delta = 0.05$ and dominant parameter $\tilde{\omega} = 6.8$, $\tau = 0.9$ to give a computation. The results are shown in Figure 2. It needs 7 inner iterations and 10 outer iterations to generate convergence.

**Remark 6.1** *To safeguard Pred is not too small, we choose $\epsilon = 1.0 \times 10^{-32}$. If Pred $\leq \epsilon$, then we regard it as zero and stop the inner iteration. However, this is not activated in our numerical test.*

**Remark 6.2** *In practical applications, the right-hand side is the observation data $y_\delta$ which contains noise or error instead of the exact data $y_{true}$. To give a reasonable simulation of the observation data, we add Gaussian white noise rand to the right-hand side $y_{true}$, i.e.,*

$$y_{noise} = y_{true} + \delta * rand, \tag{58}$$

*where rand is a vector with its components some random numbers in $[0, 1]$.*

*Note that $\delta$ should not be too small or too large. If $\delta$ is too small, according to (58), the noise will not be enough important to give interesting results. However, if $\delta$ is too large, the*

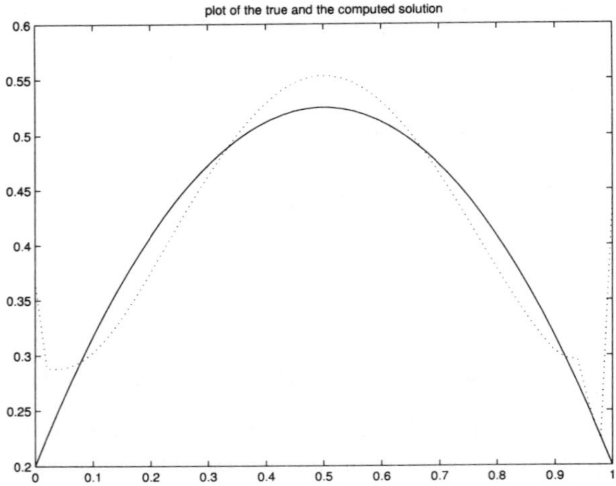


Figure 2: Solution of the inverse problem: nonlinear Fredholm equation

***observation is a poor approximation to the original problem, it will also not enough to give a reasonable results.***

# 7 Conclusion

We have establised the convergence and regularity of the trust region-cg method for nonlinear ill-posed inverse problems. It deserved pointing out that Plato in [18] had establised the regularity property of the conjugate gradient method. Later on, Hanke in [8] had established the regularity of Newon-CG method. All of the methods are stable for solving ill-posed inverse problems.

# References

[1] Bakushinsky A and Goncharsky A 1994, *Ill-Posed Problems: Theory and Applications*, Kluwer Academic Publishers.

[2] Duff I S, Nocedal J 1987, Reid J K et al., The use of linear programming for the solution of sparse sets of nonlinear equations. *SIAM J Sci Stat Comput*, **8**, 99-108.

[3] Engl H W, Hanke M and Neubauer 1996, *Regularization of Inverse Problems*, Dordrecht: Kluwer.

[4] Fletcher R 1987, *Practical Methods of Optimization*, 2nd ed., Chichester: John Wiley and Sons.

[5] Fletcher R 1982, A model algorithm for composite NDO problem, *Math Prog Study*, **17**, 67-76.

[6] Gay D M 1981, Computing optimal local constrained step, *SIAM J Sci Stat Comp*, **2**, 186-197.

[7] Groetsch C W 1984, *The Theory of Tikhonov Regularization for Fredholm Equations of The First Kind* (Boston,MA:Pitman)

[8] Hanke M, Regularizing Properties of a Truncated Newton-CG Algorithm for Nonlinear Inverse Problems, *submitted.*

[9] Hanke M 1997, A Regularizing Levenberg-Marquardt Scheme, with Applications to Inverse Groundwater Filteration Problems, *Inverse Problems* **13**, 79-95.

[10] Hanke M 1995, *Conjugate Gradient Type Methods for Ill-posed Problems*, Longman Scientific and Technical, Harlow, Essex.

[11] Kaltenbacher B 1998, On Broyden's Method for Nonlinear Ill-posed Problems, *Numer Funct Anal Opt*, **19**.

[12] Kaltenbacher B 1998, A Posteriori Parameter Choice Strategies for Some Newton Type Methods for the Regularization of Nonlinear Ill-posed Problems, *Numer.Math.* **79**.

[13] Kirsch A 1996, *An Introduction to the Mathematical Theory of Inverse Problems*, New York: Springer-Verlag.

[14] Louis A K 1992, *Inverse und Schlecht Gestellte Problems*, Teubner, Stuttgart.

[15] Moré J J 1983, Recent developments in algorithms and software for trust region methods. In: Bachem A, Grötschel, Korte B, eds. Mathematical Programming: The State of the Art. Berlin: Springer-Verlag, 258-287.

[16] Moré J J and Sorensen D C 1983, Computing a Trust Region Step, *SIAM J.Sci.Stat.Comput.* **4**, 553-572.

[17] Morozov V A 1984 *Methods for Solving Incorrectly Posed Problems* (New York: Springer)

[18] Plato R 1990, Optimal Algorithms for Linear Ill-posed Problems Yield Regularization Methods, *Numer.Funct.Anal.and Optim.* **11**, 111-118.

[19] Powell M J D 1984, Nonconvex minimization calculations and the conjugate gradient method. In: Griffiths, ed. Lecture Notes in Mathematics 1066: Numerical Analysis. Berlin: Springer-Verlag, 122-141.

[20] Powell M J D 1975, Convergence properties of a class of minimization algorithms. In: Mangasarian O L, Meyer R R, Robinson S M, eds. Nonlinear Programming. New York: Academic Press, **2**, 1-27.

[21] Sorensen D C 1982, Newton's Method with a Model Trust Region Modification, *SIAM J.Numer.Anal.* **19**, 409-426.

[22] Steihaug T 1983, The conjugate gradient method and trust regions in large scale optimization. SIAM J Numer Anal **20**, 626-637.

[23] Tikhonov A N and Arsenin V Y 1977, *Solutions of Ill-Posed Problems*, New York: Wiley.

[24] Toint Ph L 1981, Towards an efficient sparsity exploiting Newton method for minimization. In: Duff I ed. Sparse Matrices and Their Uses. Berlin: Academic Press, 57-88.

[25] Yan-fei Wang and Ya-xiang Yuan 2001, A Trust Region Method for Solving Distributed Parameter Identification Problems, *Report, AMSS, Chinese Academy of Sciences, 2001.*

[26] Yan-fei Wang, Ya-xiang Yuan and Hong-chao Zhang 2001, A Trust Region-CG Algorithm for Deblurring Problem in Atmospheric Image Reconstruction, to appear in *Science in China.*

[27] Y. Yuan 1994, Nonlinear Programming: Trust Region Algorithms, in: S.T. Xiao and F. Wu, eds., Proceedings of Chinese SIAM annual meeting (Tsinghua University, Beijing) pp. 83 –97.

[28] Y. Yuan 1998, Matrix Computation Problems in Trust Region Algorithms for Optimization, in: Q.C. Zeng, T.Q. Li, Z.S. Xue and Q.S. Cheng, eds. Proceedings of the 5th CSIAM annual meeting, (Tsinghua Unviersity Press, Beijing) pp. 52–64.

[29] Y. Yuan 1999, Problems on Convergence of Unconstrained Optimization Algorithms, in Y. Yuan ed., Numerical Linear Algebra and Optimization, (Science Press, Beijing, New York), pp. 95-107.

[30] Y. Yuan 2000, On the truncated conjugate gradient method. Math Prog, **87**, 561-571.

# Electronic Information and Publications in EMIS - An offer supervised by EMS

Bernd Wegner, Technical University Berlin
Strasse des 17. Juni 136
10623 Berlin
e-mail: wegner@math.tu-berlin.de

**Abstract**

Literature databases, scientific journals and communication between researchers on the electronic level are rapidly developing tools in mathematics having high impact on the daily work of mathematicians. They improve the availability of information on all important achievements in mathematics, speed up the publication and communication procedures and lead to enhanced facilities for the preparation and presentation of research in mathematics. The European Mathematical Information Service (EMIS) tries to bundle several of these facilities in a joint offer, to extend them by innovative components and to distribute them in a system of replicators for the benefit of mathematicians worldwide. The aim of this article is to introduce to the different offers collected in EMIS like the databases MATH and MATHDI, the preprint index MPRESS, the project to install the EULER search engine for distributed sources, the JFM-project for the indexation and digitalization of classical literature in mathematics and most importantly the Electronic Library, offering a variety of freely accessible electronic journals in mathematics.

## 1 Introduction

The increasing development of electronic devices for the publication of papers and books in mathematics lead to a drastic change in the communication facilities between authors and editors, to new ways of distributing mathematical publications to research mathematicians - like electronic journals, and to an extension of the offers of information on mathematical research to the mathematical community. Confronted with a permanently increasing number of research papers and high production and distribution costs for mathematical journals and monographs, it becomes more and more difficult to get access to these publications. Electronic publishing may provide better facilities for these purposes. It hopefully will reduce the production costs for the preparation of the articles and improve the methods to find his way through the tremendous bulk of mathematical research papers. It will enable reviewing journals to speed up and improve their procedures.

Moreover, as an extension of these offers, a lot of additional items of interest for mathematicians appear in the web already like pre-prints, educational material in mathematics, software, graphical material a others. Information databases provide a qualified access to some of th offers, but for the whole set a uniform access facility with a high deg of de-duplication will be highly desirable. The result of the EULI project may provide one tool for this.

The aim of this article is to report on these aspects related with met ods of electronic publishing and electronic communication by exhibiti offers provided by EMIS (European Mathematical Information Servic as an example which may indicate a good model for the future. Tl refers to the production and distribution of electronic publications mathematics, to electronic access to evaluated information on all math matical papers as well as to general search facilities leading to interestir documents in mathematics.

## 2 The General Concept of EMIS

The idea to develop the European Mathematical Information Servic EMIS was born at the meeting of the executive committee of the EM (European Mathematical Society) in Cortona/Italy, October 1994. I was decided to set up a system of electronic servers in Europe for Math ematics under the auspices of the EMS, and this was extended ver soon to the current version of a central server collecting mathematica information and distributing this through a world-wide system of mir- ror servers. The installation of the central server began in March 199 in co-operation with FIZ Karlsruhe at the editorial office of Zentral-blatt MATH in Berlin. In June 1995 EMIS went on-line under the URL http://www.emis.de/. The first mirrors were established very soon.

The sections of the contents of EMIS related to the topics above are "Projects", "Databases", and the "Electronic Library". World Wide Web access is regarded the primary access method. The access to the contents of EMIS is free up to the usage of some databases. In these cases a link leads directly to the corresponding system of gateways to the databases, and the user will be subject to the conditions valid for accessing these databases. In any case he will be able to do searches. But in the case his institution does not subscribe to the service, only some preliminary information will be available from the hit list.

Before giving a more detailed description of the services of EMIS which are of central interest for electronic publication and communication some important general aspect should be pointed out at the beginning. This is the idea to distribute EMIS through a world-wide system of mirrors where the full content of the service will be available and updated periodically. This improves the accessibility of EMIS, and it simultaneously is important for the safety of the data and their archiving: if the master server or one component of the system crashes down, it can be regenerated easily from the other components. In principle, every European country should have one mirror at least. The current mirrors of EMIS in Asia are in Novosibirsk, Ankara, Shanghai and Kyoto.

## 3 The Databases Section

This section contains three items: MATH - the online version of Zentralblatt MATH, MATHDI - the online version of a similar service for education in mathematics, and MPRESS - a global pre-print index.

To connect EMIS with the database of Zentralblatt MATH is one part of the increasing involvement of the European Mathematical Society in the edition of this reviewing service. In contrast to the variety of "databases" offered by commercial publishers now, the word "reviewing" is taken quite seriously and it is not "abstracting" only. Hence a lot of reviewers is employed for the evaluation of the literature. All reviewers work for Zentralblatt MATH more or less as volunteers. There are other important aspects of quality like comprehensiveness, precision of data, competent indexing, easy search menus, convenient linking with full texts and document delivery systems etc. All these aspects are important features for a reliable service, where quick and dirty offers cannot compete with, though some of them pretend to have this quality, when their promotion is considered. Mathematics as a science, where the truths detected by mathematicians do not loose their validity with time passing by, needs such a precision to maintain control over the achievements in the past and to get a reliable information on what is new.

Here Zentralblatt MATH stands for the conventional printed reviewing service as well as for the two electronic offers, one on CD-ROM for off-line use and one as a database with WWW-access. EMIS provides a link to the WWW-access which can be reached directly under the URL http://www.emis.de/ZMATH. Additional access will be directly available to mirrors of MATH which have been installed world-wide in Strasbourg, New York, Cornell, Berkeley, Mexico City, Rehovot, Rio de Janeiro, Lecce and Santiago de Compostela. Clearly, the full use of the database only will be possible for subscribers, but it should be mentioned the free component in the link from EMIS to MATH: Any user of EMIS can do searches in MATH. But non-subscribers only will get information on at most three hits from their list. These hits will be taken from the top of the list, where the most recent items are listed. Hence, if very precise information is searched which probably leads to a small hit list, then this can be reached by everyone who has access to the internet.

Some quick details on Zentralblatt MATH: It covers information on all mathematical literature, starting with 1931. The total amount of documents for which information is stored in MATH is about 1.700.000.

This increases by more than 70.000 items annually. The update of the database is made every two weeks which corresponds to the production of the print version of Zentralblatt. The scope of publications handled by this service includes in addition to mathematics mathematical applications in physics, mechanics, computer science, economics and biology.

Searches can be made using the following list of fields: authors, titles, classification, basic index, source, language, year of publication etc. A search can be formulated in logical combinations of these terms. The search will be guided by a graphic menu. The information is available in the AMSTEX source code, but several choices for a convenient formula display are available. Download of the hit list at the users site is possible, and, coming back to the other content of EMIS and other electronic journals and publications, links to the full text of the corresponding article will be arranged, if this is available electronically. Other options to get the full text of such an article consist of central document delivery services. Buttons to connect to such services and to see, if the corresponding article is available there, are installed in the menu for the search. Document delivery can be arranged by these services electronically or by sending copies by ordinary mail at reasonable rates.

MATHDI is the online version of the printed service of Zentralblatt für Didaktik der Mathematik. It provides comprehensive information on publications in education in mathematics. All aspects of quality and search facilities are the same like for Zentralblatt MATH. The editors are the European Mathematical Society and FIZ Karlsruhe. It is supervised by an Editorial Committee. The contents of the database comprise 85.000 items.

In contrast to these two charged databases, MPRESS is provided as a freely accessible service. It stores combined information on mathematics pre-prints available in the web. The gathering of information is done by robots, which are run by national brokers for harvesting of data. This procedure leads to a data structure which only allows for simple search facilities. MPRESS has no ambition to offer a pre-print server itself, only links to full texts are provided. The service is supervised by EMS among others. Countries which support the harvesting are Germany, France, Austria, and Italy. In addition to this some special servers are harvested by MPRESS. Among them are the Topology Atlas and the LANL-archive.

As a new item the link to a free offer of high-quality geometric models and animations had been arranged. This is a preliminary version, and it has to be investigated, how the data of these models could be stored in a convenient way, to make them accessible within the same menu as it is provided for searching mathematical articles.

## 4 The Projects

EMS is involved in three projects, where two of them are funded by the European Union and one is funded by Deutsche Forschungsgemeinschaft. The first two are LIMES and EULER while the other one is the Jahrbuch-Project. The one closely related to Zentralblatt MATH is LIMES (Large Infrastructures in Mathematics - Enhanced Services).

The objective of the LIMES project is to upgrade the database Zentralblatt MATH into a European based world class database for mathematics and its applications by a process of technical improvement and wide Europeanisation. Upgrading the existing database, improving the present system and developing a new, distributed system both for the input and output of the data are necessary to allow Zentralblatt MATH to use the latest developments and to anticipate on future developments of electronic technologies.

Improvements will be made in three areas: 1) Improvement of content and retrieval facilities through sophisticated further development of the current data sets and retrieval programs. 2) Broader and improved access to the database via national access nodes and new data distribution methods. In particular, improvement of access for isolated universities in regions in economic difficulties and in associated states of Central and Eastern Europe where a mathematical tradition of excellence is under economic threat. Stimulation of usage for all kinds of research as well as for funding organisations before decision making by initial support of two national test sides. 3) Improved coverage and evaluation of research literature via nationally distributed editorial units, development of technologies for efficient database production (exemplified by two further European member states).

The EULER-project is funded by the European Union within the programme "Telematics for Libraries". The goal of the project is to develop a prototype of an internet service which integrates some of the most relevant publication-related resources in the field of mathematics. This should provide a "one-stop-shopping" in heterogeneous resources for the user. The approach should be open to other participants, scalable for different kinds of user interests and it should be easily portable to other subject domains. The team co-operating in this project comprises in addition to partners in charge of software supply different providers of electronic resources available in the web: OPACs, databases, pre-print servers, electronic journals and other WWW-catalogues.

Tools have been developed facilitating the production of Dublin-Core-metadata for the different resources. Using the Z39.50 protocol the so-called EULER-engine users of the EULER-systems can carry out combined searches for mathematical information in all these resources. Different search levels are installed depending on the quality of the metadata offered by the resources. De-duplication checks are done automatically and links are arranged to full texts resp. the more detailed information given by the provider.

The EULER-project terminated in September 2000. The engine has passed the beta-test successfully and the current beta-version has been proved to be installable at new service providers. During the exploitation period a model has been developed how to establish EULER as an internet service. There will be a consortium of different providers for the future service, including some of the partners of the project. New partners already have agreed to take part in the service. The first goal is to establish a library and society based web service which includes a system of resources with a well-distributed European background. Economic viability of the EULER system was not among the concerns of the project. This has to be investigated by future measures to enable commercial publishers to join the service by making their holdings visible through EULER. To produce the meta-data will not be that problem. But to make the content accessible they will ask for an access control, and that yet has to be designed for the EULER machine.

By the way, EULER stands for "EU"ropean "L"ibraries and "E"lectronic "R"esources in Mathematical Sciences.

The aim of the Jahrbuch-Project, which officially is called ERAM (Electronic Research Archive in Mathematics), is to capture the "Jahrbuch" as a classical bibliographic service in mathematics in a database and to use this activity to select important publications from the period of the Jahrbuch for digitisation and storage in a digital archive. The database will not be just a copy of the printed bibliography. It will contain a lot of enhancements like modern subject classifications as far as possible, keywords giving ideas about the content in modern terms and comments relating classical results to modern mathematical research areas. These features will remain open for additions within a living project. The digital archive built up in connection with the database will be linked to the database and provide all facilities associated with current digitisation projects. The content will be distributed to mirrors and combined with similar archiving activities in mathematics.

## 5 The Electronic Library

The Electronic Library of EMIS has the aim to present a collection of freely accessible electronic publications which should be as comprehensive as possible. There are four sections: journals, proceedings volumes, monographs, and collected works. In order to guarantee that the electronic publications stored in the Electronic Library meet the requirements satisfied by articles in print journals, the decision on the inclusion of journals, proceedings or monographs is taken in accordance with the

Electronic Publishing Committee of the EMS. Hence no items will enter the library which have not been evaluated and recommended by a referee within the editorial procedures of the corresponding journal or series. This is in particular important in order to rule out the reservations of many mathematicians who have the opinion that electronic publishing will damage the quality of mathematical publications.

The section on Electronic Journals contains pure electronic journals as well as electronic versions of print journals. The pure electronic journals are produced elsewhere and EMIS only serves as an additional distributor. The installation of electronic versions of print journals depends on the technical facilities of the editors of these journals. We prefer that the offer of the electronic version is installed at the site of the editors, such that a mirror of the journal can be taken over by EMIS. The organisers of EMIS provide support for this purpose. Most of these print journals are published at a low-budget level, and hence the risk to loose subscribers to the print version due to the free electronic offer currently is considered as low by them. Some of them give the electronic offer with a delay to EMIS such that the earlier availability will be considered as an advantage of the print version.

Though there obviously is no chance to offer free access to electronic versions of the journals in mathematics published by commercial publishers, EMIS succeeded to get the agreement of a series of quite established journals to be made freely available in its library, as can be seen from the samples in the subsequent list. The pure electronic journals are marked by an (e): Acta Mathematica Universitatis Comenianae (Bratislava), Annales Academiae Scientiarum Fennicae Series A. Mathematica (Helsinki), Archivum Mathematicum (Brno), Beiträge zur Algebra und Geometrie / Contributions to Algebra and Geometry, Commentationes Mathematicae Universitatis Carolinae (Prague) DOCUMENTA MATHEMATICA (Bielefeld) The Electronic Journal of Combinatorics (e), The Electronic Journal of Differential Equations (e), Electronic Research Announcements of the AMS (e), Electronic Transactions on Numerical Analysis (e), Geometry and Topology (e), Journal de Théorie des Nombres de Bordeaux (Bordeaux), Matematicki Vesnik (Belgrade), Mathematical Physics Electronic Journal (e), Revista Colombiana de Matemáticas (Bogotá), Séminaire Lotharingien de Combinatoire, Theory and Applications of Categories (e). More or less all freely available electronic journals in mathematics are mirrored in EMIS.

The access to the journals in EMIS is organised quite conventionally. On the home page of EMIS a list of mirrors is provided where the site with the (probably) quickest access can be clicked. Then a choice can be made, to enter the Electronic library though the short list of journals without graphics or to use the full display of these items. The first one is preferable, if the choice of journal is clear already and if one wants to avoid the lengthy transfer of the graphical data associated with this journal. The full display contains also background information on the editorial policy of the corresponding journal and instructions how to submit an article. In cases style files for such a submission can be found on this level.

Having made the choice of journal one wants to read, the level of the contents will be reached which is organised as usual. At that level information on the offer of files is given. In all cases DVI- and Postscript-files will be available, sometimes also TEX-files can be found in addition to that, and in cases geometric material is stored in separate files. PDF is coming up rapidly and will be an obligatory offer in the near future. Clicking one of these files its content is transferred to the computer of the user and can be viewed there. Also printing or storage of these files will be possible at the site of the user, but he is requested to respect the copyright according to the rules of the corresponding journal. The access to the section of Proceedings Volumes is organised in a similar way.

At present we have only a few Proceedings Volumes in EMIS, because the organisers of conferences prefer to publish printed proceedings. They can support this publication from the conference fees or money they receive from sponsors, and the majority of participants consider a publication on paper as more valuable than an electronic one. Our hope is that the contributors to such volumes will become aware very soon, that the print versions of proceedings have a very low distribution in the average. Decreasing library budgets and growing prices for these volumes will make the distribution even worse. Hence, if these contributors really are interested to make their work public in a reasonable way, they will see that the only chance to improve the distribution lies in an electronic offer which can be accessed easily at several places. EMIS is just the appropriate system for a good distribution of electronic publications in mathematics. This is underlined by the present access numbers which are growing continuously.

Admittedly, after five years of EMIS some of these features will have to be modified. The request for PDF-files will become a must, because the preinstalled readers at electronic access facilities in libraries and desktop-computers of research mathematicians already point into that direction. For other offers special measures have to be taken for making the article readable or printable. Decisions about affordable systems of digital object identifiers (beyond DOI) have to be taken to enable a richer linking system between electronic articles. The upcoming method to offer the electronic version of an article as soon as the article will be available in final form ("online first") will have impact on the free access to the installation of the dual journals in EMIS. They will not be able to survive from what they get for the paper version, when online will be declared as "first class" and print will be still important, but second class only. But anyway, EMIS will remain non-profit during such a transition, and provide as much as possible on the free level.

## 6 Conclusion

The offers in EMIS mentioned above and their distribution through a system of mirrors provide a unique facility for quick and easy access to qualified mathematical publications. These tools can be used by the mathematical community at suitable sites simultaneously for free or at modest rates. I think that this is a future-leading way to provide an alternative to the high-priced offers of commercial publishers which cannot be afforded by the majority of potential users anymore. In particular the mathematical community never will have the financial background to be considered as a market where research literature could be sold in accordance with the interests of the share holders of commercial publishers. The initiatives of some publishers to go into that direction will damage the publication facilities for the mathematical community rather soon. Hence mathematicians have to care about their own systems to maintain a reasonable infrastructure for communicating their research achievements. EMIS is one part of this enterprise.

Last but not least, it has to be pointed out that EMIS could not survive on the current level without the big group of its supporters, who serve as volunteers for maintaining and installing electronic journals, caring about submissions and transfer of content and keeping the mirrors running. Without these activities EMIS would not have been possible. It shows that a viable service can be based on the collaboration of several volunteers, in contrast to those who argue that all these activities can only be pursued seriously on a commercial level. They should keep in their mind that at least in mathematics all the commerce hooks on voluntary services from the mathematical community: publishers do not pay the authors, editors provide their service for free, referees do not even get the mailing expenses for their contribution to the business and finally readers have to look for public funding to get access to the publications. Hence the question only is where to shift the voluntary support and where to spend the funding.

Professor Dr. Bernd Wegner
Scientific Coordinator of EMIS
Editor-in-Chief of Zentralblatt MATH
Fachbereich Mathematik
Technische Universität Berlin
Strasse des 17. Juni 135,
D-10623 Berlin, Germany
e-mail: wegner@math.tu-berlin.de

# Mathematical Modeling of Electromagnetic Sounding for a Conductive 3–D Circular Cylinder Body Embedded in a Conducting Half-Space.

S. Yooyuanyong and N. Chumchob

*Department of Mathematics, Faculty of Science, Silpakorn University, Nakorn-Pathom 73000, Thailand.*

## ABSTRACT

Integral equations that give the electromagnetic response of a 3-D circular cylinder body embedded in an inhomogeneous conductive half-space are derived. The expressions for the electric fields on each area are given. The equations obtained can be used to estimate unknown parameters such as the conductivity, the thickness and the depth of target. The estimation of the above parameters is important in the identification of subsurface structures. An inversion procedure is outlined to get the optimal values for these parameters.

## INTRODUCTION

Usually the interpretations of electric soundings are conducted by assuming the earth structure consists of horizontally stratified layers. Many authors have investigated the potential field due to a DC current source at the surface of the earth where the conductivity vary continuously with depth only. In the particular case of an exponentially varying conductivity profile, Stoyer and Wait (1977) considered the forward problem for a two layered earth models where a homogeneous overburden overlying a medium whose conductivity varies exponentially with depth. Banerjee et al. (1980) considered the forward problem of apparent resistivity of a multi-layered earth with a layer having exponentially varying conductivity. Kim and Lee (1996) considered layered earth model where each layer has exponentially varying conductivities.

For electromagnetic methods, Lee and Ignetik (1994) considered the forward problem of the transient electromagnetic response of a half-space with an exponentially varying conductivity profiles. Yooyuanyong (1999) considered the inverse problem for the electromagnetic response of a conductive thin disk embedded in a conducting half-space with an exponentially varying conductivity profiles. In inversion process, Yooyuanyong used idea in optimization theory to estimate the values of the model parameters such as the depth and the conductance of the disk. Yooyuanyong and Siew (2000) considered the inverse problem for the electromagnetic response of a conductive thin disk beneath an exponential varying conductive overburden. They gave outlined to estimate model parameters such as the thickness of the overburden, the conductance and the depth of the disk. However, the algorithms in model of Yooyuanyong (1999) and Yooyuanyong and Siew (2000) are not possible to differentiate between the conductivity and the thickness of the disk.

In this paper, we developed the exponential variation function of the subsurface difference from the exponential variation model used by Stoyer and Wait (1977),

Banerjee et al. (1980), Lee and Ignetik (1994), Kim and Lee (1996), Yooyuanyong (1999) and Yooyuanyong and Siew (2000). The objective of this paper is to present a mathematical model and techniques for finding the conductivity, the thickness, and the depth of the conductive 3-D circular cylinder body embedded in a conducting half-space with an exponentially varying conductivity profiles.

## FORMULATION OF THE PROBLEM

Consider a primary field generated by a coil of radius $a$ located at $z = h \leq 0$ above the surface of the earth ($z = 0$). The source current density, with angular frequency $\omega$, is given by $J_s e_\phi = I(\omega)a(r-a)(z+h)e_\phi/r$ where $I(\omega)$ is the current in the coil and $e_\phi$ is a unit vector in azimuthal direction. A cylindrical coordinate system $(r, \phi, z)$ is introduced with $z > 0$ are taken vertically positive downwards and origin below the center of the coil are used. A time variation $\exp(i\omega t)$ is assumed throughout. For the half-space at $z > 0$, it is a region of the ground where a conductivity is given by $\sigma_P(z) = \sigma_0 + (\sigma_1 - \sigma_0)e^{-bz}$, here $\sigma_0$, $\sigma_1$ and $b$ are positive values. At $z = z_1$, there is a circular cylinder body of radius $c$, thickness $nd$, where $n$ is a positive integer, $d$ is a positive real number and conductivity $\sigma_s$.

By construction, the horizontal loop source gives the electric field has only $E_\phi$ component, for simply denoted by $E$, where the scalar quantity $E$ is understood to be a function of $r$, $z$ and $\omega$. The magnetic fields have only the radial $H_r$ and the vertical $H_z$ components. The field quantities $E$, $H_r$ and $H_z$ are related by Maxwell's equations as

$$\frac{\partial E}{\partial z} = i\omega\mu H_r, \tag{1}$$

$$\frac{1}{r}\frac{\partial (rE)}{\partial r} = -i\omega\mu H_z, \tag{2}$$

and

$$\frac{\partial H_r}{\partial z} - \frac{\partial H_z}{\partial r} = (\sigma + i\omega\varepsilon)E + J_s, \tag{3}$$

where $\mu$ is the magnetic permeability, $\varepsilon$ is the electric permittivity, $\sigma$ is the conductivity. Eliminating $H_r$ and $H_z$ from the above equations, this gives

$$\frac{\partial^2 E}{\partial r^2} + \frac{1}{r}\frac{\partial E}{\partial r} - \frac{E}{r^2} + \frac{\partial^2 E}{\partial z^2} - \alpha^2 E = i\omega\mu_0 J_s, \tag{4}$$

where $\alpha^2 = i\omega\mu\sigma - \omega^2\mu\varepsilon$. In air (z < 0), let $E^{air} = E^{sou} + E^{res}$, where $E^{sou}$ is the primary field due to the coil and $E^{res}$ is the electric field response from the ground. $E^{sou}$ satisfies the equation

$$\frac{\partial^2 E^{sou}}{\partial r^2} + \frac{1}{r}\frac{\partial E^{sou}}{\partial r} - \frac{1}{r^2}E^{sou} + \frac{\partial^2 E^{sou}}{\partial z^2} - \alpha_{air}^2 E^{sou} = i\omega\mu_0 J_{s\phi}, \tag{4}$$

where $\alpha_{air}^2 = -\omega^2\mu_0\varepsilon_0, \sigma \approx 0, \mu \approx \mu_0,$ and $\varepsilon \approx \varepsilon_0$. Introducing the Hankel transforms defined by

$$\widetilde{E}(\lambda, z, \omega) = \int_0^\infty rE(r, z, \omega)J_1(\lambda r)dr,$$

where $J_1(\ )$ denotes the Bessel function of the first kind of order 1. Equation (4) becomes

$$\frac{d^2\widetilde{E}^{\text{sou}}}{dz^2} - m_0^2\widetilde{E}^{\text{sou}} = i\omega\mu_0 I(\omega)a\delta(z+h)J_1(\lambda a), \tag{5}$$

where $m_0 = \sqrt{\lambda^2 - \alpha_{\text{air}}^2} \approx \lambda$. The solution to (5) is obtained by using the method of variation of parameters, and the solution can be written as

$$\widetilde{E}^{\text{sou}}(\lambda, z, \omega) = C_1 e^{\lambda z} - \frac{i\omega\mu_0 I(\omega)aJ_1(\lambda a)}{2\lambda} e^{-\lambda|z+h|}, \qquad h, z \le 0. \tag{6}$$

The solution remains bounded as $z \to -\infty$, and $C_1$ is an arbitrary constant. $E^{\text{res}}$ satisfies the homogeneous form of equation (5) and the bounded solution is given by

$$\widetilde{E}^{\text{res}}(\lambda, z, \omega) = C_2 e^{\lambda z}, z \le 0, \tag{7}$$

where $C_2$ is an arbitrary constant. Thus, the total electric field in air is given by

$$\widetilde{E}^{\text{air}}(\lambda, z, \omega) = -\frac{i\omega\mu_0 I(\omega)aJ_1(\lambda a)}{2\lambda} e^{-\lambda|h+z|} + Ae^{\lambda z}, \qquad h, z \le 0, \tag{8}$$

where $A = C_1 + C_2$ is an arbitrary constant to be determined by using the boundary conditions at air–earth interface, $z = 0$.

In ground ($z > 0$), let the total electric field denoted by $E^{\text{ground}} = E^g + E^s$, where $E^g$ is the electric field response in ground without target and $E^s$ is the electric field response in ground with target. The governing equations are now

$$\frac{\partial^2 E^g}{\partial r^2} + \frac{1}{r}\frac{\partial E^g}{\partial r} - \frac{1}{r^2}E^g + \frac{\partial^2 E^g}{\partial z^2} - \alpha_p^2 E^g = 0, \tag{9}$$

and

$$\frac{\partial^2 E^s}{\partial r^2} + \frac{1}{r}\frac{\partial E^s}{\partial r} - \frac{1}{r^2}E^s + \frac{\partial^2 E^s}{\partial z^2} - \alpha_p^2 E^s = i\omega\mu_0(\sigma_s - \sigma_p)E_{tar}, \tag{10}$$

where $\alpha_p^2 = i\omega\mu_0\sigma_p$ and $E_{tar}$ is the electric field on the target. $(\sigma_s - \sigma_p)E_{tar}$ may be viewed as an equivalent current distribution on the target.

**Exponentially decreasing conductivity profile ($\sigma_0 < \sigma_1$)**

Using Hankel transform to (9) and (10), we have

$$\frac{d^2\widetilde{E}^g}{dz^2} - \left(\alpha_p^2 + \lambda^2\right)\widetilde{E}^g = 0, \tag{11}$$

and

$$\frac{d^2\widetilde{E}^s}{dz^2} - \left(\alpha_p^2 + \lambda^2\right)\widetilde{E}^s = i\omega\mu_0 \int_0^c s(\sigma_s - \sigma_p)E_{tar}(s, z, \omega)J_1(\lambda s)ds. \tag{12}$$

Following Lee & Ignetik (1994), we will use the exponential substituting $\zeta = e^{-bz/2}$ which transform equation (11) into the modified Bessel differential equation. Equation (11) becomes

$$\zeta^2 \frac{d^2\widetilde{E}^g}{d\zeta^2} + \zeta \frac{d\widetilde{E}^g}{d\zeta} - \left(\alpha^2\zeta^2 + \nu^2\right)\widetilde{E}^g = 0, \tag{13}$$

where $\alpha^2 = 4i\omega\mu_0\,[\sigma_1 - \sigma_0]/b^2$ and $\nu^2 = 4[\lambda^2 + i\omega\mu_0\sigma_0]/b^2$. The bounded solution for equation (13) is given by

$$\widetilde{E}^g(\lambda,\zeta,\omega) = C_3 I_\nu(\alpha\zeta), \tag{14}$$

where $C_3$ is an arbitrary constant. $I_\nu(\,)$ is the modified Bessel function of the first kind of order $\nu$. Similarly, equation (12) becomes

$$\zeta^2 \frac{d^2\widetilde{E}^s}{d\zeta^2} + \zeta \frac{d\widetilde{E}^s}{d\zeta} - \left(\alpha^2\zeta^2 + \nu^2\right)\widetilde{E}^s = \frac{4i\omega\mu_0}{b^2}\int_0^c s(\sigma_s - \sigma_p)E_{tar}(s,\zeta,\omega)J_1(\lambda s)ds. \tag{15}$$

The solution to (15) is obtained by using the method of variation of parameters, and the bounded solution can be written in form of Fredholm's integral equation as

$$\widetilde{E}^s(\lambda,\zeta,\omega) = C_4 I_\nu(\alpha\zeta) + \frac{4i\omega\mu_0}{b^2}\int_0^1 G(\zeta;\tau)\left[\int_0^c s(\sigma_s - \sigma_p)E_{tar}(s,\zeta,\omega)J_1(\lambda s)ds\right]d\tau, \tag{16}$$

where $C_4$ is an arbitrary constant and $G(\zeta;\tau)$ is the Green's function defined by

$$G(\zeta;\tau) = \begin{cases} \dfrac{I_\nu(\alpha\zeta)K_\nu(\alpha\tau)}{(-\tau)}, & \zeta \le \tau, \\[2ex] \dfrac{K_\nu(\alpha\zeta)\,I_\nu(\alpha\tau)}{(-\tau)}, & \tau \le \zeta. \end{cases} \tag{17}$$

$(\sigma_s - \sigma_p)E_{tar} \neq 0$, where $e^{-b(z_1+nd)/2} = \zeta_{n+1} \le \zeta \le \zeta_1 = e^{-bz_1/2}$. The integral equation (16) can be determined numerically by dividing target into $n$ disks having the thickness equal to $d$ and assume that the disk is very thin, so the current distribution and Green's function for each disk may be approximated to be constant at each position $z$. Expanding $e^{-bd/2}$ by using Maclaurin series, we obtain

$$\widetilde{E}^s(\lambda,\zeta,\omega) = C_4 I_\nu(\alpha\zeta) + \frac{2i\omega\mu_0}{b}\sum_{j=1}^{n} G(\zeta;\zeta_j)\left\{\int_0^c sS_j E_{tar}^j(s,\zeta_j,\omega)J_1(\lambda s)ds\right\}(\zeta_j), \tag{18}$$

where $\zeta_j = e^{-b(z_1+\{j-1\}d)/2}$, $E_{tar}^j(s,\zeta_j,\omega)$ is the electric field on $j^{\text{th}}$–disk, and $S_j$ is the conductance of $j^{\text{th}}$–disk defined as $S_j = d(\sigma_s - [\sigma_0 + (\sigma_1 - \sigma_0)\zeta_j^2])$. The term $S_j E_{tar}^j(s,\zeta_j,\omega) = p(s, \zeta_j, \omega)$ can be interpreted as the integrated surface current on $j^{\text{th}}$–disk. Hence, the electric field response in ground can be presented by

$$\widetilde{E}^{\text{ground}}(\lambda,\zeta,\omega) = BI_\nu(\alpha\zeta) + \frac{2i\omega\mu_0}{b}\sum_{j=1}^{n} G_j(\zeta;\zeta_j)P(\lambda,\zeta_j,\omega)(\zeta_j), \tag{19}$$

where $B = C_3 + C_4$ is an arbitrary constant to be determined. The arbitrary constants $A$ and $B$ are now determined from the boundary conditions at air-earth interface. We require that the electric field and the tangential component of electric field to be continue at the interface. This leads the electric field in air to

$$E^{\text{air}}(r,z,\omega) = -i\omega\mu_0 I(\omega) a \int_0^\infty \frac{\lambda I_\nu(\alpha) J_1(\lambda a) J_1(\lambda r) e^{-\lambda|z+h|}}{\left(\lambda - \sqrt{i\omega\mu_0\sigma_0 + \lambda^2}\right) I_\nu(\alpha) + \frac{\alpha b}{2} I_{\nu-1}(\alpha)} d\lambda$$

$$-i\omega\mu_0 \sum_{j=1}^{n} \int_0^c sp(s,\zeta_j,\omega) ds \int_0^\infty \lambda \left\{ \frac{I_\nu(\alpha e^{-\frac{bz_j}{2}}) J_1(\lambda s)\, e^{\lambda z}}{\left(\lambda - \sqrt{i\omega\mu_0\sigma_0 + \lambda^2}\right) I_\nu(\alpha) + \frac{\alpha b}{2} I_{\nu-1}(\alpha)} \right\} J_1(\lambda r)\, d\lambda, \quad (20)$$

where $h$ and $z \le 0$. In ground, the electric field is given by

$$E^{\text{ground}}(r,\zeta,\omega) = -i\omega\mu_0 I(\omega) a \int_0^\infty \frac{\lambda J_1(\lambda a)\, J_1(\lambda r) I_\nu(\alpha\zeta) e^{-\lambda|h|}}{\left(\lambda - \sqrt{i\omega\mu_0\sigma_0 + \lambda^2}\right) I_\nu(\alpha) + \frac{\alpha b}{2} I_{\nu-1}(\alpha)} d\lambda$$

$$+ \frac{2i\omega\mu_0}{b} \sum_{j=1}^{n} \int_0^c sp(s,\zeta_j,\omega) ds \int_0^\infty \lambda\gamma(\lambda,\zeta,\omega) I_\nu(\alpha\zeta_j) J_1(\lambda s) J_1(\lambda r) d\lambda, \quad (21)$$

where $\zeta_1 \le \zeta \le 1$ and

$$\gamma(\lambda,\zeta,\omega) = \frac{\lambda K_\nu(\alpha) I_\nu(\alpha\zeta) - \lambda I_\nu(\alpha) K_\nu(\alpha\zeta)}{\left(\lambda - \sqrt{i\omega\mu_0\sigma_0 + \lambda^2}\right) I_\nu(\alpha) + \frac{\alpha b}{2} I_{\nu-1}(\alpha)}$$

$$+ \frac{\frac{\alpha b}{2}\left\{ \frac{\nu}{\alpha}\left[I_\nu(\alpha) K_\nu(\alpha\zeta) - K_\nu(\alpha) I_\nu(\alpha\zeta)\right] - \left[K_{\nu-1}(\alpha) I_\nu(\alpha\zeta) + I_{\nu-1}(\alpha) K_\nu(\alpha\zeta)\right]\right\}}{\left(\lambda - \sqrt{i\omega\mu_0\sigma_0 + \lambda^2}\right) I_\nu(\alpha) + \frac{\alpha b}{2} I_{\nu-1}(\alpha)}. \quad (22)$$

**Exponentially increasing conductivity profile ($\sigma_0 > \sigma_1$)**

The results for the case where $\sigma_0 > \sigma_1$ can now be obtained in a similar manner to previous subsection. Instead of (14), we now have

$$\widetilde{E}^{g}(\lambda,\zeta,\omega) = C_5 K_\nu(\alpha\zeta), \quad (23)$$

where $C_5$ is an arbitrary constant. $K_\nu(\ )$ is the modified Bessel function of the second kind of order $\nu$. Instead of (16), we now have

$$\widetilde{E}^{s}(\lambda,\zeta,\omega) = C_6 K_\nu(\alpha\zeta) - \frac{4i\omega\mu_0}{b^2} \int_0^1 G(\zeta;\tau) \left[ \int_0^c s(\sigma_s - \sigma_p) E_{tar}(s,\zeta,\omega) J_1(\lambda s) ds \right] d\tau, \quad (24)$$

where $C_6$ is an arbitrary constant and $G(\zeta;\tau)$ is the Green's function defined by

$$G(\zeta;\tau) = \begin{cases} \dfrac{K_\nu(\alpha\zeta) I_\nu(\alpha\tau)}{(-\tau)}, & \zeta \le \tau, \\[2ex] \dfrac{I_\nu(\alpha\zeta)\ K_\nu(\alpha\tau)}{(-\tau)}, & \tau \le \zeta. \end{cases} \quad (25)$$

Instead of (19), we now have

$$\widetilde{E}^{\text{ground}}(\lambda,\zeta,\omega) = CK_\nu(\alpha\zeta) - \frac{2i\omega\mu_0}{b} \sum_{j=1}^{n} G(\zeta;\zeta_j) P(\lambda,\zeta_j,\omega)(\zeta_j), \quad (26)$$

where $C = C_5 + C_6$ is an arbitrary constant to be determined. Similarly, matching the solutions at air-earth interface to determine the unknown constants $A$ and $C$, the electric field in air is given by

$$E^{\text{air}}(r,z,\omega) = -i\omega\mu_0 I(\omega) a \int_0^\infty \frac{\lambda K_\nu(\alpha) J_1(\lambda a) J_1(\lambda r) e^{-\lambda|z+h|}}{\left(\lambda - \sqrt{i\omega\mu_0\sigma_0 + \lambda^2}\right) K_\nu(\alpha) - \frac{\alpha b}{2} K_{\nu-1}(\alpha)} d\lambda$$
$$-i\omega\mu_0 \sum_{j=1}^{n} \int_0^c sp(s,\zeta_j,\omega) ds \int_0^\infty \lambda \left\{ \frac{K_\nu(\alpha e^{\frac{bz_j}{2}}) J_1(\lambda s)\, e^{\lambda z}}{\left(\lambda - \sqrt{i\omega\mu_0\sigma_0 + \lambda^2}\right) K_\nu(\alpha) - \frac{\alpha b}{2} K_{\nu-1}(\alpha)} \right\} J_1(\lambda r)\, d\lambda, \quad (27)$$

where $h$ and $z \le 0$. In ground, the electric field is given by

$$E^{\text{ground}}(r,\zeta,\omega) = -i\omega\mu_0 I(\omega) a \int_0^\infty \frac{\lambda J_1(\lambda a) J_1(\lambda r) K_\nu(\alpha\zeta) e^{-\lambda|h|}}{\left(\lambda - \sqrt{i\omega\mu_0\sigma_0 + \lambda^2}\right) K_\nu(\alpha) - \frac{\alpha b}{2} K_{\nu-1}(\alpha)} d\lambda$$
$$- \frac{2i\omega\mu_0}{b} \sum_{j=1}^{n} \int_0^c sp(s,\zeta_j,\omega) ds \int_0^\infty \lambda \gamma(\lambda,\zeta,\omega) K_\nu(\alpha\zeta_j) J_1(\lambda s) J_1(\lambda r) d\lambda, \quad (28)$$

where $\zeta_1 \le \zeta \le 1$ and

$$\gamma(\lambda,\zeta,\omega) = \frac{\lambda I_\nu(\alpha) K_\nu(\alpha\zeta) - \lambda K_\nu(\alpha) I_\nu(\alpha\zeta)}{\left(\lambda - \sqrt{i\omega\mu_0\sigma_0 + \lambda^2}\right) K_\nu(\alpha) - \frac{\alpha b}{2} K_{\nu-1}(\alpha)}$$
$$+ \frac{\frac{\alpha b}{2}\left\{ \frac{\nu}{\alpha}\left[K_\nu(\alpha) I_\nu(\alpha\zeta) - I_\nu(\alpha) K_\nu(\alpha\zeta)\right] + \left[I_{\nu-1}(\alpha) K_\nu(\alpha\xi) + K_{\nu-1}(\alpha) I_\nu(\alpha\zeta)\right]\right\}}{\left(\lambda - \sqrt{i\omega\mu_0\sigma_0 + \lambda^2}\right) K_\nu(\alpha) - \frac{\alpha b}{2} K_{\nu-1}(\alpha)}. \quad (29)$$

## INVESION PROCESS

In this subsection, we now concentrate on the primary field generated by a dipole source of moment $m = \pi a^2 I(\omega)$, $a \to 0$, located on air-earth interface, $h = 0$. The inversion scheme makes use of the electric field at air-earth interface, $z = 0$, which may be obtained from the equations (21) or (28) depending on whether $\sigma_0 < \sigma_1$ or $\sigma_0 > \sigma_1$. At first, we will consider on the case of an exponentially decreasing ground profile, $\sigma_0 < \sigma_1$.

### Exponentially decreasing conductivity profile

For the case $\sigma_0 < \sigma_1$, the expression for the electric field in the ground, $\zeta_1 \le \zeta \le 1$ is given by

$$E^{\text{ground}}(r,\zeta,\omega) = -\frac{i\omega\mu_0 m}{2\pi} \int_0^\infty \frac{\lambda^2 J_1(\lambda r) I_\nu(\alpha\zeta)}{\left(\lambda - \sqrt{i\omega\mu_0\sigma_0 + \lambda^2}\right) I_\nu(\alpha) + \frac{\alpha b}{2} I_{\nu-1}(\alpha)} d\lambda$$
$$+ \frac{2i\omega\mu_0}{b} \sum_{j=1}^{n} \int_0^c sp(s,\zeta_j,\omega) ds \int_0^\infty \lambda \gamma(\lambda,\zeta,\omega) I_\nu(\alpha\zeta_j) J_1(\lambda s) J_1(\lambda r) d\lambda, \quad (30)$$

where $\gamma(\lambda, \zeta, \omega)$ is presented by (22). From relations $I_\nu(\alpha) = e^{-i\nu\pi/2} J_\nu(\alpha e^{-i\pi/2})$ and $I_{\nu-1}(\alpha) = e^{-i(\nu-1)\pi/2} J_{\nu-1}(\alpha e^{-i\pi/2})$, where $-\pi < \arg\alpha \leq \pi/2$, then, we can see that $I_\nu(\alpha)$ and $I_{\nu-1}(\alpha)$ never vanishes. The zero of $J_\nu(w)$ and $J_{\nu-1}(w)$ in complex $w$-plane are all real (see Abramowitz, M. & Stegun, I. A., (1972), pp. 372 and 375). Therefore, the integral terms of the equation (30) are all well-defined. In general, the magnitude of measured electric field tend to be small compared to unity, and it is expedient to scale it against the magnitude of the electric field in free space, that is, $E_0 = -i\omega\mu_0 m/4\pi r^2$. Defining $L$ to be length scale and $\Omega$ is a typical frequency, the non-dimensional quantities are given by

$$r^* = r/L,\ c^* = c/L,\ \omega^* = \omega/\Omega,\ \lambda^* = \lambda L,\ s^* = s/L,\ b^* = bL,\ S_j^* = \Omega\mu_0 L S_j,$$
$$E_{tar}^{j*} = E_{tar}^j / E_0,\ E^* = -4\pi L^2 E^{\text{ground}} / (i\omega\mu_0 m). \tag{31}$$

Applying the non-dimensional quantities to (30), and dropping the asterisks, the equations (30) becomes

$$E(r,\zeta,\omega) = 2\int_0^\infty \frac{\lambda^2 J_1(\lambda r) I_\nu(\alpha\zeta)}{\left(\lambda - \sqrt{i\omega\Omega\mu_0\sigma_0 L^2 + \lambda^2}\right) I_\nu(\alpha) + \frac{\alpha b}{2} I_{\nu-1}(\alpha)} d\lambda$$
$$+ \frac{2i\omega}{b}\sum_{j=1}^{n}\int_0^{c'} sp(s,\zeta_j,\omega)ds \int_0^\infty \lambda\gamma(\lambda,\zeta,\omega) I_\nu(\alpha\zeta_j) J_1(\lambda s) J_1(\lambda r) d\lambda, \tag{32}$$

where $c' = c^* L$, $\zeta_1 \leq \zeta \leq 1$ and

$$\gamma(\lambda,\zeta,\omega) = \frac{\lambda K_\nu(\alpha) I_\nu(\alpha\zeta) - \lambda I_\nu(\alpha) K_\nu(\alpha\zeta)}{\left(\lambda - \sqrt{i\omega\Omega\mu_0\sigma_0 L^2 + \lambda^2}\right) I_\nu(\alpha) + \frac{\alpha b}{2} I_{\nu-1}(\alpha)}$$
$$+ \frac{\frac{\alpha b}{2}\left\{\frac{\nu}{\alpha}\left[I_\nu(\alpha) K_\nu(\alpha\zeta) - K_\nu(\alpha) I_\nu(\alpha\zeta)\right] - \left[K_{\nu-1}(\alpha) I_\nu(\alpha\zeta) + I_{\nu-1}(\alpha) K_\nu(\alpha\zeta)\right]\right\}}{\left(\lambda - \sqrt{i\omega\Omega\mu_0\sigma_0 L^2 + \lambda^2}\right) I_\nu(\alpha) + \frac{\alpha b}{2} I_{\nu-1}(\alpha)}. \tag{33}$$

The electric field at air-earth interface, $z = 0$ or $\zeta = 1$, is given by

$$E(r,1,\omega) = 2\int_0^\infty \frac{\lambda^2 J_1(\lambda r) I_\nu(\alpha)}{\left(\lambda - \sqrt{i\omega\Omega\mu_0\sigma_0 L^2 + \lambda^2}\right) I_\nu(\alpha) + \frac{\alpha b}{2} I_{\nu-1}(\alpha)} d\lambda$$
$$- i\omega\sum_{j=1}^{n}\int_0^{c'} sp(s,\zeta_j,\omega)ds \int_0^\infty \frac{\lambda I_\nu(\alpha\zeta_j) J_1(\lambda s) J_1(\lambda r)}{\left(\lambda - \sqrt{i\omega\Omega\mu_0\sigma_0 L^2 + \lambda^2}\right) I_\nu(\alpha) + \frac{\alpha b}{2} I_{\nu-1}(\alpha)} d\lambda, \tag{34}$$

where the Wronskian for $I_\nu(\alpha)$ and $K_\nu(\alpha)$ has been used. On each disk, we can use the fact that $p(s,\zeta_j,\omega) = S_j E_{tar}^j(s,\zeta_j,\omega)$ to write (32) as

$$\frac{p(r,\zeta_j,\omega)}{S_j} = 2\int_0^\infty \frac{\lambda^2 J_1(\lambda r) I_\nu(\alpha\zeta_j)}{\left(\lambda - \sqrt{i\omega\Omega\mu_0\sigma_0 L^2 + \lambda^2}\right) I_\nu(\alpha) + \frac{\alpha b}{2} I_{\nu-1}(\alpha)} d\lambda$$
$$+ \frac{2i\omega}{b}\sum_{j=1}^{n}\int_0^{c'} sp(s,\zeta_j,\omega)ds \int_0^\infty \lambda\gamma(\lambda,\zeta_j,\omega) I_\nu(\alpha\zeta_j) J_1(\lambda s) J_1(\lambda r) d\lambda. \tag{35}$$

For a given frequency $\omega$, let $r_k$ is the distance between the receiver and the center of the source. We note that $\{E^{obs}(r_k,1,\omega)\}$, of size N, is a set of the electric field on the air-earth interface which obtained from the receiver. $\{E^{model}(r_k,1,\omega)\}$, of size N, is a set of the electric field on the air-earth interface which calculated by (34). If $z_1$ is known then equation (34) can be used to estimate value for $p(s,\zeta_j,\omega)$, $1 \le j \le n$. Equation (35) is used to estimate the true value of the conductance $S_j$, $1 \le j \le n$.

However, in practical, we do not know the true value of model parameters such as the conductivity, the depth and the thickness of the target under the ground. In our inverse algorithm, we will use the idea of unconstrained nonlinear optimization to minimize objective function

$$f(\sigma_s,z_1,d)=\sum_{k=1}^{N}\left\|E^{obs}(r_k,1,\omega)-E^{model}(r_k,1,\omega)\right\|, \tag{36}$$

to search unknown parameters of the model, where 2-norm is taken. Applying the downhill simplex method for finding the optimum value for $\sigma_s$, $z_1$ and $d$. The method employs four points for computing the value of $f(\sigma_s, z_1, d)$. The method locates the point with best and worst values, and then attempts to replace the worst-value point with a better point. The replacement process involves expanding and contracting the tetrahedron near the worst point to determine a better replacement point. The iterations of the method shrink the size of the tetrahedron around the optimum solution.

**Exponentially increasing conductivity profile**

For the case $\sigma_0 > \sigma_1$, the ground profile is an exponentially increasing and instead of equation (30), we now have

$$E^{\text{ground}}(r,\zeta,\omega)=-\frac{i\omega\mu_0 m}{2\pi}\int_0^\infty\frac{\lambda^2 J_1(\lambda a)\,J_1(\lambda r)K_\nu(\alpha\zeta)}{\left(\lambda-\sqrt{i\omega\mu_0\sigma_0+\lambda^2}\right)K_\nu(\alpha)\;-\frac{\alpha b}{2}K_{\nu-1}(\alpha)}d\lambda$$

$$-\frac{2i\omega\mu_0}{b}\sum_{j=1}^{n}\int_0^c sp(s,\zeta_j,\omega)ds\int_0^\infty\lambda\gamma(\lambda,\zeta,\omega)K_\nu(\alpha\zeta_j)J_1(\lambda s)J_1(\lambda r)d\lambda, \tag{37}$$

where $\gamma(\lambda, \zeta, \omega)$ is presented by (29). Since, $K_\nu(\alpha)$ and $K_{\nu-1}(\alpha)$ has no zeros in $|\arg\alpha|\le \pi/2$ (Abramowitz & Stegun, (1972), p. 377), hence, the integral terms of the equation (37) are all well-defined. Similarly, instead of (32), we now have

$$E(r,\zeta,\omega)=2\int_0^\infty\frac{\lambda^2\, J_1(\lambda r)K_\nu(\alpha\zeta)}{\left(\lambda-\sqrt{i\omega\Omega\mu_0\sigma_0 L^2+\lambda^2}\right)K_\nu(\alpha)\;-\frac{\alpha b}{2}K_{\nu-1}(\alpha)}d\lambda$$

$$-\frac{2i\omega}{b}\sum_{j=1}^{n}\int_0^{c'} sp(s,\zeta_j,\omega)ds\int_0^\infty\lambda\gamma(\lambda,\zeta,\omega)K_\nu(\alpha\zeta_j)J_1(\lambda s)J_1(\lambda r)d\lambda, \tag{38}$$

where $\zeta_1 \le \zeta \le 1$ and

$$\gamma(\lambda,\zeta,\omega) = \frac{\lambda I_\nu(\alpha)K_\nu(\alpha\zeta) - \lambda K_\nu(\alpha)I_\nu(\alpha\zeta)}{\left(\lambda - \sqrt{i\omega\Omega\mu_0\sigma_0 L^2 + \lambda^2}\right)K_\nu(\alpha) - \frac{\alpha b}{2}K_{\nu-1}(\alpha)}$$
$$+ \frac{\frac{\alpha b}{2}\left\{\frac{\nu}{\alpha}[K_\nu(\alpha)I_\nu(\alpha\zeta) - I_\nu(\alpha)K_\nu(\alpha\zeta)] + [I_{\nu-1}(\alpha)K_\nu(\alpha\xi) + K_{\nu-1}(\alpha)I_\nu(\alpha\zeta)]\right\}}{\left(\lambda - \sqrt{i\omega\Omega\mu_0\sigma_0 L^2 + \lambda^2}\right)K_\nu(\alpha) - \frac{\alpha b}{2}K_{\nu-1}(\alpha))}. \tag{40}$$

Instead of equations (35) and (36), we now have

$$E(r,1,\omega) = 2\int_0^\infty \frac{\lambda^2 J_1(\lambda r)K_\nu(\alpha)}{\left(\lambda - \sqrt{i\omega\Omega\mu_0\sigma_0 L^2 + \lambda^2}\right)K_\nu(\alpha) - \frac{\alpha b}{2}K_{\nu-1}(\alpha)}d\lambda$$
$$- i\omega\sum_{j=1}^{n}\int_0^{c'} sp(s,\zeta_j,\omega)ds\int_0^\infty \frac{\lambda K_\nu(\alpha\zeta_j)J_1(\lambda s)J_1(\lambda r)}{\left(\lambda - \sqrt{i\omega\Omega\mu_0\sigma_0 L^2 + \lambda^2}\right)K_\nu(\alpha) - \frac{\alpha b}{2}K_{\nu-1}(\alpha)}d\lambda, \tag{41}$$

and

$$\frac{p(r,\zeta_j,\omega)}{S_j} = 2\int_0^\infty \frac{\lambda^2 J_1(\lambda r)K_\nu(\alpha\zeta_j)}{\left(\lambda - \sqrt{i\omega\Omega\mu_0\sigma_0 L^2 + \lambda^2}\right)K_\nu(\alpha) - \frac{\alpha b}{2}K_{\nu-1}(\alpha)}d\lambda$$
$$- \frac{2i\omega}{b}\sum_{j=1}^{n}\int_0^{c'} sp(s,\zeta_j,\omega)ds\int_0^\infty \lambda\gamma(\lambda,\zeta_j,\omega)K_\nu(\alpha\zeta_j)J_1(\lambda s)J_1(\lambda r)d\lambda. \tag{42}$$

Similarly, we use downhill simplex method to estimate the optimum value of the model parameters by using objective function in (37).

## INVERSION EXAMPLES

The results obtained can be used to estimate the model parameters such as the conductivity, the thickness and the depth of a conductive 3-D circular cylinder body embedded in a conducting half-space with an exponentially varying conductivity profiles. In our inversion examples, we will consider firstly in case of an exponentially decreasing ground profile, $\sigma_0 < \sigma_1$.

### Exponentially decreasing ground profile ($\sigma_0 < \sigma_1$)

We consider the conductivity of the subsurface given by $\sigma_p(z)$ = 0.0000001 + (0.1 – 0.0000001)exp(–1.51793$z$). In forward problem, a conductive 3-D circular cylinder body of radius 100 metres, thickness 10 metres and conductivity 100 Siemens is assumed to be located at a depth of 100 metres. Fortran90 programming is used under Windows98 system running on a PC computer (Pentium II, 533 MHz) at the department of Mathematics, faculty of Science, Silpakorn university. The Hankel transform can be computed by using Chave's algorithm (Chave, 1983), which is an efficient direct integration scheme giving numerical results of high accuracy. The determination of the special functions are computed by using IMSL libraries. The data set $\{E^{obs}\}$, which is normally obtained from the receiver on the air-earth interface, is generated using values of $r_k$ from 10 metres to 200 metres in step of 10 metres. The random error is made up to 3% to the generated data. In the inversion process, we start with initial guess values for the four points, $P_1$(125.00,125.00,1.25), $P_2$(75.00, 75.00, 0.75), $P_3$(50.00,50.00,0.50) and $P_4$(25.00, 25.00, 0.25). Our algorithm generates a sequence of the new point that converges to the optimum solution, which base on the size of the tetrahedron. The results of algorithm are shown in Table 1–A. In addition, we may use the downhill simplex method to minimizes the model objective function, which are given by

$$g(\sigma_s, z_1, d) = \sum_{k=1}^{N}\left[\mathrm{Re}\left(E^{obs}(r_k,1,\omega) - E^{model}(r_k,1,\omega)\right)\right]^2, \tag{43}$$

and

$$h(\sigma_s, z_1, d) = \sum_{k=1}^{N}\left[\mathrm{Im}\left(E^{obs}(r_k,1,\omega) - E^{model}(r_k,1,\omega)\right)\right]^2, \tag{44}$$

to estimate the optimum values of the model parameters, where N = 20. The results are shown in Table 1–B and 1–C, respectively.

**Exponentially increasing ground profile ($\sigma_0 > \sigma_1$)**

For the case of an exponentially increasing ground profile, $\sigma_0 > \sigma_1$, we consider the conductivity of the subsurface given by $\sigma_p(z)$ = 0.0000001 + (0.0000000001 – 0.0000001)exp(–1.51793$z$). We start with initial guess values for the four points, $P_1$(115.00, 115.00, 1.15), $P_2$(90.00,90.00,0.90), $P_3$(85.00,85.00,0.85) and $P_4$(75.00, 75.00, 0.75). The result of algorithm is shown in Table 2–A. Similarly, we may use the model objective functions in equations (43) and (44) to estimate the optimum values of the model parameters. The results are shown in 2–B and 2–C, respectively.

| $i$ | $\sigma_s$ | $z_1$ | $d$ | $f(\sigma_s, z_1, d)$ | Size of the tetrahedron |
|---|---|---|---|---|---|
| 1 | 83.345000 | 83.345000 | 0.833450 | 0.424106 | 1.105526 |
| 2 | 86.226149 | 86.226149 | 0.862261 | 0.622936 | 0.857613 |
| 3 | 98.200021 | 98.200021 | 0.982000 | 0.192156 | 1.057008 |
| 4 | 90.772510 | 90.772510 | 0.907725 | 1.071780 | 1.391879 |
| 5 | 100.6730189 | 100.6730189 | 1.006730 | 0.086512 | 1.297077 |
| 6 | 96.882995 | 96.882995 | 0.968829 | 0.202206 | 1.121643 |
| 7 | 98.585663 | 98.585663 | 0.985856 | 0.235047 | 0.923842 |
| 8 | 99.629341 | 99.629341 | 0.996293 | 0.240859 | 1.084353 |
| 9 | 100.151180 | 100.151180 | 1.001511 | 0.018663 | 1.714681 |
| 10 | 100.342347 | 100.342347 | 1.003423 | 0.056830 | 1.290186 |
| **Optimum** | **100.151181** | **100.151181** | **1.001511** | **0.018663** | **$1.176991\times10^{-6}$** |

Table 1–A : Results for the case of an exponentially decreasing ground profile using the model objective function $f$

| $i$ | $\sigma_s$ | $z_1$ | $d$ | $g(\sigma_s, z_1, d)$ | Size of the tetrahedron |
|---|---|---|---|---|---|
| 1 | 83.916666 | 83.916666 | 0.839166 | 0.094551 | 0.996744 |
| 2 | 95.085277 | 95.085277 | 0.950852 | 0.067190 | 0.860960 |
| 3 | 101.339248 | 101.339248 | 1.013392 | 0.004114 | 1.833204 |
| 4 | 107.373757 | 107.373757 | 1.073737 | 0.041511 | 1.767666 |
| 5 | 104.356502 | 104.356502 | 1.043565 | 1.888379 | 1.991304 |
| 6 | 102.847875 | 102.847875 | 1.028478 | 0.001545 | 1.985251 |
| 7 | 101.320663 | 101.320663 | 1.013206 | 0.004139 | 1.856673 |
| 8 | 101.311815 | 101.311815 | 1.013118 | 0.004151 | 1.857074 |
| 9 | 100.293821 | 100.293821 | 1.002938 | 0.001116 | 1.515551 |
| 10 | 100.034814 | 100.034814 | 1.000348 | 0.000075 | 1.407450 |
| **Optimum** | **100.121236** | **100.121236** | **1.001212** | **0.000022** | **$2.766934\times10^{-6}$** |

Table 1–B : Results for the case of an exponentially decreasing ground profile using the model objective function $g$

| $i$ | $\sigma_s$ | $z_1$ | $d$ | $h(\sigma_s, z_1, d)$ | Size of the tetrahedron |
|---|---|---|---|---|---|
| 1 | 83.345000 | 83.345000 | 0.833450 | 0.007259 | 1.933907 |
| 2 | 86.226149 | 86.226149 | 0.862261 | 0.055211 | 1.739036 |
| 3 | 98.200021 | 98.200021 | 0.982000 | 0.001840 | 1.909973 |
| 4 | 88.899627 | 88.899627 | 0.888996 | 0.012005 | 1.870939 |
| 5 | 90.772510 | 90.772510 | 0.907725 | 0.312786 | 1.976597 |
| 6 | 94.693128 | 94.693128 | 0.946931 | 0.008741 | 1.535041 |
| 7 | 95.035486 | 95.035486 | 0.950354 | 0.013608 | 1.523409 |
| 8 | 95.035411 | 95.035411 | 0.950354 | 0.013608 | 1.523407 |
| 9 | 96.446574 | 96.446574 | 0.964465 | 0.004768 | 1.517439 |
| 10 | 97.106919 | 97.106919 | 0.971069 | 0.000840 | 1.402572 |
| **Optimum** | **97.550483** | **97.550483** | **0.975504** | **0.018663** | **$5.634810\times10^{-6}$** |

Table 1–C : Results for the case of an exponentially decreasing ground profile using the model objective function $h$

| $i$ | $\sigma_s$ | $z_1$ | $d$ | $f(\sigma_s, z_1, d)$ | Size of the tetrahedron |
|---|---|---|---|---|---|
| 1 | 108.340000 | 96.670999 | 0.966709 | $1.007029\times10^{-7}$ | 1.962803 |
| 2 | 102.5578000 | 98.979236 | 0.989792 | $1.226392\times10^{-7}$ | 1.828059 |
| 3 | 98.468925 | 93.335579 | 0.933355 | $1.242594\times10^{-7}$ | 1.673740 |
| 4 | 103.125866 | 96.330871 | 0.963308 | $1.223657\times10^{-7}$ | 1.534645 |
| 5 | 104.796403 | 97.404327 | 0.974043 | $1.225183\times10^{-7}$ | 0.016091 |
| 6 | 105.509177 | 96.817671 | 0.968176 | $1.222884\times10^{-7}$ | 0.002967 |
| 7 | 105.712941 | 96.647324 | 0.966473 | $1.223993\times10^{-7}$ | 0.001981 |
| 8 | 105.717524 | 96.643539 | 0.966435 | $1.223579\times10^{-7}$ | 0.001009 |
| 9 | 105.821732 | 96.556445 | 0.965564 | $1.223541\times10^{-7}$ | 0.000670 |
| 10 | 104.510505 | 96.542706 | 0.965427 | $1.223452\times10^{-7}$ | 0.001482 |
| **Optimum** | **103.375881** | **96.448421** | **0.964484** | **$1.223850\times10^{-7}$** | **0.001981** |

Table 2–A : Results for the case of an exponentially increasing ground profile using the model objective function $f$

| $i$ | $\sigma_s$ | $z_1$ | $d$ | $g(\sigma_s, z_1, d)$ | Size of the tetrahedron |
|---|---|---|---|---|---|
| 1 | 108.340000 | 96.670999 | 0.966709 | $8.403183\times10^{-15}$ | 1.999995 |
| 2 | 102.557800 | 98.979236 | 0.989792 | $8.423709\times10^{-15}$ | 1.962078 |
| 3 | 98.629326 | 93.545788 | 0.935457 | $8.397473\times10^{-15}$ | 1.867348 |
| 4 | 103.303464 | 96.441891 | 0.964418 | $8.413720\times10^{-15}$ | 0.333749 |
| 5 | 103.378926 | 95.447859 | 0.954478 | $8.374569\times10^{-15}$ | 0.023692 |
| 6 | 103.396875 | 95.138575 | 0.951385 | $8.363416\times10^{-15}$ | 0.022506 |
| 7 | 101.682325 | 94.620368 | 0.946203 | $8.367051\times10^{-15}$ | 0.021252 |
| 8 | 100.075625 | 93.977974 | 0.939779 | $8.428843\times10^{-15}$ | 0.024302 |
| 9 | 100.108914 | 93.988227 | 0.939882 | $8.379535\times10^{-15}$ | 0.019381 |
| 10 | 99.822932 | 93.901856 | 0.939018 | $8.357163\times10^{-15}$ | 0.018435 |
| **Optimum** | **99.350158** | **93.707671** | **0.937076** | $\mathbf{8.361082\times10^{-15}}$ | **0.016658** |

Table 2–B : Results for the case of an exponentially increasing ground profile using the model objective function $g$

| $i$ | $\sigma_s$ | $z_1$ | $d$ | $g(\sigma_s, z_1, d)$ | Size of the tetrahedron |
|---|---|---|---|---|---|
| 1 | 108.340000 | 96.670999 | 0.966709 | $2.065378\times10^{-18}$ | 1.999995 |
| 2 | 102.557800 | 98.979236 | 0.989792 | $2.564837\times10^{-18}$ | 1.999983 |
| 3 | 98.468925 | 93.335579 | 0.933355 | $1.081946\times10^{-17}$ | 1.999973 |
| 4 | 103.125866 | 96.330871 | 0.963308 | $1.675132\times10^{-18}$ | 1.463726 |
| 5 | 104.674865 | 97.327235 | 0.973272 | $1.846705\times10^{-18}$ | 0.419675 |
| 6 | 105.408468 | 96.754302 | 0.967543 | $1.983508\times10^{-18}$ | 0.208659 |
| 7 | 104.363697 | 96.805480 | 0.968054 | $1.687042\times10^{-18}$ | 0.168574 |
| 8 | 104.054539 | 96.821209 | 0.968212 | $2.821105\times10^{-18}$ | 0.110238 |
| 9 | 104.202485 | 96.813672 | 0.968136 | $1.858607\times10^{-18}$ | 0.837904 |
| 10 | 104.065452 | 96.791956 | 0.967919 | $1.859304\times10^{-18}$ | 0.449196 |
| **Optimum** | **104.087147** | **96.815263** | **0.968156** | $\mathbf{1.691179\times10^{-18}}$ | **0.042790** |

Table 2–C : Results for the case of an exponentially increasing ground profile using the model objective function $h$

## DISCUSSION

Lee & Ignetik (1994), Yooyuanyong (1999) and Yooyuanyong & Siew (2000) used the exponential variation function of the subsurface as given by $\sigma_P(z) = \sigma_1 . e^{-bz}$, where $\sigma_1 > 0$ and $b$ may be either positive or negative. It can be shown that as $z \to \infty$, in case of an exponentially increasing ground profile, $b < 0$, we obtain $\sigma_P \to \infty$. In case of an exponentially decreasing ground profile, $b > 0$, we obtain $\sigma_P \to 0$ as $z \to \infty$. This is obviously true for a shot range of depth. In this paper, we develop a more realistic model as $\sigma_P(z) = \sigma_0 + (\sigma_1 - \sigma_0)e^{-bz}$, where $\sigma_0$, $\sigma_1$and $b$ are positive values. It can be shown that as $z \to \infty$, in case of an exponentially increasing ground profile, $\sigma_0 > \sigma_1$, we obtain $\sigma_P \to \sigma_0$. In case of an exponentially decreasing ground profile, $\sigma_0 < \sigma_1$, we obtain $\sigma_P \to \sigma_0$ as $z \to \infty$. If the ground without target and $\sigma_0 << 1$, the equations (22) and (29) become

$$E^{\text{ground}}(r,\zeta,\omega) = -\frac{2i\omega\mu_0 I(\omega)a}{\alpha b}\int_0^\infty \frac{\lambda J_1(\lambda a)J_1(\lambda r)I_\nu(\alpha\zeta)e^{-\lambda|h|}}{I_{\nu-1}(\alpha)}d\lambda, \tag{45}$$

and

$$E^{\text{ground}}(r,\zeta,\omega)=\frac{2i\omega\mu_0 I(\omega)a}{\alpha b}\int_0^\infty \frac{\lambda J_1(\lambda a)\,J_1(\lambda r)K_\nu(\alpha\zeta)e^{-\lambda|h|}}{K_{\nu-1}(\alpha)}d\lambda, \tag{46}$$

respectively, where $\alpha^2 = (4i\omega\mu_0\sigma_1)/b^2$, $\nu^2 = (4\lambda^2)/b^2$, $\zeta \leq 1$ and $h \leq 0$. In this case, we recover the results obtained by Lee & Ignetik. In addition, for the case of an exponentially decreasing ground profile, if the target is a conductive thin disk, the equation (22) becomes

$$E^{\text{ground}}(r,\zeta,\omega)=-i\omega\mu_0 I(\omega)a\int_0^\infty \frac{\lambda J_1(\lambda a)\,J_1(\lambda r)I_\nu(\alpha\zeta)e^{-\lambda|h|}}{\left(\lambda-\sqrt{i\omega\mu_0\sigma_0+\lambda^2}\right)I_\nu(\alpha)+\frac{\alpha b}{2}I_{\nu-1}(\alpha)}d\lambda$$

$$+\frac{2i\omega\mu_0}{b}\int_0^c sp(s,\zeta_1,\omega)ds\int_0^\infty \lambda\gamma(\lambda,\zeta,\omega)I_\nu(\alpha\zeta_1)J_1(\lambda s)J_1(\lambda r)d\lambda, \tag{47}$$

where $\zeta_1 \leq \zeta \leq 1$ and $\gamma(\lambda, \zeta, \omega)$ is given by the equation (23). In the special case $\sigma_0 \ll 1$, we can rewrite the equation (47) as

$$E^{\text{ground}}(r,\zeta,\omega)=-\frac{2i\omega\mu_0 I(\omega)a}{\alpha b}\int_0^\infty \frac{\lambda J_1(\lambda a)\,J_1(\lambda r)I_\nu(\alpha\zeta)e^{-\lambda|h|}}{I_{\nu-1}(\alpha)}d\lambda$$

$$-\frac{2i\omega\mu_0}{b}\int_0^c sp(s,\zeta_1,\omega)ds\int_0^\infty \lambda I_\nu(\alpha\zeta_1)J_1(\lambda s)J_1(\lambda r)\times$$

$$\left[\frac{I_\nu(\alpha\zeta)K_{\nu-1}(\alpha)+K_\nu(\alpha\zeta)I_{\nu-1}(\alpha)}{I_{\nu-1}(\alpha)}\right]d\lambda, \tag{48}$$

where $\zeta_1 \leq \zeta \leq 1$, which is exactly as same as the work done by Yooyuanyong (1999). In the inverse problem, if the primary field generated by a magnetic dipole at air-earth interface, and using the non-dimensional quantities as mention in (32) to (48), we obtain

$$E(r,\zeta,\omega)=\frac{4}{\alpha b}\int_0^\infty \frac{\lambda^2 J_1(\lambda r)I_\nu(\alpha\zeta)}{I_{\nu-1}(\alpha)}d\lambda$$

$$-\frac{2i\omega}{b}\int_0^{c'} sp(s,\zeta_1,\omega)ds\int_0^\infty \lambda\left[\frac{I_\nu(\alpha\zeta)K_{\nu-1}(\alpha)+K_\nu(\alpha\zeta)I_{\nu-1}(\alpha)}{I_{\nu-1}(\alpha)}\right]$$

$$\times I_\nu(\alpha\zeta_1)J_1(\lambda s)J_1(\lambda r)d\lambda. \tag{49}$$

At the air-earth interface, $\zeta = 1$, the expression for the electric field on the ground surface is given by

$$E(r,1,\omega)=\frac{4}{\alpha b}\int_0^\infty \frac{\lambda^2 J_1(\lambda r)I_\nu(\alpha)}{I_{\nu-1}(\alpha)}d\lambda-\frac{2i\omega}{b\alpha}\int_0^{c'} sp(s,\zeta_1,\omega)ds\int_0^\infty \lambda\frac{I_\nu(\alpha\zeta_1)J_1(\lambda s)J_1(\lambda r)}{I_{\nu-1}(\alpha)}d\lambda. \tag{50}$$

On the disk we can use the fact that $p(s, \zeta_1, \omega) = SE_{tar}(s, \zeta_1, \omega)$ to rewrite (49) as

$$\frac{p(s,\zeta_1,\omega)}{S}=\frac{4}{\alpha b}\int_0^\infty \frac{\lambda^2 J_1(\lambda r)I_\nu(\alpha\zeta_1)}{I_{\nu-1}(\alpha)}d\lambda$$

$$-\frac{2i\omega}{b}\int_0^{c'} sp(s,\zeta_1,\omega)ds\int_0^\infty \lambda\left[\frac{I_\nu(\alpha\zeta_1)K_{\nu-1}(\alpha)+K_\nu(\alpha\zeta_1)I_{\nu-1}(\alpha)}{I_{\nu-1}(\alpha)}\right]\times$$

$$I_\nu(\alpha\zeta_1)J_1(\lambda s)J_1(\lambda r)d\lambda. \tag{51}$$

Thus, we recover the work done by Yooyuanyong (1999). In our inversion examples, the optimum values of the model parameters are presented at the last row of each table in pages 10-12. It is clearly that our inverse algorithm converges in 10 iterations. Using $f$ is the model objective function, in case of an exponentially decreasing ground profile and case of an exponentially increasing ground profile, the algorithm gives the maximum percentage errors in the optimum values of the model parameters less than 0.16% and 3.5%, respectively, of their true values. This is much better than using the model objective function from equations (43) and (44), where the maximum percentage errors are less than 0.12% and 6.3%, respectively, of their true values for the case of an exponentially decreasing ground profile and 0.12% and 4.08%, respectively, of their true values for the case of an exponentially increasing ground profile.

## CONCLUSION

We have outlined an efficient method and techniques for the inverse problem of determining the model parameters such as the conductivity, the thickness and the depth of the conductive 3-D circular cylinder body. Some examples are given to perform the efficient of algorithm. The results from inversion examples are shown in Table 1–A, 1–B, 1–C, 2–A, 2–B and 2–C. We can see that our algorithm is robust and it can obviously be adopted to deal with more complicated structures.

## REFERENCES


Abramowitz, M. and Stegun, I. A., 1972, *Handbook of mathematical functions*, New York: Dover.

Banerjee, B., Sengupta, B. J. and Pal, B. P., 1980, Apparent resistivity of a mutilayered earth with a layer having exponential conductivity, *Geophysical Prospecting* **28**, 435-452.

Chave, A. D., 1983, Numerical integration of related Hankel transforms by quadrature and continued fraction expansion, *Geophysics* **48**, 1671-1686.

David, M. H.,1972, Applied Nonlinear Programming, McGraw-Hill, Inc.

Kim, H. S., and Lee, K., 1996, Response of a multilayered earth with layers having exp onentially varying resistivities, *Geophysics* **61**, 180-191.

Lee, T. J. and Ignetik, R., 1994, Transient Electromagnetic Response of a half-space with Exponential Conductivity Profile and its Applications to Salinity Mapping, *Exploration Geophysics* **25**, 39-51.

Stoyer C. H. and Wait J. R., 1977, Resistivity probing of an exponential earth with a homogeneous overburden, *Geoexploration* **15**, 11-18.

Yooyuanyong, S., 1999, Inversion by EM sounding for a disk embedded in a conducting half-space, *Songklanakarin J. Sci. Technol.* **21(2)**, 97-205.

Yooyuanyong, S.,and Siew, P. F., 2000, The electromagnetic response of a disk beneath an exponentially varying conductive overburden, *J. Australian Mathematical Society.*, **Series B, 41**, E1-E28.

# $(l)$-Green's Relations and Perfect rpp Semigroups *

K.P.Shum
*Department of Mathematics,*
*The Chinese University of Hong Kong, Hong Kong.*
X.J.Guo
*Department of Mathematics, Jiangxi Normal University,*
*Nanchang, Jiangxi 330027, China*
X.M.Ren
*Department of Mathematics,*
*The Chinese University of Hong Kong, Hong Kong.*

**Abstract**

We characterize the perfect rpp semigroups by using $(l)$-Green's relations. We show that perfect rpp semigroups share an analogous property of Green's relations on completely regular semigroups and Green' star relations on superabundant semigroups, respectively. In particular, we prove an interesting result that $\mathcal{D}^{(l)} = \mathcal{J}^{(l)}$ holds on a perfect rpp semigroup. A new characterization for perfect rpp semigroups is consequently given.

**AMS** Mathematics Subject Classification (2000): 20M15
**Keywords**: $(l)$-Green's Relations, Perfect rpp Semigroups, Strongly rpp Semigroups.

# 1 Introduction and Background

A semigroup $S$ is called an *rpp* semigroup if all its principal right ideals $aS^1(a \in S)$, regarded as a right $S^1$-systems, are projective. This class of semigroups and its special subclasses have been extensively studied by J.B.Fountain ([3],[4] and [5]). On a semigroup $S$, the Green's star relation $\mathcal{L}^*$ is defined by $(a,b) \in \mathcal{L}^*$ if and only if the elements $a, b$ of $S$ are related by the usual Green's relation $\mathcal{L}$ on some oversemigroup of $S$. Fountain then showed that a monoid $S$ is *rpp* if and only if every $\mathcal{L}^*$-class contains an idempotent. Thus, a semigroup

*This research is partially supported by a RGC direct grant #2060196

$S$ is *rpp* if and only if every $\mathcal{L}^*$-class of $S$ contains at least one idempotent. *lpp* semigroups can be dually defined and a semigroup which is both rpp and lpp is called abundant [5]. Thus, all rpp semigroups and abundant semigroups form two kinds of generalized regular semigroups.

It is noted that rpp semigroups with central idempotents have similar structure as Clifford semigroups. We call this kind of rpp semigroups $C$-rpp semigroups. New varities of semigroups generated by $C$-rpp semigroups and abundant semigroups have been extensively investigated by El-Qallali, Fountain, Y.Q.Guo, the authors and others (see [2], [3], [4],[5], [9], [10],[16] and [17]).

Let $a$ be an arbitrary element of a semigroup $S$ and $E(S)$ the set of all idempotents of $S$. Then we call the set $\mathcal{M}_a = \{\, e \in E(S) \mid S^1a \subseteq Se$ for all $x, y \in S^1, ax = ay \Longrightarrow ex = ey\}$ the E-envelope of $a$. Also, an rpp semigroup $S$ is called strongly rpp if there exists a unique idempotent $e \in \mathcal{M}_a$ such that $a = ea$. In order to specify that this unique idempotent $e$ is closely related with $a \in S$, we usually denote the unique idempotent $e$ in the E-envelope $\mathcal{M}_a$ by $a^+$. Thus, in this case, we always have $a = a^+a = aa^+$. Although the idempotent $a^+$ in the strongly rpp semigroup is always commutative with $a \in S$, $S$ is not necessarily a $C$-rpp semigroup unless $E(S)$ is contained in the center of $S$ (see [8] and [16]).

In a recent paper, Shum, X.J.Guo and Y.Q.Guo [8] have considered the perfect rpp semigroup which is a strongly *rpp* semigroup whose idempotents form a normal band and satisfying $(ab)^+ = a^+b^+$ for all $a, b \in S$. It has been proved by them in [8] that a semigroup $S$ is a perfect *rpp* semigroup if and only if $S$ is a strong semilattice of left cancellative planks. In view of this nice structure, it can be easily seen that the perfect semigroup can be regarded as a generalized $C$-rpp semigroup because Fountain has proved that a $C$-rpp semigroup can be described as a strong semilattice of left cancellative monoids [3].

It is well known that the usual Green's relations $\mathcal{L}, \mathcal{R}, \mathcal{H}, \mathcal{D}$ and $\mathcal{J}$ on a semigroup play a fundamental role in the algebraic theory of semigroups, especially in the structure of regular semigroups. These relations and their properties have been thoughtfully described by Clifford-Preston [1] and Howie [11].

As a corresponding to the Green's relations on a semigroup, the Green's star relations on a semigroup $S$ defined by Pastijn and Fountain can be formulated in the following format:

$$a\mathcal{L}^*b \Longleftrightarrow \operatorname{Ker} a_l = \operatorname{Ker} b_l;$$
$$a\mathcal{R}^*b \Longleftrightarrow \operatorname{Ker} a_r = \operatorname{Ker} b_r;$$
$$\mathcal{D}^* = \mathcal{L}^* \vee \mathcal{R}^*;$$
$$\mathcal{H}^* = \mathcal{L}^* \wedge \mathcal{R}^*;$$
$$a\mathcal{J}^*b \Longleftrightarrow \mathcal{J}^*(a) = \mathcal{J}^*(b),$$

where $a_l(a_r)$ is the inner left (right) translation determined by $a \in S, \mathcal{J}^*(a)$ is the smallest ideals including $a$ which is both $\mathcal{L}^*$-saturated and $\mathcal{R}^*$-saturated. In general, for an equivalent relation $\rho$ on the semigroup $S$, a subset $A$ of $S$ is called $\rho$-saturated if $A$ is the union of some $\rho$-classes. It was observed by Fountain [5], the set of Green's star relations extends the set of the usual Green's relations in the sense $\mathcal{L} \subseteq \mathcal{L}^*, \mathcal{R} \subseteq \mathcal{R}^*, \mathcal{D} \subseteq \mathcal{D}^*, \mathcal{J} \subseteq \mathcal{J}^*$ and $\mathcal{H} \subseteq \mathcal{H}^*$. The Green's star relations are particularly useful in the study of abundant semigroups. However, for rpp semigroups, the Green's star relations are usually unfit due to the fact that an abundant semigroup is only a special semigroup. Under such circumstance, Pastijn and Y.Q.Guo have redefined and modified the Green's star relations for rpp semigroups in the following format:

The following five relations are named by Pastijn and Y.Q.Guo the $(l)$-Green's relations defined on a semigroup $S$ ( we once called them $\widetilde{Green's}$ relations in [10]).

$$\begin{gathered}
\mathcal{L}^{(l)} = \mathcal{L}^*; \quad i.e.\ a\mathcal{L}^{(l)}b \Longleftrightarrow \operatorname{Ker} a_l = \operatorname{Ker} b_l; \\
\mathcal{R}^{(l)} = \mathcal{R}; \quad i.e.\ a\mathcal{R}^{(l)}b \Longleftrightarrow \operatorname{Im} a_l = \operatorname{Im} b_l; \\
\mathcal{D}^{(l)} = \mathcal{L}^{(l)} \vee \mathcal{R}^{(l)} = \mathcal{L}^* \vee \mathcal{R}; \\
\mathcal{H}^{(l)} = \mathcal{L}^{(l)} \wedge \mathcal{R}^{(l)}; \\
a\mathcal{J}^{(l)}b \Longleftrightarrow J^{(l)}(a) = J^{(l)}(b)
\end{gathered}$$

where $\mathcal{J}^{(l)}(a)$ is the smallest ideal containing $a \in S$ and is $\mathcal{L}^{(l)}$-saturated.

Apparently, they are all equivalence relations on a semigroup $S$ and $\mathcal{L}^{(l)}(\mathcal{R}^{(l)})$ is a right (left) congruence on $S$.

Comparing with the above three types of Green's relations, we have

(i) $\mathcal{L} \subsetneq \mathcal{L}^{(l)}$; $\mathcal{D} \subsetneq \mathcal{D}^{(l)}$; $\mathcal{J} \subsetneq \mathcal{J}^{(l)}$ and $\mathcal{H} \subsetneq \mathcal{H}^{(l)}$,

(ii) $\mathcal{R}^{(l)} \subsetneq \mathcal{R}^*$; $\mathcal{D}^{(l)} \subsetneq \mathcal{D}^*$; $\mathcal{J}^{(l)} \subsetneq \mathcal{J}^*$ and $\mathcal{H}^{(l)} \subsetneq \mathcal{H}^*$.

In fact, all "$\subseteq$" are clear by definitions. To see that "$\subseteq$" in (i) is proper, we consider, for instance, the dual of Baer-Levi semigroup $S$ which is a left simple left cancellative semigroup and let $M = S^1$. Then $M$ is obviously a non-trivial left cancellative monoid in which $ab = a$ does not hold for all $a, b \in S$. Because the semigroup $S$ is left simple, we have $\mathcal{J}^{(l)} = \mathcal{D}^{(l)} = \mathcal{L}^{(l)}$-simple on $M$. However, $\mathcal{J}(1) \supsetneq \mathcal{J}(s)$ for all $s \in S$, that is, $\mathcal{J} \subsetneq \mathcal{J}^{(l)}$, $\mathcal{D} \subsetneq \mathcal{D}^{(l)}$ and $\mathcal{L} \subsetneq \mathcal{L}^{(l)}$.

To see "$\subseteq$" in (ii) is proper, we conside the set $T = A \cup B \cup \{1\}$, where $A$ is a monogenic semigroup generated by $a$ and $B$ is a monogenic monoid with identity $e$ generated by $b$. Define a multiplication on $T$ by $a^m b^n = b^{m+n}, b^n a^m = a^{m+n}, b^0 = e$ for all $m = 1, 2, \cdots, n = 0, 1, 2, \cdots$. Then $T$, equipped with the above multiplication, is a semigroup on which $\mathcal{D}^* = \mathcal{J}^*$ universally. Also, $\mathcal{L}^{(l)} = \mathcal{L}^*$ has two classes in $T$, namely $A \cup \{1\}$ and $B$;

$\mathcal{R}^*$ has also two classes in $T$, namely $\{1\}$ and $A \cup B$ (see [1] and [5]). It is now easy to see that $\mathcal{R} = \mathcal{L}$ which is the equality relation on $T$. Thereby, $\mathcal{D}^{(l)} = \mathcal{L}^{(l)} \vee \mathcal{R} = \mathcal{L}^{(l)} \subsetneq \mathcal{D}^*$; $\mathcal{R}^{(l)} = \mathcal{R} \subsetneq \mathcal{R}^*$. on $T$ Examples of other cases can be also constructed and we omit the details.

It was obversed by Y.Q.Guo that a similar Green's theorem and its related results of the usual Green's relations can be correspondingly extended to $(l)$-Green's relations. Because the $(l)$-Green's relations are particularly defined for rpp semigroups and the strongly rpp semigroup defined in [10] is respectively an analogue for rpp semigroups of completely regular semigroups and superabundant semigroups, one would naturally ask whether the strongly rpp semigroups have the same property on $(l)$-Green's relations, similar to completely regular semigroups and superabundant semigroups, that is, whether $\mathcal{D}^{(l)} = \mathcal{J}^{(l)}$ holds on strongly rpp semigroups, same as $\mathcal{D} = \mathcal{J}$ holds for completely regular semigroups and same as $\mathcal{D}^* = \mathcal{J}^*$ holds for superabundant semigroups ? This question is not easy to answer and in fact, it is much harder in this case than the cases of completely regular semigroups and superabundant semigroups.

In this paper, we shall answer the above question by showing that a strongly rpp semigroup $S$ whose idempotents form a normal band is a perfect rpp semigroup if and only if $\mathcal{D}^{(l)} = \mathcal{J}^{(l)}$ holds on $S$. We shall also provide an example to illustrate that $\mathcal{D}^{(l)} = \mathcal{J}^{(l)}$ does not generally hold on strongly rpp semigroups. Thus, $\mathcal{D}^{(l)} = \mathcal{J}^{(l)}$ only holds on a special strongly rpp semigroup and this notable result differs from the corresponding results on completely regular semigroups and superabundant semigroups, respectively.

For notations and terminologies which are not explained in this paper, the reader is referred to the text [11] and the paper [8].

## 2 $(l)$-Green's Relations on Strongly rpp Semigroups

In this section, we first study the properties of the $(l)$-Green's relations on strongly rpp semigroups and then we give a characterization theorem of perfect rpp semigroups by using the $(l)$-Green's relations.

We start with the following lemmas.

**Lemma 2.1** ([8]) *Let $S$ be a strongly rpp semigroup whose set of idempotents $E$ forms a normal band. Then, $S$ is a perfect rpp semigroup if and only if for all $a, b \in S$, $(ab)^+ = a^+b^+$ .*

**Lemma 2.2** *Let $S = [Y; S_\alpha, \varphi_{\alpha,\beta}]$ be a strong semilattice of semigroups $S_\alpha$. If every $S_\alpha$ is an rpp semigroup and the structure mapping $\varphi_{\alpha,\beta}$ is an $\mathcal{L}^*$-homomorphism, that is , the homomorphism preserving the $\mathcal{L}^*$-relation, then, $S$ is an rpp semigroup.*

**Proof:** Since $S_\alpha$ is an rpp semigroup, for any $a \in S_\alpha$, there exists $e \in E(S_\alpha)$ such that $e\mathcal{L}^*(S_\alpha)a$. Suppose that $ax = ay$, for some $x, y \in S^1$. Then, there exist $\beta, \gamma \in Y$ such that $x \in S^1_\beta, y \in S^1_\gamma$ and $a\varphi_{\alpha,\alpha\beta}x\varphi_{\beta,\alpha\beta} = a\varphi_{\alpha,\alpha\gamma}y\varphi_{\gamma,\alpha\gamma}$ with $\alpha\beta = \alpha\gamma$. Because $\varphi_{\alpha,\beta}$ is an $\mathcal{L}^*$-homomorphism, we have $a\varphi_{\alpha,\alpha\beta}\mathcal{L}^*(S_{\alpha\beta})e\varphi_{\alpha,\alpha\beta}$. Consequently, $e\varphi_{\alpha,\alpha\beta}x\varphi_{\beta,\alpha\beta} = e\varphi_{\alpha,\alpha\gamma}y\varphi_{\gamma,\alpha\gamma}$, and thereby $ex = ey$. On the other hand, if $ex = ey$, then, by $e\mathcal{L}^*(S_\alpha)a$, we immediately have $a = ae$ and hence $ax = ay$. This shows that $a\mathcal{L}^*(S)e$ and by definition, $S$ is an rpp semigroup. $\square$

**Lemma 2.3** *Any bicyclic semigroup is not a strongly rpp semigroup.*

**Proof:** Let $S$ be a bicyclic semigroup. Then, by Howie [11], $S$ is isomorphic to the semigroup $T = (N \times N, *)$ with $(m,n) * (p,q) = (m - n + max\{n,p\}, q - p + max\{n,p\})$, where $N$ is the set of non-negative integers. It is known that $T$ is an inverse semigroup.

If $T$ is a strongly rpp semigroup, then by definition, there exists a unique idempotent $(m,n)^+ \in T$ for any $(m,n) \in T$ such that $(m,n)^+\mathcal{L}^*(m,n)$ and hence $(m,n)^+ * (m,n) = (m,n)$. Because $(m,m), (n,n)$ are idempotents, it is trivial to see that $(m,m)\mathcal{R}^*(m,n)$ and $(n,n)\mathcal{L}^*(m,n)$. Thus, by the uniquences of the idempotents in the $\mathcal{L}^*$-class of $T$, we have $(m,n)^+ = (n,n)$. Therby, we have $(n,n) * (m,m) = (m,m)$, by $(m,m)\mathcal{R}^*(m,n)$. However, because $T$ is an inverse semigroup, we have $(n,n) * (m,m) = (m,m) * (n,n)$. This induces that $(m,m) \leqslant (n,n)$ by the natural partial order "$\leqslant$" endowed in $E(T)$ of the rpp semigroup $T$. This leads to $n \leqslant m$, which contradicts to the choice of the element $(m,n)$. This shows that $T$ is not a strongly rpp semigroup. $\square$

**Lemma 2.4** *Let $S$ be a strongly rpp semigroup whose idempotents $E(S)$ form a semilattice. Then, every regular element in $S$ is completely regular.*

**Proof:** Since $E(S)$ is a semilattice, the set of all regular elements, namely, $Reg(S)$, is an inverse subsemigroup of $S$. If $Reg(S)$ is not a completely regular subsemigroup of $S$, then there exists $a \in Reg(S)$ which is not completely regular. But $S$ is a strongly rpp semigroup, $Reg(S)$, as a subsemigroup of $S$, is also strongly rpp. Hence, for $a \in Reg(S)$, there exist idempotents $a^+$ and $e$ such that $a^+\mathcal{L}^*a$ and $a\mathcal{R}^*e$. In fact, because $a^+$ and $e$ are idempotents, according to Fountain [5], we have $a^+\mathcal{L}a$ and $a\mathcal{R}e$. Thus, there exists an inverse element $a'$ of $a$ such that $a'a = a^+$ and $aa' = e$ with $aa' \leqslant a'a$ [11]. If $a'a = aa'$, then $a$ is clearly completely regular, this contradicts to our assumption. Hence, we have $a = aa'a = aa^+ = a^+a$, and $a' = a'aa' = a^+a' = a'aa'a^+ = a'a^+$. By Clifford and Preston ([1], Lemma 1.31), it follows that $Reg(S)$ contains a bicyclic subsemigroup $T$ which is strongly rpp, since the semigroup $S$ itself is strongly rpp. On the other hand, by our Lemma 2.3, the bicyclic semigroup

$T$ is not a strongly semigroup, a contradiction. This shows that the element $a \in S$ must be completely regular. $\square$

Now, we consider the $\mathcal{L}^*$-congruence on the strongly rpp semigroup $S$.

Let $S$ be a strongly rpp semigroup whose idempotents $E(S)$ form a normal band. Define $a\pi b \iff (\exists\, e, f \in E(b^+))(a = ebf)$, where $E(b^+)$ is the $\mathcal{D}$-class of $E(S)$ containing the idempotent $b^+ \in S$. Then, by a result of X.J.Guo, Shum and Y.Q.Guo in [8], the relation $\pi$ is an $\mathcal{L}^*$-congruence on the strongly rpp semigroup $S$.

Equipped with the above notion, we establish the following lemma.

**Lemma 2.5** *Let $S$ be a strongly rpp semigroup whose set of idempotents is a normal band, Then $\mathcal{D}^{(l)}|_E = \mathcal{D}^E$ on $S$.*

**Proof:** Obviously, $\mathcal{D}^E \subseteq \mathcal{D}^{(l)}|_E$. Now, let $(e, f) \in \mathcal{D}^{(l)}|_E$. Then, by the definition of $\mathcal{D}^{(l)}$, there exists $x \in S$ such that $e\mathcal{L}^{(l)}x\mathcal{R}^{(l)}f$. Because $\mathcal{R}^{(l)} = \mathcal{R}$ and the idempotent $f$ is trivially a regular element, $x$ is a regular elment of $S$. By [8], $S/\pi$ is a strongly rpp semigroup whose idempotents form a semilattice. Now, by Lemma 2.4, the image $x\pi$ of $x$ in $S/\pi$ is a completely regular element. This means that there exists $g \in E(S/\pi)$ such that $x\pi\mathcal{H}g$, and consequently, $e\pi = g = f\pi$ since $e\pi\mathcal{L}x\pi\mathcal{R}f\pi$ and $E(S/\pi)$ is a semilattice. In other words, $e\mathcal{D}^E f$ and so $\mathcal{D}^{(l)}|_E \subseteq \mathcal{D}^E$. Thus, $\mathcal{D}^{(l)}|_E = \mathcal{D}^E$. The proof is completed. $\square$

The following lemma is similar to a result of Fountain [5] for Green's star relation. For the sake of completeness, we provide here the proof.

**Lemma 2.6** *Let $S$ be a semigroup. Then for any $a, b \in S, b \in J^{(l)}(a)$ if and only if there are elements $a_0, a_1, \cdots, a_n \in S$ with $a_0 = a, a_n = b$ and $x_1, x_2, \cdots, x_n; y_1, y_2, \cdots, y_n \in S^1$ such that $a_i\mathcal{L}^{(l)}(x_i a_{i-1} y_i)$, for all $i = 1, 2, \cdots, n$.*

**Proof:** Let $B$ be the collection of all the elements satisfying the given conditions. Then, we only need to prove that $J^{(l)}(a) = B$ . Since for any $b \in B$, there exist $a_0 = a, a_1, \cdots, a_n = b$ in $S$ and sequences $x_1, x_2, \cdots, x_n$ and $y_1, y_2, \cdots, y_n \in S^1$ such that $a_i\mathcal{L}^{(l)}(x_i a_{i-1} y_i), i = 1, 2, \cdots, n$. If $a_{i-1} \in J^{(l)}(a)$, then because $J^{(l)}$ is an ideal of $S$, we have $x_i a_{i-1} y_i \in J^{(l)}(a)$. This leads to $a_i \in J^{(l)}(a)$ because $J^{(l)}(a)$ is $\mathcal{L}^{(l)}$-saturated. By $a_0 = a \in J^{(l)}(a)$, we have $a_i \in J^{(l)}, i = 0, 1, 2, \cdots, n$, and in particular, $b = a_n \in J^{(l)}(a)$. Hence, $B \subseteq J^{(l)}(a)$. To see the converse inclusion, we only need to show that $B$ is an $\mathcal{L}^{(l)}$-saturated ideal of $S$. In fact, $a$ is apparently in $B$ and $J^{(l)}(a)$ is the smallest ideal of $S$ containing $a$, also, by our conditions, $J^{(l)}$ is $\mathcal{L}^{(l)}$-saturated ideal containing $a$. If $b \in B$, then by the properties of $B$, we know that for any $s, t \in S^1$, $sbt \in B$. Thus, $B$ is an ideal of $S$. Now, by the properties of $B$, for any $b \in B$, we know that the $\mathcal{L}^{(l)}$-class containing $b$ is still in $B$. This shows that $B$ is $\mathcal{L}^{(l)}$-saturated. In particular, $J^{(l)}(a) \subseteq B$. Hence $J^{(l)}(a) = B$ and the lemma follows. $\square$

By using the above Lemmas, we can answer the question proposed in the introduction and in fact, we establish a characterization theorem for perfect rpp semigroup.

**Theorem 2.7** *Let $S$ be a strongly rpp semigroup whose idempotents form a normal band. Then the following statements are equivalent:*

*(i) $S$ is a perfect rpp semigroup;*

*(ii) $S$ is the spined product of a right $C$-rpp semigroup and a left normal band;*

*(iii) $\mathcal{D}^{(l)} = \mathcal{J}^{(l)}$;*

*(iv) $\mathcal{D}^{(l)}$ is a semilattice congruence on $S$;*

*(v) $\mathcal{D}^{(l)}$ is a congruence on $S$.*

**Proof:** $(i) \Longrightarrow (ii)$. Let $S$ be a perfect semigroup. Then, by [8], $S$ is isomorphic to $E(S) \ltimes S/\pi$, the spined product of $E(S)$ and $S/\pi$, where $\pi$ is a $\mathcal{L}^*$-congruence on $S$. Because $E(S)$ is a normal band, $E(S)$ can be expressed as a strong semilattice of rectangular band $E_\alpha$, that is, $E(S) = [Y; E_\alpha, \varphi_{\alpha,\beta}]$ (see [11]). Since $E_\alpha$ is a rectangular band, we can express $E_\alpha = L_\alpha \times R_\alpha$ where $L_\alpha$ and $R_\alpha$ are left zero band and right zero band, respectively. Now, we form $L = \bigcup_{\alpha\in Y} L_\alpha$ and $R = \bigcup_{\alpha\in Y} R_\alpha$, also define the mappings $\varphi^l_{\alpha,\beta} : L_\alpha \longrightarrow L_\beta$ and $\varphi^r_{\alpha,\beta} : R_\alpha \longrightarrow R_\beta$ such that for any $(l_\alpha, r_\alpha) \in E_\alpha$, $(l_\alpha, r_\alpha)\varphi_{\alpha,\beta} = (l_\alpha\varphi^l_{\alpha,\beta}, r_\alpha\varphi^r_{\alpha,\beta})$ (see [11]). Define the multiplications "$*$"and "$\circ$" on $L$ and $R$ by $l_\alpha * l_\beta = l_\alpha\varphi^l_{\alpha,\alpha\beta}$ and $r_\alpha \circ r_\beta = r_\beta\varphi^r_{\alpha,\alpha\beta}$, respectively. Then, it can be easily seen that $(L, *)$ and $(R, \circ)$ are respectively left normal and right normal band.

On the other hand, since $S/\pi$ is now a strongly rpp semigroup whose idempotents form a semilattice, we can express $S/\pi$ by a strong semilattice of left cancellative monoids $M_\alpha$, that is, $S/\pi = [Y; M_\alpha, \varphi_{\alpha,\beta}]$ (see [3]). Now, we let $T = \bigcup_{\alpha\in Y}(M_\alpha \times R_\alpha)$ and define

$$(m_\alpha, r_\alpha) * (m_\beta, r_\beta) = (m_\alpha\psi_{\alpha,\alpha\beta}m_\beta\psi_{\beta,\alpha\beta}, r_\alpha \circ r_\beta)$$

on $T$. Then we can verify that "$*$" is associative and hence $T$ is a semigroup, and in fact, it is a strong semilattice of the semigroups $T_\alpha = M_\alpha \times R_\alpha$, that is, $T = [Y; T_\alpha, \Phi_{\alpha,\beta}]$ where $\Phi_{\alpha,\beta} : T_\alpha \longrightarrow T_\beta$ is defined by $\Phi_{\alpha,\beta}(m_\alpha, r_\alpha) = (m_\alpha\psi_{\alpha,\beta}, r_\alpha\varphi_{\alpha,\beta^r})$, for $\alpha \geqslant \beta$. Clearly, $\Phi_{\alpha,\beta}$ is a structure homomorphism. Obviously, for the identity element $e_\alpha$ of the monoid $M_\alpha$, we have $(m_\alpha, r_\alpha)\mathcal{L}^*(T_\alpha)$ $(e_\alpha, r_\alpha)$ on $T_\alpha$, and thereby, $(m_\alpha\psi_{\alpha,\beta}, r_\alpha\varphi^r_{\alpha,\beta})\mathcal{L}^*(T_\alpha)(e_\beta, r_\alpha\varphi^r_{\alpha,\beta})$. This shows that $\Phi_{\alpha,\beta}$ is a strong semilattice homomorphism which preserves the $\mathcal{L}^*$-relation. Thus, by Lemma 2.2, $T$ is an rpp semigroup. Invoking a result in [8], we know that $T$ is a right $C$-rpp semigroup. Our last step is to show that the spined product $E \ltimes S/\pi$ is isomorphic to the spined product $L \ltimes T$, but this part is straightforward and we omit the details. Hence $S$ is the spined product of a left normal band and a right $C$-rpp semigroup, and so (i) $\Longrightarrow$ (ii).

*(ii)* $\Longrightarrow$ *(iii)* Suppose that $S = L \ltimes T$, where $L$ is a left normal band and $T$ is a right $C$-rpp semigroup. Then, it is easy to see that $\mathcal{D}^{(l)} \subseteq \mathcal{J}^{(l)}$. For the converse inclusion, we let $(s,l)\mathcal{J}^{(l)}(t,r)$. Then by Lemma 2.6, there exists a sequence such that $(s,l) = (s_0,l_0),(s_1,l_1),\cdots,(s_n,l_n) = (t,r) \in L \ltimes T$ and a sequence $(x_1,k_1),(x_2,k_2),\cdots,(x_n,k_n),(y_1,i_1),(y_2,i_2),\cdots,(y_n,i_n) \in (L \ltimes T)^1$ such that $(s_j,l_j)\mathcal{L}^{(l)}(x_j,k_j)(s_{j-1},l_{j-1})(y_j,i_j)$ for $j = 1,2,\cdots,n$, that is, $(s_j,l_j)\mathcal{L}^{(l)}(x_js_{j-1}y_j,k_jl_{j-1}i_j)$ for $j = 1,2,\cdots,n$. Hence, $s_j\mathcal{L}^{(l)}x_js_{j-1}y_j$ and $l_j\mathcal{L}^{(l)}k_jl_{j-1}i_j$. Consequently, we have $s \in J^{(l)}(t)$ and $E(l_j) = E(k_jl_{j-1}i_j) \subseteq E(l_{j-1})$ and, in particular, $E(l_n) \subseteq E(l_0)$. Dually, we have $t \in J^{(l)}(s)$ and obtain $E(l_0) \subseteq E(l_n)$. Thus, $E(l_0) = E(l_n)$ and $J^{(l)}(s) = J^{(l)}(t)$. These facts lead to $s\mathcal{J}^{(l)}t$ and $l_0\mathcal{L}l_n$. By [8], we can also deduce that $s\mathcal{D}^{(l)}t$. This means that there exists an element $u \in T$ such that $s\mathcal{L}^{(l)}u\mathcal{R}^{(l)}t$, and so $(s,l)\mathcal{L}^{(l)}(u,l_n)\mathcal{R}^{(l)}(t,l_n)$. Hence, $(s,l)\mathcal{D}^{(l)}(t,r)$, and thereby, $\mathcal{D}^{(l)} \supseteq \mathcal{J}^{(l)}$. This proves that $\mathcal{D}^{(l)} = \mathcal{J}^{(l)}$.

*(iii)* $\Longrightarrow$ *(iv)* Let $\mathcal{D}^{(l)} = \mathcal{J}^{(l)}$. We only need to prove that $\mathcal{J}^{(l)}$ is a semilattice congruence on $S$. Because $S$ is a strongly rpp semigroup, for any $a \in S$, there exists a unique idempotent $a^+$ corresponding to $a$ with respect to $\mathcal{L}^*$ such that $a\mathcal{L}^*a^+$ and $a = a^+a$. Since $\mathcal{L}^{(l)} = \mathcal{L}^*$ and $\mathcal{L}^*$ is a right congruence on $S$, this leads to $a\mathcal{L}^{(l)}a^2$ and so $a\mathcal{J}^{(l)}a^2$. Since $a$ is arbitrarily chosen, we have $J^{(l)}(ab) = J^{(l)}((ab)^2) = J^{(l)}(a(ba)b) \subseteq J^{(l)}(ba)$, by the definition of $J^{(l)}(ba)$. Similarly, we have $J^{(l)}(ba) \subseteq J^{(l)}(ab)$. Thus $J^{(l)}(ab) = J^{(l)}(ba)$ and so, for any $a,b \in S$, we have $ab\mathcal{J}^{(l)}ba$. It remains to show that $\mathcal{J}^{(l)}$ is a congruence on $S$. Now, suppose that $a\mathcal{J}^{(l)}b$. Then, it is obvious that $a^+\mathcal{J}^{(l)}b^+$. Now, by our hypothesis, we have $a^+\mathcal{D}^{(l)}b^+$ and consequently we obtain that $a^+\mathcal{D}b^+$. This means that there exist $u,v \in S^1$ such that $a^+ = ub^+v$. Also, it is clear that for any $c \in S, ac\mathcal{L}^{(l)}a^+c$, and thereby $J^{(l)}(ac) = J^{(l)}(a^+c) = J^{(l)}(ub^+vc) \subseteq J^{(l)}(b^+vc) = J^{(l)}(vcb^+) \subseteq J^{(l)}(cb^+) = J^{(l)}(b^+c) = J^{(l)}(bc)$. Similarly, we have $J^{(l)}(bc) \subseteq J^{(l)}(ac)$. Thus, we obtain $J^{(l)}(bc) = J^{(l)}(ac)$, that is, $ac\mathcal{J}^{(l)}bc$. This shows that $\mathcal{J}^{(l)}$ is a right congruence on $S$. Dually, we can prove that $\mathcal{J}^{(l)}$ is a left congruence on $S$. Hence, $\mathcal{J}^{(l)}$ is indeed a semilattice congruence on $S$.

*(iv)* $\Longrightarrow$ *(v)*. This part is obvious, and we omit the proof.

*(v)* $\Longrightarrow$ *(i)*. Since $S$ is a strongly rpp semigroup, for every $a \in S$ and $b \in S$ there exist unique idempotents $a^+$ and $b^+$ such that $a\mathcal{L}^*a^+$ and $b\mathcal{L}^*b^+$, respectively. Since $\mathcal{L}^* = \mathcal{L}^{(l)}$, we have $a\mathcal{L}^{(l)}a^+$ and $b\mathcal{L}^{(l)}b^+$. Thereby, $a\mathcal{D}^{(l)}a^+$ and $b\mathcal{D}^{(l)}b^+$. Since $\mathcal{D}^{(l)}$ is a congruence on $S$, we have $ab\mathcal{D}^{(l)}a^+b^+$. Because $ab\mathcal{L}^{(l)}(ab)^+$, we can easily deduce that $ab\mathcal{D}^{(l)}(ab)^+$. This leads to $ab\mathcal{D}^{(l)}a^+(ab)^+$, and whence $a^+(ab)^+\mathcal{D}^{(l)}(ab)^+$. Recall that the $\mathcal{D}^E$-class is a rectangular band and by Lemma 2.5, we have $(ab)^+\mathcal{L}a^+(ab)^+$. Since $S$ is a strongly rpp semigroup and $ab = a^+(ab)^+ab$, we have $(ab)^+ = a^+(ab)^+$, by the uniqueness of

the idempotent in the $E$-envelope $\mathcal{M}_{ab}$. On the other hand, by the definition of $\mathcal{L}^{(l)}$, it is obvious that $ab\mathcal{L}^{(l)}(ab)^+$ and $ab \cdot 1 = ab \cdot b^+$. Consequently, by the definition of $\mathcal{L}^{(l)}$, we obtain $(ab)^+ \cdot 1 = (ab)^+ \cdot b^+$. Because $E(S)$ is assumed to be a normal band, $E(S)$ can be represented as a strong semilattice of rectangular bands, that is, $E(S) = [Y; E_\alpha, \varphi_{\alpha,\beta}]$. Now, we let $a^+ \in E_\alpha$, and $b^+ \in E_\beta$. Then, we have $(ab)^+ = a^+(ab)^+b^+ = a^+\varphi_{\alpha,\alpha\beta} \cdot (ab)^+ \cdot b^+\varphi_{\beta,\alpha\beta} = a^+\varphi_{\alpha,\alpha\beta}b^+\varphi_{\beta,\alpha\beta} = a^+b^+$, by recalling that $E_{\alpha\beta}$ is a rectangular band. By using Lemma 2.1, we know that $S$ is a perfect rpp semigroup. The proof is now completed. □

In closing this paper, we give an example which indicates that $\mathcal{D}^{(l)} \neq \mathcal{J}^{(l)}$ may possibly hold on strongly rpp semigroups.

**Example 2.8**

Let $N$ be the set of non-negative integers and $S = \{(m,n) \in N \times N \mid m \geqslant n\}$. Define the following operation "$*$" on $S$ by

$$(m,n) * (p,q) = (m - n + max\{n,p\}, q - p + max\{n,p\})$$

Then, it can be easily verified that $S$, under the multiplication "$*$", is a semigroup. In fact, by Lemma 2.3, we can see that our semigroup $S$ is a full subsemigroup of the bicyclic semigroup $T$ in Lemma 2.3. Obviously, $S$ is an adequate semigroup, by Fountain [4]. Therefore, $S$ itself is an rpp semigroup.

Now, for any element $(m,n) \in S$, there is a unique idempotent $(n,n)$ satisfying $(m,n)\mathcal{L}^{(l)}(n,n)$ and $(n,n)(m,n) = (m,n)$, thereby $(m,n)^+ = (n,n)$. This shows that $S$ is a strongly rpp semigroup. In addition, we can observe that $E(S)$ of $S$ is a semilattice. However, $S$ is not a perfect rpp semigroup. This is because we can find elements $a, b$ in $S$ such that $a^+b^+ \neq (ab)^+$, for instance, if we let $a = (3,3), b = (2,1)$, then $((3,2)(2,1)) = (3,2)$. Since $(3,2)^+ = (2,2), (3,3)^+ = (3,3), (2,1)^+ = (1,1)$, we can see that $(ab)^+ = (2,2) \neq a^+b^+ = (3,3)(1,1) = (3,3)$. By Lemma 2.1 and Theorem 2.7, we therefore conclude that $\mathcal{D}^{(l)} \neq \mathcal{J}^{(l)}$ on $S$. This illustrates that $\mathcal{D}^{(l)} = \mathcal{J}^{(l)}$ may not hold on an arbitrary strongly rpp semigroup.

**Acknowledge** The authors would like to thank Professor Y.Q.Guo for his valuable suggestions and comments contributed to this paper, in particular, we thank him to propose this interesting question for our research.

# References

[1] Clifford, A.H. and G.B.Preston, *The algebraic theory of semigroups,* Vol I and Vol II, Monograph, American Math.Soc.,R.I., 1961 and 1967.

[2] El-Qallali,A., On the construction of a class of abundant semigroups, *Acta.Math.Hung.* (1-2) **56** (1990), 77–91.

[3] Fountain,J.B., Right pp monoids with central idempotents, *Semigroup Forum* **13** (1977), 229–237.

[4] Fountain,J.B., Adequate semigroups, *Proc.Edinburgh Math.Soc.* **22** (1979), 113-125.

[5] Fountain,J.B., Abundant semigroups, *Proc.Lond.Math.Soc.* (3) **44** (1982), 103-129.

[6] Guo,X.J., The structure of PI-strongly rpp semigroups, *Chinese Sci. Bull.* **18** (1996), 1647–1650.

[7] Guo,X.J., Y.Q.Guo and K.P.Shum, Semispined product structure of left C-a semigroups, *Semigroup,Kunming 95, Springer* (1998).

[8] Guo,X.J., K.P.Shum and Y.Q.Guo, Perfect C-rpp semigroups, *Communication in Algebra*, **29**(6), 2447–2459 (2001).

[9] Guo,Y.Q., The right dual of left C-rpp semigroups, *Chinese Sci. Bull.* (9) **42** (1997), 1599–1603.

[10] Guo,Y.Q., K.P.Shum and P.Y.Zhu, The structure of left C-rpp semigroups, *Semigroup Forum* **50** (1995), 9–23.

[11] Howie,J.M., *An introduction to semigroup theory*, Academic Press, London, 1976.

[12] Lallement,G., *Semigroups and combinatorical applications*, John Wiley & Sons, N. Y., 1979.

[13] Lawson,M.V., Abundant Rees matrix semigroups, *J.Austral.Math.Soc.* **42** (1987), 132–142.

[14] Pastijn,F., A representation of a semigroup by a semigroup of matrices over a group with zero, *Semigroup Forum* **10** (1975), 238-249.

[15] Petrich,M., *Introduction to semigroups,* Merrill, Columbus, 1973.

[16] Ren,X.M. and K.P.Shum, Structure theorems for right pp semigroups with left central idempotents, *Discussiones Math. General Algebra and Applications* **20** (2000), 63–75.

[17] Shum,K.P. and X.M.Ren, Abundant semigroups with left central idempotents, *Pure Math. Applications (Hung).* **10** (1999), 109–113.

